TRAPPED CHARGED PARTICLES AND FUNDAMENTAL PHYSICS

International Advisory Committee

I. Bergström, Sweden,
J. Byrne, United Kingdom,
T. Erickson, Switzerland,
H. Kluge, Germany,
C. Nordling, Sweden,
D. Pritchard, U.S.A.,
R. van Dyck, Jr., U.S.A.,
H. Walther, Germany,
G. Werth, Germany,
D. Wineland, U.S.A.

Local Organizing Committee
B. Beck, D. Church, J. McDonald, J. Steiger

Conference Chair
Dieter Schneider, L.L.N.L.

Co-Chair
Daniel Dubin, U.C.S.D.

TRAPPED CHARGED PARTICLES AND FUNDAMENTAL PHYSICS

Asilomar, California August–September 1998

EDITORS
Daniel H. E. Dubin
University of California, San Diego

Dieter Schneider
Lawrence Livermore National Laboratory

American Institute of Physics

AIP CONFERENCE
PROCEEDINGS 457

Woodbury, New York

Editor:

Daniel H. E. Dubin
Department of Physics
University of California, San Diego
9500 Gilman Drive
La Jolla, CA 92093-0319

E-mail: dhdubin@ucsd.edu

Dieter Schneider
Lawrence Livermore National Laboratory
P.O. Box 808, L-414
Livermore, CA 94551-0808

E-mail: schneider2@llnl.gov

Articles on pp. 143–147, 284–289, 295–304, 309–318, 337–342, and 378–387 were authored by U. S. Government employees and are not covered by the below mentioned copyright.

Authorization to photocopy items for internal or personal use, beyond the free copying permitted under the 1978 U.S. Copyright Law (see statement below), is granted by the American Institute of Physics for users registered with the Copyright Clearance Center (CCC) Transactional Reporting Service, provided that the base fee of $15.00 per copy is paid directly to CCC, 222 Rosewood Drive, Danvers, MA 01923. For those organizations that have been granted a photocopy license by CCC, a separate system of payment has been arranged. The fee code for users of the Transactional Reporting Service is: 1-56396-776-6/ 99/$15.00.

© 1999 American Institute of Physics

Individual readers of this volume and nonprofit libraries, acting for them, are permitted to make fair use of the material in it, such as copying an article for use in teaching or research. Permission is granted to quote from this volume in scientific work with the customary acknowledgment of the source. To reprint a figure, table, or other excerpt requires the consent of one of the original authors and notification to AIP. Republication or systematic or multiple reproduction of any material in this volume is permitted only under license from AIP. Address inquiries to Office of Rights and Permissions, 500 Sunnyside Boulevard, Woodbury, NY 11797-2999; phone: 516-576-2268; fax: 516-576-2499; e-mail: rights@aip.org.

L.C. Catalog Card No. 98-89576
ISBN 1-56396-776-6
ISSN 0094-243X
DOE CONF- 980843

Printed in the United States of America

CONTENTS

Preface ix
Program of the Conference xi

SECTION 1: PRECISION SPECTROSCOPY

Relativistic and QED Effects in Few-Electron High-Z Systems 3
    J. Sapirstein

Reducing the Systematics in the Electron g-2 Measurement 13
    R. K. Mittleman, I. I. Ioannou, and H. G. Dehmelt

QED and Nuclear Effects in Highly Charged Ions 22
    V. M. Shabaev, A. N. Artemyev, and V. A. Yerokhin

Radiative Corrections in Highly Charged Ions 32
    G. Plunien, R. Schützhold, S. Zschocke, and G. Soff

Numerical Lamb Shift Calculations for Low-Z Systems 40
    U. D. Jentschura, P. J. Mohr, and G. Soff

The g-factor of Hydrogen-Like Ions 43
    M. Diederich, H. Häffner, N. Hermanspahn, M. Immel, H. J. Kluge, R. Ley, R. Mann, S. Stahl,
    W. Quint, J. Verdú, and G. Werth

First Results from the New Muon (g-2) Experiment 52
    A. Grossmann, H. N. Brown, G. Bunce, R. M. Carey, P. Cushman, G. T. Danby, P. T. Debevec,
    H. Deng, W. Deninger, S. K. Dhawan, V. P. Druzhinin, L. Duong, W. Earle, E. Efstathiadis,
    F. J. M. Farley, G. V. Fedotovich, S. Giron, F. Gray, M. Grosse Perdekamp, U. Haeberlen,
    M. Hare, E. S. Hazen, D. W. Hertzog, V. W. Hughes, M. Iwasaki, K. Jungmann, D. Kawall,
    M. Kawamura, B. I. Khazin, J. Kindem, F. Krienen, I. Kronkvist, R. Larsen, Y. Y. Lee,
    W. Liu, I. Logashenko, R. McNabb, W. Meng, J.-L. Mi, D. Miller, J. P. Miller, W. M. Morse,
    P. Neumayer, G. Onderwater, Y. Orlov, C. Pai, C. Polly, J. Pretz, R. Prigl, G. zu Putlitz,
    S. I. Redin, O. Rind, B. L. Roberts, N. Ryskulov, R. Sanders, S. Sedykh, Y. K. Semertzidis,
    S. Serednyakov, Y. M. Shatunov, E. Solodov, M. Sossong, A. Steinmetz, L. R. Sulak, M. Tanaka,
    C. Timmermans, A. Trofimov, D. Urner, P. V. Walter, D. Warburton, D. Winn, Q. Xu,
    A. Yamamoto, and D. Zimmerman

Precise g-factor Measurements on Ba$^+$ Ions in a Penning Trap 57
    G. Marx, G. Tommaseo, and G. Werth

SECTION 2: SPECIAL TOPICS

Antihydrogen for Tests of CPT and Lorentz Invariance 65
    M. H. Holzscheiter

Hydrogen and Antihydrogen Spectroscopy for Studies of CPT and Lorentz Symmetry 70
    R. Bluhm, V. A. Kostelecký, and N. Russell

**Non-Destructive, Absolute Mass Determination of Sub-Micrometer Sized Particles
in a Paul-type Trap** 80
    S. Schlemmer, J. Illemann, S. Wellert, and D. Gerlich

SECTION 3: PRECISION MASS SPECTROMETRY

Cooling of Radioactive Isotopes for Schottky Mass Spectrometry 87
    M. Steck, K. Beckert, H. Eickhoff, B. Franzke, F. Nolden, H. Reich, B. Schlitt, and T. Winkler

Accurate Mass Measurements of Short-Lived Isotopes with the *Mistral* RF Spectrometer 95
    C. Toader, G. Audi, C. Borcea, H. Doubre, M. Duma, M. Jacotin, S. Henry, J.-F. Képinski,
    G. Lebée, G. Le Scornet, D. Lunney, C. Monsanglant, M. de Saint Simon, C. Thibault,
    and the ISOLDE collaboration

**High-Precision Penning Trap Mass Spectroscopy and a New Measurement of the
Proton's "Atomic Mass"** 101
    R. S. van Dyck, Jr., D. L. Farnham, S. L. Zafonte, and P. B. Schwinberg

Mass Measurements on Radioactive Isotopes with a Penning Trap Mass Spectrometer............. 111
    G. Bollen, F. Ames, G. Audi, D. Beck, F. Herfurth, H.-J. Kluge, A. Kohl, D. Lunney, R. B. Moore,
    M. de Saint Simon, E. Schark, S. Schwarz, J. Szerypo, and the ISOLDE Collaboration

Beam Cooling Using a Gas-Filled RFQ Waveguide 120
    S. Henry, I. Martel-Bravo, M. de Saint Simon, M. Jacotin, J.-F. Képinski, and D. Lunney

Helium and Deuterium Mass Ratios in a Room Temperature Penning Trap....................... 125
    S. Brunner, T. Engel, A. Schmitt, and G. Werth

SECTION 4: WEAK INTERACTION STUDIES

Testing CPT and Lorentz Symmetry with Protons and Antiprotons in Penning Traps 133
    R. Bluhm, V. A. Kostelecký, and N. Russell

Testing CPT and Lorentz Symmetry with Electrons and Positrons in Penning Traps 138
    R. Bluhm, V. A. Kostelecký, and N. Russell

Trapping ^{82}Rb for β-Decay Parity Violation Measurements.. 143
    D. J. Vieira, S. J. Brice, S. G. Crane, A. Goldschmidt, R. Guckert, A. Hime, D. Tupa, and X. Zhao

Search for Scalar Contributions to the ^{38m}K β^+-ν Correlation in a Magneto-Optic Trap 148
    J. A. Behr, A. Gorelov, D. Melconian, M. Trinczek, P. Dubé, O. Häusser, U. Giesen, K. P. Jackson,
    T. Swanson, J. M. D'Auria, M. Dombsky, G. Ball, L. Buchmann, B. Jennings, J. Dilling, J. Schmid,
    J. Deutsch, W. P. Alford, D. Asgeirsson, and W. Wong

Spectroscopy of Francium... 155
    J. E. Simsarian, J. S. Grossman, L. A. Orozco, M. Pearson, G. D. Sprouse, and W. Z. Zhao

Neutron Decay Using an Ion Trap.. 163
    J. Byrne and P. G. Dawber

An Electromagnetic Ion Trap for Studies in Nuclear Beta Decay....................................... 172
    D. Beck, M. Beck, G. Bollen, J. Deutsch, J. Dilling, T. Phalet, P. Schuurmans, R. Prieels,
    W. Quint, N. Severijns, B. Vereecke, S. Versyck, and the EUROTRAPS Collaboration

Parity Nonconservation in Relativistic Hydrogenic Ions.. 175
    M. Zolotorev and D. Budker

Applications of Nonlinear Magneto-Optic Effects with Ultra-Narrow Widths 177
    V. Yashchuk, D. Budker, and M. Zolotorev

SECTION 5: STORAGE RING PHYSICS

Molecular Structure by Coulomb Explosion Imaging of Stored Molecular Ions 185
    J. Levin, L. Knoll, M. Lange, M. Scheffel, R. Wester, A. Wolf, and D. Schwalm

Longitudinal Dynamics of Laser-Cooled Fast Ion Beams: Square-Well Buckets,
Space-Charge Effects, and Anomalous Beam Behaviour 194
    M. Weidemüller, B. Eike, U. Eisenbarth, M. Grieser, R. Grimm, I. Lauer, P. Lenisa, V. Luger,
    M. Mudrich, U. Schramm, and D. Schwalm

Storage of keV Ion Beams.. 203
    D. Zajfman, O. Heber, M. Rappaport, and K. G. Bushan

Storage Rings at RIKEN RI Beam Factory 210
    M. Wakasugi, Y. Batygin, N. Inabe, T. Katayama, K. Maruyama, K. Ohtomo, T. Ohkawa,
    M. Takanaka, T. Tanabe, I. Tanihata, S. Watanabe, Y. Yano, and K. Yoshida

Clusters in Storage Rings 220
    P. Hvelplund, J. U. Andersen, and K. Hansen

Negative Ion Spectroscopy with Stored H$^-$ Ions... 227
    T. Andersen, H. H. Andersen, P. Balling, and V. V. Petrunin

SECTION 6: STUDIES OF LOW ENERGY TRAPPED IONS

RETRAP: An Ion Trap for Laser Spectroscopy of Highly-Charged Ions 235
    D. A. Church, J. Steiger, B. R. Beck, L. Gruber, J. P. Holder, J. McDonald, and D. Schneider

A Quantum Mechanical Model of Rabi Oscillations Between Two Interacting Harmonic
Oscillator Modes and the Interconversion of Modes in a Penning Trap 242
    M. Kretzschmar
Quantum Measurement and Nonclassical Vibration of an Ion in a Trap 252
    R. Huesmann, C. Balzer, B. Appasamy, Y. Stalgies, and P. E. Toschek
Enhanced-Micromotion Reduction and Elimination 261
    N. Yu and H. Dehmelt
Towards Crystalline Ion Beams—the PALLAS Ring Trap 269
    T. Schätz, D. Habs, C. Podlech, J. Wei, and U. Schramm

SECTION 7: PLASMA AND COLLECTIVE BEHAVIOR

Steady-State Confinement of Electron Plasmas Using Trivelpiece-Gould Modes
Excited by a "Rotating Wall" 277
    F. Anderegg, E. M. Hollmann, and C. F. Driscoll
Coulomb Clusters in RETRAP 284
    J. Steiger, B. R. Beck, L. Gruber, D. A. Church, J. P. Holder, and D. Schneider
Chaos and Order in Ion Traps and Storage Rings 290
    R. Blümel
Crystalline Order in Strongly Coupled Plasmas 295
    J. J. Bollinger, T. B. Mitchell, X.-P. Huang, W. M. Itano, J. N. Tan, B. M. Jelenković,
    and D. J. Wineland
Sympathetic Cooling and Crystallization of Ions in a Linear Paul Trap 305
    M. Drewsen, P. Bowe, L. Hornekær, C. Brodersen, J. P. Schiffer, and J. S. Hangst
Mode and Transport Studies of Laser-Cooled Ion Plasmas in a Penning Trap 309
    T. B. Mitchell, J. J. Bollinger, X.-P. Huang, and W. M. Itano
Measurement of Cross-Field Heat Transport in a Nonneutral Plasma 319
    E. M. Hollmann, F. Anderegg, and C. F. Driscoll
The Decay Instability of Langmuir Waves in the Non-Neutral Electron Plasma Column 324
    H. Higaki
Fractional Frequency Parametric Resonances in a Paul Trap 329
    M. A. N. Razvi, X. Z. Chu, R. Alheit, R. Blümel, and G. Werth

SECTION 8: FREQUENCY STANDARDS

Lasers for an Optical Frequency Standard Using Trapped Hg^+ Ions 337
    B. C. Young, F. Z. Cruz, D. J. Berkeland, R. J. Rafac, J. C. Bergquist, W. M. Itano,
    and D. J. Wineland
Optical Frequency Standard Based Upon Single Laser-Cooled Indium Ion 343
    W. Nagourney, J. Torgerson, and H. Dehmelt
Frequency Measurement of Visible Light 348
    F. Riehle, H. Schnatz, B. Lipphardt, G. Zinner, T. Trebst, and J. Helmcke
Hg^+ Frequency Standards 357
    J. D. Prestage, R. L. Tjoelker, and L. Maleki
Probing Ca^+ Ions in a Miniature Trap 365
    M. Knoop, M. Vedel, M. Houssin, T. Schweizer, T. Pawletko, and F. Vedel

SECTION 9: COHERENT QUANTUM CONTROL

Measurement and Control of Single Atom Motions in the Quantum Regime 371
    J. Ye, C. J. Hood, T. Lynn, H. Mabuchi, D. W. Vernooy, and H. J. Kimble
Quantum Logic with a Few Trapped Ions 378
    C. Monroe, W. M. Itano, D. Kielpinski, B. E. King, D. Leibfried, C. J. Myatt, Q. A. Turchette,
    D. J. Wineland, and C. S. Wood
The Quantum Zeno Effect in Trapped Ions 388
    R. C. Thompson, J.-L. Hernandez-Pozos, J. Höffges, D. M. Segal, and J. R. Vincent

Spatial Separation of Atomic States in a Laser-Cooled Ion Crystal 393
    W. Alt, M. Block, P. Seibert, and G. Werth

APPENDIX

List of Participants 399
Author Index 407

Preface

The "Trapped Charged Particles and Fundamental Physics" conference was held from August 31 to September 4, 1998, at the delightfully rustic seaside setting of Asilomar, on the Monterey peninsula in northern California. The conference was hosted by the Lawrence Livermore National Laboratory (LLNL), and the local organization was provided by the EBIT program at LLNL. The conference was attended by 124 delegates from 14 countries, and consisted of 63 invited talks, and a continuous poster session. Special lectures were given by Charles Alcock of LLNL, and Lars Ingmar Bergström of Stockholm University, on Wednesday evening. The previous conference in this series, the Nobel Symposium Nr. 91, on "Trapped Charged Particles and Related Fundamental Physics", took place in Lysekil, located on the western coast of Sweden in August 1994. This conference series presents the latest research on charged particle trapping and related fundamental physics, and includes selected topics in neutral particle trapping.

The various talks and discussions showed that physics research, with trapped charged particles in general, is a very active and attractive area of innovative research, and provides the basis for research efforts in new areas. The research requires the application and development of state-of-the-art experimental techniques required in spectroscopy, ion confinement, and manipulation. Thus, it also serves as an excellent training ground for young scientists. It promises new research results towards the growth of our understanding of basic physics phenomena in atomic, nuclear, plasma, and particle physics. The high precision and high quality of the data that can be achieved with ion traps has been shown in excellent presentations. The unique possibilities to study most fundamental physics phenomena have been demonstrated at the conference.

In addition to the many instructive talks and posters, conference participants were treated to several enjoyable extracurricular activities. These included a whale watching tour in the Monterey Bay, a visit to the famous Monterey Aquarium where the conference dinner was held, and kayak tours of the nearby coastline.

The editors would like to express their warmest appreciation to the conference secretaries Diane Rae, Lynda Allen, and Candace Lewis, without whom the conference could not have been organized so successfully.

Daniel H. E. Dubin
Dieter Schneider

PROGRAM

International Conference on "Trapped Charged Particles and Fundamental Physics"

Asilomar Conference Center
Pacific Grove / Monterey, California, USA

August 31 - September 4, 1998

SUNDAY, August 30, 1998

3:00 p.m.	Registration Begins	Administration Building
6:00 p.m.	*Dinner Served (Asilomar)*	*Crocker Dining Hall*
7:00 p.m.- 9:00 p.m.	Welcome Reception	Merrill Hall Auditorium

MONDAY, August 31, 1998

7:30 a.m. *Breakfast Served (Asilomar)*   *Crocker Dining Hall*

9:00 a.m. Welcome   Merrill Hall Auditorium
Dieter Schneider (LLNL)

9:10 a.m. Opening Remarks
Richard Fortner (LLNL)

SESSION #1: Precision Spectroscopy (Eva Lindroth, Chair)

9:20 a.m. Jonathan Sapirstein (University of Notre Dame)
"Relativistic and QED Effects in Few-Electron High Z Systems"

9:50 a.m. Thomas Stöhlker (GSI Darmstadt)
"Lamb Shift Experiments on High-Z Hydrogenlike Ions at the ESR Storage Ring"

10:10 a.m. Ulrich Jentschura (National Institute of Standards and Technology)
"Numerical Lamb Shift Calculations for Low-Z Systems"

10:30 a.m. *Coffee Break*

11:00 a.m. Gerhard Soff (TU Dresden)
"Radiative Corrections in Highly Charged Ions"

11:20 a.m. Peter Beiersdorfer (Lawrence Livermore National Laboratory)
"Accurate QED Measurements at the LLNL EBIT"

11:40 a.m. Vladimir Shabaev (St. Petersburg State University)
"QED and Nuclear Effects in Highly Charged Ions"

12:00 p.m.	*Lunch Served (Asilomar)*	*Crocker Dining Hall*

1:30 p.m. Thomas Kühl (GSI Darmstadt)
"Ground State Hyperfine Structure in H-like Heavy Ions"

1:50 p.m. Richard Mittleman (University of Washington)
"Eliminating the Systematics in the g-2 Electron Experiment"

2:10 p.m. Wolfgang Quint (GSI Darmstadt)
"The g-Factor of the Bound Electron in Hydrogen-like Ions: A Precision Test of QED"

2:30 p.m. Alex Grossmann (University of Heidelberg)
"First Results from the New Muon (g-2) Experiment"

2:50 p.m. *Coffee Break*

SESSION #2: Special Topics (Dan Dubin, Chair)

3:15 p.m. Announcement: Proceedings
Dan Dubin (University of California, San Diego)

3:20 p.m. Alan Kostelecky (Indiana University)
"Theory of CPT Violation for Antihydrogen Studies"

3:40 p.m. Gerald Gabrielse (Harvard University)
"Antihydrogen Studies"

4:00 p.m. Michael Holzscheiter (Los Alamos National Laboratory)
"Antihydrogen at Rest for Tests of CPT and WEP"

4:20 p.m. Theodor Hänsch (Max-Planck-Institute for Quantenoptik)
"Laser Spectroscopy of Hydrogen and Antihydrogen"

4:50 p.m. Hans Miesner (Massachusetts Institute of Technology)
"Bose Einstein Condensation Review"

5:20 p.m. *Break for the day*

6:00 p.m.	*Dinner Served (Asilomar)*	*Crocker Dining Hall*

TUESDAY, September 1, 1998

7:30 a.m.	*Breakfast Served (Asilomar)*	*Crocker Dining Hall*
9:00 a.m.	Program Begins	Merrill Hall Auditorium

SESSION #3: Precision Mass Spectrometry (David Pritchard, Chair)

9:00 a.m. Conny Carlberg (Stockholm University)
"A Precision Determination of the Proton Mass Using Highly Charged Ions"

9:20 a.m.	Marielle Chartier (NSCL / Michigan State University) "Direct Mass Measurements by Use of a Cyclotron"
9:40 a.m.	Marcus Steck (GSI Darmstadt) "Cooling of Radioactive Isotopes for Schottky Mass Spectrometry"
10:00 a.m.	David Lunney (CSNSM Orsay) "MISTRAL - A Radio Frequency Mass Spectrometer for Very Short-lived Isotopes"
10:20 a.m.	*Coffee Break*
10:50 a.m.	Robert Van Dyck, Jr. (University of Washington) "Precision Mass Spectroscopy"
11:20 a.m.	Georg Bollen (CERN) "Mass Measurements on Radioactive Isotopes with a Penning Trap Mass Spectrometer"
11:40 a.m.	Simon Rainville (Massachusetts Institute of Technology) "Two Ions in Two Traps: Towards $\Delta m/m = 10^{-12}$"
12:00 p.m.	*Lunch Served (Asilomar)* *Crocker Dining Hall*
1:00 p.m.	No program scheduled for this afternoon *Optional Kayak Boat Ride, Kayak Bay Cruise, or Whale Watching Tour*
6:00 p.m.	*Dinner Served (Asilomar)* *Crocker Dining Hall*

WEDNESDAY, September 2, 1998

7:30 a.m.	*Breakfast Served (Asilomar)* *Crocker Dining Hall*
9:00 a.m.	Program Begins Merrill Hall Auditorium SESSION #4: Weak Interaction Studies (Stuart Freedman, Chair)
9:00 a.m.	Robert Bluhm (Colby College) "Testing CPT in Penning-Trap Experiments"
9:20 a.m.	James Byrne (University of Sussex) "Neutron Decay Study Using an Ion Trap"
9:40 a.m.	David Vieira (Los Alamos National Laboratory) "Weak Interaction Studies in a Laser Trap"
10:00 a.m.	Luis Orozco (State University of New York, Stony Brook) "Spectroscopy of Francium"
10:20 a.m.	*Coffee Break*

10:50 a.m.	Michael Schacht (University of Washington) "A Parity Violation Experiment with a Single Trapped Ion"
11:10 a.m.	John Behr (TRIUMF) "Search for Scalar Contributions to the ^{38m}K Beta-Neutrino Correlation in a Magneto-Optic Trap"
11:40 a.m.	Harvey Gould (Lawrence Berkeley National Laboratory) "Laser Trapping and Cooling of Francium from a Radioactive Source"
12:00 p.m.	*Lunch Served (Asilomar)* *Crocker Dining Hall*

SESSION #5: Storage Ring Physics (Ingmar Bergström, Chair)

1:30 p.m.	Reinhold Schuch (Stockholm University) "Recombination of Cold Electrons with Stored Ions"
1:50 p.m.	Torkild Andersen (University of Aarhus) "Negative Ion Spectroscopy with Stored Ions"
2:10 p.m.	Preben Hvelplund (University of Aarhus) "Clusters in Rings"
2:30 p.m.	Andreas Wolf (Max-Planck-Institute for Kernphysik) "Life-time Measurements in Storage Rings"
2:50 p.m.	*Coffee Break*
3:20 p.m.	Masanori Wakasugi (RIKEN Cyclotron Laboratory) "Storage Rings at RIKEN Beam Factory"
3:40 p.m.	Jacob (Yasha) Levin (Max-Planck Institute for Kernphysik) "Molecular Structure Studies by Coulomb Explosion Imaging"
4:10 p.m.	Matthias Weidemüller (Max-Planck Institute for Kernphysik) "Laser Cooling in Storage Rings"
4:30 p.m.	Daniel Zajfman (Weizmann Institute of Science) "Storage of keV Ion Beams"
4:50 p.m.	Alex Hamza (Lawrence Livermore National Laboratory) "Surface Physics with Highly Charged Ions from Electron Beam Ion Traps"
5:10 p.m.	*Break for the day*
6:30 p.m.	*First bus departs from the West side of Asilomar's Administration Building for the Monterey Bay Aquarium*
7:00 p.m.	*Reception and Aquarium visit*

7:45 p.m.	*Dinner begins* After Dinner Speakers: Ingmar Bergström (Stockholm University) and Charles Alcock (Lawrence Livermore National Laboratory)	
9:30 p.m.	*First bus departs the Aquarium for Asilomar*	

THURSDAY, September 3, 1998

7:30 a.m.	*Breakfast Served (Asilomar)*	*Crocker Dining Hall*
9:00 a.m.	Program Begins	Merrill Hall Auditorium
	SESSION #6: Studies of Low-Energy Trapped Ions (Herbert Walther, Chair)	
9:00 a.m.	David Church (Texas A&M University) "RETRAP - A Trap for Highly Charged Ion Laser Spectroscopy"	
9:30 a.m.	Peter Toschek (University of Hamburg) "Quantum Measurement and Non-Classical Vibration of an Ion in a Trap"	
10:00 a.m.	Nan Yu (Jet Propulsion Laboratory) "Towards a Micromotion - Free rf Trap"	
10:20 a.m.	*Coffee Break*	
10:50 a.m.	Martin Kretzschmar (Johannes Gutenberg University) "A Quantum Mechanical Model of Rabi Oscillations Between Two Interacting Harmonic Oscillator Modes"	
	SESSION #7: Plasma and Collective Behavior (Bret Beck, Chair)	
11:10 a.m.	Reinhold Blümel (University of Freiburg) "Chaos in Ion Traps"	
11:40 a.m.	Travis Mitchell (National Institute of Standards and Technology) "Crystalline Order and Modes in Strongly Coupled Plasmas"	
12:10 p.m.	*Lunch Served (Asilomar)*	*Crocker Dining Hall*
1:30 p.m.	Clifford Surko (University of California, San Diego) "Positron Plasmas: Trapping and Applications"	
1:50 p.m.	Joachim Steiger (Lawrence Livermore National Laboratory) "Highly Charged Ion Coulomb Clusters in RETRAP"	
2:10 p.m.	Fred Driscoll (University of California, San Diego) "Collisional Transport in Non-Neutral Plasmas"	

2:30 p.m.	Francois Anderegg (University of California, San Diego) "Steady-State Confinement of Non-Neutral Plasmas Using Trivelpiece-Gould Modes Excited by Rotating Wall"	
2:50 p.m.	*Coffee Break*	
3:15 p.m.	Jeffrey Hangst (University of Aarhus) "Ion Crystals in a Linear Paul Trap"	
3:35 p.m.	Luigi Moi (University of Siena) "White Light Laser Cooling"	
3:55 p.m.	Thomas O'Neil (University of California, San Diego) "Thermodynamics of Trapped Non-neutral Plasmas"	
4:25 p.m.	*Break for the day*	
6:00 p.m.	*Beach-side Barbecue Dinner & Concert*	*Meadow Area at Asilomar*

FRIDAY, September 4, 1998

7:30 a.m.	*Breakfast Served (Asilomar)*	*Crocker Dining Hall*
9:00 a.m.	Program Begins	Merrill Hall Auditorium
	SESSION #8: Frequency Standards (Gunter Werth, Chair)	
9:00 a.m.	Warren Nagourney (University of Washington) "Optical Frequency Standard Based Upon Single Indium Ion"	
9:20 a.m.	Jürgen Helmcke (PTB Braunschweig) "Frequency Measurement of Visible Light"	
9:40 a.m.	John Prestage (Jet Propulsion Laboratory) "Frequency Standards"	
10:00 a.m.	Brent Young (National Institute of Standards and Technology) "Optical Frequency Standard Using Trapped Hg^+ Ions"	
10:20 a.m.	*Coffee Break*	
10:50 a.m.	Hugh Klein (National Physical Laboratory) "Trapped Ions for Optical Frequency Standards"	
11:10 a.m.	Peter Mohr (National Institute of Standards and Technology) "The Fundamental Constants as of 1998"	
11:40 a.m.	Lutz Schweikhard (Johannes-Gutenberg University) "Clusters in Ion Cyclotron Resonance Traps"	

12:00 p.m.	*Lunch Served (Asilomar)*	*Crocker Dining Hall*

SESSION #9: Coherent Quantum Control (David Wineland, Chair)

1:30 p.m. Herbert Walther (Max-Planck-Institute for Quantenoptik)
"Cavity Quantum Electrodynamics with Trapped Ions"

2:00 p.m. Hans Briegel (University of Innsbruck)
"Quantum Computing with Trapped Atoms and Ions (Including Applications in Quantum Communication)"

2:30 p.m. Michael Holzscheiter (Los Alamos National Laboratory)
"Prospect for Quantum Computation with Trapped Ions"

2:50 p.m. *Coffee Break*

3:20 p.m. Christopher Monroe (National Institute of Standards and Technology)
"Quantum Logic with Trapped Ions"

3:50 p.m. Jun Ye (California Institute of Technology)
"Measurements and Control of Single Atom Motions in the Quantum Regime"

4:10 p.m. Richard Thompson (Imperial College, London)
"Quantum Zeno Effect in Trapped Ions"

4:40 p.m. Concluding Remarks
Dieter Schneider (LLNL)

6:00 p.m.	*Dinner Served (Asilomar)*	*Crocker Dining Hall*

SATURDAY, September 5, 1998

7:30 a.m.	*Breakfast Served*	*Crocker Dining Hall*
12:00 p.m.	*Lunch Served* *(for those not attending the EBIT tour)*	*Crocker Dining Hall*

Schedule for optional tour of the EBIT facility at LLNL
Times shown are approximate:

9:00 a.m. Bus departs from the West side of Asilomar's Administration Building

11:00 a.m. Arrive at LLNL's Westgate Badge Office and receive badges

12:30 p.m. Box lunch at the EBIT facility and tour

4:00 p.m. Bus departs LLNL for the Asilomar Conference Center

6:00 p.m. Bus arrives at the Asilomar Conference Center

SECTION 1

PRECISION SPECTROSCOPY

Relativistic and QED Effects in Few-Electron High-Z Systems

J. Sapirstein

Department of Physics, 225 Nieuwland Science Hall, University of Notre Dame, Notre Dame, IN 46556

Abstract. A QED treatment of the spectra of highly-charged ions analogous to the QED treatment of the electron g factor is advocated. In both cases a well-defined, rapidly converging set of Feynman diagrams exists that need be considered only to relatively low order in order to provide highly accurate theoretical predictions. We demonstrate this for lithiumlike bismuth, considering diagrams involving one, two, and three photons. One set of two-photon diagrams involves the two-loop Lamb shift, and we show how the size of this effect can be inferred from a recent experimental measurement.

Quantum Electrodyamics (QED) is an intrinsically many-body theory. While its classic tests involve one-electron systems, there is nothing in the machinery of the theory that prohibits one from testing it in many-electron systems. However, particularly for neutral atoms, the fact that analytic solutions to the Schrödinger equation do not exist outside hydrogenic systems has made testing QED more difficult in the many-electron case. Balancing this drawback is the fact that far more data is of course available from many-electron systems. If only for this reason, it would be important to examine the question of how to apply QED to such systems. In addition, from a purely theoretical standpoint, the many-electron QED effects one must deal with in that case are of considerable interest in their own right. It is the purpose of this talk to demonstrate the use of a form of QED, S-matrix theory [1], in calculating the spectra of highly charged few-electron ions.

By far the most stringent test of QED comes from the electron g factor. The very precise experimental determination of the anomaly $a_e = (g-2)/2$ [2],

$$a_e = 0.001\,159\,652\,219\,3(10), \tag{1}$$

can be compared to a similarly precise theoretical prediction arising from QED [3],

$$a_e = \sum_{n=1}^{\infty} C_n \left(\frac{\alpha}{\pi}\right)^n \tag{2}$$

where the constants C_n have been calculated up to $n = 4$. The close agreement between theory and experiment shows that new physics cannot enter at a level much greater than 10^{-12}, or alternatively can be used to determine the fine structure constant with orders of magnitude more accuracy than available from other methods. The simultaneous presence of a completely predictive theory and extremely precise experiment is what has made the electron g factor a paradigm of how physics at its best can work.

The main point of this talk is that this same situation exists for highly-charged ions, even when more than one electron is present. The basic reason for this is that when the attraction to the nuclear Coulomb field dominates the repulsion with the other electrons, that repulsion, which in the neutral case is the source of the difficulty of the many-body problem, can be treated with a rapidly converging perturbation theory, with the expansion governed by the small parameter $1/Z$. As with the electron g factor, it is essential to have accurate experiments to compare with, and in this case we will concentrate on a very high accuracy measurement in lithiumlike bismuth by Beiersdorfer *et. al.* [4], which has determined the splitting

$$E_{2p_{3/2}} - E_{2s_{1/2}} = 2788.139(39) \text{ eV}. \tag{3}$$

To compare with such a precise measurement, a rapidly convergent perturbation scheme is needed. For the electron g factor, the convergence is guaranteed by the factors of α/π in Eq. 2. The index n in the coefficients C_n in that equation is equal to the number of photons in the associated Feynman diagrams. For the highly charged ion case, we will also deal with the number of photons, and will see that by the time three photons are present the size of the diagrams is under experimental uncertainty, so that only one-photon and two-photon diagrams need be considered. Of course, for neutral atoms, where the $1/Z$ expansion converges slowly, expansion in the number of photons is not adequate, and different methods, such as the Bethe-Salpeter equation [5], must be used. Because one-photon diagrams are well understood, and three-photon diagrams are small, the focus of interest in lithiumlike bismuth is on 'two-photon physics', which will be described in some detail below.

There are three differences of note with the electron g factor case. The first is that for that case, since only a single electron is present, the number of photons corresponds to the number of closed loops in the Feynman diagrams: the expansion is in fact sometimes referred to as the loop expansion. For many-electron ions, while one set of two-photon diagrams indeed involves two loops, namely the two-loop Lamb shift, the presence of more than one electron allows other two-photon diagrams with one loop or no loops. The second difference is that the loop diagrams are considerably more complicated in the atomic case because one must deal with electron propagators in an external field rather than free propagators. This difficulty is partly balanced by there being a smaller number of graphs, and less severe renormalization issues. The final difference is that a freedom exists in atomic physics associated with the fact that one can use an arbitrary potential to create the starting wave functions. We will exploit this freedom to gauge the accuracy of the photon expansion, showing that when one includes up to two photons, the final answer is independent of the starting potential to a high degree of accuracy.

Before beginning the description of the QED calculation, we mention that there is a considerable literature on the structure problem for atomic physics [6]. Here we use the word 'structure' to refer to the solution of the no-pair Hamiltonian, the relativistic generalization of the Schrödinger equation with negative energies excluded. This problem can be solved to high accuracy, and one can define QED effects as the difference of experiment and structure. However, to interpret these QED effects, the somewhat subtle connection of QED and structure must be carefully addressed. While this is an interesting topic in its own right, in this work we choose to work purely in a QED framework, which we now describe.

I S-MATRIX THEORY

In S-matrix theory, the basic idea is to divide the complete Hamiltonian of QED into a noninteracting part H_0 and an interaction term H_I that is treated perturbatively. The division used here is

$$H_0 = \int d^3x \psi^\dagger(x)[\vec{\alpha}\cdot\vec{p} + \beta m - \frac{Z_{\text{nuc}}(\vec{x})\alpha}{|\vec{x}|} + U(\vec{x})]\psi(x) \tag{4}$$

and

$$H_I = -\int d^3x \psi^\dagger(x) U(\vec{x}) \psi(x) + e\int d^3x \bar{\psi}(x)\gamma_\mu \psi(x) A^\mu(x). \tag{5}$$

In the above $Z_{\text{nuc}}(\vec{x})$ is the nuclear charge, modified at small distances to account for the finite size of the nucleus. The potential $U(\vec{x})$ is introduced to account for the effect of electron repulsion in an approximate way. If it is set equal to zero, we will refer to the resulting perturbation scheme as the Coulomb potential case. The potential part of H_I is represented graphically by a circle with a cross, as in Fig. 1b, which we will refer to as a 'counterterm' graph because of its similarity to the electron self-mass counterterm. It is convenient to further define

$$U(\vec{x}) = \frac{\alpha Z_s(\vec{x})}{|\vec{x}|}, \tag{6}$$

in which case H_0 can be written in terms of an effective charge

$$H_0 = \int d^3x \psi^\dagger(x)[\vec{\alpha}\cdot\vec{p} + \beta m - \frac{\alpha Z_{\text{eff}}(\vec{x})}{|\vec{x}|}], \tag{7}$$

where
$$Z_{\text{eff}}(\vec{x}) = Z_{\text{nuc}}(\vec{x}) - Z_s(\vec{x}). \tag{8}$$

We illustrate the use of the effective charge with one of the potentials we will use, the core-Hartree (CH) potential, defined by

$$Z_s^{CH}(r) = 2r \int dr' \frac{1}{r_>}(g_{1s}^2(r') + f_{1s}^2(r')). \tag{9}$$

Here g_{1s} and f_{1s} are the upper and lower radial components of the ground state Dirac wavefunction, defined self-consistently, and $r_> = \max(r,r')$. This function can be shown to have the asymptotic value of 2, which leads to the physically sensible picture of an electron at large distances from the nucleus in a lithiumlike ion seeing the nuclear charge screened to $Z-2$ by the ground state electrons.

A well-defined formalism exists that relates Feynman diagrams to energy shifts with Sucher's extension [7] of the Gell-Mann-Low formalism [8]. (Identical results can also be obtained using Green's functions techniques [9]). In this S-matrix approach an adiabatic damping factor is multiplied into H_I,

$$H = H_0 + e^{-\epsilon|t|}H_I. \tag{10}$$

At large positive or negative times the interactions drop out, and one has only H_0 to deal with, which leads to the lowest order energy

$$E^{(0)} = 2\epsilon_{1s} + \epsilon_v \tag{11}$$

for lithiumlike ions with a filled K shell and a valence electron v, in our case either a $2s_{1/2}$ or $2p_{3/2}$ state. The energies ϵ_{1s} and ϵ_v are obtained from the numerical solution of the Dirac equation in the potential $-\alpha Z_{\text{eff}}(r)/r$, and the wave function is a Slater determinant of the associated wavefunctions.

Energy shifts can now be calculated with the formula

$$\Delta E = \lim_{\epsilon \to 0} \frac{i\epsilon}{2} \lim_{\lambda \to 1} \frac{\partial}{\partial \lambda} \ln <S_{\epsilon,\lambda}>, \tag{12}$$

where the S-matrix $S_{\epsilon,\lambda}$ is given by

$$S_{\epsilon,\lambda} = T(e^{i\lambda \int d^4x e^{-\epsilon|x_0|}H_I(x)}) \tag{13}$$

At this point one can automatically generate a set of Feynman graphs. When only the second part of the interaction Hamiltonian in Eq. 5 is expanded, the terms we will consider involve either two such terms, which we call one-photon physics, four terms, which we call two-photon physics, or six terms, which we call three-photon physics. This arises because one must contract two photon fields to create a photon propagator. However, the first term in Eq. 5 is equivalent to a photon propagator by itself, and we will define a single interaction involving it as a one-photon effect, two interactions as two-photon, and so on. Now that an unambiguous calculational scheme to describe the spectra of highly charged ions has been set up, we illustrate its use on lithiumlike bismuth.

II LOWEST ORDER RESULTS

We use four potentials in this work. We have already mentioned the Coulomb and core-Hartree potentials. We will also use the modified core-Hartree potential (MCH) and the Kohn-Sham (KS) potentials. The former is obtained from the CH potential by replacing the factor $2r$ by r in Eq. 9, and the latter by using

$$Z_s^{KS}(r) = r \int_0^\infty dr' \frac{1}{r_>}\phi(r') - \frac{2}{3}(\frac{81}{32\pi^2})^{1/3}(r\phi(r))^{1/3} \tag{14}$$

where

$$\phi(r) \equiv 2(g_{1s_{1/2}}^2(r) + f_{1s_{1/2}}^2(r)) + g_{2s_{1/2}}^2(r) + f_{2s_{1/2}}^2(r). \tag{15}$$

The associated asymptotic charges are $Z-1$ for MCH and $Z-3$ for KS. In all cases we use a Fermi distribution for bismuth with the parameters $c = 6.6842$ fm and $t = 2.3$ fm.

It is important to stress that none of these potentials is in any sense exact. They simply represent starting points for QED perturbation theory, which theory should build in the actual physics arising from inter-electron interactions perturbatively.

It is a simple matter to solve the Dirac equation in all four potentials, and the results for the $2p_{3/2} - 2s_{1/2}$ transition in lithiumlike bismuth are collected in the first row of Table 1. We also give the $2s_{1/2}$ and $2p_{3/2}$ energies in Tables 2 and 3 respectively. We note that we will always drop any term contributing to the energy that affects only the core states. These cancel out of the transition we are studying, and also do not affect the valence removal energies.

While the MCH result happens to be very close to the experimental result, that is clearly accidental, as there is a variation of up to 37 eV between the different potentials. In other fields of physics, one might be content with a one percent agreement with experiment, but for precision QED work it is clear that more physics needs to be considered, and we begin with one-photon physics.

III ONE-PHOTON PHYSICS

By far the easiest one-photon diagram to evaluate is photon exchange between the electrons, Fig. 1a, taken together with the counterterm diagram in Fig. 1b. The associated formula is

$$\Delta E^E_{1\gamma} = \sum_a ((g_{avav}(0) - g_{avva}(\delta E)) - U_{vv}, \qquad (16)$$

where

$$g_{ijkl}(E) \equiv \alpha \int d^3r_1 d^3r_2 \frac{e^{i\sqrt{E^2+i\delta}|\vec{r}_1-\vec{r}_2|}}{|\vec{r}_1 - \vec{r}_2|} \bar{\psi}_i(\vec{r}_1)\gamma_\mu \psi_k(\vec{r}_1)\bar{\psi}_j(\vec{r}_2)\gamma^\mu \psi_l(\vec{r}_2) \qquad (17)$$

and $\delta E = \epsilon_v - \epsilon_a$. We use the convention here and in the following that summations over a, b, etc. refer to summing over the two spin states of the ground state electrons. The exponential factor is associated with the fact that interactions in QED are not in general instantaneous. It leads to significant complications for the two-photon calculation, as that factor leads to a cut in the complex E plane. This factor is in general complex, and while we are interested in the real part of the energies, care must be taken when more than one such factor is present to keep both the real and imaginary parts. Minor alterations in the code would allow the study of the imaginary part, associated with decay rates.

More difficult are the self-energy and vacuum-polarization diagrams of Fig. 1c and 1d. The self-energy (SE) can be written as $E^{SE}_{1\gamma} = \Sigma_{vv}(\epsilon_v)$, where (the self-mass counterterm is understood to be included)

$$\Sigma_{mn}(\epsilon) \equiv -ie^2 \int d^3x \int d^3y \int \frac{d^n k}{(2\pi)^n} \frac{e^{i\vec{k}\cdot(\vec{x}-\vec{y})}}{k^2 + i\delta} \bar{\psi}_m(\vec{x})\gamma_\mu S_F(\vec{x},\vec{y};\epsilon - k_0)\gamma^\mu \psi_n(\vec{y}). \qquad (18)$$

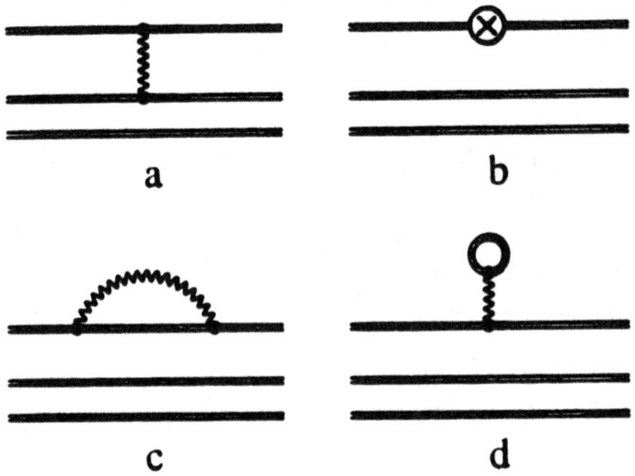

FIGURE 1. One-photon graphs

Note, as usual, that while larger, the self-energy of the core states is not included, as it affects only the core energy. The vacuum polarization term in Uehling approximation is given by

$$E_{1\gamma}^{VP} = \frac{\alpha^2}{4\pi^2} \int_0^1 dy \frac{y^2(1-y^2/3)}{1-y^2} \int d^3r \psi_v^\dagger(\vec{r})\psi_v(\vec{r}) \int d^3x \frac{e^{-\frac{2m|\vec{x}-\vec{r}|}{\sqrt{1-y^2}}}}{|\vec{x}-\vec{r}|} \vec{\nabla}_x^2 \left(\frac{Z_{\text{eff}}(x)}{|\vec{x}|}\right), \tag{19}$$

to which Wichmann-Kroll [11] corrections must be added.

The most accurate calculations of the self-energy have been carried out in the point-nucleus Coulomb case by Mohr and collaborators [12]. Techniques that work for the general, non-Coulomb case have also been developed [13], and it is now possible to carry out a complete one-photon calculation for any potential with relative ease. We present the overall effect of the one-photon diagrams on the $2p_{3/2} - 2s_{1/2}$ splitting in the second row of Table 1, and give a breakdown for the individual states in Tables 2 and 3.

When the individual contributions are considered, one can see that the radiative correction terms are relatively stable under change of potential, while the photon exchange term varies widely. This is largely because of the factor U_{vv}, which is acting to compensate the large lowest order variation. Inclusion of one-photon physics is seen to bring all potential results for the splitting into agreement to about 0.7 eV: if the Coulomb potential is excluded, the agreement is improved to about 0.4 eV. This variation is still an order of magnitude larger than the experimental error, and we must turn to two-photon physics to meaningfully interpret experiment. However, it is already clear that no large breakdown of QED is taking place, and that the radiative corrections are large, and dominated by the one-loop terms.

IV TWO-PHOTON PHYSICS

The two-photon diagrams that we will discuss here are shown in Figure 2. For simplicity, vacuum-polarization diagrams are suppressed, and only a representative two-loop Lamb shift diagram is shown. We consider them in order of increasing numbers of loops. Figs. 2a-2c, with no loops, are the simplest computationally to evaluate. As they are generalizations of the one-photon exchange terms, we denote them as $E_{2\gamma}^E$. They are given by a relatively complicated expression, which we break into two parts. The first is

$$\begin{aligned}\Delta E_{2\gamma}^E = &\sum_{abi}^{i\neq v} \frac{(g_{bvbi}(0)-g_{vbbi}(\delta E))(g_{iava}(0)-g_{iaav}(\delta E))}{\epsilon_v - \epsilon_i} + \sum_i^{i\neq v} \frac{U_{vi}U_{iv}}{\epsilon_v - \epsilon_i} \\ &- \sum_{ai}^{i\neq v}\left[\frac{(g_{avai}(0)-g_{vaai}(\delta E))U_{iv}}{\epsilon_v-\epsilon_i} + \frac{U_{vi}(g_{aiav}(0)-g_{iaav}(\delta E))}{\epsilon_v-\epsilon_i}\right] \\ &+ \sum_{abi}^{i\neq a} \frac{(g_{vbvi}(0)-g_{vbiv}(-\delta E))(g_{iaba}(0)-g_{iaab}(0))}{\epsilon_a-\epsilon_i} \\ &+ \sum_{abi}^{i\neq a} \frac{(g_{vivb}(0)-g_{ivvb}(-\delta E))(g_{baia}(0)-g_{bia}(0))}{\epsilon_a-\epsilon_i} \\ &- \sum_{ai}^{i\neq a} \frac{(g_{aviv}(0)-g_{vaiv}(-\delta E))U_{ia}}{\epsilon_a-\epsilon_i} - \sum_{ai}^{i\neq a} \frac{U_{ai}(g_{ivav}(0)-g_{ivva}(-\delta E))}{\epsilon_a-\epsilon_i} \\ &- \sum_{abi} \frac{(g_{avbi}(0)-g_{avib}(\delta E))(g_{ibva}(0)-g_{ibav}(\delta E))}{\epsilon_v-\epsilon_i} \\ &+ \sum_{abi} \frac{(g_{bavi}(\delta E)-g_{abvi}(\delta E))g_{ivba}(\delta E)}{2\epsilon_a-\epsilon_v-\epsilon_i}. \end{aligned} \tag{20}$$

Note that while $g_{ijkl}(E)$ is an even function of E, we have kept the correct signs in the above. This is because there is a second kind of contribution coming from these graphs known as derivative terms, which are sensitive to that sign. They are associated with the $i=v$ and $i=a$ terms excluded in the above, and are given by

$$\Delta E_{2\gamma}^{E'} = U_{vv} \sum_a g'_{vaav}(\delta E) + U_{aa} \sum_a g'_{vaav}(-\delta E) + \sum_{ab} g'_{avvb}(\delta E)[g_{vbva}(0) - g_{vbav}(\delta E)]$$
$$- \sum_{ab} g'_{vbav}(-\delta E)[g_{vaba}(0) - g_{vaab}(0)] - \sum_{ab} g'_{vbbv}(\delta E)[g_{vava}(0) - g_{vaav}(\delta E)]. \tag{21}$$

They are purely QED effects, inasmuch as they do not exist for instantaneous interactions.

The above summations over the intermediate states i, which arise from making a spectral decomposition of the electron propagator in Figs. 2a-c, are complete, involving both positive and negative energy states. When positive, these terms are closely related to the structure problem, and when negative are associated with the three-particle interactions studied by Zygleman [14]. However, here we treat them as a single QED contribution.

While the one-loop diagram of Fig. 2d is also associated with the structure problem, here we take it together with Fig. 2e and evaluate them as a unit. The loop is associated with an integration over the fourth component of photon momentum z, and it is straightforward to derive for Fig. 2d, which we call the ladder (L)

$$\Delta E_L = \frac{i}{2\pi} \sum_{amn} \int dz \frac{g_{avmn}(z)[g_{mnav}(z) - g_{mnva}(z - \delta E)}{[\epsilon_a + z - \epsilon_m(1 - i\delta)][\epsilon_v - z - \epsilon_n(1 - i\delta)]} \tag{22}$$

and for 2e, the crossed ladder (X)

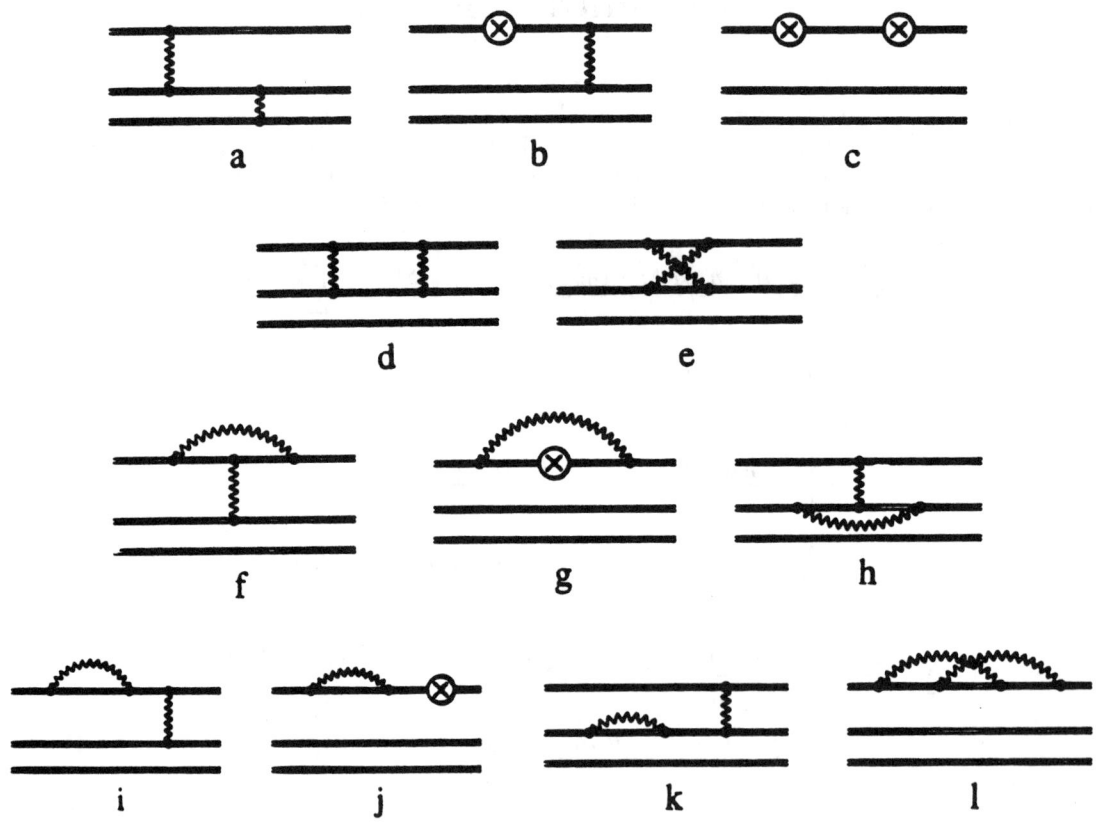

FIGURE 2. Two-photon graphs

$$\Delta E_X = \frac{i}{2\pi} \sum_{amn} \int_{-\infty}^{\infty} \left[\frac{g_{anmv}(z)g_{mvan}(z)}{(z+\epsilon_a - \epsilon_m(1-i\delta))(z+\epsilon_v - \epsilon_n(1-i\delta))} - \frac{g_{anma}(z)g_{mvvn}(z-\delta E)}{(z+\epsilon_a - \epsilon_m(1-i\delta))(z+\epsilon_a - \epsilon_n(1-i\delta))} \right]. \tag{23}$$

Analogous expressions for ground-state helium were evaluated in Ref. [15]. The technique used was that of a Wick rotation, in which the transformation $z \to i\omega$ was carried out and the ω integration carried out numerically. In the present case, while this basic approach is used, significant complications from the photon cuts are encountered, which requires the use of a somewhat complicated contour integration. The results of a preliminary calculation are tabulated as $E_{2-\gamma}^{LX}$ in Tables 2 and 3. Certain terms associated with reference state singularities, which were found to be quite small in Ref. [15], have not yet been included in the results.

At this point it is of interest to combine all terms involving exchange with the lowest order results in the different potentials. When this is done, the four transition energies are 2814.302 eV, 2814.312 eV, 2814.306 eV, and 2814.318 eV for the CH, MCH, KS, and Coulomb potentials respectively. After including a -0.04 eV contribution from a finite nuclear mass effect, one can use this clustering, which is typical of structure calculations, to infer the 'QED' value -26.13(4) eV from the experiment, if QED is defined to be the remaining one and two-photon diagrams and if, as will be shown below, three-photon physics is negligible. The one-loop radiative correction diagrams account for the bulk of this effect, contributing -26.47 eV, -26.96 eV, -26.56 eV, and -27.53 eV for the same set of potentials. As with the first order energy, the variation among the potentials is smaller if only those incorporating screening are considered, but is still larger than the experimental uncertainty. To reduce this variation, the diagrams shown in Figs. 2f through 2k must be considered.

Diagrams of this sort can be treated as a generalization of recent calculations of radiative corrections in the presence of perturbing potentials [16]. They are more complicated because instead of an instantaneous potential, one must deal with the in general noninstantaneous interaction arising from the presence of the other electrons. We begin by giving the formulas for the 'perturbed orbital' part of the calculation, which is the only part of the calculation so far completed. Essentially the same formulas have been used by Blundell [17] in his treatment of highly charged ions, which differs from the present approach only in his treating the diagrams of Fig. 2a-2e in structure approximation.

When a spectral decomposition of the propagator separating the self-energy and the exchanged photon or counterterm is made, and the degenerate state treated separately, diagrams 2i-k can be written as

$$\Sigma^{PO} = \Sigma_{v\tilde{v}} + \Sigma_{\tilde{v}v} + \sum_a (\Sigma_{a\tilde{a}} + \Sigma_{\tilde{a}a}), \tag{24}$$

where the perturbed orbitals are given by

$$\psi_{\tilde{v}}(\vec{y}) \equiv \alpha \sum_{m \neq v, a} \int \frac{d^3z d^3w}{|\vec{z}-\vec{w}|} \frac{\psi_m(\vec{y})}{\epsilon_v - \epsilon_m} [\bar{\psi}_m(\vec{z})\gamma_\mu \psi_v(\vec{z})\bar{\psi}_a(\vec{w})\gamma^\mu \psi_a(\vec{w}) - e^{i\delta E|\vec{z}-\vec{w}|}$$
$$\bar{\psi}_m(\vec{z})\gamma_\mu \psi_a(\vec{z})\bar{\psi}_a(\vec{w})\gamma^\mu \psi_v(\vec{w})] - \alpha \sum_{m \neq v} \int d^3z \frac{\psi_m(\vec{y})}{\epsilon_v - \epsilon_m} \psi_m^\dagger(\vec{z})\psi_v(\vec{z}) \frac{Z_s(z)}{z} \tag{25}$$

and

$$\psi_{\tilde{a}}(\vec{y}) \equiv \alpha \sum_{m \neq a} \int \frac{d^3z d^3w}{|\vec{z}-\vec{w}|} \frac{\psi_m(\vec{y})}{\epsilon_a - \epsilon_m} [\bar{\psi}_m(\vec{z})\gamma_\mu \psi_a(\vec{z})\bar{\psi}_v(\vec{w})\gamma^\mu \psi_v(\vec{w}) - e^{i\delta E|\vec{z}-\vec{w}|} \bar{\psi}_m(\vec{z})\gamma_\mu \psi_v(\vec{z})\bar{\psi}_v(\vec{w})\gamma^\mu \psi_a(\vec{w})]. \tag{26}$$

We note the absence of a potential term in the core perturbed orbital: the graph that would give such a term contributes only to the energy of the core, and does not contribute to the transition energy. We tabulate the perturbed orbital terms as $E_{2-\gamma}^{scr}$ in Tables 2 and 3. Their inclusion tends to reduce the spread between the different potentials, but the remaining parts of the calculation are known in the potential case [16] to also play an important role, so no conclusions about the two-loop Lamb shift can yet be drawn.

As with the exchange terms, the graphs just considered also have derivative terms associated with them, specifically

$$E^D = E_v^{(1)} \Sigma'_{vv}(\epsilon_v) + \sum_{m_a} (g_{vava}(0) - g_{vaav}(\delta E)) \Sigma'_{aa}(\epsilon_a). \tag{27}$$

These are most naturally grouped with the diagrams of Fig. 2f-h. We give the expressions for diagrams 2f and 2g, including exchange terms in the former:

$$\Delta E_{2f}^{dir} = -4i\pi\alpha^2 \sum_a \int d^3x d^3y d^3z d^3w \frac{1}{|\vec{y}-\vec{w}|} \int \frac{d^n k}{(2\pi)^n} \frac{e^{i\vec{k}\cdot(\vec{x}-\vec{z})}}{k^2} \bar{\psi}_v(\vec{x})\gamma_\mu$$
$$S_F(\vec{x},\vec{y};\epsilon_v - k_0)\gamma_\nu S_F(\vec{y},\vec{z};\epsilon_v - k_0)\gamma^\mu \psi_v(\vec{z})\bar{\psi}_a(\vec{w})\gamma^\nu \psi_a(\vec{w}), \qquad (28)$$

$$\Delta E_{2f}^{exc} = 4i\pi\alpha^2 \sum_a \int d^3x d^3y d^3z d^3w \frac{e^{i\delta E|\vec{y}-\vec{w}|}}{|\vec{y}-\vec{w}|} \int \frac{d^n k}{(2\pi)^n} \frac{e^{i\vec{k}\cdot(\vec{x}-\vec{z})}}{k^2} \bar{\psi}_v(\vec{x})\gamma_\mu$$
$$S_F(\vec{x},\vec{y};\epsilon_v - k_0)\gamma_\nu S_F(\vec{y},\vec{z};\epsilon_a - k_0)\gamma^\mu \psi_a(\vec{z})\bar{\psi}_a(\vec{w})\gamma^\nu \psi_v(\vec{w}), \qquad (29)$$

and

$$\Delta E_{2g} = 4i\pi\alpha^2 \int d^3x d^3y d^3z \int \frac{d^n k}{(2\pi)^n} \frac{e^{i\vec{k}\cdot(\vec{x}-\vec{z})}}{k^2} \bar{\psi}_v(\vec{x})\gamma_\mu S_F(\vec{x},\vec{y};\epsilon_v - k_0)\gamma_0 \frac{Z_s(y)}{y}$$
$$S_F(\vec{y},\vec{z};\epsilon_v - k_0)\gamma^\mu \psi_v(\vec{z}). \qquad (30)$$

At this point we note that the part of the first term in which only the timelike parts of the $\gamma_\nu..\gamma^\nu$ summation are kept has a particularly simple interpretation. In that approximation the d^3w integration in ΔE_{2f}^{dir} can be carried out to give

$$\sum_a \int d^3w \frac{\psi_a^\dagger(\vec{w})\psi_a(\vec{w})}{|\vec{y}-\vec{w}|} = \frac{Z_s^{CH}(y)}{y}. \qquad (31)$$

If the core-Hartree potential is used, ΔE_{2g} can be seen to cancel entirely the timelike part of ΔE_{2f}^{dir}. Thus, by using this potential, a significant part of the two-photon physics is automatically accounted for, which is why this potential is a particularly good choice for alkalilike ions. However, the exchange term, ΔE_{2g}^{ex}, along with Fig. 2h cannot be eliminated in the same way, and a direct evaluation is necessary.

Finally, two-loop diagrams, a typical one of which is shown in Fig. 2l, are by far the most difficult to evaluate. They have been treated in some detail in Ref. [18], but remain incompletely evaluated. At present, the known contributions contribute about 0.04 eV to the transition being considered here: however, one term remains uncalculated, and the final result may be larger. If that term, known as the P term [18], turns out to be small, the two-photon physics that will be tested in lithiumlike bismuth will involve exclusively many-electron effects.

V THREE-PHOTON PHYSICS

There are a very large number of three-photon diagrams, which however we argue are all very small. For example, the three-loop Lamb shift is a factor of α smaller than the already small two-loop Lamb shift, and is entirely negligible. Screening corrections to the two-loop Lamb shift are likewise unlikely to be detectable. However, three-photon exchange, being connected with structure, could in principle be important. However, just as in the two-photon case a large part of the calculation was associated with structure, specifically second-order many-body perturbation theory (MBPT), the size of a large part of the three-photon calculation will be associated with third-order MBPT. When the MBPT calculations are carried out, very small results are found, typically of order 0.02 eV. Thus three-photon physics can be entirely neglected at the present level of experimental precision. If the experimental accuracy continues to improve in highly charged ions, while one may begin to detect these very small three-photon effects, it is likely that strong interaction physics uncertainties will begin to play quite an important role.

VI CONCLUSIONS

One of the reasons that QED has had such success in one-electron atoms and the electron g factor is the unambiguity of the perturbation expansion. All researchers agree, for example, on the mathematical

TABLE 1. Cumulative lowest order, one-photon, and two-photon contributions to the $2p_{3/2} - 2s_{1/2}$ transition energy of lithiumlike bismuth in different potentials: units eV.

	CH	MCH	Kohn-Sham	Coulomb
$E^{(0)}$	2784.202	2788.417	2821.375	2792.173
$E_{1-\gamma}$	2788.182	2787.825	2787.741	2788.458
$E_{2-\gamma}$	2788.014	2787.938	2787.929	2787.765

TABLE 2. Breakdown of contributions to $E^{(0)}$, $E_{1-\gamma}$ and $E_{2-\gamma}$ for the $2s$ state in lithiumlike bismuth in different potentials: units eV.

	CH	MCH	Kohn-Sham	Coulomb
$E^{(0)}$	-25639.380	-26210.445	-25508.807	-26787.995
$E^{E}_{1-\gamma}$	-48.972	525.322	-178.709	1111.571
$E^{SE}_{1-\gamma}$	39.517	40.443	39.697	41.448
$E^{VP}_{1-\gamma}$	-8.102	-8.299	-8.148	-8.498
$E_{1-\gamma}$	-25656.937	-25652.979	-25655.967	-25643.474
$E^{E}_{2-\gamma}$	6.648	3.311	5.785	-5.499
$E^{E'}_{2-\gamma}$	-0.090	0.008	-0.042	0.109
$E^{LX}_{2-\gamma}$	-6.349	-6.364	-6.379	-6.367
$E^{scr}_{2-\gamma}$	-0.382	-0.907	-0.394	-1.427
$E_{2-\gamma}$	-25657.110	-25656.931	-25656.997	-25656.658

TABLE 3. Breakdown of contributions to $E^{(0)}$, $E_{1-\gamma}$ and $E_{2-\gamma}$ for the $2p_{3/2}$ state in lithiumlike bismuth in different potentials: units eV.

	CH	MCH	Kohn-Sham	Coulomb
$E^{(0)}$	-22855.178	-23422.028	-22687.433	-23995.822
$E^{E}_{1-\gamma}$	-18.524	551.693	-185.787	1135.391
$E^{SE}_{1-\gamma}$	4.996	5.234	5.043	5.474
$E^{VP}_{1-\gamma}$	-0.049	-0.053	-0.049	-0.058
$E_{1-\gamma}$	-22868.755	-22865.154	-22868.226	-22855.016
$E^{E}_{2-\gamma}$	6.819	3.194	6.250	-6.971
$E^{E'}_{2-\gamma}$	-0.198	0.046	-0.071	0.301
$E^{LX}_{2-\gamma}$	-6.760	-6.761	-6.805	-6.762
$E^{scr}_{2-\gamma}$	-0.202	-0.318	-0.216	-0.445
$E_{2-\gamma}$	-22869.096	-22868.993	-22869.068	-22868.893

expressions that lead to the coefficients C_n in Eq. 2. If disagreements are found, and they have many times, as these calculations can be extraordinarily complex, they have to date always been resolved. For many-electron atoms, there is less agreement on what perturbation expansion to use. There are a wide variety of methods that have been introduced to solve the non-relativistic Schrödinger equation, and a number of relativistic extensions have been proposed for highly charged ions, which incorporate QED in different ways, frequently by simply interpolating the hydrogenic values using different prescriptions. What I want to strongly advocate for highly charged ions is the S-matrix approach. As with the electron g factor, the diagrams are entirely unambiguous. While some of them can be thought of as purely structure related, others purely QED, and still others a mixture, as long as one deals with the graphs themselves everyone should be able to agree on the net result.

By far the most difficult two-photon diagrams are those associated with the two-loop Lamb shift. While still challenging, the other two-photon diagrams are relatively simple. For this reason, along with the fact that much greater precision is available from many-electron ions, it seems likely that the two-loop Lamb shift will first be detected in these ions, despite the fact that fewer diagrams need be considered in hydrogenic ions. As at least at first it is likely to be measured only very approximately, the Coulomb potential is probably adequate for the first calculations. Once screening becomes important, the core-Hartree potential should, for the reasons described above, allow for the great bulk of the screening to be incorporated. If experimental precision continues to advance, and nuclear uncertainties can be sufficiently controlled, eventually the extremely challenging task of the exact evaluation of the large set of diagrams contributing to three-photon physics will have to be faced by theory.

VII ACKNOWLEDGEMENTS

This work was supported in part by NSF grant 95-13179, and is being carried out in collaboration with K.T. Cheng. Closely related work on the two-photon exchange graphs in helium is being carried out in collaboration with S. Blundell and P. Mohr. I would like to thank P. Beiersdorfer, A.E. Livingston, and K.-H. Schartner for useful conversations.

REFERENCES

1. Mohr, P.J., Phys. Rev. A **32**, 1949 (1985).
2. Van Dyck, R.S. Jr., Schwinberg, P.H. and Dehmelt, H.G., Phys. Rev. Lett. **59**, 26 (1987).
3. Kinoshita, T. Rep. Prog. Phys. **59**, 1459 (1996).
4. Beiersdorfer, P. Osterheld, A.L., Scofield, J.H., J.R. Crespo López-Urrutia, and J. Widmann, Phys. Rev. Lett. **80**, 3022 (1998).
5. Araki, H., Prog. Theor. Phys. **17**, 619 (1957); Sucher, J., Phys. Rev. **107**, 1448 (1957).
6. References to this literature along with a discussion of the relation of the no-pair Hamiltonian to S-matrix methods are given by Sapirstein, J., Cheng, K.T., and Chen, M.H., submitted to Phys. Rev. A. (1998).
7. Sucher, J., Phys. Rev. **107**, 1448 (1957).
8. Gell-Mann, M. and Low, F., Phys. Rev. **84**, 350 (1951).
9. Shabaev, V.M., J. Phys. B **26**, 4703 (1993).
10. Furry, W.H., Phys. Rev. **81**, 115 (1951).
11. Soff, G. and Mohr, P.J., Phys. Rev. A **38**, 5066 (1988).
12. Mohr, P.J., Phys. Rev. A **46**, 4421 (1992); Mohr. P.J. and Kim, Y.K., Phys. Rev. A **45**, 2727 (1992).
13. Blundell, S.A. and Snyderman, N., Phys. Rev. A **44**, R 1427 (1991); Cheng, K.T., Johnson, W.J. and Sapirstein, J., Phys. Rev. A **47**, 1817 (1993); Persson, H., Lindgren, I, Salomonson, S. and Ynnerman, A., Phys. Rev. A **47**, R4555 (1993).
14. Zygleman, B., in *Relativistic, Quantum Electrodynamic, and Weak Interaction Effects in Atoms*, AIP Conference Proceedings 189, eds. Walter Johnson, Peter Mohr, and Joseph Sucher (AIP Press, N.Y), pg. 408.
15. Blundell, S.A., Johnson, W.R., Mohr, P.J. and Sapirstein, J., Phys. Rev. A **48**, 2615 (1993).
16. Indelicato, P. and Mohr, P.J., Theor. Chem. Acta **80**, 207 (1991); Blundell, S.A., Cheng, K.T., and Sapirstein, J., Phys. Rev. A **55**, 1857 (1997).
17. Blundell, S.A., Phys. Rev. A **47**, 1790 (1993).
18. Mallampalli, S. and Sapirstein, J., Phys. Rev. A **80**, 1234 (1997).

Reducing the Systematics in the Electron g-2 Measurement

R.K. Mittleman, I.I. Ioannou and H.G. Dehmelt

*Department of Physics,
University of Washington,
Seattle, Washington 98195-1560*

"*You know, it would be sufficient to really understand the electron*".
Albert Einstein

Abstract. The measurement of the anomalous magnetic moment of the electron provides the highest precision test of QED, or the most precise measurement of the fine structure constant. In order to improve this measurement beyond the 1ppb level a number of systematics have to be investigated and reduced. We report on the current status of this effort.

INTRODUCTION

Experimental measurements of the properties of an electron are as important as the measurements of the other known fundamental particles, quarks and leptons. A single permanently trapped electron provides an ideal system in which to test QED and allows for probing distances much smaller than those available from other experiments.

The measurement of the anomalous moment of the electron is currently the most precise test of QED. The hydrogenic Lamb shift measurement [1,2] currently has a higher precision, but does not test QED to as great a degree, due to uncertainties in the nuclear radius. With the recent improvement of the QED calculation to a part in 10^9 [3] it provides by far the most precise measurement of the fine structure constant. The reigning best measurement of the electron anomalous moment, by Van Dyck, Schwinberg and Dehmelt [4], has an accuracy of 4×10^{-9}, which was dominated by experimental systematics. With the improvement in resolution of almost an order of magnitude in the measured quantities, experimental systematics are all that remains to be overcome in order to achieve a significant improvement in this measurement. The gyromagnetic ratio of the electron (or any other charged body) in a uniform magnetic field is defined as twice the ratio of the spin precession frequency to the orbital frequency

$$g \equiv 2(\frac{\omega_s}{\omega_c}). \tag{1}$$

Dirac theory predicts that a point fermion will have a gyromagnetic ratio of exactly 2. QED modifies this prediction slightly [5], and this difference is defined as the anomalous moment of the electron

$$a \equiv \frac{g-2}{2} = \frac{\omega_s - \omega_c}{\omega_c}. \tag{2}$$

The experimentally measured quantities are the difference, or anomaly, frequency $\omega_a = \omega_s - \omega_c$ and the orbital frequency ω_c, which gives the simple formula

$$a = \frac{\omega_a}{\omega_c}. \tag{3}$$

The three remaining systematics which may limit the experimental accuracy to less then the current resolution of a few parts in 10^{10} are (in order of magnitude);

- the shift in ω_a due to the axial drive needed to create spin transitions,
- the shift in ω_c due to the conducting trapping cavity,
- the shift in the measured ω_c due to the thermal background and drive power.

We will discuss the progress and future plans which have been made in reducing each of these systematics.

EXPERIMENTAL APPARATUS

The Penning trap is well documented and has already been described in many different journals [6,7]. We will restrict ourselves to a brief overview and provide references where the interested reader can find in depth information.

We use a cryogenically cooled asymptotically symmetric Penning trap to permanently confine one isolated electron. Three electrodes form the basic Penning trap (Fig. 1), two end caps and one ring electrode. These are carefully machined hyperbolas of revolution. When a voltage, $V_0 = 10$ Volts, is applied to the ring, an almost perfect quadrupole potential is generated at the center of the trap.

$$V = V_0 \frac{z^2 - \frac{\rho^2}{2}}{2d^2} \tag{4}$$

where d is the characteristic trap dimension (d= 0.335cm). This potential provides the axial trapping force and results in an oscillation of the electron, along the z-axis at

$$\omega_z = \sqrt{eV_0/md^2} \simeq 2\pi \times 62.5 MHz. \tag{5}$$

The radial confinement is provided by a large uniform axial magnetic field ($B_0 \sim 58,000$ Gauss). This combination of fields results in a slightly shifted cyclotron frequency

$$\omega_c' = \omega_c - \omega_m, \quad \omega_c = eB_0/mc \tag{6}$$

and an $E \times B$ drift frequency $\omega_m = \frac{\omega_z^2}{2\omega_c'} \sim 12 kHz$, (where ω_c' stands for the observed trap eigenfrequency).

The axial motion is the only one which we monitor directly. This is done by connecting a tuned amplifier to one of the end-caps and observing the image current of the 60MHz oscillation. All other resonances must be measured via weak coupling channels which shift the axial resonance frequency. The energy in the cyclotron motion couples to the axial motion via two different channels. The first is through a relativistic mass gain which causes the axial oscillation frequency to decrease as energy is added to the cyclotron motion

$$\omega_z = \omega_{z_0}(1 - \frac{E_c}{2mc^2}). \tag{7}$$

The second channel is through a small inhomogeneity (magnetic bottle) which is added to the large uniform field

$$\Delta B = \frac{B_2}{2}(z^2 + \frac{\rho^2}{2})\hat{z} + B_2 z\rho\hat{\rho} \tag{8}$$

which couples to the magnetic moment of the trapped particle (u_z), creating an additional restoring force

$$\omega_z = \omega_{z_0} - \frac{B_2 u_z}{m\omega_{z_0}} = \omega_{z_0}(1 + E_c(\frac{B_2}{B_0})(\frac{1}{m\omega_z^2})) \tag{9}$$

The reverse coupling also exists, and both of these mechanisms couple the axial energy into the magnetic resonances, resulting in broadened lines.

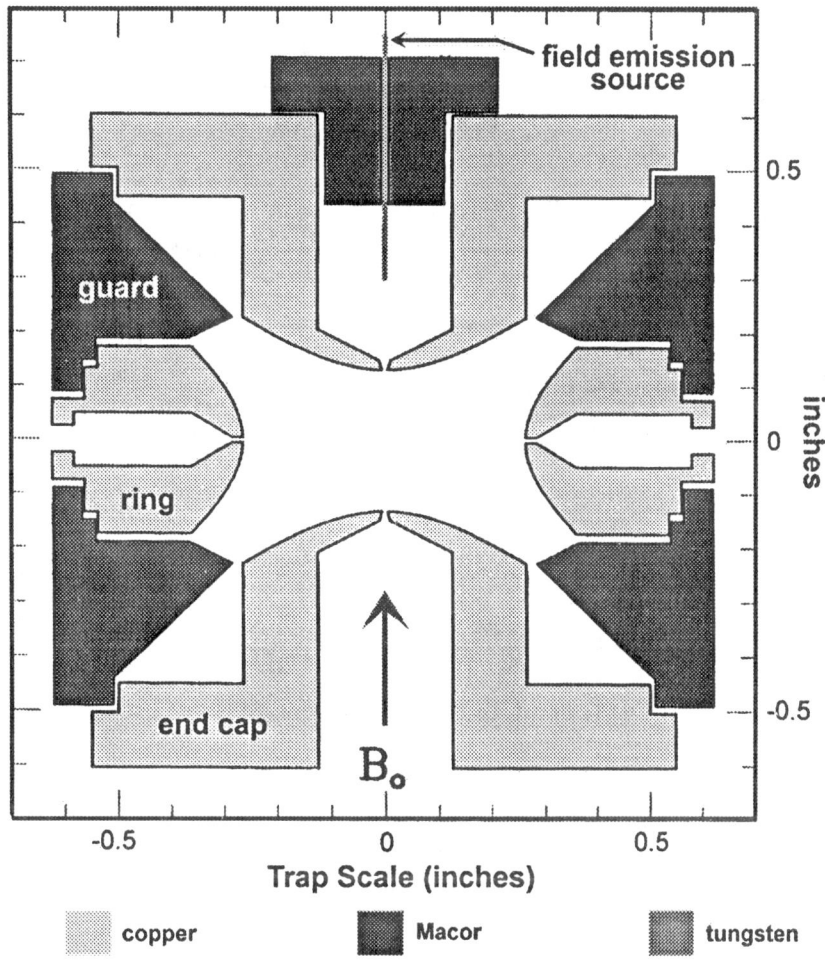

FIGURE 1. Schematic of our axially symmetric Penning trap. The ring is biased positive relative to the end caps creating a quadrapole field in the center of the trap. The guards are used to tune out residual anharmonic components in the potential.

The magnetic bottle is an adjustable quantity and can be adjusted to counteract the relativistic term and reduce the line broadening, and unhappily, the detection coupling. A second method to reduce this width is to reduce the temperature of the axial resonance below 1K. [8]. We typically have adjusted the bottle to slightly over compensate the relativistic term $\Delta\omega_{z_{bottle}}/\Delta\omega_{z_{Rel}} \sim 1.4$, and will in the future probably tune it to cancel the relativistic term to an even greater degree. With this weak a coupling we need to excite the cyclotron resonance to high energies, typically to $n > 50$, where $E_c = n\hbar\omega_c$, to detect the magnetic resonance. The relativistic perturbation also affects the cyclotron motion, making the resonance weakly anharmonic.

$$\delta\omega_c = -\omega_{c_0}(\frac{E_c}{mc^2}) \sim 200 Hz/quantum\ level \qquad (10)$$

These effects determine the method which we use to track the cyclotron frequency over time (the magnetic field typically drifts 1ppb/hr) [9]. The microwave drive power is turned on at a sample frequency, near ω_{c_0}, and then swept down in frequency, while ramping up the power. If the start frequency is above ω_{c_0}, then the electron can get accelerated, forming a mini-synchrocyclotron. At the end of the sweep, with microwave drive

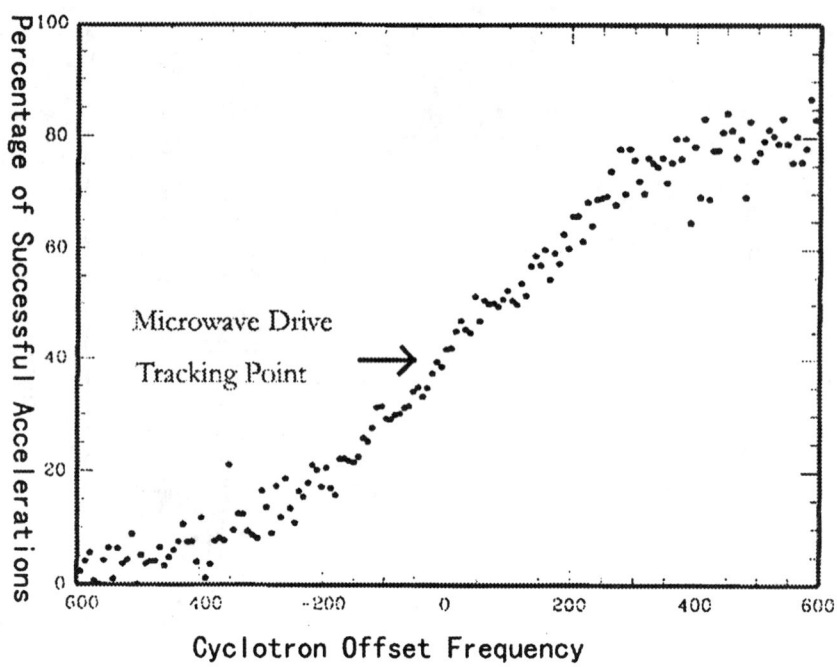

FIGURE 2. The probability of accelerating the electron to a large cyclotron energy is plotted as a function of the starting frequency of the swept microwave drive.

still on, the axial detection drives are turned on and it is determined whether the electron is in a highly excited cyclotron state ($n \sim 100$) or in the ground state. This process is repeated every three seconds, resulting in the excitation probability curve of Fig. 2. The steepest part of the curve, at approximately 40% excitations per sweep, is designated as the tracking frequency, ω_t. Fig. 3 shows a plot of ω_t vs time, exhibiting a width of $\sim 50Hz$.

To find the cyclotron frequency, ω_{c_0}, relative to ω_t, a "$\pi-pulse$" is applied immediately before the acceleration process begins and the change in acceleration probability vs pulse frequency is plotted, as in Fig. 4. The response is fitted to a Gaussian, with the center of the line fitted to $\pm 5.5Hz$ (0.03ppb).

The spin state of the electron is monitored by measuring ω_t. The energy of the particle changes by $\hbar\omega_s$ when the spin state changes which results a relativistic shift of approximately 200Hz in ω_c and ω_t when the electron changes spin state(Fig. 3). The frequency of the applied anomaly drive is then binned and the spin flipping probability is plotted (Fig 5).

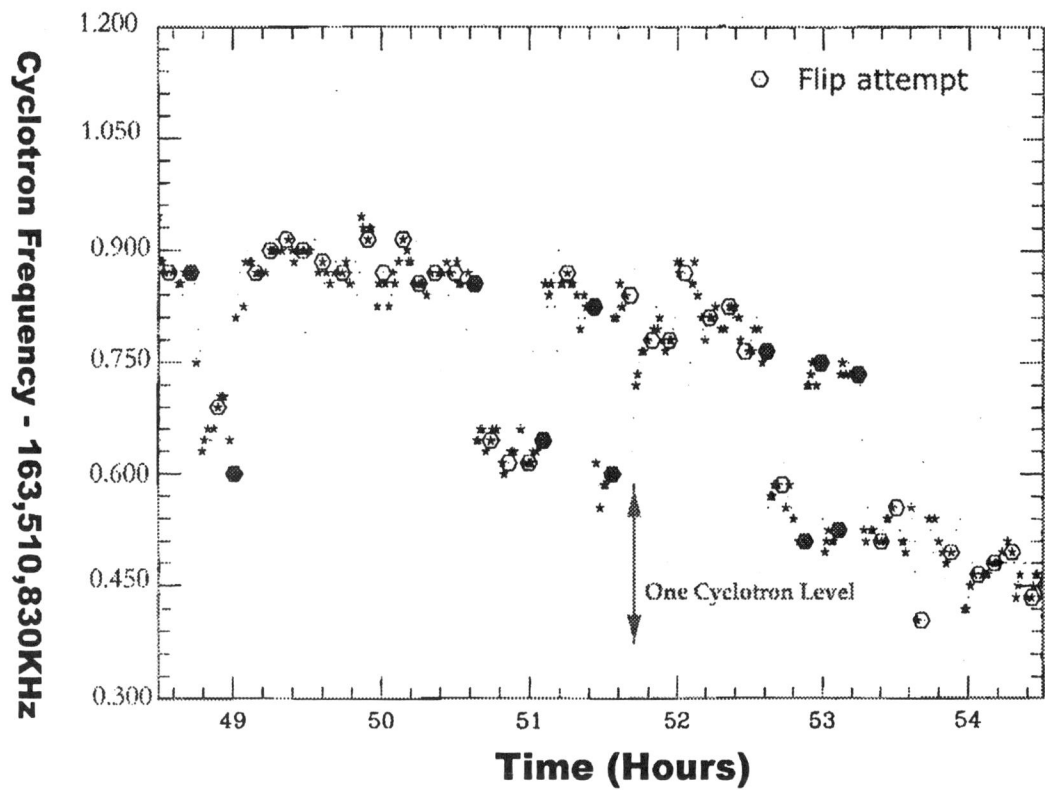

FIGURE 3. The microwave tracking frequency is plotted vs time. Each point represents 20 microwave sweeps. When the spin state of the electron is changed the tracking frequency (ω_t) changes by approximately 200Hz. There are eleven spin state changes in this data.

SYSTEMATICS

The largest systematic, that of the shift in anomaly frequency with anomaly drive has been calculated to be as large as 13 ppb [10]. This is the sum of three separate channels. The largest component comes from the change in well depth, when the anomaly drive modulates the potential, creating a Paul type trap deepening the axial well. The second channel is that of the second order Doppler effect, where the relativistic time dilation lowers the anomaly frequency. The weakest effect that is large enough to be of interest is due to the inhomogeneous magnetic field (the B_2 term); the driven trapped particle is in a different average field from the one at rest.

In order to reduce this shift we first notice that the radial component of the bottle field is proportional to both z and ρ (Eq. 8). This is the component which couples to the spin state of the electron, and when modulated by the motion of the electron in both z and r can create spin transitions. If the radius of the free precession of the cyclotron orbit can be increased, then the energy in the axial motion can be decreased while maintaining the same spin transition rate. It is important to first investigate the systematics associated with energy in the cyclotron motion to make sure that we are not trading one set of problems for another. Luckily, in a well known [11] effect that has long been used in the high energy experiments to measure the gyromagnetic ratio of the muon, the anomaly frequency is relativistically independent of the cyclotron energy. The microwave drive is much too weak to create a Paul trap effect, so the only channel left to produce shifts is the positional shift in the magnetic field. The cyclotron radius is very small, and an excitation to $<n> = 7 \times 10^4$ will only have a $\rho = 4.0 \times 10^{-4} cm$ and produce a shift in the magnetic resonances of 0.01 ppb, well below our resolution.

A much weaker axial anomaly drive has a second benefit; it will not interfere with the detection of the other

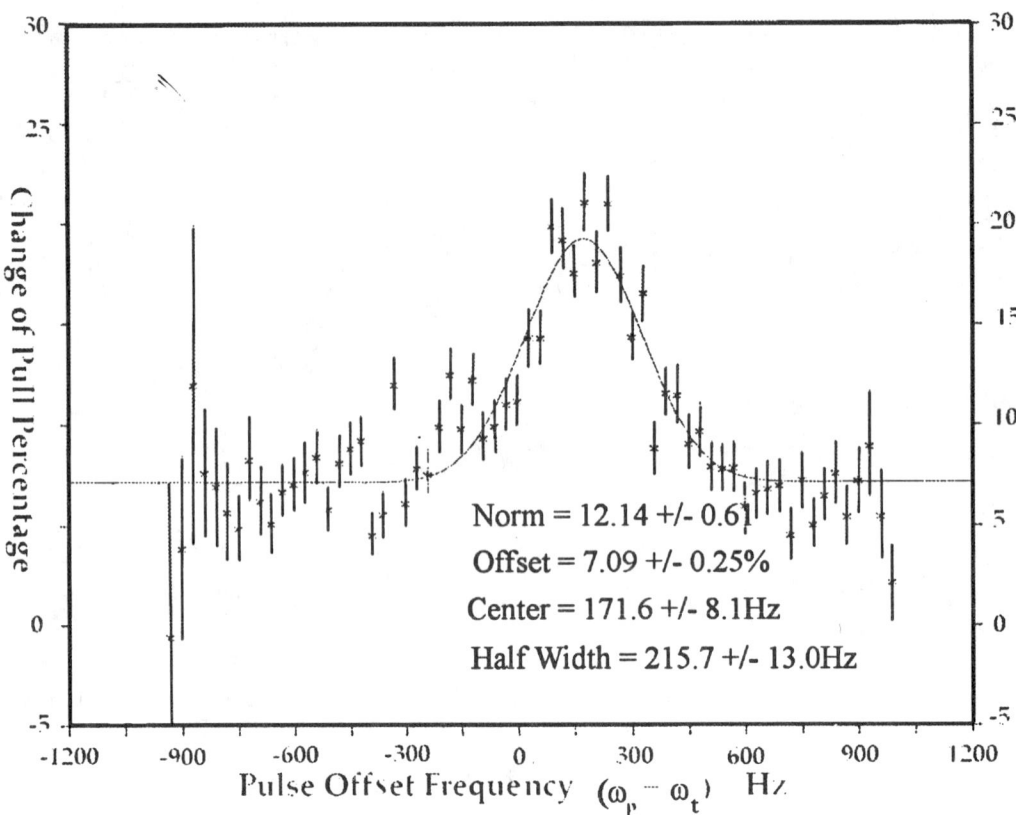

FIGURE 4. The change in acceleration probability is plotted as a function of the applied microwave pulse frequency.

resonances and can be detuned from the anomaly resonance and left on continuously [4]. This means that the well depth will not change when the spin transition is being attempted and the largest component of this systematic is reduced and eliminated.

When increasing the cyclotron radius it is important to maintain the free orbital frequency and not add any coherent component from the microwave drive. To achieve this end we drive the cyclotron resonance with a pseudo-white noise drive that is flat in the vicinity of the anharmonic resonance, and wide compared to the resonance. In order to check that no coherent components are left in the drive, producing shifts, we have taken three spin resonances with identical, weak axial anomaly drive, but differing radial drive conditions (Fig. 5). The average cyclotron radius has been increased by approximately a factor of twenty, with an equal reduction in the axial amplitude, and a reduction in the shift by a factor of 400. All three curves give the same anomaly frequency, to within 1 sigma. So to within 0.5 ppb we can say that no systematics are being created by the noise microwave drive.

The problem of the effects of the cavity are more difficult. While a very simple physical picture exists, translating this into a calculation or measurement has proven complicated. In the simplest sense the effect may be thought of as the interaction between the trapped electron and it's mirror image in the conducting surfaces of the trap. Moving with the electron the image provides a force resonant with the motion, shifting the orbital frequency by as much as 0.08ppb [12]. This shift is not significant in the measurement of ω_c, but since the spin precession frequency is not affected by the cavity [13], this translates to a shift in the anomaly frequency, $\omega_a = \omega_c - \omega_s$, of 70ppb. While a complete calculation of the mode structure of the cavity is possible, and has been performed very successfully [12,14] in a Penning trap with a cylindrical geometry, it has not been as promising in a hyperbolic trap. The mode frequencies and strengths are very sensitive to small perturbations in the trap structure. Machining imperfections and the slits and holes which are necessary to turn a microwave cavity into an electron trap provide ample perturbations and eventually make the correlation

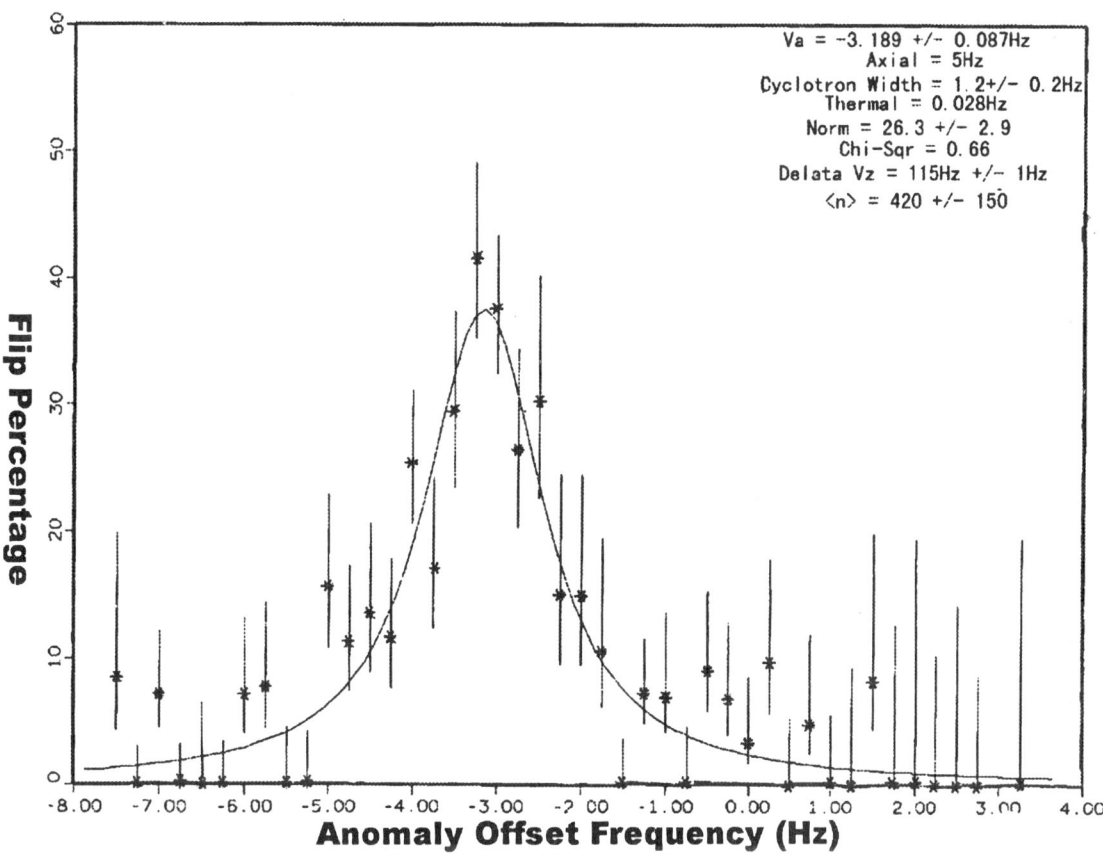

FIGURE 5. An anomaly resonance is mapped out by histograming the flip success rate. This resonance has been generated with a white noise background added to increase the average cyclotron occupation number to over 400 without affecting the width or position of the line.

between the calculated and measured spectrum exceedingly difficult. The wider spacing between modes due to the degeneracy of the cylindrical geometry reduces the problem provided by the perturbations.

Since in our experiment we are unable to theoretically calculate the shifts due to the cavity, as can be done in the cylindrical geometry [15,16], we have chosen to attempt to measure the shift directly using a small cloud as a probe. This has the additional advantage of eliminating uncertainties associated with patch effects and trap perturbations which can cause the electron not to be in the geometrical center of the trap.

To measure the cavity shift we use the fact that the force between the trapped charge and its image is proportional to the square of the charge. This means that putting in more than one electron, and measuring the center of mass frequency as a function of number should result in a direct measurement of the cavity shift. Before this can be done it must be verified that the cloud only occupies the volume of interest, or that the size of the cloud is small compared to λ_{cyc}. We have accomplished this by measuring the rotation frequency of the cloud. Knowing this frequency and the number of particles in the cloud, both the density and aspect ratio of the cloud can be calculated [17]. A plot of the cyclotron frequency and the upper and lower rotation frequency sidebands is shown in Fig. 6. The cloud of 1400 electrons when initially loaded is almost spherical, which would correspond to a rotation frequency of 36kHz, or three times the magnetron frequency. Over a period of 9 hrs the rotation frequency decreases, as the cloud spreads out to form a flattened ellipsoid [18]. For this cloud the initial measured radius, one half-hour after loading, was $1.7 \times 10^{-2} cm$ and after 3 hours it had increased to $2.0 \times 10^{-2} cm$. While this is 20-100 times larger than the magnetron orbit of a single trapped electron, it is only 10% of λ_c, and represents the cavity shifts experienced by the single electron quite well.

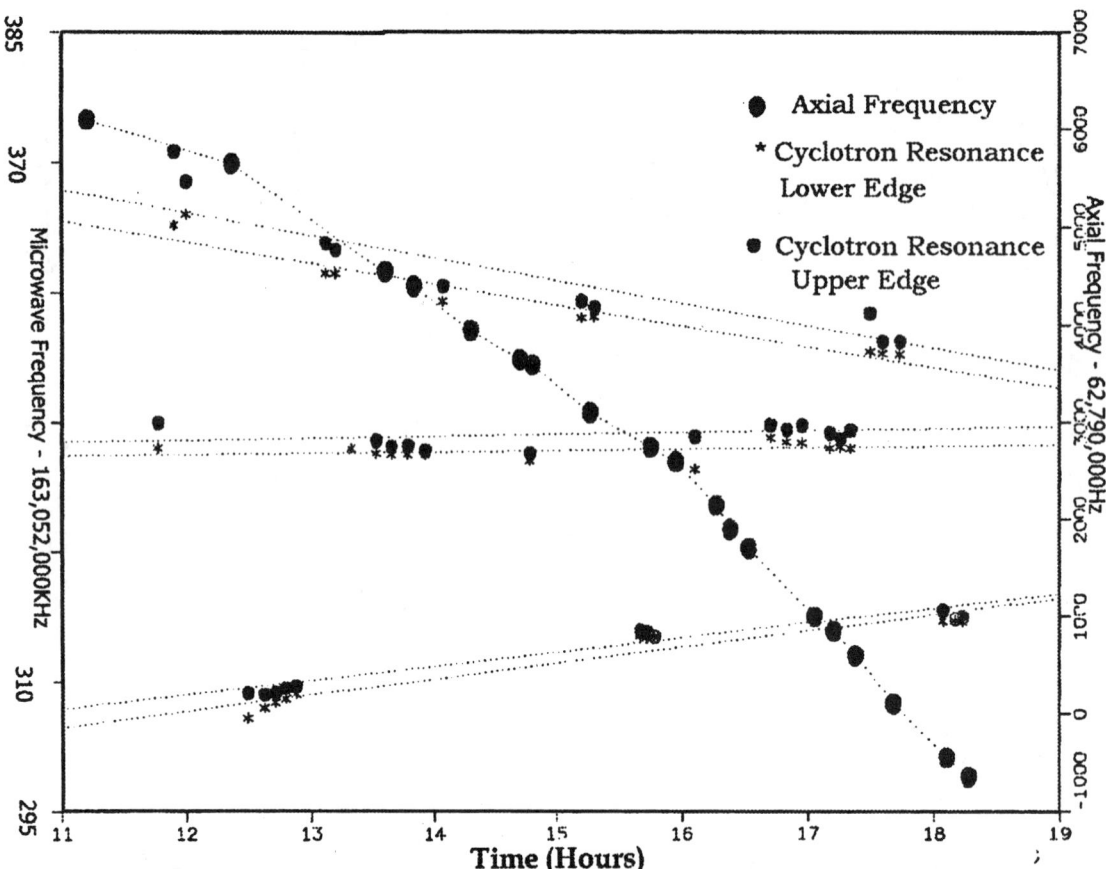

FIGURE 6. The center of mass cyclotron resonance and the first rotation sidebands are plotted over time. As the cloud relaxes from an initial sphere to a flattened ellipsoid, the rotation frequency decreases. The axial frequency is also changing as a function of the average electron radius, showing either effect can be used to track the cloud size.

There is a second way to measure the cloud size, which measures the average radius of the cloud. This is a result of the intentionally large anharmonic term which has been introduced in the trapping potential. The guard electrodes in Fig. 1 are normally used to minimize the fourth order component of the potential

$$V = V_0 \left(\frac{z^2 - \frac{\rho^2}{2}}{2d^2} + \frac{C_4}{2} \frac{z^4 - 3z^2\rho^2 + \frac{3}{8}\rho^4}{d^4} \right). \quad (11)$$

Where when working with single electrons $C_4 < 10^{-5}$. When working with clouds, in order to detect the electrons without applying any axial drives, we use the guards to increase C_4, to as large as 0.01. This introduces a radial dependence into the axial frequency. The axial frequency of the cloud is plotted in Fig. 6. The calculated average radius, agrees well with the data gathered from the rotation frequency. Smaller clouds have a longer spreading time constant, so a measurement of the center of mass cyclotron frequency of clouds of less than 2000 electrons, which is made within a few hours of loading is adequate.

The last systematic which could affect our precision is that of the cyclotron line-shape. There was an early suggestion [19] that the center of the cyclotron pulse resonance should be a function of the of the pulse power, and that a shift as large as 1ppb could be observed. This is caused by a combination of the anharmonicity in the cyclotron resonance and the 4.2K thermal bath, which excites the electron out of the ground state 15% of the time. Using a combination of classical techniques [20] and quantum calculations [21,19] to simulate the acceleration process, we should be able to generate a theoretical line shape, good enough to resolve this to a part in 10^{10}. Then, by analyzing the line as a function of power and start frequency of the microwave sweep we should be able to achieve the required accuracy.

This research is supported by the SEPARIS grant from the National Science Foundation

REFERENCES

1. Bourzeix S., de Beauvoir B., Nez F., Plimmer M.D., de Tomasi F., Julien L., and Biraben F., Phys. Rev. Lett. **76** 384 (1996)
2. Udem Th., Huber A., Gross B., Reichert J., Prevedelli M., Weitz M., and Hansch T.W., Pys. Rev. Lett. **79** 2646 (1997)
3. Kinoshita T. 1996 Rep. Prog. Phys. **59** 1459-1492
4. Van Dyck Jr. R.S., Schwinberg P.B., and Dehmelt H.G., Phys. Rev. Lett. **59** 26 (1987)
5. Nafe J.E., Nelson E.B. and Rabi I.I., Phys. Rev. **71** 914 (1947) J. Schwinger, Phys. Rev. **73**, 416 (1948) Phys. Rev. **76**, 790 (1949)
6. Brown L.S., and Gabrielse G., Rev. of Mod. Phys. **50**, 223 (1986).
7. Van Dyck Jr. R.S., Schwinberg P.B., and Dehmelt H.G., Phys. Rev. D **34**, 722 (1986).
8. Peil S. and G. Gabrielse, DAMOP 1998
9. Mittleman R.K., Palmer F.L., and Dehmelt H.G., Hyperfine Interactions **81**, 105 (1993).
10. Palmer F.L., Phys. Rev. A **47**, 2610 (1993).
11. Rich, A., and Wesley, J.C., Revs. of Mod. Phys. **44**, 250 1972
12. Brown L.S., Gabrielse G., Tan J., and Chan K.C.D., Phys. Rev. A. **37** 4163 (1987)
13. Boulware D.G., Brown L.S, Lee T., Phys. Rev. D **32** 729 (1985)
14. Brown L.S., Gabrielse G., Helmerson K., and Tan J., Phys. Rev. A **32**, 3204 (1985).
15. Gabrielse G., Tan J., *Cavity Quantum Electrodynamics* ed. P.R. Berman (Academic, NY 1994) p267
16. Tan J., and Gabrielse G., Phys. Rev. Lett. **67** 3090 (1991) Phys. Rev. A **48** 3105 (1993)
17. Bollinger J.J, Heinzen D.H., Moore F.L., Itano W.M., Wineland D. J., and Dubin D.H.E., Phys. Rev. A. **48**, 525 (1993)
18. Mittleman R.K., Dehmelt H.G., and Kim S. Phys. Rev. Lett. **75**, 2839 (1995).
19. Palmer F.L., Hyperfine Interactions **81**, 115 (1993).
20. Dehmelt H.G., Palmer F.L., and Mittleman R.K., Proc. Natl. Acad. Sci. USA **89**, 5203 (1992).
21. Enzer D., and Gabrielse G., Phys. Rev. Lett. **78**, 1211 (1997), Enzer D., Harvard 1996 PhD Thesis

QED and Nuclear Effects in Highly Charged Ions

Vladimir M. Shabaev*†, Anton N. Artemyev* and Vladimir A. Yerokhin‡

*Department of Physics, St.Petersburg State University,
Oulianovskaya 1, Petrodvorets, St.Petersburg 198904, Russia
† GSI, Postfach 11 05 52, D-64223 Darmstadt, Germany
‡ Institute for High Performance Computing and Data Bases,
Fontanka 118, St.Petersburg 198005, Russia

Abstract. Theory of quantum electrodynamic and of nuclear effects in highly charged ions is reviewed. The currently available theoretical results for the Lamb shift, the hyperfine structure splitting, and the bound-electron g-factor in high-Z few-electron atoms are compared with recent experiments. A special attention is focused on testing quantum electrodynamics in domains which were not available before.

INTRODUCTION

Precision measurements in highly charged ions have initiated accurate calculations of energy levels and transition probabilities in these systems. Since in highly charged ions the number of the atomic electrons is much less than the nuclear charge, the accurate calculations in these systems are more manageable than in many-electron atoms. In particular, the quantum electrodynamic (QED) and nuclear effects become important. As a consequence, it allows to probe theoretical predictions for the QED effects in the domain where, in contrast to low-Z atoms, the calculations based on expansion in αZ are no longer applicable. In this paper we discuss the present status of the all orders αZ calculations of the Lamb shift and the hyperfine splitting in highly charged ions.

Relativistic units ($\hbar = c = 1$) are used in the paper.

THEORETICAL BACKGROUND

The starting point for the description of a high-Z few-electron atom is the Furry picture of QED. We assume that in the zeroth approximation the electrons of the atom interact only with the Coulomb field of the nucleus. The interaction with the quantized electromagnetic field is accounted by perturbation theory. To formulate the QED perturbation theory for the calculations of energy levels and transition amplitudes it is convenient to use the two-time Green function method proposed in [1] and described in details in [2]. This method is equally suitable for calculation of single, degenerate, and quasidegenerate levels and does not demand any additional analysis of the renormalization procedure. In addition, in comparison with other methods, it simplifies considerably the derivation of formal expressions for energy shifts in the second and higher orders in α (see [1-2]). It allows also to derive formal expressions for the QED corrections to transition and scattering amplitudes.

The expressions for the QED corrections derived by the perturbation theory contain infinite summations over intermediate electron states. These sums are generally evaluated by using analytical formulas for the relativistic Coulomb Green function [3-6] or by using the relativistic finite basis set methods [7-10]. In some cases the summation can be done analytically by employing the generalized virial relations for the Dirac equation [11].

LAMB SHIFTS

Hydrogenlike ions

For the point nucleus case, the Dirac energy of a one-electron atom is given by

$$E_{nj} = \frac{mc^2}{\sqrt{1 + \frac{(\alpha Z)^2}{[n-(j+1/2)+\sqrt{(j+1/2)^2-(\alpha Z)^2}]^2}}}, \qquad (1)$$

where n is the principal quantum number and j is the total angular momentum of the electron. The nuclear and QED corrections to the Dirac energy are discussed below. The values of the theoretical contributions to the ground state energy of hydrogenlike ^{238}U are given in Table 1. The uncertainty of the Dirac binding energy indicated in the table comes from the uncertainty of the Rydberg constant. The theoretical Lamb shift prediction is in good agreement with the experimental values measured in [12,13].

TABLE 1. Theoretical contributions to the ground state energy of ^{238}U^{91+}, without higher order QED corrections. The contribution of the higher order QED corrections is expected to be on the level of a few eV.

Contribution	Value [eV]	Uncertainty [eV]	Reference
Dirac binding energy	-132279.98	± 0.04	
Nuclear size	198.81	± 0.38	
Nuclear polarization	-0.20	± 0.05	[16,17]
First order SE+VP	266.45		[19,22]
Nuclear recoil	0.46		[29]
Lamb shift theory	465.52	± 0.39	
Lamb shift experiment	470	± 16	[12]
	469	± 13	[13]

Finite nuclear size correction

The finite nuclear size correction to the Dirac energy can be evaluated by numerical solving the Dirac equation for an extended nucleus (see, e.g., [14]) or by using analytical formulas from [15]. For Z=1-100, with relative accuracy of $\sim 0.2\%$, this correction can easily be found using the following approximate expressions [15]

$$\Delta E_{ns} = \frac{(\alpha Z)^2}{10n}[1 + (\alpha Z)^2 f_{ns}(\alpha Z)]\left(2\frac{\alpha Z}{n}\frac{R}{(\hbar/mc)}\right)^{2\gamma} mc^2, \qquad (2)$$

$$\Delta E_{np_{1/2}} = \frac{(\alpha Z)^4}{40}\frac{n^2-1}{n^3}[1+(\alpha Z)^2 f_{np_{1/2}}(\alpha Z)]\left(2\frac{\alpha Z}{n}\frac{R}{(\hbar/mc)}\right)^{2\gamma} mc^2, \qquad (3)$$

where $\gamma = \sqrt{1-(\alpha Z)^2}$,

$$f_{1s}(\alpha Z) = 1.380 - 0.162\alpha Z + 1.612(\alpha Z)^2,$$

$$f_{2s}(\alpha Z) = 1.508 + 0.215\alpha Z + 1.332(\alpha Z)^2,$$

$$f_{2p_{1/2}}(\alpha Z) = 1.615 + 4.319\alpha Z - 9.152(\alpha Z)^2 + 11.87(\alpha Z)^3,$$

and R is an effective radius defined by

$$R = \left\{\frac{5}{3}\langle r^2 \rangle \left[1 - \frac{3}{4}(\alpha Z)^2\left(\frac{3}{25}\frac{\langle r^4 \rangle}{\langle r^2 \rangle^2} - \frac{1}{7}\right)\right]\right\}^{1/2}. \qquad (4)$$

For the Fermi model of the nuclear charge distribution

$$\rho(r) = \frac{N}{1 + \exp[(r-c)/a]} \qquad (5)$$

one can obtain with very high precision

$$N = \frac{3}{4\pi c^3}\left(1 + \frac{\pi^2 a^2}{c^2}\right)^{-1}, \qquad (6)$$

$$\langle r^2 \rangle = \frac{4\pi N c^5}{5}\left(1 + \frac{10}{3}\frac{\pi^2 a^2}{c^2} + \frac{7}{3}\frac{\pi^4 a^4}{c^4}\right), \qquad (7)$$

$$\langle r^4 \rangle = \frac{4\pi N c^7}{7}\left(1 + 7\frac{\pi^2 a^2}{c^2} + \frac{49}{3}\frac{\pi^4 a^4}{c^4} + \frac{31}{3}\frac{\pi^6 a^6}{c^6}\right). \qquad (8)$$

Nuclear polarization correction

The other nuclear correction arises from the nuclear polarization effect which is caused by exciting the intermediate states of the nucleus. This correction was estimated by Plunien and Soff [16] and by Nefiodov et al. [17].

Self-energy and vacuum-polarization corrections

The self energy (SE) is the dominant QED correction in the first order in α. It is defined by the diagram represented in Figure 1. The most accurate calculations of this correction were done by Mohr [4,18] for the point nucleus case and by Mohr and Soff [19] for the extended nucleus case.

FIG. 1. The first-order self-energy diagram.

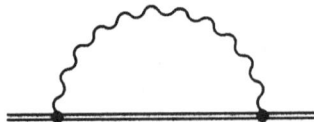

The dominant contribution to the vacuum-polarization (VP) correction (Figure 2) arises from the Uehling potential. The remaining part of the VP correction is called by the Wichmann-Kroll contribution [3]. Calculation of the Uehling contribution causes no problem and was done by many authors. The Wichmann-Kroll part was first calculated by Soff and Mohr [20] for the extended nucleus case and by Manakov et al. [21] for the point nucleus case. Recently Persson et al. [22] calculated this effect for some specific ions to higher precision.

FIG. 2. The first-order vacuum-polarization diagram.

Second-order QED corrections

The QED corrections of the second order in α have not yet been calculated completely. To date only some VP-VP and SE-VP contributions have been calculated (see [6,23], and references therein). There is some progress in the calculation of the SE-SE diagrams [24-25] but the complete value of this correction, which contributes on the level of a few eV, remains unknown.

Nuclear recoil effect

Relativistic theory of the nuclear recoil effect can be formulated only within QED. In contrast to the QED corrections discussed above, calculation of the recoil effect demands using QED beyond the external field approximation. Complete αZ-dependence formulas for the recoil effect in hydrogenlike atoms were derived in [26]. The relativistic recoil correction for a state a of a hydrogenlike atom is the sum of a low-order term ΔE_L and a higher-order term ΔE_H [26], where

$$\Delta E_L = \frac{1}{2M}\langle a|[\mathbf{p}^2 - (\mathbf{D}(0)\cdot\mathbf{p} + \mathbf{p}\cdot\mathbf{D}(0))]|a\rangle, \tag{9}$$

$$\Delta E_H = \frac{i}{2\pi M}\int_{-\infty}^{\infty} d\omega\, \langle a|\Big(\mathbf{D}(\omega) - \frac{[\mathbf{p},V]}{\omega+i0}\Big)G(\omega+\varepsilon_a)\Big(\mathbf{D}(\omega) + \frac{[\mathbf{p},V]}{\omega+i0}\Big)|a\rangle. \tag{10}$$

Here, $|a\rangle$ is the unperturbed state of the Dirac electron in the Coulomb potential $V(r) = -\alpha Z/r$, $\mathbf{p}$ is the momentum operator, $G(\omega) = (\omega - H(1-i0))^{-1}$ is the relativistic Coulomb Green function, $H = (\boldsymbol{\alpha}\cdot\mathbf{p}) + \beta m + V$, α_l ($l = 1,2,3$), β are the Dirac matrices, $D_m(\omega) = -4\pi\alpha Z\alpha_l D_{lm}(\omega)$, and $D_{ik}(\omega,r)$ is the transverse part of the photon propagator in the Coulomb gauge. In equation (10), the scalar product is implicit. The term ΔE_L contains all the recoil corrections within the $(\alpha Z)^4 m^2/M$ approximation. Its calculation yields [26]

$$\Delta E_L = \frac{m^2 - \varepsilon_a^2}{2M}, \tag{11}$$

where ε_a is the Dirac electron energy. ΔE_H contains the contribution of order $(\alpha Z)^5 m^2/M$ and all contributions of higher order in αZ which are not included in ΔE_L. The calculation of this term to all orders in αZ was performed in [27].

Recently [28] the formulas (9)-(10) were generalized to partially include the finite size effect on the recoil correction. The corresponding calculation was carried out in [29].

Heliumlike ions

In [30] the two-electron contribution to the ground state energy of helium-like ions was measured directly by comparing the ionization energies of heliumlike and hydrogenlike ions. In this way the one-electron contributions are completely eliminated. The dominant two-electron contribution results from the one-photon exchange diagram depicted in Figure 3. The contribution of this diagram, taken from [31], is given in the second column of the Table 2. The uncertainty of the one-photon exchange contribution comes from the uncertainty of the nuclear charge distribution parameters.

FIG. 3. The one-photon exchange diagram.

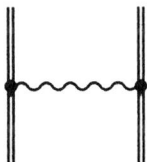

In the second order in α we have the following three types of diagrams: the two-photon exchange diagrams (Figure 4), the self energy screening diagrams (Figure 5), and the vacuum polarization screening diagrams (Figure 6).

FIG. 4. The two-photon exchange diagrams.

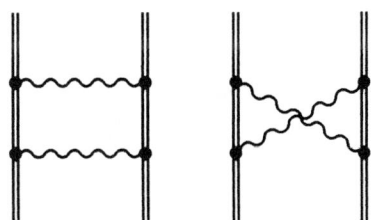

FIG. 5. The self-energy screening diagrams.

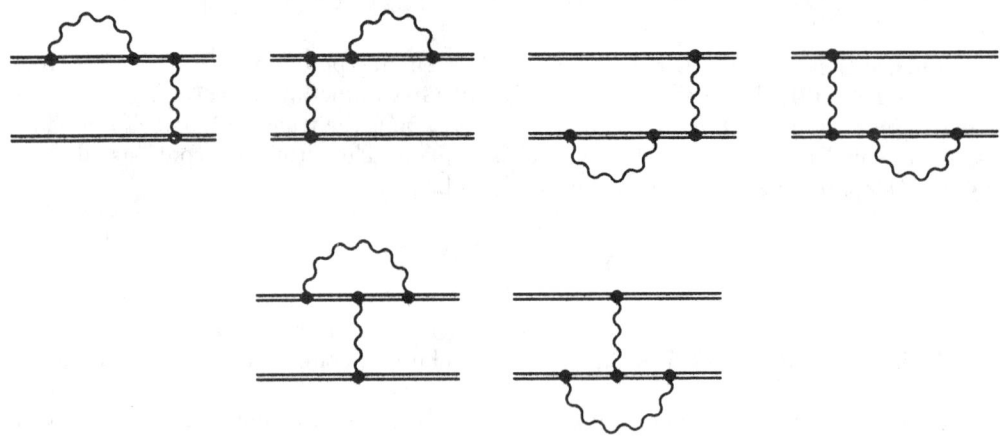

FIG. 6. The vacuum-polarization screening diagrams.

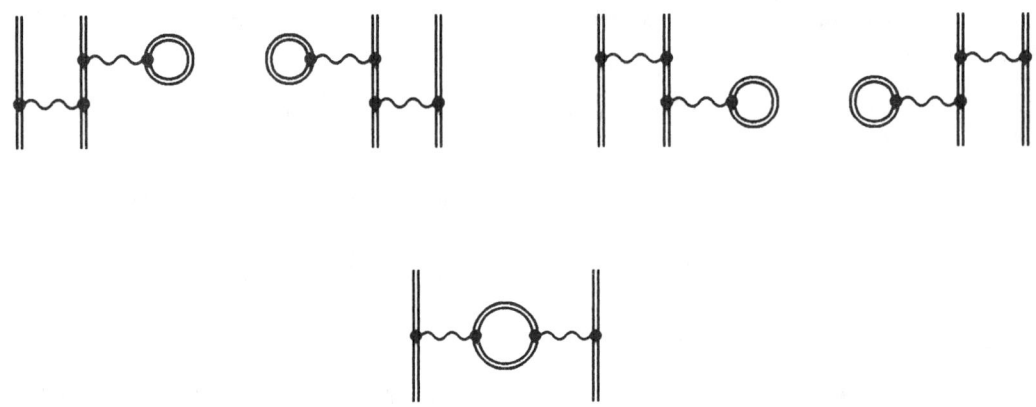

The two-photon exchange contribution was evaluated in [32,33] (the complete formal expression for this contribution was derived in [1]). This contribution is conveniently divided into two gauge invariant parts. The first part ("non-QED part"), which can be derived from the Breit equation, is given by [34]

$$\Delta E^{(2)}_{\text{"non-QED"}} = \alpha^2[-0.15766638 - 0.6356(\alpha Z)^2]mc^2. \tag{12}$$

The second part ("QED part") is the residual one. These contributions are given in the third and fourth columns of the table, respectively. The self-energy screening and vacuum polarization screening contributions were calculated in [31,35,36]. The sum of these corrections, taken from [31], is given in the fifth column. In the sixth coulumn we give an estimate of the three and more photon contribution (see [31]). The total theoretical two-electron contributions are presented in the seventh column. The corresponding experimental values are given in the last column of the table. As it follows from the table, improvement of the experimental precision by an order of magnitude will allow us to probe the QED effects in the second order in α.

TABLE 2. The two-electron contribution to the ground-state energy of helium-like ions (in eV).

Z	1-photon exchange	2-ph. exch. "non-QED"	2-ph. exch. "QED"	SE + VP screening	≥ 3 photon	Total theory	Exp. [30]
20	345.76	-4.66	0.01	-0.14	0.03(1)	341.00(1)	
32	567.61	-5.23	0.04	-0.43	0.03(1)	562.02(1)	562.5(1.6)
54	1036.56	-6.98	0.10	-1.56	0.04(3)	1028.16(3)	1027.2(3.5)
66	1347.45(1)	-8.30	0.06	-2.66	0.05(4)	1336.58(4)	1341.6(4.3)
74	1586.93(2)	-9.33	-0.04(1)	-3.68	0.05(5)	1573.92(6)	1568(15)
83	1897.56(1)	-10.64	-0.30(1)	-5.18	0.06(7)	1881.50(7)	1876(14)
92	2265.88(1)	-12.09	-0.79	-7.15	0.06(9)	2245.92(9)	

Lithiumlike ions

Measurements of the $2p_{1/2} - 2s$ and $2p_{3/2} - 2s$ transitions in high-Z lithiumlike ions [37,38] look most promising for testing QED effects in the second order in α. The experimental accuracy of these experiments is by an order of magnitude smaller than the second order QED contribution to the transition energy. The theoretical predictions for these values calculated by partial including the second order QED corrections [39-42] are in reasonable agreement with experiment. However, complete accurate calculations of the α^2 corrections are required in order to probe actually the second order QED effects in these systems. A first step in this direction was recently done in [43], where the complete gauge invariant sets of the self-energy and vacuum polarization screening diagrams for the $(1s)^2 2s$ and $(1s)^2 2p_{1/2}$ states were calculated. In Table 3 we give the theoretical contributions to the $2p_{1/2} - 2s$ transition energy in $^{238}\text{U}^{89+}$. The two and more photon exchange contribution is estimated by using the results of the relativistic configuration interaction calculations of Ref. [40]. The uncertainty caused by the uncalculated "QED" part of the two-photon exchange diagrams and by the second order one-electron QED corrections is about a few tenths eV.

TABLE 3. The theoretical contributions to the $2p_{1/2} - 2s$ transition energy in $^{238}\text{U}^{89+}$, without second order one-electron QED corrections. The contribution of the uncalculated second order QED corrections is expected to be on the level of a few tenths eV.

Contribution	Value [eV]	Reference
One-photon exchange + one-electron nuc. size	335.49(8)	This work, [23]
First order SE+VP	-42.93	[19,22]
≥ 2 photon exchange	-13.19(?)	[40]
SE screening	1.52	[43]
VP screening	-0.36	[43]
Nuclear recoil	-0.07	[27]
Nuclear polarization	0.03(1)	[16,17]
Total theory	280.49(?)	
Experiment	280.59(9)	[37]

It should be noted also that the experimental uncertainty (0.04 eV) of the $2p_{3/2} - 2s$ transition energy in Bi^{80+} [38] is smaller than the relativistic recoil contribution which constitutes -0.07 eV [27]. It gives good perspectives for testing QED in the domain of strong coupling ($\alpha Z \sim 1$) beyond the external field approximation.

HYPERFINE SPLITTING

Hydrogenlike ions

High precision measurements of the hyperfine splitting on high-Z hydrogenlike ions [44-47] have initiated calculations of various contributions to this effect (see [48,49] and references therein). The ground state hyperfine splitting of hydrogenlike ions is conveniently written in the form [50]

$$\Delta E_\mu = \frac{4}{3}\alpha(\alpha Z)^3 \frac{\mu}{\mu_N} \frac{m}{m_p} \frac{2I+1}{2I} mc^2$$
$$\times \{A(\alpha Z)(1-\delta)(1-\varepsilon) + x_{\rm rad}\}. \quad (13)$$

Here m_p is the proton mass, μ is the nuclear magnetic moment, μ_N is the nuclear magneton, I is the nuclear spin. $A(\alpha Z)$ denotes the relativistic factor

$$A(\alpha Z) = \frac{1}{\gamma(2\gamma-1)} = 1 + \frac{3}{2}(\alpha Z)^2 + \frac{17}{8}(\alpha Z)^4 + \cdots, \quad (14)$$

where $\gamma = \sqrt{1-(\alpha Z)^2}$. δ is the nuclear charge distribution correction, ε is the nuclear magnetization distribution correction (the Bohr-Weisskopf correction), and $x_{\rm rad}$ is the QED correction. In Table 4 we give the hyperfine splitting values, based on calculations of Refs. [48,49]. Except for Ho, the magnetic moment values are taken from [51]. In the case of Ho we use the value of Nachtsheim [52] corrected for the diamagetic factor by Gustavsson and Mårtensson-Pendrill [53]. The theoretical uncertainty indicated in the table is caused by the uncertainty of the Bohr-Wesskopf effect evaluated within the single particle nuclear model. Note, that this uncertainty is only an estimate of the order of magnitude of the real error [48]. So, we can say that the theoretical values are in satisfactory agreement with the experimental ones. However, the large uncertainties of the theoretical values due to the Bohr-Weisskopf effect strongly restrict a possible test of QED corrections by these experiments. In addition, as it follows from a detailed analysis of Ref. [53], remeasurements of the nuclear magnetic moments by using the modern experimental technique are needed. More elaborated calculations of the Bohr-Weisskopf effect, based on the many-particle nuclear models, are also urgent. In case of Bi, such a calculation was considered in [54].

TABLE 4. Various contributions to the ground state hyperfine splitting in hydrogenlike ions (in eV).

Ion	μ/μ_N	Rel. value (point nuc.)	Nuc. size	Bohr-Weisskopf	QED	Total theory	Experiment
^{165}Ho^{66+}	4.177(5)	2.326(3)	-0.106(1)	-0.020(6)	-0.011	2.189(7)	2.1645(5) (Ref. [45])
^{185}Re^{74+}	3.1871(3)	3.010	-0.213(2)	-0.034(10)	-0.015	2.748(10)	2.719(2) (Ref. [46])
^{187}Re^{74+}	3.2197(5)	3.041	-0.215(2)	-0.035(10)	-0.015	2.776(10)	2.745(2) (Ref. [46])
^{207}Pb^{81+}	0.592583(9)	1.425	-0.149	-0.053(5)	-0.007	1.215(5)	1.2159(2) (Ref. [47])
^{209}Bi^{82+}	4.1106(2)	5.839	-0.649(2)	-0.061(27)	-0.030	5.100(27)	5.0840(8) (Ref. [44])

Lithiumlike ions

The energy difference between the ground state hyperfine splitting components of a lithiumlike ion is conveniently written in the form [55]

$$\Delta E_{(1s)^2 2s} = \frac{1}{6}\alpha(\alpha Z)^3 \frac{m}{m_p}\frac{\mu}{\mu_N}\frac{2I+1}{2I}mc^2 \Big\{[A^{(2s)}(\alpha Z)(1-\delta^{(2s)})(1-\varepsilon^{(2s)}) + x_{\rm rad}^{(2s)}]$$
$$+ \frac{1}{Z}B(\alpha Z) + \frac{1}{Z^2}C(\alpha Z) + \cdots\Big\}. \quad (15)$$

Here $A^{(2s)}(\alpha Z)$ denotes the one-electron relativistic factor for the 2s state

$$A^{(2s)}(\alpha Z) = \frac{2[2(1+\gamma) + \sqrt{2(1+\gamma)}]}{(1+\gamma)^2 \gamma(4\gamma^2-1)} = 1 + \frac{17}{8}(\alpha Z)^2 + \frac{449}{128}(\alpha Z)^4 + \cdots. \quad (16)$$

$\delta^{(2s)}$ is the one-electron nuclear charge distribution correction, $\varepsilon^{(2s)}$ is the one-electron nuclear magnetization distribution correction, and $x_{\rm rad}^{(2s)}$ is the one-electron radiative correction. The terms $B(\alpha Z)/Z$ and $C(\alpha Z)/Z^2$ correspond to the interelectronic interaction corrections of first and second orders in $1/Z$, respectively. All these corrections were calculated in [56]. As in the hydrogenlike ions, the uncertainty of the theoretical values is mainly defined by the uncertainty of the Bohr-Weisskopf effect evaluated within the single particle nuclear model. However, in Ref. [56] it was found that the Bohr-Weisskopf effect for the lithiumlike ions can be calculated with high precision by using the experimental values of the ground state hyperfine splitting in the corresponding hydrogenlike ions. Since the Bohr-Weisskopf effect gives a domimant contribution to the uncertainty of the theoretical value, this method allows significant improvement of the theoretical prediction. The results of such a calculation for Bi and Ho are presented in the Table 5. In the case of ^{209}Bi^{80+}, the total theoretical value amounts to 0.7969(2) eV and agrees with the experimental value (0.820(26) eV [38]). For comparison, a calculation, based on using the single particle nuclear model for the Bohr-Weisskopf effect (see [56]), gives 0.800(4) eV.

TABLE 5. The individual contributions (in eV) to the ground state hyperfine splitting in ^{209}Bi^{80+} for $\Delta E_{\text{exp}}^{(1s)} =$ 5.0840(8) eV [44], $\mu = 4.1106(2)\mu_N$ and in ^{165}Ho^{64+} for $\Delta E_{\text{exp}}^{(1s)} = 2.1645(6)$ eV [45], $\mu = 4.177(5)\mu_N$. The Bohr-Weisskopf effect is found by using the experimental values of the 1s hyperfine splitting.

Contribution	^{209}Bi^{80+}	^{165}Ho^{64+}
Nonrelativistic one-electron value	0.34349(2)	0.1888(2)
Relativistic one-electron value	0.95850(5)	0.3432(4)
Nuclear size	-0.1138(3)	-0.0165(1)
Bohr-Weisskopf	-0.0139(3)	-0.0070(4)
One-electron QED	-0.0046	-0.0015
Interelectronic interaction	-0.02945(4)	-0.01345(4)
Interelectronic interaction-QED	0.00016(8)	0.00006(3)
Total theoretical value	0.7969(2)	0.3049(1)
Experiment [38]	0.820(26)	

Transition probability and bound-electron g-factor

In a recent work [57] it was shown that the transition probability between the ground state hyperfine structure components of a hydrogenlike ion, including the first order QED and nuclear corrections, is given by

$$w = \frac{\alpha}{3} \frac{\omega^3}{m^2} \frac{I}{2I+1} \left[g^{(e)} - g_I^{(n)} \frac{m}{m_p} \right]^2 , \qquad (17)$$

where ω is the transition frequency, $g^{(e)}$ is the bound-electron g-factor, and $g_I^{(n)}$ is the nuclear g-factor (both g-factors are defined to be positive). The bound-electron g-factor is

$$g^{(e)} = g_0 + \Delta g_{\text{rad}} + \Delta g_{\text{NS}} , \qquad (18)$$

where

$$g_0 = \frac{2}{3}\left[1 + 2\sqrt{1 - (\alpha Z)^2}\right] \qquad (19)$$

is the zeroth order relativistic contribution, Δg_{rad} is the QED correction, and Δg_{NS} is the nuclear size correction. Formulas (17)-(19) allow for a simple calculation of the QED and nuclear corrections to the transition probability if the corresponding corrections to the bound-electron g-factor are known. Such a calculation was done in [57]. To evaluate the QED correction, an interpolation of the recent results for Δg_{rad} from [58] was done. It was found that in the experimentally interesting cases of Pb and Bi, QED and nuclear corrections increase the transition probability by about 0.3%. Of course, these corrections can not be responsible for the discrepancy between theory and experiment for the transition probability in Bi [59]. However, they will be important if the experimental precision for the transition probability will be improved by two orders of magnitude.

Using the formula (17) and the experimental values of the hyperfine splitting and the transition probability in ^{207}Pb^{81+} [47], we find the experimental value of the bound-electron g-factor to be 1.78(12). The corresponding theoretical value calculated by formula (18) is 1.7383.

CONCLUSION

In this paper we have reviewed the present status of calculations of the QED and nuclear effects in highly charged ions. In the case of hydrogenlike uranium, the recent experiment [13] on the ground state Lamb shift provides a test of the first-order (in α) QED contribution on the level of about 5%. Improvement of the experimental precision by an order of magnitude and accurate calculations of the complete gauge invariant set of the SE-SE diagrams are necessary to probe the QED effects in the second order in α.

At present time the two-electron contribution to the ground state energy of the heliumlike ions is the only measured value [30] which has been calculated to the second order in α [31,35]. Improvement of the experimental precision by an order of magnitude would provide testing the QED effects in the second order in α.

The high precision measurements of the transition energies in high-Z lithiumlike ions [37,38] demand of accurate calculations of all QED corrections in the second order in α. The experimental uncertainty of the $2p_{3/2} - 2s$ transition energy in lithiumlike bismuth [38] is about two times smaller than the nuclear recoil contribution. It gives good perspectives for testing QED beyond the external field approximation.

The uncertainty of the Bohr-Weisskopf correction strongly restricts testing the QED effects in the hyperfine splitting investigations. However, this uncertainty can be eliminated in a combination of the hyperfine splitting values for hydrogenlike and lithiumlike ions. The latter has allowed us to improve significantly the theoretical predictions for the hyperfine splitting values in lithiumlike ions by using the experimental hyperfine splitting values in the corresponding hydrogenlike ions. High precision measurements of the hyperfine splitting in lithiumlike ions would be very important for a further development of the theory. Remeasurements of the nuclear magnetic moments by employing the modern experimental technique are also needed.

Theory of the bound-electron g-factor can be probed by measurements of the transition probabilities between the hyperfine splitting components.

ACKNOWLEDGEMENTS

Valuable conversations with Thomas Stöhlker are gratefully acknowledged. This work was supported in part by Grant No. 98-02-18350 from the Russian Foundation for Basic Research and by the programme "Russian Universities. Basic Research" (project No. 3930).

REFERENCES

1. Shabaev, V.M., *Sov. Phys. J.* **33**, 660-670 (1990).
2. Shabaev, V.M., and Fokeeva, I.G., *Phys. Rev. A* **49**, 4489-4501 (1994); Shabaev, V.M., *Phys. Rev. A* **50**, 4521-4534 (1994).
3. Wichmann, E.H., and Kroll, N.M., *Phys. Rev.* **101**, 843-859 (1956).
4. Mohr, P.J., *Ann. Phys.* **88**, 26-51, (1974); **88**, 52-87 (1974).
5. Zapryagaev, S.A., Manakov, N.L., and Palchikov, V.G., *Theory of One- and Two-Electron Multicharged Ions*, Moscow: Energoatomizdat, 1985, ch. 2, pp. 40-70.
6. Mohr, P.J., Plunien, G., and Soff, G., *Physics Reports*, **293**, 227-372 (1998).
7. Drake, G.W.F., and Goldman, S.P., *Phys. Rev. A*, **23**, 2093-2098 (1981).
8. Grant, I.P., *Phys. Rev. A*, **25**, 1230-1232 (1982).
9. Johnson, W.R., Blundell, S.A., and Sapirstein, J., *Phys. Rev. A* **37**, 307-315 (1988).
10. Salomonson, S., and Öster, P., *Phys. Rev. A* **40**, 5548-5558 (1989).
11. Shabaev, V.M., *J. Phys. B* **24**, 4479-4488 (1991).
12. Beyer, H.F., *IEEE Trans.Instrum. Meas.* **44**, 510-513 (1995); Beyer, H.F., Menzel, G., Liesen, D., Gallus, A., Bosch, F., Deslattes, R., Indelicato, P., Stöhlker, Th., Klepper, O., Moshammer, R., Nolden, F., Eickhoff, H., Franzke, B., and Steck, M., *Z. Phys. D* **35**, 169-175 (1995).
13. Stöhlker, Th., Mokler, P.H., Bosch, F., Dunford, R.W., Kozhuharov, C., Menzel, G., Rymuza, P., Stachura, Z., Swiat, P., Warczak, A., Franzke, B., Klepper, O., Krämer, A., Ludziejewski, T., Prinz, H.T., Reich, H., and Steck., M., *GSI scientific report 1997*, 99 (1998).
14. Franosch, T., and Soff, G., *Z. Phys. D* **18**, 219-222 (1991).
15. Shabaev, V.M., *J. Phys. B* **26**, 1103-1108 (1993).
16. Plunien, G., and Soff, G., *Phys. Rev. A* **51**, 1119-1131 (1995); **53**, 4614 (1996).
17. Nefiodov, A.V., Labzowsky, L.N., Plunien, G., and Soff, G., *Phys. Lett. A* **222**, 227-232 (1996).
18. Mohr, P.J., *Phys. Rev. A* **46**, 4421-4424 (1992).
19. Mohr, P.J., and Soff, G., *Phys. Rev. Lett.* **70**, 158-161 (1993).
20. Soff, G., and Mohr, P.J., *Phys. Rev. A* **38**, 5066-5075 (1988).
21. Manakov, N.L., Nekipelov, A.A., and Fainshtein, A.G., *Sov. Phys. JETP* **68**, 673-678 (1989).
22. Persson, H., Lindgren, I., Salomonson, S., and Sunnergren, P., *Phys. Rev. A* **48**, 2772-2778 (1993).
23. Beier, T., Mohr, P.J., Persson, H., Plunien, G., Greiner, M., and Soff, G., *Phys. Lett. A* **236**, 329-338 (1997).

24. Mitrushenkov, A., Labzowsky, L.N., Lindgren, I., Persson, H., and Salomonson, S., *Phys. Lett. A* **200**, 51-55 (1995).
25. Mallampalli, S., and Sapirstein, J., *Phys. Rev. A* **57**, 1548-1564 (1998).
26. Shabaev, V.M., *Theor. Math. Phys.* **63**, 588-596 (1985).
27. Artemyev, A.N., Shabaev, V.M., and Yerokhin, V.A., *Phys. Rev. A* **52**, 1884-1894 (1995); *J. Phys. B* **28**, 5201-5206 (1995).
28. Shabaev, V.M., *Phys. Rev. A* **57**, 59-67 (1998).
29. Shabaev, V.M., Artemyev, A.N., Beier, T., Plunien, G., Yerokhin, V.A., and Soff, G., *Phys. Rev. A* **57**, 4235-4239 (1998).
30. Marrs, R.E., Elliott, S.R., and Stöhlker, T., *Phys. Rev. A* **52** 3577-3585 (1995).
31. Yerokhin, V.A., Artemyev, A.N., and Shabaev, V.M., *Phys. Lett. A* **234**, 361-366 (1997).
32. Blundell, S.A., Mohr, P.J., Johnson, W.R., and Sapirstein, J., *Phys. Rev. A* **48** 2615-2626 (1993).
33. Lindgren, I., Persson, H., Salomonson, S., and Labzowsky, L.N., *Phys. Rev. A* **51**, 1167-1195 (1995).
34. Drake, G.W., *Can. J. Phys.* **66**, 586-611 (1988).
35. Persson, H., Salomonson, S., Sunnergren, P., and Lindgren, I., *Phys. Rev. Lett.* **76** 204-210 (1996); Persson, H., Salomonson, S., Sunnergren, P., Lindgren, I., and Gustavsson, M.G.H., *Hyperfine Interaction* **108**, 3-17 (1997).
36. Artemyev, A.N., Shabaev, V.M., and Yerokhin, V.A., *Phys. Rev. A* **56**, 3529-3534 (1997).
37. Schweppe, J., Belkacem, A., Blumenfeld, L., Claytor, N., Feinberg, B., Gould, H., Costram, V.E., Levy, L., Misawa, S., Mowat, J.R., Prior, M.H., *Phys. Rev. Lett.* **66**, 1434-1437 (1991).
38. Beiersdorfer, P., Osterheld, A.L., Scofield, J.H., Crespo-López-Urrutia, J.R., and Widmann, K., *Phys. Rev. Lett.* **80**, 3022-3025 (1998).
39. Indelicato, P., and Mohr, P.J., *Theor. Chem. Acta* **80**, 207-214 (1991); Mohr, P.J., *Phys. Scr. T* **46**, 44-51 (1993).
40. Cheng, K.T., Johnson, W.R., and Sapirstein, J., *Phys. Rev. Lett.* **66**, 2960-2963 (1991); Chen, M.H., Cheng, K.T., Johnson, W.R., and Sapirstein, J., *Phys. Rev. A* **52**, 266-273 (1995).
41. Blundell, S.A., *Phys. Rev. A* **47**, 1790-1803 (1993).
42. Persson, H., Lindgren, I., and Salomonson, S., *Phys. Scr. T* **46**, 125-130 (1993); Lindgren, I., Persson, H., Salomonson, S., and Ynnerman, A., *Phys. Rev. A* **47**, R4555-R4558 (1993).
43. Yerokhin, V.A., Artemyev, A.N., Beier, T., Shabaev, V.M., and Soff, G., *J. Phys. B*, in press; *Phys. Scr. T*, in press.
44. Klaft, I., Borneis, S., Engel, T., Fricke, B., Grieser, R., Huber, G., Kühl, T., Marx, D., Neumann, R., Schröder, S., Seelig, P., Völker, L., *Phys. Rev. Lett.* **73**, 2425-2427 (1994).
45. Crespo Lopez-Urrutia, J.R., Beiersdorfer, P., Savin, D., and Widmann, K., *Phys. Rev. Lett.* **77**, 826-829 (1996).
46. Crespo Lopez-Urrutia, J.R., Beiersdorfer, P., Widmann, K., Birket, B., Mårtensson-Pendrill, A.-M., and Gustavsson, M.G.H., *Phys. Rev. A***57**, 879-887 (1998).
47. Seelig, P., Borneis, S., Dax, A., Engel, T., Faber, S., Gerlach, M., Holbrow, C., Huber, G., Kühl, T., Marx, D., Meier, K., Merz, P., Quint, W., Schmitt, F., Tomaselli, M., Völker, L., Würtz, M., Beckert, K., Franzke, B., Nolden, F., Reich, H., Steck, M., Winkler, T., *Phys. Rev. Lett.*, in press.
48. Shabaev, V., Tomaselli, M., Kühl, T., Artemyev, A.N., and Yerokhin, V.A., *Phys. Rev. A* **56**, 252-255 (1997).
49. Sunnergren, P., Persson, H., Salomonson, S., Schneider, S.M., Lindgren, I., and Soff, G., *Phys. Rev. A* **58**, 1055-1069 (1998).
50. Shabaev, V.M., *J. Phys. B* **27**, 5825-5832 (1994).
51. Raghavan, P., *At. Data Nucl. Data Tables* **42**, 189-291 (1989).
52. Nachtsheim G., *Präzisionsmessung der Hyperfeinstruktur-Wechselwirkung von ^{165}Ho im Grundzustand*, Ph. D. Thesis (Bonn 1980), unpublished.
53. Gustavsson, M.G.H., and Mårtensson-Pendrill, A.-M., *Phys. Rev. A*, in press.
54. Tomaselli, M., Schneider, S.M., Kankeleit, E., Kühl, T., *Phys. Rev. C* **51**, 2989-2997 (1995).
55. Shabaeva, M.B., and Shabaev, V.M., *Phys. Rev. A* **52**, 2811-2819 (1995).
56. Shabaev, V.M., Shabaeva, M.B., Tupitsyn, I.I., Yerokhin, V.A., Artemyev, A.N., Kühl, T., Tomaselli, M., and Zherebtsov, O.M., *Phys. Rev. A* **57**, 149-156 (1998); **58**, 1610 (1998).
57. Shabaev, V.M., to be published.
58. Persson, H., Salomonson, S., Sunnergren, P., and Lindgren, I., *Phys. Rev. A* **56**, R2499-2502 (1997).
59. Schneider, S.M., Greiner, W., and Soff, G., *Z. Phys. D* **31**, 143-144 (1994).

Radiative Corrections in Highly Charged Ions

Günter Plunien, Ralf Schützhold, Sven Zschocke and Gerhard Soff

Institut für Theoretische Physik, TU Dresden, Mommsenstr. 13
D-01062 Dresden, Germany

Abstract. We are aiming at predictions for the Lamb shift in hydrogen-like systems with a relative accuracy of 10^{-6}. This necessitates calculations of all radiative corrections of first and second order in α but exact in the coupling constant $Z\alpha$ to the external Coulomb field. Results for all known corrections including unclear structure effects are presented for the 1S-ground state in lead $^{208}_{82}\text{Pb}^{81+}$ and uranium $^{238}_{92}\text{U}^{91+}$.

I INTRODUCTION

Bound-state QED provides a relativistic describtion of highly charged few-electron atomic systems. Its predictions can be tested to very high precision by measurements of the Lamb shift of electron levels in highly charged ions. Finite nuclear-size effects, the self energy and vacuum polarization of order α represent the dominant corrections to the energy spectrum. Both radiative corrections have to be evaluated to all orders in $Z\alpha$ in the interaction with the external Coulomb potential to achieve agreement with Lamb shift data for hydrogen-like system measured with the relative precision of about 10^{-4}. However, at the level of 10^{-5} even the influence of nuclear size on these QED corrections have to be examined. The natural limitation for testing QED is set by nuclear polarization effects and by the uncertainties of nuclear parameter. In heavy systems nuclear structure becomes non-negligible at the level of relative precision of about 10^{-6}. To provide predictions for the Lamb shift taking into account this ultimate standard requires the exact evaluation of all QED-radiative corrections of order α^2. For hydrogen-like lead and uranium we summarize the current status of Lamb shift predictions for the 1S-ground state.

II FINITE NUCLEAR SIZE

The theory of energy levels and QED effects in high-Z atoms is based primarily on the Dirac equation for electrons in a spherically symmetric external Coulomb potential V_{ex}:

$$\left[-i\vec{\alpha}\cdot\vec{\nabla}+V_{\text{ex}}(r)+\beta-E_A\right]\phi_A(\vec{r})=0 \quad . \tag{1}$$

The difference between the energy eigenvalue obtained in the presence of the Coulomb potential of extended nuclei and the Bohr-Sommerfeld value for point-like nuclei is per definition considered as the effect of the finite nuclear size. The present experimental accuracy necessitates to employ general potentials V_{ex} in any calculations of energy levels generated by an extended nuclear charge density ρ_{ex}. A characteristic empirical parameter is the root mean square (rms) radius. Simple models for the nuclear charge distribution used frequently is that of a uniformly sphere and the Fermi-distribution. The rms radius is always kept fixed when comparing different model distributions. The difference in the calculated binding energies employing a Fermi distribution and the uniform sphere model with the same rms radius might well serve as an estimate of the "nuclear shape uncertainty". Any uncertainty in the knowledge of the rms radius also induces a "size uncertainty" in the binding energy. Predictions for the influence of the finite nuclear size on the Lamb shift based on a Fermi distribution are usually presented together with an estimate of the shape uncertainty. Nuclear size effects represent an important contribution to the Lamb shift which in heavy systems becomes as important as the most dominant radiative correction, i.e. the self energy of order α.

III QED CORRECTIONS OF ORDER α

Performing perturbation theory (in the Furry picture) with respect to the interaction with the radiation field we obtain the following formal expression for the self-energy correction

$$\Delta E_A^{\text{SE}} = i\alpha \int d^3r\, d^3r'\, \phi_A^\dagger(\vec{r}) \int_{C_F} \frac{dE}{2\pi}\, \alpha^\mu\, G(\vec{r},\vec{r}',E)\, \alpha^\nu$$
$$\times D_{\mu\nu}(\vec{r},\vec{r}',E_A-E)\phi_A(\vec{r}') - \delta m \int d^3r\, \phi_A^\dagger(\vec{r})\beta\phi_A(\vec{r}) \tag{2}$$

and for the vacuum-polarization correction

$$\Delta E_A^{\text{VP}} = -i\alpha \int d^3r\, d^3r'\, \phi_A^\dagger(\vec{r})\alpha^\mu \phi_A(\vec{r}) D_{\mu\nu}(\vec{r},\vec{r}',E=0)$$
$$\times \int_{C_F} \frac{dE}{2\pi}\, \text{Tr}[\alpha^\nu G(\vec{r},\vec{r}',E)] \quad , \tag{3}$$

respectively. The corresponding Feynman diagrams are depicted in Fig. 1. The self energy of the electron arises due to the emission and absorption of a virtual photon by the bound electron. The vacuum-polarization correction can be viewed as additional interaction of the bound electron with virtual electron-positron pairs induced by the Coulomb potential of the nucleus which modify the external field.

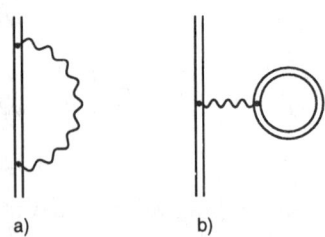

FIGURE 1. Feynman diagrams for the self energy (a) and the vacuum polarization (b) of a bound electron of order α. The double lines indicate wave functions and propagators in the external Coulomb potential of the nucleus.

For the evaluation of the self energy we employ *Mohr's* method [1–3], which is based on the expansion of the electron Green's function in terms of eigenfunctions of angular momentum. Since the unrenormalized self energy is infinite, a method of subtracting off the infinite mass renormalization term must be employed which is suitable for a numerical calculation as well. Details to the rather involved derivations of expressions can be found in [2,4]. The dependence of the self energy on the nuclear radius indicates that for heavy systems ($Z > 50$) also the influence of the finite nuclear size on the self-energy correction has to be taken into account [2].

The vacuum polarization correction needs to be calculated to all orders in $Z\alpha$ as well. It is usually performed by considering the Uehling- and Wichmann-Kroll contributions separately. The evaluation of this contribution for a high-Z finite-size nucleus avoids a perturbative calculation by making use of explicit solutions of the Dirac equation in a Coulomb potential together with the partial wave decomposition of the Green function. A detailed discussion of this type of higher-order vacuum polarization effects is presented in Ref. [9]. The calculations performed in [5,8,9] underline the importance of including finite-size effects in any state-of-the-art calculations of the vacuum polarization correction for high-Z systems. As in the case of the self energy its examination is essential before calculating radiative effects of higher order in α. However, since the vacuum polarization and the self energy correction carry opposite sign the influence of finite nuclear extension on both radiative effects nearly cancel. An intensive elaboration of this issue can be found in [8].

IV NUCLEAR POLARIZATION

Nuclear polarization represents a natural limitation for any high-precision tests of QED in heavy atoms. Due to the exchange of virtual photons the nucleus can undergo virtual electromagnetic transitions giving rise to additional contributions to the energy shifts of bound electrons. The field-theoretical treatment of nuclear polarization has been developed by *Plunien* et al. [10,11] introducing the concept of effective photon propagators with nuclear polarization insertions. It allows a systematic treatment of the influence of nuclear polarization to any atomic processes. In the case of one-electron systems the dominant lowest-order contributions of nuclear polarization involving one effective photon propagator (order α^2) appear as effective self energy (NPSE) and vacuum polarization (NPVP). Explicit expressions for the effective self-energy and vacuum-polarization term of electron bound-states ϕ_A are obtained from Eqs. (2) and (3) by the replacing the free photon propagator by the modified propagator. The NPSE contribution of nuclear polarization to the total Lamb shift carries an overall minus sign, i.e. thus tends to increase the binding energy of electronic bound states. The effective self-energy shift depends parametrically on the nuclear excitation energies, the corresponding reduced transition strengths and on the radial profile of the nuclear transition current densities associated with the collective nuclear multipole excitations under consideration.

The mixed NPVP correction has been evaluated in Uehling approximation [13]. It is dominated by a logarithmic dependence on the nuclear radius. This feature allows to distinguish the combined NPVP effect from ordinary finite size corrections. The corresponding energy shift carries opposite sign as the ordinary NPSE correction. In the lead system nuclear monopole excitations contribute and the sum of both nuclear polarization corrections almost cancel completely [13]. This may indicate that it is a most preferable system for tests of QED in strong fields.

V QED CORRECTIONS OF ORDER α^2

The recent experimental progress made in the measurements of the Lamb shift in highly charged few-electron ions may indicate that calculations of higher-order QED corrections become relevant. In this regard we point out that recent measurements of the $1s_{1/2}$-ground-state Lamb shift in hydrogen-like uranium have increased the precision from 520 ± 130 eV [14] to 429 ± 63 eV, which has been achieved by *Stöhlker* et al. [15], respectively, 470 ± 16 eV measured by *Beyer* et al. [16,17]. For lithium-like uranium the splitting between the $2s_{1/2}$ and the $2p_{1/2}$ levels was determined to be 280.59 ± 0.09 eV corresponding to a precision of 3×10^{-4} [18].

The Feynman diagrams of the various α^2-corrections for one-electron atoms are depicted in Fig. 2. At a first glance we may distinguish between diagrams involving closed fermion loops and those representing the two-photon self energy SESE. The first subset embraces the second-order vacuum polarization contributions VPVPa,b,c, the mixed self energy - vacuum polarization SEVPa,b,c and the effective self energy - vacuum polarization contribution S(VP)E. Progress has been made towards the evaluation of all second-order corrections given above (see e.g. [4] for a review and [19] for recent numerical results). While the calculation of the VPVP, SEVP and S(VP)E contributions are almost completed the two-photon self energy is far more elaborate to obtain and remains subject of current research. Instead of trying to evaluate the individual energy shifts as they stand most of the diagrams involving closed fermion loops are easier to compute employing the modified potential approach. The general idea behind is to generate first the complete spectrum of orbitals $\widetilde{\phi}_A$ and corresponding energy eigenvalues $\widetilde{E}_A$ by solving the Dirac eigenvalue problem (1) including the exact one-loop vacuum polarization

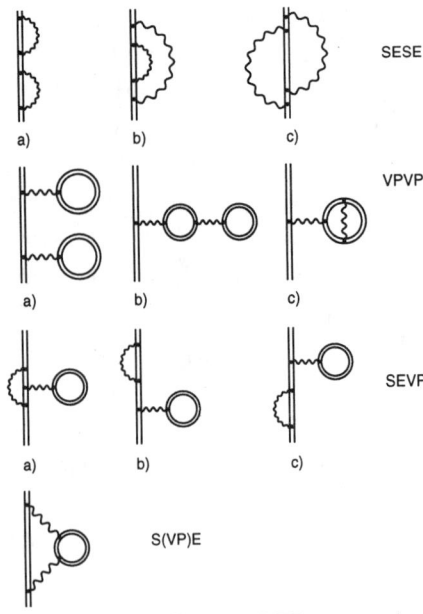

FIGURE 2. Feynman diagrams corresponding to QED corrections of order α^2 in hydrogen-like ions. Notations are the same as in Fig. 1.

potential $\mathcal{U}^{\text{VP}}$ induced by the external nuclear Coulomb potential. The resulting modified wave functions and shifted energy eigenvalues account for an infinite sequence of VP loops.

Until now the major theoretical uncertainty concerning the evaluation of the complete α^2-corrections arises from missing contributions of the two-photon self energy SESE. Renormalization schemes have been developed recently [24–27]. Although progress has been made in numerical evaluation [28] its final completion remains as challenge.

VI LAMB-SHIFT PREDICTIONS

So far we briefly discussed all relevant QED and nuclear corrections involved in current prediction of the Lamb shift in hydrogen-like heavy systems aiming at the limiting relative precision of 10^{-6}. A summary of all presently known contributions to the Lamb shift of K- and L-shell electrons in various one-electron ions has been given in Ref. [19]. In Table 1 the current status is exemplified for the Lamb shift of the $1s_{1/2}$-ground state in lead $^{208}_{82}\text{Pb}^{81+}$ and uranium $^{238}_{92}\text{U}^{91+}$. It also contains references to details of the calculation of the α^2-corrections. The numbers presented here are slightly different both due to new calculations for the two-loop vacuum polarization diagram [21] as well as improved numerical results for the nuclear polarization [12,13].

A rigorous describtion of nuclear recoil effects within the framework of QED has been developed by *Shabaev* et al. (see e.g. [29] and references therein). The updated

TABLE 1. One-electron Lamb-shift contributions (eV) for the $1s_{1/2}$-ground state in Lead and Uranium ions. The nuclear shape uncertainties are estimated to be 0.36 eV for Uranium and 0.14 eV for Lead.

	^{208}Pb^{81+}	^{238}U^{91+}	
Binding energy E_B (point nucleus)	−101581.37	−132279.96	
Correction:			Reference
Finite nuclear size	67.25	198.82	
Self energy (order α)	226.33	355.05	
VP: Uehling contribution	−50.70	−93.58	
VP: Wichmann-Kroll contribution	2.29	4.99	
Total vacuum polarization (order α)	−48.41	−88.60	
SESE a) b) c)	not complete		[26,28]
VPVP a) (ladder diagrams)	−0.09	−0.22	[5,20]
VPVP b) (Källén-Sabry cont. + h.o.)	−0.07	−0.15	[21]
VPVP c) (Källén-Sabry cont.)	−0.34	−0.60	[19,22]
SEVP a) b) c)	0.53	1.14	[6,7]
S(VP)E	0.07	0.13	[6,23]
Reduced mass	0.27	0.30	
Relativistic recoil	0.09	0.16	
Total recoil	0.36	0.46	[29,30]
Nuclear polarization	0.00	-0.19	[12,13]
Sum of corrections	245.63	465.84	
Resulting total binding energy	−101335.74	−131814.11	
Lamb Shift (theory)	245.36	465.54	

numbers for the total recoil and for the relativistic recoil respectively involve also corrections due to the finite nuclear size [30]. The predicted value for the Lamb shift of the $1s_{1/2}$-ground state in hydrogen-like Uranium amounts to 465.84(36) eV, which has to be compared with the experimental value of about 470 eV. The zero value presented for the nuclear polarization in lead arises due to a different aspect of nuclear polarization, where the nucleus also interacts with a virtual electron-positron pair [13]. For Lead it practically cancels that of the ordinary nuclear polarization [12] due to excisting collective monopole vibrations of this particular nucleus. Therefore the nuclear polarization effects which otherwise limit very precise Lamb shift predictions are almost completely neglible for $^{208}_{82}$Pb^{81+}, making this ion especially suitable for the most precise theoretical predictions. Apart from the missing contribution of the two-photon self energy, currently all QED corrections to the Dirac binding energy value are expected to be known to a precision of about

0.01 eV or better. As soon as the missing two-photon self-energy values become available the goal to provide Lamb-shift predictions for all of the systems under consideration with an accuracy of about 0.1 eV will be achieved.

ACKNOWLEDGEMENTS

Financial support from BMBF, DAAD, DFG and GSI is greatefully acknowledged.

REFERENCES

1. P. J. Mohr, Ann. Phys. (NY) **88**, 26 (1974)
2. P. Indelicato, P. J. Mohr, Phys. Rev. **A46**, 172 (1992)
3. P. J. Mohr, G. Soff, Phys. Rev. Lett. **70**, 158 (1993)
4. P. Mohr, G. Plunien, G. Soff, Phys. Rep. **293**, 227 (1998)
5. H. Persson, I. Lindgren, S. Salomonson, P. Sunnergren, Phys. Rev. **A48**, 2772 (1993)
6. H. Persson, I. Lindgren, L. N. Labzowsky, G. Plunien, T. Beier, G. Soff, Phys. Rev. **A54**, 2805 (1996)
7. I. Lindgren, H. Persson, S. Salomonson, V. Karasiev, L. Labzowsky, A. Mitrushenkov, M. Tokman, J. Phys. B **26**, L503 (1993)
8. T. Beier, P. J. Mohr, H. Persson, G. Soff, Phys. Rev. **A58**, 954 (1998)
9. G. Soff, P. Mohr, Phys. Rev. **A38**, 5066 (1988)
10. G. Plunien, B. Müller, W. Greiner and G. Soff, Phys. Rev. **A43**, 5853 (1991)
11. G. Plunien, G. Soff, Phys. Rev. **A51**, 1119 (1995);
    G. Plunien, G. Soff, Phys. Rev. **A53**, 4614 (1996)
12. A. V. Nefiodov, L. N. Labzowsky, G. Plunien, G. Soff, Phys. Lett. **A222**, 227 (1996)
13. L. N. Labzowsky, A. V. Nefiodov, G. Plunien, T. Beier, G. Soff, J. Phys. **B29**, 3841 (1996)
14. J. P. Briand, P. Chevallier, P. Indelicato, K. P. Ziock, D. D. Dietrich, Phys. Rev. Lett. **65**, 2761 (1990)
15. Th. Stöhlker, P. H. Mokler, K. Beckert, F. Bosch, H. Eickhoff, B. Franzke, M. Jung, T. Kandler, O. Kleppner, C. Kuzhuharov, R. Moshammer, F. Nolden, H. Reich, P. Rymuza, P. Spädtke, M. Steck, Phys. Rev. Lett. **71**, 2184 (1993)
16. H. F. Beyer, IEEE Trans. Instrum. Meas. **44**, 510 (1995)
17. H. F. Beyer, G. Menzel, D. Liesen, A. Gallus, F. Bosch, R. Deslattes, P. Indelicato, Th. Stöhlker, O. Klepper, R. Moshammer, F. Nolden, H. Eickhoff, B. Franzke, and M. Steck, Z. Phys. **D 35**, 169 (1995)
18. J. Schweppe, A. Belkacem, L. Blumenfield, N. Claytor, B. Feinberg, H. Gould, V. E. Koster, L. Levy, S. Misawa, J. R. Mowat, M H. Prior, Phys. Rev. Lett. **66**, 1434 (1991)
19. T. Beier, P. Mohr, H. Persson, G. Plunien, M. Greiner, G. Soff, Phys. Lett. **A 236**, 329 (1997)
20. T. Beier, G. Plunien, M. Greiner, G. Soff, J. Phys. **B28**, 2761 (1997)
21. G. Plunien, T. Beier, G. Soff, H. Persson, Eur. Phys. J. D **1**, 177 (1998)

22. S. M. Schneider, W. Greiner, G. Soff, J. Phys. **B26**, L529 (1993)
23. S. Mallampalli, J. Sapirstein, Phys. Rev. **A54**, 2714 (1996)
24. L. N. Labzowsky, A. O. Mitrushenkov, Phys. Lett. **A198**, 333 (1995)
25. A. O. Mitrushenkov, L. N. Labzowsky, I. Lindgren, H. Persson, S. Salomonson, Phys. Lett. **A200**, 51 (1995)
26. L. N. Labzowsky, A. O. Mitrushenkov, Phys. Rev. **A53**, 3029 (1996)
27. I. Lindgren, H. Persson, S. Salomonson, P. Sunnergren, Phys. Rev. **A58**, 1001 (1998)
28. S. Mallampalli, J. Sapirstein, Phys. Rev. **A57**, 1548 (1998)
29. V. M. Shabaev, Phys. Rev. **A57**, 59 (1998)
30. V. M. Shabaev, T. Beier, G. Plunien, V. A. Yerokhin, G. Soff, Phys. Rev. **A57**, 4235 (1998)

Numerical Lamb Shift Calculations for low-Z Systems

U. D. Jentschura[1,2] P. J. Mohr[1] and G. Soff[2]

[1]*National Institute of Standards and Technology (NIST), Gaithersburg, Maryland MD 20899-0001, USA*
[2]*Institut für Theoretische Physik, TU Dresden, Mommsenstraße 13, 01062 Dresden, Germany*

Abstract. For bound systems with a small atomic number Z, numerical evaluations of self-energy corrections, which are non-perturbative in the binding field, entail severe numerical cancellations at intermediate stages of the calculation. This paper reports on a result for the non-perturbative self-energy remainder function $G_{\rm SE}$ in atomic hydrogen with a relative accuracy of 10^{-5}. We discuss consistency checks on the results of numerical Lamb shift calculations in systems with a small atomic number. The precise determination of radiative corrections in low-Z bound systems is of crucial importance for the interpretation of precision measurements in atoms, for tests of quantum electrodynamics and for the determination of fundamental constants.

PACS numbers: 12.20.Ds, 31.30.Jv, 06.20.Jr. **Keywords**: Quantum electrodynamics, bound state calculations, quantum electrodynamic effects in atoms and molecules, atomic hydrogen, Lamb shift, determination of fundamental constants.

PROPERTIES OF NUMERICAL LAMB-SHIFT CALCULATIONS

Consistency checks are important for numerical evaluations of self-energy corrections in atomic systems. This applies in particular to low-Z atoms where the theoretical data is intended to be used for the interpretation of precision measurements and for the determination of fundamental constants. Traditionally, theoretical Lamb shift data was obtained for low-Z systems by analytic calculations (see e.g. [1–3] and the comprehensive list of references in [4]). The problem arises that in general, numerical calculations cannot be checked as rigorously as analytic evaluations.

A numerical calculation defies the consistency checks which were applied to analytic evaluations of higher-order binding corrections to the Lamb shift [5,6]. In the analytic calculations, an auxiliary parameter was introduced which is related to a cut-off in the photon energy. Cancellation of this auxiliary parameter constitutes an important consistency check.

Other checks of analytic calculations include the verification of certain limits of matrix elements of the atomic state as a function of the energy of the virtual photon. The limit of vanishing photon energy $\omega \to 0$ can usually be evaluated independently from a complete evaluation of the matrix element for arbitrary photon energy. The two independently obtained results can be compared for consistency in the limit $\omega \to 0$. Note that the complete result for the matrix elements, at the level of precision reached in current analytic calculations, may comprise rather complex mathematical entities (cf. Eq. (79) in [5]). This mathematical complexity suggests that additional consistency checks on the result should be performed.

The question arises if similar consistency checks can be devised for numerical calculations. It should be noted that it is possible to split up numerical calculations of self-energy effects into distinct individual contributions. This separation can be done with regard to the energy of the virtual photon or with regard to other integration variables (see e.g. [7]). For each one of the individual contributions, the low-Z limit can be calculated analytically. The numerical data can be compared to the independently evaluated low-Z limit for consistency.

Another important consistency check, which should be performed for *any* numerical calculation, is variation of the input parameters. For self-energy calculations, the only input parameter (except for the atomic state quantum numbers) is the coupling of the electron to the central binding field $\gamma = Z\alpha$, where Z is the atomic number and α is the fine structure constant.

A final check for the consistency of numerical data is the comparison of the numerical results to known lower-order analytic coefficients. If the analytic results are correct and the procedure for estimating unknown higher-order terms in the expansion is consistent, then the main dependence on the input parameter is expected to be absorbed in the analytically known lower-order coefficients. This leads to a non-perturbative remainder function $G_{\rm SE}(\gamma)$, which is evaluated numerically. The remainder function is expected to vary smoothly with γ.

We therefore identify as important consistency checks for numerical evaluations of radiative corrections in low-Z electronic systems,

- a variation of the input parameter, which is the coupling $\gamma = Z\alpha$,

- a comparison of numerical results for distinct individual contributions to their independently evaluated low-Z limit, and

- a comparison of the final numerical all-order result to the contribution of lower-order coefficients known from analytic calculations.

In our numerical evaluation of the one-loop self-energy correction in hydrogen, several consistency checks are performed. In order to address the variation of the input parameter, we repeat the evaluation of the correction with the following values of the fine structure constant,

$$\alpha_< = 1/137.036\,000\,5\,, \qquad (1)$$
$$\alpha_0 = 1/137.036\,000\,0 \quad \text{and} \qquad (2)$$
$$\alpha_> = 1/136.035\,999\,5\,. \qquad (3)$$

This set of values of α is chosen for consistency with current bounds on the value of the fine structure constant as obtained by Mohr and Taylor [8].

CURRENT STATUS OF THE EVALUATION

Evaluations of the self energy of the $1S_{1/2}$-ground state are currently being performed for hydrogenlike ions with the atomic numbers $Z = 1, 2, 3, 4, 5, 10$ and for each of the values $\alpha_<$, α_0 and $\alpha_>$ of the fine structure constant. We plan to obtain numerical values for the scaled self-energy function $F(\gamma)$, which is defined and discussed for example in [4]. The current result for the dimensionless scaled self-energy function of the ground state of atomic hydrogen is

$$F(1 \times \alpha_0) = 10.316\,793\,65(1)\,. \qquad (4)$$

The relative uncertainty of this result is 10^{-9}. This result, which has been obtained by assuming a value $\alpha_0 = 1/137.036$ for the fine structure constant, is expected to vary at the level of 10^{-9}, i.e. at the level of the current precision, with a variation of the fine structure constant in the interval $(\alpha_<, \alpha_>)$. However, the result for the self-energy remainder function $G_{\rm SE}(\alpha)$ for atomic hydrogen presented below in Eq. (5) is expected to remain *stable* with a variation of the fine structure constant in the interval $(\alpha_<, \alpha_>)$ at the current level of precision (because the main dependence on the input parameter is absorbed in known lower-order coefficients). First results seem to be consistent with this fact and will be described in more detail in [9]. The variation of α is an important check for the numerical stability of the evaluation. It is our ultimate objective to obtain numerical values of the dimensionless self-energy function F with a relative accuracy of the order of 10^{-10}, which in frequency units corresponds to an uncertainty at the 1 Hz-level.

Comparisons of the numerical data to analytically evaluated low-Z limits of distinct individual contributions to F have been performed and lead to consistent results. These consistency checks will be described in [9].

In relation to the one-loop self energy, the lower-order coefficients A_{41}, A_{40}, A_{50}, A_{62} and A_{61} are known from analytical calculations of Bethe [1], Baranger, Bethe and Feynman [2] and Erickson and Yennie [3]. These coefficients are defined and their numerical values are listed e.g. in [4]. A subtraction of the lower-order coefficients leads to the result

$$G_{\rm SE}(1 \times \alpha_0) = -30.2902(2) \qquad (5)$$

for the non-perturbative self-energy remainder function $G_{\rm SE}$ (with a relative accuracy of 10^{-5}). This result is listed in Table 1, together with values for the remainder function $G_{\rm SE}$ for $Z = 5$ and $Z = 10$ obtained from numerical data presented in [10]. Note that for $Z = 5$ and $Z = 10$, more significant digits are known than presented in Table 1. The relevant data for $Z = 5$ and $Z = 10$ can be found in [10], a recent more precise evaluation for $Z = 5$ and $Z = 10$, which confirms the results in [10], will be presented in [9]. Here it is our objective to show the apparent monotonic dependence of the remainder function $G_{\rm SE}(\gamma)$ on its argument γ.

TABLE 1. Values of the remainder functions $G_{\rm SE}$ for hydrogen-like ions with various atomic numbers Z, obtained with a value of $\alpha_0 = 1/137.036$ for the fine structure constant. For $Z = 5$ and $Z = 10$, not all known significant digits are shown (see text).

Atomic number	$G_{\rm SE}$	Ref.
0	$-30.92890(1)$	[6]
1	$-30.2902(2)$	this work
2	calculation in progress	
3	calculation in progress	
4	calculation in progress	
5	$-28.4435\ldots$	[10]
10	$-26.6041\ldots$	[10]

We note that an error in the evaluation of any of the subtracted analytically known lower-order coefficients A_{41}, A_{40}, A_{50}, A_{62} and A_{61} would be likely to manifest itself in a grossly inconsistent dependence of $G_{\rm SE}(\gamma)$. However, such a behaviour is not observed. A detailed comparison to the result of Pachucki for $G_{\rm SE}(0) = A_{60}$, which is listed in the first row of Table 1, and to the result of Karshenboim for A_{71} [11] will require the availability of numerical data for $Z = 2, 3, 4$. Work in this direction is currently in progress.

The current and projected level of precision of the high-precision spectroscopic experiments in low-Z systems requires non-perturbative treatments of the self-energy. The difference between the perturbative result $G_{\rm SE}(0) = A_{60} = -30.92890(1)$ given in [6] and the non-perturbative result $G_{\rm SE}(\alpha) = -30.2902(2)$ reported in this work is 28 kHz for atomic hydrogen.

ACKNOWLEDGEMENTS

U. D. J. gratefully acknowledges helpful conversations with J. Baker, J. Devaney and J. Sims and support by DAAD and DFG (contract no. SO333/1-2). P. J. M. acknowledges continued support by the Alexander v. Humboldt foundation, and G. S. acknowledges continued support by DFG and GSI.

REFERENCES

1. H. A. Bethe, Phys. Rev. **72**, 339 (1947).
2. M. Baranger, H. A. Bethe, and R. P. Feynman, Phys. Rev. **92**, 482 (1953).
3. G. W. Erickson and D. R. Yennie, Ann. Phys. (N.Y) **35**, 271, 447 (1965). I
4. P. J. Mohr in *Atomic, Molecular and Optical Physics Handbook*, ed. by G. W. F. Drake, p. 341 (AIP, Woodbury (N.Y.), 1996). I
5. U. D. Jentschura, G. Soff, and P. J. Mohr, Phys. Rev. A **56**, 1739 (1997).
6. K. Pachucki, Ann. Phys. (NY) **226**, 1 (1993).
7. P. J. Mohr, Ann. Phys. (NY) **88**, 26, 52 (1974).
8. P. J. Mohr and B. N. Taylor, *private communication*.
9. U. Jentschura, P. J. Mohr and G. Soff, *manuscript in preparation*.
10. P. J. Mohr, Phys. Rev. A **46**, 4421 (1992).
11. S. Karshenboim, Z. Phys. D **39**, 109 (1997).

The g-Factor of Hydrogen-like Ions

M.Diederich*, H.Häffner†, N.Hermanspahn*, M.Immel*, H.J.Kluge†, R.Ley*, R.Mann†,
S.Stahl*[1], W.Quint†, J.Verdú*, G.Werth*

Johannes Gutenberg Universität, 55099 Mainz, Germany
† *Gesellschaft für Schwerionenforschung, 64291 Darmstadt, Germany*

Abstract. We report on the first direct measurement of the g-factor of a highly charged ion. The experimental determination of the magnetic moment (g-factor) of the bound electron in hydrogen-like ions is an important test of the theory of Quantum Electrodynamics in strong nuclear Coulomb fields. For this purpose a single hydrogen-like ion is stored in the magnetic field of a Penning trap. The g-factor is measured by inducing spin flip transitions with a microwave field at the Larmor precession frequency of the bound electron. The magnetic field is calibrated by measuring the cyclotron frequency of the stored ion. The first results were obtained for a hydrogen-like carbon ion (C^{5+}). The experimental precision is high enough to verify the relativistic contribution to the g-factor on the 10^{-3} level.

INTRODUCTION

The development of modern Quantum Electrodynamics (QED) was initiated by experiments on the fine-structure (Lamb-shift) [1] and hyperfine-structure [2] of atomic hydrogen. Experiments, like the measurement of the g-factor of the free electron [3] to an accuracy of a few parts in 10^{-12}, and their agreement with theoretical calculations have clearly demonstrated the success of QED. So far differences between experiment and theoretical prediction have been resolved. Presently limitations of theoretical calculations are due to uncertainties in the used parameters. In the case of atomic hydrogen nuclear size and structure, in the case of the g-factor of the free electron our knowledge of the fine structure constant α limit the comparison between theory and experiment. In weak external or nuclear fields, QED can (at present) be regarded as a self-contained theory.

Strong electric fields, of up to 10^{18}V/m (for U^{91+}), exist in highly charged hydrogen-like ions. Presently, a number of experiments on Lamb-shift, hyperfine-structure and electronic g-factors are carried out to test QED in highly charged hydrogen-like systems [4–7]. Theoretical calculations for highly charged ions are finally limited because of uncertainties in nuclear parameters as in the case of atomic hydrogen. The g-factor of the bound electron seems to be a good system for testing bound-state QED (BS-QED), because nuclear parameters influence the g-factor only through changes in the electron propagator. We report on the measurement of the g-factor of the bound electron for Z=6 (carbon). In the future we plan to extend the measurements up to Z=92 (uranium).

THEORETICAL ASPECTS

The g-factor is defined as a dimensionless constant connecting the magnetic moment μ (measured in units of the Bohr magneton μ_B) to the spin s (measured in units of Planck's constant $\hbar$):

$$\frac{\mu}{\mu_B} = g\frac{s}{\hbar}. \tag{1}$$

According to the Dirac theory a free particle at rest should have a g-factor of exactly 2. The g-factor of the bound electron can be expressed as (see fig. 1)

$$g = 2 + \Delta g_{Breit} + \Delta g_{free} + \Delta g_{BS-QED} + \Delta g_{nuc}. \tag{2}$$

---

[1] Present address: Inst. of Physics and Astronomy, University of Aarhus, Denmark

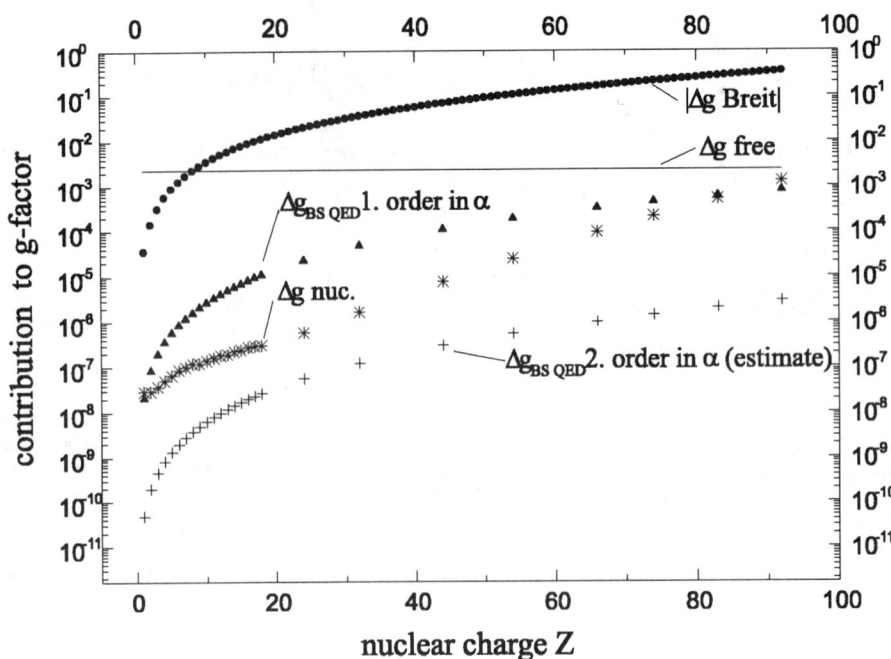

FIGURE 1. Theoretical contributions to the g-factor of the bound electron. $g = 2 - |\Delta g_{Breit}| + \Delta g_{free} + \Delta g_{BS-QED} + \Delta g_{nuc.}$. The first-order radiative corrections for the bound state Δg_{BS-QED} include self-energy and vacuum-polarisation contributions. The plotted second-order corrections are only an estimation.

- The **Dirac** theory predicts the g-factor in the bound system to be $2 + \Delta g_{Breit}$ [8]. This contribution can be expressed analytically for a point-like nucleus and was calculated by Breit in 1928 to be

$$\Delta g_{Breit} = -\frac{4}{3}\left(1 - \sqrt{1 - (Z\alpha)^2}\right). \tag{3}$$

The Breit term is due to a relativistic effect decreasing the magnetic moment. For most hydrogen-like ions it constitutes the biggest correction to the g-factor of the bound electron. In the case of hydrogen-like uranium it reduces the g-factor by 15%.

- Theoretical evaluations of the **QED** contributions to the g-factor of the free electron (Δg_{free}) are extensive with the calculation up to the eighth order terms. For the bound electron additional QED contributions due to the presence of the nuclear Coulomb field have to be taken into account. These bound-state QED corrections were first calculated by Grotch and Hegstrom [9] in a $Z\alpha$ expansion.

$$\Delta g_{QED} = \frac{\alpha}{\pi}\left(1 + \frac{(Z\alpha)^2}{6} + \ldots\right) = \Delta g_{free} + \Delta g_{BS-QED}. \tag{4}$$

Recently, the bound-state self-energy corrections to first order in α have been evaluated numerically by two groups in a non-perturbative calculation (all orders in $Z\alpha$) [10,11]. Their results are in good agreement. The Gothenburg group additionally calculated the first-order vacuum polarisation contribution [11].

- The g-factor of the bound electron is an interesting system for QED tests because the influence of **nuclear size and nuclear structure** ($\Delta g_{nuc.}$), even for heavy systems, is small enough to ensure a test of the radiative corrections with a relative precision of 10^{-4}. The theoretical uncertainty in the g-factor of hydrogen-like uranium, due to nuclear size and recoil, is presently in the 10^{-7} range.

The g-factor of an atomic state is determined by measuring the Zeeman splitting in a known magnetic field.

$$\Delta E = \hbar\omega_L = g\mu_B B. \tag{5}$$

The magnetic field can be measured via the cyclotron frequency

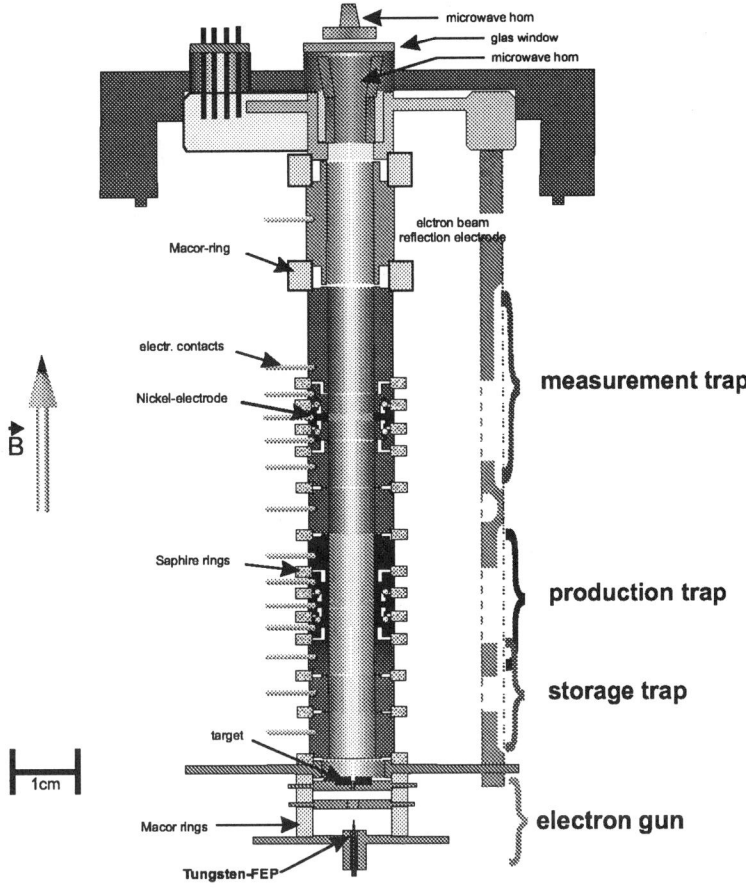

FIGURE 2. Ions can be stored at three different locations in the electrode assembly.

$$\omega_c = \frac{q}{m} B \qquad (6)$$

of the stored ion. The g-factor can be calculated using these two measured frequencies

$$g = 2 \frac{Q/M}{q/m} \frac{\omega_L}{\omega_c} \qquad (7)$$

and the ratio of the charge-to-mass ratios of electron (q/m) and ion (Q/M). In our experiment we plan to measure the g-factor of the bound electron for a number of different nuclear charge numbers to validate the theoretical predictions and finally to test QED in strong electric fields. In our first measurement we determined the g-factor of hydrogen-like carbon (C^{5+})[2].

EXPERIMENTAL SET-UP

Highly charged ions are to be confined in a small spatial volume (an ion trap) and stored for long periods of time (several days or longer). The spin direction of a single ion and its motional frequencies are to be measured with high precision.

Ion Storage

Confinement of ions at low energies is achieved in ion traps. We store a single hydrogen-like ion in the electric and magnetic fields of a **Penning trap**. An ideal Penning trap consists of a homogeneous magnetic field and an

---

[2] The charge-to-mass ratios of electron and carbon are known with sufficient accuracy [13,14].

electrostatic quadrupole field. The ion's trajectory in the trap is composed of three independent oscillatory motions:

1. The axial oscillation along the magnetic field lines has the frequency

$$\omega_z = \sqrt{C_2 \frac{qU}{md^2}}, \qquad (8)$$

where U is the trapping voltage, q and m the charge and mass of the ion, d the characteristic trap dimension and C_2 the relative strength of the electric quadrupole field ($C_2 = 1$ for hyperbolic trap electrodes).

2. The cyclotron motion in the magnetic field is slightly modified due to the presence of the electric field and orbits at the reduced cyclotron frequency

$$\omega_+ = \frac{\omega_c}{2} + \sqrt{\left(\frac{\omega_c}{2}\right)^2 - \frac{\omega_z^2}{2}}. \qquad (9)$$

3. The electric field pulls the ion constantly away from the trap axis resulting in the magnetron motion with frequency

$$\omega_- = \frac{\omega_c}{2} - \sqrt{\left(\frac{\omega_c}{2}\right)^2 - \frac{\omega_z^2}{2}}. \qquad (10)$$

To measure the magnetic field strength one needs to determine the free cyclotron frequency (see equ.6) which is given by the trap frequencies through

$$\omega_c = \sqrt{\omega_+^2 + \omega_z^2 + \omega_-^2}. \qquad (11)$$

The magnetic field of the Penning trap is produced by a commercial superconducting magnet operated at a field strength of B = 3.8 T. Situated in the center of the magnetic field is the electrode structure generating the electrostatic quadrupole field. The electrode assembly consists of 13 cylindrical elements (see fig. 2). This way three Penning traps are stacked on top of each other and microwaves can enter the trap region through the open end. Two of these traps are compensated traps with five electrodes [12]. The ions are created by electron impact ionisation in the "production trap" in the homogeneous part of the magnetic field. Electrons are emitted from a field emission point and accelerated to an energy of approx. 1 keV. The electron beam is reflected into itself after traversing the ion trap structure. After multiple reflections the beam finally hits the anode which is covered with a carbon coating acting as a target material. Apart from carbon also oxygen ions are created (see fig. 3) resulting from residual oxygen atoms in the target.

After production contaminant ions are removed and the ion number is reduced to one [15]. The single hydrogen-like ion is then transported to the "measurement trap". Ion transport is performed by moving the potential minimum adiabatically along the electrode structure taking the ion(s) along with it. A single ion was transported more than 1000 times back and forth between the two traps without any effect on the ion's motional amplitudes. To perform spin flip measurements on single hydrogen-like ions it is necessary to couple the spin magnetic moment to the ion's motional frequency. For this purpose the "measurement trap" is fitted with a ferromagnetic ring to form a quadratic magnetic field component (a so-called magnetic bottle).

Storage Time

The loss mechanism for stored highly charged ions is charge exchange collisions with background gas atoms or molecules. The loss rate is therefore proportional to the background gas pressure. To reduce the gas pressure the trap structure is located inside a completely sealed copper vessel. This enclosure is evacuated and cooled down to a temperature of T = 4 K. Cryopumping reduces the pressure below the range of vacuum gauges. So far no ion loss was observed in our trap, setting a lower limit for the storage time of one year and an upper limit for the vacuum pressure of 10^{-16} mbar. Cooling is achieved by thermally coupling the vacuum chamber (as well as front end electronics) to a liquid helium reservoir. The cryostat is inserted into the room-temperature bore of the magnet.

A background gas pressure of 10^{-16} mbar He results in a storage time of 20 days for U^{91+}. This is sufficient for performing high precision measurements.

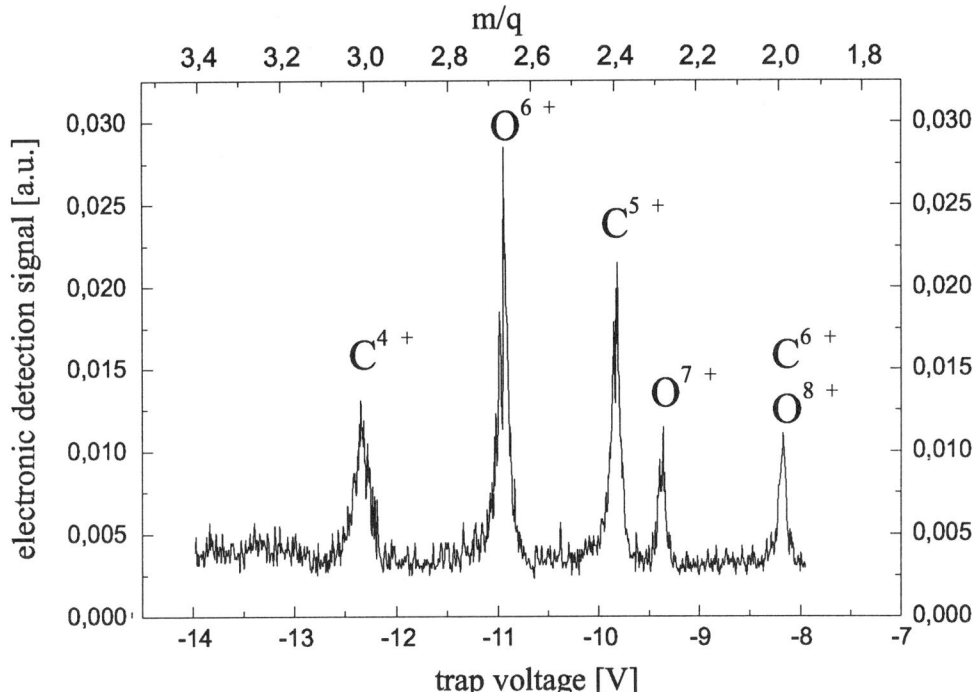

FIGURE 3. Mass spectrum of an ion cloud in the production trap. Only higher charge states are shown.

Frequency Measurements

Resonant circuits are connected to the trap electrodes to cool and detect the ions. Axial and cyclotron motion are cooled to a temperature of 4 K with the method of resistive cooling with cooling time constants of typically 130 ms and 30 s respectively. The magnetron motion is centered using a sideband drive at $\nu_z + \nu_-$. The ions are detected by performing a Fourier transformation of the induced image current flowing over a resonant circuit. The motion of the ions is excited with a drive applied to one of the electrodes to record mass spectra (fig. 3) or to measure the cyclotron frequency (see fig. 4). With this method it is also possible to detect a single highly charged ion while the ion is in thermal equilibrium with the resonant circuit (see fig. 5). In this case a characteristic deformation of the resonance shape of the external resonant circuit is observed (instead of the signal peak as in fig. 4). At the ion's oscillation frequency (here axial frequency) the ion acts as a shunt decreasing the thermal noise of the attached circuit.

EXPERIMENTAL RESULTS

Detection of Spin Direction

To detect the spin orientation of a single hydrogen-like ion we use Dehmelt's method utilising a magnetic bottle [3]. The magnetic bottle is produced by a ring of ferromagnetic material positioned at the center of the ring electrode of the ion trap. Along the axis of symmetry the magnetic field can be well described by a parabola superimposed to the homogeneous component of the magnetic field. The position-dependent potential energy $V_{mag} = \mu \cdot B_2 z^2$ of the magnetic moment of the bound electron adds a small binding or anti-binding (depending on the spin orientation) harmonic force to the axial electrostatic binding. This leads to a coupling between the magnetic moment and the axial motion. The axial frequency in the inhomogeneous magnetic field thus depends on the spin direction. The change in the axial frequency ($\nu_z = 360$ kHz) due to a spin flip is $\Delta\nu_z = 0.7 Hz$ for C^{5+} with the present set-up. This change in frequency can be easily detected as can be seen in fig. 6.

To drive the spin-flip transition a microwave field at 105 GHz is coupled into the trap. The microwaves are generated by multiplication ($\times 6$) of the output of a microwave synthesizer at 17.5 GHz. Application of microwaves

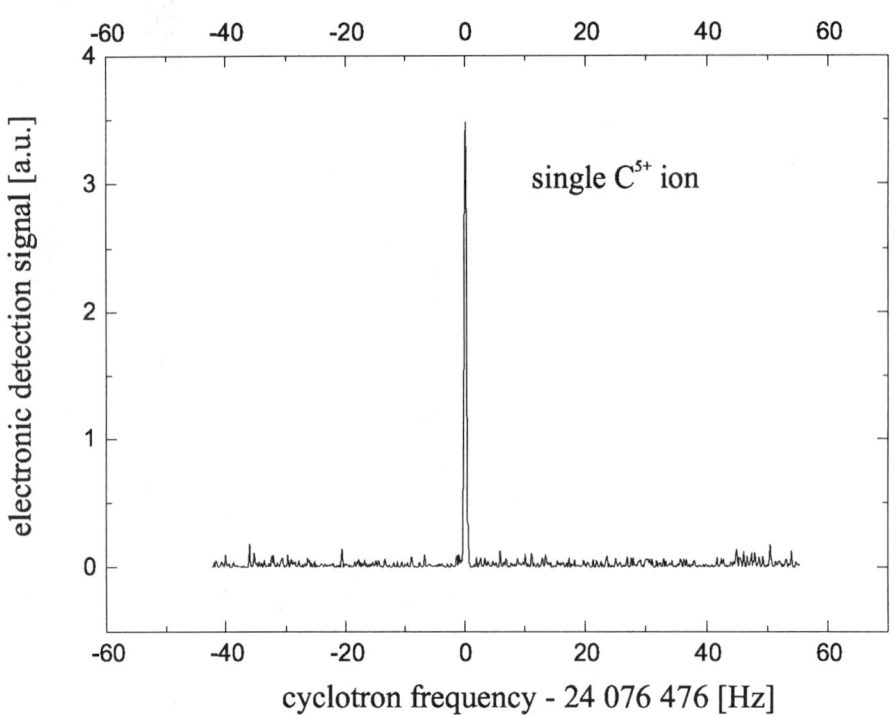

FIGURE 4. Fourier transform of current induced by cyclotron motion of single ion.

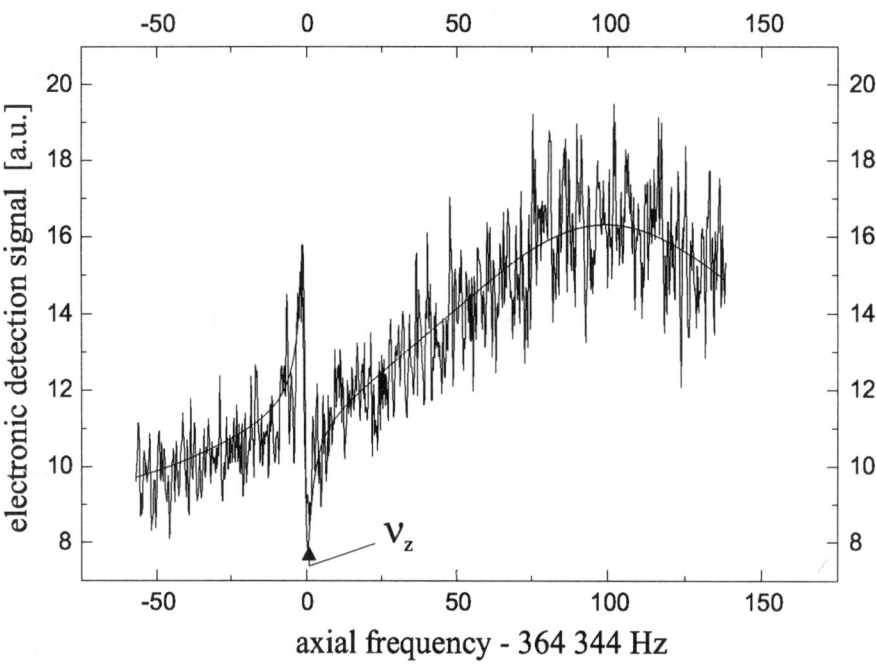

FIGURE 5. Measurement of the axial frequency (ν_z) of a single hydrogen-like carbon ion - named Carola. The broad feature is the thermal noise of the tuned circuit.

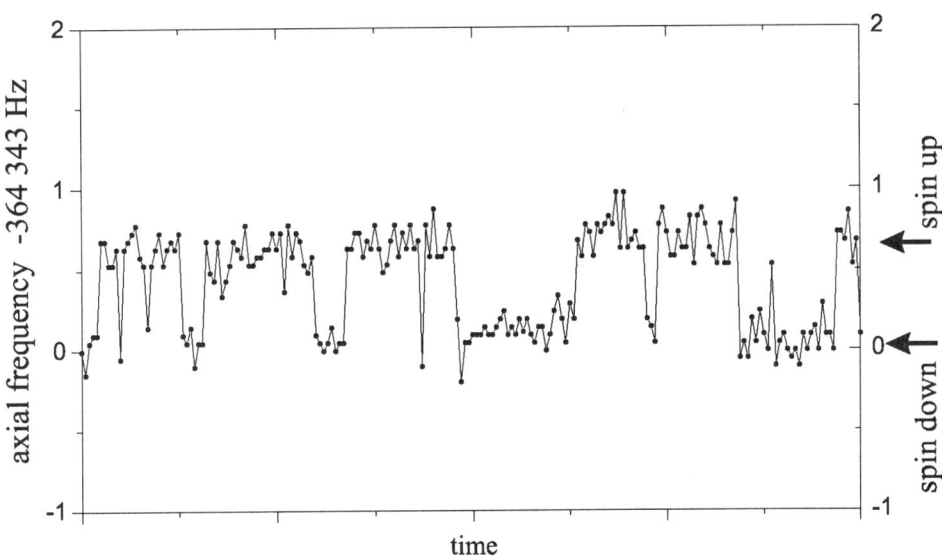

FIGURE 6. Statistical change of spin direction of ion *Carola* with application of microwaves at the Larmor precession frequency. The frequency noise on the two quantum states is mainly produced by the voltage noise of our voltage supply.

at the Larmor precession frequency leads to statistical quantum jumps which are observed as transitions between the two axial frequency levels (see fig. 6).

Resulting Lineshape

The magnetic field in the Penning trap with the magnetic bottle can be described on the z-axis as $B(z) = B_0 + B_2 z^2$. The ion experiences a modulated magnetic field because of its axial motion

$$B(t) = B_0 + B_2 z_0^2 \sin^2(\omega_z t) = B_0 + 1/2\, B_2 z_0^2 \left(1 - \cos(2\omega_z t)\right). \tag{12}$$

Here z_0 denotes the amplitude of the axial oscillation. Both Larmor frequency as well as cyclotron frequency are proportional to the magnetic field strength. Therefore they are shifted as well as frequency modulated. The value of the frequency shift and modulation depth, $\Delta\omega = \omega_0 \frac{B_2}{2B_0} z_0^2$, depends linearly on the axial motional energy. The axial energy is not fixed but fluctuates because of the coupling to the resonant circuit, which acts as a thermal reservoir. Both frequencies are therefore broadened with a Boltzmann distribution. For the reduced cyclotron frequency, the frequency modulation depth is small, resulting in sidebands that are well suppressed. For the Larmor frequency the sidebands have to be taken into account to explain the resulting lineshape. The relative amplitude of the carrier (n=0) and the sidebands (of order n) is given by the Bessel functions, $J_n\left(\frac{\Delta\omega_L}{2\omega_z}\right)$. The relative amplitude depends on the axial energy through $\Delta\omega_L$ and is therefore connected to the frequency shift of the carrier (Boltzmann shape).

The g-Factor of C^{5+}

The Larmor resonance line is obtained by determining the spin flip rate at a given microwave frequency and subsequent variation of this frequency. The resulting lineshape agrees well with the theoretical fit (see fig. 7). The resonance frequency can be determined with a fractional precision of $2 \cdot 10^{-6}$. The magnetic field strength is measured with the same ion. The reduced cyclotron motion is excited with a rf-signal. In resonance the cyclotron orbit will increase thus changing the orbital magnetic moment of the ion. This causes an increase of the axial frequency (see fig. 8) because of the presence of the magnetic bottle. It is important to note that the cyclotron motion in this measurement is excited to an energy of only 5 meV. The resulting change in motional radius does not change the cyclotron frequency significantly. The free cyclotron frequency was calculated using equ. 11. The axial frequency is measured directly during this measurement (see fig. 8).

Inserting the values of the measured Larmor and cyclotron frequency into equ. 7, yields

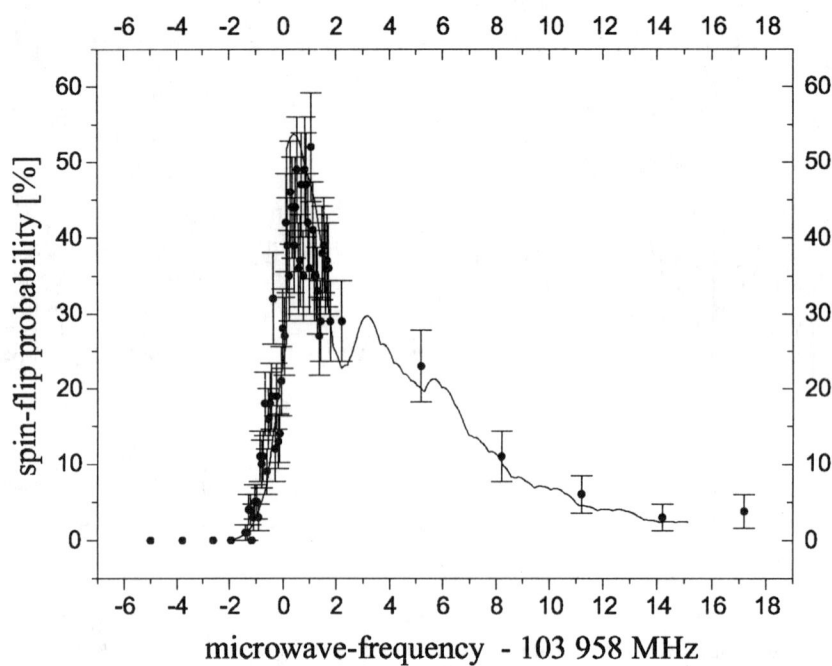

FIGURE 7. Larmor resonance of a single hydrogen-like carbon ion in the inhomogeneous magnetic field. The theoretical lineshape (solid line) results from the convolution of a Boltzmann distribution with the frequency dependence of the relative amplitudes of the carrier and sidebands due to motional frequency modulation.

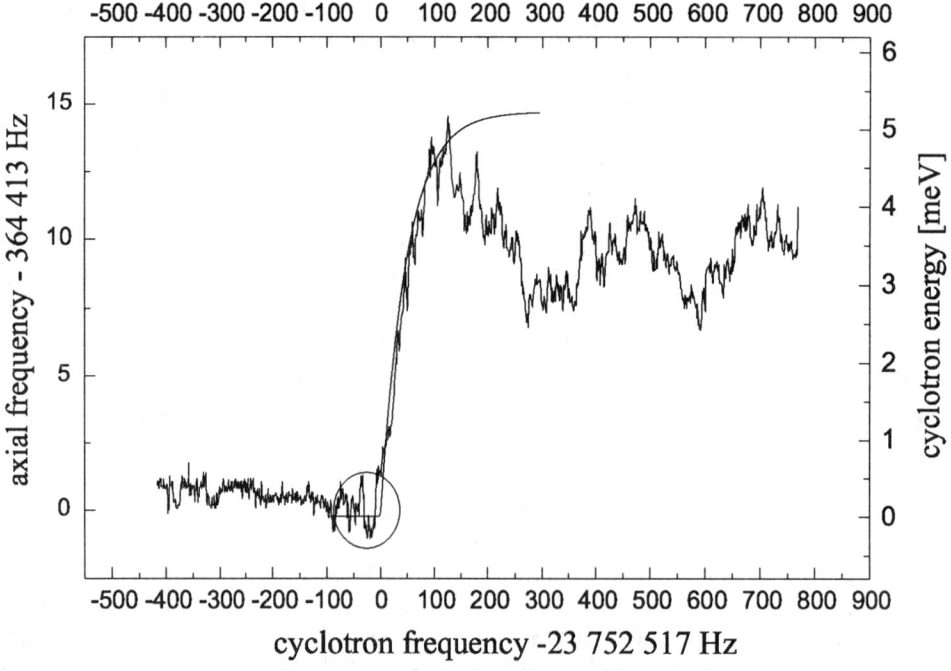

FIGURE 8. Excitation of the cyclotron motion results in an increase of the axial frequency due to the coupling of the cyclotron energy with the magnetic bottle. The second solid line shows the expected shape of the resonance.

$$g(exp.) = 2.001\,040\,(6) \tag{13}$$

which is in good agreement with the theoretical prediction [11]

$$g(theo.) = 2.001\,041\,59. \tag{14}$$

These measurements constitute the first direct determination of the g-factor of the bound electron for a highly charged ion.

SUMMARY AND OUTLOOK

First measurements of the g-factor of the bound electron in hydrogen-like carbon (C^{5+}) have been performed. The method is applicable to any nuclear charge number. The open trap structure makes loading of ions from external ion sources possible.

Presently, the precision of the frequency measurements is limited by the ion motion in the inhomogeneous magnetic field at thermal energies (T = 4 K), resulting in line shifts and line broadening. Improvement can be achieved by performing the measurement in a homogeneous magnetic field. Therefore it is planned to spatially separate the functions of inducing and detecting the spin-flip transition. Irradiation with microwaves and measurement of the cyclotron frequency will be performed in the trap located in the homogeneous part of the magnetic field. The ion is to be transported to the magnetic bottle trap to detect the spin direction. With this method an improvement in accuracy of one or two orders of magnitude should be achievable. With improved accuracy it will be possible, even for light ions, to compare recent numerical calculations with the old calculations using a $Z\alpha$ expansion. For Ne^{9+} one could compare the calculations performed by Blundell et al. [10] with those performed by Persson et al. [11]. Additional measurements with light ions could be performed with oxygen (O^{7+}) and magnesium (Mg^{11+}). Later measurements on medium-heavy ions like Sn^{49+} could be performed to verify the Z-dependence of the g-factor before changing to heavier ions. Measurements on very heavy ions (like U^{91+}) will pose a stringent test of QED in strong electric fields.

REFERENCES

1. Lamb W.E. and Retherford R.C., *Phys.Rev.* **72**, 971, 1947
2. Nafe J.E. et al., *Phys.Rev.* **71**, 914, 1947
3. Dehmelt H.G., *Rev.Mod.Phys.* **62**, 525, 1990
4. Beyer H.F. et al., *Z.Phys.D* **35**, 169, 1995
5. Crespo J.R. et al., *Phys.Rev.Lett.* **77**, 826, 1996
6. Quint W., *Physica Scripta* **T59**, 203, 1995
7. Hermanspahn K. et al., *Acta Phys. Pol.* **B27**, 357, 1996
8. Breit G., *Nature*, **122**, 649, 1928
9. Grotch H. and Hegstrom R.A., *Phys.Rev.* **A4**, 59, 1971
10. Blundell S.A.,et al., *Phys. Rev.* **A55**, 1857, 1997
11. Persson H. et al., *Phys.Rev.* **A56**, R2499, 1997
12. Gabrielse G. et al., *Int.Jour. of Mass Spect. and Ion Proc.* **88**, 319, 1989
13. Farnham D.L. et al., *Phys.Rev.Lett.* **75**, 3598, 1995
14. Van Dyck R.S. et al., *Physica Scripta* **T59**, 134, 1995
15. Diederich M. et al., *Hyperfine Int. (accepted)*

First Results from the New Muon (g-2) Experiment

A. Grossmann[e], H.N. Brown[b], G. Bunce[b], R.M. Carey[a], P. Cushman[h], G.T. Danby[b],
P.T. Debevec[g], H. Deng[l], W. Deninger[g], S.K. Dhawan[l], V.P. Druzhinin[i], L. Duong[h],
W. Earle[a], E. Efstathiadis[a], F.J.M. Farley[l], G.V. Fedotovich[i], S. Giron[h], F. Gray[g], M. Grosse
Perdekamp[l], U. Haeberlen[f], M. Hare[a], E.S. Hazen[a], D.W. Hertzog[g], V.W. Hughes[l],
M. Iwasaki[k], K. Jungmann[e], D. Kawall[l], M. Kawamura[k], B.I. Khazin[i], J. Kindem[h],
F. Krienen[a], I. Kronkvist[h], R. Larsen[b], Y.Y. Lee[b], W. Liu[l], I. Logashenko[i], R. McNabb[h],
W. Meng[b], J.-L. Mi[b], D. Miller[h], J.P. Miller[a], W.M. Morse[b], P. Neumayer[e], G. Onderwater[g],
Y. Orlov[c], C. Pai[b], C. Polly[g], J. Pretz[l], R. Prigl[b], G. zu Putlitz[e], S.I. Redin[l], O. Rind[a],
B.L. Roberts[a], N. Ryskulov[i], R. Sanders[b], S. Sedykh[g], Y.K. Semertzidis[b], S. Serednyakov[i],
Yu.M. Shatunov[i], E. Solodov[i], M. Sossong[g], A. Steinmetz[l], L.R. Sulak[a], M. Tanaka[b],
C. Timmermans[h], A. Trofimov[a], D. Urner[g], P.V. Walter[e], D. Warburton[b], D. Winn[d], Q. Xu[b],
A. Yamamoto[j], D. Zimmerman[h].

[a]*Department of Physics, Boston University, Boston, MA 02215, USA,* [b]*Brookhaven National Laboratory, Upton, NY 11973, USA,* [c]*Newman Laboratory, Cornell University, Ithaca NY 14853, USA,* [d]*Fairfield University, Fairfield, CT 06430, USA,* [e]*Physikalisches Institut der Universität Heidelberg, 69120 Heidelberg, Germany,* [f]*MPI für Med. Forschung, 69120 Heidelberg, Germany,* [g]*Department of Physics, University of Illinois, Urbana, IL 61820, USA,* [h]*Department of Physics, University of Minnesota, Minneapolis, MN 55455, USA,* [i]*Budker Institute of Nuclear Physics, Novosibirsk, Russia,* [j]*KEK, Japan,* [k]*Tokyo Institute of Technology, Tokyo, Japan,* [l]*Department of Physics, Yale University, New Haven, CT 06511, USA.*

Abstract. A new and improved experiment for measuring the muon magnetic anomaly at the Brookhaven National Laboratory (BNL) was successfully started and has yielded first results. New major components of the experiment include a superferric storage ring, a superconducting inflector, electrostatic quadrupoles, lead-scintillating fiber electron calorimeters and a high precision NMR magnetic field measurement and control system. The first measurement of the ratio R of the spin precession frequency of the positive muon relative to that of a free proton gives $R = (3.707\,219 \pm 0.000\,048) \times 10^{-3}$. It is similar in accuracy and in good agreement with previous CERN measurements for μ^+ and μ^-. A muon kicker has been installed to boost the number of stored particles in the storage ring magnet and was successfully commissioned recently. The data acquired so far is expected to lower significantly the uncertainty in R. First extensive data-taking will start soon.

INTRODUCTION

The anomalous magnetic moment of leptons can be calculated to very high precision within the framework of standard theory. In precision experiments the validity of the theory can be verified, respectively possible, yet speculative, extensions to it can be explored. To the current level of experimental precision (3.7 ppb), the anomalous magnetic moment for the electron,

$$a_e \equiv \frac{(g-2)_e}{2} \qquad (1)$$

arises almost entirely from virtual photons and electron-positron pairs [1] and can be calculated exclusively within the framework of quantum electrodynamics (QED). All contributions of particles with higher masses including the muon only amount to less than 4 ppb. The good agreement between the experimental and theoretical values of a_e provides one of the most decisive tests of pure QED [2]. This is limited in accuracy to about 25 ppb to which the fine structure constant α is known from other experiments. Since the theoretical accuracy is higher, α can be extracted with the best precision from this experiment. For the muon, the relative contributions to a_μ of heavier particles or new contact interactions are much larger than for the electron because they usually scale as the square of the muon/electron mass ratio $(m_\mu/m_e)^2 \simeq 4 \times 10^4$. In particular, the contribution from diagrams including virtual hadrons is about 60 ppm, and was verified for the first time in the so far most accurate measurement of a_μ at CERN [3] which reached a precision of 7.3 ppm. The design goal of the new BNL muon $g-2$ experiment is to measure a_μ to 0.35 ppm. With this precision the electroweak contribution [4] of 1.3 ppm, which arises mostly from virtual processes involving the Z and W vector bosons, should be observed, thus offering an important clean test of electroweak renormalization. Whereas all other contributions to $(g-2)$ are at present known at least one order of magnitude more precisely than the experimental goal the hadronic contribution is available to 0.64 ppm [5]. It has been evaluated using experimental data of the cross section $e^+e^- \to hadrons$ over a large energy range. Currently there are ongoing efforts in Novosibirsk, Frascati and Bejing to improve the accuracy of this cross section. At the same time the theoretical approach to include already measured hadronic τ decay data from ALEPH and CLEO is pushed further.

Already with the present knowledge of a_μ from theory a sensitive and powerful test for contributions beyond the standard model will be provided. These new models include μ and W substructure or anomalous couplings, but also supersymmetry and leptoquarks [4,6–8].

THE BROOKHAVEN EXPERIMENT

The principle of the measurement is similar to the third and latest CERN experiment [3]. Polarized muons are stored in a uniform dipole magnetic field of 1.45 T with electrostatic quadrupoles providing weak vertical focussing. The muon spin precesses in an external magnetic field $\vec{B}$ relative to the momentum vector in an experimentally observable frequency

$$\vec{\omega}_a = -\frac{e}{m_\mu} \left[a_\mu \vec{B} - \left(a_\mu - \frac{1}{\gamma_\mu^2 - 1} \right) \vec{\beta} \times \vec{E} \right], \qquad (2)$$

assuming that $\frac{E}{c} \ll B$ and $\vec{\beta} \cdot \vec{B} \approx 0$, where $\vec{\beta} = \vec{v}/c$, $\gamma = \sqrt{1 - (v/c)^2}$ and $\vec{E}$ any electric field. The dependence of ω_a on the electric field can be eliminated by storing muons with the "magic" $\gamma_\mu = 29.3$, corresponding to a muon momentum $p_\mu = 3.09$ GeV/c. In this ideal case $a_\mu - 1/(\gamma_\mu^2 - 1) = 0$, and the focussing electric field does not affect the spin precession frequency. a_μ is then extracted from $\omega_a \approx 2\pi \cdot 230 kHz$ through

$$a_\mu = \frac{\omega_a/\omega_p}{\mu_\mu/\mu_p - \omega_a/\omega_p} \qquad (3)$$

where $\omega_p \approx 2\pi \cdot 62 MHz$ is the free proton Larmor precession frequency in the same magnetic field which is seen by the muons. The ratio of muon to proton magnetic moments, μ_μ/μ_p, is currently known to 0.15 ppm from the hyperfine structure interval in muonium and muon spin rotation in liquid bromine [9], and is expected to be improved by a recently completed experiment [10].

The source of the stored muons is the AGS proton beam, which delivered 6 bunches every 2.5 s or 8 bunches every 2.6 s with a total of up to $\approx 50\times 10^{12}$ protons at 24 GeV/c onto a nickel production target. From each bunch about $(4-6) \times 10^7$ pions at ≈ 3.1 GeV/c are transported from the production target along a 116 m beam line. About 50% of the pions decay along the transport line. A momentum slit and a bending magnet near the downstream end select either pions or forward decay muons for injection into the storage ring. After passing through a channel in the back of the storage ring magnet yoke and a field free region supplied by a superconducting inflector magnet [11], the pion or muon beam enters nearly tangentially the toroidal storage region which has a radius of 7.112 m and a 9 cm diameter cross section. With pion injection, our mode of operation in the first experimental test run, a small fraction of muons from pion decay, $\pi^+ \to \mu^+ \nu_\mu$, are launched onto stable orbits and are stored.

One of the major improvements over the last CERN experiment is the use of direct muon injection. In this mode, a total kick of ≈ 11 mrad during the first one or two turns is needed to store the muons born in the pion decay channel. The muon kicker is a pulsed magnetic device consisting of three sections of pairs of current sheets, each 1.7m long and separated by 0.1m. The peak current through the plates during a 300 ns wide pulse

integrated field contour plots

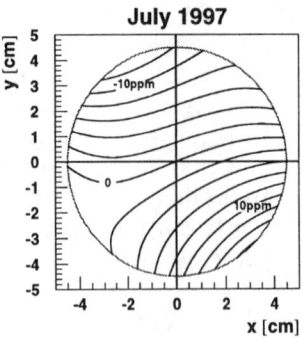 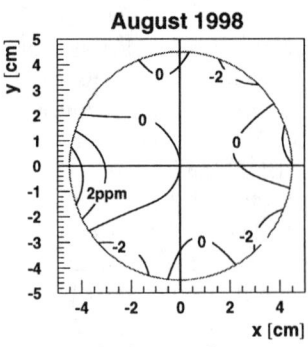

FIGURE 1. Contour plot of the magnetic field integrated over azimuth during the 1997 run (a) and in August 1998 after additional shimming and applying correction currents(b). x denotes the radial and y the vertical direction. Each contour line represents a 2ppm change. The skew quadrupole, dominant in (a), and all other multipoles except for the normal quadrupole, have been reduced to less than 2ppm amplitude at r=4.5cm.

is about 4200 A, providing a vertical field of 16 mT superimposed on the storage ring magnetic field. The residual field following a pulse was measured with an optical polarimeter using the Faraday rotation effect in a TGG crystal. Its contribution to the integral of the storage ring field decays to less than 0.1 ppm within $25\mu s$ or 0.4 muon lifetimes. The continuous superferric 'C'-shaped storage ring magnet is excited by superconducting coils which carry a current of 5177 A. The yoke consists of twelve 30 degree sections bolted together at the four corners, with azimuthal gaps of less than 1mm. The pole pieces are 10 degrees long and aligned with the yoke sectors. The azimuthal gap between adjacent pole pieces of about $75\,\mu m$ is filled with insulating Kapton foils to avoid irregular eddy current effects. The vertical gap between pole and yoke decouples the yoke and pole pieces, which are fabricated from high quality steel, and allows the insertion of iron wedges to compensate for the C-magnet quadrupole. The 864 wedges, each $10\,cm$ wide, between pole and yoke are radially adjustable to improve the field homogeneity in azimuth. The four edge shims, $5\,cm$ wide and about $3\,mm$ high, are the main tool for reducing field variations over the beam cross section. A first adjustment of the shims in preparation for the 1997 run resulted in the field distribution shown in Figure 1 (a). Since then, the 144 shims have been ground individually and the field homogeneity locally as well as integrated over azimuth (Figure 1 (b)) has improved significantly. Continuous current shims on the poles, which were installed but not operational in 1997, have been employed to reduce the inhomogeneity in the integrated field to below 4ppm over the 9cm diameter aperture.

The magnetic field is monitored by a novel NMR magnetometer. It has 375 NMR probes embedded in the top and bottom plates of the twelve vacuum chambers [12]. The field inside the storage region is mapped several times every week using a trolley precision magnetometer containing a matrix of 17 NMR probes which operates in vacuum. The probes in the magnetometer are calibrated in place against a standard spherical water probe [13].

The decay positrons from $\mu^+ \to e^+ \nu_e \bar{\nu}_\mu$, which constitute our signal, range in energy from 0 to 3.1 GeV, and are detected with 24 Pb-scintillating fiber calorimeters placed symmetrically around the inside of the storage ring. Because of parity violation in the weak muon decay, the high energy positrons are preferentially emitted along the muon spin direction. The muon spin precession is reflected in the decay positron time spectrum as determined through observing positrons at each detector station exceeding a threshold energy, N(t), where we expect

$$N(t) = N_0 e^{-t/\tau_\mu} \left[1 - A\cos\left(\omega_a t + \phi\right)\right]. \qquad (4)$$

The asymmetry parameter, A, is energy dependent, and for a positron energy threshold of 1.8 GeV is ~ 0.4. The arrival times of the positrons are recorded in multi-hit TDCs, and the calorimeter pulses are sampled by a custom 400 MHz waveform digitizer (WFD). A laser and LED system are used to monitor potential time shifts, as well as gain shifts. Several detector stations are outfitted with a finely segmented hodoscope array of 20×32 small scintillating elements connected to a multianode phototube, which provide position sensitive information on the muon decay positron. Additional spatial and temporal event information is derived from five stations each equipped with five scintillator paddles oriented horizontally. Finally, a traceback chamber provides information on the stored muon phase space. The latter has to be convoluted with the magnetic field

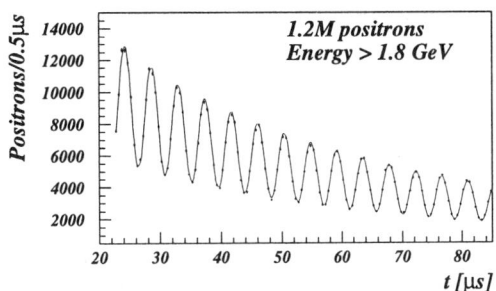

FIGURE 2. Analysis of decay positron counts as a function of time. The exponential muon decay is modulated at the $(g-2)$ precession period. The solid line represents a fit to the data.

to determine the average field seen by the muons. The mean radial distribution of the stored muons can also be obtained from the fast rotation signal from the initial time structure of the injected beam. The results of the analysis were consistent with a uniformly filled phase space, as expected in pion injection mode. Vertically, the stored beam was approximately 1mm high due to a nonzero average of the radial field component B_x of about 20 ppm, which was measured with Hall probes prior to the run.

THE RESULT FROM THE 1997 FIRST CHECKOUT EXPERIMENT

The exponential muon decay, modulated by the $(g-2)$ precession is shown in Figure 2. It contains a subset of the data taken in 1997. An 8-parameter function was fitted to the data

$$F(t) = N_0 e^{-\frac{t}{\gamma\tau}} \frac{1}{(t-t_0)^\alpha}[1 + (A_1 t + A_2)cos(\omega_a t + \Phi)] + B, \tag{5}$$

where $t_0 = 5\,\mu s$ after injection. The function includes a time dependent energy threshold due to a particular operating mode of the WFDs and a resulting time dependent asymmetry parameter. The power law was found to best describe the time dependent background. Preliminary analysis of the first muon-injection data recently taken shows that the experimental data can now be described using eq(4). The experiment measures the frequency ratio $R = \omega_a/\omega_p$, where ω_p is the proton NMR frequency measured with the beam tube magnetometer. Including corrections of about 1 ppm from the electric quadrupole field for vertical focusing, we obtain

$$R = 3.707220(47)(11) \times 10^{-3} \tag{6}$$

where the first error is statistical and the second systematic. Systematic errors are listed in table 1. Adding the two errors in quadrature results in a 13 ppm relative uncertainty. For a_μ we obtain when using fundamental constants from [9]

$$a_\mu = \frac{R}{\lambda - R} = 1165925(15) \times 10^{-9} \tag{7}$$

where $\lambda = \mu_\mu/\mu_p = 3.18334547(47)$ [14], in good agreement with the last CERN measurement, $a_{\mu,CERN} = 1165924(8.5) \times 10^{-9}$ [3].

Table 1: Systematic Errors in ppm.

Systematic Effect	ϵ (ppm)
Magnetic Field B	1.0
Muon Distribution and $<B>$	0.9
WFD Energy Threshold	1.5
Gain and Pedestal Shift	0.6
Muon Losses	0.2
Timing shifts	0.1
Radial E field, Pitch Correction	0.05
Fitting Start Time	2.0
Binning Effects	0.2
Total Systematic Error	2.9

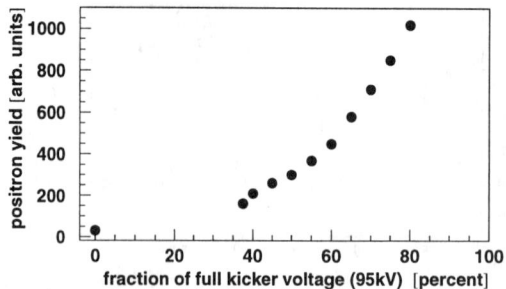

FIGURE 3. The number of decay positrons seen by an online scaler plotted against the kicker high voltage. Note the point taken with pion injection at 30 on the vertical scale. The kicker can not be operated at less than 35 percent of its nominal voltage.

The 8-parameter function does not describe the signal completely at early times due to a high rate at the detectors arising from pion reactions in the apparatus. Therefore the fitted frequency depends on the fitting start time and there is an uncertainty associated with the choice of the start time, which in fact is the largest contribution to the systematic error. This uncertainty as well as the WFD energy threshold, gain and pedestal shift contributions are expected to be absent or smaller by at least one order of magnitude in the upcoming runs due to the significantly reduced background in the muon injection mode and the absence of the WFD problem. The only other significant entries in Table 1 are associated with the uncertainty in the knowledge of the average field seen by the muon beam. The improved field homogeneity, the use of active shimming, the thermal insulation of the magnet yoke and the installation of two plunging probes that allow continuous cross calibration of the beam tube magnetometer, should reduce these errors by about a factor of ten as well.

IV. OUTLOOK

After successfully commissioning all major components in 1997 and 1998, the muon $(g-2)$ collaboration is preparing for a muon injection high statistics experiment. The muon kicker increased the number of detected decay positrons by more than one order of magnitude (Fig 3). Other major improvements include the reduced magnetic field inhomogeneity, substantial increases in the light output of several detector stations for better separation between decay electron pulses and background, and additional beam instrumentation to maximize the number of stored muons per proton on target. The collaboration aims to reduce the uncertainty in R and a_μ by a factor of ten or more relative to their first result achieved with pion injection, and is confident that the proposed measurement accuracy of $0.35\,ppm$ in a_μ can be achieved for both charges of the muon with additional beamtime, thus providing a sensitive test of standard theory and of possible extensions to it.

REFERENCES

1. R. van Dyke et al., Phys. Rev. Lett., Vol. **59**, 26, (1987).
2. T. Kinoshita, Rep. Prog. Phys., Vol. **59**, 1459, (1996).
3. J. Bailey et al., Nucl. Phys. B., Vol. **B150**, 1, (1979).
4. T. Kinoshita and W.J. Marciano in Quantum Electrodynamics, ed. T. Kinoshita, Singapore, World Scientific, 419, (1990).
5. R. Alemany, M. Davier and A. Höcker, Eur. Phys. J., Vol. **C2**, 123, (1990).
6. P. Méry et al., Z. Phys. C., Vol. **46**, 229, (1990).
7. J. Lopez et al., Phys. Rev. D., Vol. **49**, 366, (1994); G. Couture and H. Konig, Phys. Rev. D., Vol. **53**, 555, (1996); U. Chattopadhyay and P. Nath, Phys. Rev. D., Vol. **53**, 1648, (1996); T. Moroi, Phys. Rev. D., Vol. **53**, 6565, (1996).
8. F.M. Renard et al., Phys. Lett. B., Vol. **409**, 398, (1997).
9. E. R. Cohen and B. N. Taylor, Rev. Mod. Phys., Vol. **59**, 1121, (1987).
10. M. G. Boshier et al., Phys. Rev. A., Vol. **52**, 1948, (1995).
11. F. Krienen et al., Nucl. Instr. Meth. A., Vol. **283**, 5, (1989).
12. R. Prigl et al., Nucl. Instr. Meth. A., Vol. **374**, 118, (1996).
13. X. Fei et al., Nucl. Instr. Meth. A., Vol. **394**, 349, (1997).
14. C. Caso et al. (Particle Data Group), The European Physical Journal **C3**, 1, (1998).

Precise g-factor measurements on Ba⁺ Ions in a Penning Trap

G.Marx, G. Tommaseo and G. Werth

Institut für Physik, Johannes Gutenberg Universität, D-55099 Mainz, Germany

Abstract. We have performed Laser-microwave double and triple resonance experiments on clouds of Ba⁺ ions confined in a Penning ion trap to induce and detect electronic and nuclear spin flip transitions. We used collisions with buffer gas molecules in the trap to cool the ions to the ambient temperature and to quench a long lived metastable state, in which population trapping might occur. Loss of ions from the trap by collisions was prevented by coupling the magnetron and reduced cyclotron motions by an additional r.f. field at the sum frequency of the two motions. Electronic Zeeman transitions in ^{138}Ba⁺ and ^{135}Ba⁺ were observed at a transition frequency of 80 GHz in a 3 Tesla B-field with a fractional uncertainty of $3 \cdot 10^{-9}$. From the magnetic field calibration by the cyclotron resonance of electrons stored in the same trap the g_J-factor for both isotopes could be determined to $2 \cdot 10^{-8}$. From radio frequency induced $\Delta m_I = 1$ transitions in ^{137}Ba⁺ the nuclear g-factor could be determined to $5 \cdot 10^{-6}$. Both measurements improve earlier results by about one order of magnitude.

INTRODUCTION

The determination of g-factors has always been an important subject in atomic physics. Electronic g_J-factors serve as sensitive test of electronic wave functions, particularly for relatively simple systems such as the $S_{1/2}$ ground states of alkali atoms or isoelectronic ions. Progress in computational methods has made it possible to calculate the deviation of the g-factor from the value of the free electron, which is of completely relativistic origin, with great precision. g_I factors are basic properties of atomic nuclei and their knowledge is required for the interpretation of measured hyperfine splittings of atomic energy levels.

In this contribution we report about a new measurement of the ground state g_J-factor of Ba⁺ ions using laser-microwave double resonance techniques in a Penning trap. We also present a precise value of the nuclear g_I-factor in ^{137}Ba⁺ obtained by direct nuclear Zeeman transitions. Ba⁺ is of particular interest since accurate wave functions are required for the investigation of parity violation effects in these ions. An experiment to search for such an effect is presently under way [1]. Calculations of the $6S_{1/2}$ g_J-factor have been performed by Lindroth and Ynnerman [2] using accurate relativistic wave functions obtained in the Coupled-Cluster Singles and Doubles (CCSD) approximation, including correlations due to the Coulomb as well as the Breit interaction.

EXPERIMENT

Our experimental setup is described in ref. [3]: A Penning ion trap of 13 mm radius of the hyperbolic shaped ring electrode (fig.1) was placed at the center of the horizontal room-temperature bore of a superconducting solenoid of 2.89 Tesla magnetic field strength. Guard electrodes were placed between the ring and the endcap electrodes to compensate partially for possible trap imperfections. Ions were created by surface ionisation of a sample of Ba isotopes on a Rhenium filament, placed in a slot in one of the endcap electrodes. The ions were excited at their $6S_{1/2}$ - $6P_{1/2}$ resonance transition at 493.4 nm by a nitrogen pumped pulsed dye laser operated at 20 Hz repetition frequency. The laser beam was guided parallel to the solenoid´s axis, reflected into the trap through a hole in the ring electrode by a mirror and back reflected onto itself by a second mirror placed opposite to the entrance hole. Laser induced fluorescence light was collected by a two-lens optics into a light guide and detected by a photomultiplier tube outside the solenoid at a distance of about 1.5 m from the trap. The overall detection efficiency including solid angle, transmission losses and detector quantum efficiency was approximately $4 \cdot 10^{-3}$. The decay of the excited $6P_{1/2}$ state into the metastable $5D_{3/2}$ level at 649.6 nm was detected and the laser stray light was blocked by an interference filter. Microwaves were guided into the trap by a waveguide through a hole in the ring electrode perpendicular to the laser beam direction to induce electronic Zeeman transitions at 80 GHz. To induce the nuclear Zeeman transitions we used the Ba-filament as antenna for the 2 GHz radio frequency field.

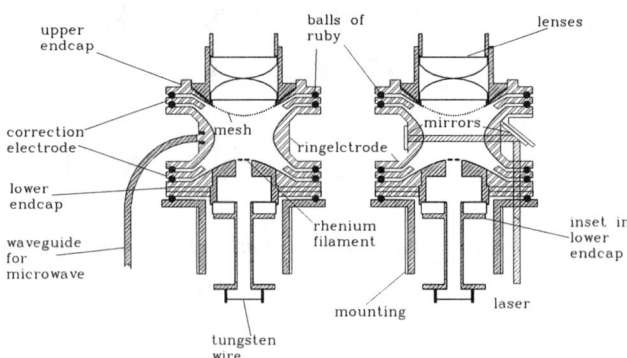

FIGURE 1. Penning ion trap used in our experiment. The two sketches are at 90° to each other showing the laser beam direction and the microwave guide, respectively. The magnetic field is directed along the trap axis.

A problem arises due to the fact that the radiative decay time of the state metastable $5D_{3/2}$ is about 80 s [4] and population trapping into this state occurs. In order to reduce the effective lifetime we used collisional quenching of the metastable state at high background pressures. Under the influence of collisions, however, the ion motion in a Penning trap becomes unstable through an increase of the magnetron radius. Ion loss can be overcome by coupling of the magnetron motion at frequency

$$\omega_m = \omega_c/2 - [\omega_c^2/4 - \omega_z^2/2]^{1/2} \quad (1)$$

to the reduced cyclotron frequency

$$\omega_c' = \omega_c/2 + [\omega_c^2/4 - \omega_z^2/2]^{1/2} \quad (2)$$

by an additional radio frequency field at their sum frequency [5]

$$\omega = \omega_m + \omega_c' = \omega_c \quad (3)$$

$\omega_c = (e/m)B$ is the free ions cyclotron frequency and ω_z is the axial oscillation frequency in the trap. The cooling effect of collisions on the cyclotron motion supercedes the increase in magnetron radius and as a result the ions concentrate near the trap center. Details of this technique are published elsewhere [6]. We benefit in our experiment from this method in several ways: Besides the increase in ground state population density by quenching of the metastable state, the concentration of the ions near the trap center increases the spatial overlap with the exciting laser and improves the fluorescence signal. The same holds for the spectral overlap, since the ions are cooled. The maximum benefit, however, comes from the fact that by cooling the amplitude of the ion oscillation is reduced to a value which is smaller than the wavelength of the microwave radiation for the Zeeman transition. We then operate in the Dicke regime where the microwave spectrum consist of a central narrow line, unbroadened by first order Doppler effect, and sidebands at the ion oscillation frequency. As shown below, this reduces the spectral linewidth in the electronic Zeeman transitions by about two orders of magnitude compared to our previous experiment [10]. Finally, another advantage of the collisional cooling of the ions is the extended storage time from previously 20 min under UHV conditions to several hours at 10^{-5} mbar of N_2 buffer gas.

The magnetic field at the ions position was measured by the cyclotron frequency of electrons stored in the same trap after inversion of the trapping potential. Excitation of the motional eigenfrequencies of stored electrons by a radio frequency field increases the oscillation amplitude. This, in our experiment, is detected in the following way: The trap endcaps are connected by an outer inductance to form a tank circuit, which is weakly excited at its resonance frequency. The axial ion oscillation frequency of a stored particle of charge e and mass m in a trap of radius r_0

$$\omega_z^2 = (eV)/(mr_0^2) \quad (4)$$

depends on the square root of the potential V seen by the particles. This potential is the sum of the applied trapping voltage and the space charge potential of the electron cloud. When we sweep the trapping potential by a linear ramp voltage, ω_z coincides at a certain instant with the resonance frequency of the tank circuit. Then the ions absorb energy and dampen the circuit, which can be detected by a sensitive amplifier. Upon excitation of the electron cloud the space charge potential decreases and consequently the axial resonance appears at a different value of the applied trapping potential. We plot the applied voltage at which the axial electron resonance appears and obtain a resonance curve as in fig.2. It fits well to a Lorentzian lineshape and we determine the center frequency with an uncertainty of $3 \cdot 10^{-8}$.

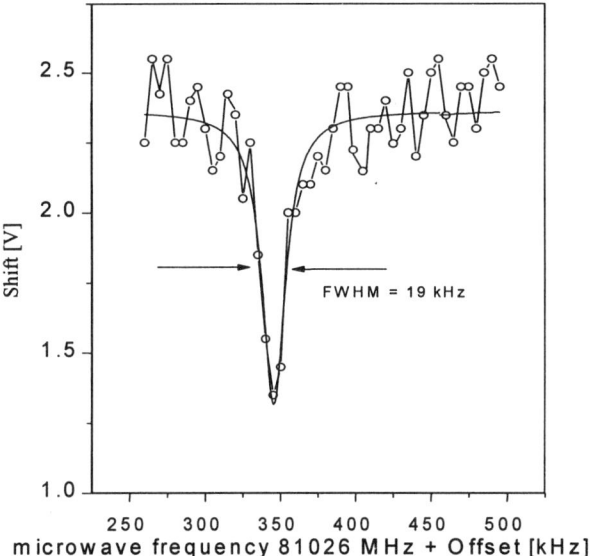

FIGURE 2. Cyclotron resonance of stored electrons for magnetic field calibration. The voltage at which the axial resonance appears is plotted versus the applied excitation frequency. At resonant excitation the space charge in the stored electron cloud is reduced, which leads to a shift in the position of the resonance.

To account for a temporal variation of the magnetic field we measured the electron cyclotron frequency before and after a Zeeman resonance measurements for the Ba$^+$ ions. The measured drift of the field was $\delta B/B = 1 \cdot 10^{-9}$/h. By linear interpolation of the measured field strength from the electron cyclotron resonance, we determined the field strength at the time of the Zeeman resonance measurement.

We determined the g_J-factor of the two Ba$^+$ isotopes with mass 138 and 135 in laser-microwave double resonance experiments. After storage of about 10^5 ions from isotope enriched samples, cooling and confinement of the ions in the center of the trap as described above, one of the electronic ground state Zeeman sublevels was selectively excited by a broadband ($\delta\nu = 7$ GHz) nitrogen pumped pulsed dye laser operated at a repetition rate of 20 Hz. After a few seconds the observed fluorescence decreases to the scatter level by depletion of the pumped state. When the microwave frequency is swept across the $m_J = +1/2 - m_J = -1/2$ Zeeman resonance, we observe an increase in the fluorescence intensity. At higher microwave powers a central peak occurs alongwith strong sidebands at combinations of the ions axial and radial oscillation frequencies. At lower microwave powers, the amplitude of the sidebands decreases. When we scan with higher resolution over the central resonance, which is unaffected by first order Doppler effect, the resonance shows a Lorentzian lineshape. The minimum linewidth which we obtained was 3 kHz at a transition frequency of 81 GHz for the even isotope ^{138}Ba$^+$ (fig.3). The line center could be determined by a least squares fit of a Lorentzian to the data points with an uncertainty of 0.22 kHz, corresponding to a fractional uncertainty of $2.5 \cdot 10^{-9}$. Further reduction of the uncertainty was limited by the temporal stability of the magnetic field produced by the superconducting solenoid. Furthermore the uncertainty of the magnetic field calibration could be performed with a precision of 10^{-8} only and hence improvements on the Zeeman resonance would not improve our results.

In a similar way a $\Delta m_J = 1$, $\Delta m_I = 0$ transition was induced in the odd isotope ^{135}Ba$^+$. Here the transition frequency is shifted to 79.6 GHz compared to ^{138}Ba$^+$ by hyperfine interaction and nuclear Zeeman effect. The experimentally obtained linewidth was of the same order as for the even isotope. From the transition frequency in ^{138}Ba$^+$ and the magnetic field strength, we derived the value for the g_J-factor in the $6S_{1/2}$ ground state of Ba$^+$ as:

$$g_J = 2.002\,491\,92\,(3)$$

The quoted standard deviation ($1.5 \cdot 10^{-8}$) is entirely due to the uncertainty in the calibration of the the magnetic field.

For the odd isotope ^{135}Ba$^+$ we obtain a value for the g_J-factor using the Breit-Rabi formula. For the hyperfine structure coupling constant A we used the value $A = 3\,591670117.45\,(29)$ Hz [7] and for the nuclear g_I factor $g_I = 0.557\,67$ [8]. The g_J-value for ^{135}Ba$^+$ agrees within the limits of error with that of ^{138}Ba$^+$ and we can quote an upper limit for the isotope dependence of the g_J-factor:

$$g_J(^{135}\text{Ba}^+) - g_J(^{138}\text{Ba}^+) = 1\,(5) \cdot 10^{-8}$$

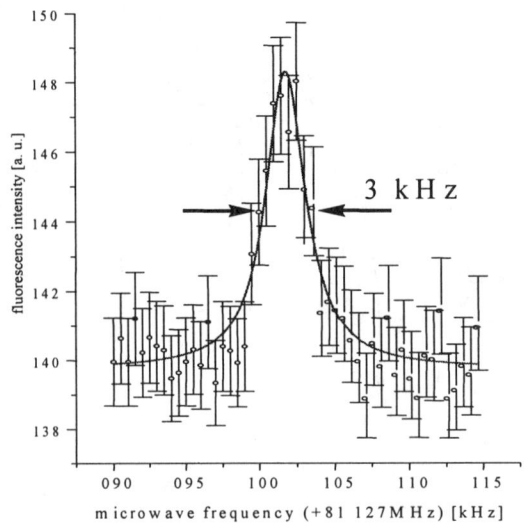

FIGURE 3. High resolution scan of the central part of the $\Delta m_J=1$ resonance in $^{138}Ba^+$. The experimental points are least-squares fitted by a Lorentzian lineshape. The statistical uncertainty of the line center is $3 \cdot 10^{-9}$

In a triple resonance experiment on $^{137}Ba^+$ we determined the nuclear g-factor of this isotope. The $m_j = ½$ manifold of the nuclear Zeeman structure was depleted by selective excitation with a broadband dye laser in a similar way as in the experiments described above on an even isotope. A second transition between the $m_J = 1/2$, $m_I = -1/2$ and $m_J = -1/2$, $m_I = -1/2$ levels depletes one of the nuclear Zeeman levels in the $m_J = -1/2$ manifold. Finally a radio frequency field between the depleted $m_I = -1/2$ and the adjacent $m_I = -3/2$ or $m_I = +1/2$ levels is detected by an increase in fluorescence intensity. Fig. 4 shows the two resonances. The linewidth of 4.7 kHz at a transition frequency of about 2.1 GHz is limited by power broadening. Since the line intensity was rather weak, averaging times of about one hour were required to record the data.

From the two transition frequencies and a magnetic field calibration by a Zeeman resonance in $^{138}Ba^+$, we obtain two independent values for the g_I-factor, which agree with each other within the statistical error. The average value from both the measurements is

$$g_I = 0.620\ 235\ (3).$$

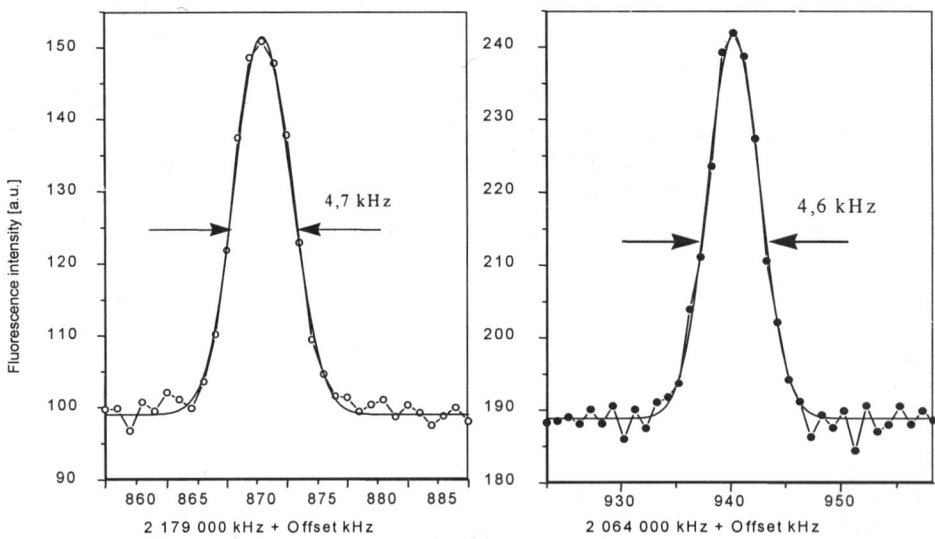

FIGURE 4. Microwave induced nuclear Zeeman transitions in a triple resonance experiment on $^{137}Ba^+$.
left: $m_I = -1/2 \rightarrow m_I = -3/2$, right: $m_I = -1/2 \rightarrow m_I = +1/2$. The experimental data are least squares fitted by a Lorentzian. The relative uncertainty of the line centers is $3 \cdot 10^{-8}$.

Table 1 compares the measured g_J value with results from earlier experiments and with a theoretical calculation. All values agree within their quoted uncertainties. Compared to earlier experiments in Penning traps, we have improved the experimental value by a factor of 35. This was made possible by significant signal enhancement by collisional quenching of a long lived metastable state and ion cooling and confinement near the trap center. Further reduction of the uncertainty was limited by calibration of the magnetic field. The theoretical value is two orders of magnitude less accurate than the experimental value and considering the complexity of calculations in a multi-electron system like Ba^+ it may be a difficult task for the theory to match the experimental precision.

TABLE 1 Published values of g_J-factors in the $6S_{1/2}$ ground state of Ba^+

	g_J-factor	Reference
experimental	2.002 491 92 (3)	this work
	2.002 490 6 (11)	3
	2.002 492 2 (10)	11
theoretical	2.002 491 1 (30)	2

Table 2 lists available values for the g_I-factor of $^{135}Ba^+$. We have applied a diamagnetic correction $(1 - \sigma)^{-1} = 1.00590$ [9] to our experimental value of $g_I = 0.620235$. All the quoted values agree with each other. Compared to the most accurate previous measurements on neutral Ba atoms by optical pumping in a buffer gas cell, our measurements using Penning trap have one order of magnitude lower uncertainty. Further improvement, if required, is possible by reducing the power broadening.

TABLE 2: Nuclear g_I-factors of ^{137}Ba

g_I-factor	Method	Reference
0.623 876 (3)	Penning ion trap	this work
0.623 82 (4)	Opt. Pumping in gas cell	8
0.623 88 (1)	Opt. Pumping in gas cell	12
0.623 8 (5)	Nuclear magnetic Resonance	13

In conclusion, we have demonstrated that the Penning trap technique can yield very accurate values of electronic and nuclear g-factors. Since it has been shown earlier [6] that radioactive isotopes of minute quantities can be successfully investigated in ion traps, the extension of the method described above to unstable isotopes could result in improved values for nuclear magnetic moments which are not derived from hyperfine structure measurements, but measured by direct nuclear Zeeman transitions. This would in combination with measured hyperfine coupling constants of the same isotopes, result in values of the hyperfine anomaly in unstable isotopes, which is desirable for a better understanding of the distribution of magnetisation over extended nuclei.

ACKNOWLEDGMENTS

Our experiments were supported by the Deutsche Forschungsgemeinschaft.

REFERENCES

[1] E.N. Fortson, Phys. Rev. Lett. **70**, 2383 (1993)

[2] E. Lindroth and A. Ynnerman, Phys. Rev. **A47**, 961 (1993)

[3] H. Knab, K.H. Knöll, F. Scheerer and G. Werth, Z. Phys. **D25**, 205 (1993)

[4] N. Yu, W. Nagourney and H.G. Dehmelt, Phys. Rev. Lett **78**, 4898 (1997)

[5] G. Savard et al., Phys. Lett **A158**, 247 (1991)

[6] Ch. Lichtenberg, G. Marx, G. Tommaseo, P.N. Ghosh and G. Werth, Europ. Phys. Journ.**D** (in press)

[7] W. Becker and G. Werth, Z. Phys. **A311**, 41 (1983)

[8] L. Olschewski and E. Otten, Z. Phys. **196**, 77 (1966)

[9] H. Kopfermann, Kernmomente, Akad. Verlagsanstalt, Frankfurt (1956)

[10] H.Knab, M.Schupp and G.Werth, Europhys. Lett. 4, 1361 (1987)

[11] K.H. Knöll et al., Phys. Rev. **A54**, 1199 (1996)

[12] L. Olschewski, Z. Phys. **249**, 205 (1972)

[13] O.Lutz and H. Oehler, Z. Phys. **A288**, 11 (1978)

SECTION 2

SPECIAL TOPICS

Antihydrogen for Tests of CPT and Lorentz Invariance

Michael H. Holzscheiter[1]
for the ATHENA collaboration

Los Alamos National Laboratory, P-23, MS H803, Los Alamos, NM 87545

Abstract.
Antihydrogen atoms, produced near rest, trapped in a magnetic well, and cooled to the lowest possible temperature (kinetic energy) could provide an extremely powerful tool for the search of violations of CPT and Lorentz invarianz. We describe our plans to form a significant number of cold antihydrogen atoms for comparative precision spectroscopy of hydrogen and antihydrogen.

I INTRODUCTION

CPT invariance is a fundamental property of quantum field theories in flat space-time, which results from the basic requirements of locality, Lorentz invariance and unitarity [1–5]. Principal consequences include the predictions that particles and their antiparticles have equal masses and lifetimes, and equal and opposite electric charges and magnetic moments. It also follows that the fine structure, hyperfine structure, and Lamb shifts of matter and antimatter bound systems should be identical.

A number of experiments have tested some of these predictions with impressive accuracy [6], e.g. with a precision of 10^{-12} for the difference between the moduli of the magnetic moment of the positron and the electron [7] and of 10^{-9} for the difference between the proton and antiproton charge-to-mass ratio [8]. The most stringent CPT test to date comes from a mass comparison of neutral kaon and antikaon, where an accuracy of 10^{-18} has been reached, albeit in a theoretically dependent manner.

Recent years have seen a steady increase in discussions of possible mechanism for, and implications of, CPT violation [9–12]. Specifically, a model based on an extension of the Standard Model (SM) and Quantum-Electrodynamics (QED) has been formulated and used to quantitatively analyze specific experiments for their sensitivity to CPT and Lorentz violations [13]. Spontaneous breaking of CPT and Lorentz symmetry has been suggested as a possible source of observable experimental effects in conventional four-dimensional spacetime. In the framework of this theoretical model existing and proposed experiments have been studied and new, more meaningful figure-of-merits have been established. It is found that current g-2 experiments on electrons and protons test CPT at a level of 2×10^{-17} [14] and could be improved to 1×10^{-20} and that a proposed experiment on measuring the ratio of the magnetic moments of protons and antiprotons [15,16] could yield a figure of merit of 1×10^{-23} [17]. Similar work analyzing the sensitivity of specific spectroscopic measurements in hydrogen and antihydrogen to CPT and Lorentz violation show that the highest sensitivity may be achieved in studies of the hyperfine interaction in antihydrogen [18].

The formation of antihydrogen in flight has been demonstrated to date by two experiments [19,20]. Antihydrogen was formed by collisions between high energy antiprotons and a gas jet, creating electron-positron pairs. In kinematic favourable cases the antiproton could capture a positron and continue its flight path as a neutral antihydrogen atom. While this was sufficient to identify the formed antihydrogen, the extremely low production rate and the relativistic energy of the particles prohibited any measurements at a level of accuracy necessary for meaningful tests of CPT and Lorentz invarianz. Such precision can only be reached by capturing antihydrogen in a magnetic trap and cooling it to the lowest possible temperatures.

In the following we give a brief overview of the ATHENA (**A**ppara**T**us for **H**igh precision **E**xperiments on **N**eutral **A**ntimatter) experiments and discuss some of the physics issues relevant to antihydrogen formation.

---

[1]) e-mail: mhh@lanl.gov

II EXPERIMENTAL OVERVIEW

In order to form antihydrogen atoms at rest one must start with both constituents, antiprotons and positrons, stored in electromagnetic field configurations known as Penning traps and cooled to the ambient temperature of the cryogenic environment. When antihydrogen atoms are then formed by appropriately overlapping the two oppositely charged particle plasmas, the product atom will carry the kinetic energy of the heavier particle, the antiproton, and therefore will be "cold" as well.

The technique of capturing antiprotons into traps and cooling them to milli-eV energies has been developed at LEAR over the last 10 years [8,21]. To reduce the kinetic energy of the incoming beam from 5.9 MeV (the lowest energy available at LEAR, and the output energy anticipated for the AD) to several tens of keV (the energy where electromagnetic trapping of particles has been demonstrated), energy loss in thin foils [22] is being used. To capture and confine the antiprotons once the energy has been reduced to ≤ 30 keV, we employ a modified Penning trap [23]. The trap structure typically consists of seven electrodes: the entrance foil, a central region comprising five cylinders (two endcaps, two compensation electrodes, and the central ring), and a cylindrical high voltage exit electrode. The trap system is situated in the cryogenic bore of a superconducting solenoidal magnetic field of 3 to 6 Tesla for radial confinement, while the axial confinement is given by the electrostatic potentials applied to the trap electrodes.

The initial kinetic energy of antiprotons after capture is in the keV range. Electron cooling is used to reduce the antiproton energy to values below 1 meV. For this purpose, a dense electron cloud is preloaded into the central region of the trap. These electrons cool to equilibrium with their cryogenic environment via synchrotron radiation with a time constant of ≤ 0.4 s at 3 Tesla. The antiprotons oscillate through the cold electron cloud and lose their energy via Coulomb collisions with a time constant of a few minutes. The efficiency observed for this process is better than 90%.

Our previous experiment at LEAR, PS200, has set the world record in collecting and cooling one million antiprotons from a single shot from LEAR [24]. It has also demonstrated that subsequent pulses can be "stacked" to increase the overall number of antiprotons in the trap. Using this method we plan to accumulate 10^7 cold antiprotons from the Antiproton Decelerator (AD) [25] currently under construction at CERN.

Large numbers of positrons have also been accumulated in similar field configurations. Our collaboration will use a system based upon the positron accumulator presently operated at the University of California in San Diego, in which 10^8 low energy positrons are routinely accumulated in a few minutes. With minor modifications in the source design and in the vacuum system we anticipate to accumulate 10^{10} positrons in an intermediate ultra-high vacuum storage trap, from which the positrons will be transferred into the main apparatus on demand.

One of the major challenges in the formation of antihydrogen will consist of bringing the oppositely charged antiprotons and positrons in close contact for a time sufficiently long to allow the recombination process to take place. For this we will use a nested Penning trap [26] which consists of a sequence of axial electrostatic wells in a common magnetic field. These wells are arranged in such a way that particles of opposite charge are stored in separate locations in close proximity. Mixture of the plasmas can be achieved by adjusting the potential wells or by heating the particles in one well so they can leak over the barrier into an adjacent well. Latter method has been used to generate ultra-low energy beams from Penning traps [27] and appears to be a promising scheme to mix dense antiproton and positron clouds at low relative velocity.

The experimental program of ATHENA can be broken down into two distinct phases. Phase one will entail the capture and cooling of antiprotons, the injection of positrons into the recombination region, and the formation of low energy antihydrogen atoms. In this phase we will sytematically characterize the formation process and the distribution of energy and initial state resulting from the specific recombination process under a varying parameters like plasma densities, plasma temperature, etc. While no attempt will be made to actually capture the neutral atoms in a magnetic well, studies are planned to systematically address the question of stability of charged plasmas in azimuthally assymetric fields as presented by a three dimensional magnetic well.

Only after this research is completed we will embark into phase 2, the actual capture of antihydrogen and spectroscopy on this system. Once antihydrogen atoms have been formed they can be confined using the force produced by the interaction of a magnetic gradient field with the magnetic moment of the atoms. Typically a combination of quadrupole coils (Ioffe bars) for radial confinement and Helmholtz coils for the axial confinement is used [30]. Such a field can be superimposed onto the homogeneous field generated by the superconducting magnet needed for the Penning traps by either adding additional superconducting magnet coils to the solenoid or by placing appropriately shaped permanent magnets in the gap between the magnet bore and the vacuum system.

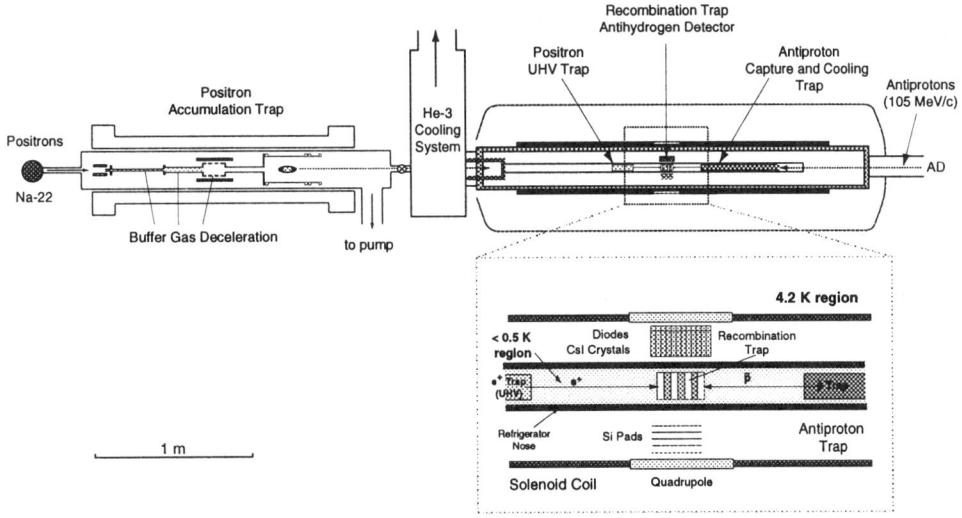

FIGURE 1. *Overview of the ATHENA apparatus showing the superconducting solenoid with the antiproton capture trap, the positron storage trap, and the recombination trap surrounded by the magnetic gradient trap and the particle detector.*

Figure 1 shows a general lay-out of the central portion of the apparatus containing the antiproton capture trap, the recombination trap with superimposed neutral trap, the particle detector system to study the dynamics of the recombination processes, and the final positron storage trap. Also shown is the primary positron accumulator, which is designed as a stand-alone system.

III ANTIHYDROGEN FORMATION

In order to meet the requirements of precision spectroscopy of antihydrogen the recombination technique used in the ATHENA experiment must

- provide sufficient numbers of antihydrogen atoms for spectroscopy,
- produce the atoms at very low temperatures (T $\leq$ 1 K) to allow trapping within achievable magnetic well depths,
- form antihydrogen atoms in the ground state or in low lying excited states, and
- achieve all of the above within a reasonably short time period.

To form a bound state of antiproton and positron starting from free particles, excess energy and momentum has to be carried away by a third particle. Various schemes for producing antihydrogen have been proposed and discussed in the literature in some detail [31–37], with the first mentioning of the possible production of antihydrogen in traps by Dehmelt and co-workers [38].

The simplest process is spontaneous radiative recombination:

$$e^+ + \overline{p} \Rightarrow \overline{H} + h\nu \qquad (1)$$

(see ref. [31,32]). The rate for this process can be increased by laser stimulation [33]:

$$e^+ + \overline{p} + nh\nu \Rightarrow \overline{H} + (n-1)h\nu \ . \qquad (2)$$

An entirely different approach is based on three-body collisions: [26,34–37]:

$$e^+ + e^+ + \overline{p} \Rightarrow \overline{H} + e^+ \ , \qquad (3)$$

$$Ps + \bar{p} \Rightarrow \overline{H} + e^- \text{ , and} \tag{4}$$

$$Ps^* + \bar{p} \Rightarrow \overline{H^*} + e^- . \tag{5}$$

Reaction 1 - 3 require two plasmas of opposite charge (antiprotons and positrons) trapped and brought into contact, while reactions 4 and 5 need only one charged species to be confined.

The most critical issues to be considered in the analysis of a specific reaction for the purpose of our experimental program are the total recombination rate and the distribution of states in which the antihydrogen atoms are produced. For this reason, the two most promising schemes are the spontaneous radiative recombination (SRR) (possibly with laser enhancement), and the three-body recombination (TBR) using dense positron plasmas.

The cross-section for spontaneous radiative recombination [39] is related by time-reversal invariance to photo-ionization, and depends only on the kinetic energy E of the electron in the center-of-mass (c.m.) system of the proton, and the capture level n:

$$\sigma_{SRR}(n,E) = 2.1 \cdot 10^{-22} cm^2 \frac{1}{nx(1+n^2x)} \quad x = E/E_0, \quad E_0 = 13.6 eV, \quad E = \frac{1}{2}mv^2 . \tag{6}$$

This cross-section decreases with high n and predominately low-lying n states are populated ($\approx$ 60% of the atoms are produced in states n $\leq$ 10). The total cross-section is obtained by summing over all n up to a "cut-off" level n_{cut}, which is reached when antihydrogen atoms are ionized in collisions with neighboring atoms or by external electric fields. For example, a temperature of the antihydrogen atoms of 4 K (or an ambient electric field of 1 V/cm) would lead to a cut-off at $n_{cut} \sim 200$. For a center-of-mass energy of $E_{c.m.} \sim 0.1$ meV the cross-section for this process is $1.5 \cdot 10^{-16}$ cm^2. By integration over the three-dimensional velocity distribution and assuming the relative velocity to be equal in all three dimensions as well as a perfect geometric overlap of the two plasmas we obtain an order of magnitude estimate for $\alpha(v_r) = \langle \sigma(v)v \rangle = 0.9 \cdot 10^{-10}$ This value agrees within a factor 2 or better with more elaborate calculations [40] and with experimental results from storage ring experiments [41]. With the parameters for the charged plasmas $N_e = 10^8$, $N_p = 10^7$ anticipated for the ATHENA apparatus, we obtain an upper limit for the spontaneous recombination rate R = 90.000 atoms/sec .

An alternative route to recombination has been proposed [26], where three particles (two positrons and an antiproton) collide simultaneously. This reaction, the three-body recombination (TBR), plays a role predominantly at high positron densities and very low temperatures. The rate $\alpha_{TBR}(n)$ as a function of the capture level n has been calculated [42] by considering the time-reversed process, i.e. electron-impact ionization of hydrogen, which is well known, yileding:

$$\alpha_{TBR}(n) = 1.96 \cdot 10^{-29} cm^6 s^{-1} n_e \left(\frac{1}{kT/eV}\right) n^6 \tag{7}$$

The steep dependence on the principal quantum number n indicates that mostly very high Rydberg states close to the "cut-off" level $n^* \sim \sqrt{R/2kT}$, $R = 13.6$ eV, are populated. Summing up all contributions from $n=1$ to n^*, the total three-body recombination rate for a Maxwellian positron velocity distribution of temperature T becomes:

$$\alpha_{TBR}(n^*) = 2.7 \cdot 10^{-27} cm^6 s^{-1} n_e \left(\frac{1}{kT/eV}\right)^{4.5} \tag{8}$$

which highlights the strong temperature dependence, in excellent agreement with previously quoted results [43].

A comparison of the recombination coefficients for spontaneous radiative recombination and three body recombination shows, that for a certain positron density and energy the rate for three body recombination becomes equal to the rate for spontaneous radiative recombination, and then increases by 4.5 orders of magnitude per factor 10 of decreasing temperature. But due to the population of intrinsically unstable, high-lying Rydberg states it is not clear that this increase in rate can be fully exploited experimentally. At the very least, this process requires introduction of a laser to pump the produced atoms down to stable, low-lying states, making this reaction look very similar to the laser induced stimulated radiative recombination.

IV SUMMARY

We have described the plans of the ATHENA collaboration to form ultra-low energy antihydrogen atoms for precision spectroscopy. While all individual steps have been demonstrated in separate experiments, combining them in a single experiment is posing a formidable challenge. But once the formation and capture of antihydrogen atoms has been achieved a powerful new probe for fundamental physics will be available, undoubtedly leading to interesting physics results.

REFERENCES

1. G. Lüders, Kong. Danske Vidensk. Selsk. Mat.-Fys. Medd. 28 No. 5 (1954) 1; Ann. Phys. 2 (1957) 1.
2. W. Pauli, in: *Niels Bohr and the Development of Physics*, ed. by W. Pauli (Pergamon, New York, 1955), p. 30.
3. J. S. Bell, Proc. Roy. Soc. A 231 (1955) 479.
4. R. Jost, Helv. Phys. Acta 30 (1957) 409;
5. J. J. Sakurai, *Invariance Principles and Elementary Particles* (Princeton University Press, Princeton, 1964); R. F. Streater and A. S. Wightman, *PCT, Spin & Statistics, and All That* (Benjamin, New York, 1964).
6. Particle Data Group, Phys. Lett. B204 (1988) 46
7. R. S. Van Dyck, P. B. Schwinberg, and H. G. Dehmelt, Phys. Rev. Lett. 59 (1987) 26.
8. G. Gabrielse, D. Phillips, W. Quint, H. Kalinowsky, and G. Rouleau, Phys. Rev. Lett. 74 (1995) 3544.
9. P. Huet and M. Peskin; Nucl. Phys. B434 (1995) 3
10. A. Kostelecky and R. Potting; Phys. Rev. D51 (1995) 3923
11. D. Colladay and A. Kostelecky; Phys. Lett. B344 (1995) 259;
12. J. Ellis, J. Lopez, N. Mavromatos and D. Nanopoulos, CERN-TH.95-99;
13. D. Colladay and A. Kostelecky; Phys. Rev. D 52 (1995) 6224
14. R. Bluhm, V. A. Kostelecky, and N. Russell; Phys. Rev. Lett. 79 (1997) 1432
15. D. J. Heinzen and D. J. Wineland; Phys. Rev. A42 (1990) 2977
16. W. Quint and G. Gabrielse; Hyperfine Int. 76 (1993) 379
17. R. Bluhm, V. A. Kostelecky, and N. Russell; Phys. Rev. D 57 (1998) 3932
18. R. Bluhm, V. A. Kostelecky, and N. Russell; to be published
19. G. Baur et al.; Phys. Lett. B311 (1993) 343
20. G. Blanford et al.; Phys. Rev. Lett. 80 (1998) 3037
21. M. H. Holzscheiter; Physica Scripta T59 (1995) 326
22. M. H. Holzscheiter; Physica Scripta 46 (1992) 272
23. H. G. Dehmelt; Adv. At. Mol. Phys. 3 (1967) 53 and 5 (1969) 109
24. M. H. Holzscheiter et al.; Phys. Lett. A 214 (1996) 279
25. S. Maury, *et al.*; CERN/PS 96-43 (AR)
26. G. Gabrielse, S. L. Rolston, L. Haarsma, and W. Kells; Phys. Lett. A129 (1988) 38
27. X. Feng, M. H. Holzscheiter, R. A. Lewis, R. Newton, and M. M. Schauer; Hyperfine Int. 100 (1996) 103
28. Guo-Zhong Li, R. Poggiani, G. Testera, G. Torelli, and G. Werth; Hyperfine Int. 76 (1993) 343
29. J. W. Humberston, M. Charlton, F. M. Jacobsen, and B. I. Deutch; J. Phys. B 20 (1987) L25
30. Y. V. Gott, M. S. Ioffe, V. G. Tel'kovskii; Nucl. Fusion, 1962 suppl., Pt. 3 (1962) 1045
31. G. Budker and A. Skrinsky; Sov. Phys.-Usp. 21, 277 (1978)
32. H. Herr, D. Möhl, and A. Winnacker; in *Proc. 2nd Workshop on Physics with Cooled Low Energy Antiprotons at LEAR*, Erice, May 9-16, 1982, (eds. U. Gastaldi and R. Klapisch) p. 659, Plenum, New York 1984
33. R. Neumann, H. Poth, A. Wolf, and A. Winnacker; Z. Phys. A313, 253 (1984)
34. B. I. Deutch, F. M. Jacobsen, L. H. Andersen, P. Hvelplund, H. Knudsen, M. H. Holzscheiter, M. Charlton, G. Laricchia, Phys. Scrip. T22, 288 (1988)
35. B. I. Deutch et al; Hyperfine Int. 44, 271 (1988)
36. M. Charlton; Phys. Lett. A143, 143 (1990)
37. B. I. Deutch et al.; Hyperfine Int. 76 (1993) 153
38. H. Dehmelt, R. Van Dyck, P. Schwinberg, and G. Gabrielse; Bull. Am. Phys. Soc. 24, 757 (1979)
39. H. A. Bethe and E. E. Salpeter; *Quantum Mechanics of One- and Two- Electron Atoms*, Springer, Berlin (1957)
40. M. Bell, J. S. Bell, Part. Acc. 12 (1982) 49
41. A. Wolf et al.; Z. Phys. D21 (1991) 69
42. M. Pajek and R. Schuch; Hyperfine Int. 108 (1997) 185
43. P. Mansbach, B. Keck, Phys. Rev. 181 (1969) 275

Hydrogen and Antihydrogen Spectroscopy for Studies of CPT and Lorentz Symmetry[1]

Robert Bluhm[a], V. Alan Kostelecký[b], and Neil Russell[b]

[a] *Physics Department, Colby College, Waterville, ME, 04901 U.S.A.*
[b] *Physics Department, Indiana University, Bloomington, IN, 47405 U.S.A.*

Abstract. A theoretical study of possible signals for CPT and Lorentz violation arising in hydrogen and antihydrogen spectroscopy is described. The analysis uses a CPT- and Lorentz-violating extension of quantum electrodynamics, obtained from a general Lorentz-violating extension of the minimal standard model with both CPT-even and CPT-odd terms. Certain 1S-2S transitions and hyperfine Zeeman lines exhibit effects at leading order in small CPT-violating couplings.

INTRODUCTION

At presently accessible energy scales, which are determined by the electroweak scale m_W and are small relative to the Planck mass M_P, the predictions of the minimal $SU(3) \times SU(2) \times U(1)$ standard model appear to be in agreement with nature. However, at scales closer to M_P this model is expected to be superseded by a fundamental theory that also combines quantum mechanics and gravitation in a consistent way. The fundamental theory is likely to involve qualitatively new physics as, for example, occurs in string (M) theory at the Planck scale. Associated low-energy signals may exist. However, approximately 17 orders of magnitude separate m_W from M_P, so effects specific to the fundamental theory and accessible via existing techniques are likely to be heavily suppressed. Experiments that search for effects forbidden in the usual renormalizable gauge theories and that are of high precision are therefore of particular interest.

In this talk, the idea is considered that the new physics includes a spontaneous violation of Lorentz symmetry [1]. If a theory with Lorentz-covariant dynamics involves Lorentz-tensor interactions acting to destabilize the naive vacuum, some finite Lorentz-tensor expectation values may arise. This can occur in some string theories, for instance. In the low-energy theory at the level of the standard model, apparent Lorentz violations would ensue if the orientation of the tensor expectation values includes the physical four spacetime dimensions.

The CPT theorem connects Lorentz transformations to the discrete charge-conjugation (C), parity-reflection (P), and time-reversal (T) transformations [2]. It implies that all local relativistic quantum field theories satisfying mild technical assumptions are invariant under CPT. This suggests that both CPT and Lorentz violations represent unconventional and potentially observable effects emerging from a fundamental theory. However, the heavy suppression expected from the hierarchy between m_W and M_P implies that detection of these effects would be feasible only in particularly sensitive experiments.

EXTENDED QUANTUM ELECTRODYNAMICS

At the level of the minimal standard model, the consequences of spontaneous Lorentz and CPT breaking can be investigated by incorporating possible terms that would represent violations of these symmetries. There exists a general Lorentz-violating extension of the minimal $SU(3) \times SU(2) \times U(1)$ standard model [3]. It includes both CPT-even and CPT-odd terms. To date, it appears to be the sole existing candidate for a consistent standard-model extension based on a microscopic description of CPT and Lorentz violation. In any

---

[1] Presented by V.A.K.

event, this theory is necessarily the low-energy limit of any fundamental theory that contains the standard model and incorporates spontaneous CPT and Lorentz violation.

The standard-model extension is theoretically attractive for several reasons. For one, the usual structure of the gauge invariances and the spontaneous gauge-symmetry breaking are unaffected, and energy and momentum are conserved provided the Lorentz symmetry breaking produces position-independent expectation values. Also, standard quantization methods apply, and the extension is hermitian and power-counting renormalizable. Even though Lorentz symmetry is spontaneously broken, various desirable features of Lorentz-covariant theories such as positivity of the energy and microcausality are expected to persist [3]. This is largely a result of the Lorentz covariance of the underlying fundamental theory and the conventional quantum description. In fact, invariance under rotations or boosts of the observer's inertial frame (*observer* Lorentz transformations) is retained even at the level of the standard-model extension. Only rotations or boosts of particles and localized field distributions (*particle* Lorentz transformations) introduce Lorentz breaking, as a result of couplings to the tensor vacuum expectation values.

Details of the construction and the specific form of the standard-model extension, including both CPT-even and CPT-odd terms, are provided in the literature [3]. Various limits of this theory are of direct relevance to experiments testing aspects of quantum electrodynamics (QED). In this talk, attention is primarily given to the special limit that produces a CPT- and Lorentz-violating theory for a charged fermion interacting via the electromagnetic force [3]. As an explicit example, here are the terms appearing in the lagrangian extension of the usual quantum theory of photons, electrons, and positrons. In units with $\hbar = c = 1$, the standard QED lagrangian is

$$\mathcal{L}^{\text{QED}} = \overline{\psi}\gamma^\mu(\tfrac{1}{2}i\overleftrightarrow{\partial}_\mu - qA_\mu)\psi - m\overline{\psi}\psi - \tfrac{1}{4}F_{\mu\nu}F^{\mu\nu} \quad . \tag{1}$$

In the fermion sector, there are two CPT-breaking terms:

$$\mathcal{L}_e^{\text{CPT}} = -a_\mu \overline{\psi}\gamma^\mu \psi - b_\mu \overline{\psi}\gamma_5 \gamma^\mu \psi \quad , \quad , \tag{2}$$

while there is one possibility in the photon sector:

$$\mathcal{L}_\gamma^{\text{CPT}} = \tfrac{1}{2}(k_{AF})^\kappa \epsilon_{\kappa\lambda\mu\nu} A^\lambda F^{\mu\nu} \quad . \tag{3}$$

The possible Lorentz-violating but CPT-preserving terms in the fermion sector are:

$$\mathcal{L}_e^{\text{Lorentz}} = c_{\mu\nu}\overline{\psi}\gamma^\mu(\tfrac{1}{2}i\overleftrightarrow{\partial}^\nu - qA^\nu)\psi + d_{\mu\nu}\overline{\psi}\gamma_5\gamma^\mu(\tfrac{1}{2}i\overleftrightarrow{\partial}^\nu - qA^\nu)\psi - \tfrac{1}{2}H_{\mu\nu}\overline{\psi}\sigma^{\mu\nu}\psi \quad , \tag{4}$$

while the only possibility in the photon sector is:

$$\mathcal{L}_\gamma^{\text{Lorentz}} = -\tfrac{1}{4}(k_F)_{\kappa\lambda\mu\nu}F^{\kappa\lambda}F^{\mu\nu} \quad . \tag{5}$$

In the above expressions, the unconventional coupling coefficients govern the magnitude of the CPT- and Lorentz-violating effects and are expected to depend on the small ratio m_W/M_P. Note that all the extra couplings are hermitian. It can be shown using field redefinitions that some coupling-coefficient components are physically unobservable. The reader is referred to the literature [3] for details about this and other issues, and for more information about the notation used above.

OVERVIEW OF SOME EXPERIMENTAL TESTS

Most experiments testing Lorentz invariance or CPT symmetry are likely to be insensitive to the extra couplings in the standard-model extension due to the expected heavy suppression factors. A few experiments of exceptional sensitivity could bound or in principle detect these effects despite the suppression. In such cases, the standard-model extension can be used as a quantitative theoretical guide to potential experimental signals. It also offers the possibility of analyzing and comparing bounds on CPT and Lorentz violation arising from different experiments.

At present, implications of the standard-model extension have been studied for CPT and Lorentz tests that involve: observations of neutral-meson oscillations [4-7], measurements of particle and antiparticle properties in Penning traps [8,9], spectroscopic comparisons of hydrogen and antihydrogen [10,11], determination of photon properties [3], and baryon-number generation [12]. A variety of additional studies are in progress, notably one [13] establishing the implications for the standard-model extension of high-precision clock-comparison experiments [14].

This section of the talk provides a short summary of a subset of the results obtained. The use of hydrogen and antihydrogen spectroscopy to test CPT and Lorentz symmetries is described in the following three sections.

The flavor oscillations of certain neutral-meson systems provide a valuable interferometric tool for studying CP violation. The effective hamiltonian for the time evolution of a P-meson state, where P represents one of the neutral K, D, B_d, or B_s mesons, depends on two kinds of (indirect) CP violation. The first involves T violation with CPT invariance and is conventionally described with a complex parameter ϵ_P. The second involves CPT violation with T invariance and is described with a complex parameter δ_P. The standard-model extension can be used to derive an expression for δ_P [7].

It turns out that flavor oscillations in neutral-P systems are sensitive to only one type of CPT-violating term in the standard-model extension, $-a_\mu^q \bar{q}\gamma^\mu q$, where q is a quark field and a_μ^q is a spacetime-constant coupling coefficient with value dependent on the quark flavor q. None of the other experiments discussed in this talk involve flavor changes, and it has been shown that as a result these other experiments are insensitive to a_μ^q-type coupling coefficients. In this respect, the bounds on CPT violation from neutral-meson tests of CPT are entirely disjoint from those of other experiments.

In the observer frame in which the Lorentz-violating coupling coefficients are defined, denote the P-meson four-velocity by $\beta^\mu \equiv \gamma(1, \vec{\beta})$. Then, at leading order in all the standard-model coupling coefficients, the expression for δ_P is [7]

$$\delta_P \approx \frac{\gamma(\Delta a_0 - \vec{\beta} \cdot \Delta \vec{a})}{\Delta m} i \sin \hat{\phi} e^{i\hat{\phi}} \quad . \tag{6}$$

In this equation, $\Delta a_\mu \equiv a_\mu^{q_2} - a_\mu^{q_1}$, where q_1 and q_2 represent the valence-quark flavors in the P meson. Also, $\hat{\phi} \equiv \tan^{-1}(2\Delta m/\Delta\gamma)$, where the mass and decay-rate differences between the P-meson eigenstates are, respectively, Δm and $\Delta \gamma$.

The expression for δ_P implies a proportionality between the real and imaginary parts of δ_P [5]. Note that the magnitude of δ_P can vary with P because the couplings a_μ^q are flavor dependent [5], so the magnitude of CPT-violating effects may differ in distinct neutral-meson systems. For instance, the magnitude of CPT violation might grow with the mass of the quarks involved, as the Yukawa couplings do in the standard model. Also, the explicit dependence in Eq. (6) of δ_P on the boost magnitude and orientation implies several types of potentially observable effect including, for instance, larger CPT-violating effects in boosted mesons [7]. Experiments involving mesons with different momenta may therefore have different CPT reaches. The best reported bounds to date come from the kaon system [4]. Recently, two CERN experiments [6] have obtained results for the B_d system, following the observation [5] that existing data already suffice to yield CPT limits. Other studies are ongoing.

A number of experiments that test CPT and Lorentz symmetries in a different way have been performed with the goal of comparing particle and antiparticle properties. An important technique is the use of a Penning trap to confine single particles over relatively large time scales while high-precision measurements are taken of properties such as anomaly and cyclotron frequencies [8]. Experiments of this type can constrain, for example, the coupling coefficients in the fermion sector of the extended QED. Possible observable signals in the context of this theory, the corresponding relevant figures of merit, and the associated CPT and Lorentz reaches have been obtained [9]. As just one example, using existing technology and implementing a relatively minor change in experimental procedure, Penning-trap experiments comparing the anomalous magnetic moments of electrons and positrons could place a bound of roughly 10^{-20} on a figure of merit involving the spatial components of the coefficient b_μ.

The extra terms (3) and (5) in the QED extension represent modifications to photon properties. It turns out that the ensuing generalized Maxwell equations describe two independent propagating degrees of freedom as in the conventional case [3]. Typically, however, each has a distinct dispersion relation, which implies several interesting effects. For example, the vacuum becomes birefringent, so that in the presence of the CPT and Lorentz violation an electromagnetic wave propagating in the vacuum exhibits properties similar to those displayed by conventional radiation traveling in an optically anisotropic and gyrotropic transparent crystal having spatial dispersion of the axes. Behavior of this type can be constrained from the observed absence of birefringence on radio waves propagating over cosmological distances. The components of the CPT-odd coefficient $(k_{AF})_\mu$ are presently bounded to $\lesssim 10^{-42}$ GeV [15,16], although a disputed claim [17,18] exists for a nonzero effect with $|\vec{k}_{AF}| \sim 10^{-41}$ GeV. The rotation-invariant irreducible component of the CPT-even coefficient $(k_F)_{\kappa\lambda\mu\nu}$ is bounded to $\lesssim 10^{-23}$ by cosmic-ray existence [19] and other experiments. The rotation-violating irreducible components of $(k_F)_{\kappa\lambda\mu\nu}$ could in principle be bounded to about 10^{-27} with existing techniques seeking cosmological birefringence [3], but no actual limit has been obtained to date.

The CPT-even term in Eq. (5) introduces no theoretical difficulties. However, the CPT-odd term in Eq. (3) can generate negative contributions to the energy [15]. This may represent a theoretical difficulty and indicates $(k_{AF})^\kappa$ vanishes [3], which would be in agreement with the tight experimental bound from cosmological birefringence. It can be argued that a zero value of $(k_{AF})_\mu$ is acceptable theoretically despite the possibility of radiative corrections from diagrams involving the CPT-violating couplings in the fermion sector because the one-loop effects are finite.

1S-2S SPECTROSCOPY IN FREE HYDROGEN AND ANTIHYDROGEN

The remainder of this talk addresses the possibility of searching for CPT and Lorentz violations by making high-precision comparisons of the spectra of hydrogen and antihydrogen [11]. The feasibility of the idea of comparative tests [10] has received a boost following the recent production and observation of antihydrogen [20,21], and several proposals for antihydrogen spectroscopy have been advanced. In the near future, the antihydrogen fine structure and Lamb shift may be obtained within a few percent by observations on a relativistic antihydrogen beam [22]. A more ambitious goal is to measure the two-photon 1S-2S transition in antihydrogen, which is expected to have a natural linewidth of only 1.3 Hz and is therefore a promising candidate for high-precision spectroscopy. Proposed experiments [23] would provide a comparison of the 1S-2S transitions in spin-polarized hydrogen and antihydrogen confined within a magnetic trap. For hydrogen, a cold atomic beam has been used to measure the 1S-2S transition frequency to 3.4 parts in 10^{14} [24], while trapping techniques have yielded a frequency precision of about 10^{-12} [25]. A limiting accuracy of about 10^{-18} may be attainable [26].

A theoretical analysis of signals for CPT and Lorentz violations in hydrogen and antihydrogen spectroscopy is feasible [11] in the context of the QED extension described in the second part of this talk. In this section, possible effects on the free-atom 1S-2S transition are considered. These are relevant, for example, to the experiments with cold atomic beams of hydrogen [24]. The next section treats the trapped-atom case. A detailed theoretical treatment of the proposed experiments with relativistic beams [22], which are expected to have significantly poorer frequency resolutions than those based on other techniques, remains to be performed and is not discussed here. Note, however, that all the experimental situations discussed below are sensitive only to spatial or mixed spatio-temporal components of the CPT- and Lorentz-violating couplings in the comoving Earth frame, whereas a boost can induce sensitivity to purely timelike components and can enhance CPT- and Lorentz-violating effects [7].

To calculate effects on the free-atom 1S and 2S energy levels, the modified Dirac equation for a four-component electron field ψ in the proton Coulomb potential

$$A^\mu = \frac{|e|}{4\pi r}(1, \vec{0}) \tag{7}$$

is needed. The desired equation is found from Eqs. (1), (2), and (4) to be

$$\left(i\gamma^\mu D_\mu - m_e - a^e_\mu \gamma^\mu - b^e_\mu \gamma_5 \gamma^\mu - \tfrac{1}{2} H^e_{\mu\nu} \sigma^{\mu\nu} + i c^e_{\mu\nu} \gamma^\mu D^\nu + i d^e_{\mu\nu} \gamma_5 \gamma^\mu D^\nu \right)\psi = 0 \quad , \tag{8}$$

where m_e is the electron mass and the covariant derivative is

$$iD_\mu \equiv i\partial_\mu - qA_\mu \tag{9}$$

with the electron charge being $q = -|e|$. Both a free electron and a free proton have distinct CPT- and Lorentz-violating coupling coefficients in the typical case [3,9], so superscripts e have been added to the couplings in Eq. (8). In what follows, the corresponding couplings for a free proton are denoted by a^p_μ, b^p_μ, $H^p_{\mu\nu}$, $c^p_{\mu\nu}$, $d^p_{\mu\nu}$. Note that, as mentioned following Eq. (5), certain combinations of the electron and proton couplings can be shown on general grounds to be physically unobservable [3]. This is true, for example, of the coefficients a^e_μ and a^p_μ. Although all couplings are kept explicitly in the derivations that follow, it is to be expected that the ensuing possible spectroscopic signals in hydrogen and antihydrogen are independent of the unobservable couplings.

Since the coupling coefficients are expected to be highly suppressed, it is reasonable to calculate the dominant effects on the hydrogen and antihydrogen spectra via perturbation theory in relativistic quantum mechanics. The relevant unperturbed hamiltonians and the corresponding eigenstates are identical for hydrogen and antihydrogen, as are all perturbative effects from conventional quantum electrodynamics. The unconventional coupling coefficients introduce hermitian perturbations that can differ for hydrogen and antihydrogen. For the electron and positron, the explicit forms of these perturbations follow from Eq. (8) after application of suitable field redefinitions to obtain the hamiltonian and, for the positron, a standard charge-conjugation procedure [9]. The CPT and Lorentz violations from the proton sector also produce energy perturbations, which at leading order can be derived using relativistic two-fermion techniques [27].

In what follows, the uncoupled angular-momentum quantum numbers for the S-state electron/positron and for the proton/antiproton are denoted by $J = 1/2$ and $I = 1/2$, respectively. Their components along the spin-quantization axis are m_J, m_I, and the corresponding basis states are denoted $|m_J, m_I\rangle$. Note that distinct real experiments are likely to involve different spin-quantization axes relative to any single specified inertial frame, so comparisons between various experiments may require care in allowing for possible geometrical factors.

The result of the perturbative calculation is that the 1S and 2S levels in hydrogen are shifted by identical amounts ΔE^H [11]:

$$\Delta E^H(m_J, m_I) \approx (a_0^e + a_0^p - c_{00}^e m_e - c_{00}^p m_p) + (-b_3^e + d_{30}^e m_e + H_{12}^e)\frac{m_J}{|m_J|}$$
$$+ (-b_3^p + d_{30}^p m_p + H_{12}^p)\frac{m_I}{|m_I|} \quad , \tag{10}$$

where m_p is the proton mass.

A similar calculation for antihydrogen also yields equal 1S and 2S level shifts $\Delta E^{\overline{H}}$, given by Eq. (10) with the substitutions

$$a_\mu^e \to -a_\mu^e \quad , \quad d_{\mu\nu}^e \to -d_{\mu\nu}^e \quad , \quad H_{\mu\nu}^e \to -H_{\mu\nu}^e \quad ; \quad a_\mu^p \to -a_\mu^p \quad , \quad d_{\mu\nu}^p \to -d_{\mu\nu}^p \quad , \quad H_{\mu\nu}^p \to -H_{\mu\nu}^p \quad . \tag{11}$$

Note that in all these expressions the leading-order contributions from the proton/antiproton have the same mathematical form as those from the electron/positron.

The electron (positron) and proton (antiproton) angular momenta are coupled through the hyperfine interaction. The relevant basis states are thus linear combinations $|F, m_F\rangle$ of the $|m_J, m_I\rangle$ states, where F is the total angular-momentum quantum number and m_F is its projection on the quantization axis. For the two-photon 1S-2S transition, the selection rules are $\Delta F = 0$ and $\Delta m_F = 0$ [28], which allows four 1S-2S transitions in hydrogen and four in antihydrogen. These transitions involve states with identical spin configurations. However, for hydrogen the result (10) of the perturbative calculation implies that the leading-order level shifts for 1S and 2S hydrogen states with the same spin configuration are identical. The same follows from Eq. (11) for antihydrogen. Therefore, the 1S-2S frequencies are unaffected at leading order for all these transitions. Indeed, this result could have been anticipated from the discussion in Ref. [9] showing that observable CPT-violating effects must also involve spin-flip processes and CT violation.

In summary, *no leading-order 1S-2S spectroscopic signal occurs for Lorentz or CPT violation in free hydrogen or in free antihydrogen* [11].

Non-leading level shifts can produce observable signals, but these are suppressed. The dominant subleading effects from electron/positron and proton/antiproton CPT- and Lorentz-violation terms are relativistic corrections suppressed by at least $\alpha^2 \simeq 5 \times 10^{-5}$. As an explicit example, consider the coupling coefficient b_μ^e in Eq. (8). If this coupling is nonzero, the $m_F = 0 \to m_{F'} = 0$ is unaffected but a subleading-order frequency shift in the $m_F = 1 \to m_{F'} = 1$ transition appears. It is given by

$$\delta\nu_{1S-2S}^H \approx -\frac{\alpha^2 b_3^e}{8\pi} \quad . \tag{12}$$

The potential signals from subleading effects are suppressed to the extent that feasible $g - 2$ experiments could exclude their observation in free hydrogen or antihydrogen. As mentioned above, an electron-positron $g - 2$ comparison using present technology with a minor change in experimental procedure could attain a tight bound on b_3^e [9]. The effect of a nonzero b_3^e at this level on the 1S-2S frequency in free hydrogen would be to produce a nonzero frequency shift $\delta\nu_{1S-2S}^H \lesssim 5$ μHz, which is below the resolution of the 1S-2S line center.

Similarly, bounds attainable in Penning-trap experiments comparing $g - 2$ for protons and antiprotons could exclude observable signals in 1S-2S transitions. The basic reason why $g - 2$ experiments are so effective in constraining possible violations is that they involve spin-flip transitions that exhibit unsuppressed sensitivity to CPT and Lorentz breaking. The $g - 2$ experiments have an absolute frequency resolution of about 1 Hz. Although the idealized line-center resolution for free-hydrogen or free-antihydrogen 1S-2S transitions is about three orders of magnitude better, the CPT- and Lorentz-violating effects on these transitions are suppressed by about five orders of magnitude and so the net sensitivity of the $g - 2$ experiments is better. Note that it is inappropriate in this context to compare the conventional figure of merit for CPT breaking in $g-2$ experiments [29],

$$r_g = \frac{|g_{e^-} - g_{e^+}|}{g_{\mathrm{av}}} \lesssim 2 \times 10^{-12} \quad, \tag{13}$$

with the idealized resolution of the 1S-2S line,

$$\Delta\nu_{1S-2S}/\nu_{1S-2S} \simeq 10^{-18} \quad . \tag{14}$$

These two quantities are physically very different [9]. A relevant comparison would involve the same physics, such as the absolute frequency resolution and sensitivity to CPT- and Lorentz-violating effects used above.

1S-2S SPECTROSCOPY IN TRAPPED HYDROGEN AND ANTIHYDROGEN

The results in the previous section for free hydrogen and antihydrogen may be modified in the presence of external fields, which can induce transitions between states with different spin configurations. External fields are present for the class of proposed experiments [23] involving spectroscopy of hydrogen or antihydrogen confined within a magnetic trap with an axial bias magnetic field, such as an Ioffe-Pritchard trap [30]. Next, a theoretical analysis of possible signals of CPT and Lorentz violation in this context is described. In what follows, the four 1S hyperfine Zeeman levels in hydrogen are denoted by $|a\rangle_1$, $|b\rangle_1$, $|c\rangle_1$, $|d\rangle_1$, in order of increasing energy in a magnetic field B. The corresponding four 2S levels are denoted $|a\rangle_2$, $|b\rangle_2$, $|c\rangle_2$, $|d\rangle_2$. The same notation is used for the 1S and 2S hyperfine Zeeman levels in antihydrogen.

The Zeeman levels $|a\rangle_n$ and $|c\rangle_n$, $n = 1, 2$, are mixed-spin states. For hydrogen, they are given in terms of the basis states $|m_J, m_I\rangle$ by

$$\begin{aligned}|a\rangle_n &= \cos\theta_n |-\tfrac{1}{2}, \tfrac{1}{2}\rangle - \sin\theta_n |\tfrac{1}{2}, -\tfrac{1}{2}\rangle \quad, \\ |c\rangle_n &= \sin\theta_n |-\tfrac{1}{2}, \tfrac{1}{2}\rangle + \cos\theta_n |\tfrac{1}{2}, -\tfrac{1}{2}\rangle \quad . \end{aligned} \tag{15}$$

The mixing angles θ_n are given by

$$\tan 2\theta_n \approx \frac{(51 \text{ mT})}{n^3 B} \quad . \tag{16}$$

Expressions similar to (15) hold for antihydrogen, but the spin labels are reversed.

For hydrogen, in the absence of perturbations and prior to excitation, the low-field-seeker states $|c\rangle_1$ and $|d\rangle_1$ are confined in the trap. Spin-exchange collisions cause the $|c\rangle_1$ occupation to decrease with time: $|c\rangle_1 + |c\rangle_1 \to |b\rangle_1 + |d\rangle_1$. The primary states in the trap are therefore $|d\rangle_1$. Moreover, the transition $|d\rangle_1 \to |d\rangle_2$ is field independent for small magnetic fields. It might therefore seem reasonable to perform an experiment comparing the frequency ν_d^H for the 1S-2S transition $|d\rangle_1 \to |d\rangle_2$ in hydrogen with the frequency $\nu_d^{\overline{H}}$ for the corresponding transition in antihydrogen. However, the $|d\rangle_n$ states in hydrogen have no spin mixing, so the frequency is unaffected to leading order by CPT- and Lorentz-breaking effects. A similar result is true for antihydrogen. This means that [11]

$$\delta\nu_d^H = \delta\nu_d^{\overline{H}} \simeq 0 \tag{17}$$

to leading order.

Thus, *no leading-order 1S-2S spectroscopic signal for Lorentz or CPT violation occurs for unmixed-spin states in hydrogen or antihydrogen confined in a magnetic trap with an axial bias field [11]*.

It therefore appears worthwhile theoretically to examine 1S-2S transitions involving mixed-spin states. Indeed, for the $|c\rangle_1 \to |c\rangle_2$ transition in hydrogen the spin mixing induces an unsuppressed frequency shift

$$\delta\nu_c^H \approx -\frac{\kappa}{2\pi}(b_3^e - b_3^p - d_{30}^e m_e + d_{30}^p m_p - H_{12}^e + H_{12}^p) \quad , \tag{18}$$

where

$$\kappa \equiv \cos 2\theta_2 - \cos 2\theta_1 \quad . \tag{19}$$

The corresponding transition in antihydrogen in the same magnetic field exhibits a frequency shift $\delta\nu_c^{\overline{H}}$ given by an expression of the form (18) except with opposite signs for b_3^e and b_3^p.

The unsuppressed sensitivity to CPT and Lorentz breaking of the $|c\rangle_1 \to |c\rangle_2$ transition represents a theoretical advantage of a factor of about $4/\alpha^2 \simeq 10^5$ over the suppressed effects from the $|d\rangle_1 \to |d\rangle_2$ transition. Since the frequencies ν_c^H and $\nu_c^{\overline{H}}$ vary with the spatial components of the CPT-violating couplings b_μ^e and b_μ^p in the comoving Earth frame, they would exhibit diurnal variations. Moreover, frequency measurements in a given magnetic trapping field would also display a nonzero difference

$$\Delta\nu_{1S-2S,c} \equiv \nu_c^H - \nu_c^{\overline{H}} \approx -\frac{\kappa}{\pi}(b_3^e - b_3^p) \quad . \tag{20}$$

This difference varies with the amount of spin mixing according to the parameter κ, so a maximal value of κ is theoretically desirable. It is $\kappa \simeq 0.67$, and it is attained when $B_0 \simeq 0.01$ T.

From the experimental perspective, the 1S-2S transition $|c\rangle_1 \to |c\rangle_2$ is likely to be less advantageous because it is field dependent in both hydrogen and antihydrogen. An experiment would therefore have to address the issue of Zeeman broadening from the inhomogeneous trapping fields. For instance, at $B \simeq 10$ mT the 1S-2S linewidth for the $|c\rangle_1 \to |c\rangle_2$ transition is broadened to over 1 MHz for both hydrogen and antihydrogen even at a temperature of 100μK. Although present methods might reduce the impact of this effect, it would seem necessary that other techniques be developed if resolutions of the order of the natural linewidth are to be reached.

To summarize, *unsuppressed 1S-2S spectroscopic signals for Lorentz and CPT violation appear for transitions involving mixed-spin states in hydrogen or antihydrogen atoms confined in a magnetic trap with an axial bias field* [11].

HYPERFINE SPECTROSCOPY IN HYDROGEN AND ANTIHYDROGEN

The remainder of this talk addresses the issue of possible CPT- and Lorentz-violating signals in frequency measurements of hyperfine Zeeman transitions in trapped hydrogen and antihydrogen [11]. The interest in these is partially motivated by the resolution below 1 mHz that has already been attained in transitions between $F = 0$ and $F' = 1$ hyperfine levels of a hydrogen maser [31].

Perturbative calculations along the lines described above show that all four hyperfine levels in the 1S ground state of hydrogen are shifted by CPT- and Lorentz-violating effects. One contribution to the shifts, $a_0^e + a_0^p - c_{00}^e m_e - c_{00}^p m_p$, is identical for all four levels and therefore has no effect on any frequencies. There are also spin-dependent energy shifts, given by [11]

$$\begin{aligned}
\Delta E_a^H &\approx \hat{\kappa}(b_3^e - b_3^p - d_{30}^e m_e + d_{30}^p m_p - H_{12}^e + H_{12}^p) \quad , \\
\Delta E_b^H &\approx b_3^e + b_3^p - d_{30}^e m_e - d_{30}^p m_p - H_{12}^e - H_{12}^p \quad , \\
\Delta E_c^H &\approx -\Delta E_a^H \quad , \\
\Delta E_d^H &\approx -\Delta E_b^H \quad ,
\end{aligned} \tag{21}$$

where

$$\hat{\kappa} \equiv \cos 2\theta_1 \tag{22}$$

is a parameter analogous to κ of Eq. (19) that grows with B, with $\hat{\kappa} \simeq 1$ when $B \simeq 0.3$ T.

To begin, suppose the magnetic field vanishes. Then, $\hat{\kappa} = 0$ and so Eq. (21) shows that the states $|a\rangle_1$ and $|c\rangle_1$ are unchanged. However, the energies of $|b\rangle_1$ and $|d\rangle_1$ shift equally in magnitude but oppositely in sign. Thus, even for $B = 0$ the three $F = 1$ levels are split.

If instead the magnetic field is nonzero, then the energies of all four hyperfine levels are changed. Consider first the conventional hydrogen maser, which uses a small magnetic field and involves the (approximately field-independent) transition $|c\rangle_1 \to |a\rangle_1$. For this situation, the value of $\hat{\kappa}$ is roughly 10^{-4} so the spin-mixing is small. This would act as a suppression factor for CPT- and Lorentz-breaking effects in possible high-precision measurements of the maser σ line $|c\rangle_1 \to |a\rangle_1$.

In contrast, unsuppressed frequency differences appear between the field-dependent transitions $|d\rangle_1 \to |a\rangle_1$ and $|b\rangle_1 \to |a\rangle_1$. Equation (21) gives

$$|\Delta\nu_{d-b}^H| \approx \frac{1}{\pi}|b_3^e + b_3^p - d_{30}^e m_e - d_{30}^p m_p - H_{12}^e - H_{12}^p| \quad . \tag{23}$$

This difference would vary diurnally in the comoving Earth frame, as occurs with the shifts (18), so in principle a measurement of $|\Delta\nu_{d-b}^H|$ in hydrogen alone could provide a signal of CPT and Lorentz violation. However, in practice the attainable frequency resolution is likely to be affected by the broadening due to field inhomogeneities. An experiment of this type would also need to address the issue of distinguishing the signal from possible backgrounds due to residual Zeeman splittings.

Instead, one could envisage using a field-independent transition point to minimize the frequency dependence on the magnetic field and making a direct comparison of hydrogen and antihydrogen transition frequencies to avoid issues with the background splittings. Consider, for example, an experiment performing high-resolution radiofrequency spectroscopy in trapped hydrogen and antihydrogen on the $|d\rangle_1 \to |c\rangle_1$ transition at the field-independent transition point $B \simeq 0.65$ T. To avoid Doppler broadening, cooling to temperatures of 100 μK with a good signal-to-noise ratio is likely to be needed. Also, the relatively high bias field suggests potentially larger field inhomogeneities would occur, so a stiff box shape would be preferable for the trapping potential. Under these circumstances, it may be possible to attain frequency resolutions of order 1 mHz.

In a magnetic field of 0.65 T, the state $|c\rangle_1$ in hydrogen is well approximated as a spin-polarized level with $|m_J, m_I\rangle = |1/2, -1/2\rangle$. This means that the transition of interest, $|d\rangle_1 \to |c\rangle_1$, involves a proton spin flip, which in turn implies a signal dependence only on CPT- and Lorentz-violating effects for the proton. Explicit calculation shows that the frequency shift for hydrogen is

$$\delta\nu_{c\to d}^H \approx \frac{1}{\pi}(-b_3^p + d_{30}^p m_p + H_{12}^p) \quad , \tag{24}$$

while that for antihydrogen is

$$\delta\nu_{c\to d}^{\overline{H}} \approx \frac{1}{\pi}(b_3^p + d_{30}^p m_p + H_{12}^p) \quad , \tag{25}$$

confirming the expected dependence on proton coupling coefficients.

Like the quantity $|\Delta\nu_{d-b}^H|$ of Eq. (23), diurnal variations of the the frequencies $\nu_{c\to d}^H$ and $\nu_{c\to d}^{\overline{H}}$ would provide a signal for CPT and Lorentz violation. However, the difference

$$\Delta\nu_{c\to d} \equiv \nu_{c\to d}^H - \nu_{c\to d}^{\overline{H}} \approx -\frac{2b_3^p}{\pi} \tag{26}$$

between these frequencies has the potential to provide an instantaneous, clean, and accurate test of CPT-violating couplings b_3^p for the proton.

Relevant figures of merit for the diurnal and instantaneous signals in Eqs. (24), (25), and (26) (as well as ones for other signals mentioned above in this and earlier sections) can be defined following the methods developed for Penning-trap tests [9]. For the instantaneous signal in Eq. (26), an appropriate choice is

$$r_{rf,c\to d}^H \equiv \frac{|(\mathcal{E}_{1,d}^H - \mathcal{E}_{1,c}^H) - (\mathcal{E}_{1,d}^{\overline{H}} - \mathcal{E}_{1,c}^{\overline{H}})|}{\mathcal{E}_{1,\text{av}}^H}$$
$$\approx \frac{2\pi|\Delta\nu_{c\to d}|}{m_H} \quad , \tag{27}$$

where m_H is the atomic mass of hydrogen and where the relativistic energies in the ground-state hyperfine levels are denoted by $\mathcal{E}_{1,d}^H$, $\mathcal{E}_{1,c}^H$ for hydrogen and by $\mathcal{E}_{1,d}^{\overline{H}}$, $\mathcal{E}_{1,c}^{\overline{H}}$ for antihydrogen. Suppose, for instance that a 1 mHz frequency resolution could indeed be reached in an experiment of this type. This would represent an estimated upper bound on the figure of merit (27) of approximately $r^H_{rf,c\to d} \lesssim 5 \times 10^{-27}$. The associated constraint on the coefficient b^p_3 would be $|b^p_3| \lesssim 10^{-18}$ eV. This is more than four orders of magnitude better than bounds attainable from 1S-2S transitions and roughly three orders of magnitude better than estimated attainable bounds [9] from $g-2$ experiments in Penning traps.

To summarize, *unsuppressed Zeeman hyperfine spectroscopic signals for Lorentz or CPT violation appear for transitions involving spin-flip hyperfine states in hydrogen and antihydrogen atoms confined in a magnetic trap with an axial bias field [11].*

ACKNOWLEDGMENTS

V.A.K. thanks Orfeu Bertolami, Don Colladay, Rob Potting, Stuart Samuel, and Rick Van Kooten for collaborations leading to some of the results described in this talk. This work is supported in part by the Department of Energy under grant number DE-FG02-91ER40661 and by the National Science Foundation under grant number PHY-9503756.

REFERENCES

1. V.A. Kostelecký and S. Samuel, Phys. Rev. Lett. **63** (1989) 224; *ibid.*, **66** (1991) 1811; Phys. Rev. D **39** (1989) 683; *ibid.*, **40** (1989) 1886; V.A. Kostelecký and R. Potting, Nucl. Phys. B **359** (1991) 545; Phys. Lett. B **381** (1996) 89.
2. The discrete symmetries C, P, T are discussed, for example, in R.G. Sachs, *The Physics of Time Reversal* (University of Chicago Press, Chicago, 1987).
3. D. Colladay and V.A. Kostelecký, Phys. Rev. D **55** (1997) 6760; preprint IUHET 359, Phys. Rev. D in press (hep-ph/9809521).
4. B. Schwingenheuer et al., Phys. Rev. Lett. **74** (1995) 4376; L.K. Gibbons et al., Phys. Rev. D **55** (1997) 6625; R. Carosi et al., Phys. Lett. B **237** (1990) 303.
5. V.A. Kostelecký and R. Potting, in D.B. Cline, ed., *Gamma Ray–Neutrino Cosmology and Planck Scale Physics* (World Scientific, Singapore, 1993) (hep-th/9211116); Phys. Rev. D **51** (1995) 3923; D. Colladay and V. A. Kostelecký, Phys. Lett. B **344** (1995) 259; Phys. Rev. D **52** (1995) 6224; V.A. Kostelecký and R. Van Kooten, Phys. Rev. D **54** (1996) 5585.
6. OPAL Collaboration, R. Ackerstaff et al., Z. Phys. C **76** (1997) 401; DELPHI Collaboration, M. Feindt et al., preprint DELPHI 97-98 CONF 80 (July 1997).
7. V.A. Kostelecký, Phys. Rev. Lett. **80** (1998) 1818.
8. P.B. Schwinberg, R.S. Van Dyck, Jr., and H.G. Dehmelt, Phys. Lett. A **81** (1981) 119; Phys. Rev. D **34** (1986) 722; L.S. Brown and G. Gabrielse, Rev. Mod. Phys. **58** (1986) 233; R.S. Van Dyck, Jr., P.B. Schwinberg, and H.G. Dehmelt, Phys. Rev. Lett. **59** (1987) 26; G. Gabrielse et al., *ibid.*, **74** (1995) 3544.
9. R. Bluhm, V.A. Kostelecký and N. Russell, Phys. Rev. Lett. **79** (1997) 1432; Phys. Rev. D **57** (1998) 3932.
10. See, for example, M. Charlton et al., Phys. Rep. **241** (1994) 65; J. Eades, ed., *Antihydrogen* (J.C. Baltzer, Geneva, 1993).
11. R. Bluhm, V.A. Kostelecký and N. Russell, Indiana University preprint IUHET 388 (1998).
12. O. Bertolami et al., Phys. Lett. B **395** (1997) 178.
13. V.A. Kostelecký and C.D. Lane, in preparation.
14. V.W. Hughes, H.G. Robinson, and V. Beltran-Lopez, Phys. Rev. Lett. **4** (1960) 342; R.W.P. Drever, Philos. Mag. **6** (1961) 683; J.D. Prestage et al., Phys. Rev. Lett. **54** (1985) 2387; S.K. Lamoreaux et al., *ibid.*, **57** (1986) 3125; T.E. Chupp et al., *ibid.*, **63** (1989) 1541.
15. S.M. Carroll, G.B. Field, and R. Jackiw, Phys. Rev. D **41** (1990) 1231.
16. P. Haves and R.G. Conway, Mon. Not. R. Astr. Soc. **173** (1975) 53P; J.N. Clarke, P.P. Kronberg and M. Simard-Normandin, *ibid.*, **190** (1980) 205.
17. B. Nodland and J.P. Ralston, Phys. Rev. Lett. **78** (1997) 3043; Phys. Rev. Lett. **79** (1997) 1958; astro-ph/9706126.
18. S.M. Carroll and G.B. Field, Phys. Rev. Lett. **79** (1997) 2394; D.J. Eisenstein and E.F. Bunn, Phys. Rev. Lett. **79** (1997) 1957; J.P. Leahy, astro-ph/9704285; J.F.C. Wardle, R.A. Perley and M.H. Cohen, Phys. Rev. Lett. **79** (1997) 1801; T.J. Loredo, E.E. Flanagan and I.M. Wasserman, Phys. Rev. D **56** (1997) 7507.
19. S. Coleman and S. Glashow, Phys. Lett. B **405** (1997) 249.
20. G. Baur et al., Phys. Lett. B **368** (1996) 251.
21. G. Blanford et al., Phys. Rev. Lett. **80** (1998) 3037.
22. G. Blanford et al., Phys. Rev. D **57** (1998) 6649.

23. B. Brown *et al.*, Nucl. Phys. B (Proc. Suppl.) **56A** (1997) 326; M.H. Holzscheiter *et al.*, *ibid.*, 336.
24. T. Udem *et al.*, Phys. Rev. Lett. **79** (1997) 2646.
25. C.L. Cesar *et al.*, Phys. Rev. Lett. **77** (1996) 255.
26. See, for example, T.W. Hänsch, in D.J. Wineland, C.E. Wieman, and S.J. Smith, eds., *Atomic Physics 14* (A.I.P. New York, 1995).
27. See, for example, G. Breit, Phys. Rev. **34** (1929) 553.
28. B. Cagnac, G. Grynberg, and F. Biraben, J. Physique **34** (1973) 845.
29. See, for example, R.M. Barnett *et al.*, Review of Particle Properties, Phys. Rev. D **54** (1996) 1.
30. Y.V. Gott, M.S. Ioffe, and V.G. Tel'kovskii, Nucl. Fusion, Suppl. Pt. 3 (1962) 1045; D.E. Pritchard, Phys. Rev. Lett. **51** (1983) 1336.
31. N.F. Ramsey, Physica Scripta **T59** (1995) 323.

Non-destructive, Absolute Mass Determination of Sub-micrometer Sized Particles in a Paul-type Trap

Stephan Schlemmer, Jens Illemann, Stefan Wellert, and Dieter Gerlich

Institut für Physik, Gasentladungs- und Ionenphysik,
Technical University of Chemnitz, 09107 Chemnitz, Germany

Abstract. A method for non-destructive determination of the mass of a single submicron particle under ultra high vacuum conditions is presented. In an electrodynamic quadrupole trap the eigenfrequencies, ω_r and ω_z, of the particle motion and thus its q/m is measured by recording the scattered light from a laser beam. This allows for online weighing and determination of the particles charge state. Electron bombardment changes the particle charge state in integral steps of elementary charges due to secondary electron emission. The emission yield has been determined event by event at two primary electron energies. In addition the absolute number of charges has been inferred from these measurements and, therefore, the particles absolute mass as well. A relative reproducibility of $3 \cdot 10^{-4}$ over a period of several hours is achieved. Possible applications for this new technique such as the determination of sticking coefficients are mentioned.

Introduction

Localization, long term containment and isolation from disturbing surrounding influences are the prominent features of ion trapping [1]. This confinement technique is the foundation for a number of precision measurements carried out in the last four decades. Mass spectrometry and optical spectroscopy are two important examples. Besides storage of atomic and molecular ions, quadrupole ion traps have also been used for storage of single micrometer size dust particles [2] or droplets [3] since the late 1950's. Whereas in the beginning of these studies the size of the stored droplet was determined via direct observation through a microscope, application of Mie-scattering [4] allowed much more accurate determination of particle radii. In addition, electrostatic balancing of the charged particle was used as a method to measure the particles charge to mass ratio [4]. On the one hand, this technique became known as the electrodynamic balance (EDB) or picobalance, because the balance voltage is a direct measure for relative mass changes, e.g., due to evaporation, as long as the charge state remains the same throughout the experiment. On the other hand, single electron removal by UV-radiation was employed to infer the particle's absolute charge and thus its absolute mass [5]. The accuracy of this method is limited to a few percent due to imperfect determination of the localization by optical means.

Already in the early development of the quadrupole trap, detection of ions was intended to be achieved by resonant excitation at the "eigenfrequency", ω_{res}, of the particular ion in the trap. Since this secular frequency is proportional to the ions charge to mass ratio, resonant detection also serves as an accurate mass analyzer [6]. Unfortunately, this method has not been used for the mass determination of macroscopic particles until very recently [7]. The authors observe a star shape pattern of the particle trajectory in a hyperbolic quadrupole trap using a CCD camera. They determine the secular frequency, ω, and thus also q/m of the particle by adjusting the driving frequency, Ω, to an overtone of ω, since in this case the pattern stabilizes analogous to a Lissajous trajectory.

In this contribution we describe a different laser based non-destructive method of accurate mass determination of sub-micrometer particles in a quadrupole trap. As a first result of this powerful method we present the detection of individual events of secondary electron emission from a single 500 nm in diameter SiO_2 sphere. The inherent precision of the mass determination by a frequency measurement opens up a variety of applications which will be mentioned in the conclusions.

EXPERIMENTAL SECTION

Storage in a Quadrupole Trap

The inhomogeneous electric field

$$\mathbf{E}_0(\mathbf{r},t) = \mathbf{E}_0(\mathbf{r}) \cos(\Omega t) \qquad (1)$$

in an electrodynamic trap gives rise to a time-averaged pseudo-potential

$$V^*(\mathbf{r}) = q^2 \, \mathbf{E}_0(\mathbf{r})^2 / 4 \, m \, \Omega^2, \qquad (2)$$

where q is the particles charge and m its mass. For a quadrupolar field $\mathbf{E}_0^2$ is harmonic in all directions. Therefore the particle

motion along independent coordinates is separable and V* can be expressed for any direction, e.g., for the z-direction as
$$V_z^* = 1/2\, m\, \omega_z^2\, z^2. \qquad (3)$$
Here ω_z is the secular frequency of the particles motion in z-direction. For the geometry used in our trap (Paul-type trap) the motion along the x- and y-direction results in a degenerate solution and is therefore characterized by the same frequency ω_r for both coordinates. Using eqs. 2 and 3 one can determine ω_z, which can be solved for
$$q/m = \tfrac{1}{2}\sqrt{2}\,\omega_z\,\Omega\,/\,\sqrt{(E_{o,z}^2/z^2)}. \qquad (4)$$
For a Paul-trap, $\sqrt{E_{o,z}^2/z^2} = V_0/z_0^2$ is a constant, where V_0 is the amplitude of the applied voltage and $2\,z_0$ is the axial distance of the two cap electrodes. Thus for fixed Ω, the secular frequency ω_z is measured experimentally and q/m is determined. For a realistic Paul-type trap, $\sqrt{E_{o,z}^2/z^2} = |\partial_z E_{o,z}|$ is a constant only in the vicinity of the electrical center of the trap and depends on the geometry of the actual electrodes as well as the applied voltage. Numerical values for $|\partial_z E_{o,z}|$ and $|\partial_r E_{o,r}|$ have been calculated for the actual trap geometry using an ion optics simulation program, SIMION [8].

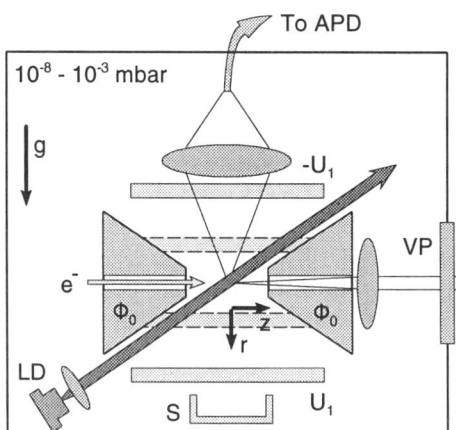

FIGURE 1: Schematic view of the experimental setup. A high voltage at audio frequency ($\Phi_0 = V_0 \cdot \cos(\Omega t)$) is supplied to the cone electrodes. DC voltages ($\pm U_1$) for compensation of the weight of the stored particle are supplied to the rod electrodes. A reservoir with membrane serves as particle source (S) and a collimated laser diode (LD) for illumination. Scattered light is focused onto a multimode fibre and detected by an avalanche photo diode (APD) outside the vacuum chamber or alternatively, observed through the view port (VP). Electrons (e-) from an electron gun are used for changing the charge state of the particle.

Trap-Design and Experimental Setup

Fig. 1 shows a schematic view of the actual trap design and additional features of the experimental setup. The trap consists of two cones, opposing each other at a distance of $2\,z_0 = 6.6$ mm with axial bored holes. Their axis of cylinder symmetry defines the z-axis used in the theoretical descritption of the system. They are surrounded by eight rods which form a cage like structure resembling the ring electrode of a Paul-trap. This open design has been chosen in order to obtain a large solid angle for light detection and also to access the trap volume through several ports with additional tools such as the particle source, the laser beam and the electron gun. The potential ($\Phi_0 = V_0 \cos(\Omega t)$) for the driving field of the quadrupole trap is applied to the cones only. Pairs of the rods are electrically connected and used to control the position in x- and y-direction by applying symmetrical low DC voltages to opposing pairs. In this way the particle can be steered into the center of the trap where the secular frequency shall be measured.

A small UHV compatible loudspeaker is filled with monodisperse 500 nm diameter SiO_2 spheres and serves as a particle source. For injection, the membrane vibrates and ejects particles into the trap. In order to overcome the problem of the particle being able to settle in the trap when injecting particles from an external source which has been described already in the early paper by E. Fischer [9], the background pressure is increased to several 10^{-4} mbar. Friction due to the buffer gas is sufficient to dissipate kinetic energy of the particle while entering the effective potential given by Eq. 2. Light from a collimated laser diode (I ≈ 2.5 mW/mm²) is directed near the center of the trap. Upon storage of a particle scattered light is observed through a view port (VP) using a CCD camera. If several particles are stored at the same time the effective trapping potential is weakened by increasing Ω until only a single particle is left. The residual secular motion of the particle leads to a modulation of the scattered light at the eigenfrequencies ω_r and ω_z. This light is collected by a lens, transferred outside the vacuum in a multimode fibre and transmitted to an avalanche photo diode (APD). With the laser intensity given above and a detection angle of 0.013 sr a maximum power of about 4 pW is scattered from a 500 nm SiO_2 sphere. The corresponding APD-signal is further amplified and filtered electronically, recorded by an AD-converter and Fourier transformed by a computer. The relevant frequency peak for the motion in z-direction, $\omega_z/2\pi$, is found manually and thus the temporal evolution of q/m (Eq. 4) can be followed over a period of hours or even days. Without interfering with the particle we found that the measured q/m remains constant over such long periods of time.

RESULTS

Precision of the q/m Determination

Fig. 2a shows the result of a series of 997 consecutive determinations of ω_z over a period of more than 30 hours. ω_z and therefore q/m remain constant over approx. the first 20 hours. Between two neighboring time steps ω_z increases by a factor of 1.132924(50) and remains again constant for the rest of the measurement. We attribute this step like increase in ω_z to a spontaneous increase in charge of the particle. The measured ratio of the secular frequencies is consistent with an increase from +15 e to +17 e for a single 500 nm in dia. SiO_2 sphere, where $+e$ is the positive elementary charge. Fig 2b illustrates the precision of the experiment. Here the probability of a deviation from the mean value binned in increments of 0.001 Hz is shown for the first 567 data points ($\approx$ 20 h). The envelope of this histogram is fitted to a Gaussian peak shape with a FWHM of 8.7 mHz which corresponds to a precision of 270 ppm for a single q/m determination. Over the period of 20 hours the q/m determination is precise to $4.4 \cdot 10^{-5}$ which results in the five significant figures given above for the frequency ratio at the step. The expected ratio (1.13333) deviates significantly ($3.6 \cdot 10^{-4}$) from the measured. This is due to imperfections of the quadrupolar field even near the center of the trap where the particle is located [10]. It becomes measurable when the particle is lifted to a slightly different location (r-direction, see Fig. 1) upon charging. Its effect is smaller in ω_z than in ω_r. During this long time measurement we observe also a small drift of about -1 ppm/minute which might be due to minor variations of the distance of the cap electrodes (nm/minute). This drift has been corrected in Fig 2b. Measuring a single step, $\Delta q = + n \cdot e$, of the charge state is not sufficient to unambiguously determine the absolute particle charge state, since an initial charge larger by a factor of 2 and a step with an assumed change in charges twice as big will lead to the same final result.

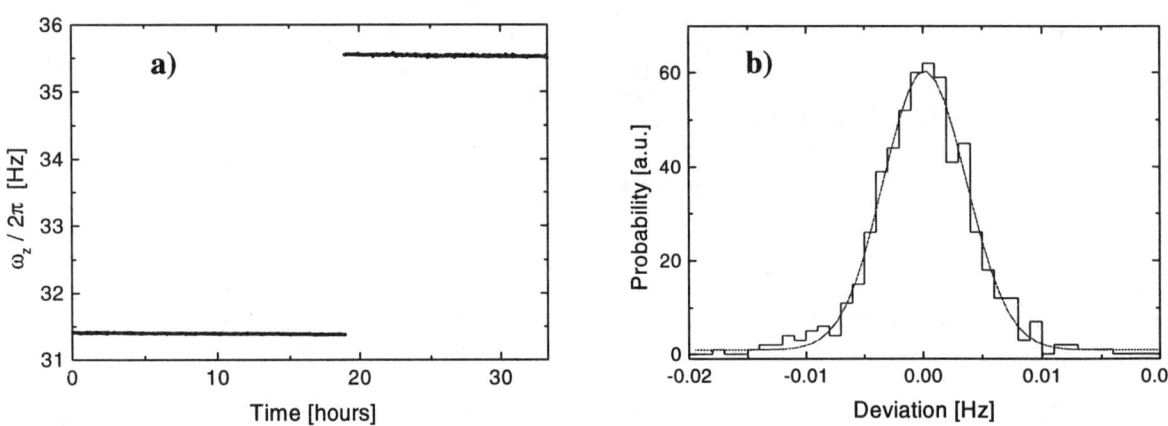

FIGURE 2 a): Repeated determination of the secular frequency ω_z of a single stored particle over a period of more than 30 hours. After approx. 20 h a step in the secular frequency ω_z is observed. The measured ratio of frequencies is consistent with the expected one for one particle carrying +15 e and +17 e before and after the change of charge, respectively. Change of charge of the particle is attributed to the emission of electrons. **b):** Precision of the experiment. The scatter of the individual measurements before the step (13.2 sec integration time, S/N = 25) is shown in a histogram as a probability of deviation from a mean value. The envelope is fitted to a Gaussian with a FWHM of $8.7 \cdot 10^{-3}$ Hz. This corresponds to a precision of 270 ppm for a single q/m determination.

Absolute Mass Determination

In order to overcome the problem of determining the absolute particle charge and mass usually the ejection of single photoelectrons is used [5], but in the case of SiO_2 the work function is too large to use a conventional UV-lamp. Therefore we employed an electron gun to charge the particle. Due to the fact that the driving voltage is supplied to the cones only, the kinetic energy of the electrons hitting the particle is modulated by the driving potential, $V_0 \cos(\Omega t)$. We use a repelling grid to allow injection of electrons only during a time window of several 10 µs. Varying the phase shift of the gate pulse with respect to the driving potential the electron energy can be selected from 0 V up to about 0.8 V_0. Fig. 3 shows ω_r as a function of time with an electron energy of about 190 eV. Here, thirteen charge steps are observed within 3 hours. It is clearly seen that the steps are at integer multiple of about 0.45 Hz. Since this is the smallest step and since we observed all step sizes between +1 and +5 elementary frequency units we relate this smallest unit to ejection of a single electron. From this we infer that at t=0 the particle carried +31 e. Using Eq. 4 we determine q/m = 30.8 mC/kg for this particle and therefore the absolute mass is m = $(1.6 \pm 0.3) \cdot 10^{-16}$ kg. A single 500 nm SiO_2 sphere ($\rho = 2$ g/cm^3 [11]) has a mass of $1.3 \cdot 10^{-16}$ kg. Taking into account an uncertainty of 10 % in the radius of the prefabricated SiO_2 spheres [11], the mass determined by the method described here matches well with the manufacturers specifications. However, while the precision of the experiment is very good, the fairly poor absolute accuracy is related to the determination of the values for $|\partial_z E_{0,z}|$ and $|\partial_r E_{0,r}|$ which derive from a numerical simulation of a model of the actual trap with a spatial resolution of only $1/25\, z_0$. More work increasing the resolution as well as taking into account possible misalignments of the electrodes is in progress to improve the accuracy of absolute mass determination.

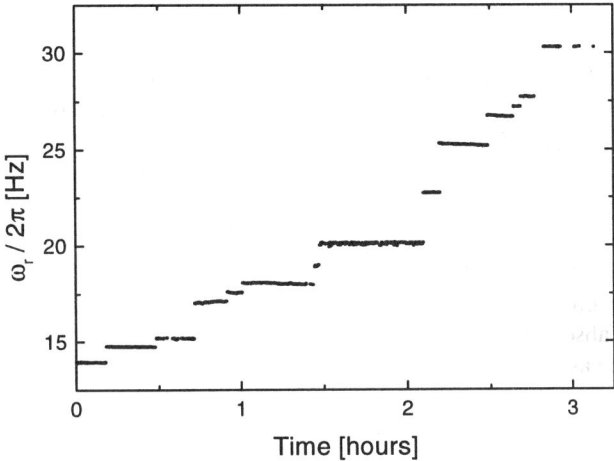

FIGURE 3: Thirteen charging steps of a single particle are induced by use of (190 ± 20) eV electrons. The height of the steps varies between +1 and +5 elementary frequency units of 0.45 Hz which corresponds to charging by +1 e. The net charge is increasing from 31 e to 64 e during this three hour measurement. At the beginning of the measurement q/m was 30.8 mC/kg. From this the absolute mass of the particle has been determined to m = $1.6 \cdot 10^{-16}$ kg.

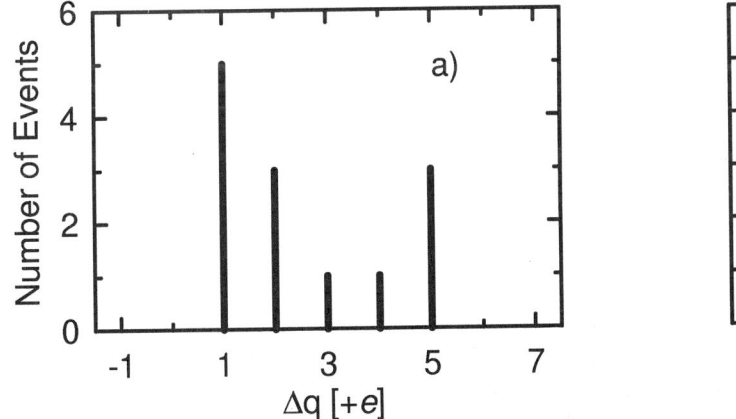 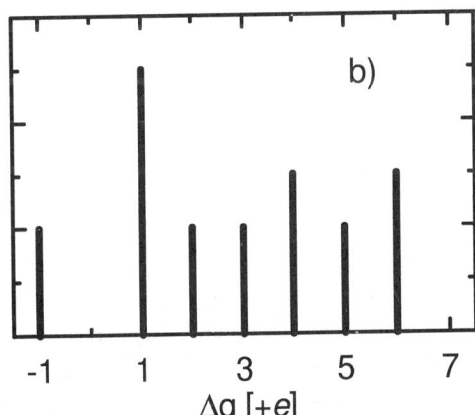

FIGURE 4: Secondary electron emission yields at two different primary electron energies. *a)* 190 eV. *b)* 570 eV. $\Delta q = 0$ events are not detectable.

Secondary Electron Emission Yields

Charging the particle using electron bombardment leads to an increase in positive charge as can be seen in Fig. 3. This charging process can be attributed to secondary electron emission from the surface of the SiO_2 particle. As shown in Fig. 3 this is a statistical process where up to 6 electrons are ejected in a single event. The energetic primary electron penetrates the surface to a depth up to about 5Å [12] and loses energy continuously due to inelastic collisions. Eventually it transfers enough energy to another electron which then might escape from the solid if it can cover the distance to the surface without losing its energy (mean free path also in the Å regime [12]). For the thirteen steps measured in the experiment shown in Fig. 3 we display the probability of secondary emission yield in Fig. 4a. Although the statistical error is still quite large due to the small total number of recorded events this histogram shows that the probability peaks at a yield of about 2 secondary electrons per primary electron. $\Delta q = 0$ events, which signify one primary electron yielding one secondary electron, are undetectable. We repeated the experiment with the same particle at an elevated primary electron energy (570 eV, see Fig. 4b). Here the average yield appears to be slightly increased as well as it is spread over a wider range. This is in agreement with the fact that higher energy electrons produce more secondary electrons. Besides the generally occurring positive charging two events of negative charging (-1 e) have been detected. They are attributed to trapping of the primary electron in the solid with no secondary electron emission. This finding is consistent with the fact that 570 eV electrons penetrate about twice as deep into the solid as 190 eV ones. Therefore the chance of escape for the secondary electron is reduced.

CONCLUSIONS

With the experiments presented here we demonstrate the capability of accurate, non-destructive mass determination of a single very well localized and isolated particle in a mass regime of 10^{-16} kg ($\approx 10^{11}$ amu $\approx 10^{9}$ SiO_2). At the present stage and for a 500 nm SiO_2 particle the mass resolution is sufficient to resolve molecular adsorption (e.g., H_2O) at a sub-monolayer level. Therefore one of the future applications of this method will be the determination of sticking coefficients at various conditions (e.g., temperatures). Since the mass determination is based on a frequency measurement there is room for further improvement of the resolution ($\leq 10^{-6}$). Especially for much smaller particles ($\varnothing \leq 50$ nm) single molecule mass resolution is feasible. In that case the secular frequencies are increasing from the Hz- towards the kHz-regime. On the one hand low frequency noise which is hard to be rejected will no longer play a role and gravitational compensation is no longer necessary. On the other hand the time to determine q/m by FFT methods will be reduced and the accuracy will be increased at the same time. In summary it might become feasible to form a macroscopic structure which mass is determined to a single atom accuracy leading to a new reference of absolute mass unit.

The first physical process studied in our new apparatus was the secondary electron emission. This process could be followed online event by event, revealing its statistical nature, and the probability of yields has been measured for two primary electron energies. Due to the small mean free path of primary and secondary electrons this process is related to the outermost shell of the particle. The investigation of further interesting phenomena such as charge transfer from adsorbed molecules can be undertaken. Since the detection scheme is laser based combination of this sensitive method of mass analysis with optical information (e.g., absorption and emission spectra) opens up a variety of additional applications in the fields of aerosol-chemistry (gas/liquid interface), dust-chemistry (gas/grain interface), physics and chemistry of colloidal systems as well as applications in biology.

ACKNOWLEDGEMENT

This work is funded by the Deutsche Forschungsgemeinschaft.

REFERENCES

1. M.H. Holzscheiter, Physica Scripta, **T59**, (1995), 69-76.
2. R.F. Wuerker, H. Shelton, R.V. Langmuir, J. Appl. Phys., **30** (1959) 342.
3. T.G.O. Berg and T.A. Gaukler, Amer. J. Phys., **37** (1969) 1013; M N.R. Whetten, J. Vac. Sci. Technol., **11** (1974) 515.
4. E.J. Davis and A.K. Ray, J. Colloid and Interface Science, **75** (1980) 566.
5. M.A. Philip, F. Gelbard, and S. Arnold, J. Colloid and Interface Science, **91** (1983) 507.
6. Today this technique is even commercially available as a Paul-trap mass analyzer, e.g. from Finnigan Inc., San Jose.
7. G. Hars and Z.Tass, J. Appl. Phys., **77**, 4245, (1994).
8. SIMION 3D, Ion Optics Simulation Program, Idaho National Engineering Laboratory, P.O. Box 1625, Idaho Falls, ID 83415.
9. E. Fischer, Z. Phys.**156**, (1959), 1.
10. S. Wellert, Diploma thesis, TU-Chemnitz, March 1998.
11. Value specified by the manufacturer, Merck Company.
12. A. Zangwill, physics at surfaces, New York, 1988, ch. 2, p. 21.

SECTION 3

PRECISION MASS SPECTROMETRY

Cooling of Radioactive Isotopes for Schottky Mass Spectrometry

M. Steck, K. Beckert, H. Eickhoff, B. Franzke,
F. Nolden, H. Reich, B. Schlitt, T. Winkler

*Gesellschaft für Schwerionenforschung,
Planckstr. 1, D-64291 Darmstadt, Germany*

Abstract. Nuclear masses of radioactive isotopes can be determined by measurement of their revolution frequency relative to the revolution frequency of reference ions with well-known masses. The resolution of neighboring frequency lines and the accuracy of the mass measurement is dependent on the achievable minimum longitudinal momentum spread of the ion beam. Electron cooling allows an increase of the phase space density by several orders of magnitude. For high intensity beams Coulomb scattering in the dense ion beam limits the beam quality. For low intensity beams a regime exists in which the diffusion due to intrabeam scattering is not dominating any more. The minimum momentum spread $\delta p/p = 5 \times 10^{-7}$ which is observed by Schottky noise analysis is considerably higher than the value expected from the longitudinal electron temperature. The measured frequency spread results from fluctuations of the magnetic field in the storage ring magnets. Systematic mass measurements have started and can be presently used for ions with half-lives of some ten seconds. For shorter-lived nuclei a stochastic precooling system is in preparation.

I INTRODUCTION

The availability of cooled beams in ion storage rings provides conditions to perform high precision spectroscopy on stored ion beams. The high beam velocity of typically 70 % of the velocity of light allows to study completely stripped or few electron ions up to the heaviest species. Electron cooling [1] plays a crucial role in the preparation of the ion beam. The reduction of the phase space volume results in ion beams with a highly precise definition of the beam parameters. For the novel method of Schottky mass spectrometry [2] particularly the reduction of the longitudinal momentum spread of a coasting ion beam allows a precise definition of the particle momentum. A relative momentum difference $\Delta p/p$ is observed as a relative frequency difference

$$\frac{\Delta f}{f} = \eta\, \frac{\Delta p}{p} = (\frac{1}{\gamma^2} - \alpha_p)\, \frac{\Delta p}{p}. \qquad (1)$$

The frequency dispersion η is dependent on the ion optical momentum compaction factor α_p of the storage ring and the relativistic beam energy γ. Usually the Schottky noise of a coasting ion beam is analyzed in frequency domain and provides precise information about the beam momentum and the momentum distribution of the ions. The mass of a particular isotope can be determined by comparison of its revolution frequency with the revolution frequency of fragments of well-known mass which are co-circulating. The frequency separation is determined by

$$\frac{\Delta f}{f} = -\alpha_p \frac{\Delta(m/q)}{(m/q)} + (\frac{1}{\gamma^2} - \alpha_p)\, \gamma^2\, \frac{\Delta v}{v} \qquad (2)$$

with the ion mass and charge m, q and the velocity difference $\Delta v/v$. To achieve the highest mass resolution one has to take into account the small variations of α_p over the acceptance of the storage ring which are mainly due to higher order components in the magnetic field of the storage ring components.

The small velocity spread of cooled beams allows a high resolution of neighboring lines in the Schottky noise frequency spectrum which originate either from a slightly different charge to mass ratio of two ions or isomeric states with small excitation energy. In this respect cooling of the ion beam to smallest velocity spread improves the accuracy of the frequency and consequently also of the mass determination.

Although electron cooling is very powerful if the ion beam is already cold the initial cooling time for the hot ion beam after injection at the typical energy of 300 MeV/u for injection into the storage ring amounts to several ten seconds. For radioactive isotopes with half-lives on the order of seconds a special cooling scheme allowing considerably shorter cooling times is foreseen. In order to reduce the initial cooling period for fragments with large emittances and momentum spread it combines stochastic precooling of the hot ion beam with final electron cooling. This offers the advantage of faster initial cooling and the achievement of the highest beam quality for the mass determination from frequency spectra.

II PRODUCTION OF RADIOACTIVE ION BEAMS

The GSI accelerator complex has been described in more detail elsewhere [3]. For the investigation of radioactive isotopes the following stages are relevant. The linear accelerator UNILAC can accelerate all ion species to an energy of 11.4 MeV/u. After stripping the ions to a higher charge state they are injected at this energy into the synchrotron SIS [4] filling the horizontal phase space by multiturn injection. Recently the synchrotron has been supplemented by an electron cooling system which supports beam accumulation at the injection energy [5]. This method results in intensities of up to 10^9 particles for the heaviest ions occupying a small phase space volume which allows high efficiency for acceleration and beam transport. Acceleration up to a maximum energy of 1 GeV/u is possible for the heaviest elements. After acceleration the ions are extracted by means of a fast kicker magnet during a single turn and injected into the storage ring ESR [6]. The beam can be transferred to the storage ring either directly passing a stripper foil of some 10 mg/cm^2 thickness to produce the primary beam in high charge states or through the fragment separator FRS [7].

For the production of radioactive isotopes the beam from the synchrotron is directed onto a 4 or 8 g/cm^2 thick beryllium target in front of the fragment separator. The fragmentation products thereafter are selected in a magnetic separation system according to their magnetic rigidity. Thus a mixture of up to 60 fragments is produced which can be accepted by the storage ring and which after injection circulate on the same orbit. Although their magnetic rigidity is nearly identical, their velocity is different. A degrader system in the FRS also offers the option to select a single isotope by a $B\rho$–ΔE–$B\rho$ separation technique.

III ELECTRON COOLING

For precision Schottky mass spectrometry the mixture of isotopes from the fragment separator which are simultaneously circulating in the storage ring is cooled by the ESR electron cooling system [8]. Electron cooling results in a multi-component beam with identical velocity of all components. If the various ions are cooled to the same velocity their revolution frequency differs due to small differences in the charge to mass ratio. The cooled isotopes circulate on slightly different orbits with corresponding revolution frequencies. The differences in revolution frequencies have to be resolved and are the basis of the mass determination. For the precise determination of revolution frequencies the properties of the cooled ion beam are most important, particularly cooling to equal velocity of all beam components. The quality of cooling determines the resolution of neighboring frequency lines in the spectrum as well as the accuracy of the mass determination.

The ion beam after injection into the storage ring from the synchrotron has a large transverse emittance on the order of 10 π mm mrad and a longitudinal momentum spread of a few permille. Even for beams which were cooled in the synchrotron at the injection energy and extracted with small emittances and momentum spread the primary beam quality is destroyed by scattering and energy loss processes in the stripper or fragmentation target. In order to restore the high beam quality the hot ion beam is merged in one straight section of the storage ring over an effective length of about 2 m with a cold electron beam. The electron beam guided by a longitudinal magnetic field of about 0.1 T strength is comoving with the ion beam at the same average velocity. By Coulomb interaction the heat of the thermal motion is transferred from the hot ions to the cold electrons.

Electron cooling is particularly powerful when the ion beam quality is already good, but the cooling power decreases strongly when the thermal velocity components in the ion beam are comparable or even higher than the thermal motion of the electrons. The typical time to cool an initially hot ion beam with a velocity $\beta \simeq 0.7$ is on the order of a few 10 s depending inversely proportionally on the electron current. During the cooling process the thermal ion motion decreases and this consequently results in a reduction of the cooling time below 1 s.

However, the cooling process does not continue until ion and electron temperature are equalized. For a cold beam of highly charged ions the increase of the phase space density causes stronger Coulomb interactions between the ions thus creating an additional heating mechanism by intrabeam scattering which increases with the phase space density of the ion beam.

IV EQUILIBRIUM WITH INTRABEAM SCATTERING

The highest quality of cooled highly charged ion beams is reached when the heating rate by intrabeam scattering and the cooling rate are balanced [9]. The equilibrium beam parameters for a constant electron current of 250 mA at a typical energy of around 300 MeV/u in the ESR were measured for bare ions. They are characterized by an increase of the transverse emittances and the longitudinal momentum spread with the number of stored ions N (Fig. 1). The transverse emittances which were measured non-destructively by detection of down-charged ions after the cooler section [10] increase approximately with the square root of the particle number. The momentum spread which is crucial for precise mass measurements by Schottky noise detection of stored radioactive beams is for particle numbers $N \geq 10^5$ larger than 10^{-5} and increases proportional to $N^{0.3}$. Extrapolating these results to smaller particle numbers a momentum spread not below 10^{-6} can be concluded, if intrabeam scattering is the dominant diffusion process.

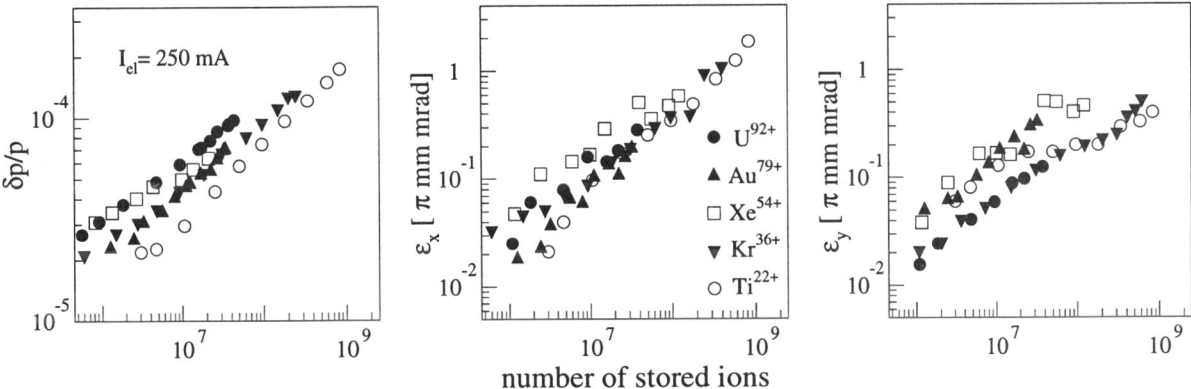

FIGURE 1. Equilibrium beam quality for bare ions cooled with an electron current of 250 mA. Intrabeam scattering in the dense ion beam causes a growth of the transverse emittances and of the longitudinal momentum spread with the number of stored ions.

The dependence of the equilibrium beam parameters on the ion charge and mass is weak. For higher charges intrabeam scattering grows stronger (heating rate proportional to q^4/m^2) than cooling (cooling rate proportional to q^2/m) and a weak increase with the ion charge results. Albeit the equilibrium is quite sensitive to the cooling conditions the general tendency in the measurements confirms the increase of the phase space volume with the charge.

An increase of the electron current I_e in order to increase the cooling power results in a reduction of the momentum spread proportional to $I_e^{-0.3}$ [9]. As the beam lifetime decreases proportional to the electron current, an electron current of a few hundred milliampere is usually chosen as a trade off between beam quality and beam lifetime.

V SUPPRESSION OF INTRABEAM SCATTERING

A more thorough investigation for small particle numbers exhibited a surprising result for the achievable momentum spread of low intensity beams of highly charged ions. At particle numbers around 10^3 a sudden reduction of the momentum spread $\delta p/p$ by a factor of 10 from about 5×10^{-6} to 5×10^{-7} was observed which indicates a suppression of intrabeam scattering. This behavior was unambiguously detected by observing the

Schottky noise of a stored beam with a known lifetime due to capture of cooler electrons and associated loss of the ions at the momentum acceptance of the storage ring (Fig. 2). The noise power exhibits the expected exponential decrease with the decay time corresponding to the beam lifetime due to radiative electron capture with free electrons in the cooling section. After a storage time of 110 min a discontinuous momentum spread reduction was observed. The intensity of the ion beam could be determined independently from the known lifetime and the initial beam intensity which was measured with a standard current transformer. It agreed with the value determined from the Schottky noise power and amounted at the transition point to about 1000 stored ions.

The reduction of the momentum spread indicates a suppression of intrabeam scattering which might be connected with an ordering effect in the cold ion beam. A possible explanation is the formation of a linear chain where the ions are at fixed longitudinal positions similar to ordering effects for laser-cooled ions at rest in a ring trap [11].

The strong Coulomb force between highly charged ions and the high potential energy between neighboring ions can support such ordering phenomena. Nearly the same reduction factor of 10 for the momentum spread was found for all ions with masses heavier than gold, but also for lighter ions a smaller but still significant reduction of the momentum spread was observed [12].

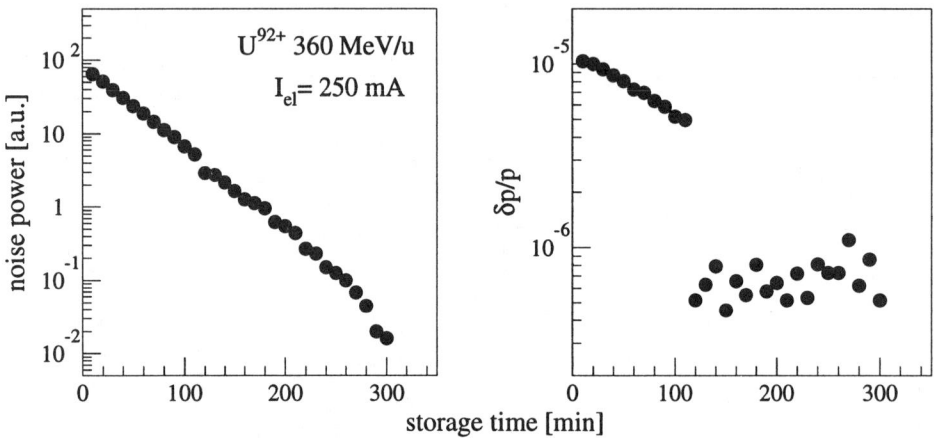

FIGURE 2. Total noise power and momentum spread determined from the Schottky noise of a cooled U^{92+} beam at 360 MeV/u. The noise power shows the expected decrease corresponding to a beam lifetime of 39 min due to radiative electron capture with free electrons. The momentum spread drops by one order of magnitude when the particle number decreases to about 1000 ions.

In Fig. 2 it is visible that at the transition point also the Schottky noise power is slightly reduced. This is common to all ion species and is caused by coherent beam interaction in the dense ion beam with small momentum spread [13]. The use of the Schottky noise signal for lifetime measurements of radioactive isotopes should exclude the vicinity of the transition point to avoid disturbances by collective Schottky noise signals.

A further confirmation for the interpretation as a manifestation of an ordering effect comes from recent measurements of the transverse beam size which represents the beam emittance. For lack of non-destructive diagnostics with sufficient position resolution the beam radius was measured with a beam scraper which is moved horizontally into the circulating beam. The center of the beam is defined by the outermost scraper position where all the stored ions are lost. Relative to this beam center the number of ions surviving for a certain scraper position r follows roughly a power law $r \propto N^{0.3}$ (corresponding to an emittance increase $\epsilon \propto N^{0.6}$) which reasonably agrees with the non-destructive measurements in the intrabeam scattering dominated regime. Below $N \simeq 1000$ the beam radius collapses from 0.25 mm to less than 0.05 mm. The longitudinal momentum spread drops nearly for the same particle number by a factor of 10. The scraper positioning is accurate to about 10 μm, therefore the measurement of the beam radius is not as reproducible as the measurement of the longitudinal momentum spread by Schottky noise. In addition the position of the closed ion orbit can change due to the finite dispersion function at the scraper position which causes a change of the horizontal position as a result of fluctuations of the field strength in the main ring magnets.

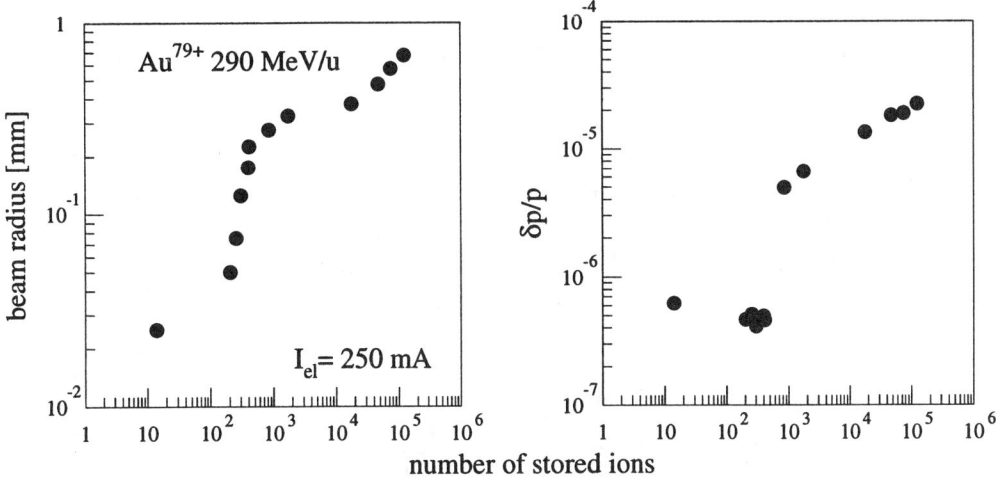

FIGURE 3. Beam radius obtained from scraper measurements for a Au^{79+} beam at 290 MeV/u cooled with an electron current of 250 mA. The momentum spread and the intensity was determined by analyzing the Schottky noise of the coasting ion beam with respect to width and total noise power. Radius and momentum spread show a similar discontinuous reduction for about 1000 stored ions.

The ion beam temperature estimated from the beam radius and the optics function of the storage ring results in a value which resembles the transverse electron temperature of 0.1 eV. The transverse ion beam temperature and the transverse potential energy $U_\perp = q^2 e^2 / 4\pi\epsilon_o \langle r_{rms} \rangle$ for the measurement with the gold beam are equal within the accuracy of the scraper technique. The plasma parameter $\Gamma_\perp = U_\perp / kT_\perp$ for the transverse degree of freedom is therefore close to unity which supports the idea of a phase transition from a gaseous to a liquid phase. The measured momentum spread results in an upper limit for the longitudinal beam temperature of 5 meV. It is large compared to the longitudinal potential energy $U_\parallel = q^2 e^2 N / 4\pi\epsilon_o C \gamma$ with the ring circumference C which has a value $U_\parallel \simeq 0.06$ meV. However, this is difficult to compare to the longitudinal temperature for which only an upper limit can be determined. Consequently the main argument for a linear ordering comes from the restriction of the betatron amplitudes of the cold particles to such low values that the thermal energy is too small for an efficient exchange of energy in a collision between particles.

VI LIMITS OF SCHOTTKY MASS SPECTROMETRY

The minimum momentum spread that has been observed for cooled ion beams is around 5×10^{-7} for all ion species [12] as also demonstrated in Figs. 2 and 3. For the heaviest ions this corresponds to an estimated longitudinal temperature of a few meV which is at least an order of magnitude higher than the longitudinal electron temperature. The momentum spread is derived from the spread of the ion revolution frequencies which depend on the field strength in the storage ring magnets. The power supplies for the ring magnets show output current fluctuations on the time scale of seconds. The frequency analysis of the circulating ions uses data acquisition and averaging over times much longer than the time scale of the power supply fluctuations. Therefore the Schottky diagnostics rather represents the long term average over these revolution frequency fluctuations caused by the power supplies than the intrinsic velocity spread connected with the finite beam temperature.

With fast Schottky scans (measuring time less than 1 s) a correlation between revolution frequency changes of the stored beam and the cycle of the synchrotron which is operated in the vicinity of the storage ring have been detected (Fig. 4). Small jumps of the revolution frequency could be observed for a manganese isotope circulating in the ground and an isomeric state which were associated with a certain part of the acceleration cycle of the synchrotron. Improved stabilization of the output currents is difficult to achieve as the current fluctuations in the main power supplies are on the level of the electronic noise. For a further reduction of the linewidth in the spectra of cooled low intensity beams a fast data acquisition system will allow a correction of

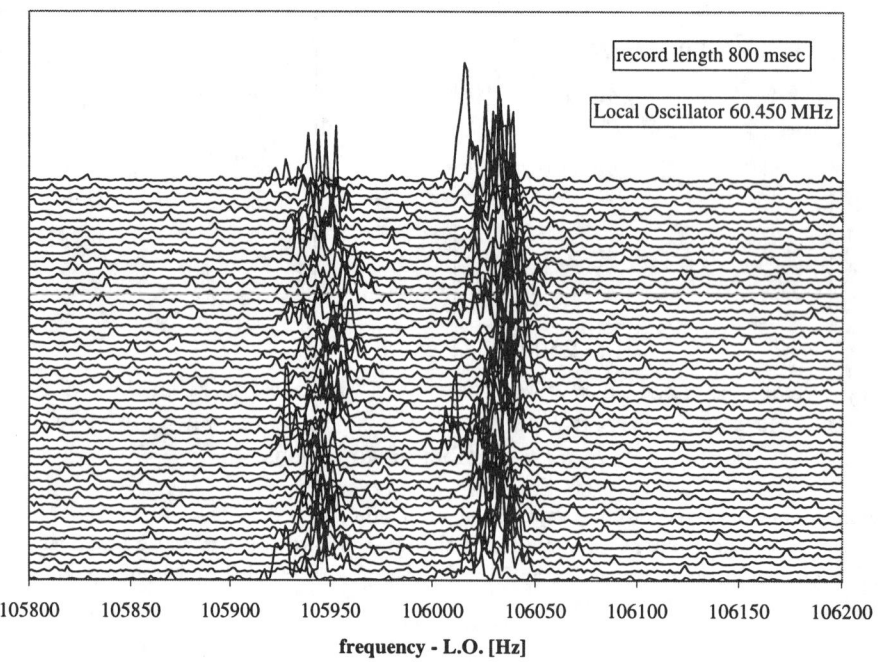

FIGURE 4. Fluctuations of the revolution frequency due to power supply ripple are demonstrated for a Mn^{25+} beam. Ground and isomeric state ions are circulating at slightly different revolution frequencies. Frequency spectra were recorded every 0.8 s. They show a periodic shift to a slightly lower frequency ($\Delta f/f = -3 \times 10^{-7}$) with a time period of 12 s caused by the ramping of the nearby synchrotron.

correlated frequency shifts in multi-component beams. After offline correction of spectra with short measuring times and averaging of several corrected spectra an improved momentum resolution is expected which should approach the final temperature determined by the longitudinal electron temperature.

VII MASS MEASUREMENTS OF RADIOACTIVE ISOTOPES

The ESR storage ring in combination with the fragment separator FRS has been used for direct mass measurements of projectile fragments (Fig. 5). The primary beam energy was matched to the energy loss in the production target in order to inject bare ions or ions with only one or two bound electrons at an energy of 300 MeV/u. This energy presently gives the best performance of the electron cooler and the optical setting of the storage ring magnets. By now the masses of more than 100 mainly proton-rich isotopes produced with primary beams of nickel and bismuth ions have been measured for the first time, and another 200 masses have been determined with improved accuracy.

A detailed analysis and the status of the mass measurements is given elsewhere [14]. The performance of the storage ring and of the cooling system can be summarized as follows. The small momentum spread of the cooled multi-component beams for intensities below 1000 ions allows to resolve mass differences down to $\Delta m/m = 1 \times 10^{-6}$ which is the present limit by the stability of the magnetic field. The stability of the accelerating voltage for the electron beam has no influence on the mass resolution for ion energies exceeding 100 MeV/u. The relative error $\delta m/m$ of the mass measurement is dependent on several experimental parameters, values down to 5×10^{-7} have been achieved. Measurements of short-lived nuclei are limited by the cooling time of the electron cooling system. An extension to nuclei with half-lives in the range of seconds will be available after commissioning of the stochastic precooling system.

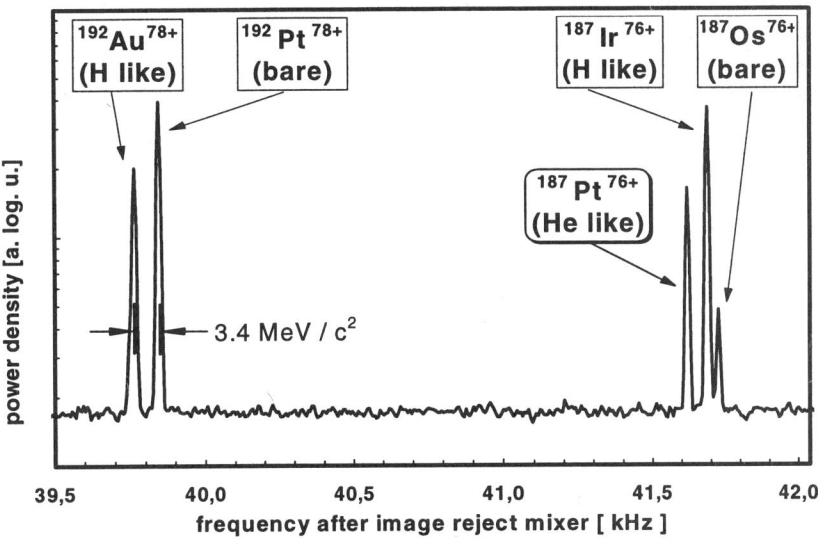

FIGURE 5. Narrow band Schottky spectrum of cooled fragments stored simultaneously in the storage ring. The fragments were produced with a primary gold beam. Frequency lines of isotopes with differences in the mass to charge ratio of less than 1×10^{-5} are well separated.

VIII STOCHASTIC PRECOOLING OF HOT ION BEAMS

As it was expected that electron cooling is weak when dealing with large relative velocities a stochastic precooling system has been designed for emittance and momentum spread reduction of hot ion beams [15]. Theoretical simulations have shown that cooling times for highly charged ions and intensities below 10^5 stored ions on the order of 100 ms can be expected [16]. The ions can be efficiently cooled from emittances of $20\,\pi$ mm mrad to $2.5\,\pi$ mm mrad and from a momentum spread of $\pm 3.5 \times 10^{-3}$ to $\pm 5 \times 10^{-4}$. This regime which is achieved after stochastic precooling is well matched for further electron cooling to the lowest temperatures with overall cooling times on the order of seconds. With the combination of the two cooling techniques mass or lifetime measurements of radioactive isotopes with half-lifes of seconds can be addressed which could not be investigated with electron cooling only.

The stochastic cooling system which has been partly commissioned in 1997 and which is presently being completed in the storage ring is designed for cooling of heavy ion beams with energies in excess of 400 MeV/u. It consists of separate subsystems for cooling the horizontal, vertical and longitudinal degree of freedom. The electrodes are installed on the injection orbit in order to allow beam stacking. A total microwave power of 2 kW in the frequency range from 0.9 to 1.7 GHz is available to achieve the expected short cooling times. Some parts of the system that had been installed in advance have been commissioned with beam. Longitudinal cooling of carbon and argon beams by the Palmer method has been demonstrated with cooling times close to the expectations for the relatively low charge of these ions [17]. The final commissioning and first experiments with stochastic precooling in combination with electron cooling will be performed in 1999.

REFERENCES

1. G.I. Budker, *Sov. J. Atom. Energy* **22**, 438 (1967).
2. B. Franzke, K. Beckert, H. Eickhoff, F. Nolden, H. Reich, A. Schwinn, M. Steck, T. Winkler, Performance of Schottky Mass Spectrometry at the ESR, *Proc. of the 6th Europ. Part. Acc. Conf.*, Stockholm, 1998.
3. N. Angert, Status and Development of the GSI Accelerator Facilities, *Proc. of the 5th Europ. Part. Acc. Conf.*, World Scientific, Singapore, 1996, 125.
4. K. Blasche, B. Franzke, *Proc. of the 4th Europ. Part. Acc. Conf.*, V. Suller and Ch. Petit-Jean-Genaz eds., World Scientific, Singapore, 1994, 133–137.

5. M. Steck, K. Blasche, H. Eickhoff, B. Franczak, B. Franzke, L. Groening, T. Winkler, Commissioning of the Electron Cooling Device in the Heavy Ion Synchrotron SIS, *Proc. of the 6th Europ. Part. Acc. Conf.*, Stockholm, 1998.
6. B. Franzke, K. Beckert, F. Bosch, H. Eickhoff, B. Franczak, A. Gruber, O. Klepper, F. Nolden, P. Raabe, H. Reich, P. Spädtke, M. Steck, J. Struckmeier, Heavy Ion Beam Accumulation, Cooling, and Experiments at the ESR, *Proc. of the 1993 Part. Acc. Conf.*, Washington D.C., 1993, 1645–1648.
7. H. Geissel, *Nucl. Instr. Methods* **B24/25**, 286 (1992).
8. M. Steck, K. Beckert, H. Eickhoff, B. Franzke, F. Nolden, P. Spädtke, Electron Cooling of Heavy Ions at GSI, *Proc. of the 1993 Part. Acc. Conf.*, Washington D.C., 1993, 1738–1740.
9. M. Steck, K. Beckert, F. Bosch, H. Eickhoff, B. Franzke, O. Klepper, R. Moshammer, F. Nolden, P. Spädtke, T. Winkler, Recent Results on Equilibrium Temperatures and Cooling Forces with Electron Cooled Heavy Ion Beams in the ESR, *Proc. of the 4th Europ. Part. Acc. Conf.*, V. Suller and Ch. Petit-Jean-Genaz eds., World Scientific, Singapore, 1994, 1197–1199.
10. M. Steck, *Nucl. Phys.* **A626**, 473c (1997).
11. I. Waki, S. Kassner, G. Birkl, H. Walther, *Phys. Rev. Lett.* **68**, 2007 (1992).
12. M. Steck, K. Beckert, H. Eickhoff, B. Franzke, F. Nolden, H. Reich, B. Schlitt, T. Winkler, *Phys. Rev. Lett.* **77**, 3803 (1996).
13. V. V. Parkhomchuk, and D. V. Pestrikov, Sov. Phys. Tech. Phys. **25(7)**, 818 (1980).
14. H. Geissel, T. Radon, F. Attallah, K. Beckert, F. Bosch A. Dolinsky, H. Eickhoff, M. Falch, B. Franczak, B. Franzke, Y. Fujita, M. Hausmann, M. Hellström, F. Herfurth, Th. Kerscher, O. Klepper, H.-J. Kluge, C. Kozhuharov, K.E.G. Löbner, G. Münzenberg, F. Nolden, Yu. Novikov, Z. Patyk, W. Quint, H. Reich, C. Scheidenberger, B. Schlitt, M. Steck, K. Sümmerer, L. Vermeeren, M. Winkler, Th. Winkler, H. Wollnik, Experiments with Stored Relativistic Exotic Nuclei, *Proc. of 2nd Int. Conf. on Exotic Nuclei and Atomic Masses (ENAM 98)*, Bellaire, Michigan, USA, 1998.
15. B. Franzke, *Nucl. Instr. Meth. in Phys. Res.* **B24/25**, 18 (1987).
16. F. Nolden, *Zur stochastischen Vorkühlung am ESR*, PhD thesis, Technical University Munich, 1996.
17. F. Nolden et al., First Experiments on Stochastic Cooling of Heavy Ion Beams at the ESR, *Proc. of the 6th Europ. Part. Acc. Conf.*, Stockholm, 1998.

ACCURATE MASS MEASUREMENTS OF SHORT-LIVED ISOTOPES WITH THE *MISTRAL*[*] RF SPECTROMETER

C. Toader[1,2], G. Audi[1], C. Borcea[2], H. Doubre[1], M. Duma[2],
M. Jacotin[1], S. Henry[1], J.-F. Képinski[1], G. Lebée[3], G. Le Scornet[1],
D. Lunney[1†], C. Monsanglant[1], M. de Saint Simon[1], C. Thibault[1],
and the *ISOLDE* collaboration[3]

[1]*CSNSM-IN2P3-CNRS*, F-91405 Orsay, France
[2]*Inst. Atomic Physics*, Bucharest, Romania
[3]*CERN*, EP Division, Geneva, Switzerland

Abstract. The *MISTRAL*[*] experiment has measured its first masses at *ISOLDE*. Installed in May 1997, this radiofrequency transmission spectrometer is to concentrate on nuclides with particularly short half-lives. *MISTRAL* received its first stable beam in October and first radioactive beam in November 1997. These first tests, with a plasma ion source, resulted in excellent isobaric separation and reasonable transmission. Further testing and development enabled first data taking in July 1998 on neutron-rich Na isotopes having half-lives as short as 31 ms.

The atomic mass is a global property that can elucidate interesting physics from the binding energy. Atomic physics requires particularly high accuracy mass measurements in order to isolate contributions beyond those accounted for by quantum electrodynamics [1,2] as does any effort at verifying fundamental symmetries between, for example, particles and their anti-particles [3].

The interest in measuring masses of radioactive isotopes comes from an ever-present need to understand the nature of the nuclear force. The topography of the mass surface gives us clues to the nature of nuclear structure. As we climb either side of the so-called valley of stability, we expect the landscape to become more and more exotic out to the point where a nuclear configuration reaches saturation at the cliffs of nuclear particle stability that are called the drip lines.

The mass difference of isotopes involved in a nuclear reaction can also provide constraints on the nature of the involved interaction, radioactive β-decay being of particular interest as it is the signature of the weak interaction [4,5].

---

[*]Mass measurements at ISolde using a Transmission RAdiofrequency spectrometer on-Line
[†]Invited speaker and corresponding author (*lunney@csnsm.in2p3.fr*)

Finally, knowledge of the nuclear binding energy is extremely important for nucleosynthesis, in particular the rapid neutron capture process, thought to occur in exploding supernovae. Masses are required to calculate a variety of physical quantities involved in this process and to reproduce the abundances of the heavy elements present in the solar system [6].

MISTRAL is one of several programs dedicated to the accurate mass measurement of radioactive isotopes. These programs are all complementary in technique and/or applicability. Cyclotrons that accelerate reaction products allow us to reach far up into the cliffs of exotic nuclei [7] and storage rings offer nuclear lifetime measurements as an added bonus to huge mass harvests from high energy fragmentation reactions [8]. Penning traps offer a kinder and gentler environment for mass measurements and consequently hold the records for accuracy [9,10]. The key here is that a single ion may be held for as long as necessary to make a measurement. *ISOLTRAP*, a Penning trap spectrometer on-line at *ISOLDE*, has its accuracy slightly compromised by the adverse on-line conditions and connection to the outside world but nevertheless provides excellent and systematically accurate measurements of radioactive isotopes [11].

The *MISTRAL* spectrometer, also at *ISOLDE*, is a sort of hybrid between a cyclotron and a Penning trap. Its rather special technique of radiofrequency excitation of the cyclotron motion at the full beam transport energy allows very rapid measurements of high accuracy thus rendering it particularly suitable for short-lived isotopes [12,13]. Thus, *MISTRAL* complements *ISOLTRAP* which must store ions for longer periods in order to make a very accurate measurement.

A schematic diagram of the *MISTRAL* spectrometer with its nominal trajectory is shown in figure 1. Ions injected at the full *ISOLDE* beam energy (60 kV) follow a two-turn helicoidal trajectory inside the annular, homogeneous magnetic field (figure 1, inset center) and are counted using a secondary electron multiplier. With an injection slit size of 0.4 mm and orbit radius of 0.5 m, a mass resolution of 2500 is obtained using no radiofrequency. In order to make a measurement, a longitudinal kinetic energy modulation is effected using two symmetric electrode structures (figure 1, inset right) located at the one-half and three-half turn positions inside the magnetic field. This way the ions make one cyclotron orbit between the two modulators. A radiofrequency voltage is applied to the central modulator electrodes. Depending on the phase of this voltage when the ions traverse the structure, the resulting longitudinal acceleration produces a larger or smaller cyclotron radius than that of the nominal trajectory (all the trajectories are isochronous). The ions are transmitted through the 0.4 mm exit slit when the net effect of the two modulations is zero. This happens when the radiofrequency voltage is an integer-plus-one-half multiple of the cyclotron frequency which means that during the second modulation the ions feel exactly the opposite of what they felt during the first. For high harmonic numbers (e.g. larger than 1000) and a radiofrequency voltage of about 200 V, the ion signal over a frequency scan shows narrow transmission peaks having resolutions of up to 100,000 evenly spaced at the cyclotron frequency (figure 1, inset left).

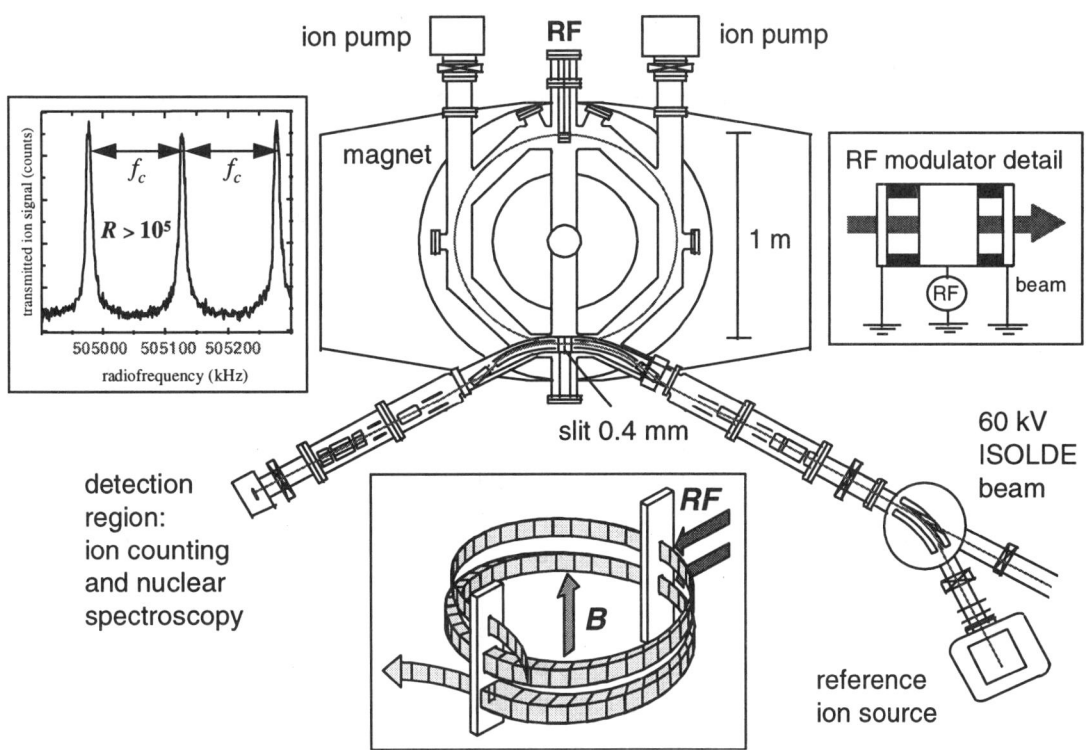

FIGURE 1. Layout of the MISTRAL spectrometer showing the nominal ion trajectory. Ions are injected from the *ISOLDE* beam line at the full transport voltage of 60 kV while the reference mass is alternately injected (without changing the magnetic field) at its required (lower) energy. Inset (right) shows the modulator electrode structure the geometry of which is selected depending on the mass range of operation. Inset (center) shows an isometric view of the trajectory envelope with the 0.4 mm injection slit followed by the first modulator at one-half turn, the phase-definition slit (up to 5 mm wide to incorporate the envelope of cyclotron radii), the second modulator at three-half turns and finally the exit slit. Inset (left) shows the transmitted ^{39}K ion signal as a function of radiofrequency spanning three harmonic numbers (around 3400). The mass resolution is greater than 100,000.

A mass measurement is made when an unknown mass is alternately injected with a reference mass. These comparisons are done in rapid succession (seconds) in order to eliminate short-term drift in the magnetic field. Comparing masses in this way, without changing the magnetic field, requires changing not only the transport energy of the reference beam but the voltages of all electrostatic elements in the spectrometer (two triplets, eight pairs of steering plates, and two benders plus the injection switchyard bender). Since, for the moment, the reference ion source does not withstand more than 60 kV, we are obliged to use a reference mass that is heavier than the ISOLDE mass in order to operate ISOLDE and its transport system at the nominal voltage. A reference source upgrade later this year should avoid this limitation for future runs.

ISOLDE uses a pulsed proton beam extracted from a set of synchrotron booster rings [14]. In the case of short-lived isotopes (as well as elements with very rapid release times from the target matrix, such as Na) it is impossible to scan the entire required frequency range in time after the impact of the proton pulse. In this case, a special acquisition mode is used (called, appropriately: point-by-point). For each radioactive beam pulse, the ion transmission signal is recorded for only one radiofrequency point (determined randomly) and the resonance peak is reconstructed at the end. This mode not only allows us to increase statistics in the peak but for each point, the ion signal is recorded with the radiofrequency switched off so that not just the intensity but the true transmission is measured.

The *MISTRAL* spectrometer was installed in the new beam hall extension of *ISOLDE* in mid-1997 and a first test run using radioactive isotopes around $A = 27$ took place at the end of that year using a UC_2 target coupled with a plasma ion source. The spectrometer was able to cleanly separate the isobaric components with relatively good sensitivity and very encouraging indications for measurement precision.

In July 1998, *MISTRAL* again took radioactive beam from a UC_2 target but this time coupled to a surface ionization source to get a clean beam of Na isotopes. This element is nevertheless very challenging since the release time from the target matrix is very fast so the measurements are made using the point-by-point mode described above.

During this run we were able to measure the masses of $^{23-30}$Na. Shown in figure 2 is a recorded (reconstructed) peak for ^{28}Na. This measurement corresponds to 64 (random) frequency steps of 320 ms each. Each step is triggered by the PS booster proton pulse with a period of (at least) 1.2 s. The center frequency (derived from a triangular fit [15]) is about 480,050 kHz corresponding to harmonic number 2342 of the cyclotron frequency of ^{28}Na in the 0.37359 T field at a beam energy of 60 keV. The mass resolution in this case is 47,860.

When we compare our preliminary values to those in the mass table we perceive an offset (proportional to ΔM, the mass-doublet difference) of about $7 \times 10^{-7}/\Delta M$: the measurements are precise (reproducible, to 0.2 ppm) but not accurate. The mass uncertainties for the isotopes further from stability are naturally dominated by statistical error. If we correct this offset, the residual difference scatters randomly about zero within a value of about ±30 keV, a fairly good accuracy; already better than the present one for ^{28,29,30}Na ($T_{1/2}$ = 31, 45, 48 ms) and which will be certainly improved. The relatively large systematic error is due to a lack of congruency between the reference ion trajectory and that of the mass being measured. Ions of differing trajectories do not experience the same magnetic field because of residual gradients ($\sim 10^{-5}$/cm). We plan to reduce these field gradients through the use of current shim coils [16].

A important point is the chronic problem of sensitivity - exotic nuclei are produced in such small quantities that it is a shame to waste a single ion! This problem can be tackled by the addition of a beam cooling device at the entrance of the spectrometer. By reducing the emittance of the incoming beams not only is the transmission through the many slits inside the spectrometer improved but both the reference and ISOLDE beam are "brainwashed" before they go through, forgetting their differing characteristic

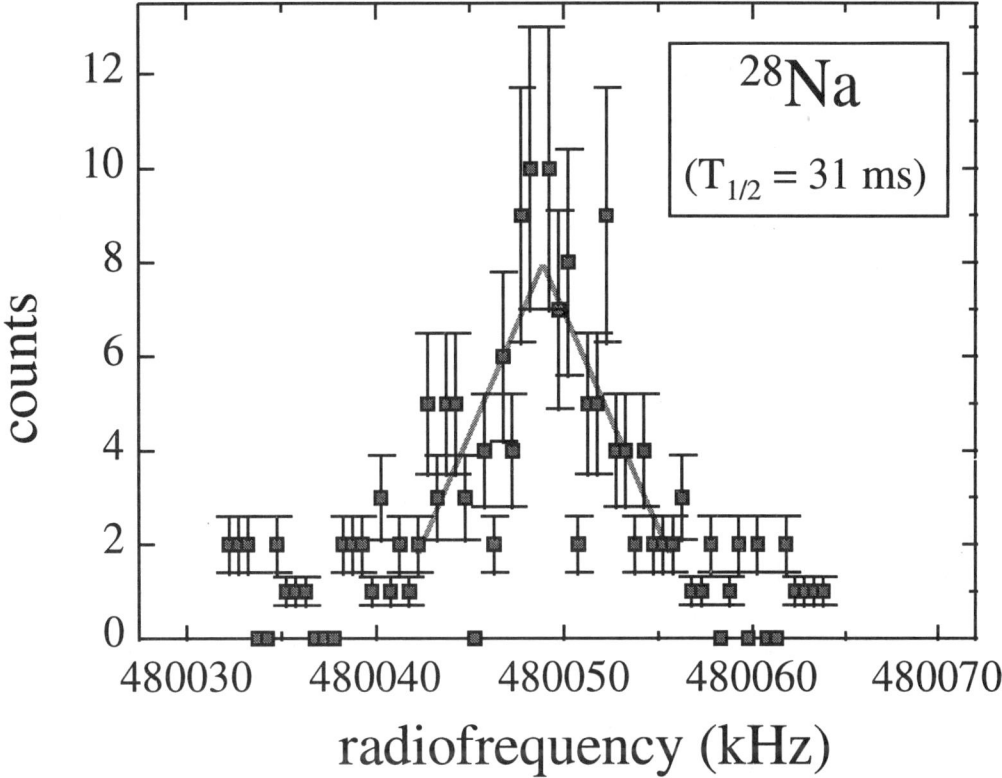

FIGURE 2. A recorded (reconstructed) peak for ^{28}Na ($T_{1/2}$ = 30.5 ms). This measurement corresponds to 64 (random) frequency steps of 320 ms each. The center frequency of about 480,050 kHz is derived from a triangular fit [15] and corresponds to harmonic number 2342 of the cyclotron frequency of ^{28}Na in the 0.37359 T field at a beam energy of 60 keV. The mass resolution is 47,860.

divergences, positions and energy spreads, so following the same path through the spectrometer. Cooling a beam can be done several ways but the one technique that seems the fastest and most universal makes use of a light, neutral buffer gas. We plan to install a gas-filled ion guide that will use the alternate focusing of a radiofrequency quadrupole field to continually refocus the ion beam onto the axis while it loses kinetic energy from collisions with the gas [17].

The potential measurement program at *ISOLDE* is quite rich. Though the calculated transmission of 1% has not yet been reached, there are some one hundred candidates for either new measurements or considerably reduced error. The next scheduled beam time for *MISTRAL* is November, 1998. In the longer term, we hope that the development of the ion cooler will extend our measurement possibilities with better sensitivity but also aid in reducing systematic error.

ACKNOWLEDGMENTS

The authors would like to thank the following researchers who gave their advice and support to the original proposal for the *ISOLDE* experiments committee: G. Bollen, D. Guillemaud-Mueller, P.G. Hansen, B. Jonson, H.-J. Kluge, R.B. Moore, A.C. Mueller, G. Nyman and H. Wollnik. We would also like to thank A. Coc, R. Le Gac and F. Touchard for their contributions to the early stages of the development of the spectrometer. One of us (David Lunney) would also like to thank CERN for a one-year Scientific Associateship.

REFERENCES

1. C. Carlberg *et al.*, this volume.
2. G. Soff, this volume.
3. G. Gabrielse *et al.*, this volume.
4. J. Byrne *et al.*, this volume.
5. J. Behr *et al.*, this volume.
6. B. Jonson, Physica Scripta T59 (1995) 53
7. M. Chartier *et al.*, this volume.
8. M. Steck *et al.*, this volume.
9. R. Van Dyke *et al.*, this volume.
10. S. Rainville *et al.*, this volume.
11. G. Bollen *et al.*, this volume.
12. D. Lunney *et al.*, Hyperfine Interactions 99 (1996) 105
13. M. de Saint Simon *et al.*, Physica Scripta T59 (1995) 406
14. J. Lettry *et al.*, Nucl. Instr. and Meth. B 126 (1997) 130
15. A. Coc *et al.*, Nucl. Instr. and Meth. A 271 (1988) 512
16. A. Coc *et al.*, Nucl. Instr. and Meth. A 305 (1991) 143
17. S. Henry *et al.*, this volume.

High Precision Penning Trap Mass Spectroscopy and a New Measurement of the Proton's "Atomic Mass"

Robert S. Van Dyck, Jr., Dean L. Farnham, Steven L. Zafonte and Paul B. Schwinberg

Department of Physics, Box 351560
University of Washington, Seattle, Washington 98195-1560

Abstract. The Penning trap mass spectrometer (PTMS) at the University of Washington has been rebuilt into a new state-of-the-art magnet/cryostat system with external Helmholtz compensation coils (controlled by a nearby flux-gate sensor). This system gives a total magnetic shielding factor of $\sim 10^4$ (which includes the effects of a passive internal flux-stabilizing coil supplied by the manufacturer). When the new magnet/cryostat is fitted with a system to control its boil-off pressure, the typical temporal field stability is $\sim -0.017(2)$ ppb/h. The ultimate resolution of this improved spectrometer is expected to exceed 0.020 ppb with 100 hours of data using a single C^{4+} ion. The comparison of a C^{4+} ion with a C^{5+} ion suggests that the spectrometer's accuracy may indeed match its resolution. To demonstrate the spectrometer's improved performance over its previous version, the cyclotron frequency of a single proton is compared to the corresponding frequency of a single C^{4+} ion, yielding a determination of the proton's atomic mass given by $M_p = 1,007,276,466.89(14)$ nu. The primary systematic error (overcome in this case) is due to a position sensitivity within the non-uniform magnetic field, enhanced by the electrostatic trapping potentials of these ions which differ by a factor of three. A residual limitation to the overall field stability appears to be due to a long-term temperature-dependent temporal wander in the magnetic field.

INTRODUCTION

The basic mass spectrometer has experienced many changes over the last several decades which have improved its resolution by many orders of magnitude. Initially conceived as a roughly uniform region of magnetic field which could bend a beam of velocity-selected ions into a semi-circular orbit onto a photographic emulsion, the mass spectrometer has now evolved into a sophisticated device which requires only a single sample ion followed by a single calibration ion, in order to determine the atomic mass of the sample particle to an accuracy that exceeds 0.1 ppb. This paper briefly describes the Penning trap mass spectrometer (PTMS) which uses a new custom-designed magnet/cryostat system that has made the device remarkably more stable than any previous spectrometer. Its immunity to both external and internal field variations now makes it possible to fully realize the inherent accuracy associated with the PTMS. To illustrate its improved performance, the measurement of the proton's atomic mass will be described.

MAGNET/CRYOSTAT SYSTEM

In the past, typical superconducting magnets were observed to have field drifts that could occasionally exceed 1 ppb/h or more (even when placed in a so called "persistent state"). Over the years, it has become evident that these field changes are not due exclusively to externally varying magnetic fields. In fact, one of the most serious sources of temporal instability is associated with the temperature-dependent susceptibility of all materials that pass through the bore of the superconducting solenoid [1]. As helium boil-off gas exits the cryostat, a flow-rate dependent temperature distribution is established over some of these offending materials. This distribution (and thus the magnetic susceptibility) changes as atmospheric pressure fluctuations disturb the boil-off rate.

In order to significantly reduce this effect, an innovative design by one of the authors (PBS) was submitted to a major manufacturer of superconducting magnet systems (Nalorac Cryogenic Corporation). In the schematic illustration shown in Fig. 1, one notes that the liquid helium used in the main reservoir to maintain the superconducting state of the solenoid is also used to provide the cryogenic environment for the Penning trap placed

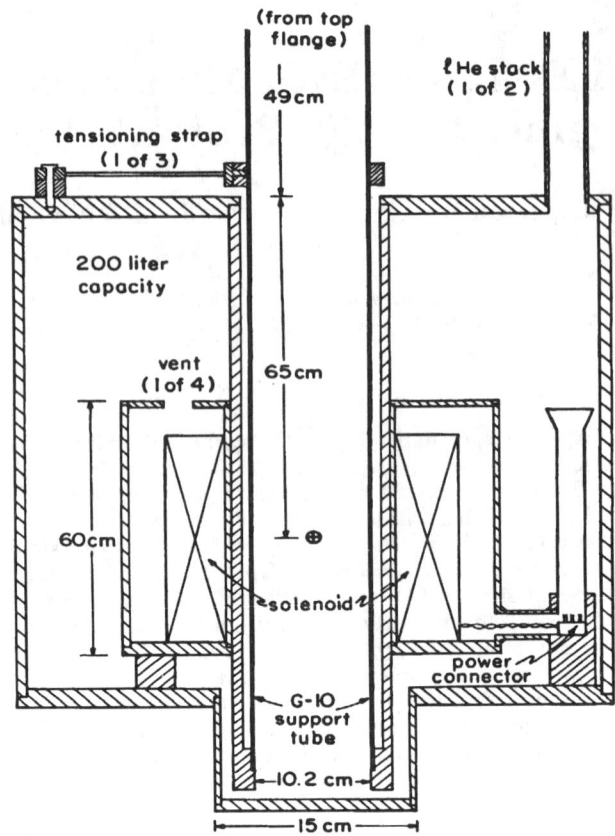

FIGURE 1. Overall schematic illustration of the new custom-designed magnet/cryostat system. Note in particular, that the main helium reservoir supplies the cryogenic environment for both the magnet and the experimental apparatus.

at the center of the solenoid. A 7–8 cm long helical resonator [2] is used to sense the level of the liquid helium in the bore. The 18 MHz frequency of this resonator changes by about 40 kHz/cm with the dielectric change of the liquid level within the device. By using a voltage controlled valve and a feedback circuit, the pressure is raised in the main liquid helium reservoir in order to force the liquid up to a height in the bore (determined by the location of the helical resonator) where the field is below 1% of its peak value. This level is controlled to ±50 μm in order to stabilize thermal-electric effects on the the conducting cables that provide the ring voltage to the Penning trap. As a result, all critical surfaces now have liquid helium providing the temperature stabilization instead of easily warmed boil-off vapor of some earlier designs.

Other notable features of this new magnet/cryostat design are the following (all of which will be described in a future article):

- The main support for the entire system of liquid helium reservoir, magnet, and experimental apparatus, is provided by the 10-cm I.D. bore tube, thus reducing relative motion between trap and magnet center.

- The use of an inner magnet reservoir provides enhanced protection of the superconducting solenoid from accidental quenching when warm gas is inadvertently injected during the liquid helium transfer process.

- The charging plug at the bottom allows the internal magnet reservoir to be filled separately and the associated vent stack can be used for an independent capacitive level sensor.

- The boil-off gas pressure is stabilized relative to an absolute pressure reference to within ±1 ppm.

- The helium fill and vent stacks are temperature stabilized on the order of $\sim 10^{-3}$ C°.

- The combined liquid helium loss rate is about two liters/day.

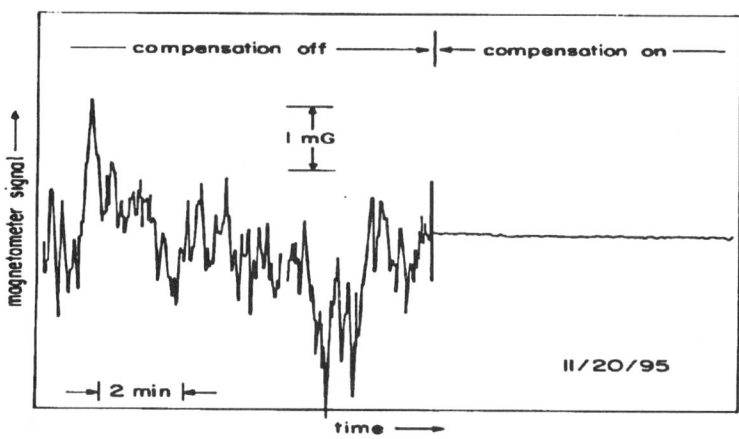

FIGURE 2. Magnetic field sensed by a commercial magnetometer located at the center of the Helmholtz pair with and without compensating field applied. The superconducting solenoid was not in place during this test.

As a further refinement, necessitated by a move of the Physics Department to a new building near an electric trolley line, the cryostat is centered inside a 1.7 meter diameter Helmholtz coil pair. To compensate for a 30× larger external magnetic noise at this location, a custom-fabricated flux-gate device is used to sense the magnetic noise at a strategic location about two meters from the magnet's center. A correction current is then generated and fed into the Helmholtz coils in order to complete the field cancelation. With sufficient care, a compensation factor of 150 can be achieved, but typical compensation is about half this large as exemplified by the data shown in Fig. 2 (measured without the superconducting magnet in place). In addition, the solenoid has been supplied by the manufacture with a superconducting flux-stabilizing coil as described by Gabrielse [3]. This coil by itself can provide substantial shielding against external field variations that typically arise in the lab because of the redistribution of the ambient magnetic field with, for example, opening steel cabinet doors or moving steel chairs. For a uniform external field change, a compensation of about 180 has been observed. When combined with the shielding of the Helmholtz coil, the shield factor for the entire system is conservatively about 10^4 for distant field disturbances which can be considered as uniform near the magnet center.

THE PENNING TRAP MASS SPECTROMETER

The PTMS at the University of Washington has been described in some detail in the literature [4–6]. However, to see its improved potential for measuring atomic masses, it is necessary to outline its basic operation. A five-electrode Penning trap (consisting of hyperbolic ring and endcaps with rotational symmetry around the magnetic field axis, plus two guard or compensation electrodes [7]) is used to isolate a single charged particle. Its motion along the symmetry axis can be excited by a very stable frequency synthesizer, causing current to be induced in the endcaps. One of these electrodes is used to observe this motion by means of an attached LC circuit, referenced to common ground, that is tuned to the axial frequency $(2\pi)\nu_z = \sqrt{qV_0/md^2}$ where V_0 is the ring-endcap potential difference, $d = 0.211$ cm is the characteristic trap dimension (related to the separation between endcaps), and q/m is the particle's charge-to-mass ratio. The resulting signal voltage is amplified by a custom-built, cryogenically cooled preamplifier and mixed with the original (appropriately phase-shifted) drive voltage to generate an error signal. If something happens to shift ν_z, the error signal becomes non-zero and is then integrated to produce a current that moves the ring voltage supply to a new value that again forces the error signal to zero. In this way, the charged particle's axial motion is kept frequency locked to the stable drive synthesizer and the resulting correction voltage (now referred to as the "frequency-shift signal") gives real time information about perturbative changes in the frequency of this easily observed resonance.

Of experimental interest are the two radial modes of the ion isolated in the Penning trap, both of which can be observed using the axial frequency shift detector. One of these is the $\vec{E} \times \vec{B}$ magnetron mode whose frequency is given by $\nu_m = \nu_z^2/2\nu_c'$. Here, the "observable" cyclotron frequency, $\nu_c' = \nu_c - \nu_m$, is the other observable radial frequency. All three normal mode frequencies can be taken in quadrature [8] to determine the free-space

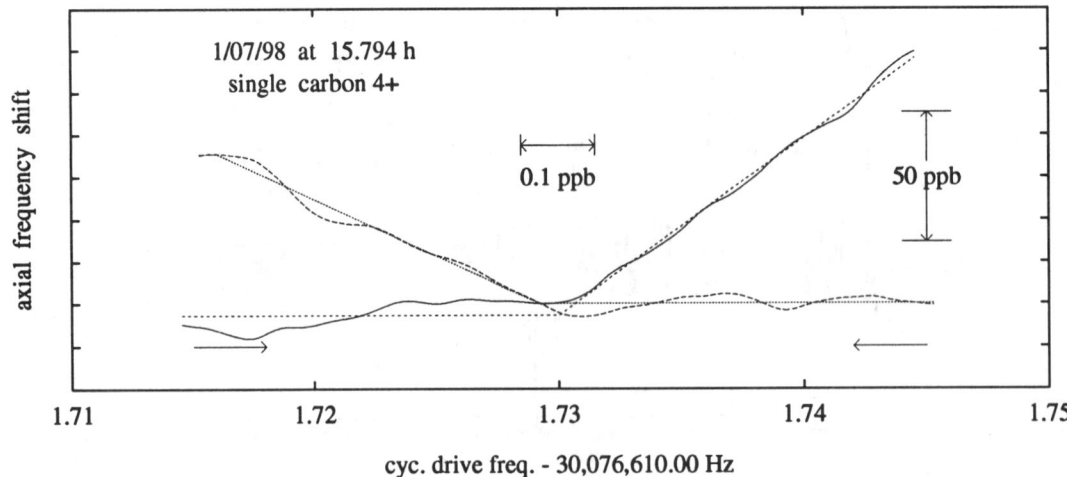

FIGURE 3. The bracketed cyclotron resonance for a single C^{4+} ion using our extrapolated trigger anharmonic detection method. The superimposed straight line segments are least-squares fitted lines and the typical "linewidth" (between corner frequencies) is 0.1–0.2 ppb. Also, full sweep range is one ppb and typical sweep time is 200–300 sec (each direction).

cyclotron frequency, $\nu_c = qB/2\pi m$, given by

$$\nu_c^2 = (\nu_c')^2 + (\nu_z)^2 + (\nu_m)^2. \qquad (1)$$

THE PROTON/CARBON COMPARISON

In order to thoroughly test the improved spectrometer, the proton's atomic mass has been determined using C^{4+} as the reference ion. For fixed ν_z, the respective well depths for this choice differ by a factor of three, which aggravates the possible relative shifts in the determination of the respective free-space cyclotron frequencies. For the case of the C^{4+} ion, it can be shown that the relative axial sensitivity to the cyclotron energy E_c is given by

$$\delta\nu_z/\nu_z = [10\,\text{ppb/eV}]E_c \qquad (2)$$

when the C_4 anharmonic potential coefficient is chosen to be $\sim 5 \times 10^{-5}$ (see reference [9] for the details concerning the anharmonic potential). For protons, the constant in brackets is about 2.5 times larger, determined in this case by the residual quadratic term in the magnetic field, $B_2 \sim 1$ G/cm^2. Figure 3 shows a typical example of the axial shifts observed when the cyclotron resonance is excited by a frequency-swept rf electric field and cooled using sideband cooling resonances [10]. In the limit that one uses long detection response times and sweeps a very narrow range of about one ppb, the energy absorption signature looks quite similar for both sweep directions. However, for a down-sweep, the absorbed energy increases the relativistic mass of the ion which then shifts the resonant frequency downward. If the sweep is slow enough, the resonance stays ahead of the swept frequency and continues to absorb energy. The up-sweep response would normally be a step function because the resonant frequency is pulled through the drive frequency, whereupon it indicates a beating between the free and driven motions. However, the long detection response time filters out the beat note and produces a similar ramp of the axial frequency. Such pairs of sweeps are used to bracket the resonance; in addition, one generally adjusts the sweep rate and the power applied such that the two corner frequencies coincide with a typical scatter of ± 0.1 ppb. The averages of consecutive corner-frequencies are then plotted versus time as shown in Fig. 4 where a linear drift of -0.016 ppb/h has been removed from the data. The precision of the linear fit for this one week's worth of data (~ 1200 points) is 5×10^{-12} which is roughly the value expected assuming Gaussian statistics.

Dependence on Field Inhomogeneity

It was anticipated early in the proton/carbon comparison, that the magnetic field inhomogeneity would produce the largest systematic shift in the cyclotron frequency ratio for these two ions. In order to determine the resulting

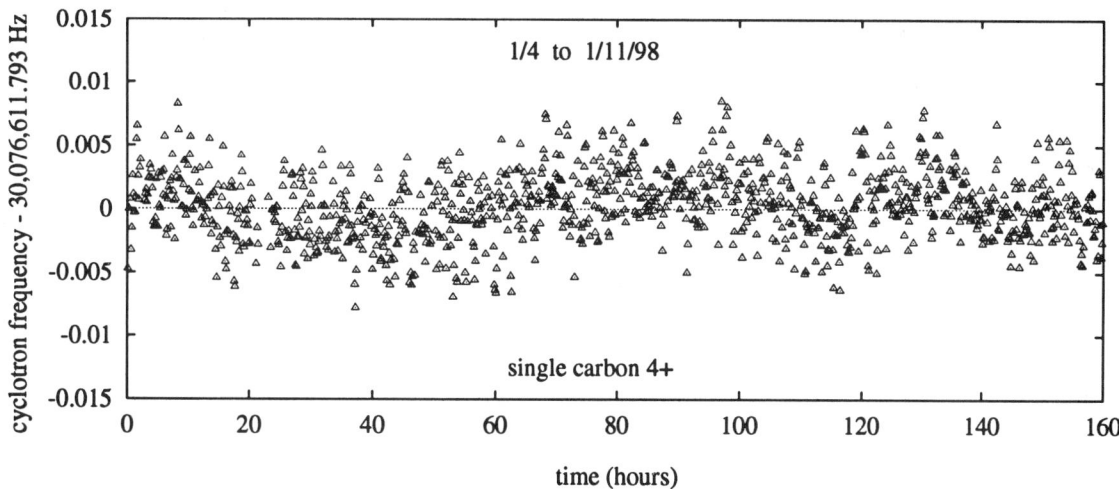

FIGURE 4. Typical cyclotron resonance data taken using a single C^{4+} ion after final modifications to the magnet/cryostat stabilization systems. Residual structure in this record is believed to be associated with temperature effects (driven by changes in the ambient atmosphere) of electronic parts in various control systems. The full vertical scale represents one ppb for residuals with a -0.016 ppb/h linear drift removed. Precision of this fit is about 5×10^{-12}.

sensitivity of such frequency ratios, three coils had been wrapped around the vacuum envelope, symmetrically located relative to the main ring electrode. The outer two were operated like a set of bucking coils to tune out the linear term in the magnetic field. With the assumption that the middle coil was centered on the ring electrode, it was expected to generate only a useful quadratic dependence in the magnetic field, and indeed, the quadratic term was measured to be $+80$ $(G/cm^2)/amp$. As an example of the data taken when $+18$ mA was flowing in this middle coil, see Fig. 5. Here, a single C^{4+} ion is used to observe the field drift for approximately one week, and then a single proton is used for three days to observe the same field drift (but now subtracted from the data). Even after scaling the proton residuals by the inverse cyclotron frequency ratio, one still notes a $3\times$ larger scatter in the proton data. This is probably due to a cryogenic preamplifier being noisier than usual, which further reduces the driven signal/noise for protons in the $3\times$ shallower potential well. The results obtained by varying the current in the middle coil from -18 to $+27$ mA are shown in Fig. 6, where a linear dependence of 0.056 ppb/mA has been removed. At this point, it became evident that this sensitivity was still coming from the linear

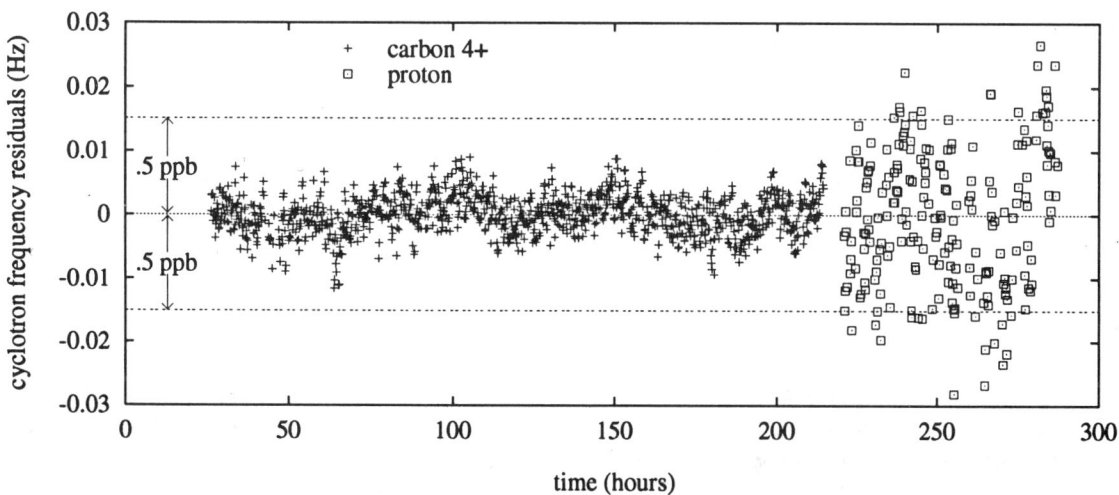

FIGURE 5. The cyclotron frequency residuals versus time for a single C^{4+} ion and a single proton with $+18$ mA flowing in the middle coil. By fitting data, obtained from the bracketed corner frequencies, to a common drift of -0.017 ppb/h, the fitted cyclotron frequency ratio $= 2.977,783,718,335(74)$ allows all data to be displayed on the same record.

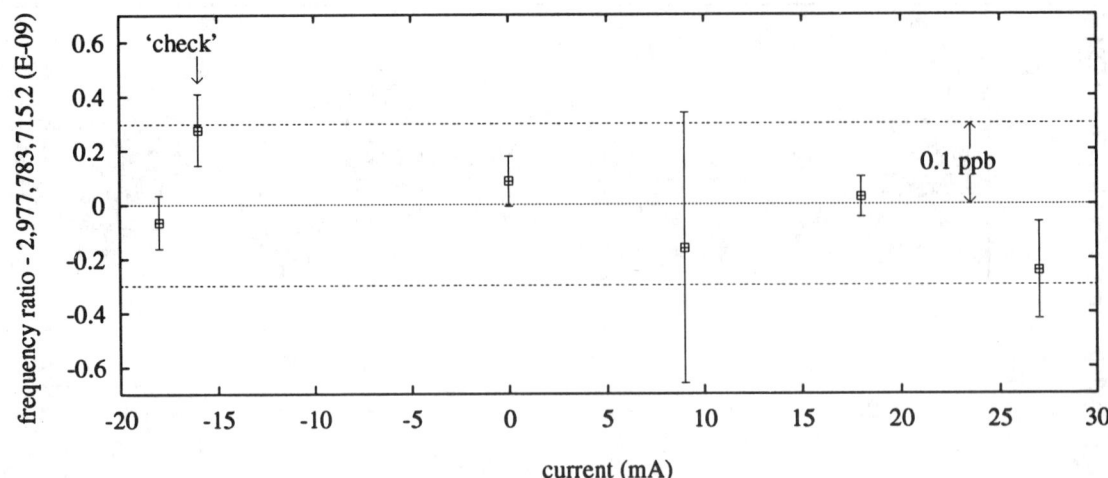

FIGURE 6. Cyclotron frequency ratio versus current in the coil that is centrally located on the vacuum envelope relative to the main ring electrode. A fitted linear dependence of 0.056 ppb/mA has been removed from the data and the "check" data point was determined with the linear gradient compensated using the bucking coils surrounding the central coil before that measurement was made.

term in the magnetic field, not from the quadratic term, and a plot of ν_c' versus axial offset potential between endcaps [11] verified the existence of a strong linear gradient = 7.5 (G/cm)/amp coming from the middle coil. This term arises because the central coil misses the center of the main ring electrode by 0.047(5) cm. To verify this conclusion, a current of -16 mA was passed through the middle coil, but with the linear term nulled out to 0.002 G/cm using the two outer bucking coils. The resulting cyclotron frequency ratio is plotted (without any further adjustments) in Fig. 6 and is referred to as the "check" point. Clearly, there is very little dependence on the B_2 term within the tolerance of 0.1 ppb. Also, when B_1 is nulled to a limit < 0.002 G/cm and the potential difference between endcaps (due to patch effects, etc.) is < 100 mV, the relative magnetic field difference for both protons and C^{4+} ions is < 0.01 ppb.

Dependence on Normal Mode Energies

In order to appreciate the residual sensitivity of the cyclotron frequency ratio of these two ions to the energies E_c, E_z and E_m which exist in the cyclotron, axial and magnetron modes respectively, first-order perturbation theory [12,13] is used to establish the relative shifts in the cyclotron frequency which are appropriate for the values of C_4 and B_2 which are needed for the detection of each ion, along with the corresponding relativistic mass effect. The dominant shifts are the following (with coefficients in units of ppb/eV):

$$\frac{\delta\nu_c'}{\nu_c'}(C^{4+}) \approx (-0.03)E_c + (4.9)E_z + (3.9)E_m \qquad (3)$$

$$\frac{\delta\nu_c'}{\nu_c'}(H^+) \approx (-1.07)E_c + (23)E_z + (46)E_m. \qquad (4)$$

Dependence on Magnetron Energy

Because of the potential impact on the relative shift due to the magnetron energy term, E_m must be estimated. This can be done for the C^{4+} ion because its detection sensitivity allows an adequate resolution of its magnetron energy. First-order perturbation theory indicates that

$$\nu_m = \nu_m(0)\{1 + (aC_4)E_z - (2aC_4)E_m\} \qquad (5)$$

where $a \approx 1.1 \times 10^{-2}$/eV for the C^{4+} ion. By using the axial frequency shift detector in the same manner as used for the measurement of the cyclotron frequency shown in Fig. 3, and measuring ν_m for $C_4 = -5.6 \times 10^{-5}$

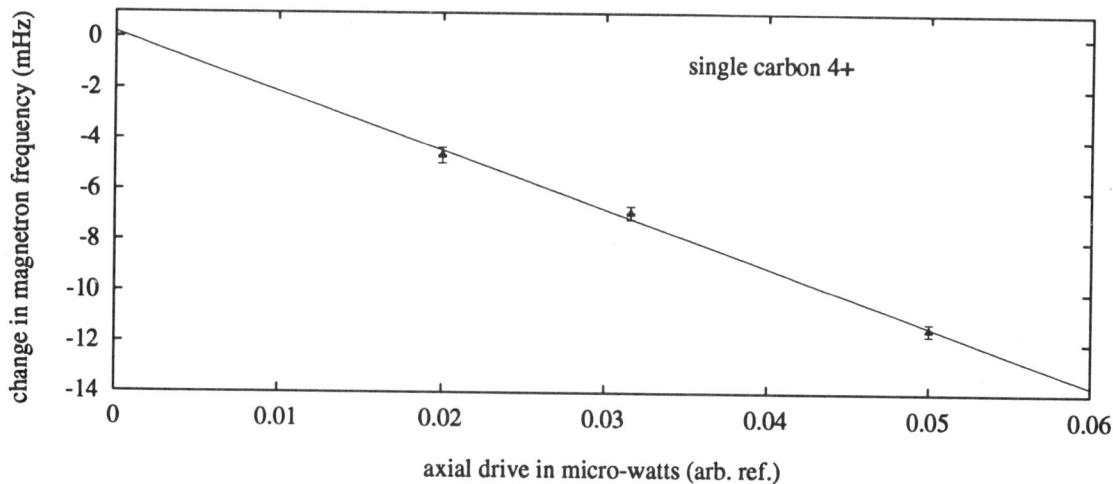

FIGURE 7. The difference in magnetron frequency for a single C^{4+} ion measured at the symmetric extremes of $\pm 5.6 \times 10^{-5}$ for the anharmonic C_4 coefficient versus axial drive power. The uncertainty of the linear fit, extrapolated to zero axial drive power, is ± 0.39 mHz.

and $C_4 = +5.6 \times 10^{-5}$, one can plot the difference of these two measurements as a function of the axial drive power as shown in Fig. 7. By extrapolating to zero axial drive power, the residual uncertainty of ± 0.39 mHz for $\nu_m \sim 209,690,020$ mHz yields

$$E_z(\text{th}) + 2|E_m| \approx 1.51 \text{ meV} \qquad (6)$$

where $E_z(\text{th})$ is the un-driven axial energy established by the equilibrium of the resonant motion with the detection amplifier near 4 K, and E_m is the magnetron energy which must be negative. Choosing a lower limit of 0.4 meV for $E_z(\text{th})$ requires $|E_m| < 0.6$ meV. However, the theoretical cooling limit expected for the magnetron motion is given by $(\nu_m/\nu_z)E_z(\text{th}) < 0.06$ meV and three times still smaller for the proton. Thus, for these two ions, it is safe to assume $E_m < 1$ meV for each ion, which implies a systematic relative shift of the cyclotron frequency ratio due to the magnetron energy < 0.04 ppb.

Dependence on Axial Drive Energy

Another potential problem noted by the perturbation relations in Eqs. (3) and (4) is the dependence on axial drive power. These relations suggest that if the axial energy is strictly thermal, there is a negligible shift due to E_z. However, cyclotron resonances are taken with axial drive applied and $E_z(\text{dr})$ is greater that $E_z(\text{th})$. Figure 8 shows the effect for the C^{4+} ion when the anharmonic C_4 coefficient is about twice that which is needed for detection. Normally, this drive is reduced until such shifts are not visible within the available resolution; however, as suggested by the data in Fig. 8, even in this case the drive power can be extrapolated to zero, yielding an uncertainty of $\sim 2 \times 10^{-11}$ for C^{4+}. The axial drive for protons is typically reduced by another factor of 2–4 in power (relative to the drive for C^{4+}) in order to keep this effect well below the 0.1 ppb level. However, since it could not be measured, it is reasonable to estimate that the possible shift due to proton axial energy could be ± 0.02 ppb times roughly the ratio of coefficients in Eqs. (3) and (4) (for the range of coupling coefficients used in the experiment), thus yielding an estimated uncertainty of ~ 0.07 ppb for this effect.

Dependence on Cyclotron Drive Energy

Finally, the perturbation relations in Eqs. (3) and (4) suggest that the available cyclotron energy could produce a sizeable systematic shift for these two ions. This fear is based on the value for the cyclotron energy which is required to see an axial frequency shift above the typical axial noise [6], given by $E_c(\text{det}) \lesssim 1$ eV. However, the frequency-shift detection method relies on the fact that the expected width of the cyclotron resonance is quite small, typically $< 10^{-11}$ due to residual external magnetic noise and thermal axial noise convolved into

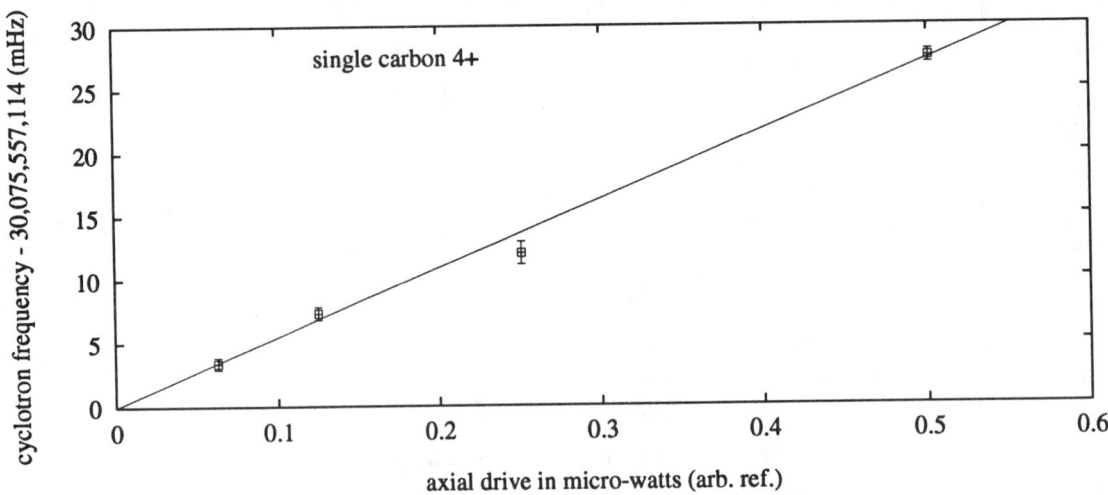

FIGURE 8. The cyclotron frequency of a single C^{4+} ion is measured as a function of the axial drive power. The strong dependence reflects the use of drive powers much larger than normally used as well as the use for detection of a larger than normal value of the anharmonic C_4 coefficient ($\sim 6.7 \times 10^{-5}$). The uncertainty of the linear fit extrapolated to zero power is $\sim 2 \times 10^{-11}$.

the cyclotron resonance through the perturbation effects due to the coupling coefficients shown in Eqs. (3) and (4). Thus, one must be essentially right on the cyclotron resonance before appreciable power is absorbed, and sweeping slowly should reduce the frequency error that occurs before sufficient energy is absorbed by the rapidly growing orbit radius that eventually allows the resonance to be detected. Figure 9 shows the shifts in the cyclotron frequency for a single C^{4+} ion when the cyclotron drive power is varied, yielding a relative error of about 2×10^{-11} for the weakest drives which are typically used for this ion. To give another perspective, one can also plot cyclotron frequency versus the anharmonic C_4 coefficient. In the case of the C^{4+} ion, the data extrapolated to zero C_4 indicates a relative error of 3×10^{-11} remains for the smallest C_4 value used in the experiment. However, the relativistic mass shift is actually the dominant perturbation for E_c. Thus, another check is required to test for the dependence on cyclotron energy. For this purpose, one compares ions with different charge states of the same element, since typically, the ratio of such cyclotron frequencies are known to a few parts in 10^{13}. However, the magnitude of error made by the relevant perturbations depends on the difference in relative shifts of the respective cyclotron frequencies. For convenience, the C^{5+} ion has been compared with a C^{4+} ion as shown

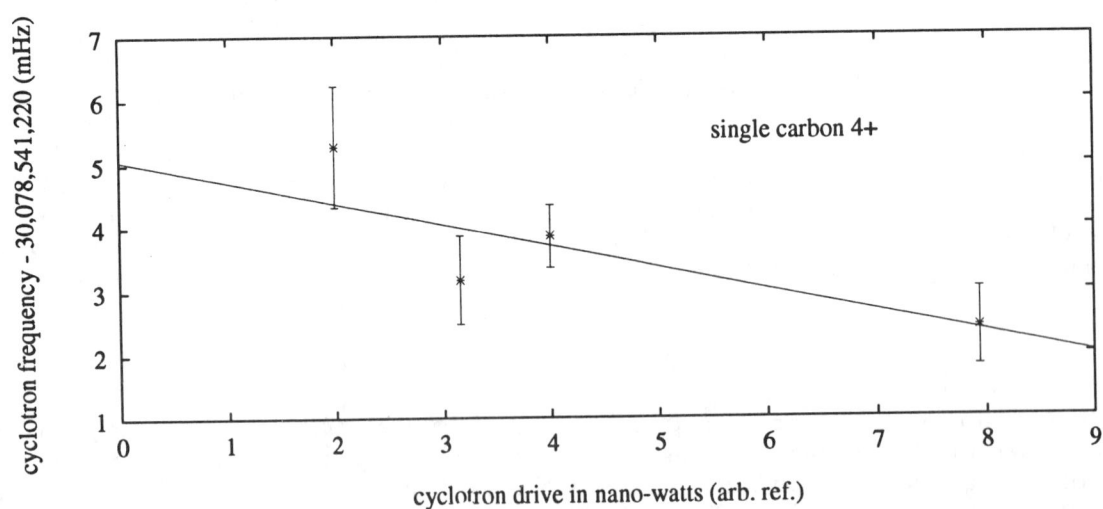

FIGURE 9. The cyclotron frequency of a single C^{4+} ion measured as a function of the cyclotron drive power. The anharmonic C_4 coefficient is -5.6×10^{-5} for this data and the fit uncertainty, extrapolated to zero power, is $\sim 2 \times 10^{-11}$.

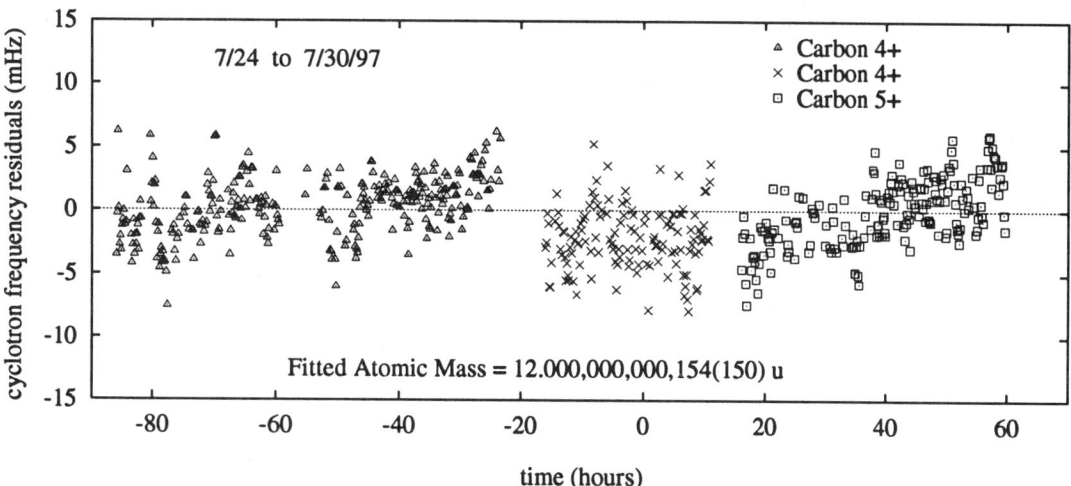

FIGURE 10. The cyclotron frequency is measured as a function of time for two separate C^{4+} ions followed by a single C^{5+} ion. A linear drift of -0.012 ppb/h has been subtracted from the data and the fitted mass ratio $= 0.799,963,449,374(10)$ yields an error of 1.3×10^{-11} relative to the known atomic mass. Note that the full vertical scale is one ppb.

TABLE 1. Error budget for the proton/carbon comparison

Source	Uncertainty[a]
Statistical	0.05
Residual temperature and/or pressure effect	0.08
Dependence on axial energy	0.07
Dependence on cyclotron energy	0.05
Dependence on magnetron energy	0.04
Image charge shifts	0.04
Residual linear gradient effect	0.01
Quadrature sum	0.14

[a] in parts per billion.

in Fig. 10 (though the C^{6+}/C^{4+} comparison would be more sensitive). For this consistency test, the fitted mass ratio differs from the "correct" value by 1.3×10^{-11}. One can also show that the perturbation relations for these two ions would predict a systematic error $\sim 7\times$ larger if the relative shift in the observed cyclotron resonances were caused by $E_c(\text{det})$. In this case, the accuracy of the C^{5+}/C^{4+} comparison is consistent with

$$E_c < \left(\frac{1}{7}\right) E_c(\text{det}), \qquad (7)$$

and from Eqs. (3) and (4), it is safe to assume that the systematic error associated with E_c is less than 0.05 ppb.

Conclusions

The current accuracy of this PTMS is $\sim 2 \times 10^{-11}$ for the measurement of the cyclotron frequency of a single C^{4+} ion but with a potential precision of 5×10^{-12}. For measurements of the lightest of elements however, the accuracy of the spectrometer decreases, assuming one does not invoke substantial modifications. Table 1 summarizes the potential errors that could remain in this measurement, many of which have already been described. The possible image charge shift is based on the measurements [14] made in a $2\times$ smaller "quadring" trap [11] which has been scaled down according to the model that this effect varies as the inverse cube of the trap size. The second item listed in the table of errors has yet to be discussed. Most of the data displayed in Fig. 6 were taken with the final version of our stabilization controls. But earlier, a strong correlation between the barometric pressure and the internal magnetic field was observed, and the fit parameters were adjusted to compensate for this effect. With the final stabilization system in place, a weak correlation still remains that appears to be related to ambient atmospheric pressure changes, but delayed by 10–20 hours, suggesting a residual temperature dependence in one of the control systems produced by changes in weather systems. The uncertainty due to this effect that remains in the least squares fitting of the data is ±0.08 ppb. Thus, our current value for the proton's atomic mass [15] is

$$M_p = 1,007,276,466.89(14)\,\text{nu} \tag{8}$$

(where 1-nu = 10^{-9} unified atomic mass units). This result is 20 times more precise than our first measurement of M_p in 1989 [16] and agrees with the recent value obtained by the Stockholm group [17]:

$$M_p = 1,007,276,466.45(70)\,\text{nu}. \tag{9}$$

By adding the "atomic mass" of the missing electron, with $M_e = 548,579.911,1(12)$ nu [13,18], and subtracting the corresponding binding energy [19], one can determine a new value for the atomic mass of hydrogen:

$$M(H) = 1,007,825,032.21(14)\,\text{nu}. \tag{10}$$

This result also agrees well with the previous most accurate determination by the MIT group [20]

$$M(H) = 1,007,825,031.60(50)\,\text{nu}. \tag{11}$$

For future work, the sensitivity to position in the trap can be removed by using the same ring potential for both ions (and thus two different tuned preamplifiers), as well as by making the magnetic field as uniform as possible. Also, with some effort, the residual wander in the temporal field stability can be further reduced. As a result, it should be possible to reduce the uncertainty of the proton's atomic mass by yet another factor of three, with the C_4 dependence extrapolated to zero.

This research is supported by the Mono-Ion Research grant from the National Science Foundation.

REFERENCES

1. G.L. Salinger and J.C. Wheatley, Rev. Sci. Instrum. **32**, 872 (1961); J.M. Lockart, R.L. Fagaly, L.W. Lombardo, and B. Muhlfelder, Physica B **165 & 166**, 147 (1990).
2. W.W. Macalpine and R.O. Schildknecht, Proc. IRE **47**, 2099 (1959).
3. G. Gabrielse, and J. Tan, J. Appl. Phys. **63**, 5143 (1988).
4. F.L. Moore, L.S. Brown, D.L. Farnham, S. Jeon, P.B. Schwinberg, and R.S. Van Dyck, Jr., Phys. Rev. A **46**, 2653 (1992).
5. R.S. Van Dyck, Jr., D.L. Farnham, and P.B. Schwinberg, Physica Scripta **T59**, 134 (1995).
6. R.S. Van Dyck, Jr., in *Atomic, Molecular, and Optical Physics: Charged Particles,* R. Hulet and B. Dunning, Eds., in Vol 29A of series on Experimental Methods in the Physical Sciences (Academic Press, New York, 1995) pp. 363–389.
7. R.S. Van Dyck, Jr. D.J. Wineland, P.A. Ekstrom, and H.G. Dehmelt, Appl. Phys. Lett. **28**, 446 (1976).
8. L.S. Brown and G. Gabrielse, Phys. Rev. **A 25**, 2423 (1982).
9. G. Gabrielse, Phys. Rev. **A 27**, 2277 (1983).
10. R.S. Van Dyck, Jr., D.L. Farnham, and P.B. Schwinberg, J. Mod. Opt. **39**, 243 (1992).
11. R.S. Van Dyck, Jr., P.B. Schwinberg, and S.H. Bailey, in *Atomic Masses and Fundamental Constants 6,* J.A. Nolen, Jr. and W. Benenson, Eds. (Plenum, N.Y., 1980) p. 173.
12. L.S. Brown and G. Gabrielse, Rev. Mod. Phys. **58**, 233 (1986).
13. D.L. Farnham, Ph.D. thesis, University of Washington, 1995.
14. R.S. Van Dyck, Jr., F.L. Moore, D.L. Farnham, and P.B. Schwinberg, Phys. Rev. **A40**, 6308 (1989).
15. R.S. Van Dyck, Jr., D.L. Farnham, S. Zafonte, and P.B. Schwinberg, *Abstracts of Contributed Papers,* for the Sixteenth International Conference on Atomic Physics at the University of Windsor, (Windsor, ON, Canada) p. 294 (1998).
16. R.S. Van Dyck, Jr., F.L. Moore, D.L. Farnham, and P.B. Schwinberg, in *Frequency Standards and Metrology,* A. De Marchi, Ed. (Springer-Verlag, Berlin, 1989)p. 349.
17. C. Carlberg *et al.*, "A Precision Determination of the Proton Mass Using Highly Charged Ions," Conference Proceedings on Trapped Charged Particles and Fundamental Physics, Daniel H.E. Dubin (editor), held in Asilomar, CA, August 31, 1998.
18. D.L. Farnham, R.S. Van Dyck, Jr. and P.B. Schwinberg, Phys. Rev. Lett. **75**, 3598 (1995).
19. R.L. Kelly. J. Phys. Chem. Ref. Data **16**, 1 (1987).
20. F. DiFilippo, V. Natarajan, M. Bradley, F. Palmer, and D.E. Pritchard, Physica Scripta **T59**, 144 (1995).

Mass measurements on radioactive isotopes with a Penning trap mass spectrometer

G. Bollen[1], F. Ames[2], G. Audi[3], D. Beck[4], F. Herfurth[4], H.-J. Kluge[4], A. Kohl[4], D. Lunney[3], R. B. Moore[5], M. de Saint Simon[3], E. Schark[2], S. Schwarz[4], J. Szerypo[7] and the ISOLDE Collaboration[1]

[1]CERN, Geneva, Switzerland, [2]Institut für Physik, Universität Mainz, Germany, [3]CSNSM-IN2P3-CNRS, Orsay, France, [4]Gesellschaft für Schwerionenforschung GSI, Darmstadt, Germany, [5]McGill University, Montreal, Canada and [6]Institute of Experimental Physics, Warsaw University, Warsaw, Poland

Abstract. Penning trap mass measurements on short-lived isotopes are performed with the ISOLTRAP mass spectrometer at the radioactive beam facility ISOLDE/CERN. In the last years the applicability of the spectrometer has been considerably extended by the installation of an RFQ trap ion beam buncher and a new cooler Penning trap, which is operated as an isobar separator. These improvements allowed for the first time measurements on isotopes of rare earth elements and on isotopes with Z = 80 - 85. In all cases an accuracy of $\delta m/m \approx 1 \cdot 10^{-7}$ was achieved.

INTRODUCTION

The binding energy of the atomic nucleus is one of the most fundamental properties of such a many-body system. Accurate mass data serve as testing grounds for nuclear models and stimulate their further improvement. Furthermore, systematic investigation of the binding energy as a function of proton and neutron number allows the direct observation of nuclear properties like pairing, shell and sub-shell closures, as well as deformation effects, and leads to a deeper understanding of nuclear structure. In addition, very precise mass differences are for example required in the context of precision weak interaction studies in nuclear β-decay. Therefore, large efforts are presently devoted to the application of classical as well as new mass spectrometric techniques, such as time-of-flight, Smith-RF or Schottky mass spectrometry, for the accurate mass determination of short-lived isotopes far from the valley of beta stability (1,2).

Penning traps have proven to be very accurate mass spectrometers (3). A large variety of mass measurements with highest accuracy has been performed on stable, mostly light particles. ISOLTRAP is so far the only spectrometer operational for the investigation of short-lived radioactive isotopes. It is installed at the on-line mass separator ISOLDE/CERN, one of the world-leading facilities for the production of radioactive isotopes. Another project, the CPT trap system (4) installed at Argonne National Laboratory will start to deliver first results very soon.

ISOLTRAP is now in operation since more than 10 years and masses of more than 100 radioactive isotopes have been investigated with an accuracy of $\delta m/m = 1 \cdot 10^{-7}$. The more recent measurements have been carried out on rare earth isotopes in the vicinity of ^{146}Gd, of neutron-deficient mercury isotopes and of isotopes with Z = 82 - 85.

THE ISOLTRAP SPECTROMETER

The basic principle of mass measurements with ISOLTRAP is the determination of the cyclotron frequency $\omega_c = q/m \cdot B$ of ions with a charge-over-mass ratio q/m stored in a Penning trap with known magnetic field B. Fig. 1 shows the present layout of the ISOLTRAP spectrometer (5, 7). The first section of the spectrometer has the task to

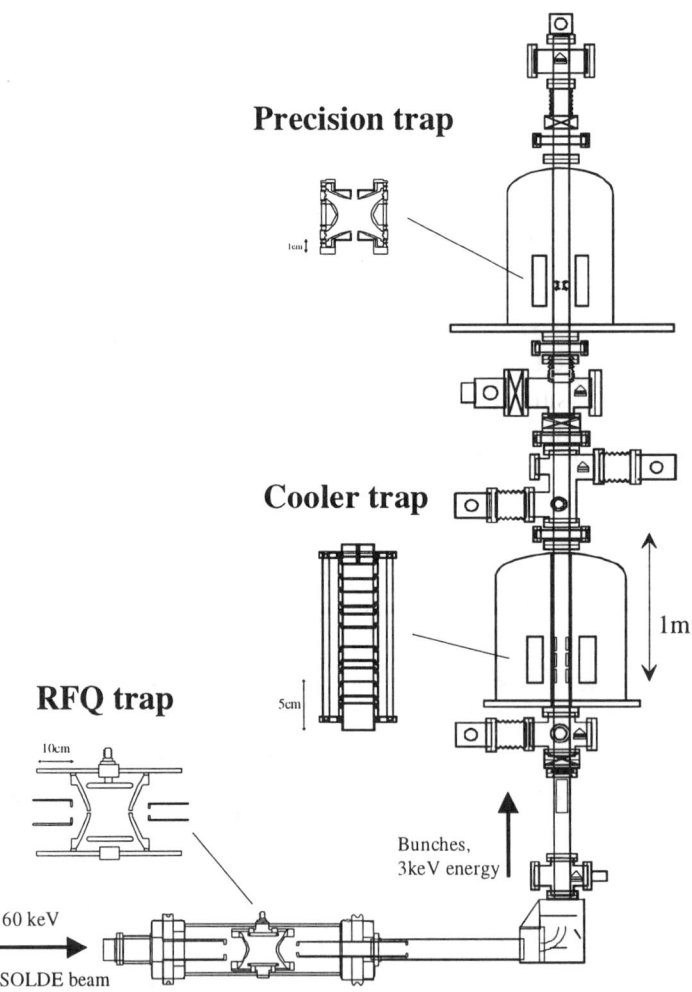

FIGURE 1. Experimental set-up of the ISOLTRAP mass spectrometer at ISOLDE/CERN

stop the 60 keV ISOLDE beam and to prepare it for an efficient transfer into the cooler trap. This is achieved by an RFQ trap ion beam buncher (6), which allows to capture the continuous ISOLDE beam in flight. The lower Penning trap (7) has the task to accumulate, cool, and mass separate the ions delivered from the ion preparation section and to bunch them for an efficient delivery to a second Penning trap. This precision trap is the actual mass spectrometer where the cyclotron frequency of the captured ions is determined. Important aspects of the set-up and the performance of ISOLTRAP will be discussed in the following.

In-flight accumulation and bunching

At ISOLDE radioactive ions are delivered as 60 keV continuous ion beams. In order to prepare a low energy ion beam that can be captured in the cooler Penning trap, a stopping re-ionization scheme was employed for many years at ISOLTRAP. ISOLDE ions were first stopped on a rhenium filament. Then, by heating it the accumulated species were released, surface-ionized and sent to the cooler trap. The main drawback of this technique was its applicability to surface-ionizable elements only.

The only way to overcome this major limitation was the implementation of an accumulation and bunching technique where no implanting and re-ionization processes are involved. For this purpose an RFQ trap ion beam buncher was developed and integrated into the ISOLTRAP system two years ago (6,8). The buncher consists of an RFQ trap connected directly to the ISOLDE beam line but installed on a high voltage platform close to the potential of the beam. Ions from ISOLDE are thereby decelerated from their initial energy to a few electron volts before entering the RFQ trap. The ions are then captured in the trap by energy loss from buffer gas collisions. The collected ion cloud is ejected as a single bunch with an average kinetic energy of about 3 keV. In order to deliver the ions to the cooler Penning trap (Fig. 1), which is just above ground potential, the ion bunch enters a cavity immediately after extraction which is brought from the trap potential to ground potential while the ions are still inside. The RFQ trap ion beam buncher was for the first time successfully used for the mass measurements on neutron-deficient mercury isotopes discussed below.

Isobar and isomer separation

Penning trap mass measurements require rather clean beams in order to avoid systematic errors in the mass determination arising from Coulomb interaction of different ion species in the trap. The resolving power of the ISOLDE general-purpose separator is by far not high enough to deliver such isobarically pure beams. A mass selective cooling technique (9, 10, 11), based on the simultaneous application of a buffer gas cooling and radio frequency excitation of the ion motion, is employed in the first "cooler" Penning trap. This trap system has been completely reconstructed (7) and optimized for a mass selectivity high enough to resolve isobars and for the delivery of

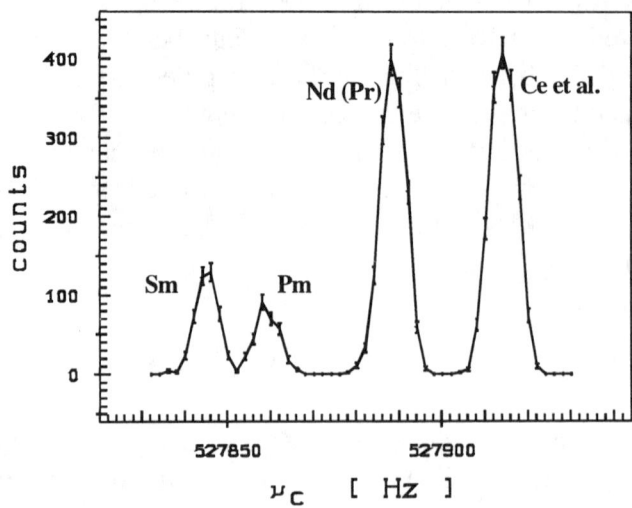

FIGURE 2. 'Mass scan' with the cooler trap for A=138 ions delivered by ISOLDE. Shown is the number of ions extracted from the trap as a function of the applied radio frequency.

clean and cooled ion bunches to the precision trap, an essential ingredient for highly accurate mass measurements. As an example, Fig. 2 shows a 'mass scan' performed with the cooler trap for an A=138 ion beam delivered by ISOLDE from a Ta-foil target. Shown is the number of ions extracted from the trap as a function of the applied radio frequency. The mass resolving power achieved here is about $R=10^5$, which is sufficient to resolve and separate isobars even close to stability.

The precision trap in which the cyclotron frequency determination of the ions takes place is in operation without major modification since several years and performs excellently (12). This trap is normally operated with a resolving power R close to one million corresponding to rf excitation times of $T_{rf} = 1$ s for A = 100 ions. If required, the resolving power $R \sim T_{rf}$ can be considerably increased by increasing T_{rf}. The maximum resolving power that has been realized in off-line tests with ^{133}Cs ions is R = 8 million using T_{rf} = 12s. This corresponds to a mass resolution of Δm_{fwhm} = 15 keV.

For the investigation of trends in nuclear binding energies an accuracy of $\delta m/m \approx 10^{-7}$ is normally sufficient and is already achieved with modest resolving powers (R < 10^6). Higher resolving powers become important in the case of long-lived isomers produced simultaneously with isotopes in their ground state. Over the nuclide chart nearly one third of the isotopes have long-lived isomeric states with (in many cases unknown) excitation energies down to < 100 keV. Only in a few cases information about the production ratio exists which may vary drastically depending on the half-lives and release times from the targets. Therefore, the resolution of isotopes in their ground or isomeric state is essential for an unambiguous determination of the mass of the isotope in one or the other state. That this can be achieved with ISOLTRAP has now been demonstrated several times. Two recent examples are shown in Fig. 3 for ^{141}Sm and ^{185}Hg.

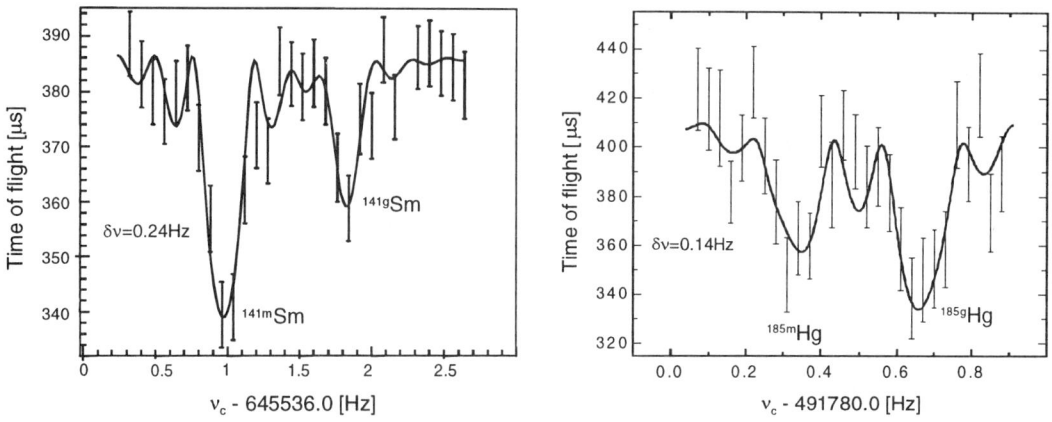

FIGURE 3. Resolved isomeric and ground states for ^{141}Sm ($\Delta E = 175$ keV) and ^{185}Hg ($\Delta E = 118$ keV).

MASS MEASUREMENTS

In total 76 isotopes and states were investigated since 1994, which are listed in table 1. An accuracy in the mass determination of $\delta m/m = 1 \cdot 10^{-7}$ was achieved for most of these isotopes. The measurements concentrated on rare earth isotopes, isotopes of mercury and of heavier elements, which became only possible by the recent improvements of the spectrometer.

Rare earth isotopes

So far direct mass measurements in this region were hampered by the fact that many isobars are delivered simultaneously by ISOLDE. Since the cooler trap can be operated as an isobar separator, clean ion samples can be prepared and sent to the precision trap. In several beam times it was possible to investigate more than 50 isotopes in the vicinity of ^{146}Gd, most of them with N ≤ 82 and Z < 64. For most of the isotopes the detailed analysis of the data is finished (13) and an atomic mass evaluation similar to the work by G. Audi et al. (14) has been performed. The evaluation shows that the ISOLTRAP measurements have a large impact on this mass region. This is illustrated in Fig. 4, which shows the trend of the two neutron separation energies. The upper and lower part of the figure show the situation before and after the ISOLTRAP data have been included.

Prior to the ISOLTRAP measurements strong discontinuities were observed in the S_{2n} trends derived from estimated but also from experimental mass values as can be seen in the upper figure. Above the N=82 shell closure, for Z=67 and Z=78, these discontinuities are now removed and the separation energies follow the regular trend observed in the neighboring isotopic chains. Most of the isotopes investigated by ISOLTRAP are in the region with N < 82 and Z < 64. Also here trends are now more clearly established. Systematic deviations from a linear trend for Z > 56 around N=76,

TABLE 1. List of isotopes investigated with ISOLTRAP since 1995.

Element	Mass Number	Element	Mass Number
Ba	123, 125, 127, 131	Ho	150
Cs	133 (reference isotope)	Tm	165
Ce	132, 133, 134	Yb	158, 159, 160, 161, 162, 163, 164
Pr	133-137	Hg	184, 185g+m, 186-190, 191m,
Nd	130, 132, 134-138		192, 193g+m, 194-196, 197g
Pm	136-141, 143	Pb	196, 198, 208 (reference isotope)
Sm	136-140, 141m, 141g, 142, 143	Bi	197
Eu	139, 141-149, 151, 153	Po	198
Dy	148, 149, 154	At	203

77 are now visible up to Z=60. They might be related to the eradication of the proton sub-shell gap at $Z \approx 64$ as one departs from N=82 and be accompanied by a change in nuclear deformation.

Since mass values of many isotopes are linked via known Q-values to other isotopes, accurate mass measurements of a few key isotopes can have a large impact on the knowledge of masses over a whole mass region. The case of ^{150}Ho will be discussed as an example. Mass differences between 19 isotopes linked to ^{150}Ho, some of them beyond the proton drip-line around $Z \approx 80$, are already known via experimental Q-values. No link existed between these nuclei and the backbone of stability, since a doubtful experimental Q-value for ^{150}Ho was rejected in the 1995 atomic mass evaluation (15). This unsatisfactory situation is now resolved by the ISOLTRAP measurement on ^{150}Ho, which justifies the early rejection of the old experimental datum, which is 810 keV away from the ISOLTRAP value. The ISOLTRAP measurement therefore not only gives an accurate experimental mass value for ^{150}Ho but also anchors the masses for all 19 isotopes linked to it.

Neutron-deficient mercury isotopes

The interest for nuclear structure investigations and mass measurements in this region arises from the appearance of shape coexistence at low excitation energies in the region around the shell closure at Z=82. The onset of rotational bands built on low-lying 0^+ states has been found (16) in even-even Pt, Hg, Pb and Po isotopes mid-shell between N = 82 and N = 126. A large staggering in the $\delta \langle r^2 \rangle$ values determined from isotopic shift measurements was observed for $A \leq 185$ for the ground-states of the light Hg isotopes, a jump from small to strong deformation in the neighboring Au isotopes at $A \leq 186$ and a smooth transition in the Pt isotopes (17, 18, 19). However, until recently

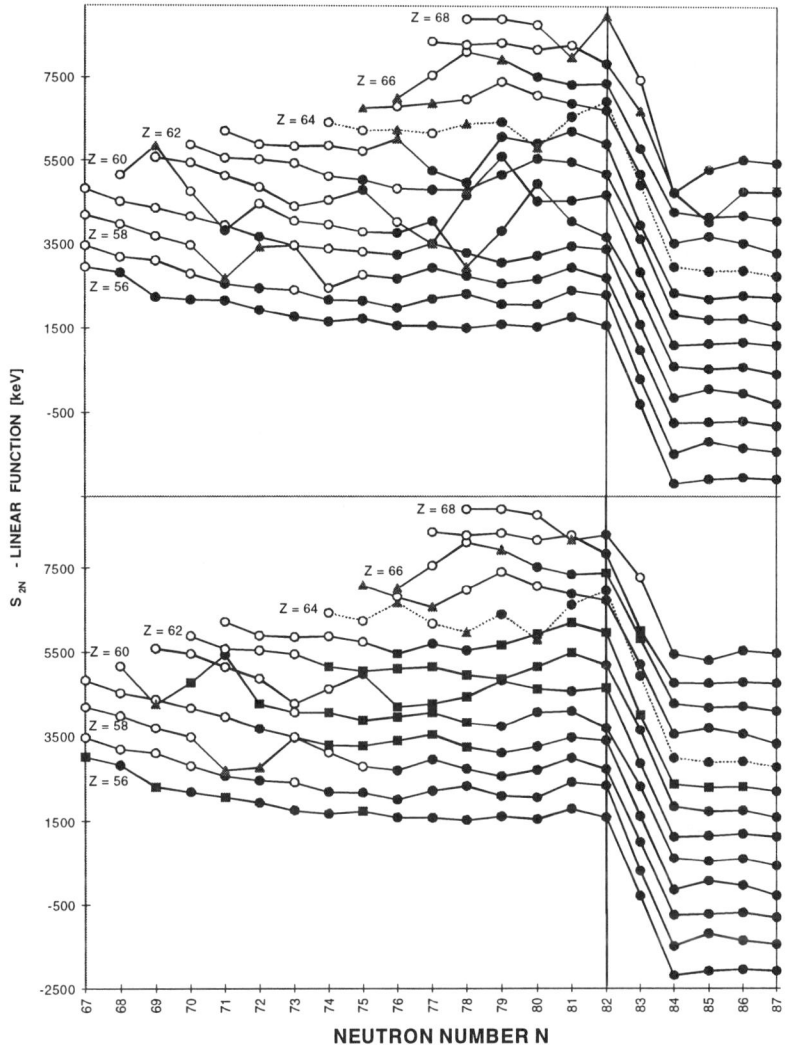

FIGURE 4. Two-neutron separation energies as a function of neutron number. Shown are S_{2n}-values excluding (top) and including (bottom) ISOLTRAP data in the atomic mass evaluation. The isotopes are marked by squares (ISOLTRAP data), filled circles (other experimental data), open circles (estimates from systematic trends), and triangles (doubtful experimental value).

no mass values were known in this mass region. Today, precise information is still lacking for $A \leq 185$, where the strongest structural changes happen.

The neutron-deficient isotopes of elements around $Z = 82$ are all members of long α-decay chains with well-known Q-values. Therefore, an accurate determination of such isotopes allows to fix these chains, making a large impact on a whole mass area starting at the upper part of the rare earth region and reaching to the border of known proton-rich isotopes.

With ISOLTRAP, a first series of mass measurements on the neutron-deficient mercury isotopes $^{185-197}$Hg was carried out in December 1996 after the installation of the RFQ trap ion beam buncher. In the case of the even isotopes where no isomeric

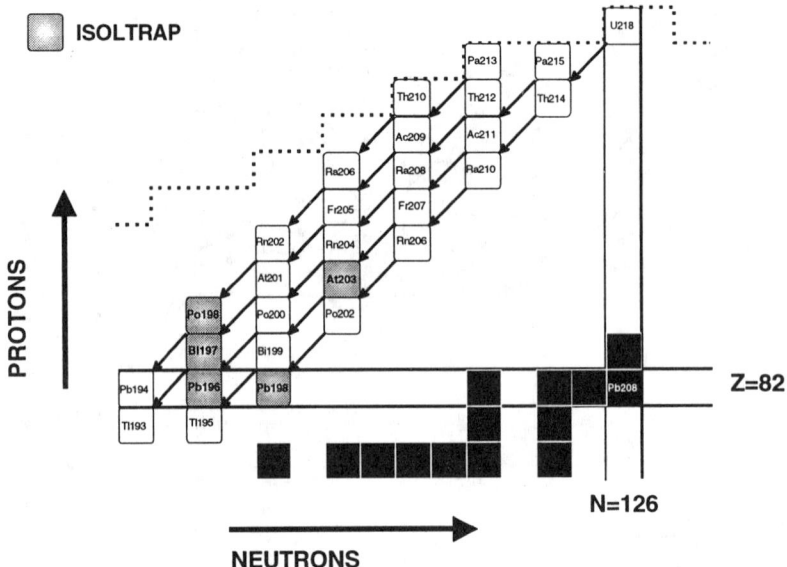

FIGURE 5. Q_α-value decay chains with isotopes investigated by ISOLTRAP (shaded). The dashed line indicates the borderline of known nuclei.

states exist, the evaluation was straightforward and an accuracy of $\delta m \approx 20$ keV can be assigned to all mass values. However, in the case of the odd isotopes long-lived isomers exist and are produced at ISOLDE. The excitation energies of typically 100 - 150 keV of these isotopes are very low. In the first measurements in December 1996 the spectrometer was operated with a resolving power of $R \approx 500000$, which corresponds to a mass resolution of 300 keV in this mass range. Therefore it was not possible to resolve isomeric and ground states.

Therefore, in a second run the attempt was made to verify the production of the isomers and to resolve them and the corresponding ground states. For this purpose the spectrometer was operated with resolving powers up to $R = 5$ million with which a mass resolution of $\delta m \approx 30$ keV was achieved. Using such a scenario it was possible to resolve isomeric and ground state in the cases of ^{185}Hg (see Fig. 3) and ^{193}Hg. Furthermore, it was verified that the ground state is dominantly produced in the case of ^{197}Hg, while for ^{195}Hg only the isomeric state has been seen, for which the excitation energy is known. Therefore for all these isotopes the ground state masses are now identified and determined with an accuracy of 20 keV. In addition, during this run it was possible to extend the measurements in the mercury chain out to ^{184}Hg.

Isotopes with Z = 82 - 85

Using the Paul trap ion beam buncher, the investigation of a new region of isotopes with $Z \geq 82$ was started very recently. In a first experiment isotopes were selected which are members of long alpha decay chains, those either not linked to an isotope with known mass or to one with a large mass uncertainty. Fig. 5 shows the decay chains and the isotopes investigated by ISOLTRAP. Due to the high accuracy of the ISOLTRAP data together with the availability of the Q_α-values accurate information on nuclear binding energies is now available even for very heavy proton-rich isotopes like ^{210}Th, ^{213}Pa, or ^{218}U, situated at the borderline of known nuclei.

CONCLUSION AND OUTLOOK

Penning trap mass spectrometry provides high accuracy mass data far from stability. The most recent investigations by ISOLTRAP have contributed significantly to our knowledge about nuclear binding of neutron-deficient rare earth isotopes and of isotopes with Z = 80 - 85. On-going measures (20) to increase the efficiency of ISOLTRAP will allow to extend the studies even farther from stability and to explore new mass regions. One example is the hardly explored region of neutron-rich isotopes at and above the magic proton number Z = 82, which is of importance in the context of nuclear astrophysics but also for the theoretical prediction of properties of super-heavy elements.

---

1. Bollen, G., Nucl. Phys. A626, 297c (1997).
2. Mittig, W. et al., Annu.Rev. Nucl. Sci. 47, 22 (1997).
3. Proc. of the Nobel Symposium 91 on Trapped Charged Particles and Related Fundamental Physics, Lysekil, Sweden, August 19-26,1994, Physics Scripta T59 (1995).
4. Savard, G., et al., Nucl Phys A626, 353c (1997).
5. Bollen, G., et al., Nucl. Instr. Meth. A368, 675 (1996).
6. Moore, R.B, et al., J. Mod. Optics 39, 361 (1992).
7. Raimbault-Hartmann, H., et al., Nucl. Instr. Meth. B126, 374 (1997).
8. Schwarz, S., PhD thesis work, Mainz 1998, and to be published
9. Bollen, G., et al., J. Appl. Phys. 68, 4355 (1990).
10. Savard, G. et al., Phys. Lett. A158, 247 (1991).
11. König, M., et al., Int. J. Mass Spec. Ion. Proc. 142, 95 (1995).
12. Beck, D., et al., Nucl. Instr. Meth. B126,378 (1997).
13. Beck, D., et al., Nucl. Phys. A626, 343c (1997).
14. Audi G., et al., Nucl. Phys. A565, 1 (1993).
15. Audi G., et al., Nucl. Phys, A595, 409 (1995).
16. Wood, J.L., Phys.Rep. 215, 101 (1992).
17. Ulm, G., et al., Z. Phys. A325, 2471 (1986).
18. Passler, G., et al., Nucl. Phys. A580, 173 (1994).
19. Hilberath, T., et al., Z. Phys. A342, 1 (1992).
20. Dilling, J., et al, A New Radio Frequency Quadrupole Ion Beam Buncher for ISOLTRAP, these proceedings.

BEAM COOLING USING A GAS-FILLED RFQ ION GUIDE

S. Henry, I. Martel-Bravo[*], M. de Saint Simon,
M. Jacotin, J.-F. Képinski, and D. Lunney[†]

CSNSM-IN2P3-CNRS, F-91405 Orsay, France

Abstract. A radiofrequency quadrupole mass filter is being developed for use as a high-transmission beam cooler by operating it in buffer gas at high pressure. Such a device will increase the sensitivity of on-line experiments that make use of weakly produced radioactive ion beams. We present simulations and some preliminary measurements for a device designed to cool the beam for the *MISTRAL* RF mass spectrometer on-line at *ISOLDE*. The work is carried out partly within the frame of the European Community research network: *EXOTRAPS*.

For most experiments using weak radioactive beams, for example the MISTRAL project at ISOLDE/CERN [1], it is desirable that the beam emittance be as small as possible. This requires that the beam be concentrated in a small geometrical area, with a small angular divergence, and a small energy spread. The way to reduce the beam emittance is to cool the beam for which several schemes exist: stochastic cooling and electron cooling are generally used in storage rings while ion trapping applications rely on resistive cooling, laser cooling, and buffer-gas cooling. The latter is of great interest as it is relatively simple and more or less universal.

Ions confined in a Paul or Penning trap can be cooled by introducing a light, neutral gas such as H_2 or He. The ion motion is viscously damped, in principle down to the temperature of the gas itself. This scheme was extended to a continuous beam traversing a radiofrequency quadrupole mass filter by Douglas and French [2] who demonstrated a dramatic gain in transmission through a small hole at the end of a tandem mass filter system by introducing buffer gas into the last quadrupole section. The mass filter continuously focuses the ion beam onto the central axis to avoid the diffusion that would otherwise occur as the ion transit the gas. Similar work using sextupole ion guides has also been reported [3,4].

---

[*]Present address: University of Seville, Spain
[†]Corresponding author (*lunney@csnsm.in2p3.fr*)

The purpose of the simulation presented here is to design a similar system capable of decelerating the 60 kV ISOLDE beam and cooling it to the buffer gas temperature before re-acceleration. Since radionuclides are generally short-lived, the cooling process must be rapid. Furthermore, any losses incurred must be minimal. In principle, simply increasing the buffer gas pressure will increase the cooling rate however the consequences of operating electric devices at high pressure are obvious. The design of such a device requires detailed simulation of a system of quadrupole rods (which we refer to as an RF ion guide) and a collisional cooling mechanism as discussed below.

Shown in figure 1 is a schematic diagram of the beam cooling system. Taking pumping speeds of 500 l/s, the end pressures are calculated using Knudsen's formula for vacuum impedance assuming a set of enclosed rods 50 cm long with 10 mm apertures and 0.033 torr of gas introduced in the center. The ion trajectories are simulated by the program described below (using the mass filter operating parameters indicated) and show that the ions reach thermal equilibrium with the gas before exiting the guide.

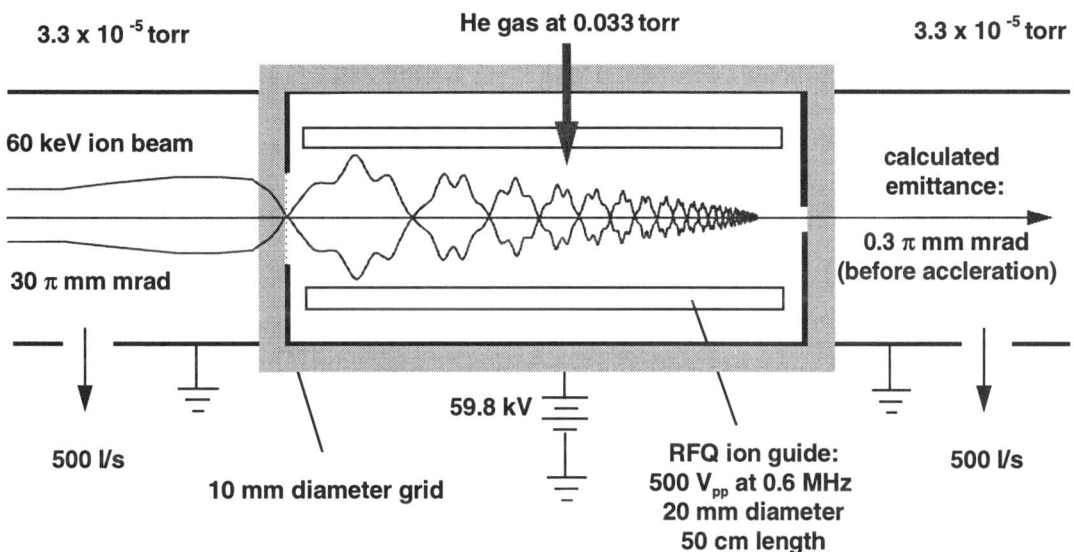

FIGURE 1. Schematic diagram of beam cooling system with optimized design parameters indicated.

The deceleration system was modeled using the SIMION program in which a 60 keV ion beam is gradually slowed to about 150 eV maintaining a moderate focus into the mass filter aperture. This is done using three electrodes: the first, at ground, with an aperture of 40 mm; the second, 40 mm downstream with a 30 mm aperture, is at 53 kV; a further gap of 40 mm separates the injection aperture grid of 10 mm diameter which floats at 59.9 kV. The grid is necessary to prevent over-focusing. The RF ion guide is located 4 mm further and floats at 59.8 kV thus creating a small potential barrier to stop cooled ions from diffusing back upstream. The beam focus, given the initial emittance of 35 π·mm·mrad at 60 kV (the measured ISOLDE value), is 7 mm in diameter with a divergence of 15° (corresponding to a transverse energy of about 8.5 eV) with a final axial energy of 130 eV, all in accordance with Liouville's theorem.

Due to the phenomenon of RF distortion, the acceptance of an RF ion guide is limited and furthermore, depends on the phase of the RF voltage at injection. Note that though the RF motion is coherent and not affect the ion temperature (in the absence of RF heating), it can cause part of the ion distribution to fall outside the confines of the ion guide. In order to minimize this effect, the operating parameters of the ion guide have to be chosen carefully.

First, reducing the effect of the RF can be done simply by lowering the RF voltage however this has the disadvantage of also lowering the ion guide confining power, or potential well depth D. Therefore, we must minimize the Mathieu parameter q and maximize D which results in choosing an RF voltage as high as possible (*caveat emptor*: high buffer gas pressure!) and an RF frequency such that q remains modest (less than about 0.6). For this simulation we have chosen an RF voltage of 500 V_{pp} and a frequency of 0.6 MHz which results in $q = 0.5$ and $D = 32$ eV. The resulting macromotion frequency ω_0 (for $A = 133$) is 108 kHz.

The effect of RF distortion was studied and indeed, some of the ions are lost before having their motion adequately damped by the buffer gas. However, if we recall that the number of ions actually present in the wings is relatively small (assuming the phase space diagram has a Gaussian distribution), the resulting loss in transmission will be of the order of less than 10%. This loss can eventually be remedied by a simple bunching voltage added to the injection electrode in order to let the ions arrive at an RF phase where RF distortion is minimized.

The viscous damping term is calculated using a parametrization of measurement data from ion mobilities in gases [5] which are tabulated versus ion drift velocity. An example of this parametrization is shown in figure 2 (left) for K^+ ions in He gas at 1×10^{-4} torr. For comparison, the mobility calculated using the gas density and the gas-kinetic cross-section for hard-ball collisions is also shown. At higher energies, one can see that the measured mobility approaches the elastic collision curve and indeed, the fit is tailored to more or less follow this curve at higher energies. There is a preponderance to treat buffer gas collisions as hard-ball collisions which is not correct for ions approaching thermal energies as can be seen from the graph. Evidence for this claim can be found from cooling times observed for ions in Paul traps which are much shorter that what would seem possible from elastic collisions. This is illustrated in figure 2 (right) where the damping time constant is plotted versus energy for mobilities derived from the two approaches discussed above along with an experimental result of the cooling time for K in a He gas-filled Paul trap [6]. Still further support for viscous damping calculated from measured mobilities can be found from measurements with buffer-gas cooled ions in Penning traps [7].

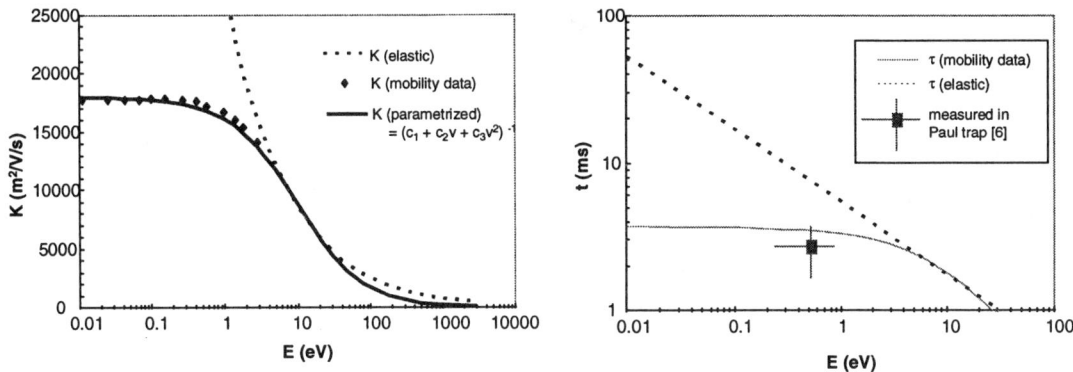

FIGURE 2. (left) K^+ ($A = 39$) ion mobility in He versus kinetic energy for a pressure of 1×10^{-4} torr (data points from [5]) shown with the mobility calculated using the gas-kinetic cross-section ($\sigma \approx 3 \times 10^{-19}$ m^{-2}). Note that the damping function is parametrized using velocity though energy is shown here for convenience (fit parameters: $c_1 = 0.0455$, $c_2 = 5\times 10^{-7}$, $c_3 = 9\times 10^{-10}$). (right) Cooling time constant versus energy as calculated from the mobilities shown (left). Also plotted is an experimental cooling time constant measured for K^+ in He gas in a Paul trap [6].

For the trajectory calculations, an acceleration term derived from the Lorentz equation is numerically integrated by a 4th order Runge-Kutta algorithm. From this term is subtracted the parametrized mobility calculation illustrated above which is recalculated using the new velocity after each time step. As the final energy of the cooled ions should be in equilibrium with the gas, the corresponding gas temperature energy is also added to the acceleration term after each step. This is done, for the transverse coordinates, using the macromotion frequency ω_0 which represents the statistical motion of the ions and for the longitudinal coordinate, just the simple thermal "diffusion" velocity.

Using the ion guide parameters in fig.1 and the action diagram calculated by SIMION, it was found that a 50 cm long ion guide operated in a pressure of 0.033 torr cools the ions to the gas temperature in all three dimensions. The calculated ion trajectory for a corner point on the action diagram is shown in fig. 3 (left). While it is interesting to see the trajectory inside the ion guide, it gives us no indication of how much the emittance is reduced as a result of the cooling. For this, we must compare action diagrams at the entrance and exit of the ion guide, as done in figure 3 (right). Here the input and output phase space diagrams - directly proportional to the emittances - are superimposed. When the emittance is calculated from these diagrams ($\pi/4 \times \delta y \times \delta \theta$) and normalized for the beam transport energy, we get a reduction factor of 100 (for each coordinate). In principle this reduction is simply determined by the ratio of the temperature corresponding to the energy spread in the ion source to that of the buffer gas. For an ISOLDE beam, this corresponds to a few eV compared to 0.026 eV if a room-temperature gas is used.

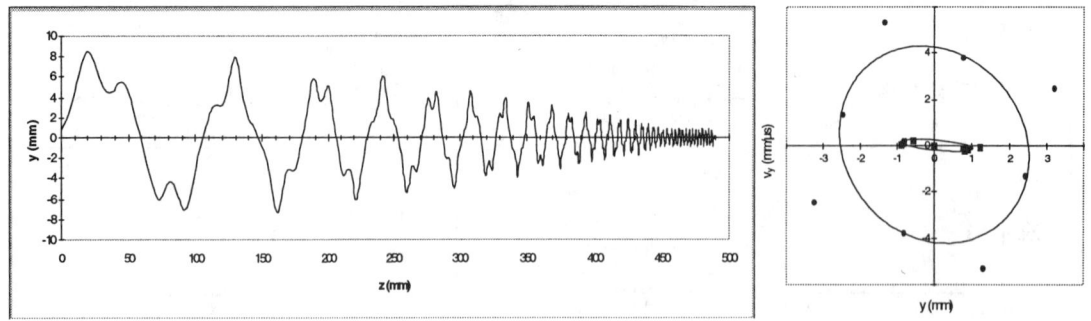

Figure 3. (left) Simulated trajectory of a single ion with an axial energy of 130 eV and a transverse energy of 8.5 eV subject to buffer gas cooling at 0.033 torr in a quadrupole ion guide. (right) Transverse phase space diagrams (for one dimension) of the ion beam before and after transit through the ion guide. The ratio of the corresponding emittance is a factor of 100. Ion guide operating parameters are as in figure 1.

These preliminary results show that such a system is definitely feasible. The critical problem seems to be the transmission loss due to RF distortion and perhaps the gas load on the system. The loss of ions due to charge exchange and other chemical reactions has also not been evaluated. This is sure to play some role as previous results have already shown [4]. Another factor not considered here is the phenomenon of RF heating that occurs in Paul traps. This effect causes an effective increase in temperature (and hence, emittance) when collisions with heavier ions cause a transfer of the coherent RF motion to the statistical (macro) motion. A test system scaled-down (by about a factor of ten) in energy is being built at the *CSNSM* using an existing mass spectrometer set-up. It is hoped that the *ISOLDE* emittance can thus be significantly reduced, corresponding to a similar gain in sensitivity for *MISTRAL* and eventually, other on-line experiments at *ISOLDE* and elsewhere. A similar device, with the added feature of bunching, is being built for the *ISOLTRAP* experiment [8], also at *ISOLDE*. Both projects are part of the European Community research network: *EXOTRAPS* [9] devoted to cooling and purifying radioactive beam.

REFERENCES

1. C. Toader *et al.*, this volume.
2. D.J. Douglas and J.B. French, J. Am. Soc. Mass Spectrom. 3 (1992) 398
3. H.J. Xu *et al.*, Nucl. Instr. and Meth. A333 (1993) 274
4. P. Van den Bergh *et al.*, Nucl. Instr. Meth. B126 (1997) 194
5. L.A. Viehland and E.A. Mason, At. Data Nucl. Data Tables 60 ((1995) 37 and references therein.
6. M.D. Lunney, Ph. D. Thesis, McGill University, Montreal, 1992, unpublished.
7. M. König *et al.*, Int. J. Mass Spectrom. Ion Proc. 142 (1995) 95
8. G. Bollen *et al.*, these proceedings.
9. http://www.jyu.fi/~armani/exotraps/frames.htm

Helium and Deuterium mass ratios in a room temperature Penning trap

S. Brunner, T. Engel, A. Schmitt and G. Werth

Johannes Gutenberg Universität, Institut für Physik, D-55099 Mainz, Germany

Abstract. The cyclotron frequencies of single He^+, H_2^+, and D_2^+ ions, stored in a room temperature Penning trap, have been measured with uncertainties in the ppb range. A time-of-flight method has been used to detect the ion´s excited motional resonances after they are kicked out of the trap. Linewidth limited by the 1 s interrogation time have been obtained and the statistical uncertainties of the line centers was several parts in 10^{-10}. The ratio of the cyclotron frequencies for different ions gives directly the mass ratio of the particles. After a careful investigation of possible systematic frequency shifts we obtain results in the ppb range in agreement with results from other authors using different techniques.

Introduction

Among different types of mass spectrometers Penning traps are at present those instruments by which the highest precision in mass determination of atoms and molecules has been obtained. They use the fact that the motional frequencies of ions in a quadrupole potential can be determined with great precision. These frequencies depend on the ions mass. Various instruments which are presently in use in different laboratories differ essentially by the way in which the ions and their motional frequencies are detected: Non-destructive techniques [1,2] measure the image charges induced in the trap electrodes by the ion motion, while destructive techniques [3,4] detect the ions outside the trap and require repetitive loading.

Experiment

Our experimental principle and details of the setup are discussed in ref. [5]. We used for the present experiments a room temperature Penning trap with hyperbolic electrodes of 7.3 mm characteristic dimension $d_0 = [z_0^2/2 + \rho_0^2/4]^{1/2}$, $2z_0$ being the closest distance between the endcap electrodes and ρ_0 the radius of the ring electrode. The trap was placed in a superconducting solenoid of 7 Tesla strength. The ring elctrode was divided into 4 segments to allow the application of a radio frequency field between opposite or adjacent electrodes to excite radial motions of the ions. Between ring and endcap electrodes guard rings were placed which could be used to compensate partially for deviations from the perfect trap geometry. Additionally we could apply electric fields to electrodes placed above and below the endcap electrodes. They served to compensate for imperfections arising from holes of 2 mm diameter at the center of the endcaps. These holes were used for entrance of electrons into the trap and for ion extraction. The trap structure was made from a Copper-Nickel alloy of low susceptibility and uses an amount of material as small as possible to minimize magnetic field inhomogeneities from the trap. Fig. 1 shows a scale drawing of the trap.

The ions were produced by an electron pulse of typically 10 ms length from the background gas which was held at a pressure of 1×10^{-10} mbar. Unwanted ion species were removed from the trap by excitation of their axial oscillation by a radio frequency field applied between the two endcap electrodes. After a storage time of 1 s the ions were kicked out from the trap at low energies. The are accelerated in the inhomogeneous part of the solenoids magnetic field by the force of the field gradient acting upon their orbital magnetic moment. After about 60 μs they arrive at a channel plate detector placed outside the solenoid at 50 cm distance from the trap and are detected.

The cyclotron frequency was excited by a radio frequency field, applied between two adjacent segments of the ring electrode. The detection of the resonance was performed in the following way: When the cyclotron motion is excited the orbital angular momentum and consequently the associated orbital magnetic moment is increased. This results in an larger force on the ions in the inhomogeneous part of the magnetic field and a reduced flight time to the detector. When we plot the average arrival time versus the frequency of the applied r.f. field, we obtain a minimum at resonant excitation. In order to obtain a sufficient signal-to-noise ratio every data point in a resonance curve was repeated 50 times. In this way we observed the perturbed cyclotron frequency ω_c', which is an eigenfrequency of the stored ions motion, $\omega_c' = \omega_c/2 + [\omega_c^2/4 - \omega_z^2/2]^{1/2}$, as well as the free ions cyclotron frequency $\omega_c = (e/m)B$, which is a

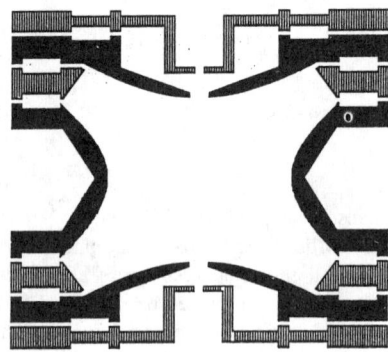

FIGURE 1. Scale drawing of our Penning trap. The magnetic field is directed along the vertical axis. Black areas are the primary trap electrodes, shaded areas the correction electrodes.

sideband at the sum of ω_c' and the magnetron frequency $\omega_m = \omega_c/2 - [\omega_c^2/4 - \omega_z^2/2]^{1/2}$. $\omega_z = (eV_0/md_0^2)^{1/2}$ is the axial oscillation frequency at a trapping voltage V_0. The observed linewidth is the Fourier limit from the interrogation time, which was usually 1 s. Figure 2 shows an example.

It is of particular importance to account for systematic frequency shifts arising from trap imperfections. These imperfections can be described by a series expansion of the potential

$$V' = V_2 + V_0 \sum C_k (\rho/d_0)^k P_k (\cos\theta) \tag{1}$$

Here V_2 is the ideal trap potential

$$V_2 = (V_0/2d_0^2) (z^2 - \rho^2/2) \tag{2}$$

and ρ the radial coordinate. The coefficients C_k denote the strength of the perturbing potential.

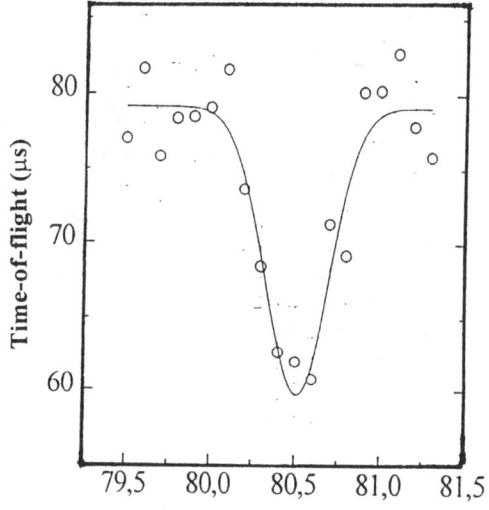

FIGURE 2. Average flight time of ions to the detector at resonant cyclotron excitation. The solid line is a least-squares fitted Lorentzian. The linewidth of 0.4 Hz corresponds to the Fourier limit of the 2 s interrogation time

In a similar way magnetic field inhomogeneities can be described by an expansion

$$B = B_0 (1+b_2 z^2 ...)\tag{3}$$

The associated shifts of the cyclotron frequency are up to k=8

$$\Delta\omega_c = \Delta\omega_c' + \Delta\omega_m$$
$$= \frac{\omega_z^2}{\omega_c' - \omega_m} \left[\begin{array}{l} \frac{3C_4}{4d^2}\left(\rho_m^2 - \rho_c'^2\right) + \frac{15C_6}{8d^4}\left(3z^2\left(\rho_m^2 - \rho_c'^2\right) + \left(\rho_c'^4 - \rho_m^4\right)\right) \\ + \frac{35C_8}{16d^6}\left(18z^4\left(\rho_m^2 - \rho_c'^2\right) - 24z^2\left(\rho_m^4 - \rho_c'^4\right) + 3\left(\rho_m^6 - \rho_c'^6\right) + 6\left(\rho_m^4\rho_c'^2 - \rho_c'^4\rho_m^2\right)\right) \end{array}\right] \tag{4}$$

Similarly for the magnetic case, the shift in frequency up to the 4th order is given by

$$\Delta\omega_c = \omega_c \frac{b_2}{2}\left(z^2 + \frac{\omega_m \rho_c'^2 - \omega_c' \rho_m^2}{\omega_c' - \omega_m}\right)$$
$$+ \omega_c \frac{b_4}{8}\left(3z^4 - 2\rho_m^2\rho_c'^2 + 12z^2 \frac{\omega_m \rho_c'^2 - \omega_c'^2 \rho_m^2}{\omega_c' - \omega_m} - 5\frac{\omega_c' \rho_c'^4 + \omega_m \rho_m^4}{\omega_c' - \omega_m} + \frac{\omega_c' \rho_m^4 - \omega_m \rho_c'^4}{\omega_c' - \omega_m}\right) \tag{5}$$

Here z, ρ_c', ρ_m are the axial oscillation amplitudes and the radii of the reduced cyclotron and magnetron motion, respectively. In our experiment the largest contribution comes from the terms containing z. z can be as large as several mm, since it is given by the position of the creation of ions by electroionisation which can be anywhere along the z-axis. The second largest part comes from terms containing ρ_m which can assume a maximal value of 0.4 mm, given by the radius of the electron entrance hole in the endcap compensation electrode. The radius of the reduced cyclotron orbit at our magnetic field strength of 7 Tesla and a radial ion energy, which increases from 1/40 eV at creation to about 1 eV at resonant excitation is only 10 μm and terms containing higher orders in this coordinate can generally be neglected.

Finally an angle ε between the traps symmetry axis and the magnetic field direction shifts the cyclotron frequency by an amount [5,6]

$$\Delta\omega_c = 9/4\ \omega_c \sin^2 \varepsilon \tag{6}$$

We took a number of measures in order to minimize systematic frequency shifts: Initial trapping of the ions was performed at low trapping voltages (150 mV). The potential was then raised adiabatically to 1850 V. The axial ion oscillation amplitude then decreases by the 4th root of the voltage ratio, in our case nearly a factor 2. The voltages at the correction electrodes were varied to optimize the trap potential. The linewidth of the cyclotron resonance shows a distinct minimum at the correct tuning voltage (Fig. 3). Furthermore at the same voltage we observed that the change in time-of-flight to the multichannel plate, which serves to detect the ions, is a maximum at resonant ion excitation (Fig. 3). The values of the coefficients for higher order perturbations and their influence on the measured mass ratios after setting the correction voltages to the proper values are listed in table 1

To avoid perturbations from ion-ion interaction we reduced the number of ions to a level that only in one out of three attempts a single ion was detected. This makes it sufficiently unlikely that more than one ion is stored in the trap at the same time. From the measured number dependency of the cyclotron frequency we set an upper limit for the contribution of ion-ion interaction to the cyclotron shift (table 1).

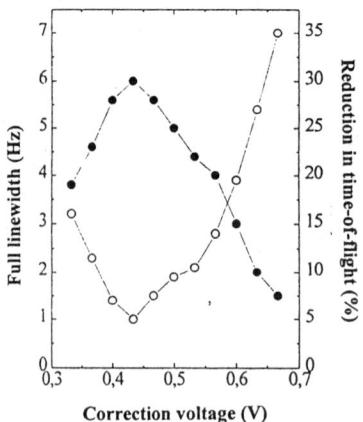

FIGURE 3. Half width of the cyclotron resonance line (full circles) and reduction in time-of-flight (open circles) at resonant cyclotron excitation at different voltages applied to the correction electrodes. The trapping potential was 3o V

The maximum size of an angle ε is derived from the size of the entrance and exit hole in the endcaps, through which we pass the electron beam. It is 0.2^0 and it's influence on the mass ratio is small (see table 1).

A serious limitation arises from the temporal stability of the magnetic field of the superconducting solenoid. Although we did not observe a long term drift at the 10^{-9} level we cannot exclude short term variations. To reduce the influence of such a drift on our results we switched between two ion species after each frequency point of a resonance curve. Considering the time for a measurement cycle of 1.2 s and a 50 times repetition of each point, this switch took place every minute. Checks on the statistical distribution of the measured frequency ratios did not show any deviation from the expectations beyond statistical scatter. So we conclude that time variations of the magnetic field do not contribute to our error significantly at our level of precision.

Results

We determined the mass ratios of $^4\text{He}^+/\text{H}_2^+$ and $^4\text{He}^+/\text{D}_2^+$ by measurements of the sideband frequency ω_c in the manner described above. An additional measurement on the $^4\text{He}^+/\text{D}_2^+$ mass ratio was performed by an excitation of the frequency $2\omega_c'$. This resonance could be observed at high excitation amplitudes. The magnetron frequency ω_m, required for the calculation of ω_c, was determined from the difference of ω_c and ω_c'.

Table 2 lists our result. The quoted error bars are the quadratic sum of the statistical and systematic uncertainties. For $^4\text{He}^+/\text{D}_2^+$ we have combined our results from the independent determinations by ω_c and $2\omega_c'$, which gave the same results within the statistical error. We compare our results to those obtained previously by van Dyck et al. and find general agreement on a level of precision in the ppb range.

TABLE 1. Origins and size of perturbations and their influence on the mass ratios

perturbation	Size	$^4\text{He}^+/\text{H}_2^+$	$^4\text{He}^+/\text{D}_2$
C_4	$2\ 10^{-5}$	$7.8\ 10^{-11}$	$4.7\ 10^{-13}$
C_6	$-6\ 10^{-4}$	$4.5\ 10^{-11}$	$2.7\ 10^{-13}$
b_2 (mm^{-2})	$4.3\ 10^{-9}$	$2.5\ 10^{-11}$	$1.1\ 10^{-11}$
ε_{max}	0.2^0	$4.1\ 10^{-10}$	$2.5\ 10^{-12}$
ion-ion		$1.9\ 10^{-9}$	$1.2\ 10^{-10}$
relativistic. mass shift		$3.2\ 10^{-10}$	$2.5\ 10^{-13}$
quadratic sum		$2.0\ 10^{-9}$	$1.2\ 10^{-10}$

TABLE 2 Final results for the measured mass ratios and comparison to other results

	$^4\text{He}^+/\text{H}_2^*$	$^4\text{He}^+/\text{D}_2^+$
This work	1.986 031 327 4 (31)	0.993 643 871 28 (55)
Ref. 7	1.986 031 325 2 (10)	0.993 643 869 94 (110)

Acknowledgments

Our experiment was supported by the Deutsche Forschungsgemeinschaft.

References

[1] F.L.Moore, D.L.Farnham, P.B.Schwinberg and R.S. Van Dyck, Jr., Nucl.Instr. Meth. B43, 425 (1989)

[2] E.A. Cornell et al., Phys. Rev. Lett. 63, 1674 (1989)

[3] G. Bollen, R.B.Moore, G.Savard and H. Stolzenberg, J. Appl. Phys. 68, 4355 (1990)

[4] R. Jertz et al., Z. Phys. D21, 179 (1991)

[5] Ch. Gerz, D. Wilsdorf and G. Werth, Nucl. Instr. Meth. B47, 453 (1990)

[6] L. Brown and G. Gabrielse, Rev.Mod.Phys. 58, 233 (1986)

[7] R.S. van Dyck, Jr., Proc 6th Intern Conf. on Nuclei far from Stability and 9th Intern. Conf. on Atomic Masses and Fundamental Constants, (R. Neugart and A. Wöhr eds) Bernkastel, Inst. Conf. Series 132 (1992)

SECTION 4

WEAK INTERACTION STUDIES

Testing CPT and Lorentz Symmetry with Protons and Antiprotons in Penning Traps[1]

Robert Bluhm[a], V. Alan Kostelecký[b], and Neil Russell[b]

[a] *Physics Department, Colby College, Waterville, ME, 04901 U.S.A.*
[b] *Physics Department, Indiana University, Bloomington, IN, 47405 U.S.A.*

Abstract. A theoretical analysis is performed of Penning-trap experiments comparing protons and antiprotons to test CPT and Lorentz symmetry through measurements of anomalous magnetic moments and charge-to-mass ratios. Possible CPT and Lorentz violations arising at a fundamental level are treated in the context of a general extension of the standard model of particle physics and its restriction to quantum electrodynamics. In a suggested experiment measuring anomaly frequencies a bound on CPT violation of 10^{-23} for a relevant figure of merit is attainable. Experiments comparing cyclotron frequencies are sensitive within this theoretical framework to different kinds of Lorentz violation that preserve CPT. Constraints could be obtained on one figure of merit at 10^{-24} and on another in a related experiment with H^- ions and antiprotons at the level of 10^{-25}.

INTRODUCTION

The standard model of particle physics is symmetric under the discrete transformation CPT [1]. A consequence of this is the equality of various experimentally measurable quantities. In particular, the charge-to-mass ratio of the proton should equal that of the antiproton, and the gyromagnetic ratios of these two particles should be equal. Experiments in Penning traps can test this symmetry to high precision. Measurements of proton and antiproton cyclotron frequencies using Penning traps allow a comparison of their charge-to-mass ratios [2], producing the bound

$$r^p_{q/m} \equiv |[(q_p/m_p) - (q_{\overline{p}}/m_{\overline{p}})]/(q/m)_{\mathrm{av}}| \lesssim 1.5 \times 10^{-9} \ . \tag{1}$$

In the present work, we analyze past and future experiments on protons, antiprotons and hydrogen ions confined within a Penning trap. The theoretical framework is a CPT- and Lorentz-violating extension of the standard model [3]. Since the dominant interactions are electromagnetic, we consider the pure-fermion sector of the extension of quantum electrodynamics emerging as a limit of the general standard-model extension.

Our primary goal is to consider the sensitivity of Penning-trap experiments to possible CPT- and Lorentz-violating effects in the extension of quantum electrodynamics. We investigate the relevance of the conventional figures of merit as measures of CPT violation. In some cases, more suitable figures of merit and corresponding experiments are suggested. Estimates are also made of bounds accessible to experiments with existing technology.

THEORY

The framework for the extension of the SU(3)×SU(2)×U(1) standard model and quantum electrodynamics originates from the idea of spontaneous breaking of CPT and Lorentz symmetry in a more fundamental model such as string theory [4,5]. The standard-model extension lies within the context of conventional quantum

---

[1] Presented by N.R.

field theory and appears to preserve desirable features of the standard model, including gauge invariance, power-counting renormalizability, and microcausality.

At the level of the standard model, protons and antiprotons are composite particles formed as bound states of quarks and antiquarks, respectively. Possible CPT- and Lorentz-violating effects in the standard-model extension appear as perturbations involving the basic fields [3]. A distinct set of quantities is assigned to each quark flavor, and suitable combinations of these determine the CPT- and Lorentz-violating features of the proton. For our present investigation of protons and antiprotons in a Penning trap, it suffices to work within the usual effective theory in which the protons and antiprotons are regarded as basic fermions described by a four-component Dirac quantum field with dynamics governed by a minimally coupled lagrangian. Based on the quantities for the fundamental particles, we introduce effective quantities controlling possible CPT- and Lorentz-breaking effects for the proton. The lagrangian is taken to be the standard one for proton-antiproton quantum electrodynamics but extended to include possible small CPT- and Lorentz-violating terms. Further details are given in Ref. [6].

For the Penning-trap experiments of interest, the dominant contributions to the energy spectrum arise from the interaction of the proton or antiproton with the constant magnetic field of the trap. The quadrupole electric and other fields generate smaller effects. In a perturbative calculation, the dominant corrections due to CPT- and Lorentz-violating effects can therefore be obtained by considering a constant uniform magnetic field only. Since the signals of interest are energy-level shifts rather than transition probabilities, it suffices to use relativistic Landau-level wave functions as the unperturbed basis set and to calculate within first-order perturbation theory. However, the unperturbed energy levels must be taken as the Landau levels shifted by an anomaly term and other corrections.

To lowest order in the fine-structure constant, the perturbative hamiltonian $\hat{H}^p_{\text{pert}}$ for a proton of mass m is

$$\hat{H}^p_{\text{pert}} = a^p_\mu \gamma^0 \gamma^\mu - b^p_\mu \gamma_5 \gamma^0 \gamma^\mu - c^p_{00} m \gamma^0 - i(c^p_{0j} + c^p_{j0})D^j + i(c^p_{00}D_j - c^p_{jk}D^k)\gamma^0\gamma^j$$
$$- d^p_{j0} m \gamma_5 \gamma^j + i(d^p_{0j} + d^p_{j0})D^j \gamma_5 + i(d^p_{00}D_j - d^p_{jk}D^k)\gamma^0 \gamma_5 \gamma^j + \tfrac{1}{2} H^p_{\mu\nu} \gamma^0 \sigma^{\mu\nu} \quad, \tag{2}$$

where D_μ is the appropriate covariant derivative. The quantities a^p_μ, b^p_μ, $H^p_{\mu\nu}$, $c^p_{\mu\nu}$, $d^p_{\mu\nu}$ are effective couplings for the proton and antiproton arising in the standard-model extension. We note that the terms involving a^p_μ, b^p_μ break CPT while those involving $H^p_{\mu\nu}$, $c^p_{\mu\nu}$, $d^p_{\mu\nu}$ preserve it, and that Lorentz invariance is broken by all five terms.

For the antiproton, the hamiltonian can be found via charge conjugation. However, it should be taken into account that experimental procedures for replacing particles with antiparticles in Penning traps typically reverse the electric field but leave unchanged the magnetic field.

We denote the proton energy levels without CPT- and Lorentz-violating perturbations by $E^p_{n,s}$, where n is the principal quantum number and s is the spin. The proton cyclotron and anomaly frequencies are defined as

$$\omega_c = E^p_{1,+1} - E^p_{0,+1} \quad, \qquad \omega_a = E^p_{0,-1} - E^p_{1,+1} \quad. \tag{3}$$

From the CPT theorem, these frequencies have the same values as those of the antiproton.

We can calculate energy corrections in first-order perturbation theory for the CPT- and Lorentz-breaking couplings. The corrected cyclotron and anomaly frequencies can then be found. Calculations at leading order in the CPT- and Lorentz-breaking quantities, in the electromagnetic fields, and in the fine-structure constant give the results

$$\omega^p_c = \omega^{\bar{p}}_c \approx (1 - c^p_{00} - c^p_{11} - c^p_{22})\omega_c \quad, \tag{4}$$

$$\omega^p_a \approx \omega_a + 2b^p_3 - 2d^p_{30}m_p - 2H^p_{12} \quad, \qquad \omega^{\bar{p}}_a \approx \omega_a - 2b^p_3 - 2d^p_{30}m_p - 2H^p_{12} \quad, \tag{5}$$

for the modified frequencies. In these expressions, ω_c and ω_a are the unperturbed frequencies of Eq. (3).

ANOMALOUS MAGNETIC MOMENTS

Currently, the antiproton magnetic moment is known to a precision of 3 parts in 10^3 from experiments with exotic atoms [7]. In principle, the anomalous magnetic moments of protons and antiprotons could be measured in Penning traps. A comparison of the experimental ratios $2\omega^p_a/\omega^p_c$ and $2\omega^{\bar{p}}_a/\omega^{\bar{p}}_c$ would then provide a sharp test of CPT and Lorentz violation. The possibility of such experiments has received some attention in the literature [8,9].

The sensitivity of possible future $g-2$ experiments to CPT and Lorentz violation can be investigated using the present theoretical framework. We find the proton-antiproton differences at leading order for the cyclotron and anomaly frequencies are

$$\Delta\omega_c^p \equiv \omega_c^p - \omega_c^{\bar{p}} = 0 \quad , \qquad \Delta\omega_a^p \equiv \omega_a^p - \omega_a^{\bar{p}} = 4b_3^p \quad . \tag{6}$$

The leading-order signal for CPT breaking is thus an anomaly-frequency difference. Denoting the exact physical energy levels with possible CPT violation by $\mathcal{E}_{n,s}^p$ and $\mathcal{E}_{n,s}^{\bar{p}}$, the corresponding figure of merit providing a well-defined measure of the violation is

$$r_{\omega_a}^p \equiv \frac{|\mathcal{E}_{n,s}^p - \mathcal{E}_{n,-s}^{\bar{p}}|}{\mathcal{E}_{n,s}^p} \quad , \tag{7}$$

where the weak-field, zero-momentum limit is understood. We find

$$r_{\omega_a}^p \approx |\Delta\omega_a^p|/2m_p \approx |2b_3^p|/m_p \tag{8}$$

within the present theoretical framework.

Assuming an experiment could measure ω_a^p and $\omega_a^{\bar{p}}$ with a resolution of the order of $2\pi \times (1 \text{ Hz})$ and assuming equality of ω_c^p, $\omega_c^{\bar{p}}$ is observed to one part in 10^8, a bound of $|b_3^p| \lesssim 10^{-15}$ eV becomes possible. The corresponding estimated bound on the figure of merit $r_{\omega_a}^p$ is

$$r_{\omega_a}^p \lesssim 10^{-23} \quad . \tag{9}$$

The estimate shows the promise held by this type of experiment to tightly bound CPT violation in a baryon system. The standard-model extension has also been applied to neutral mesons [10] and leptons [11].

Measurements of diurnal variations in the anomaly frequency could also place bounds on a combination of couplings in the standard-model extension. An estimate of one part in 10^{-21} has been made for a suitable figure of merit [6].

CHARGE-TO-MASS RATIOS

Penning-trap experiments confining single protons and antiprotons can provide precision comparisons of their cyclotron frequencies [2], yielding the limit $|\Delta\omega_c^p|/\omega_c^p \lesssim 10^{-9}$. Equation (1) gives the corresponding conventional figure of merit $r_{q/m}^p$ and its bound.

Within the present theoretical framework, the perturbed proton and antiproton cyclotron frequencies are given in Eq. (4). Both are independent of leading-order CPT-violating quantities. As the cyclotron frequencies are unshifted even if CPT is broken, a comparison of these frequencies would represent an inappropriate measure of CPT violation in the context of the present theory. For example, the figure of merit $r_{q/m}^p$ in Eq. (1), which is proportional to the frequency difference $\Delta\omega_c^p$, may vanish even though explicit CPT violation occurs in the standard-model extension.

The effect of the Lorentz-breaking but CPT-preserving couplings is to induce identical shifts in the proton and antiproton cyclotron frequencies. This indicates that the frequency difference $\Delta\omega_c^p$ would also be an inappropriate measure of Lorentz violation in the present theoretical context.

Another possible experimental signal is the occurrence of diurnal variations in the cyclotron frequencies, which could be induced by the Earth's rotation during an experiment. Such variations would arise in the present standard-model extension from the dependence of the cyclotron frequencies on the components $|c_{11}^p + c_{22}^p|$ of $c_{\mu\nu}^p$.

A suitable theoretical figure of merit can be introduced by defining for the proton

$$\Delta_{\omega_c}^p \equiv \frac{|\mathcal{E}_{1,-1}^p - \mathcal{E}_{0,-1}^p|}{\mathcal{E}_{0,-1}^p} \quad . \tag{10}$$

It is the amplitude $r_{\omega_c,\text{diurnal}}^p$ of periodic fluctuations in $\Delta_{\omega_c}^p$. We find

$$r^p_{\omega_c,\text{diurnal}} \approx |c^p_{11} + c^p_{22}|\omega_c/m_p \qquad (11)$$

in the comoving Earth frame. The appearance of ω_c implies that the value of this figure of merit depends on the magnetic field.

A crude upper bound on $r^p_{\omega_c,\text{diurnal}}$ can be obtained from the data in Ref. [2], which represent alternate measurements of proton and antiproton cyclotron frequencies ω^p_c, $\omega^{\bar{p}}_c$ over a 12-hour period. The slow drifts in these frequencies are confined to a band of width about $2\pi \times (2 \text{ Hz})$. This suggests a bound on a possible diurnal variation in $r^p_{\omega_c,\text{diurnal}}$ arising from the contribution proportional to $|c^p_{11} + c^p_{22}|$, given by

$$r^p_{\omega_c,\text{diurnal}} \lesssim 10^{-24} \quad . \qquad (12)$$

Diurnal fluctuations in the antiproton cyclotron frequency could be analyzed similarly.

EXPERIMENTS WITH HYDROGEN IONS

The precision of proton-antiproton cyclotron-frequency comparisons is limited by the need to reverse the electric field each time the other species is loaded in the trap [2]. A recent experiment by Gabrielse and coworkers [12] has addressed this issue by comparing antiproton cyclotron frequencies with those of an H^- ion instead of a proton. The electric field is fixed throughout the experiment, and the magnetic-field variation between measurements is reduced due to the rapid interchange possible between simultaneously trapped hydrogen ions and antiprotons. The expected theoretical value of the difference $\Delta\omega^{H^-}_c \equiv \omega^{H^-}_c - \omega^{\bar{p}}_c$ can be obtained in the context of conventional quantum theory using known values of the electron mass and the H^- binding energy. Comparison of this theoretical value with the experimental result for $\Delta\omega^{H^-}_c$ is expected to provide a symmetry test with a ten-fold improvement on the previous test [2].

The theoretical analysis of this experiment within the present theoretical framework requires a description of the electromagnetic interactions of the hydrogen ion in a Penning trap in the presence of possible CPT and Lorentz violation. The hydrogen ion is treated as a charged composite fermion of mass m_{H^-}, so its electromagnetic interactions can be discussed within an effective spinor electrodynamics producing a modified hamiltonian of the form (2), but with a different set of CPT- and Lorentz-violating couplings. The modified cyclotron frequency is then calculated as for the proton-antiproton case. All the effective CPT- and Lorentz-breaking couplings for a hydrogen ion are determined by appropriate combinations of the corresponding quantities for its constituent proton and electrons. Lowest-order perturbation theory can be used to find approximations to these relationships, giving expressions involving the proton-antiproton couplings as well as electron-positron couplings a^e_μ, b^e_μ, $H^e_{\mu\nu}$, $c^e_{\mu\nu}$, and $d^e_{\mu\nu}$.

Subject to the approximations above, the component $\Delta\omega^{H^-}_{c,\text{th}}$ of $\Delta\omega^{H^-}_c$ that is determined theoretically to arise purely from CPT- and Lorentz-violating effects is

$$\Delta\omega^{H^-}_{c,\text{th}} \approx (c^p_{00} + c^p_{11} + c^p_{22})(\omega_c - \omega^{H^-}_c) - \frac{2m_e}{m_p}(c^e_{00} + c^e_{11} + c^e_{22} - c^p_{00} - c^p_{11} - c^p_{22})\omega^{H^-}_c \quad . \qquad (13)$$

Again, ω_c is the proton-antiproton cyclotron frequency in the absence of CPT or Lorentz perturbations. This result implies that in the context of this theory the experiment constrains a combination of Lorentz-violating but CPT-preserving quantities, including c^e_{00} and c^p_{00}. The latter would be inaccessible through the experiments with cyclotron or anomaly frequencies. In addition, this experiment does not look for diurnal variations in the cyclotron frequency, which means potential systematics associated with diurnal field drifts are eliminated.

The definition of a model-independent figure of merit follows from considerations similar to those leading to the figures of merit defined above. We define the quantity

$$\Delta^{H^-}_{\omega_c} \equiv \frac{|\mathcal{E}^{H^-}_{1,-1} - \mathcal{E}^{H^-}_{0,-1}|}{2\mathcal{E}^{H^-}_{0,-1}} - \frac{|\mathcal{E}^{\bar{p}}_{1,-1} - \mathcal{E}^{\bar{p}}_{0,-1}|}{2\mathcal{E}^{\bar{p}}_{0,-1}} \quad . \qquad (14)$$

As given here, $\Delta^{H^-}_{\omega_c}$ is nonzero even if CPT and Lorentz symmetry is preserved. To arrive at a measure that vanishes in the exact symmetry limit, we remove from the hydrogen-ion terms in $\Delta^{H^-}_{\omega_c}$ the conventional contributions arising from the differences between the H^- ion and a proton: the masses of the two electrons and the binding energy. The result is an appropriate figure of merit for Lorentz violation, denoted by $r^{H^-}_{\omega_c}$.

Estimating a precision of one part in 10^{10} in measurements of the ratio $|\Delta\omega_c^{H^-}|/\omega_c^{H^-}$, we estimate an experimentally attainable bound of $r_{\omega_c}^{H^-} \lesssim 10^{-25}$. Indeed, the Gabrielse experiment [12] placed a bound of

$$r_{\omega_c}^{H^-} \lesssim 4 \times 10^{-26} \tag{15}$$

on this figure of merit.

CONCLUSIONS

We have used a general theoretical framework based on an extension of the standard model and quantum electrodynamics to establish and investigate possible signals of CPT and Lorentz breaking in Penning-trap experiments with protons and antiprotons. We have looked for leading-order limits arising from precision measurements of anomaly and cyclotron frequencies.

Sharp tests of CPT symmetry would be possible in experiments comparing anomaly frequencies. We have introduced appropriate figures of merit with attainable bounds of approximately 10^{-23} for a plausible experiment with protons and antiprotons.

In contrast, comparative measurements of cyclotron frequencies for protons and antiprotons are insensitive to leading-order effects from CPT breaking within the present framework. However, diurnal variations of cyclotron frequencies and comparisons of cyclotron frequencies for hydrogen ions and antiprotons are affected by different CPT-preserving Lorentz-violating quantities. These experiments could generate bounds on various dimensionless figures of merit at the level of 10^{-24} in the proton-antiproton system, and 10^{-25} using the H^--antiproton system.

ACKNOWLEDGMENTS

This work is supported in part by the Department of Energy under grant number DE-FG02-91ER40661 and by the National Science Foundation under grant number PHY-9801869.

REFERENCES

1. See, for example, R.G. Sachs, *The Physics of Time Reversal* (University of Chicago Press, Chicago, 1987).
2. G. Gabrielse et al., Phys. Rev. Lett. **74** (1995) 3544.
3. D. Colladay and V.A. Kostelecký, Phys. Rev. D **55** (1997) 6760; preprint IUHET 359, Phys. Rev. D, in press (hep-th/9809521).
4. V.A. Kostelecký and R. Potting, Nucl. Phys. B **359** (1991) 545; Phys. Lett. B **381** (1996) 389.
5. V.A. Kostelecký and S. Samuel, Phys. Rev. Lett. **63** (1989) 224; ibid. **66** (1991) 1811; Phys. Rev. D **39** (1989) 683; ibid. **40** (1989) 1886.
6. R. Bluhm, V.A. Kostelecký and N. Russell, Phys. Rev. D **57** (1998) 3932.
7. A. Kriessle et al., Z. Phys. C **37** (1988) 557.
8. D.J. Heinzen and D.J. Wineland, Phys. Rev. A **42** (1990) 2977.
9. W. Quint and G. Gabrielse, Hyperfine Int. **76** (1993) 379.
10. V.A. Kostelecký, Phys. Rev. Lett. **80** (1998) 1818.
11. R. Bluhm, V.A. Kostelecký and N. Russell, Phys. Rev. Lett. **79** (1997) 1432.
12. G. Gabrielse et al., to be published.

Testing CPT and Lorentz Symmetry with Electrons and Positrons in Penning Traps[1]

Robert Bluhm[a], V. Alan Kostelecký[b], and Neil Russell[b]

[a] *Physics Department, Colby College, Waterville, ME, 04901 U.S.A.*
[b] *Physics Department, Indiana University, Bloomington, IN, 47405 U.S.A.*

Abstract. We present a theoretical analysis of signals for CPT and Lorentz violation in $g - 2$ and charge-to-mass-ratio experiments on electrons and positrons in Penning traps. Experiments measuring anomaly frequencies are found to be the most sensitive to CPT violation. We find that the conventional figure of merit for CPT breaking, involving the difference of the electron and positron g factors, is inappropriate in this context, and an alternative is introduced. Bounds of approximately 10^{-20} are attainable.

INTRODUCTION

The CPT theorem [1] is a general and powerful result that holds for local relativistic quantum field theories of point particles in flat spacetime. Any field theory of this kind must be invariant under the combined operations of charge conjugation C, parity reversal P, and time reversal T. As a consequence of this invariance, particles and antiparticles have equal masses, lifetimes, charge-to-mass ratios, and gyromagnetic ratios. The CPT theorem has been tested to great accuracy in a variety of experiments [2]. The sharpest bound is obtained in experiments with neutral kaons, where the CPT figure of merit is

$$r_K \equiv \frac{|m_K - m_{\overline{K}}|}{m_K} \lesssim 2 \times 10^{-18} \quad . \tag{1}$$

Experiments on electrons and positrons confined in Penning traps also yield sharp bounds on CPT violation. Indeed, these experiments provide the tightest bounds on CPT in the lepton system. Two types of experimental comparisons of electrons and positrons are possible in Penning traps. They involve making accurate measurements of cyclotron frequencies ω_c and anomaly frequencies ω_a of single isolated particles confined in the trap. The first compares the ratio $2\omega_a/\omega_c$ for particles and antiparticles. In the context of conventional quantum electrodynamics, this ratio equals $g - 2$ for the particle or antiparticle. A second experiment compares values of $\omega_c \sim q/m$, where $q > 0$ is the magnitude of the charge and m is the mass, and is therefore a comparison of charge-to-mass ratios.

The conventional figure of merit adopted in $g - 2$ experiments on electrons and positrons is given as the relative difference in their g factors [3,4],

$$r_g^e \equiv \frac{|g_{e^-} - g_{e^+}|}{g_{\text{avg}}} \quad , \tag{2}$$

which is known to be less than 2×10^{-12}. The bound obtained in charge-to-mass-ratio experiments [5] is expressed as the ratio

$$r_{q/m}^e \equiv \frac{|(q_{e^-}/m_{e^-}) - (q_{e^+}/m_{e^+})|}{(q/m)_{\text{avg}}} \quad , \tag{3}$$

---

[1] Presented by R.B.

which is less than or equal to 1.3×10^{-7}.

Measurements of frequencies in atomic systems typically have experimental uncertainties four or five orders of magnitude better than the measurements made in kaon experiments. However, the figure of merit r_g^e is poorer than r_K by about six orders of magnitude. This raises some interesting questions about the Penning-trap experiments as to why they do not provide better tests of CPT despite having better experimental precision. However, it is impossible to pursue these types of questions in the context of conventional quantum electrodynamics, since CPT breaking is strictly forbidden. Instead, one would need to work in the context of a theoretical framework that incorporates CPT-violating interactions, making possible an investigation of possible experimental signatures. Only recently has such a theoretical framework in the context of the standard model been developed [6].

In this paper, we summarize the results of our analysis on CPT and Lorentz tests performed with electrons and positrons in Penning traps. A more complete description of this analysis can be found in Refs. [7,8].

THEORETICAL FRAMEWORK

The theoretical framework we use [6] is based on a general extension of the SU(3) × SU(2) × U(1) standard model in particle physics. It includes all possible leading-order CPT- and Lorentz-violating interactions that could arise from spontaneous symmetry breaking at a more fundamental level, such as in string theory. This type of CPT violation is a possibility in string theory because the usual axioms of the CPT theorem do not apply to extended objects like strings. In spontaneous symmetry breaking, the dynamics of the action remains CPT invariant, which means the framework can preserve desirable features of quantum field theory such as gauge invariance, power-counting renormalizability, and microcausality. The CPT and Lorentz violation occurs only in the solutions of the equations of motion and is similar to the spontaneous breaking of the electroweak theory in the standard model.

To analyze interactions involving electrons and positrons in a Penning trap, we use a restriction of the full particle-physics framework to quantum electrodynamics. The resulting model divides into two sectors, one that breaks CPT and one that preserves CPT, while both break Lorentz symmetry. Possible violations of CPT and Lorentz symmetry are parametrized by quantities that can be bounded by experiments. Within this framework, the modified Dirac equation describing a fermion with charge q and mass m in an electromagnetic field is given by

$$\left(i\gamma^\mu D_\mu - m - a_\mu \gamma^\mu - b_\mu \gamma_5 \gamma^\mu - \tfrac{1}{2} H_{\mu\nu} \sigma^{\mu\nu} + i c_{\mu\nu} \gamma^\mu D^\nu + i d_{\mu\nu} \gamma_5 \gamma^\mu D^\nu \right)\psi = 0 \quad . \tag{4}$$

Here, ψ is a four-component spinor, $iD_\mu \equiv i\partial_\mu - qA_\mu$ is the covariant derivative, A^μ is the electromagnetic potential in the trap, and a_μ, b_μ, $H_{\mu\nu}$, $c_{\mu\nu}$, $d_{\mu\nu}$ are the parameters describing possible violations of CPT and Lorentz symmetry. The properties of ψ under transformations imply that the terms involving a_μ, b_μ break CPT while those involving $H_{\mu\nu}$, $c_{\mu\nu}$, $d_{\mu\nu}$ preserve it, and that Lorentz symmetry is broken by all five terms.

Since there have been no experimental observations to date of CPT or Lorentz breaking, the quantities a_μ, b_μ, $H_{\mu\nu}$, $c_{\mu\nu}$, $d_{\mu\nu}$ must all be small. We can estimate the suppression scale for these parameters by taking the scale governing the fundamental theory as the Planck mass m_{Pl} and the low-energy scale as the electroweak mass scale m_{ew}. The natural suppression scale for Planck-scale effects in the standard model would then be of order $m_{\text{ew}}/m_{\text{Pl}} \simeq 10^{-17}$.

EXPERIMENTS IN PENNING TRAPS

We use this theoretical framework to analyze comparative tests of CPT and Lorentz symmetry on electrons and positrons in Penning traps. First, we note that the time-derivative couplings in (4) alter the standard procedure for obtaining a hermitian quantum-mechanical hamiltonian operator. To overcome this, we first perform a field redefinition at the lagrangian level that eliminates the additional time derivatives. We then use charge conjugation to obtain a Dirac equation and hamiltonian for the antiparticle.

In tests of CPT, experiments compare the cyclotron and anomaly frequencies of particles and antiparticles. According to the CPT theorem, electrons and positrons of opposite spin in a Penning trap with the same magnetic fields but opposite electric fields should have equal energies. The experimental relations $g - 2 = 2\omega_a/\omega_c$ and $\omega_c = qB/m$ provide connections to the quantities g and q/m used in defining the figures of merit r_g^e and

$r^e_{q/m}$. We perform calculations using Eq. (4) to obtain possible shifts in the energy levels due to either CPT-breaking or CPT-preserving Lorentz violation. In this way, the effectiveness of Penning-trap experiments on electrons and positrons as tests of both CPT-breaking and CPT-preserving Lorentz violation can be analyzed. From the computed energy shifts we determine how the frequencies ω_c and ω_a are affected and whether the conventional figures of merit are appropriate.

In experiments performed in Penning traps, the dominant contributions to the energy come from interactions of the electron or positron with the constant magnetic field of the trap, while the quadrupole electric fields generate smaller effects. In a perturbative calculation, the dominant CPT- and Lorentz-breaking effects can therefore be obtained by working with relativistic Landau levels as unperturbed states. Conventional perturbations, such as the usual corrections to the anomalous magnetic moment, are the same for electrons and positrons. Violations of CPT and Lorentz symmetry result in either differences between electrons and positrons or in unconventional effects such as diurnal variations in the measured frequencies.

RESULTS

The results of our calculations for electrons and positrons in Penning traps [7,8] show that the leading-order effects due to CPT and Lorentz breaking cause corrections to the cyclotron and anomaly frequencies:

$$\omega_c^{e^-} \approx \omega_c^{e^+} \approx (1 - c_{00}^e - c_{11}^e - c_{22}^e)\omega_c \quad , \tag{5}$$

$$\omega_a^{e^\mp} \approx \omega_a \mp 2b_3^e + 2d_{30}^e m_e + 2H_{12}^e \quad . \tag{6}$$

In our notation, ω_c and ω_a represent the unperturbed frequencies for the electron (e^-) and the positron (e^+), while $\omega_c^{e^\mp}$ and $\omega_a^{e^\mp}$ denote the frequencies including the corrections. Superscripts have also been added on the coefficients b_μ, etc., to denote that these parameters describe the electron-positron system. From these relations we find the differences in the electron and positron cyclotron and anomaly frequencies to be

$$\Delta\omega_c^e \equiv \omega_c^{e^-} - \omega_c^{e^+} \approx 0 \quad , \tag{7}$$

$$\Delta\omega_a^e \equiv \omega_a^{e^-} - \omega_a^{e^+} \approx -4b_3^e \quad . \tag{8}$$

We find that in the context of this framework, comparisons of cyclotron frequencies to leading order do not provide a signal for CPT or Lorentz breaking, since the corrections to ω_c for electrons and positrons are equal. However, comparisons of anomaly frequencies provide unambiguous tests of CPT since the CPT-violating term with b_3 results in a nonzero value for the difference $\Delta\omega_a^e$, while the CPT-preserving coefficients do not appear.

We also find that to leading order there are no corrections due to CPT or Lorentz violation to the g factors for either electrons or positrons. This leads to some interesting and unexpected results concerning the figure of merit r_g in Eq. (2). With $g_{e^-} \approx g_{e^+}$ to leading order, we find that r_g vanishes, which would seem to indicate the absence of CPT breaking. However, this conclusion would be incorrect because the model contains explicit CPT violation. In addition, our calculations show that with $\vec{b} \neq 0$ the experimental ratio $2\omega_a/\omega_c$ depends on the magnetic field and is undefined in the limit of a vanishing B field. Therefore, the usual relation $g - 2 = 2\omega_a/\omega_c$ does not hold in the presence of CPT violation. For these reasons, we conclude that the figure of merit r_g in Eq. (2) is inappropriate in the context of our framework. An alternative is suggested next.

Since a prediction of the CPT theorem is that electron and positron states of opposite spin in the same magnetic field have equal energies, we propose as a model-independent figure of merit

$$r_{\omega_a}^e \equiv \frac{|E_{n,s}^{e^-} - E_{n,-s}^{e^+}|}{E_{n,s}^{e^-}} \quad , \tag{9}$$

where $E_{n,s}^{e^\mp}$ are the energies of the relativistic states labeled by their Landau-level numbers n and spin s. Our calculations show $r_{\omega_a}^e \approx |\Delta\omega_a^e|/2m_e \approx |2b_3^e|/m_e$. Assuming frequency resolutions on the order of 1 Hz, we estimate as a bound on this figure of merit,

$$r_{\omega_a}^e \lesssim 10^{-20} \quad . \tag{10}$$

This definition of the figure of merit $r^e_{\omega_a}$ is compatible with the corresponding figure of merit r_K arising from experiments with the neutral-kaon system. This is because both figures of merit involve ratios of energy scales, and therefore comparisons across experiments are more meaningful. This is not the case for the figures of merit r^e_g and r_K, since each involves different physical quantities. Our estimate suggests that a somewhat tighter bound for $r^e_{\omega_a}$ is attainable in Penning-trap experiments than that for the corresponding figure of merit r_K arising from experiments with the neutral-kaon system. This result is more in line with the greater precision that is experimentally accessible in frequency measurements in a Penning trap. However, performing the CPT tests in the kaon system remains essential because neutral-meson CPT violation is controlled by distinct CPT-violating parameters that appear only in the quark sector [10].

In Ref. [8], we describe additional possible signatures of CPT and Lorentz violation. These include possible diurnal variations in the anomaly and cyclotron frequencies. Tests for these effects would provide bounds on some of the components of the parameters $c^e_{\mu\nu}$, $d^e_{\mu\nu}$, and $H^e_{\mu\nu}$.

One type of experiment looking for diurnal variations involves the electron alone or the positron alone. In the standard-model extension, these variations would occur because the components of the couplings in Eq.(6) would change as the Earth rotates. Consider the following quantities for the electron and positron:

$$\Delta^e_{\omega_a^{e^-}} \equiv \frac{|\mathcal{E}^{e^-}_{0,+1} - \mathcal{E}^{e^-}_{1,-1}|}{\mathcal{E}^{e^-}_{0,-1}} \quad , \quad \Delta^e_{\omega_a^{e^+}} \equiv \frac{|\mathcal{E}^{e^+}_{0,-1} - \mathcal{E}^{e^+}_{1,+1}|}{\mathcal{E}^{e^+}_{0,+1}} \quad . \tag{11}$$

Suitable figures of merit $r^e_{\omega_a^-,\text{diurnal}}$ and $r^e_{\omega_a^+,\text{diurnal}}$ can be defined as the amplitude of the diurnal variations in $\Delta^e_{\omega_a^{e^-}}$ and $\Delta^e_{\omega_a^{e^+}}$, respectively. In the context of our framework, we find

$$r^e_{\omega_a^{\mp},\text{diurnal}} \approx \frac{2|\mp b^e_3 + d^e_{30}m_e + H^e_{12}|}{m_e} \quad . \tag{12}$$

The experimental issues involved in obtaining a bound on $r^e_{\omega_a^{\mp},\text{diurnal}}$ include maintaining stability in the magnetic field. For example, limiting variations in the magnetic field to a level of about 5 parts in 10^9 over the duration of the experiment would keep any drift in the 200 MHz anomaly frequency within a 1 Hz margin. The data would also need to be suitably binned according to the orientation of the magnetic field as a function of star time. A more elaborate approach to such diurnal experiments would be to mount the apparatus on a suitable rotating platform and thereby to investigate any geometrical dependence more directly.

An experiment of this nature on electrons alone or positrons alone would bound the combination $\mp b^e_3 + d^e_{30}m_e + H^e_{12}$ of couplings in the standard-model extension. It would involve searching for leading-order corrections to the anomaly and cyclotron frequencies which exhibit periodicities of approximately 24 hours. Subleading order corrections involving tensor couplings might exhibit 12-hour periodicities. However, these effects would be suppressed relative to the leading-order effects in Eq. (11). All three of these quantities in Eq. (11) break Lorentz symmetry, but only the coupling b^e_3 breaks CPT. If a signal were detected, it would indicate Lorentz violation but not necessarily CPT violation. It would provide strong motivation for a subsequent experiment comparing anomaly frequencies of electrons and positrons, which would bound the CPT-breaking parameter b^e_3 in isolation.

Data for this type of experiment on electrons alone already exist, and a preliminary analysis has been performed [11]. Assuming a precision of approximately 1 Hz in detecting diurnal variations, we estimate a bound on Lorentz breaking of

$$r^e_{\omega_a^{\mp},\text{diurnal}} \lesssim 10^{-20} \quad . \tag{13}$$

CONCLUSIONS

We find that the use of a general theoretical framework incorporating CPT and Lorentz breaking allows a detailed investigation of possible experimental signatures in Penning-trap experiments on electrons and positrons. Our results indicate that the best tests of CPT symmetry in Penning traps emerge from comparisons of anomaly frequencies in $g-2$ experiments. Our estimated bound on CPT from a variety of signals is approximately 10^{-20} in electron-positron experiments. A table showing these estimated bounds is presented in Ref. [8]. We also find that experiments searching for diurnal variations in electrons alone can provide bounds on Lorentz breaking at a level of approximately 10^{-20}.

ACKNOWLEDGMENTS

This work was supported in part by the National Science Foundation under grant number PHY-9801869.

REFERENCES

1. See, for example, R.G. Sachs, *The Physics of Time Reversal* (University of Chicago Press, Chicago, 1987).
2. See, for example, R.M. Barnett et al., Review of Particle Properties, Phys. Rev. D **54** (1996) 1.
3. R.S. Van Dyck, Jr., P.B. Schwinberg, and H.G. Dehmelt, Phys. Rev. Lett. **59** (1987) 26; Phys. Rev. D **34** (1986) 722.
4. L.S. Brown and G. Gabrielse, Rev. Mod. Phys. **58** (1986) 233.
5. P.B. Schwinberg, R.S. Van Dyck, Jr., and H.G. Dehmelt, Phys. Lett. A **81** (1981) 119.
6. D. Colladay and V.A. Kostelecký, Phys. Rev. D **55** (1997) 6760; Indiana University preprint IUHET 359, Phys. Rev. D, in press.
7. R. Bluhm, V.A. Kostelecký and N. Russell, Phys. Rev. Lett. **79** (1997) 1432.
8. R. Bluhm, V.A. Kostelecký and N. Russell, Phys. Rev. D **57** (1998) 3932.
9. V.A. Kostelecký and S. Samuel, Phys. Rev. Lett. **63** (1989) 224; *ibid.*, **66** (1991) 1811; Phys. Rev. D **39** (1989) 683; *ibid.*, **40** (1989) 1886; V.A. Kostelecký and R. Potting, Nucl. Phys. B **359** (1991) 545; Phys. Lett. B **381** (1996) 89.
10. V.A. Kostelecký, Phys. Rev. Lett. **80** (1998) 1818.
11. R. Mittleman, private communication.

TRAPPING ^{82}Rb FOR β-DECAY PARITY VIOLATION MEASUREMENTS

D. J. Vieira[1], S. J. Brice[1], S. G. Crane[1,2], A. Goldschmidt[1], R. Guckert[1], A. Hime[1], D. Tupa[1] and X. Zhao[1]

[1]*Los Alamos National Laboratory, Los Alamos, NM 87545*
[2]*Physics Department, Utah State University, Logan, UT 84322*

Abstract. We report on the Los Alamos effort to measure the β-nuclear spin correlation function in the Gamow-Teller decay of polarized ^{82}Rb atoms confined in a time orbiting potential (TOP) magnetic trap as a means of probing the origin of parity violation in the electroweak interaction. As a first step in this experiment, we have recently trapped 6 million ^{82}Rb ($t_{1/2}$=75 s) atoms in a magneto-optical trap (MOT) that is coupled to a mass separator. The hyperfine structure of the D_1 transition has been measured for the first time and the D_2 line has been remeasured with higher precision. Moreover, we have recently transferred radioactive atoms from one MOT to another MOT where they have been retrapped and accumulated due to the improved vacuum in the second MOT. We are currently working to optically pump these retrapped atoms into the desired weak-field-seeking, stretched magnetic substate and to load them into the pure magnetic moment TOP trap which will serve as a rotating beacon of spin-polarized nuclei. Using a single (or small set of) positron detectors, we will measure the β-nuclear spin asymmetry as a continuous function of both beta energy and β-nuclear spin angle. Precise measurements initially at the 1% level are planned.

INTRODUCTION

With the advancement of magneto-optical and pure magnetic traps for neutral atoms, there has been a growing interest in exploiting this technology in nuclear physics. Trapped radioactive atoms will enable a new generation of fundamental symmetry experiments including nuclear beta-decay, atomic parity nonconservation, and the search for parity and time-reversal-violating electric dipole moments. In particular, trapped β-decaying species will enable a new set of high-precision measurements that will elucidate our understanding of the helicity structure of the electroweak interaction and aid in the search of physics beyond the standard model. In many ways trapped radioactive atoms make an ideal source for β-decay correlation measurements since relatively intense sources can be harnessed which are effectively massless, point-like, and highly spin polarized. Consequently, systematic effects which have long plagued these measurements that arise from electron scattering and polarization uncertainties can be greatly reduced if not eliminated all together.

It is well known that pure Gamow-Teller transitions, such as those available in ^{82}Rb, are useful candidates to study parity violation since these transitions are driven solely by the axial vector coupling between leptons and quarks. Given the good match between the 75 second half-life of ^{82}Rb and typical trap lifetimes, this species is an excellent choice for a magneto-optical trap (MOT) based experiment. Moreover, since ^{82}Rb is fed by the long-lived ^{82}Sr ($t_{1/2}$=25 d) parent, this experiment can be performed off-line from an accelerator. Although several radioactive species have been trapped to date, the number of trapped atoms (up to 40,000 atoms) has, in most cases, been too small to attempt such a β-decay experiment. Herein, we report on the trapping of several million atoms of ^{82}Rb (a sufficiently large number to obtain 1% counting statistics in just an hour) and summarize our progress in mounting a high precision β-asymmetry measurement of ^{82}Rb.

THE ^{82}Rb β-ASYMMETRY EXPERIMENT

An overview of the ^{82}Rb β-asymmetry experiment is shown in Fig. 1. It involves a mass separator to selectively implant ^{82}Rb into a catcher foil which is located within the trapping cell of a MOT. The ^{82}Rb is fed from a ~9 mCi sample of ^{82}Sr that is placed inside the ion source of the mass separator. Upon heating, the ^{82}Rb diffuses out of the sample and is selectively ionized using an electron-bombardment-heated, thermal ion source. Singly charged ions are extracted from the source, accelerated to 20 keV, mass separated, and focussed through a 5-mm diameter opening of the trapping cell and implanted into a thin catcher foil of yttrium-coated tungsten. The foil is subsequently or continuously heated to temperatures of 700-850 °C using an inductive-heating coil located outside the trapping cell.

Upon heating the implanted ^{82}Rb atoms diffuse out of the catcher foil and are trapped in a standard MOT composed of three orthogonal, circularly-polarized laser beams that are retro-reflected back through the cell. Anti-Helmholtz coils are used to provide a quadrupole field gradient of ~7 G/cm. A Ti:sapphire laser tuned to the D_2 line of Rb at 780 nm provides the trapping laser beams. Acoustic-optical modulators are used to shift the trapping frequency from the D_2 transition in stable ^{85}Rb using a FM sideband locking technique. An electro-optic modulator (EOM) is used to provide repumping light. To improve the trapping efficiency, the inside surface of the trapping cell is coated with a nonstick coating of octadecyltrichlorosilane (OTS) dryfilm. Once trapped the fluorescing cloud of atoms can be seen with a simple CCD camera. By chopping the EOM, the fluorescence from the trapped cloud can be modulated. Improved detection sensitivity is obtained by using a photomultiplier tube with a lock-in amplifier to demodulate the trapping signal.

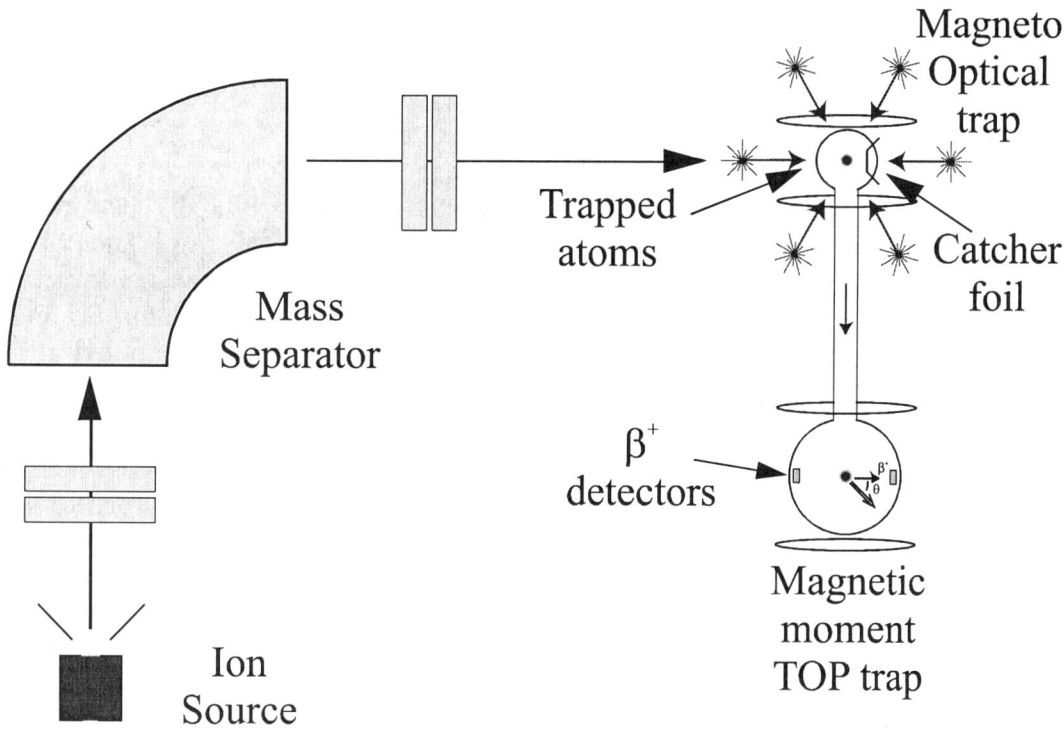

FIGURE 1. General layout of the ^{82}Rb beta-asymmetry experiment.

In our initial work (1), we were able to trap 6 million ^{82}Rb atoms in pulsed heating of the catcher foil and 3 million atoms in the continuous heating mode with trap lifetimes of 90 s and 30 s, respectively. This represents a two orders of magnitude improvement in the number of trapped atoms over previous radioactive atom trapping work. Using gamma-ray counting we were able to determine that 35% of the original activity in the source was ionized and implanted into the catcher foil and of that 30% was released from the foil upon heating to 750 °C. Based on a comparison of the number of trapped atoms determined from the fluorescene trapping signal and the number of atoms in the source, we determined an overall efficiency of ~3 x 10^{-4}. This gives a trapping efficiency of ~0.3% which is a factor of 20 below our expectations based on previous dryfilm-coated cells (2-4). Through gamma counting we learned that the quality of the dryfilm coating was poor and future coatings should help to improve this situation. In large part we attribute the success of this work to the development of a more efficient method of introducing the sample into the MOT with minimal gas loading using the ion implantation and release method with a catcher foil located inside the trapping cell.

Taking advantage of this large number of trapped atoms, we used an additional probe laser to measure the hyperfine structure of the $5P_{1/2}$ and $5P_{3/2}$ atomic states as well as the isotope shift of the D_1 transition in ^{82}Rb (see Ref. 5 for details). This was accomplished by scanning the frequency of the probe laser across each transition which produced a modulation in the trapping signal. However because these measurements were done while the trap was on, careful measurements of the light shift (AC Stark shift) as a function of trapping light intensity were required. By extrapolating these data to zero trapping light intensity, we determined the $5P_{1/2}$

hyperfine constant to be A = 122.7 (1.3) MHz and the D_1 transition isotope shift of $\delta\nu_{82\text{-}85}$ = -150.8 (2.5) MHz. We also remeasured the $5P_{3/2}$ F=5/2 to F'=1/2 and 3/2 hyperfine splitting to be 90.3 (2.0) MHz in agreement with previous results (6) of 89.3 (9.0) MHz. (Note that because the splitting between the $5P_{3/2}$ F=1/2 and F=3/2 hyperfine levels is only 0.1 MHz (Ref. 6), we were unable to resolve these two levels.) The accuracy of this trap and probe method was further verified by performing analogous measurements with trapped ^{85}Rb whose atomic structure is well know from saturated absorption measurements. Not only do these measurements enhance our understanding of the ^{82}Rb atomic structure, but this spectroscopic information is needed for the optical pumping (polarizing) step and as part of the nuclear polarization determination in the β-asymmetry experiment.

After trapping the radioactive atoms in the MOT, the next step in the β-asymmetry experiment (see Fig. 1) is to transfer the atoms to a second MOT using a laser push beam and magnetic guide approach (7). Typical results from this trap, transfer, and retrap sequence are shown in Fig. 2. After optimization we obtained a single-shot transfer efficiency of ~50%. Because the second MOT has a much better vacuum than the first MOT, the lifetime in the second MOT is ~500 s. This long lifetime makes it possible to accumulate atoms in the second MOT by running in a multi-shot transfer mode.

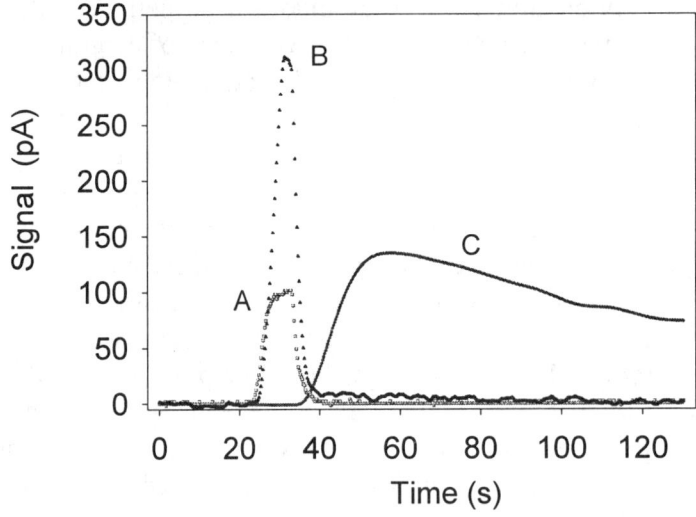

FIGURE 2. Pulsed release, trapping and transfer of ^{82}Rb in a double MOT system. Trace A gives the optical pyrometer readout of the foil temperature which is rapidly heated to a temperature of 700 °C for ~10 s. Trace B shows the lock-in trapping signal from the first MOT. At the ~35 s mark, the first MOT is switched off and the atoms are rapidly "pushed" using another laser beam over to the second MOT where they are retrapped. Trace C shows the MOT2 lock-in trapping signal.

Presently we are working on the next steps in the β-asymmetry experiment. These involve: (a) the optical pumping (polarization) of the atoms into the desired weak-field-seeking, spin-aligned magnetic substate; (b) the loading of the time-orbiting-potential (TOP) magnetic trap (8); and (c) the detection of the positrons. For the most part, the first two steps will employ techniques already developed for the loading of rubidium into TOP traps for atomic physics experiments, however, in our case a premium will be placed on optimizing the efficiency of the loading process and in maintaining a high degree of nuclear polarization once trapped. After extensive Monte Carlo modeling of the TOP trap performance as rotating source of spin–polarized nuclei, we have constructed our own TOP trap and first operation is imminent. A first-generation positron telescope is in place and we are looking forward to undertake our first prototype experiments soon.

The key idea behind using a TOP trap is that the spin vector is aligned with the rotating bias field of the TOP trap. By exploiting the point-like source geometry and rotating spin feature of the TOP trap, we will measure the β-particle – nuclear spin asymmetry function as a continuous function of β energy and angle between the β particle and the nuclear spin alignment vector using a single (or small set of) positron detector(s). Not only is the symmetry of this system attractive from the standpoint of reduced systematic errors, but it will enable the complete mapping of the parity-violating correlation. The initial goal of the ^{82}Rb experiment is a 1% measurement of the β-asymmetry, but ultimately we hope to achieve a ~0.1% determination. If realized, this and other high-precision β-decay correlation measurements will provide a new generation of fundamental symmetry experiments to test and advance our understanding of electroweak interactions.

ACKNOWLEDGMENTS

We wish to thank D. W. Preston, V. D. Sandberg, M. J. Smith, and D. Steele for their earlier contributions to this work. We also thank L. D. Benham, E. P. Chamberlin, F. O. Valdez, and LANL glass and machine shops for their excellent technical support. We gratefully acknowledge helpful discussions with Z.-T. Lu, M. Stephens, and C. Wieman. This work is supported by the Laboratory Directed Research & Development program at the Los Alamos National Laboratory and by the U.S. Dept. of Energy.

REFERENCES

1. R. Guckert *et al.*, Phys. Rev. A **58**, R1637 (1998).
2. M. Stephens and C. Wieman, Phys. Rev. Lett. **72**, 3787 (1994); M. Stephens, R. Rhodes and C. Wieman, J. Appl. Phys. **76**, 3479 (1994).
3. R. Guckert *et al.* Nucl. Instrum. Methods Phys. Res. B **126**, 383 (1997).
4. Z.-T. Lu et al., Phys. Rev. Lett. **79**, 994 (1997).
5. X. Zhao *et al.*, to be published.
6. C. Thibault *et al.*, Phys. Rev. C **23**, 2720 (1981).
7. C. J. Myatt *et al.*, Opt. Lett. **21**, 290 (1996).
8. W. Petrich et al., Phys. Rev. Lett. **74**, 3352 (1995).

Search for Scalar Contributions to the ^{38m}K β^+-ν Correlation in a Magneto-Optic Trap

J.A. Behr[‡], A. Gorelov[*], D. Melconian[*], M. Trinczek[†], P. Dubé[*], O. Häusser[*] U. Giesen[‡], K.P. Jackson[‡], T. Swanson[*], J.M. D'Auria[†], M. Dombsky[‡], G. Ball[‡], L. Buchmann[‡], B. Jennings[‡], J. Dilling[*,§], J.Schmid[*,§], J. Deutsch[¶], W.P. Alford[††], D. Asgeirsson[*], W. Wong[**]

[*] *Dept. of Physics, Simon Fraser University, Burnaby, British Columbia, Canada V5A 1S6*
[†] *Dept. of Chemistry, Simon Fraser University, Burnaby, British Columbia, Canada V5A 1S6*
[‡] *TRIUMF, 4004 Wesbrook Mall, Vancouver, British Columbia, Canada V6T 2A3*
[§] *Physikalisches Institut der Universität Heidelberg, 69120 Heidelberg, Germany*
[¶] *Université Catholique de Louvain, B-1348 Louvain-la-Neuve, Belgium*
[††] *U. Western Ontario, London, Ontario, Canada*
[**] *University of British Columbia, Vancouver, British Columbia, Canada*

Abstract. We have begun a program to test weak interaction symmetries by trapping in a Magneto-optic trap (MOT) isotopes produced at the TRIUMF cyclotron's on-line separator TISOL. We are searching for non-Standard Model scalar contributions to the $\beta^+ - \nu$ correlation in the $0^+ \rightarrow 0^+$ Fermi decay of ^{38m}K. We have trapped ^{38m}K and ^{37}K in a vapor-cell MOT, and transferred them to a second MOT which houses the nuclear detectors. The recoiling nuclei freely escape the trap and are detected in a microchannel plate; simultaneous detection of the β^+ momentum allows reconstruction of the ν momentum. Approximately 500,000 β^+-recoil coincidences have been observed in June 1998; analysis is proceeding.

INTRODUCTION

Using lasers, neutral atoms can be cooled to very low energies and trapped. The pioneering development of this technology won the 1997 Nobel Prize for S.Chu, W.Phillips, and C. Cohen-Tannoudji. TRIUMF, along with several other labs represented at this conference, is harnessing this technology to trap radioactive atoms to do precision low-energy experiments to test the Standard Model. We have coupled our magneto-optic trap to the output of radioactive alkali atoms from TRIUMF's existing on-line isotope separator TISOL [1]. In winter 1998 we will begin using the 10-1000 times more intense ion source being built for ISAC, TRIUMF's new radioactive beam facility.

For β-decay studies, the magneto-optic trap (MOT) [2] provides a sample of atoms in a localized volume with negligible source thickness, so unperturbed nuclear recoils from β-decay can be detected, and therefore the ν momentum can be deduced. In addition, the atoms (and hence nuclei) can be fully polarized using laser techniques, allowing improvements in measurements of the degree to which parity is maximally violated in the weak interaction.

Being a purely weak process, nuclear β-decay is inherently sensitive to physics at the energy scales characterized by the mass of the vector boson that mediates the weak interaction, M_W=82 GeV/c^2. Small deviations of experimental results from Standard Model predictions translate into limits on the existence of new bosons. β-ν correlation experiments in the $0^+ \rightarrow 0^+$ decay of ^{38m}K at accuracy 0.01 are sensitive to masses of new scalar exchange bosons of $\sim (0.01)^{-\frac{1}{4}} M_W \sim 250$ GeV/c^2 (for standard electroweak coupling), complementary, e.g., to direct searches at high energy colliders like HERA [3] and Fermilab [4]. β-asymmetry measurements in ^{37}K to $\leq 10^{-3}$ accuracy are necessary to set limits on right-handed W's complementary to direct searches at Fermilab [5].

This paper is primarily a progress report on our present experiment, a search for scalar contributions to the β-ν correlation of the superallowed $0^+ \rightarrow 0^+$ Fermi decay of ^{38m}K. We also have plans for spin-correlation measurements in the mixed Fermi/Gamow-Teller decay of ^{37}K. In the β^+-ν correlation of polarized ^{37}K, the β-ν correlation coefficient a, the β asymmetry A, the neutrino asymmetry B, and the tensor term c [6] all contribute and can in principle be extracted; their simultaneous measurement would allow both the Fermi/Gamow-Teller mixing ratio and limits on new physics to be determined. For the present experiment, measuring the β-ν correlation for unpolarized ^{37}K provides a test of the apparatus, as the Q-value is almost identical to ^{38m}K (5.1268(4) MeV vs. 5.0215(6) MeV). Some details of our plans to optically pump ^{37}K can be found in the contributed poster of J. Schmid.

SEARCH FOR SCALAR CURRENTS WITH β^+-ν CORRELATION

Our first experiment utilizes our MOT to provide a well-localized, suspended sample of ^{38m}K. We deduce the β^+-ν correlation from measurements of the β^+ momentum and the recoiling nucleus momentum in coincidence. In the $0^+ \to 0^+$ Fermi decay of ^{38m}K the leptons carry away no net angular momentum. Unlike the Standard Model (V-A) interaction, mediated by the W vector boson exchange, a scalar interaction would demand like helicities for both leptons and antileptons. Thus for vector bosons back-to-back emission of the leptons is forbidden (their spins would add up to 1), while back-to-back emission is maximal for scalar boson exchange. Hence, if we write the angular distribution $W(\theta) = 1 + a\frac{v_\beta}{c}\cos(\theta)$, then the β-ν coefficient $a = +1$ for the Standard Model W, and $a=-1$ for a new scalar boson. We measure back-to-back coincidences between β^+ and ^{38}Ar neutral atom recoils. The recoils will have lower energy–hence longer time-of-flight– if the leptons are emitted back-to-back. Ideally, the greatest sensitivity is achieved by comparing the rate of the slowest recoils to the fastest, but in practice there are very few slow events. So in addition, we collect charged Ar recoils with high efficiency with a uniform electric field, and reconstruct the entire angular distribution.

A scalar term could be produced fundamentally by the exchange of scalar bosons found in many standard model extensions, such as scalar leptoquarks found (for example) in GUT's [7], or the charged Higgs needed in supersymmetric models [10]. An induced scalar, which would be indistinguishable in β-decay from a new scalar boson [8], would violate CVC and would in addition be a a second-class current (which in the first generation is equivalent to strong isospin violation) [9]. Limits on the scalar interaction are poor, both from beta decay [7] and from particle physics. From the detailed measurements of the energy distribution of β-delayed protons from ^{32}Ar, a can now be determined with precision 0.005 [11], with the largest systematic error now the knowledge of the Q-value; a measurement to ≈ 0.01 of the β-ν correlation coefficient a in ^{38m}K would thus provide complementary information with an entirely different technique.

Collecting radioactives in the first MOT

The present experiment (see schematic in Fig. 1) utilizes the ^{38m}K$^+$ and ^{37}K ion beams from TISOL. A 1μA, 500 MeV proton beam from the main TRIUMF cyclotron bombards a target made of many thin CaO pressed disks with total thickness 20g/cm^2. The mass-separated ion beam is converted to neutral potassium atoms by stopping in a hot (900 °C) Zr conical foil. Using the back-to-back annihilation radiation to localize the activity, we have deduced that more than 80% of the $\approx$ 1 second half-life ^{37}K and ^{38m}K is released as neutral atoms. Zr has a much lower vapor pressure than Yttrium, which has similar release times. The magneto-optic trap is very shallow, trapping a small percentage of the atoms from the low-velocity tail of the thermal Maxwell-Boltzmann distribution on a given pass through the trap beams; we use the standard "vapor-cell" technique, letting the untrapped atoms rethermalize by bouncing on a hollow Pyrex cube surrounding the trap [12]. The cube is coated with SC-77 Dryfilm to help prevent permanent chemisorption of the alkalis [13].

The Zr foil is at the back of the cube, and the ion beam passes through the trap region, similar to the design which Los Alamos used to trap ^{82}Rb with 3x10^{-3} efficiency ([14] and D. Vieira paper in this session). We tested a number of potentially lower-temperature catchers (liquid Li, Dryfilm-coated stainless steel, Al, In) with either ^{41}K trapping or ^{37}K activity, with limited or no success. Our vapor cell collection efficiency is $\approx$ 1x10^{-3} for stable ^{41}K or for ^{37}K or ^{38m}K. By measuring the trap efficiency for stable ^{41}K with and without the cube, we have determined that our trapping efficiency is limited by our rather open cell geometry and trap depth, not the Dryfilm quality. The designed emittance of the ISAC beam is much smaller, hopefully allowing us to use smaller cube holes in the future. JILA has demonstrated 50% trapping for Fr that entered a more compact geometry [12] (see following paper of H. Gould).

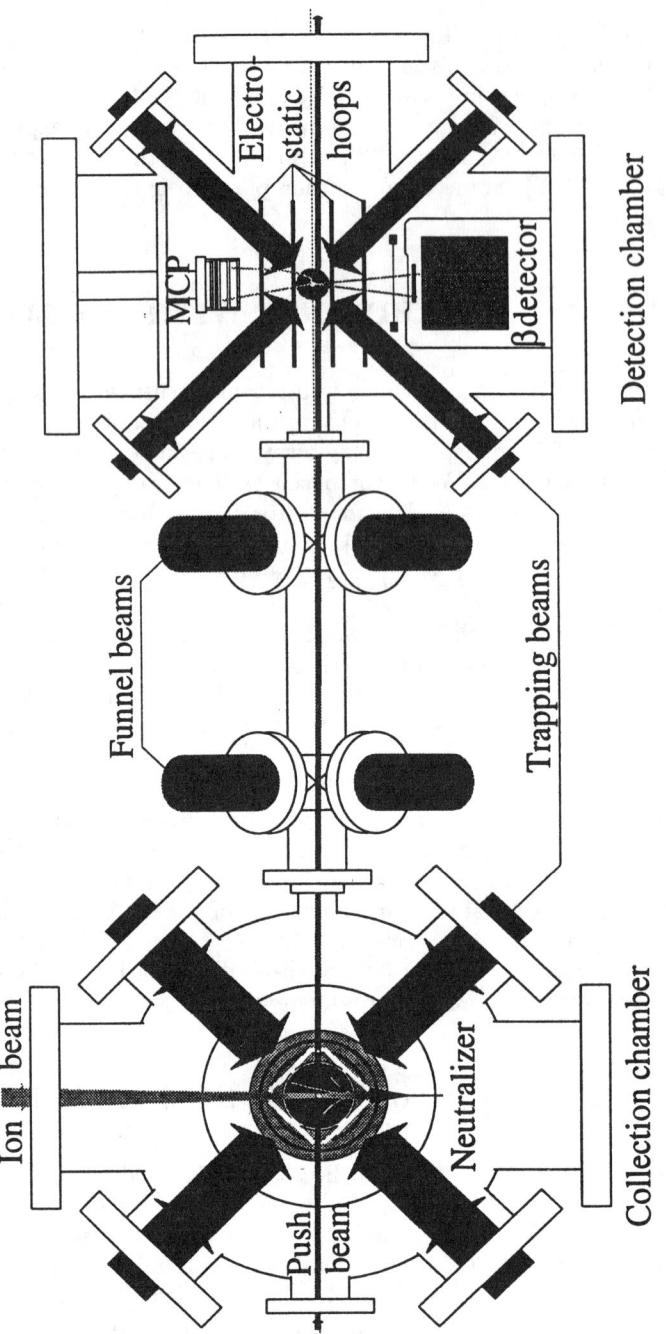

FIGURE 1. Top view of the two-MOT apparatus. The TISOL beam is neutralized and trapped in the first MOT. Push beam moves atoms to the 2nd MOT with nuclear detectors; atoms are collimated during transfer by two 2-d magneto-optic funnels. The 2nd MOT chamber is 15 cm in diameter.

Nuclear Detection trap

The mass 38 beam also includes a flux of $\approx 3 \times 10^8$/sec ^{38}K ground state, which has a half-life of 7.6 minutes and produces a 2.17 MeV γ-ray. Because of the proximity to the hot neutralizer and this much larger ^{38}K ground state activity, any detectors near the capture MOT would count at a very high background rate. So we have moved the atoms, using a push laser beam [15], to a clean, high-vacuum measurement MOT, where the nuclear detectors are housed back-to-back.

The duty cycle for the transfer of the radioactive atoms is as follows. For 20 ms, the 1st trap laser beams are lowered in power and shifted closer to resonance to make the atoms colder, and the push beam is applied. The 2nd trap beams are kept at a frequency and power to optimize collection for an additional 50ms while the atoms transfer and are trapped. Then the 2nd trap beams are shifted closer to resonance and lower in power to make the 2nd trap cloud smaller in size, and the detectors count for 150 msec. The short time between transfers makes the process relatively independent of the vacuum in the first trap, which is as high as 2×10^{-8} Torr with the Zr at 900°C.

In June 1998, we trapped on average $\approx$ 2,000 atoms of ^{37}K or 5,000 atoms of ^{38}K continuously for 10 days total to take the present data set; after $\sim$ 1 week the trap yield monotonically decreased, and the Dryfilm was found to have gone bad. The half-life of trapped ^{41}K atoms in the detection trap was $\approx$ 15 seconds, limited by collisions with the 6×10^{-10} Torr of residual gas; thus 14/15 atoms decay while in the trap. Two CCD cameras view the resonance fluorescence photons scattered from the MOT beams; we obtain quantitative information on the position and spatial distribution of the atoms in the cloud this way. The cloud size is the limitation on the timing resolution; e.g. the 0.7 mm FWHM of the cloud spatial distribution limits the timing for neutral Ar recoils to 15 nsec, and the charged (accelerated) Ar recoils to 5 nsec.

β-ν correlation measurements: First data

We have used for the β^+ telescope a 0.46mm thick double-sided Si strip detector with 24x24 1mm wide strips for position sensitivity, backed by a 6.5cm diameter x 5.5 cm long BC408 plastic scintillator with energy resolution 9.2% at 1.6 MeV.

The recoil detector, placed on the opposite side of the trap, is a microchannel plate (MCP) of 2.5 cm diameter. Position readout is done by a resistive anode, with position resolution of 0.25 mm. Relative timing of 1.0 nsec FWHM between the plastic scintillator and the MCP has been achieved. The efficiency of detection depends on the final state charge distribution of the Ar daughter products, which we have now measured. A uniform electric field of 800 V/cm applied to the entire region is used to accelerate the charged Ar recoils into the MCP to accept most or all angles (at the cost of recoil angular and momentum resolution). The original recoil angle and momentum can still be uniquely reconstructed.

The trap and the push process are completely isotopically selective, and $\approx$ 15 cm of Pb shielding is placed between the traps; nevertheless, approximately 50% of the singles rate in the MCP is from the ^{38}K ground state γ-rays, producing a finite but small rate of false coincidences with real β^+'s in the scintillator.

We have demonstrated the feasibility of these experiments by measuring β^+-Ar recoil coincidences. (We have learned to transfer radioactives to the 2nd trap with the >75% efficiency seen for stable ^{41}K.) The measured charge state distribution of Ar ions is favorable to their detection, and the MCP efficiency for Ar^0 was greater than expected. So there exists the possibility of doing the experiment 3 ways simultaneously; back-to-back angles with Ar^0, a uniform electric field to collect a larger fraction (or all) of Ar$^+$, and also Ar^{2+} (effectively twice the electric field, testing our knowledge of the field uniformity).

Fig. 2 shows scatter plots of Ar recoil TOF vs. β^+ energy, showing Ar^0, as well as Ar$^{+1,+2,+3,+4,+5,+6}$ accelerated by the uniform electric field. These can be seen more easily in the TOF projection (Fig. 3). Detailed comparison with a Monte Carlo, including the detector response function, is necessary to extract the angular distribution. Qualitatively, the slow-going Ar^0 recoils are due to neutrinos leaving at close to back-to-back angles, while the TOF of the Ar charged recoils is determined primarily by their initial direction. The position information is still under analysis and is not shown here.

The events at TOF $\approx$ 1000-1200 are kinematically forbidden, and constitute $\sim$ 1% of the total. They are clearly separated from the Ar^{+1} and Ar^0 events of interest. By making auxilary measurements—such as deliberately releasing the trapped atoms to the walls before counting— we can reproduce this background rate, with negligible additional events in the kinematically allowed regions. By these methods we can attribute this background to badly pushed atoms that were never trapped, and can set good upper limits on how much similar backgrounds could affect the extracted angular correlation.

From literature data, the accelerated Ar$^{+1,+2}$ recoils will be detected with equal probability ($\approx$ 60% of the solid angle, determined by the area density coverage of MCP channels) to within a few percent. The large ($\approx$ 20%) microchannel plate (MCP) efficiency for neutral Ar atoms, given their very low kinetic energies of 0-450 eV, is probably due to metastable Ar atomic states.

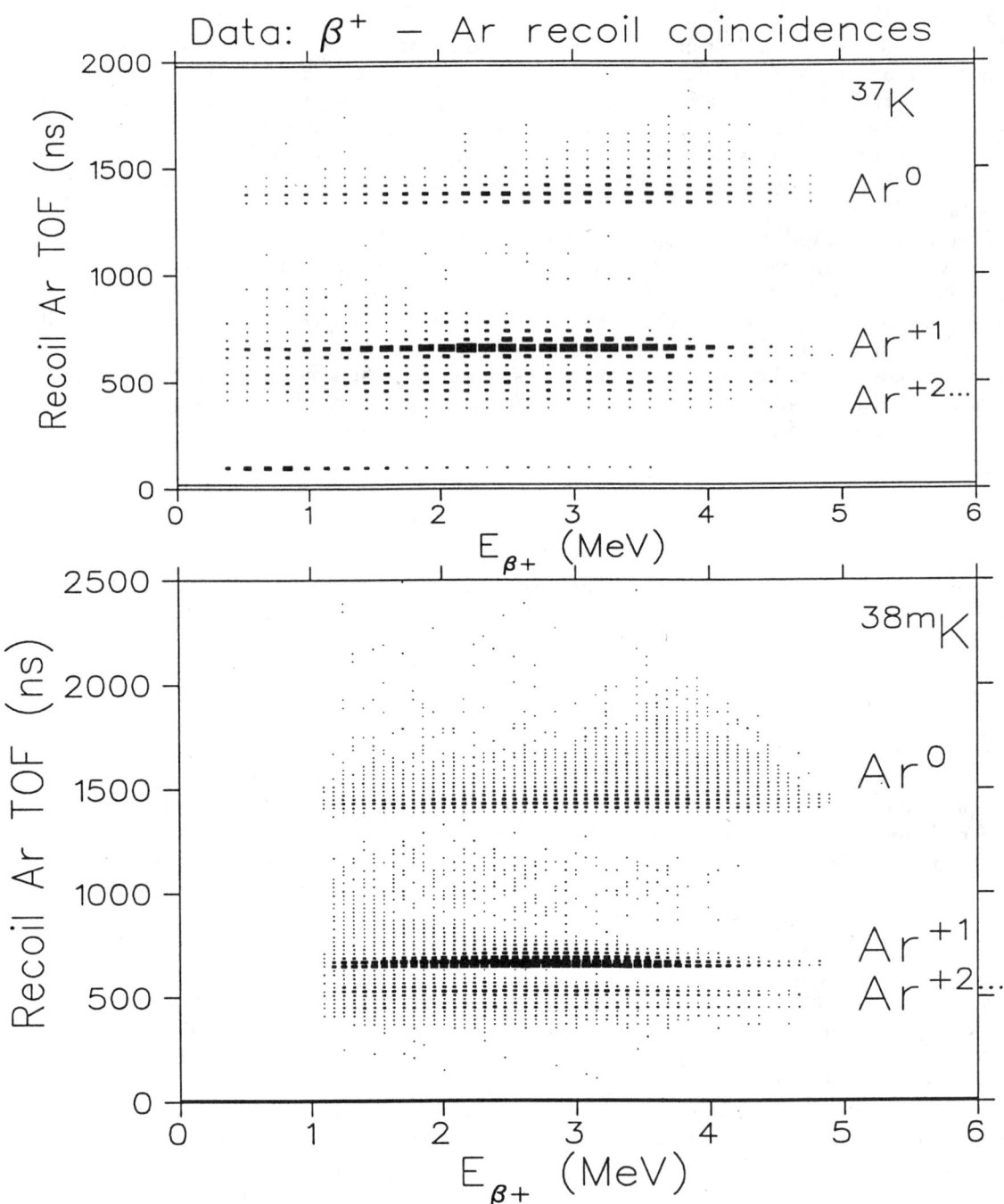

FIGURE 2. Scatter plot of Ar recoil TOF vs. kinetic $E_{\beta+}$, $\approx 20\%$ of the data. $|\vec{E}|=800$ V/cm cleanly separates Ar^0, from $Ar^{+1,+2,+3,+4,+5,+6}$ in TOF.

Ar charge state distributions

Ours are the first measurements of charge state distributions in β^+ decay (Fig. 4). Previous experimental measurements of the charge state distribution in the β^- decay of noble gases at ORNL [16] and phenomenological extrapolations of calculations [17] suggested that 80% of the Ar daughter would be produced as Ar^-, 15% as Ar^0, and the remaining 5% Ar^+ or higher charge states. Fortunately, these estimates for charged recoils were low by factors of 3. Our charge state distribution for β^+ decay is compared with the ORNL work in Fig. 4

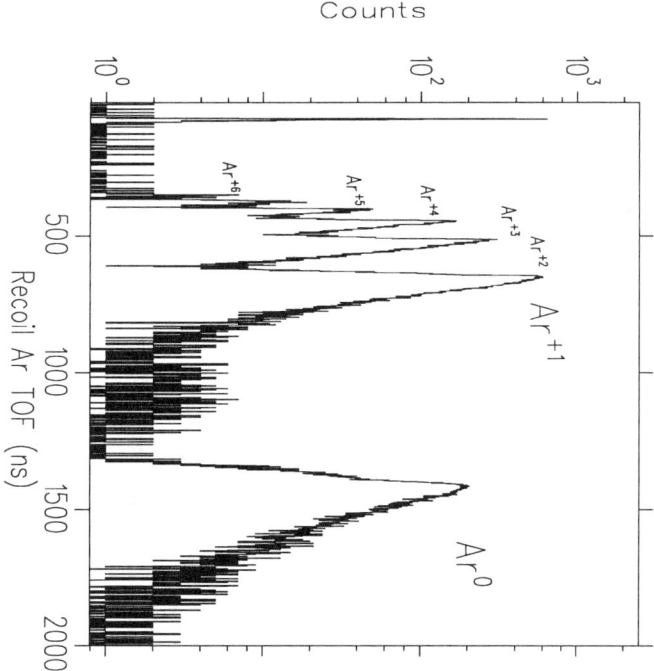

FIGURE 3. TOF of Ar recoils from ^{38m}K decay, for ≈20% of the data. The high-velocity edges are determined almost entirely by the trap size, and quantitatively agree with the spatial distribution as measured in the CCD camera. Note the log scale.

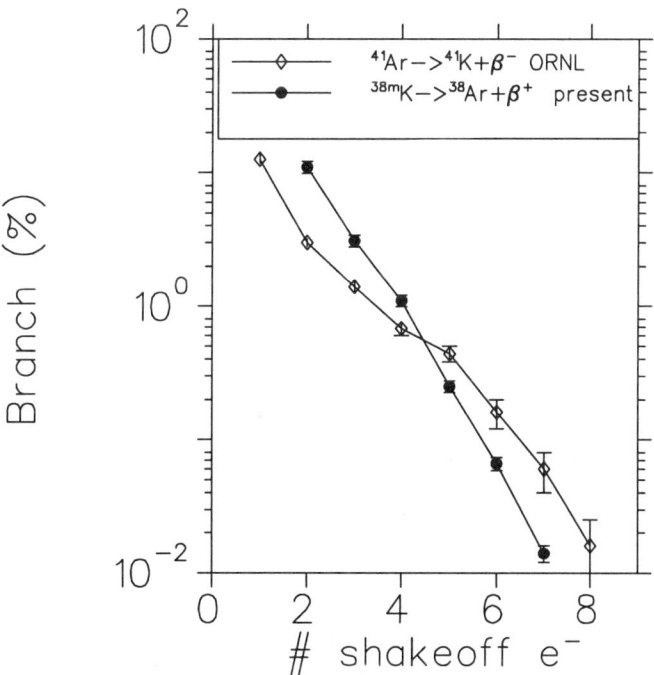

FIGURE 4. Charge state distribution measured in β^+ decay, compared to Oak Ridge β^--decay data.

as a function of number of shakeoff electrons. Sudden-approximation calculations exist for β^+ decay, but not for an alkali decaying to a noble gas, and the atomic configuration is known to be important [17].

Ar neutral metastables

The MCP is detecting Ar^0 with 20% efficiency. This was a surprise, possibly a very useful one if the MCP efficiency as a function of recoil energy can be determined.

Two of the first four excited states of Ar (sharing $3p^54s$ configuration) are metastable (with lifetimes 38 sec [19] and 0.1 sec), the other two E1 decay to the ground state with lifetimes 2 and 8 nsec. The excitation energy is 11.5 eV. Many excited states of Ar decay through these four states. Statistical population of these states would yield $\approx$50% population of the metastables.

The MCP is of standard lead-glass with a standard nichrome coating; data from the literature for secondary electron emission from similar work function surfaces [18] suggested very small efficiencies; e.g., the threshold for secondary electron emission from Mo is 600 eV for Ar^0. This strongly suggests a large fraction of metastables is being produced. Theoretically (with data at 1 energy on several surfaces), secondary electron emission from metastables is similar to Ar^+ and should have a similar roughly constant dependence on Ar energy down to zero energy [18]. We hope to determine this from the ^{37}K slow branch, although the question of whether hyperfine structure affects the state feeding remains open. If the simplest physics expectation is true – to the extent that the MCP only detects Ar^{0*}– then the Ar^0 may turn out to be quantitatively useful.

It is clear that the charged Ar will not have these complications. However, the possibility of doing simultaneous β^+-recoil experiments with the neutral Ar–with a quite different set of systematic errors– is possibly very powerful.

I CONCLUSION

We have measured 500,000 Ar-β^+ coincidences from the decay of ^{38m}K trapped in a MOT, and 75,000 from ^{37}K for detector calibration purposes. Analysis of the data to set limits on scalar interactions in β^+ decay is proceeding.

I would like to acknowledge my collaborators, in particular Otto Haüsser, who was originally invited to give this talk. Otto died in March after a long struggle with cancer. His numerous contributions to this project from its conception live on in the form of his graduate students, his November 1997 Monte Carlo simulation programs and ^{37}K optical pumping programs, and hopefully his vigorous physics style.

REFERENCES

1. J.A.Behr et al., Phys.Rev.Lett. **79** 375 (1997).
2. E.L. Raab et al., Phys. Rev. Lett. **59** 2631 (1987).
3. S. Aid et al. (H1 collaboration), Phys. Lett. **B369** (1996) 173.
4. B. Abbott et al., Phys. Rev. Lett. **80** 2051 (1998).
5. F. Abe et al., Phys. Rev. Lett. **74** 2900 (1995); S. Abachi et al, Phys. Rev. Lett. **76** 3271 (1996).
6. J.D. Jackson, S. B. Treiman and H.W. Wyld, Phys. Rev. **106** (1957) 517; Nucl. Phys. **4** (1957) 206.
7. E.G. Adelberger, Phys. Rev. Lett. **70**, 2856 (1993); Erratum, Phys. Rev. Lett. **71**, 469 (1993).
8. W.E. Ormand, B.A. Brown, and B.R. Holstein, Phys. Rev. C **40** 2914 (1989).
9. L. Grenacs, Annual Rev. of Nucl. Part. Sci. 1985 **35**, 455 (1985).
10. P. Herczeg, in *Precision Tests of the Standard Electroweak Model*, P. Langacker, ed. p.786; H.E. Haber et al., Nucl. Phys. **B161**, 493 (1979).
11. E. G. Adelberger and A. Garcia, WEEIS workshop, 1997.
12. Z.-T. Lu et al., Phys. Rev. Lett. **79** 994 (1997); Z.-T. Lu et al. Phys. Rev. Lett. **77** 3331 (1996).
13. D.R. Swenson and L.W. Anderson, NIM **B29** 627 (1988).
14. R. Gückert et al., Phys. Rev. A **58** (Sept. 1998)
15. T. Swanson et al., Journal of the Optical Society of America B, accepted for publication.
16. A. Snell in *Alpha-, Beta-, and Gamma-Ray Spectroscopy*, K. Siegbahn, Ed. (1964).
17. T. A. Carlson, C.W. Nestor, T.C. Tucker, and F.B.Malik, Phys. Rev. **169**, 27 (1968).
18. M. Kaminsky, "Atomic and Ionic Impact Phenomena on Metal Surfaces", Springer-Verlag, 1965.
19. F.Shimizu et al., Optics Lett. **16** 339 (1991).

Spectroscopy of Francium

J. E. Simsarian, J. S. Grossman, L. A. Orozco, M. Pearson, G. D. Sprouse, W. Z. Zhao

Department of Physics and Astronomy State University of New York at Stony Brook, Stony Brook, NY 11794-3800

Abstract. Francium is the least studied of the alkali atoms because it has no stable isotopes. We have performed precision spectroscopy on cold Fr atoms in a magneto optical trap. We have determined the location of the first two excited states of the S series by two-photon spectroscopy. We have measured the lifetimes of the $7p$ levels with a precision better than 0.5%. Our measurements test the many-body perturbation theory *ab initio* calculations of the dipole matrix element to very high accuracy in this relativistic alkali.

INTRODUCTION

Francium is the heaviest of the alkali atoms and has no stable isotopes. It occurs naturally from the α decay of actinium or artificially from fusion or spallation nuclear reactions in an accelerator. Its longest lived isotope has a half-life of 22 minutes. Previously, experiments to study the atomic structure of francium were possible only with the very high fluxes available at a few facilities in the world [1], or by use of natural sources [2]. We have recently captured francium atoms [3] in a magneto optical trap (MOT), opening the possibility for extensive studies of its atomic properties. Because of its large number of constituent particles, electron correlations and relativistic effects are important, but its structure is calculable with many-body perturbation theory (MBPT). Its more than two hundred nucleons and simple atomic structure make it an attractive candidate for a future atomic parity non-conservation (PNC) experiment. The PNC effect is predicted to be 18 times larger in Fr than Cs [4].

Since the pioneering work of Bouchiat [5] focusing on heavy elements for atomic PNC, efforts have centered in those with atomic number greater than fifty [6–10]. Recent measurements of PNC in Cs [6] and Tl [7] put constraints on physics beyond the Standard Model. The present francium spectroscopy serves to test the theoretical calculations in a heavier alkali to ensure that the Cs structure is well understood and serves to pave the way towards a PNC experiment in a series of isotopes of Fr.

We have been studying the spectroscopy of francium in a magneto optical trap on-line with an accelerator. The captured atoms are confined for long periods of time moving at low velocity in a small volume, an ideal environment for precision spectroscopy. Our investigations have included the location of the $8S$ and $9S$ energy levels. We have also made the first measurements of any radiative lifetime in Fr. The precisions of our lifetime measurements of the D_1 and D_2 lines are comparable to those achieved in stable atoms. They test atomic theory in a heavy atom where relativistic and correlation effects are large. We have reached sensitivity to the nuclear magnetization in measurements of the hyperfine splittings in a chain of Fr isotopes.

Atomic properties are sensitive to different ranges of the electron wave function. The energy levels are the direct eigenvalues of the wavefunctions, but do not give detailed information about the **r** dependence of the wavefunction in a particular range. The radiative lifetime, τ, depends on the matrix element of the dipole moment operator, $e\mathbf{r}$, spatially integrated with the wave functions of the two connected levels. It is most sensitive to large **r** properties. The hyperfine interaction probes the wave functions at the nucleus ($\mathbf{r} \approx 0$).

There has been considerable interest in measuring radiative lifetimes of the alkali lowest atomic levels to test the *ab initio* calculations. A result of the activity is the resolution of a prior discrepancy between the theoretical and experimental lifetimes in Li [12] and Na [13–15]. Recent measurements are now in agreement with the calculations. Dzuba *et al.* [4] and Johnson *et al.* [16] have calculated the dipole matrix elements for Fr using MBPT. The matrix element calculations have an accuracy of $\pm 1.0\%$ to $\pm 2.0\%$. The same methods used for Fr have been applied to Cs. Our comparison of the measured lifetimes to the calculations test their accuracy in a more complex system.

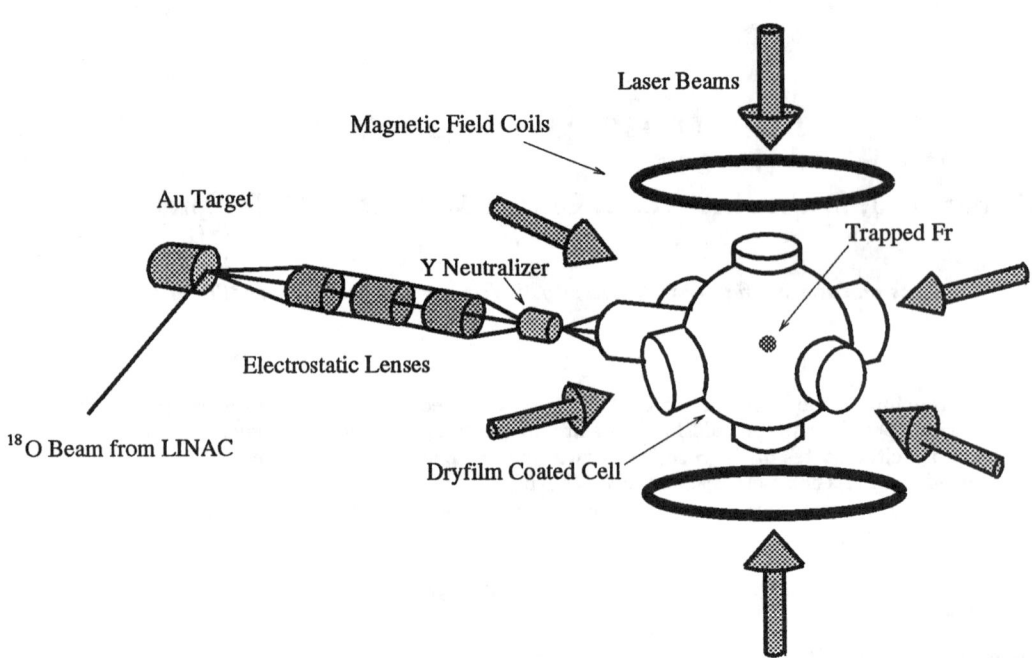

FIGURE 1. Schematic view of target, ion transport system, and magneto optical trap.(Figure adapted from Ref. [22]).

FRANCIUM PRODUCTION

Heavy-ion fusion reactions can, by proper choice of projectile, target and beam energy, provide selective production of the neutron deficient francium isotopes. Gold is an ideal target because it is chemically inert, has clean surfaces, and a low vapor pressure. The ^{197}Au(^{18}O,xn) reaction at 100 MeV produces predominantly ^{210}Fr, which has a 3.2 min half-life. Changing the energy and the isotope of the oxygen beam maximizes the production of isotopes 208, 209 or 211. The reaction ^{198}Pt(^{19}F,5n) produces ^{212}Fr.

The apparatus shown in Fig. 1 is based on the same principles of our earlier work [17]. A beam of 10^{12} ^{18}O ions/s on the Au produces ^{210}Fr in the target, with less than 10% of other isotopes. The target is heated to $\approx$ 1200 K by the beam power and by an auxiliary resistance heater. The elevated temperature is necessary for the alkali elements to rapidly diffuse to the surface and be surface ionized.

Separation of the production and the trapping regions is critical in order to operate the trap in a UHV environment. Extracted at 800 V, the $\approx 1 \times 10^6$/s ^{210}Fr ions travel about one meter where they are deposited on the inner surface of a cylinder coated with yttrium which is heated to 1000 K and located 0.3 cm away from the entrance of the cell. Neutral Fr atoms evaporate from the Y surface and form an atomic beam directed towards an aperture into the vapor cell MOT.

THE FRANCIUM TRAP

The physical trap consists of a 10 cm diameter Pyrex bulb with six 5 cm diameter windows and two viewing windows 3 cm in diameter. The MOT is formed by six intersecting laser beams each with $1/e^2$ (power) diameter of 4 cm and power of 150 mW, with a magnetic field gradient of 6 G/cm. The glass cell is coated with a non-stick Dry-film coating [18,19] to allow the atoms multiple passes through the trapping region after thermalization with the walls [20]. The trapping laser operates in the D_2 line of francium, while the repumper may operate in the D_1 or in the D_2 lines depending on the measurement. The ground state hyperfine splitting of ^{210}Fr is 46.7 GHz.

We have successfully trapped francium in a MOT [3]. We estimate the number of atoms in the trap by comparing the measured fluorescence signal from the trap with the expected fluorescence/atom. The most recently observed signals correspond to at least 10^4 atoms. The ratio of trapping rate to the ion extraction rate

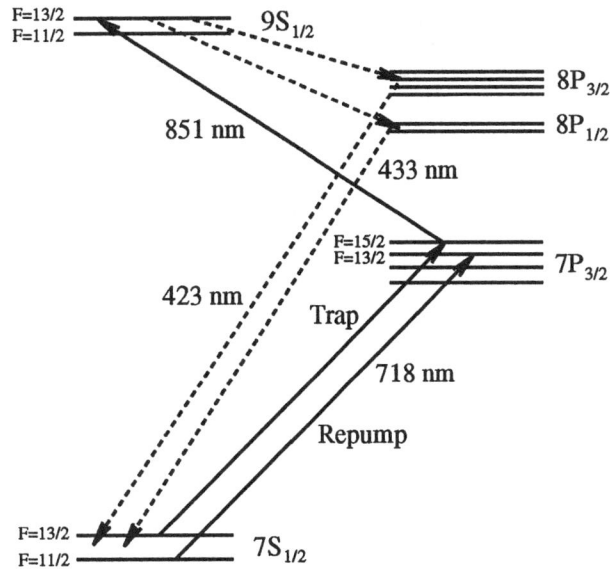

FIGURE 2. Relevant energy levels for the observation of the 9S level in francium. (Figure from Ref. [22]).

is 10^{-3}. The collaboration of Lawrence Berkeley Laboratory and The University of Colorado used a radioactive source to produce the neutron rich ^{221}Fr isotope and trapped francium in a MOT [21].

SPECTROSCOPY OF S LEVELS

Although many energy levels of francium [1] have been studied, until recently the position of the first two excited levels ($8S$ and $9S$) had not been observed. Our spectroscopic measurements have located them permitting the extraction of very accurate quantum defect parameters.

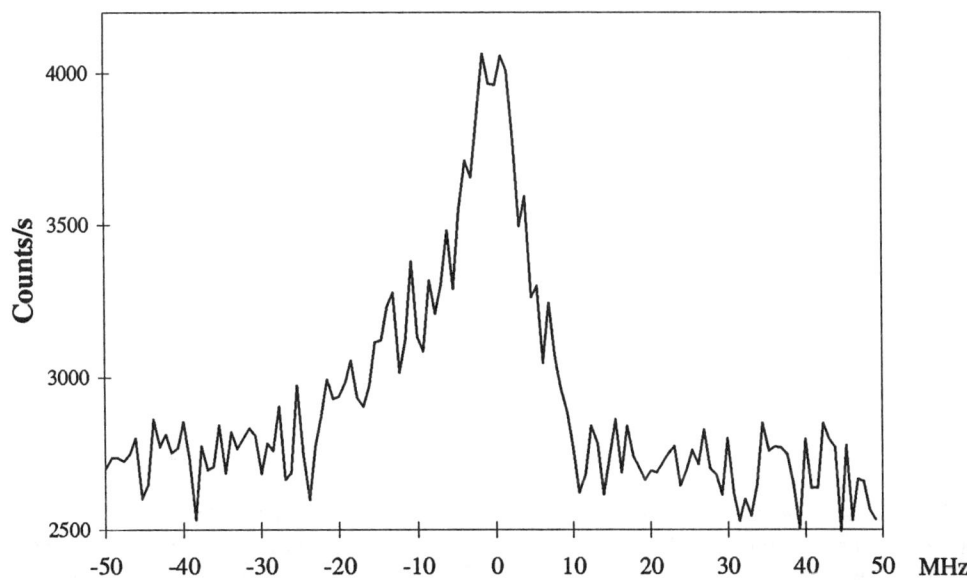

FIGURE 3. 9S resonance in francium. The zero in the horizontal axis is the sum of the two laser frequencies involved 25670.358 ± 0.005 cm^{-1}. (From Ref. [22]).

We used a two photon process to excite the atom from the $7S_{1/2} \rightarrow 7P_{3/2} \rightarrow 9S_{1/2}$ state [22](See Fig 2). The trapping laser provided the first photon and a narrow-linewidth diode laser operating at 851 nm provided the second. We looked for blue photons at 433 nm and 423 nm as the atom decayed from the $9S$ state via the $8P_{3/2}$ and $8P_{1/2}$ levels, (see Fig. 3). A photon counting system with very good discrimination from any background detected the blue photons. We monitored the decrease in the fluorescence on the cycling transition when the atoms got excited to the $9S$ level. We observed a correlation between the two signals. They showed the asymmetries expected from the Autler-Townes splitting [23] of the $7P_{3/2}$ level due to the intense trapping laser.

The calculations by Dzuba et al. [4] using *ab initio* wave-functions for the location of the $9S$ level were within the quoted uncertainty of ± 10 cm^{-1}; they were accurate to four digits. In the search for this transition we were guided by the quantum defect fit to the S series with an accuracy of 1 cm^{-1} in the location of the $9S$ level. The energy difference between the centers of gravity of the $9S_{1/2}$ level and the $7S_{1/2}$ ground state is 25671.021 ± 0.006 cm^{-1}.

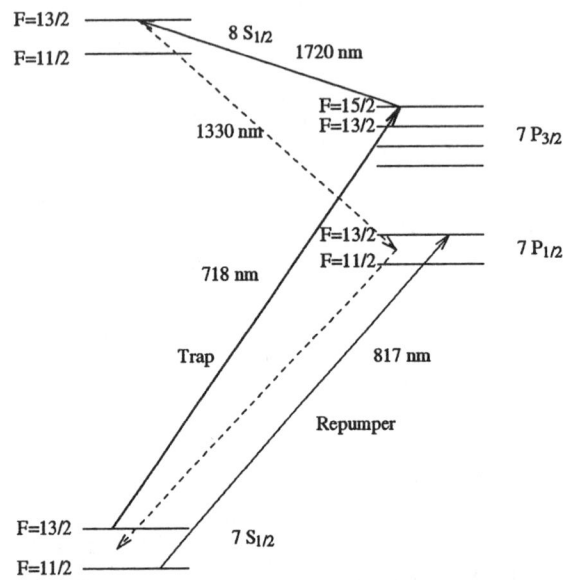

FIGURE 4. Energy levels relevant for the observation of the $8S$ state in ^{210}Fr.

We found the $8S$ level using a similar technique to that used for the $9S$ level: Two photon excitation from the $7S_{1/2} \rightarrow 7P_{3/2} \rightarrow 8S_{1/2}$ level (See Fig. 4). The first photon came from the trapping laser and the second from a diode laser operating at 1.7 μm. We detected the resonance in two different ways. First we looked for photons at 817 nm indicating the atom decayed from the $8S$ level via the $7P_{1/2}$ states. We also observed the change because of the loss of atoms from the cycling transition that produces the trap. The laser resonant to the $8S_{1/2}$ pumps a fraction of the atoms out of the cycling transition decreasing the trap fluorescence accordingly. We measured the hyperfine splitting of the $8S_{1/2}$ state we used the $7S_{1/2} \rightarrow 7P_{1/2} \rightarrow 8S_{1/2}$ two photon transition. The second step, with a diode laser at 1.3 μm could probe both hyperfine levels of the $8S_{1/2}$ state.

LIFETIME OF THE ELECTRONIC LEVELS

One of the most stringent tests for the atomic theory is the calculation of the lifetime of an electronic state. For the case of francium and the other alkalis the lifetime of the D_2 line linking the $nP_{3/2}$ excited level with the $nS_{1/2}$ ground state is directly proportional to the square of the dipole matrix element between the two levels. The calculation relies on having very good wavefunctions for both the S and the P levels.

Great effort has been devoted to the cesium D_2 line lifetime both experimentally [24,25] and theoretically [26,27]. The results bring credibility to the atomic calculations and strengthen the knowledge of the atom. The experiments and theory agree at a little better than 0.5 % accuracy.

We have measured the atomic lifetime of the Fr $7P_{3/2}$ level as it decays into the $7S_{1/2}$ ground state [28]. This is the first measurement of an atomic radiative lifetime in Fr. We used a time-correlated single photon counting technique with a cold sample of ^{210}Fr atoms in a magneto-optic trap. Figure 5a shows the time sequence required for the technique. We turn off the trapping laser for 500 ns and start a timer that stops on the detection of a fluorescence photon. We repeat the cycle with a rate of 100 kHz and histogram the delay times of the fluorescence. The lifetime for the $7p\ ^2P_{3/2}$ level of 21.02(16) ns gives a value for the reduced transition matrix element between the levels $7s\ ^2S_{1/2} \to 7p\ ^2P_{3/2}$ of 5.898(22) a_∞ atomic units. The final precision is better than the accuracy reported by state of the art MBPT calculations.

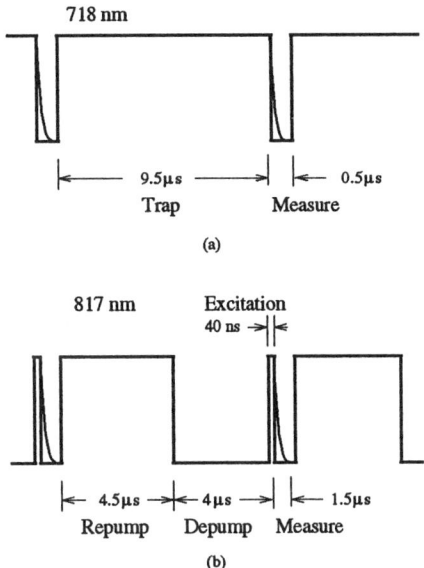

FIGURE 5. Timing for the lifetime measurements. Trace a is the chopping of the 718 nm trapping and repumping lasers for the $7P_{3/2}$ experiment. Trace b is the chopping of the 817 nm repumping laser for the $7P_{1/2}$ experiment. (Figure from Ref. [29]).

We have also measured the lifetime of the $7P_{1/2}$ level (see Fig. 6). The lifetime measurement required a more elaborate procedure for the transfer of population to the appropriate level (see Fig. 5b). We carried out extensive tests for systematic effects, and performed measurements of the lifetimes of the D_1 and D_2 lines in Rb using the exact same techniques as in Fr. The accuracy in the determination of the matrix elements for both states is better than 0.5% [29].

From the two measurements of the lifetimes of the P levels we get the atomic line strength ratio. This reveals the large relativistic effects in the heavy Fr atom. The line strength ratio, $\mathbf{S}_{1/2}/\mathbf{S}_{3/2}$ is the ratio of the reduced matrix elements and is independent of the transition energies. For the nonrelativistic case of the light alkali elements, the ratio is 0.5. As the relativistic effects become more important in heavier elements the ratio increases. Fig 8 plots the line strength ratio for the alkali. Our result of .526(3) for $\mathbf{S}_{1/2}/\mathbf{S}_{3/2}$ in Fr shows a dramatic increase in relativistic effects over Cs. The values of the line strength ratio other than for Fr come from Refs. [14,25].

According to the simple scaling of the PNC effect as Z^3, it should be a factor of four larger in Fr than in Cs but the predictions of Dzuba et al. give a factor of eighteen. The extra enhancement comes from relativistic effects. A measure of the relativistic effects is the difference of $\mathbf{S}_{1/2}/\mathbf{S}_{3/2}$ from 0.5. For Cs the difference is 0.0047(16) [24] while in Fr we have measured 0.026(3) [29]. The ratio of these two numbers accounts for the predicted factor of 18, indicating the importance of relativistic effects for a PNC experiment.

The calculations of the groups of New South Wales and Notre Dame [4,16] agree with our measurements, testing their ability to generate appropriate Fr wavefunctions. They use MBPT to calculate the atomic structure of francium to a high accuracy. To achieve precise results in the heavy alkali atoms, the interaction between the valence and the core electrons must be calculated. This is the most complicated part of the method. The accuracy of the calculated radial integrals in Fr by Dzuba et al. is expected to be ±1%.

The RHF calculation of Johnson et al. [16] is in good agreement with that of Dzuba et al. The matrix elements of the two groups differ slightly, mostly from a difference in the Brueckner-type corrections. Johnson et al. calculated the corrections to third order and included fourth- and higher-order terms empirically. Since these terms play a significant role, it is expected that the result will not be as accurate as their "all-order" calculation in Cs [26]. The accuracy of the calculated matrix elements in Fr by Johnson et al. is expected to be ±2%. (See Fig. 8).

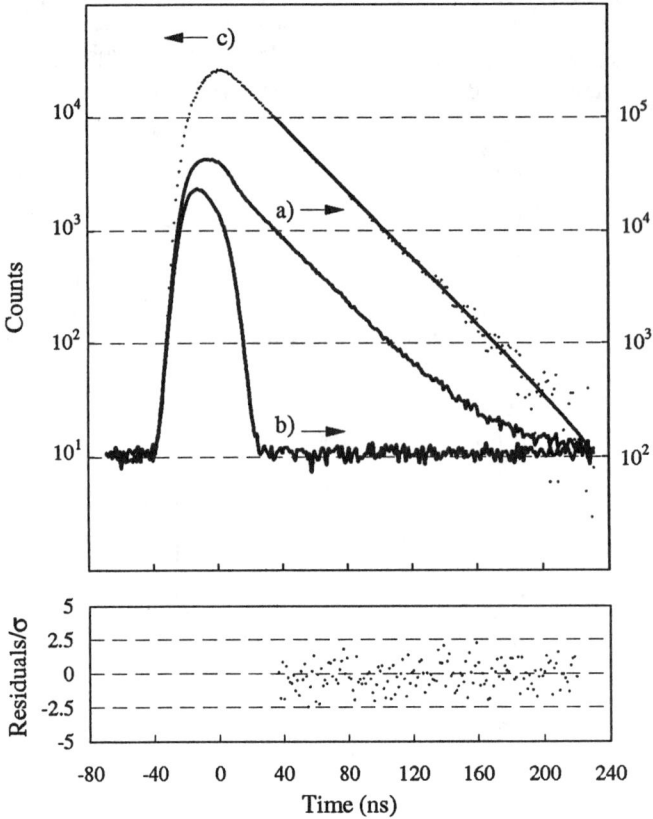

FIGURE 6. Decay curves of Fr $7P_{1/2}$ level. Trace a is the raw data with Fr in the trap and trace b is the background. Trace c is the subtraction of b from a and the straight line is a pure exponential fit to trace c. The lower trace shows the residuals of the fit to an exponential. (Figure from Ref. [29]).

The MBPT calculations predict the location of energy levels, hyperfine structure, E1 transition amplitudes, and parity non-conserving transition amplitudes. These quantities are calculated from the MBPT atomic wave functions. Comparisons with experiments test the accuracy of the wave functions over different ranges of **r**.

HYPERFINE INTERVAL MEASUREMENTS

Although there is much information about the distribution of protons within the atomic nucleus, very little is known about the distribution of neutrons in nuclei, and one has to rely heavily on theory. A unique experimental probe of the nuclear magnetization distribution is precision measurements of the magnetic hyperfine constants (A) with laser spectroscopy. The magnetic hyperfine interaction can be viewed as arising from an effective magnetic field from the electron interacting with the magnetization of the nucleus. Different atomic states have different radial wavefunctions, and will sample the nuclear magnetization distribution with different weighting. This is the origin of the hyperfine anomaly (Bohr Weisskopf effect). A possible way to get at the neutron positions in nuclei is to look at the radial dependence of the magnetization generated by the neutrons.

We are currently measuring the hyperfine splitting of the $7P_{1/2}$ atomic level for five different isotopes $^{208-212}$Fr. We have achieved sensitivity to differences in the radial distribution of the nuclear magnetization.

To obtain a high signal-to-noise ratio out of the few thousand trapped atoms we excite the atoms to the appropriate hyperfine state, and use photon counting techniques. To avoid systematic errors in the calibration of a frequency marker as well as possible slow shifts (tens of minutes) in the laser frequency, we FM modulate the probe laser to generate sidebands separated at about the hyperfine splitting of the $7P_{1/2}$ level (≈ 6 GHz). We are currently investigating systematic effects such as laser power, number of atoms and magnetic field gradient dependence.

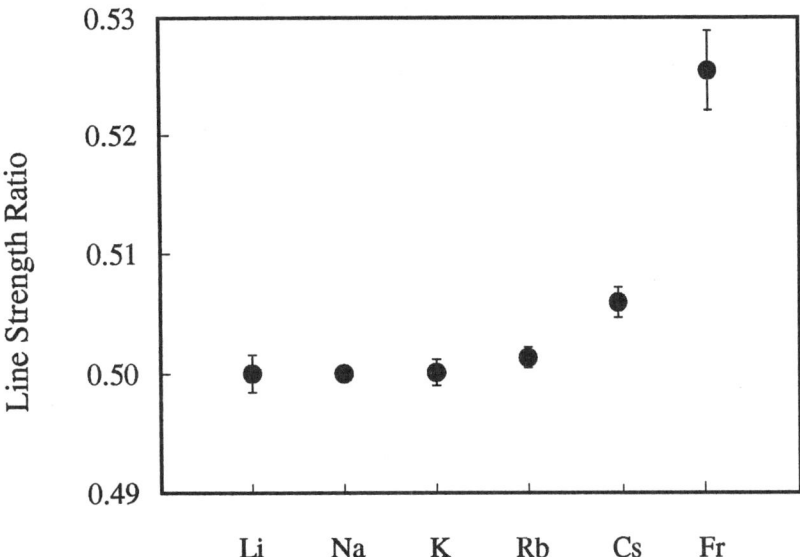

FIGURE 7. Ratio of line strength of the lowest P levels for alkali atoms. Deviations from 0.5 indicate relativistic effects. The line strength ratios for the stable elements are from Refs. [14,15,25]. (Figure from Ref. [29])

The quantitative interpretation of these effects is currently in progress. Neutron distributions are needed for the interpretation of future atomic-parity nonconservation measurements in chains of isotopes. Nuclear theories have to correctly describe both the charge distributions determined by isotope shift measurements, and the magnetization distributions from hyperfine anomaly measurements. We expect that these high precision spectroscopic measurements in Fr contribute to the limited knowledge of neutron distributions in nuclei.

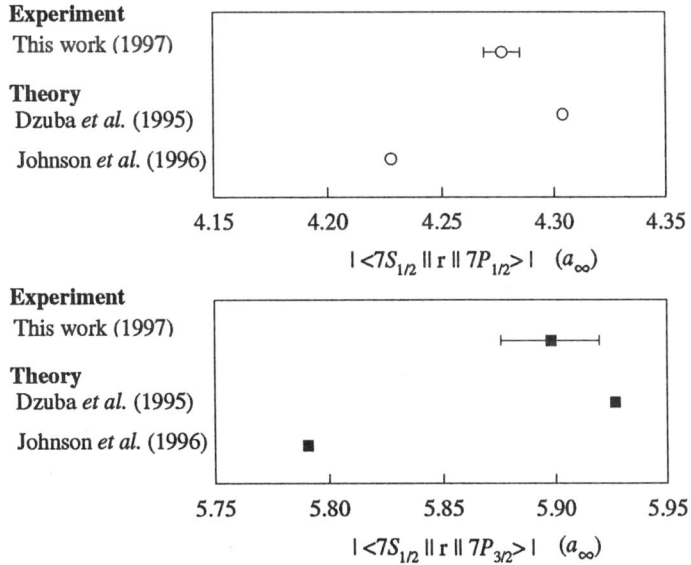

FIGURE 8. Comparison of the absolute value or the $7S_{1/2} \to 7P_{1/2,3/2}$ transition radial matrix element with *ab initio* calculations from Ref. [27,26]. (Figure from Ref. [29]).

ACKNOWLEDGMENT

We thank the National Institute of Standards and Technology and the National Science Foundation for support to carry out these studies.

REFERENCES

1. E. Arnold, W. Borchers, H. T. Duong, P. Juncar, J. Lermé, P. Lievens, W. Neu, R. Neugart, M. Pellarin, J. Pinard, J. L. Vialle, K. Wendt, and the ISOLDE Collaboration, J. Phys. B **23**, 3511 (1990).
2. S. V. Andreev, V. I. Mishin, and V. S. Letokhov, J. Opt. Soc. Am. B **5**, 2190 (1988).
3. J. E. Simsarian, A. Ghosh, G. Gwinner, L. A. Orozco G. D. Sprouse, and P. A. Voytas, Phys. Rev. Lett. **76**, 3522 (1996).
4. V. A. Dzuba, V.V. Flambaum, and O. P. Sushkov, Phys. Rev. A **51**, 3454 (1995).
5. See for example M. A. Bouchiat in *Atomic Physics 12* Ed. by R. R. Lewis and J. C. Zorn, AIP, New York, p. 399 (1991).
6. C. S. Wood, S. C. Bennett, D. Cho, B. P. Masterson, J. L. Roberts, C. E. Tanner, and C. E. Wieman, Science **275**, 1759 (1997).
7. P. A. Vetter, D. M. Meekhof, P. K. Majumder, S. K. Lamoreaux, and E. N. Fortson, Phys. Rev. Lett. **74**, 2658 (1995).
8. D. Budker, D. DeMille, E. D. Commins, and M. S. Zoltorev, Phys. Rev. Lett. **70**, 3019 (1994).
9. R. B. Warrington, C. D. Thompson, D. N. Stacey, Europhys. Lett. **24**, 641 (1993).
10. D. DeMille Phys. Rev. Lett. bf 74 4165 (1995).
11. V. A. Dzuba, V. V. Flambaum, and I. B. Khriplovich, Z. Phys. D **1**, 243 (1986).
12. W. I. McAlexander, E. R. I. Abraham, and R. G. Hulet, Phys. Rev. A **54**, R5 (1996).
13. C. W. Oates, K. R. Vogel, J. L. Hall, Phys. Rev. Lett **76**, 2866 (1996).
14. U. Volz, M. Majerus, H. Liebel, A. Schmitt, and H. Schmoranzer, Phys. Rev. Lett. **76**, 2862 (1996); U. Volz and H. Schmoranzer, Physica Scripta **T65**, 48 (1996).
15. K. M. Jones, P. S. Julienne, P. D. Lett, W. D. Phillips, E. Tiesinga, and C. J. Williams, Europhys. Lett. **35**, 85 (1996).
16. W. R. Johnson, Z. W. Liu, and J. Sapirstein, At. Data Nucl. Data Tables **64**, 279 (1996).
17. G. Gwinner, J. A. Behr, S. B. Cahn, A. Ghosh, L. A. Orozco, G. D. Sprouse, and F. Xu, Phys. Rev. Lett. **72**, 3795 (1994).
18. D. R. Swenson and L. W. Anderson, Nucl. Instr. and Methods B **29**, 627 (1988).
19. M. Stephens, R. Rhodes, and C. Wieman, J. Appl. Phys. **76**, 3479 (1994).
20. C. Monroe, W. Swann, H. Robinson, and C. Wieman, Phys. Rev. Lett. **65**, 1571 (1990).
21. Z.-T. Lu, K. L. Corwin, K. R. Vogel, C. E. Wieman, T. P. Dinneen, J. Maddi, and H. Gould, Phys. Rev. Lett. **79**, 994 (1997).
22. J. E. Simsarian, W. Shi, L. A. Orozco, G. D. Sprouse, W. Z. Zhao, Opt. Lett. **21**, 1939 (1996).
23. P. L. Knight, P. W. Milonni, Phys. Rep. **66** 21 (1980).
24. C. Tanner in *Atomic Physics 14* Ed. by D. J. Wineland, C. E. Wieman and S. J. Smith. AIP Conference Proceedings 323, p. 130, New York 1995.
25. L. Young, W. T. Hill III, S. J. Sibener, Stephen D. Price, C. E. Tanner, C. E. Wieman, and Stephen R. Leone. Phys. Rev. A. **50**, 2174 (1994).
26. S. A. Blundell, W. R. Johnson, and J. Sapirstein, Phys. Rev. A **43**, 3407 (1991).
27. V. A. Dzuba, V. V. Flambaum, A. Ya. Kraftmakher, and O. P. Sushkov, Phys. Lett. A **141**, 147 (1989).
28. W. Z. Zhao, J. E. Simsarian, L. A. Orozco, W. Shi, and G. D. Sprouse, Phys. Rev. Lett. **78**, 4169 (1997).
29. J. E. Simsarian, L. A. Orozco, G. D. Sprouse, and W. Z. Zhao, Phys. Rev. A. **57** 2448 (1998).
30. V. A. Dzuba, V. V. Flambaum, and O. P. Sushkov, J. Phys. B: At. Mol. Phys. **17**, 1953 (1984).

Neutron Decay Study Using an Ion Trap

J. Byrne and P.G. Dawber

*Physics and Astronomy Subject Group,
School of Chemistry, Physics and Environmental Science,
University of Sussex, Brighton, Sussex, BN1 9QH, UK*

Abstract. We discuss the application of the ion trapping technique to measure both the neutron lifetime and the spectrum of recoil protons in unpolarized neutron β-decay.

1. THE WEAK DECAY OF THE NEUTRON

In the β-decay of the free neutron, $n \to p e^- \bar{\nu}_e$, the endpoint kinetic energy of the electron spectrum is 0.78 MeV while the recoiling proton carries off a maximum kinetic energy of 0.75 keV. In the language of nuclear physics free neutron decay is described as a superallowed mirror transition within an isospin doublet. Since there is no spin change $\Delta I = 0$ (but not $0 \to 0$) the transition is allowed by both Fermi and Gamow-Teller selection rules with strengths governed by vector and axial vector coupling constants denoted by G_V and G_A respectively [1]. The very long lifetime for the decay [2]

$$\tau_n = 887.0 \pm 2.1 \text{ sec.}$$

is a consequence of the low energy release, reflecting the fact that the recoil parameter $\delta = (m_n - m_p)/(m_n + m_p)$ has a value $\sim 10^{-3}$.

In the language of particle physics proton and neutron are the lightest members of the lowest flavour-SU(3) baryon octet, and neutron decay is one of 16 semi-leptonic decays within this octet of which 6 are electronic decays. Neutron decay is a strangeness conserving ($\Delta S = 0$) transition as indeed are $\Sigma^+ \to \Lambda^0 e^+ \nu_e$, $\Sigma^- \to \Lambda^0 e^- \bar{\nu}_e$ and $\Xi^- \to \Xi^0 e^- \bar{\nu}_e$. The remaining two electronic decays within the octet, $\Lambda^0 \to p e^- \bar{\nu}_e$ and $\Sigma^- \to n e^- \bar{\nu}_e$, are strangeness violating ($\Delta S = 1$) transitions. All these decays are conventionally interpreted according to the universal V-A theory of weak interactions with a conserved hadronic vector current deriving from a global gauge symmetry of the QCD Lagrangian[3]. The vector coupling constants are expressed in terms of the weak Fermi coupling constant G_F according to the relations

$$G_V (\Delta S=0) = G_F \{1+\Delta_\beta - \Delta_\mu\}^{1/2} V_{ud} \; ; \quad G_V (\Delta S=1) = G_F \{1+\Delta_s - \Delta_\mu\}^{1/2} V_{us},$$

where the Δ_i ($i=\beta,\mu,s$) are radiative corrections for beta, muon and strange particle decay respectively, and V_{ud} and V_{us} are the largest elements of the CKM quark mixing matrix.[2] The matrix element V_{ud} is derived from the measured values of the ft-values of the pure Fermi ($0^+ \to 0^+$) β-transitions and independently from the study of neutron decay.[4]. The Fermi coupling constant G_F has the value

$$G_F = 1.16639(2) \times 10^{-5} \text{ GeV}^{-2}$$

as determined from the measured lifetime of the muon with appropriate radiative corrections[2].

2. THE HADRONIC MATRIX ELEMENT FOR NEUTRON DECAY

The matrix element for the hadronic weak current can be written in the form

$$<p| J_\lambda^h(0) |n> = v_p [f_1(q^2)\gamma_\lambda + i f_2(q^2)\sigma_{\lambda\mu}q^\mu + g_1(q^2)\gamma_\lambda\gamma_5 + g_3(q^2)q_\lambda\gamma_5] v_n$$

where v_p and v_n are proton and neutron Dirac spinors respectively, and second class form factors have been omitted corresponding to G-parity invariance of the strong interactions[3]. Since the momentum transfer in neutron decay is ~1 MeV the q^2-dependence of the vector ($f_i(q^2)$, i=1,2) and axial vector ($g_i(q^2)$, i=1,3) form factors can be neglected. Also since the induced pseudo-scalar form factor g_3 makes a negligible contribution to all electronic decays and the weak magnetism form factor f_2 only makes a contribution of order δ and has yet to be detected in neutron decay, it follows that the hadronic matrix element is a function of the form factors $f_1(0)$ and $g_1(0)$ alone. Since the conserved vector current requires that the vector coupling to the nucleon be identical with the coupling to the quark and therefore $f_1(0)=1$, it follows that the decay can be characterized by G_V together with the ratio

$$\lambda = G_A/G_V = g_1(0)/f_1(0)$$

In the convention whereby the operator $(1-\gamma_5)/2$ projects out left-handed fields, the parameter λ is negative with a best estimate [2]

$$\lambda = -1.2601 \pm 0.0025$$

3 DECAY OF A POLARIZED NEUTRON

We can now write down the standard expression[5] for the transition rate of a polarized neutron

$$dW(\mathbf{p_e}, \mathbf{p}_{\bar{\nu}} | <\sigma>) = dW(p_e)d\Omega_e d\Omega_{\bar{\nu}} [1 + a (\mathbf{p_e}/E_e)\cdot(\mathbf{p}_{\bar{\nu}}/E_{\bar{\nu}}) + <\sigma> \cdot \{ A(\mathbf{p_e}/E_e)$$
$$+ B(\mathbf{p}_{\bar{\nu}}/E_{\bar{\nu}}) + D((\mathbf{p_e}/E_e) \times (\mathbf{p}_{\bar{\nu}}/E_{\bar{\nu}}))\}]$$

where

$$A = -2(|\lambda|^2 + \text{Re}\{\lambda\})/(1 + 3|\lambda|^2), \qquad B = 2(|\lambda|^2 - \text{Re}\{\lambda\})/(1 + 3|\lambda|^2)$$

are P-violating, T-conserving electron and neutrino spin asymmetry coefficients respectively, and

$$D = 2 \text{Im}\{\lambda\} / (1 + 3|\lambda|^2)$$

is the P-conserving, T-violating triple correlation coefficient. Experimentally D is zero[2] showing that λ is real and neutron β-decay is a T-invariant process.

The electron-neutrino angular correlation coefficient

$$a = (1 - |\lambda|^2)/(1 + 3|\lambda|^2)$$

is a P-and T-conserving parameter which does not depend on vector-axial vector interference. The same is true of the neutron lifetime $\tau_n = t_n \ln(2)$ which is given by the comparative half-life formula

$$f t_n = (2 \ln(2) \hbar^7 / m_e^5 c^4) / (G_V^2 |f_1(0)|^2 \{1 + 3|\lambda|^2\})$$

where f is the integrated phase space factor which, including outer radiative corrections, has the value[6]

$$f = 1.71465 \pm 0.00015$$

The conventional method for determining G_V and $G_A = \lambda G_V$ from neutron decay data is to combine measurements of A and τ_n[7].

4 MEASURING THE NEUTRON LIFETIME USING AN ION TRAP

4.1 Principle of the Method

Two quite distinct techniques have been developed to measure the neutron lifetime. The modern method is based on observation of the exponential decay of ultracold neutrons, i.e. neutrons of energy $\leq 2 \times 10^{-7}$ eV, which are trapped in a fixed volume of space by material barriers[8] or inhomogeneous magnetic fields[9]. Attractive as this method sounds, in practice it is beset with difficulties [10] associated with the technical problem of identifying all possible mechanisms for neutron loss other than β-decay. However in recent years a large number of ingenious techniques have been developed to deal with these problems[8] and projects are now under way which aim to achieve precisions at the level of 0.1% and better[11].

The traditional method, which includes the ion trap variant[12,13] is based on the equation describing the rate of decay of neutrons in a beam emerging from a cold source in a high flux nuclear reactor:

$$dn(t)/dt = -n(t)/\tau_n$$

where dn(t)/dt is determined by counting the number of electrons or protons emerging from a known volume of beam and n(t) is found from absolute measurements of the integrated density of neutrons in that volume. In carrying out this kind of measurement we may identify four difficulties which must be addressed: (i) Should the experiment aim to detect electrons or protons and which detectors should be used ? (ii) How are the genuine events associated with the weak interaction to be distinguished from the potentially overwhelming background associated with the strong and electromagnetic interactions? (iii) How is the source volume to be defined and (iv) how is the neutron density to be measured?

In the ion trap technique these problems are solved in the following manner : (i) protons are detected because, having energies ≤ 0.75 keV, they can be collected and stored in an ion trap and accelerated to energies of order 30 keV after release. They can then be counted with 100% efficiency in a silicon surface barrier detector. (ii) Since electrons and γ-rays are not trapped, the background due to neutron-nucleus interactions can be suppressed in the ratio of counting time to trapping time which can be set at a level of order 10^{-3}. (iii) The source volume is that entire length of neutron beam which traverses the ion trap whose length can be varied over a pre-selected range to eliminate uncertainties associated with end effects. Thus the ion trap behaves as a 4π- counter. (iv) The neutron density is measured by counting α-particles emitted in the reaction $^{10}B(n,\alpha)^7Li$. Since ^{10}B has a neutron capture cross-section which varies as $1/v_n$, the counting rate measures the neutron density rather than the flux [14]. It is important to recognise that that the combined uncertainty in the capture cross-section and the precision with which the amount of ^{10}B in the target can be assayed sets a lower limit on the error with which the neutron lifetime can be determined by this or any other version of the neutron beam technique. Currently this limit is around 0.3%.

4.2 Experimental Apparatus

The layout of the experimental apparatus is shown in figure 1. The beam of cold neutrons is transported from left to right along an axis which coincides with the axis of the proton trap, which itself is positioned at the centre of the superconducting magnet and aligned along the magnetic field at that location. The magnetic field is uniform at a value of 5 tesla to better than 1% within the trap and the maximum cyclotron orbit radius of the trapped protons is about 0.8mm. The whole system is constructed to ultra-high-vacuum standards and a very low pressure of residual gas is maintained by cryopumping onto the liquid helium cooled interior surfaces. The protons are trapped for periods up to 50 msec and on release exit in the backwards direction where they are accelerated and counted. The counting part of the cycle is competed within about 50 μsec. At the exit end of the trap where the 'gate' electrode is situated, the magnetic field reduces slowly to a value of about 4.75 tesla at the position of the detector to ensure that there are no magnetic mirror effects to hinder the exit of protons from the trap or to generate background from magnetically trapped low energy decay electrons.

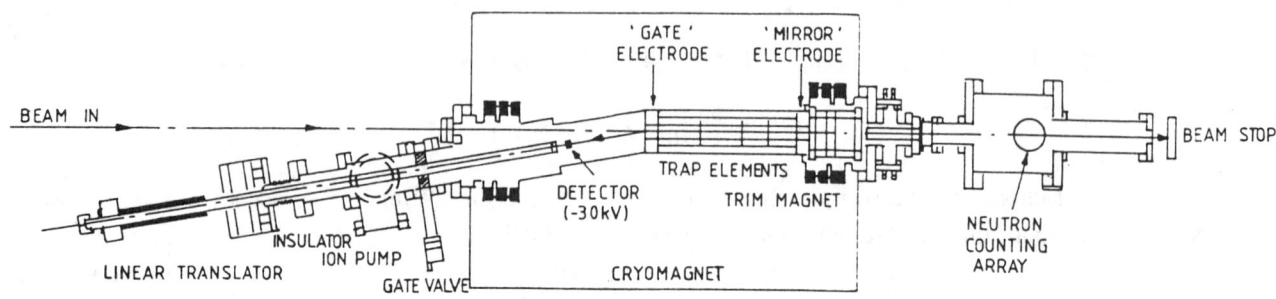

FIGURE 1. Cryomagnetic ion trap used for storing protons from neutron decay. The trim magnet used to shape the diverging magnetic field at the beam exit has been added to the original neutron lifetime apparatus for the proton spectrum measurement.

One should also be aware that, since the 'mirror" electrode, located near the point where the neutron beam leaves the system, is maintained at a potential of about 1 kV, it behaves as a potential well for electrons and is therefore lowered to ground at the end of each trap-emptying cycle in order to release any trapped electrons. Such a trapped electron can produce spurious events by ionization of residual hydrogen and this effect was observed.

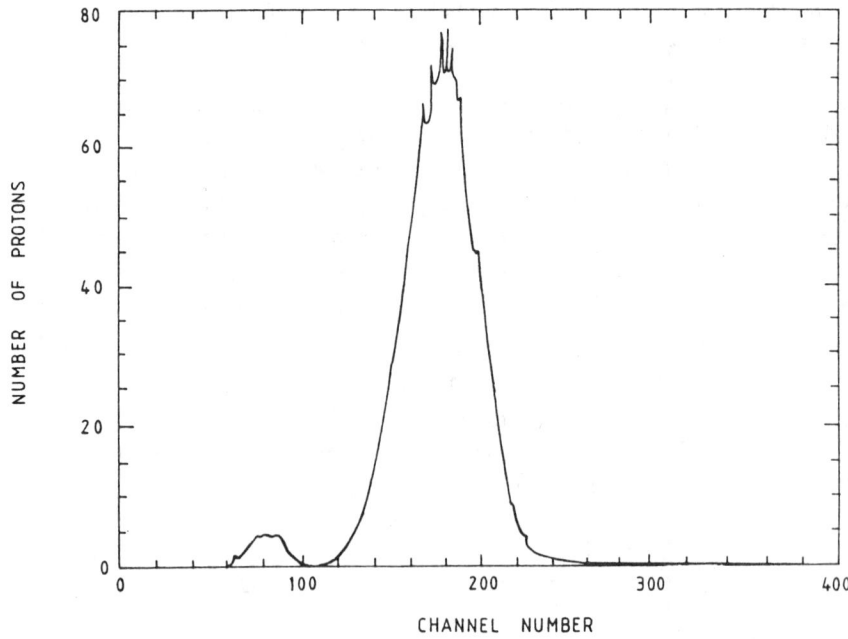

FIGURE 2. Spectrum of protons from neutron decay accelerated to 30 keV and observed in a silicon surface barrier detector.

The protons are detected in a 300 mm² silicon surface barrier detector with a 300μ depletion depth, which is necessarily placed outside the neutron beam and the guiding centres of the proton cyclotron orbits are channelled to it by bending the magnetic field through an angle of 90° away from the symmetry axis at the exit point. This detector and its associated preamplifier are maintained at a potential variable between -25 kV and -35 kV, and are enclosed in a beryllium oxide insulator to protect against electrical breakdown in the system of strong 'crossed' electric and magnetic fields which exists in this region. The signal is transmitted down to ground via an optic fibre link and the observed spectrum of trapped protons following post acceleration to an energy about 30 keV is shown in figure 2. The small peak in the extreme low energy region is the residue of the thermal noise in the detector. Because there was always some uncertainty as to where an event associated with the simultaneous trapping of two protons would appear in the spectrum the absolute number of decay events was determined, not from the energy spectrum, but from the time of arrival spectrum for which the deadtime correction could be performed unambiguously [12].

After leaving the cryomagnetic the neutrons traverse a single crystal of silicon which supports an assayed thin layer of separated ^{10}B isotope and the α-particles generated by neutron capture are counted in a system of four 500 mm² detectors for each of which the collection solid angle has been measured to an accuracy of 0.1%. The α-particle counting rate provides a measure of the neutron density n which corresponded to a capture flux of about 4×10^9 neutrons cm^{-2}sec^{-1}. This number is defined as the product of the neutron density times the velocity 2.2 km sec^{-1} of a thermal neutron, i.e. a neutron of wavelength 1.8 A° for which the capture cross section in ^{10}B has been measured to an accuracy of 0.125% [15]. In the experiments which were performed the cold neutron flux centred at a wavelength of about 4.5 A°.

4.3 Value of the Neutron Lifetime

In the neutron lifetime experiment measurements were carried out for a range of accelerating potentials varying between -24.5 kV and -34.5 kV and for trap lengths between 9 and 22 cms. The numbers of recorded trapped protons per detected neutron were then corrected for Rutherford scattering in the protective gold layer covering the counter, which itself was varied between 20 and 80 μgm cm^{-2}, and plotted as function of trap length. The slope of the resultant straight line then gave a value [12]

$$\tau_n = 893.5 \pm 5.3 \text{ sec.}$$

However, because it was decided to carry out a further experiment to measure the spectrum of decay protons, using a variant of the trapping technique incorporating the principle of adiabatic focusing[16] a very precise measurement of the profile of the magnetic field was carried out. As a result a small correction was inserted to take account of the weak variation of the magnetic field in the trapping region[13], the effect of which is to introduce a small difference between the effective trap length and the geometric trap length. The net result was a reduction of the neutron lifetime to a value which is now determined to be:

$$\tau_n = 889.2 \pm 4.8 \text{ sec.}$$

This value is in agreement with, but is less accurate than, the most recent values based on the observation of decay rates of ultracold neutrons confined in bottes [2]. However the method employed has the advantage that the systematic errors are completely different and are, on the whole, better understood [10].

5 MEASUREMENT OF THE PROTON SPECTRUM

5.1 Motivation for the Experiment

It has been observed in section 1 that the prime motive for carrying out precise measurements on the decay of the neutron is to allow an accurate determination of the coupling constants G_A and G_V. Although G_A is itself of great practical and theoretical significance, because, e.g., it determines the rate at which the proton-proton-cycle of thermonuclear reactions proceeds in the sun[17] and, via the parameter λ, it provides an important test of models of nucleon structure[18], it is the determination of G_V that has attracted the most attention[19]. This is because its value is critical to the measurement of V_{ud}. In this context two questions arise: (i) are the current measured values of V_{ud} consistent with unitarity of the CKM matrix within the quoted errors? To this question the answer appears to be no (ii) To what degree are the values of G_V determined from the ft-values of pure Fermi transitions and from neutron β-decay mutually consistent and what, if any, experimental uncertainties or ambiguities remain to be resolved in these two experimental techniques? Although it does appear to be the case that the results of both techniques are consistent, nevertheless in both methods there remain some outstanding problems [19].

In this paper we shall however be concerned with neutron decay only where the question arises as to why the measured values of the electron asymmetry coefficient A have fluctuated so dramatically in recent years[20]. This is vividly illustrated in reference [2] where the apparently inexorable rise in the measured values of the parameter $|\lambda|$ mirrors the almost continuous fall in the measured values of τ_n which was a notable problem during the period up to about 1986 [21]. We have noted that the determination of A requires that the decaying neutrons be polarized and some question must arise whether this polarization is a function of wavelength and geometry, and whether the method of its determination conceals some unsuspected systematic errors.

There is however another route towards the determination of $|\lambda|$ which has not been fully explored, namely the measurement of the electron-neutrino angular correlation coefficient a which is defined in section 3. Since this is a parity conserving correlation which does not contain interference terms proportional to λ, its measurement therefore does not require that the neutrons be polarized. Also it shares the common feature with the coefficient A that, rather than being proportional $|\lambda|$, it measures the anomaly $(|\lambda|^2 - 1)$ due to the renormalization of G_A by the strong interactions. The most accurate method for determining a relies on a measurement of the proton kinetic energy spectrum g(E) in neutron decay which has the form [22]

$$g(E) = g_1(E) + a.g_2(E), \quad 0 \leq E \leq 0.75 \text{ keV},$$

where the function $g_1(E) \geq 0$ rises to a maximum near the middle of the spectrum in a region where the function $g_2(E)$ changes sign from negative to positive. Unfortunately the current value $a = -0.1017\pm0.0051$ [23] leading to the value $|\lambda| = 1.259\pm0.017$, indicates a range of error which encompasses all the current values of $|\lambda|$ derived from measured values of A. It is also unclear whether the authors of reference [23] were aware of a misprint which has been noted [24] in formula (4.5) of reference [22].

5.2 Ion Trap method for Measuring *a*

We have investigated two possible ways of measuring a using an ion trap[16]. In the simplest method the electrostatic potential on the mirror electrode is set at different values of V_m and the number $N(V_m)$ of protons trapped at this setting is recorded. This gives the integrated one-dimensional proton spectrum since protons are trapped if the energy in their longitudinal degree of freedom is less that eV_m. Of course over many cycles of oscillation in the trap the energy will be reduced to some degree due to collisions with residual gas atoms. However this is not important because once a particle is trapped on its first cycle of longitudinal motion it remains trapped and of much greater significance are the changes in the longitudinal component of the kinetic momentum due to inhomogeneities in the magnetic field. However even allowing for these effects the real problem is that there is no maximum in the one-dimensional spectrum which is concentrated at low energies and the sensitivity to the value of a is low.

In the second variant we proposed to exploit the conservation of the adiabatic invariant [25] associated with the cyclotron motion of the trapped proton to convert the energy in the transverse degrees of freedom into longitudinal energy, an adiabatic focusing action which is the exact inverse of the magnetic mirror effect. The conversion of transverse energy $E_T(0)$ at the origin, into longitudinal energy $E_L(z)$ at the point z on axis, is given by the usual magnetic mirror formula:

$$E_L(z) = E_L(0) - E_T(0)\{B(z) - B(0)\}/B(0),$$

so that this change is positive and consistent with overall energy conservation when $B(z) < B(0)$. This energy transfer is brought about by locating the mirror electrode in a region of space where the magnetic field has a value approximately equal to 14% of the uniform field in the centre of the trap. In principle of course one could place the mirror in zero magnetic field, in which case $E_L(z) = E_L(0) + E_T(0)$, but one must bear in mind that the proton beam diameter grows to a size determined by the conservation of the adiabatic invariant associated with magnetron drift, and a suitable compromise is necessary. In the system which was eventually decided on some 93% of the total energy appears in the longitudinal mode and this leads to a dramatic increase in the dependence of the quasi-three dimensional spectrum on $|\lambda|$. The generation of a flat field in the vicinity of the mirror electrode is effected by the insertion of a permanent magnet in the position shown in figure 1.

The resultant measured field profile over the whole of the active volume of the trap is shown in figure 3. In the region of high magnetic field close to the gate electrode protons generated from neutron decay can travel down to the mirror electrode where they may or may not hit the barrier depending essentially on their total energy. Protons generated in the region of low magnetic field near the mirror electrode may be trapped behind the barrier, but a negligible proportion reach the detector because they cannot climb the huge magnetic barrier presented to them by the rapidly rising magnetic field. In the intermediate region where the field is changing, most neutron decays will not be recorded but protons emitted into a small solid angle around the magnetic axis may reach the detector.

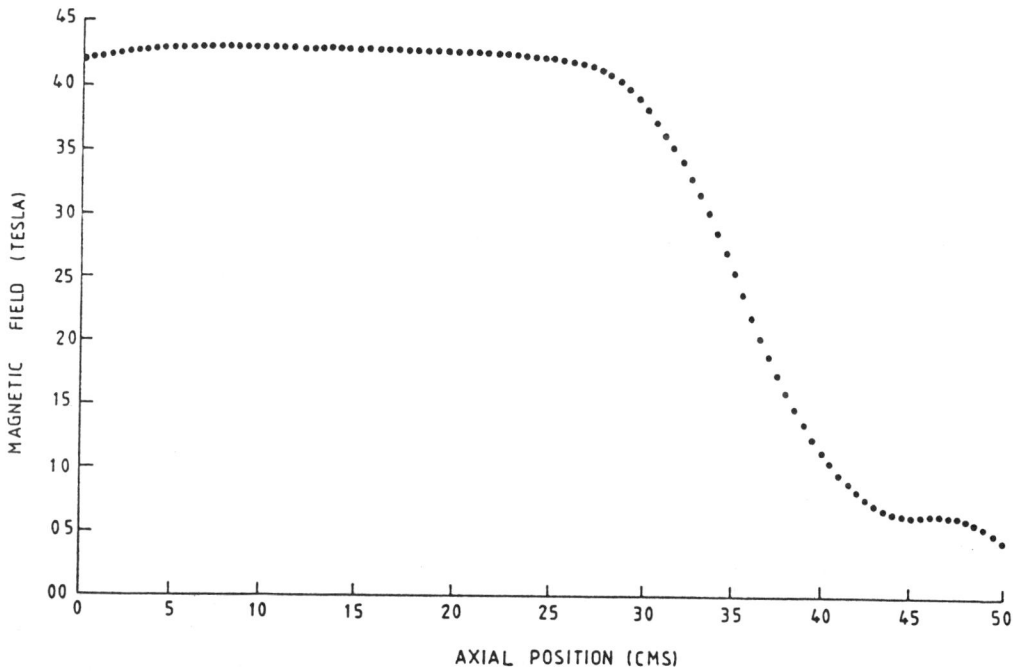

FIGURE 3. Distribution of magnetic field in the proton trap. The 'mirror' electrode is located in the region of low uniform magnetic field generated by the addition of a permanent trim magnet.

To avoid the necessity of computing the effect of varying efficiency of detection over the whole trap we have measured two spectra; the spectrum of trapped protons collected from the full trapping region from high field to low field (the 'long' trap) and the spectrum of trapped protons, from a region defined by moving the gate electrode so that only a very small fraction of the high field region has been sampled (the 'short' trap). The difference between these spectra then measures the integrated spectrum of protons generated in the region of high constant magnetic field.

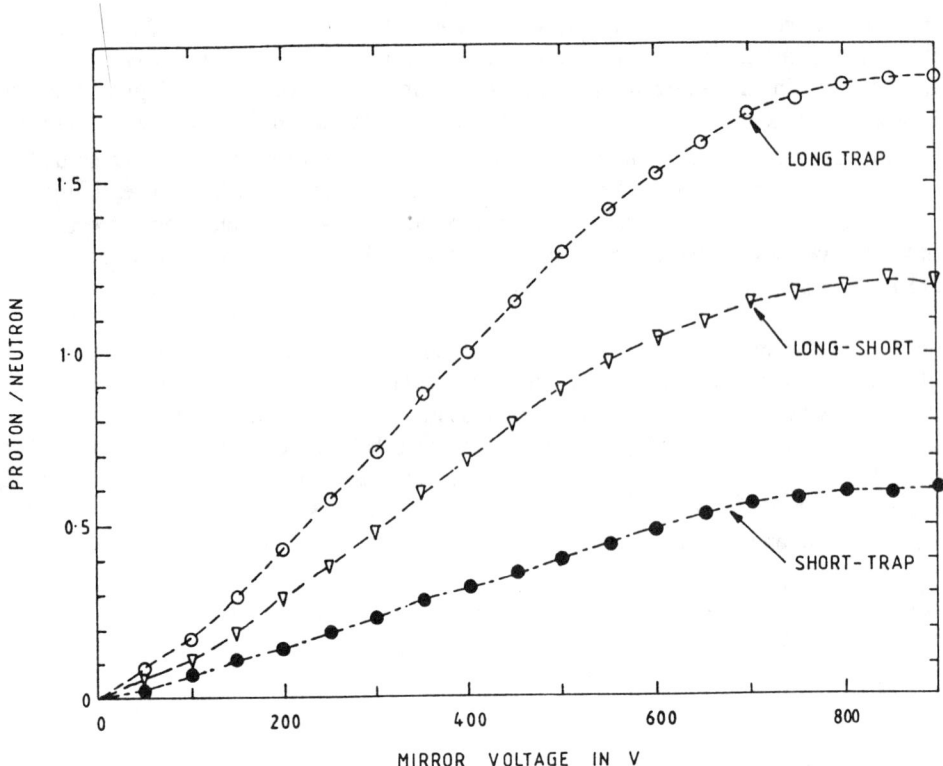

FIGURE 4. Quasi-three dimensional integrated proton spectra observed for 'long' and 'short' traps respectively. The difference spectrum is generated by events occurring in the region of uniform high magnetic field.

5.3 Preliminary Results

The experiment was first performed on beam PF1 at the Institut Laue-Langevin in Grenoble, France, during two cycles of running (~100 days) in the Summer and Autumn of 1997. Data were accumulated for both 'one-dimensional' and 'three dimensional' spectra, over a range of trapping times between 1 msec and 50 msec, and comparison of the results with Monte Carlo calculations rapidly established that the adiabatic focusing principle worked well. However there were problems associated with both short and long trapping times. With long trapping times (>30 msec) dead time effects associated with trapping of one or more protons become important and with short trapping times (< 5 msec) the number of events sampled from outside the fiducial volume when the trap is open became significant.

The results for both long and short traps, accumulated for a trapping time of 10 msec, and the difference between these, are shown in figure 4. The main problem encountered was the very low count rate recorded which reached a maximum value of about 3 sec^{-1} and correspondingly smaller values at the lowest values of V_m. Thus the statistics were insufficient to provide a useful result. However the experiment has been restarted at Grenoble during the past week using an improved collimating system which we hope will provide a significant increase in the number of transmitted neutrons, leading to a result with a precision in $|\lambda|$ at the level of 1%.

ACKNOWLEDGEMENTS

The project to measure the proton spectrum is supported by grants from the U.K. Particle Physics and Astronomy Research Council and is a collaboration with M. van der Grinten, C.G. Habeck, F. Shaikh, J.A. Spain (Sussex), R.D. Scott (Glasgow), C.A. Baker, K. Green (RAL), and O. Zimmer (ILL).

REFERENCES

[1] Holstein B.R., *Weak Interactions in Nuclei* (Princeton University Press 1989)
[2] Review of Particle Physics, *Phys.Rev.D* **54**, 1 (1996)
[3] Dubbers D., *Prog.Part. Nucl. Phys.* **26**, 173 (1991)
[4] Wilkinson D.H., *Z.Phys.A* **348**, 129 (1994)
[5] Jackson J.D., Treiman S.B. and Wyld H.W., *Phys.Rev.* **106**, 517 (1957)
[6] Wilkinson D.H., *Nucl.Phys.A* **377**, 474 (1982)
[7] Dubbers D. et al., *Europhys.Lett.* **11**, 195 (1990)
[8] Kostvintsev Y.Y. et al., *JETP Lett.* **44**, 571 (1986);
  Mampe W. et al., *Phys.Rev.Lett.* **52**, 373 (1990)
  Nesvizhevskii V.V. et al., *Sov.Phys. JETP* **75**, 405 (1992)
  Mampe W. et al., *JETP Lett.* **57**, 82 (1993)
  Arzumanov S. et al., *JINR, Dubna* (1997) Report E3-97-213
[9] Kugler K.J. et al, *Phys.Lett. B* **72**, 421 (1978)
  Anton F. et al., *Z.Phys.C* **45**, 373 (1989)
[10] Schreckenbach K. and Mampe W., *J.Phys.G* **18**, 1 (1992)
  Yerozolimsky B., *Contemp.Phys.* **35**, 191 (1994)
[11] Doyle J.M. and Lamoreaux S.K., *Europhys.Lett.* **26**, 253 (1994)
[12] Byrne J. et al., *Phys.Rev.Lett.* **65**, 289 (1990)
  Werth G., *J.Phys.G* **20**, 1865 (1994)
  Byrne J., *Phys.Script.*, **T59**, 311 (1995)
[13] Byrne J. et al., *Europhys.Lett.* **33**, 187 (1996)
[14] Pauwels J. et al., *Nucl.Instr.and Meth. A* **303**, 133 (1991)
  Scott R.D. et al., *Nucl.Instr. and Meth. A* **314,** 163 (1992)
[15] Holden N.E., '*Neutron Capture Cross Section Standards for BNL 325*' 4th Edition. BNL-NCS-51388- UC-346 (1981)
[16] Byrne J., Dawber P.G. and Lee S.R., *Nucl.Instr.and Meth. A* **349**, 454 (1994)
[17] Bahcall J.N. et al., *Rev.Mod.Phys.* **54**, 767 (1982)
[18] Thomas A.W., *Adv. Nucl.Phys.***13**, 1 (1983)
  Myhrer F. and Wroldsen J., *Rev.Mod.Phys.* **60**, 629 (1988)
[19] Deutsch J.P. *Proc.WEIN Conf. Santa Fe* (1998) (in press)
[20] Bopp P. et al., *Phys.Rev.Lett.* **56**, 919 (1986)
  Schreckenbach K. et al., *Phys.Lett.B* **349**, 427 (1995)
  Yerozolimsky B. et al., *Phys.Lett. B* **412**, 240 (1997)
  Abele H. et al., *Phys.Lett. B* **417**, 212 (1997)
[21] Freedman S, *Comm. Nucl.Part.Phys.***19**, 209 (1990)
[22] Nachtmann O., *Z.Phys.* **215**, 505 (1968)
[23] Stratowa C. et al., *Phys.Rev.D* **18**, 3970 (1978)
[24] Habeck C.G., *University of Sussex D.Phil.Thesis* (1997) (unpublished)
[25] Northrup T.G. and Teller E., *Phys.Rev.* **117**, 215 (1960)

An Electromagnetic Ion Trap for Studies in Nuclear Beta Decay

D. Beck[a], M. Beck[a], G. Bollen[c], J. Deutsch[b], J. Dilling[d], T. Phalet[a], P. Schuurmans[a], R. Prieels[b], W. Quint[d], N. Severijns[a], B. Vereecke[a], S. Versyck[a], and the EUROTRAPS[1] Collaboration[d]

[a] *Instituut voor Kern- en Stralingsfysica, KU Leuven, Celestijnenlaan 200 D, B-3001 Leuven, Belgium*

[b] *Institut de Physique Nucleaire, UCL, Chemin du Cyclotron 2, B-1348 Louvain-la-Neuve, Belgium*

[c] *CERN, CH-1211 Geneva 23, Switzerland*

[d] *GSI-Darmstadt, Postfach 110552, D-64220 Darmstadt, Germany*

Abstract. Traps are especially well suited for the investigation of weak interactions as they allow the detection of the decay products without interference from any supporting material. Here we report on the development of an electromagnetic ion trap for correlation experiments in nuclear beta decay. An ion trap was chosen over an optical trap because it is not as limited in the choice of elements that can be contained. At present, we are studying different types of traps and their possible configurations in the context of the set of observables that will give the highest sensitivity for a test of the electroweak standard model.

INTRODUCTION

Scalar currents in the weak interaction would occur if a charged scalar boson or a scalar leptoquark were exchanged instead of a W boson. They enter linearly in the Fierz interference term but the constraints which can be deduced from this quantity depend critically on the helicity-structure of the scalar coupling [1,2]. The limits from nuclear beta decay experiments on possible scalar currents are still rather poor. Their coupling constants C_S and C'_S can be as large as 17 % of the vector coupling constant C_V [2]. This is mainly due to the fact that scalar couplings must in general be inferred from observables in which they enter quadratically, e.g. the beta-neutrino correlation [1]. The experimental determination of the latter is very difficult. Firstly, the neutrino can not be detected directly. Hence, one must study the beta particles and the recoil ions, where the kinetic energy of the latter is usually very small (of the order of a few hundred electron-Volts only). Secondly, one must be sure that the recoil particle does not form a molecular state with other atoms. Indeed, this would alter the recoil energy as severely as when the decaying nucleus was embedded in a solid state matrix. Most beta-neutrino correlation experiments were therefore performed with a gaseous source. Recently, two new approaches were introduced. The first determines the kinematics of the recoil ion by measuring the Doppler shift of a gamma ray that is emitted by the recoiling nucleus and detected in coincidence with the beta particle [3]. The second one measures the recoil broadening of the beta-delayed protons emitted after pure Fermi beta decay [4]. The enhanced sensitivity of beta-delayed particle emitters was already stressed by [5]. The most precise result up to now is obtained from the recent measurement of the beta-neutrino correlation in the decay of ^{32}Ar [4], where the second approach was used and a precision of about 1 % was obtained on the beta-neutrino

---

[1] Eurotraps is a European Network for Ultra-High Precision Spectroscopy of Highly Charged, Stored, and Cooled Ions

correlation coefficient. Here we report on still another approach that is being developed now and which will use electromagnetic ion traps to further improve on the precision for beta-neutrino correlation measurements.

WHY USE TRAPS FOR WEAK INTERACTION STUDIES

As proven over the last decade with the ISOLTRAP mass spectrometer [6,7] at ISOLDE/CERN [8], electromagnetic ion traps are a powerful tool for studies with unstable nuclei. They permit to store and cool ions for an extended period of time, providing very well localized sources of radioactive ions with almost zero thickness. One of the major systematic uncertainties in most nuclear beta decay experiments is scattering in the source and the intermediate medium between the source and the detectors. Both components may significantly be reduced by the use of traps. Since they store ions free of any solvent or host material, scattering in the source is avoided to a large extent. Scattering in the intermediate medium can be limited by an open construction with very little material around the trapped ion cloud and by having the trap electrodes as little massive as possible. Ion traps thus eliminate or at least significantly reduce several systematic errors that in general limit the precision of correlation measurements in nuclear beta decay. This, of course, also holds for the laser-driven atom traps, but ion traps are not restricted to the chemical or atomic properties of the species to be trapped. In addition, ion traps provide a higher capture efficiency. However, in the case of ion traps special care has to be taken about a possible perturbation of the charged particle trajectories by the electromagnetic fields.

POSSIBLE EXPERIMENTS USING ION TRAPS

We are currently investigating the possibilities for three different experimental set-ups using an RFQ or Penning trap for beta-neutrino correlation experiments.

The first option makes use of the angular correlation between the beta and the recoil particle. This correlation originates from the beta-neutrino angular correlation and thus depends on the coupling constants too. By selecting either the beta or the recoil, a forward-backward asymmetry is created and measured for the other particle. In this integrating experiment the energy of the two coincident particles does not need to be determined. The Ions could be stored and cooled in a RFQ trap with an open structure. A multiple-detector system covering a rather large solid angle, thereby optimizing statistics, could be constructed quite easily. Simulations are being carried out in order to investigate whether the rf-field can be switched off for a sufficiently long time. This would allow to limit the detection to this field-free period during which the recoil ion trajectories are not disturbed by the field.

In the second set-up the Doppler shift of a gamma ray emitted by the recoiling nucleus is determined in coincidence with the detection of the beta-particle [3]. From the point of view of detection this is easier than the first type of experiment since a direct detection of the recoil particle is not needed. In addition, a multiple detector system can again be used in order to increase statistics. One option here is the use of a Penning trap. Although a part of the angular information gets lost, the magnetic field can advantageously be used to guide the beta particles to their detectors. Thus a large overall efficiency is obtained. The drawback of the Doppler shift method is its limited applicability to a few specific cases like ^{18}Ne.

Finally, a third option being envisaged is to trap ions in a Penning trap and then transport the recoil ions to a detector using a solenoid retarding spectrometer of the type used by the Mainz group in the neutrino mass experiment with tritium [9]. This would allow to precisely measure the shape of the recoil spectrum, that depends on the coupling constants [10]. For recoil ions such a set-up could be much smaller than the one that is being used in the tritium experiment. The transmission of such an apparatus is in general quite high and scattering effects are negligible since the particles spiral along a particular magnetic field line within a single flux tube from the source to the detector. In addition, only a single detector is needed for the recoil ions. However, the transmission characteristics of such a spectrometer have to be be known very precisely.

SUMMARY

At present, we are simulating the trajectories of the radioactive decay products for the different suitable experiments. The ion trap set-up which we will finally choose for the beta-neutrino correlation experiments will in a later stage be extended/modified so that it can be used for the measurement of other observables/correlations in nuclear beta decay as well. If, for instance, the ions could be polarized using e.g. optical pumping, a whole range of experiments with spin-polarized nuclei would become possible. In this way different types of experiments in nuclear beta decay could be carried out yielding complementary information on the various weak interaction coupling constants.

REFERENCES

1. J. D. Jackson, S. B. Treiman, and H. W. Wyld, Nucl. Phys. 4 (1957) 206.
2. E. G. Adelberger, Phys. Rev. Lett. 70 (1993) 2856.
3. V. Egorov et al., Nuclear Physics A621 (1997) 745.
4. A. Garcia et al., Proc. of the Int. Symp. on Weak and Electromagnetic Interactions in Nuclei (Santa Fe, NM, U.S.A., June 1998), to be published.
5. E. T. H. Clifford et al., Nucl. Phys. A 493 (1989) 293.
6. G. Bollen et al., Nucl. Instr. and Meth. A 368 (1996) 675
7. G. Bollen et al., these proceedings.
8. E. Kugler et al., Nucl. Instr. and Meth. B 70 (1992) 41.
9. A. Picard et al., Nucl. Inst. and Meth. B 63 (1992) 345.
10. O. Kofoed-Hansen, Phys. Rev. 74 (1948) 1785.

ERRATUM: The following are corrected pages 175 and 176.

Parity Nonconservation in Relativistic Hydrogenic Ions

Max Zolotorev[a] and Dmitry Budker[a,b]

[a] E. O. Lawrence Berkeley National Laboratory, Berkeley, California 94720
[b] Department of Physics, University of California, Berkeley, California 94720-7300

Abstract. A technique is proposed for measuring parity nonconservation with ultra-relativistic hydrogenic ions in a storage ring. The sensitivity of such measurements could be sufficient for probing physics beyond the standard model.

It has long been a dream of atomic physicists to measure parity nonconservation (PNC) in hydrogen (for a review of hydrogen experiments, see (1)), or a hydrogenic system. It appears that recent developments in relativistic ion colliders, high-brightness ion sources, and laser cooling methods of ions in storage rings, may open such a possibility for relatively light (Z~10-40) hydrogenic ions (2). Due to their simple atomic structure, high precision theoretical calculations can be carried out in these ions. In addition, neutron distribution uncertainties ([3]) will not present a serious problem in relatively light ions considered here, both because the structure of light nuclei is much better understood than that of the heavy nuclei, and because the electron wavefunction gradient at the nucleus is relatively small for light nuclei.

Consider the conceptually simplest variant of a PNC experiment in a hydrogenic system: circular dichroism (i.e. the difference in transition rates for right- and left-circularly polarized light) on the 1S→2S transition in the absence of external electric and magnetic fields. Dichroism arises due to interference between the $M1$ and the PNC-induced $E1$ amplitudes of the transition.

Due to the PNC interaction, the 2S state acquires an admixture of the $2P_{1/2}$ state; the magnitude of the PNC admixture is probed by tuning the laser in resonance with the highly forbidden 1S→2S $M1$-transition and observing circular dichroism. Tracing Z-dependences of various atomic parameters, one finds that while the weak interaction matrix element increases $\propto Z^5$, the PNC asymmetry, i.e. the relative difference in absorption for the two circular polarizations, decreases $\propto Z^{-2}$, while the statistical sensitivity increases $\propto Z^4$ (2).

The frequencies of the 1S→2S transitions for the hydrogenic ions lie outside the range directly accessibly to laser sources. This problem can be solved by using relativistic Doppler tuning. For an ion with relativistic factor $\gamma = \dfrac{1}{\sqrt{1-\beta^2}} \gg 1$ colliding head-on with a photon of frequency ω_{lab}, the frequency in the ion's rest frame is given by:

$$\omega_{ion\,frame} = \gamma(1+\beta)\omega_{lab} \approx 2\gamma\omega_{lab}. \qquad (1)$$

In order to tune to the 1S→2S resonance for a hydrogenic ion, it is necessary to satisfy the condition:

$$\Delta E_{2S-2P} \approx Z^2 \cdot 10.2 \text{ eV} = 2\gamma\hbar\omega_{lab}. \qquad (2)$$

For example, using the Relativistic Heavy Ion Collider (RHIC, $\gamma \approx 100$), with visible and near-UV lasers, it is possible to access 1S→2S transitions for ions with Z up to ≈ 11 (Na).

Evaluating the statistical sensitivity of the experiment, one arrives at the following expression:

$$\delta H_w = \frac{1}{4}\sqrt{\frac{\Gamma_D \Gamma_{2P}}{\dot{N}_{ions} T \chi_{E1}}}. \qquad (3)$$

Here $\Gamma_D \approx \omega_{ion\,frame} \cdot \frac{\Delta\gamma}{\gamma}$ is the Doppler width, $\dot{N}_{ions}$ is the average number of ions entering the interaction region per unit time, T is the overall measurement time, and χ_{E1} is a dimensionless saturation parameter. This shows that for an optimally designed PNC experiment, the statistical sensitivity is completely determined by the total number of available ions and by the transition widths. In order to obtain a certain sensitivity to weak interaction parameters, e.g. to $sin^2\theta_w$, where θ_w is the Weinberg angle, it is necessary to have exposure $\dot{N}T$ which can be represented in units of particle-Amperes×year:

$$Exposure[part.\,Amp \times year] \geq \frac{\Delta\gamma}{\gamma} \cdot \frac{0.1}{Z^4 \cdot (\delta \sin^2\theta_w)^2 \cdot \chi_{E1}}. \qquad (4)$$

As an example, for $\delta\sin^2\theta_w = 10^{-3}$, using Ne ions (Z=10) in RHIC, and substituting $\chi_{E1} = 6 \cdot 10^{-2}$ (which is found to be an optimal value limited by laser photoionization), one obtains the necessary running time ~ 1 week. In this estimate we assumed $\Delta\gamma/\gamma=10^{-6}$, which is possible to achieve using laser cooling (4,2).

Many technical problems would have to be addressed before a PNC experiment could be carried out. This includes development of a hydrogenic ion source for the accelerator, implementation of laser cooling, design of an efficient detection scheme for ions excited to the 2S state, etc. However, all of these problems appear, at least in principle, tractable, and the proposed technique may offer sensitivity sufficient for testing physics beyond the standard model (2).

REFERENCES

1. Hinds, E., in *The Spectrum of Atomic Hydrogen: Advances*, ed. G. W. Series, Singapore: World Scientific, 1988, ch. 4, pp. 278-287.
2. Zolotorev, M., and Budker, D., Phys. Rev. Lett. **78**(25), 4717 (1997).
3. Fortson, E. N., Pang, Y., and Wilets, L., Phys. Rev. Lett. **65**, 2857 (1990); Pollock, S. J., Fortson, E. N., and Wilets, L., Phys. Rev. C **46**, 2587 (1992).
4. Habs, D., Balykin, V., Grieser, M., Grimm, R., Jaeschke, E., Music, M., Petrich,W., Schwalm, D., Wolf, A., Huber, G., and Neumann, R., in *Electron Cooling and New Cooling Techniques*, R. Calabrese and L. Tecchio, eds, Singapore: World Scientific, 1991.

Applications of Nonlinear Magneto-Optic Effects with Ultra-Narrow Widths

Valeriy Yashchuk[a,b], Dmitry Budker[a,c], and Max Zolotorev[c]

[a] *Department of Physics, University of California, Berkeley, California 94720-7300*
[b] *B. P. Konstantinov Petersburg Nuclear Physics Institute, Gatchina, Russia 188350*
[c] *E. O. Lawrence Berkeley National Laboratory, Berkeley, California 94720*

Abstract. A three-axis magnetometer based on the nonlinear magneto-optic effect is described. The magnetometer was used to measure the average (over the cell volume) residual magnetic field and to determine the shielding ratio of a multi-layer magnetic shield. The shot-noise-limited sensitivity of the magnetometer is estimated to be $\approx 10^{-11}$ *Gs/Hz$^{1/2}$*. A possible application to parity and time reversal invariance violation experiments is also considered.

INTRODUCTION

In this paper we discuss applications of the recently observed nonlinear (in light power) magneto-optic effects (NMOE) with effective resonance width $\gamma = g\mu\Delta B_Z \approx 2\pi \cdot 1$ Hz (1). Here g is the Lande factor, μ is the Bohr magneton, and ΔB_Z is the peak-to-peak width of the dispersion-like magnetic field dependence of NMOE. The small value of γ corresponds to an enhancement of small-field optical rotation ($\varphi_s \propto g\mu B_Z/\gamma$), making it useful in low-magnetic field measurements.

NMOE related to the interaction of near-resonant light with atomic vapor in the presence of a magnetic field have been the subject of a number of recent investigations reviewed in (2). Most of the work (see e.g. (3,4,5,6,7)) addresses NMOE in the case of linearly-polarized light propagating along the magnetic field known as nonlinear Faraday rotation. The schematic of an experiment to observe this effect shown in Figure 1 is similar to that of M. Faraday (8), in which he discovered the linear magneto-optical rotation (with non-resonant light), named after him as the Faraday effect. In the experiment, one observes rotation of the linear polarization plane of light as it passes through a medium exposed to a longitudinal magnetic field. In our case, the medium is Rb vapor contained in a paraffin-coated cell with no buffer gas. The light frequency dependence of magneto-optic rotation in the vicinity of a resonance absorption line has a characteristic resonant profile (the Macaluso-Corbino effect (9)). In a complementary picture, the magnetic field dependence of the Faraday effect with a fixed light frequency, one observes several nested dispersion-like features with vastly different widths. This is illustrated in Figure 2 (1). The feature represented in Figure 2 by the overall slope and reaching a maximum at several Gauss is due to hole burning in the velocity distribution of ground state atoms induced by velocity selective optical pumping. The Faraday rotation with such velocity distribution can be thought of as rotation produced by the Maxwell-distributed atoms without the hole (with the Macaluso-Corbino spectral profile) *minus* the rotation that would have been produced by the pumped out atoms. While the effective width of the Macaluso-Corbino effect corresponds to the Doppler-broadened line width (300 *MHz*), the hole burning effect has effective width ~6 *MHz* determined by the natural width of the excited state. Significantly narrower features arise due to long-lived light-induced alignment of the atomic ground state (10) resulting from the process known as coherent population trapping (11). An ensemble of aligned atoms constitutes a medium with linear dichroism. In the presence of a magnetic field, the dichroism axis precesses around the direction of the field with the Larmor frequency. The effect of this precession on the light polarization can be understood if one thinks of the atomic medium as a layer of dichroic polarizing material rotating in the magnetic field, see e. g. (5,1). According to this picture, the effective width γ is determined by the alignment relaxation rate. For sufficiently low light power, this model allows one to achieve quantitative description of dispersion-like shaped Faraday rotation due to the coherence effect as well as of the characteristic behavior of nonlinear optical rotation in the presence of transverse magnetic fields (1,12,13). In Figure 2, the coherence effect is represented by two dispersion-like features corresponding to different processes leading to the relaxation of the ground state alignment. The $\Delta B_Z \approx 120$ *mGs* width of the broader structure is determined by the atoms' transit time through the laser beam. The aligned atoms fly out of the laser beam, while "fresh" atoms from the volume of the cell replace them. This effectively provides for the alignment relaxation. The inserts in Figure 2 show the narrowest NMOE features observed so far (1). Their effective

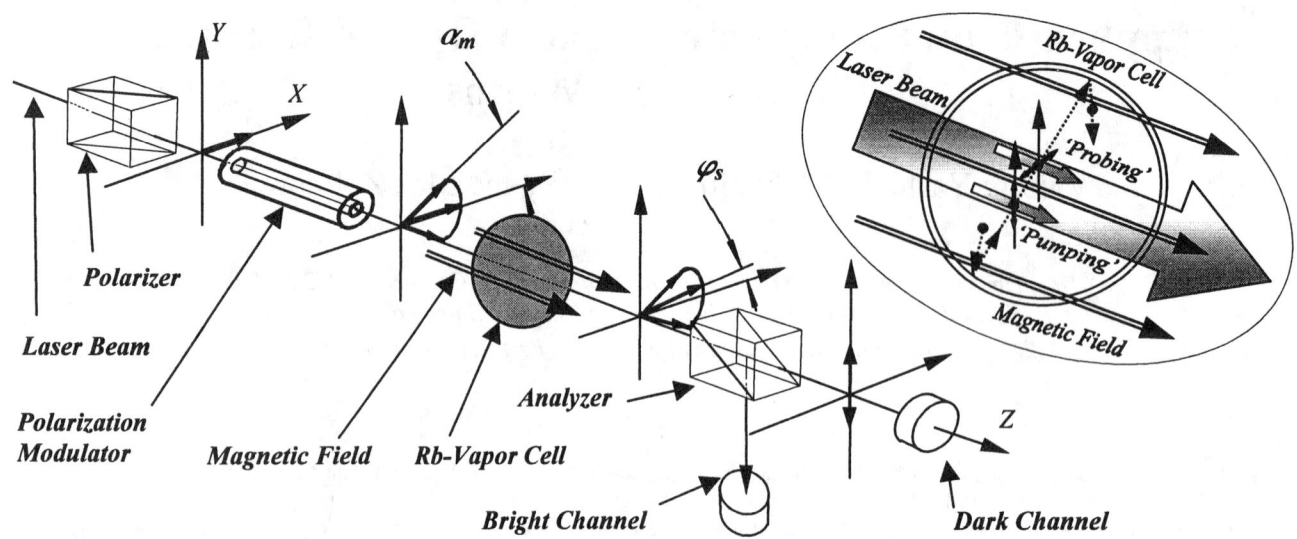

FIGURE 1. Schematic of an experiment to observe NMOE with Rb vapor. The Rb atoms aligned by linearly polarized light represent a magneto-optically active medium in a magnetic field. The measure of the polarization rotation angle is the intensity of light passing through the analyzer crossed with the polarizer. The details of the modulation polarimeter are described in the text. The inset illustrates the mechanisms for the coherence effect.

width is $\gamma = g\mu\Delta B_Z/\hbar \approx 2\pi \cdot 1.3$ Hz, where $\Delta B_Z \approx 2.8$ μGs. The width is determined by spin-exchange collisions among Rb atoms in the cell. The corresponding relaxation time, 250 ms, shows that the alignment is preserved over several thousand wall-collisions.

In this paper, we consider an application of nonlinear magneto-optic effects with ultra-narrow (~1 Hz) widths for three-axis magnetometry. The sensitivity of a possible experiment to search for a parity (P) and time reversal invariance (T) violating electric dipole moment (EDM) in cesium is also estimated based on the sensitivity to the magnetic field.

THE LOW-FIELD MAGNETOMETER

The experimental set-up was described in (1). It incorporates a paraffin-coated cell with ^{85}Rb-vapor at room temperature and a modulation polarimeter - Figure 1. The cell (diameter d=10 cm) has high quality paraffin coating (14), ensuring negligible relaxation of alignment in wall collisions compared to spin-exchange relaxation (1). The cell was surrounded with three mutually perpendicular (3-D) magnetic coils. The coils allow compensation of the residual magnetic field and application of an arbitrarily directed well-controlled field to the cell. An external cavity diode laser is used. In the following we describe experimental results and estimate the magnetometer sensitivity for the $F_g=3 \rightarrow F_e=2,3,4$ transition of the D_2 ($^2S_{1/2} \rightarrow {}^2P_{3/2}$; λ=780.2 nm) resonance line of ^{85}Rb. The optical thickness of the vapor at a temperature of 20.5 °C was measured to be $d/l_0 \approx 1.4$ for the center of the absorption line. Here l_0 is the unsaturated absorption length. The modulation polarimeter (see Figure 1) incorporates a crossed Glan prism polarizer and a polarizing beam splitter used as an analyzer (extinction ratio $\varepsilon < 5 \cdot 10^{-5}$). A Faraday glass element modulates the direction of the linear polarization of the light at a frequency $\omega_m \approx 2\pi \cdot 1$ kHz with an amplitude $\alpha_m \approx 5 \cdot 10^{-3}$ rad. The first harmonic of the signal is detected with a lock-in amplifier. Its amplitude, normalized to the transmitted light intensity detected in the bright channel of the analyzer, is a measure of the magneto-optical rotation in the vapor cell, φ_s.

In a magnetometric measurement using this technique, we record the dependences of φ_s on longitudinal magnetic field B_Z, scanning an appropriate current in the 3-D coils. Depending on the value of measured field, one of the nested dispersive features, like those shown in Figure 2, is observed. The feature shape and horizontal offset depend strongly on the value and direction of the magnetic field being measured. Fitting a series of experimental curves at different bias currents in the magnetic coils with the model developed in (1) and briefly described in the Introduction we determine all three Cartesian components of magnetic field. The highest sensitivity of the magnetometer is achieved by realizing the effect of alignment preservation in collisions of Rb atoms with paraffin-coated cell walls. To observe this effect, the cell surrounded with 3-D coils is placed inside a four-layer magnetic shield, shown in Figure 3. The shield was manufactured from 0.040" thick CONETIC-AA sheets. The three outer layers of the shield are cylinders with conical lids, while the innermost layer is a cube with rounded edges. The nearly spherical shape of the three outer layers is intended to provide nearly isotropic shielding of an external DC field. Indeed, the shielding ratio for multi-layer shield depends strongly on the shield shape as well as on the

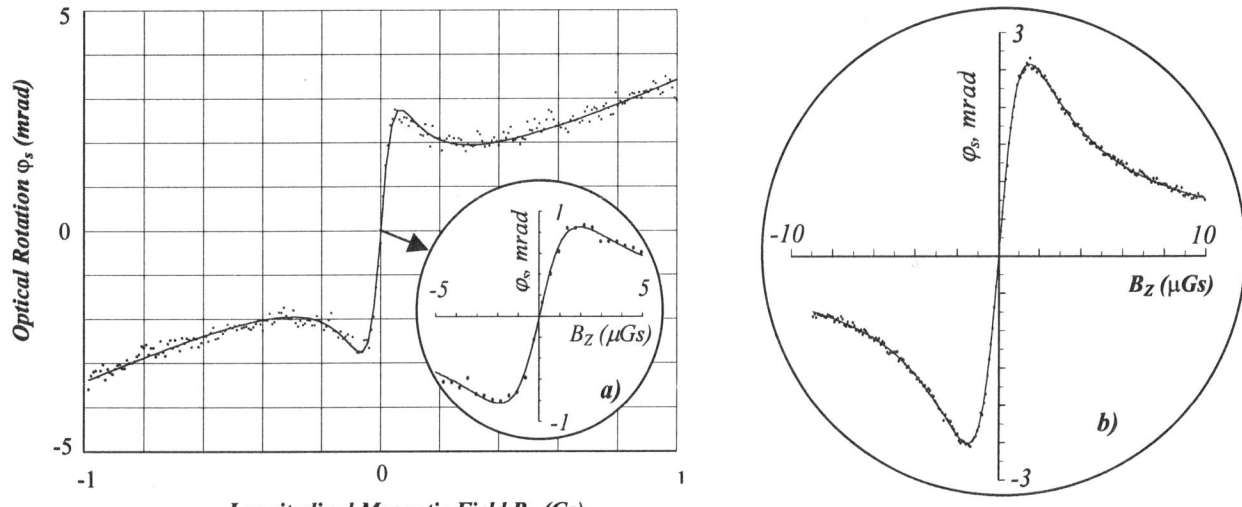

FIGURE 2. Optical rotation dependence on the longitudinal magnetic field (1). *a)* - a detailed scan of the near-zero B_Z-field region at a $2 \cdot 10^5 \times$ magnification of the horizontal scale. The experiments were performed with laser tuned on the center of Doppler profile of $F_g=3 \rightarrow F_e=2,3,4$ transition of ^{85}Rb D_2 resonance line. Light intensity: $W_p \approx 100$ $\mu W/cm$. *b)* - a near-zero B_Z-field scan with laser tuned ≈ 460 *MHz* towards high frequencies from the line-center. $W_p \approx 520$ $\mu W/cm$.

magnetic field orientation with respect to the shield. It can be estimated as (see, e. g. (15,16)):

$$S_{tot} \equiv \frac{B_{in}}{B_0} \approx S_n \cdot \prod_{i=1}^{n-1} S_i \left(1 - \left(\frac{X_{i+1}}{X_i}\right)^k\right); \quad S_i \approx \frac{X_i}{\mu_i \cdot t_i}. \quad (1)$$

In equations (1), B_0 is the homogeneous magnetic field before introducing the shield, B_{in} is the field inside the shield due to B_0; S_i is the shielding factor of a separate *i*-th layer; X_i are the layer's radius or length (depending on the relative orientation of the magnetic field and the layer); we assume $X_i > X_{i+1}$; t_i and μ_i are the thickness and magnetic permeabilities; n is the number of layers. For estimates, good approximations of the power k are: $k=3$ for a spherical shield; $k=2$ and $k=1$ for the transverse and axial shielding factors of a cylindrical shield with flat lids, respectively. Therefore spherical shells are preferable providing the best shielding properties (for shields of comparable dimensions). In the design of the three outer layers of our shield, we found a compromise between trying to approximate a spherical shape and retaining relative simplicity of manufacturing. The innermost shield has cubic shape (with rounded edges). This allows application of relatively homogeneous fields with a simple system of nested 3-D coils.

As a first application of the described magnetometer, we measured the residual magnetic fields and shielding factors of the shield. The typical components of the residual magnetic field averaged over the volume of the vapor cell were found to be $B_X \approx 53$ μGs, $B_Y \approx 14$ μGs, and $B_Z \approx 6$ μGs. (These values change after each reassembly of the shield.) The residual magnetic field at the cell is mostly attributed to the residual magnetization of the innermost-shielding layer. The inhomogeneity of this magnetic field is one of possible mechanisms decreasing the relaxation time of atomic alignment. However, in the present experiment it is less important than spin-exchange relaxation (1). In order to reach the highest sensitivity of the magnetometer, the transverse magnetic fields should be compensated to a level corresponding to a small fraction of the observed widths ΔB_Z. Zeroing of transverse magnetic fields is accomplished by finding such currents in the 3-D magnetic coils, for which the observed B_Z-dependence of NMOE has symmetric dispersive shape with minimal effective relaxation width. (This was found to be $\gamma = g\mu \Delta B_Z/\hbar \approx 2\pi \cdot 1.3$ Hz, where $g=1/3$ for the $F_g=3$ state.) The zeroed magnetometer was used to measure shielding ratio for a dc magnetic field produced by a set of six magnetic coils surrounding the shield. Unfortunately, due to the space limit, the set (the coils' diameters and separations are nearly the same, ≈ 30") is not able to produce a homogeneous field on the shield dimensions. Therefore, estimating the shielding ratio, we took the field produced by the set at the cell location without the shield, as B_0 field. It was found to be $\approx 1 \cdot 10^{-6}$, being approximately independent of field direction. The sensitivity δB_Z of the magnetometer to a magnetic field B_Z can be expressed as:

$$\delta B_Z = \left(\frac{\partial \varphi_s}{\partial B_Z}\right)^{-1}_{B_Z=0} \cdot \delta \varphi_s. \quad (2)$$

The first factor represents the slope of the dependence of the rotation angle φ_s on the magnetic field. It is known from the experiment. The second factor is the sensitivity of the polarimeter to φ_s. We can estimate it for shot-noise limited detection using the parameters of the current set-up. In order to perform this estimation, let us consider an expression for the counting

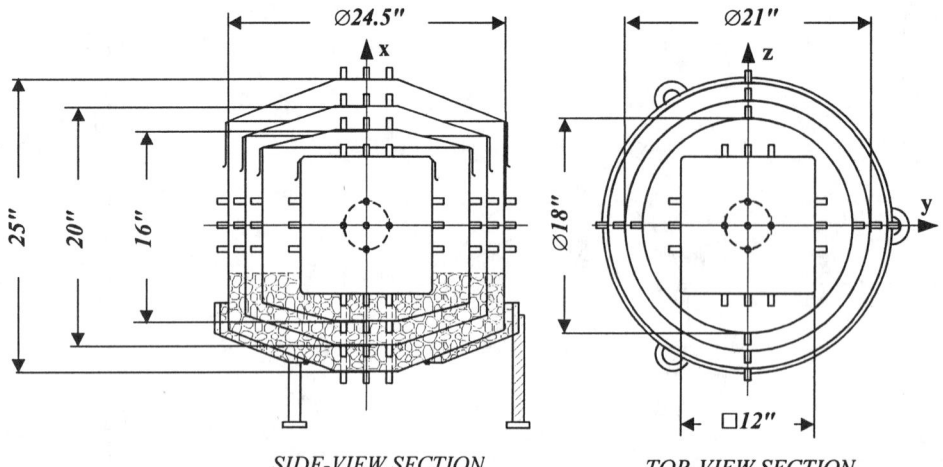

FIGURE 3. Geometrical configuration of the four-layer magnetic shield. The layers are spaced with polyurethane foam to reduce mechanical stress. All CONETIC parts were annealed in a hydrogen atmosphere upon manufacturing. No degaussing was used in the investigations described here. The cell position is shown by dashed-line circle in the center. The shield is designed to allow optical access through a number of 1/2" diameter holes. Each of the holes has an associated 1" stem preserving the shielding properties.

rate from the photodiode placed in the dark channel of the analyzer:

$$N(t) \cong \chi I_p (\varepsilon + \frac{1}{2}\alpha_m^2 + \varphi_s^2) + 2\chi I_p \alpha_m \varphi_s Sin(\omega_m t) - \frac{1}{2}\chi I_p \alpha_m^2 Cos(2\omega_m t). \qquad (3)$$

Here χ is the coefficient defined by the absorption and scattering of light by the atomic vapor cell, $\chi \approx 0.2$; I_p is the intensity of linearly polarized light (in photons/sec) transmitted by the polarizer. From the expression (3), for shot-noise-limited detection of the first harmonic signal in the case of an "ideal polarimeter" ($\alpha_m^2 >> \varepsilon + \varphi_s^2$), one obtains a relation describing the experimental sensitivity of the polarimeter with data accumulation time T:

$$\delta\varphi_s \approx \frac{1}{2 \cdot \sqrt{\chi I_p T}}. \qquad (4)$$

As it was found in (1), the light broadening effect is considerably reduced (without attenuating the light power) by working on the slope of the resonance line rather than on the line center, see Figure 2 b). This allows to improve the statistics of the measurement due to increasing both the light intensity and the slope of the magnetic field dependence of φ_s. Therefore, with pumping light intensity $W_p=520$ $\mu W/cm^2$ (pumping light cross-section area $S_p=0.1$ cm^2) one finds:

$$\frac{\partial \varphi_s}{\partial B_Z} \cong 3.3 \cdot 10^3 \, rad/Gs \quad \text{and} \quad \delta\varphi_s \approx 3 \cdot 10^{-8} \, rad/Hz^{1/2}, \qquad (5)$$

and, consequently, the achievable shot-noise-limited sensitivity δB_Z is

$$\delta B_Z \approx 10^{-11} \, Gs/Hz^{1/2}. \qquad (6)$$

This sensitivity is comparable to or surpasses the best devices based on other techniques. A more detailed comparison to other magnetometers will be given elsewhere.

SENSITIVITY OF A POSSIBLE EDM SEARCH

Application of NMOE in the search for P- and T- violating permanent electric dipole moment (EDM) was first suggested in (17), and later discussed in (6,18). The sketch of a possible EDM experiment is, in general, very similar to the one used to observe the NMOE dependence on B_Z. The only principal difference consists in the detection of alignment precession caused by interaction of atom's P- and T- violating EDM d_A with applied electric field E. The shot-noise-limited sensitivity δd_A to the measurement of EDM d_A may be found from the smallest detectable change of the magnetic field, δB_Z, according to the relation:

$$\delta d_A = g_J \mu \delta B_Z / E, \qquad (7)$$

here $g_J=2$. With the sensitivity δB_Z given by expression (6), one obtains:

$$\delta d_{Cs} \approx 10^{-26} \, e \cdot cm, \qquad (8)$$

assuming electric field $E=10$ kV/cm and data accumulation time $T=10^6$ sec. This corresponds to a sensitivity to the electron's EDM, d_e that is, according to theory (see, e.g. (19) and references therein) approximately 120 times smaller:

$$\delta d_e \approx 10^{-28} \, e \cdot cm. \qquad (9)$$

To obtain an EDM limit at the level of statistical sensitivity (9), one has to control possible systematic effects at least at a similar level. We are considering an experiment with a coated cell containing vapors of both Cs and Rb atoms. While the EDM measurements are performed on Cs atoms, Rb will be used as a "co-magnetometer."

CONCLUSIONS

In conclusion, we have considered some applications of NMOE with ultra-narrow widths observed recently in experiments with a ^{85}Rb-vapor cell with a high quality anti-relaxation coating (1). A three-axis magnetometer based on the strong dependence of NMOE on the longitudinal and transverse magnetic fields was developed. The magnetometer was used to determine the shielding ratio of a four-layer magnetic shield shaped to be close to a sphere but retaining relative simplicity of manufacturing. It was found to be $\approx 10^{-6}$ for shielding of an external dc magnetic field in accord with design expectation. The magnetometer was also used to measure and compensate the average (over the cell volume) residual magnetic field. This provided near-zero transverse magnetic field condition at the cell, which was found to be very important in order to achieve the ultra-narrow NMOE features. It was shown that the shot-noise-limited sensitivity of the magnetometer 10^{-11} Gs (for 1-sec data accumulation time) is achieved using the advantages of tuning the laser to the line-slope. Possible application of NMOE in searching for P- and T- violating EDM in atoms has been considered also. We estimated the shot-noise-limited sensitivity of EDM experiment with cesium based on the sensitivity to magnetic field achievable with our set-up. Assuming that it is possible to apply a 10 kV/cm electric field to a paraffin coated cell, this provides statistical sensitivity of $\delta d_{Cs} \sim 10^{-26} e \cdot cm$ and $\delta d_e \sim 10^{-28} e \cdot cm$ for data accumulation time 10^6 sec. These are more than two orders of magnitude better than the current limit for d_{Cs} (20) and about 40 times better than the best published experimental limit on electron EDM, $|d_e| \leq 4 \cdot 10^{-27} e \cdot cm$, established in an experiment with ^{205}Tl (21), respectively.

In the nearest future, we plan to investigate application of high voltage electric fields to a paraffin-coated cell, and the effects of the spin-exchange collisions between rubidium and cesium in the cell in the proposed EDM experiment.

We are grateful to L. R. Hunter for useful comments. This research is supported by ONR, grant # N00014-97-1-0214.

REFERENCES

1. Budker, D., Yashchuk, V., and Zolotorev, M., *Preprint LBNL-42066*, 1998, submitted for publication.
2. Gawlik, W., in: *Modern Nonlinear Optics*, Edited by M. Evans and S. Kielich, New York: Wiley, 1994, Part 3, pp.733-773.
3. Barkov, L. M., Melik-Pashayev, D., and Zolotorev, M., *Opt. Commun.* **70**(6), 467-472 (1989).
4. Zetie, K. P., Warrington, R. B., MacPherson, M. J. D., Stacey, D. N., and Schuller, F., *Opt. Commun.* **91**(3-4), 210-214, (1992).
5. Weis, A., Wurster, J., and Kanorsky, S. I., *J. Opt. Soc. Am. B* **10**(4), 716-724 (1993).
6. Schuh, B., Kanorsky, S. I., Weis, A., and Hänsch, T. W., *Opt. Commun.* **100**(5-6), 451-455, (1993).
7. Kanorsky, S. I., Weis, A., and Skalla, J., *Appl. Phys. B-Lasers and Optics* **60**(2-3), S165-S168 (1995).
8. Faraday, M., *Experimental Research*, **III**, 2164, London, 1855.
9. Macaluso, D., e Corbino, O. M., *Nuovo Cimento* **8**, 257 (1898); *Ibid.* **9**, 384-389 (1899).
10. We use the convention of Alexandrov, E. B., Chaika, M. P., and Kvostenko, G. I., *Interference of Atomic States*, Berlin, Heidelberg, New York: Springer-Verlag, 1993, in which alignment designates the second (quadrupole) polarization moment.
11. Arimondo, E., in *Progress in Optics*, Edited by E. Wolf, Amsterdam: Elsevier, 1996, Vol. 35, pp. 257-354; Brandt, S., Nagel, A., Wynands, R., and Maschede, D., *Phys. Rev. A* **56**(2), R1063-R1066 (1997).
12. Budker, D., Yashchuk, V., and Zolotorev, M., "Study of resonant magneto-optical rotation in the presence of arbitrarily-directed magnetic field," Abstracts of contributed papers, ICAP XVI, Windsor, Canada, August 3-7, 1998, pp. 356-357.
13. Budker, D., Yashchuk, V., and Zolotorev, M., *Preprint LBNL-41149*, 1997, to be published in *Sib. J. Phys.*, 1998.
14. Alexandrov, E. B., Balabas, M. V., Pasgalev, A. S., Vershovskii, A. K., and Yakobson, N. N., *Laser Physics* **6**(2), 244-251 (1996).
15. T. Rikitake. *Magnetic and Electromagnetic Shielding*. Tokyo: TERRAPUB, 1987.
16. Summer, T. J., Pendlebury, J. M., Smith, K.F., *J. Phys. D* **20**(9), 1095-1101 (1987).
17. Barkov, L. M., Zolotorev, M. S., and Melik-Pashayev, D., *JETP Letters* **48**(3), 144-147 (1988). The idea of using optical rotation in longitudinal electric field was suggested in (Sushkov, O. P., Flambaum, V. V., *Zh. Eksp. Teor. Fiz.* **75**(4), 1208-1213 (1978)).
18. Hunter, L. R., *Science* **252**, 73-79 (1991).
19. Khriplovich, I. B., and Lamoreaux, S. K., *CP Violation without Strangeness. The Electric Dipole Moments of Particles, Atoms and Molecules*. Berlin, Heidelberg, New York: Springer-Verlag, 1997.
20. Murthy, S. A., Krause, D. Jr., Li, Z. L., and Hunter, L. R., *Phys. Rev. Lett.* **63**(9), 965-968 (1989).
21. Commins, E. D., Ross, S. B., DeMille, D., Regan, B. C., *Phys. Rev. A* **50**(4), 2360-2377 (1994).

SECTION 5

STORAGE RING PHYSICS

Molecular Structure by Coulomb Explosion Imaging of Stored Molecular Ions

J. Levin, L. Knoll, M. Lange, M. Scheffel, R. Wester, A. Wolf, D. Schwalm

Max-Planck-Institut für Kernphysik, 69029 Heidelberg, Germany

A. Baer, Z. Amitay, D. Zajfman, Z. Vager

Department of Particle Physics, Weizmann Institute of Science, 76100 Rehovot, Israel

Abstract. An experimental scheme, which combines Coulomb explosion imaging (CEI) with storage of fast molecular ions, has been introduced recently at the TSR heavy ion storage ring facility in Heidelberg. CEI is an experimental technique that provides direct observation of the nuclear conformations within small molecules. The combination of CEI with the storage ring technique enables the control of the internal excitation of the measured molecules, which is an essential condition to the interpretation of CEI results in terms of "structure" assigned to specific molecular states. This structure is measured as a function of storage time, thus enabling one to study processes of slow intramolecular dynamics such as isomerization, metastable states, etc. Moreover in this scheme, CEI can be used as a diagnostic tool for the intramolecular excitation, while other molecular interactions (e.g. with electrons or photons) are investigated. In this report, the CEI principle and the new experimental setup are described with an emphasis on the new prospects for studies in molecular physics. CEI measurements of stored CH_2^+ and NH_2^+ molecular ions are presented. The study of the angular distribution in these molecules as a function of their vibrational relaxation to the ground state, reveals unexpected behavior near the linear conformation which is inconsistent with the current adiabatic theories.

INTRODUCTION

Most of our experimental knowledge of the structure of small molecules in the gas-phase comes from spectroscopic measurements. The measured spectra, which are differences between molecular eigenvalues are analyzed by a model Hamiltonian providing fitted structural parameters. This traditional method has been highly successful in describing the structure of many stable and rigid molecules where only small excursions of the nuclei from the equilibrium geometry occur. Yet, the structure of floppy molecules, which present large amplitude motion far from the geometrical equilibrium, as well as vibrationally excited molecules and molecular ions, is much less amenable to be studied by spectroscopy. Another approach to the study of molecular structure is the direct observation of molecular dissociation products which contain information on the nuclear geometry, such as in the Coulomb explosion imaging (CEI) experiment. In this case, a direct probe of the nuclear density function (the distribution of the nuclear conformations) is obtained, independently of the rotation-vibration model of a molecule. Thus, CEI applies equally well to rigid and floppy (or vibrationally excited) molecules.

It is important at this point, to eliminate certain confusion concerning the definition of "molecular structure". Since spectroscopy mainly provides structural parameters associated with the equilibrium position of nuclei in a molecule, this single equilibrium geometry is often referred to as "the structure". Yet, a more general and precise definition of structure can be given. In the quantum mechanical sense molecular structure is: *the projection of the molecular density function on the subspace of nuclear coordinates.* Note that this definition is probabilistic and is constrained to a pure molecular eigenstate [1]. For rigid molecules in the vibrational ground state, the nuclear density peaks at the equilibrium geometry and can be well characterized by a small number

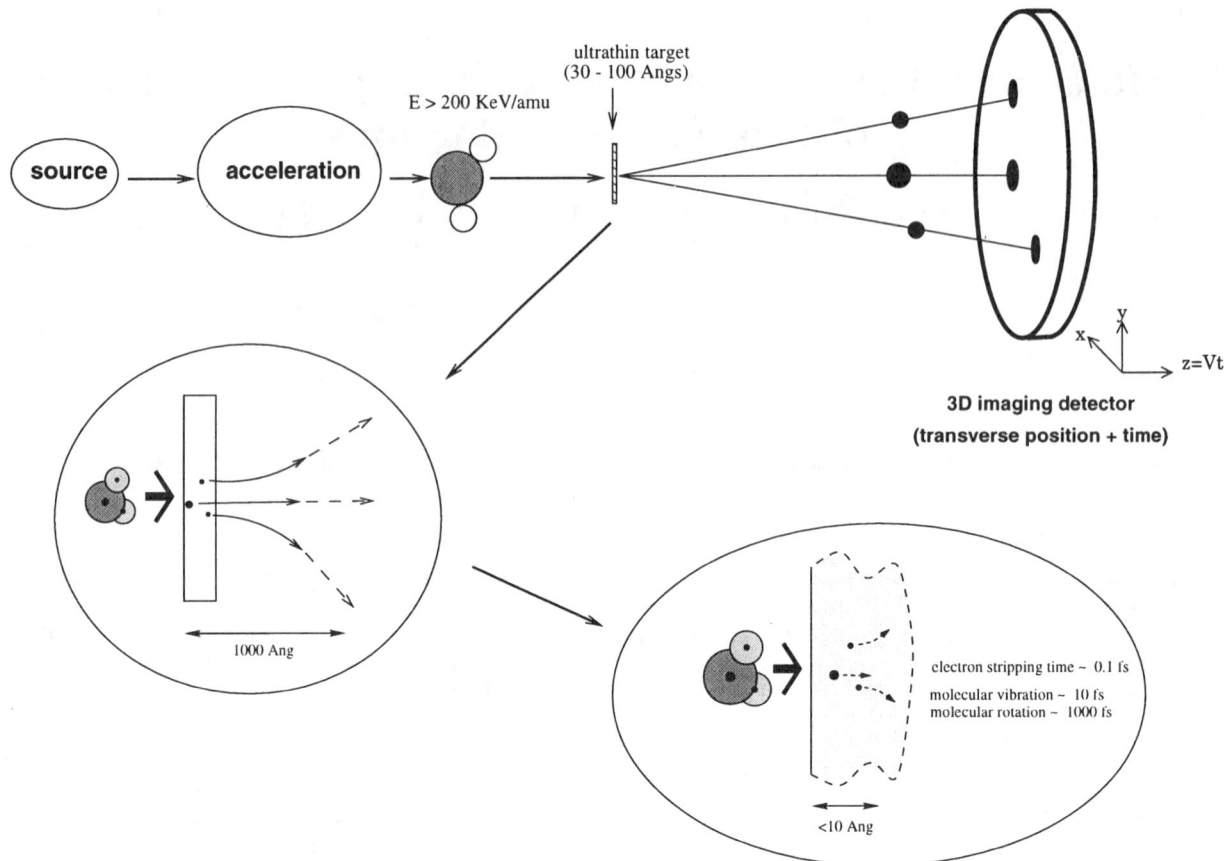

FIGURE 1. Schematic illustration of the CEI experiment. The two inserts give an enhanced microscale view of the Coulomb explosion process.

of parameters that define this equilibrium and the near equilibrium behavior. Thus, the strict definition of structure coincides with the approximate classical notion, referred to the equilibrium structure.

The data of CEI is a sampling of the nuclear density function in the ensemble of measured molecules. Therefore, it can be related to the general definition of molecular structure *if* the internal excitation of the measured ensemble can be fully characterized. Since CEI is insensitive to the internal energy of the molecules, it depends on initial preparation of the molecules in a pure state. Storage of molecules in a heavy ion storage ring is the suitable tool for preparing such a well defined molecular ensemble. Vibrational relaxation to the ground state can usually be achieved within the storage time and the structure of cold molecules, as well as well-defined excited states can be obtained. This and other applications of CEI at storage rings are described in the next section.

In the last section of this report we present CEI measurements of the structure of vibrationally cold linear triatomic molecules. The results demonstrate the profound importance of directly observing density functions. The main feature observed in these experiments is the deviation of the density function of CH_2^+ from the expected behavior when the molecule approaches the linear conformation. This deviation cannot be be explained by the current theoretical treatment of the Renner-Teller effect. Notably, in this case, spectroscopy is much less sensitive to the specific shape of the wavefunctions near linearity. The effect is localized in a small region of phase space (away from equilibrium) whereas energy eigenvalues involve integrals over the whole space.

CEI OF STORED MOLECULAR IONS

The principle of the CEI technique [2–4] and the relevant time- and length scales are illustrated in Fig. 1. In a typical experiment, molecules accelerated to more than 200 Kev/amu collide with an ultrathin (<100 Å)

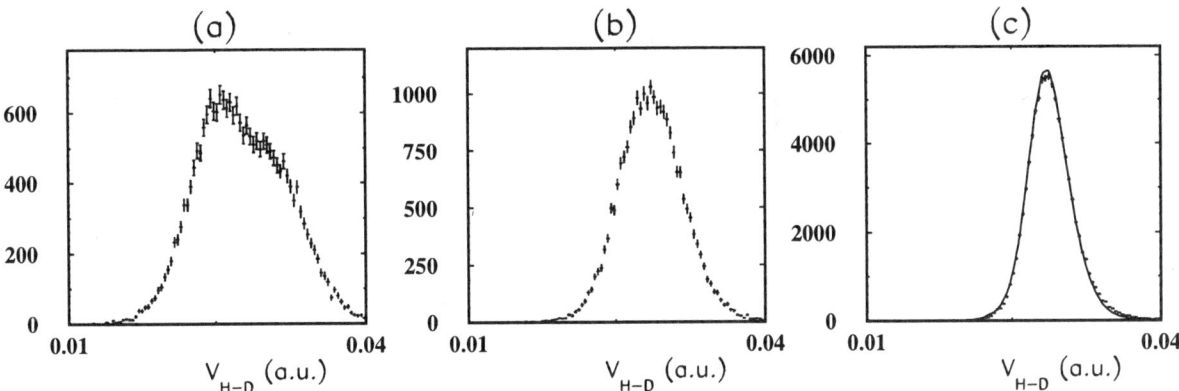

FIGURE 2. CEI data for stored HD$^+$ molecular ions. Histograms representing the distribution of the relative asymptotic velocity $V = |\vec{V}_H - \vec{V}_D|$ are given for different storage time intervals: (a) $t_s = 0 - 10$ ms, (b) $t_s = 70 - 80$ ms, and (c) $t_s > 400$ ms. The data presented in (c) corresponds to a vibrationally cold molecular ensemble. The solid line is a theoretical density function of the vibrational ground state (calculated from ref. [7]).

target. In such a collision, the valence electrons of the molecule are stripped off within the first few atomic layers of the target on a time scale of 10^{-16} sec, which is by orders of magnitude shorter than the time scale of intramolecular nuclear motion (see Fig. 1 lower-right insert). Thus, to a very good approximation, the electron stripping can be considered as being instant, producing a system of fragment ions that retain the geometrical structure of the molecule before the stripping. This system dissociates due to the mutual Coulomb repulsion, eventually converting all the potential Coulomb energy into relative kinetic energy of the fragments. Typically, a few hundred Å away from the target the Coulomb interaction becomes negligible and the fragments follow in a free drift towards the CEI detectors that are positioned at a distance of a few meters from the target. The mass and charge of the molecular fragments are analyzed by a magnetic field (not shown in Fig. 1). When the fragments of a single molecular dissociation reach the detection system, their typical separation is on a centimeter scale, and the corresponding time separation is a few nanoseconds. The fragments are detected separately, the position and time of arrival of each fragment is recorded with a resolution better than 100 μ and 100 ps respectively. Given the velocity of the molecular beam and the target-detector distance, the asymptotic velocities of the fragments in the molecular center of mass frame can be obtained. The information on the initial geometry of each exploding molecule is conserved in the Coulomb explosion kinematics and, in principle, can be retrieved by reverse calculation of the Coulomb trajectories. In practice, the situation is complicated by the interaction of the exploding fragments with the target atoms. A part of the initial information is lost in those stochastic processes, but the effect can be reduced by using higher beam energies and thinner targets. It is important to emphasize that under the current experimental conditions the retarding effect of the target on the molecular fragments is negligible. The main interaction is small-angle multiple scattering which contributes to only a minor smearing of the geometrical features. In a quantum mechanical description the instant electron stripping is approximated by a sudden replacement of the molecular Hamiltonian by a Hamiltonian of the Coulomb interaction between the fragment ions. The nuclear density of an N-atomic molecule, defined in a 3N dimensional space of molecular conformations evolves under this dissociative Hamiltonian and eventually, its Coulomb transform on the 3N dimensional space of internuclear velocities is sampled by the CEI experiment. Traditionally, the space of molecular conformations is termed "R space", and the space of the asymptotic fragment velocities - "V space". Detecting a large number of coincident molecular fragments provides sampling of the V space density function, which enables the reconstruction of the distribution in R space. The target effects are taken into account numerically in a Monte Carlo simulation that propagates an initial R space distribution through Coulomb transformation and convolutes it with the detection resolution [5]. Methods for solving the inverse problem of retrieving the R space distribution from the data with propagated error have been devised [6].

As stated in the introduction, the interpretation of CEI data as "molecular structure" relies on the ability to characterize the internal excitation of the measured ensemble of molecules. The measured density has to be associated with specific molecular states. If, for example, all the measured molecules belong to the same

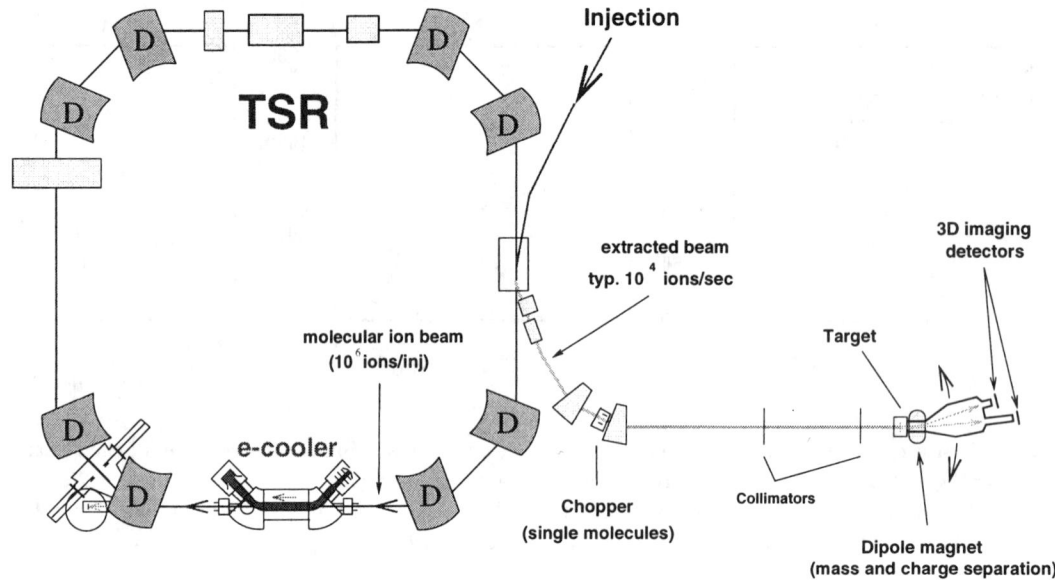

FIGURE 3. A schematic drawing of the Test Storage Ring (TSR) located at the Max-Planck Institute für Kernphysik in Heidelberg. The extraction beamline leading to the CEI target and detectors is shown.

eigenstate, then the density sampled by the measurement is the square of the wave function of this state. It is therefore essential to control the state of the measured molecules. In the past, a partial control has been achieved by cooling the molecules in the molecular ion source. This presented an experimental challenge since the standard sources of intense molecular beams produce vibrationally excited molecules. Therefore special "cold" sources were developed, based on the principle of vibrational cooling by supersonic expansion [8].

Lately, the development of the heavy-ion storage ring technique provided the means for controlling intramolecular excitation: storage of fast molecular ions for as long as 10–20 seconds. This time is long enough to enable relaxation of internal degrees of freedom by spontaneous emission to the limit of thermal equilibrium with the environment (300K). This usually assures pure population of the vibrational ground state. The simplest experimental scheme is therefore to employ the storage ring as a source of fast cold molecular ions for structure analysis by the CEI technique. An example of CEI data for stored HD^+ molecules is presented in Fig. 2. The histograms represent the distribution of the internuclear asymptotic velocity. The effect of vibrational cooling is clearly seen by comparing data from different storage times. For short storage times the distribution is very wide as a result of broad distribution of vibrational states at the molecular ion source. Vibrational cooling is manifested by the narrowing of the measured distribution. After ~ 400 ms the distribution becomes independent of storage time, which indicates that most of the stored molecules are in the ground state. The solid line in Fig. 2(c) represents the theoretical density function of HD^+ at $\nu = 0$ ($\Psi_0^2(R)$), convoluted with the experimental resolution. This yields a very good agreement with the experimental results.

This storage-time dependent CEI technique, can also be used simultaneously with other experiments on molecular ions in the storage ring. In fact, the HD^+ data shown in Fig. 2 has been used to *probe* the vibrational excitation of the stored molecules at given storage time intervals. For simple and well studied systems such as HD^+, where the structure of the low states is known *a priori*, the argument of the structure measurements can be inverted. Then the CEI output can be used as a diagnostic tool for the degree of vibrational excitation providing the relative contribution of each vibrational state to the measured distribution at a certain storage time. In the case of HD^+, a simultaneous measurement of the dissociative recombination (DR) rate of HD^+ allowed the extraction of DR cross sections for particular (excited) vibrational states [9].

The CEI setup at the heavy-ion storage ring TSR

Experiments on CEI of stored molecular ions (CEISMI) are conducted at the TSR storage ring at the Max-Planck Institut für Kernphysik - Heidelberg in collaboration with the Weizmann Institute of Science - Rehovot.

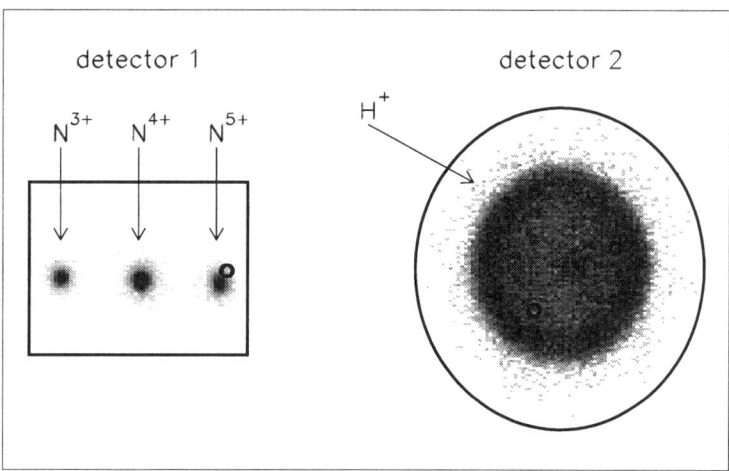

FIGURE 4. A two dimensional projection of 31500 accumulated NH_2^+ detection events. Each event is a coincident hit of a nitrogen ion on detector 1 and two protons on detector 2. The different charge states of the nitrogen (N^{3+}, N^{4+} and N^{5+}) are spatially separated. A single arbitrary event is indicated by the black circles.

A schematic drawing of the experimental setup is given in Fig. 3. Positive molecular ions are injected into the ring from a 2-MV van-de-Graaf or a 12-MV tandem accelerator. Ions with a maximum magnetic rigidity of 1.5 Tm can be stored, which sets a maximum of 32 amu for singly charged molecular ions with energies > 100 Kev/amu. Typically, using multi-turn injection $10^5 - 10^6$ molecules are stored with revolution frequency of few hundred kilohertz. Molecular storage lifetimes of 10-30 s have been observed, at residual gas pressure of about $3 \cdot 10^{-11}$ mbar. A new extraction system has been installed at the TSR which allows for continuous slow extraction of molecular ions towards the CEI setup which consists of a stripping foil, fragment analyzing magnet and two 3D imaging detectors. In order to operate the extraction, the ions are stored close to a third order resonance, such that the stable horizontal phase space region can be made smaller than the horizontal acceptance of the ring. An unstable motion is driven in a controlled way by applying a resonant RF field which excites horizontal betatron oscillations. The RF excitation method is known to lead to a small emittance of the extracted beam. The ions that have left their stable orbits are extracted by the same electrostatic septum which is used for injection. Typically $10^2 - 10^3$ ions are extracted per second which is more than sufficient for the CEI experiment that measures one molecule per detection cycle at a repetition rate of ≤ 50 Hz. The CEI setup at the extraction line has been described in details elsewhere [4]. Briefly, the extracted beam is chopped to allow for single molecule detection, then collimated and directed towards a target holder with several ultrathin target foils made of Formvar or diamond like carbon (DLC) [4]. A dipole magnet located behind the CEI targets, is used to separate molecular fragments according to their mass to charge ratio. Two imaging detectors are positioned at a distance of ~ 2.5 meters from the target foils. The detectors chamber is movable such that one detector can collect several charge states of heavy ion fragments and the other one simultaneously measures light fragments (e.g. protons or deutrons), which are deflected further away. An example of raw CEI data is shown in Fig. 4 where an accumulated distribution of the fragments of NH_2^+ on both detectors is shown. Each detected event consisted of a nitrogen ion in one of three charge states (N^3+, N^4+, N^5+) and two protons in coincidence. The position and the time of arrival were recorded for each such fragment. Such an arbitrary single event is indicated in Fig. 4 by the black circles. The shaded plot represents the distribution of all 31500 NH_2^+ events accumulated during the measurement.

ANGULAR STRUCTURE OF QUASILINEAR MOLECULES: CH_2^+, NH_2^+

The Renner-Teller effect

The CH_2^+ and NH_2^+ molecular ions are typical examples of "quasilinear" molecules, which exhibit a large amplitude bending motion in the vicinity of the linear conformation. The term "quasilinear" stems from the fact that the "equilibrium" geometries for these molecules, defined as the minima of the *ab initio* ground state potential curves, are not linear, but the barrier to linearity is low when compared to the zero-point energy of the bending vibration. There is a vast amount of literature dealing with linear and quasi-linear triatomics and the associated Renner-Teller effect (e.g. the classical paper by Jungen and Merer [10]); here we shall only state the underlying principles and quote the relevant theoretical predictions for the expected density functions.

Within the Born-Oppenheimer (BO) approximation a potential surface is defined by calculating the energy of the electronic degrees of freedom for each nuclear conformation. For a linear geometry of the nuclei, the symmetry of this conformation induces degeneracy of electronic states with non-vanishing angular momentum. At bent conformations this degeneracy is lifted which results in potential curves of the kind shown in Fig. 5(b) for CH_2^+. The breakdown of the BO approximation near linearity (due to the near degeneracy of electronic levels) has been predicted by Renner in 1934 [11]. In the case of NH_2^+, however, the electronic angular momentum along the linear axis vanishes at linearity (the ground electronic level correlates to a Σ state) and no degeneracy occurs as shown in Fig. 5(a).

Let J be the total angular momentum and S the total molecular spin operator. Λ is the projection of the electronic angular momentum on the linear axis (z) and K is defined as the z component of $N = J - S$. The coordinate related to the bending vibration is the H-X-H angle θ or the complementary angle $\rho = 180 - \theta$. The two Renner-Teller electronic states are coupled to the nuclear rotations around the symmetry axis, and the (Curiolis) coupling coefficients are of the form [10]

$$\hbar^2 \mu_{zz}(R) K \Lambda \qquad (1)$$

where μ_{zz} is the inverse moment of inertia at the conformation R. μ_{zz} is proportional to $1/\rho^2$ and therefore diverges at linearity. For the case $K = 0$ (the one treated in ref. [12]) the problem can be decoupled, but the BO potentials are replaced by the so called adiabatic potentials which contain an additional term

$$A(R) = \frac{1}{2}\hbar^2 \mu_{zz}(R)\Lambda^2. \qquad (2)$$

This term, which has the form of centrifugal energy, can be given a simple interpretation as representing the nuclear rotation which compensates for the electronic angular momentum to give in total $K = 0$ (near linearity

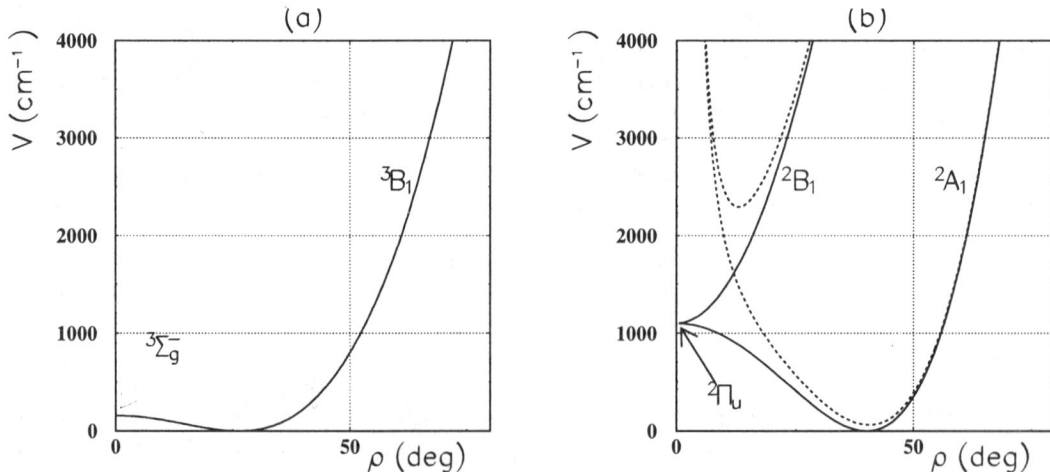

FIGURE 5. Born-Oppenheimer potential curves for the lowest electronic states of NH_2^+ (a) and CH_2^+ (b). The dashed lines in (b) represent the adiabatic potential curves for K=0.

K is a good quantum number). The adiabatic potentials for the case of CH_2^+ are illustrated in Fig. 5(b) as dashed lines. Evidently, the corresponding wavefunctions for all vibrational states have to be zero at linearity. Moreover, in their rigorous treatment of the Renner-Teller effect for all K, Jungen and Merer show that by a suitable transformation of the BO basis functions one obtains diagonal terms of the form [10]

$$V(R) = V_0(R) + \frac{1}{2}\hbar^2 \mu_{zz}(R)(K \pm \Lambda)^2. \tag{3}$$

while the new coupling terms are independent of μ_{zz} and are small near linearity. It follows from Eq. 3 that the wave function solutions near linearity are superpositions of terms: $\propto \rho^{|K \pm \Lambda| + 1/2}$. Thus, the density $\mathcal{D}(\cos\theta)$, which is the quantity measurable by CEI, is obtained by squaring the wave functions and transformation of coordinates to $\cos\theta$:

$$\mathcal{D}(\cos\theta) \propto (\cos\theta)^{|K \pm \Lambda|} \tag{4}$$

The main point is that among all terms only the terms with $|K \pm \Lambda| = 0$ can contribute to the density $\mathcal{D}(\cos\theta)$ at linearity. For ensembles at temperatures of 300K or more, where the lowest angular momenta have a small probability to occur, this contribution is expected to be small. Moreover, this contribution should decrease at higher excitations of the molecules due to the decrease in their statistical weight. Thus, the decrease in the density near linearity should be more pronounced at higher excitations.

Experimental results for NH_2^+ and CH_2^+

As follows from the discussion above, it has been expected from the conventional treatment of the Renner-Teller effect in quasilinear triatomics, that the measured angular distribution of CH_2^+ ($\mathcal{D}(\cos\theta)$) will show a decrease to zero when the bending angle θ approaches 180° ($\cos\theta = -1$). On the other hand, the angular density of NH_2^+ was expected to follow the BO prediction.

Fig. 6 presents the results of CEI measurements of the angular distribution of NH_2^+ (a) and CH_2^+ (b). Shown are the already deconvoluted distributions in the molecular R space. In both cases the data represents molecular ions that were stored in the TSR storage ring long enough to assure relaxation of internal degrees of freedom to 300K. The measured density of NH_2^+ (Fig. 6(a)) is compared to two recent theoretical calculations of the ground state density [13,14]. The comparison shows a fair agreement of the measurement to the theory.

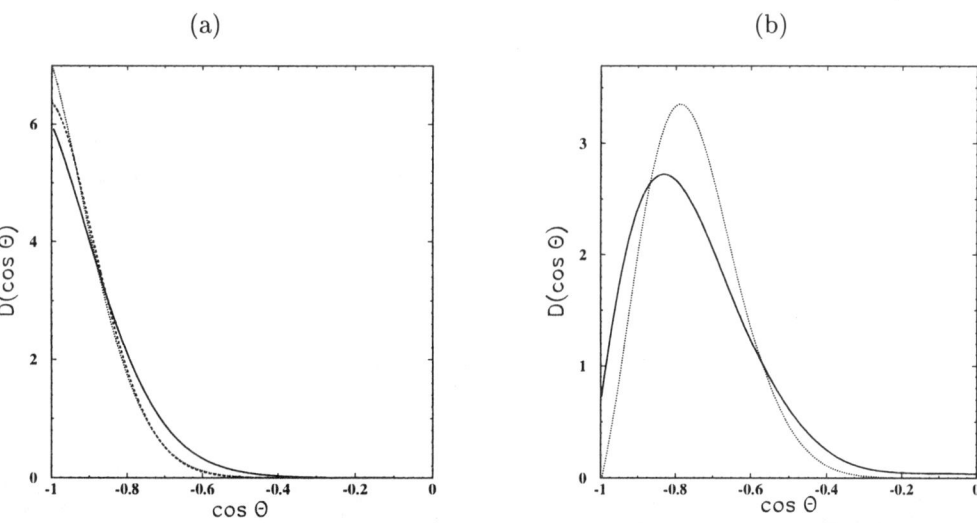

FIGURE 6. CEI results in R space for vibrationally cold NH_2^+ (a) and CH_2^+ (b) are compared to theories. The measured density functions are plotted as solid lines. Two recent theoretical predictions for the ground state density are shown in (a) for NH_2^+ as a dashed line (ref. [13]) and a dotted line (ref. [14]). The prediction of the Renner-Teller treatment for the ground state of CH_2^+ (at $K = 0$) [12] is given by a dotted line in (b).

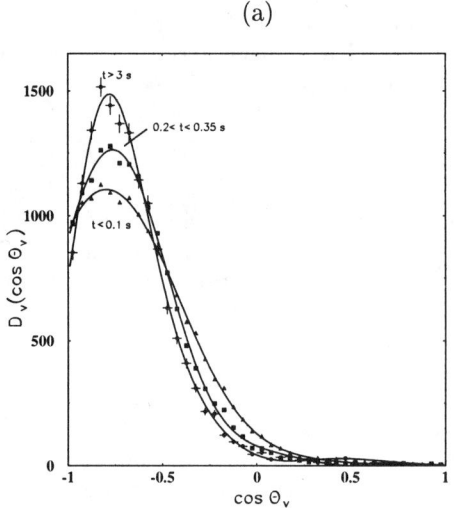 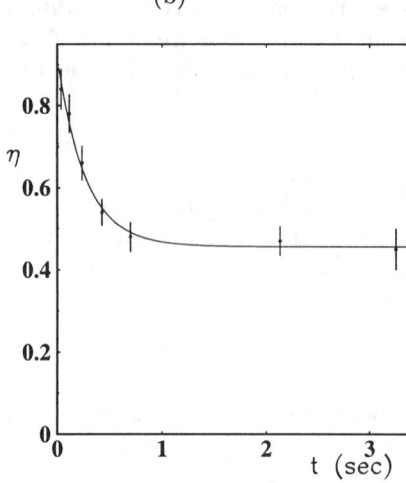

FIGURE 7. V space angular distributions of CH_2^+ at different storage times. The distributions at three different storage time intervals are shown in (a). For clarity, only the coldest distribution ($t > 3s$) is shown within error bars. The time dependence of the density at linearity $\mathcal{D}(-1)$ is represented in (b) where the ratio $\eta = \mathcal{D}(-1)/\mathcal{D}^{max}$ is plotted as a function of mean storage times.

The measured distribution seems to be wider than the prediction for the ground state, which may be attributed to the small amount of rotational excitation at 300 K.

Adiabatic bending wave functions for CH_2^+ were calculated in ref. [12] for $N = 0$. The adiabatic ground state bending density from is shown in Fig. 6(b) as a dotted line. The disagreement with the data is evident especially when contrasted by the result for the non Renner-Teller molecule NH_2^+ in (a). Notice, that the position of the peak of the theoretical distribution at $\theta = 146°$ is significantly lower than the experimental peak position at $\theta = 142°$. The finite density near the linear conformation clearly disagrees with the general considerations concerning Renner-Teller molecules. This effect is even more striking when one considers CEI measurements of CH_2^+ molecular ions after shorter storage times, when vibrational excitation is present. The trend is illustrated in Fig. 7 where the measured distributions for different storage times are plotted. Clearly, for higher internal excitations (shorter storage times) the relative population of the near linear conformation in increased. This is explicitly shown in Fig. 7(b) where the ratio $\eta = \mathcal{D}(-1)/\mathcal{D}^{max}$ is plotted as a function of storage time. The ratio decreases exponentially with a lifetime of ∼240 ms which is the vibrational cooling time. These results are in conflict with the theory that allows population of the near linear region only for a single rotational state. We suggest that non-adiabatic corrections beyond the two-state Renner-Teller treatment are needed to account for the shape of the density function near the linear configuration.

CONCLUSIONS AND PROSPECTS

We have presented a new experimental setup which combines storage of molecular ions with continuous extraction and structure measurement by the CEI technique. The novelty of this scheme is in the introduction of a new dimension to the CEI data - the storage time, which accounts for the intramolecular dynamics of the stored molecules. This enables the measurement of the structure of vibrationally cold molecules as well as monitoring the process of vibrational excitation. In fact any dynamical process which affects molecular structure and happens on a time scale of milliseconds to tens of seconds can be studied. Isomerization, slow decay of excited metastable states, redistribution of energy between different vibrational modes and other slow processes are now amenable for research. The continuous extraction scheme allows to combine CEI with other experimental probes of the stored molecular ions. CEI can be used to identify the population of different vibrational levels and allow for state selective measurements. Application of laser techniques can allow an active control of the internal excitation in the stored molecules. For example, vibrationally selective dissociation of molecular ions stored in the ring, either by laser fields or by reaction with a merged electron beam can be used for creating a pure ensemble of molecules in a single excited vibrational state. Probing the structure of such states by CEI will have a major impact on our understanding of molecules in excited states.

The example of linear triatomic molecules, given in this report, emphasizes the importance of the direct measurements of molecular structure. New, unexpected, behavior of the nuclear density far from the equilibrium configuration has important implications to theory as well as to the nature of chemical reactions.

The presented scheme for CEI of stored molecules is not restricted to the measurement of positive or negative molecular ions. The existing CEI setup at the Weizmann Institute, for example, enables studies of neutral molecules, by applying laser induced photodetachment of negative ions. In such a scheme negative molecular ions are produced and accelerated, and after acceleration they interact with a laser beam which neutralizes most of them by electron photodetachment. The resulting neutral molecules proceed towards the CEI arrangement. Certain control of the molecular excitation is achieved by the photodetaching laser frequency. Yet, it is still very important to control the internal state of the negative ions, a task easily achieved by molecular storage.

ACKNOWLEDGMENTS

JL aknowledges the fellowship and travel support of the MINERVA foundation.

REFERENCES

1. for discussion and a more general definition of molecular structure in terms of the density matrix see J. Levin, Ph.D. thesis, Weizmann Institute of Science, Rehovot (unpublished).
2. Z. Vager, R. Naaman, and E. P. Kanter, Science **244**, 426 (1989).
3. D. Kella, M. Algranati, H. Feldman, O. Heber, H. Kovner, E.Malkin, E. Miklazky, R. Naaman, D. Zajfman, J. Zajfman, and Z. Vager, Nucl. Instrum. Methods **A329**, 440 (1993).
4. R. Wester, F. Albrecht, A. Baer, M. Grieser, L. Knoll, J. Levin, R. Repnow, D. Schwalm, Z. Vager, A. Wolf, and D. Zajman Nucl. Instrum. Methods **A413**, 379 (1998).
5. D. Zajfman, T. Graber, E. P. Kanter, and Z. Vager, Phys. Rev. **A46**, 194 (1992).
6. J. Levin, D. Kella, Z. Vager Phys. Rev. **A53**, 1469 (1996).
7. T. E. Sharp, At. Data **2**, 119 (1971).
8. T. Graber, D. Zajfman, E. P. Kanter, R. Naaman, Z. Vager, and B. J. Zabransky, Rev. Sci. Instrum. **63**, 3569 (1992).
9. Z. Amitay, A. Baer, M. Dahan, L. Knoll, M. Lange, J. Levin, I. F. Schneider, D. Schwalm, A. Suzor-Weiner, Z. Vager, R. Wester, A. Wolf, and D. Zajfman, Science **281**, 75 (1998).
10. C. Jungen and A. J. Merer Mol. Phys., **40**, 1 (1980).
11. R. Renner, Z. Phys. **92**, 172 (1934).
12. W. P. Kraemer, P. Jensen, and P. R. Bunker, Can. J. Phys. **72**, (1994).
13. G. Chambaud, W. Gabriel, T. Schmelz, P. Rosmus, A. Spielfiedel, and N. Feautrier Theor. Chim. Acta **87**, 5 (1993).
14. G. Osmann, P. R. Bunker, P. Jensen, and W. P. Kraemer, J. Mol. Spec. **186**, 319 (1996).

Longitudinal dynamics of laser-cooled fast ion beams: square-well buckets, space-charge effects, and anomalous beam behaviour

M. Weidemüller,* B. Eike, U. Eisenbarth, M. Grieser, R. Grimm, I. Lauer, P. Lenisa, V. Luger, M. Mudrich, U. Schramm,[†] and D. Schwalm

Max-Planck-Institut für Kernphysik, 69029 Heidelberg, Germany

Abstract. We present recent results of our experiments on laser cooling of fast stored ion beams at the Heidelberg Test Storage Ring. The longitudinal motion of the ions is directly cooled by the light pressure force, whereas efficient transverse cooling is obtained indirectly by longitudinal-transverse coupling mechanisms. Laser cooling in novel bunch forms consisting of square-well buckets leads to longitudinally space-charge dominated beams. The observed longitudinal ion density distributions can be well described by a self-consistent mean-field model based on a thermodynamic Debye-Hückel approach. When applying laser cooling in square-well buckets over long time intervals, hard Coulomb collisions suddenly disappear and the longitudinal temperature drops by about a factor of three. The observed longitudinal behaviour of the beam shows strong resemblance with the transition to an Coulomb-ordered ion string.

I INTRODUCTION

Laser cooling is the method-of-choice for the crystallization of ion clouds confined at rest in ion traps [1–4]. The resonant light pressure acting on the ions provides an extremely strong damping force resulting in very low temperatures. At these low temperatures, the plasma parameter Γ, defined as the ratio of the Coulomb potential energy between nearest neighbor ions to the thermal kinetic energy per ion, becomes larger than one already at moderate densities. For a one-component plasma with $\Gamma > 1$, a phase transition into a Coulomb ordered state takes place. At $\Gamma \simeq 1$, a one-dimensional Coulomb string of ions can be formed, whereas for two- and three-dimensional system, the phase transition to the crystalline state occurs at $\Gamma \gg 1$.

Although storage rings confining ions with large speed share many common features with ion traps (except their size, of course), there is, at present, no unambiguous proof for ion beam crystallization [5,6]. One may wonder why laser cooling of fast ion beams in storage rings, as introduced about ten years ago [7–10], did not immediately lead to the observation of beam crystallization. Two main reasons have so far prevented the attainment of crystalline ion beams by laser cooling: the forces exerted on the ions by the ring lattice in co-operation with the huge beam energy lead to extreme heating rates [11], and direct transverse laser cooling of a fast ion beam is practically impossible. In addition, destructive effects by the ring lattice, such as shear in the bending sections of the ring, will break any kind of Coulomb-ordered structure except the linear string and possibly zig-zag bands [a].

In the course of the last year, we have made important progress in laser cooling towards crystalline beams at the Heidelberg Test Storage Ring (TSR) which has led us to new intriguing observations. On the one hand, we have introduced a novel method for indirect transverse cooling based on a single-particle interaction of

---

*) E-mail: m.weidemueller@mpi-hd.mpg.de
[†] Permanent address: Ludwig-Maximilians-Universität München, Sektion Physik, 85748 Garching, Germany
[a] For specific ring lattices, this restriction can be overcome by special cooling schemes designed to provide constant angular velocity (see, e.g. Refs. [12] and [13]). First experimental efforts into this direction are reported in Ref. [14], but the conditions necessary for constant angular velocity seem hard to be met experimentally.

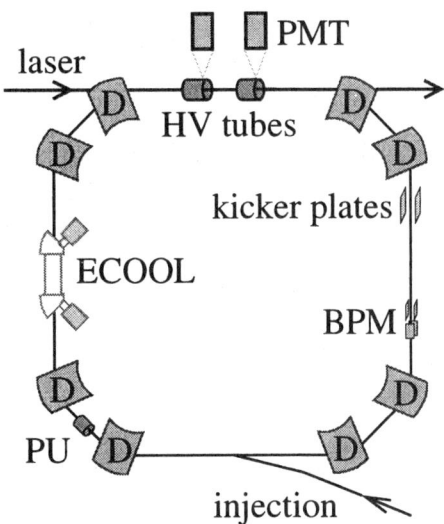

FIGURE 1. Schematic view of the Heidelberg Test Storage Ring (TSR) with its basic elements relevant for laser cooling. The laser beam is merged at one of the straight sections of the TSR. The ion beam is bunched by applying an RF voltage to the kicker plates. D: bending dipole magnets; not shown are the twenty focusing quadrupole magnets. ECOOL: electron cooler; BPM: beam profile monitor; PU: electrostatic pickup; HV tubes: high-voltage biased drift tubes; PMT: photomultiplier tubes.

ions with the laser light [14]. The method provides true 3D laser cooling, and transverse cooling rates are significantly improved as compared to indirect transverse cooling methods used earlier [15], especially when applied to dilute ion beams. On the other hand, we have developed novel bunch forms based on square-well buckets for very efficient laser cooling over long times only limited by the beam lifetime. In Sect. II, we review the essential features of longitudinal and transverse laser cooling at the TSR. In Sect. III, we present results of our experiments on laser cooling in square-well buckets yielding longitudinally space-charge dominated beams. In Sect. IV we report the sudden disappearance of hard Coulomb collisions of a dilute laser-cooled ion beam. Our observations show that we are now entering a qualitatively new, interesting regime. Sect. V finally gives a brief discussion of our results in view of future developments.

II LASER COOLING AT THE TSR

Our experiments towards Coulomb ordering are performed with ^{9}Be$^+$ ions which are injected into the TSR at a beam energy of 7.3 MeV. The initial beam current after multiturn injection is typically 0.3 μA corresponding to about 10^7 ions in a beam of several centimeters in diameter. The $1/e$ storage lifetime is typically 25 s at a residual pressure in the ring chamber of about 5×10^{-11} mbar. The TSR with its basic elements is schematically depicted in Fig. 1. The ion beam is precooled by electron cooling. Within a few seconds, the electron cooling leads to a decrease of the longitudinal temperature from typically 10^4 K to about 300 K (relative momentum spread $\delta p/p \sim 10^{-5}$). The transverse equilibrium emittance after electron cooling is about $10^{-2}\,\pi$ mm mrad which corresponds to an ion beam diameter of about 1 mm. Electron precooling thus provides excellent starting conditions for subsequent laser cooling.

A Longitudinal laser cooling

Laser cooling relies on the radiation pressure exerted by resonant laser light on the ions [16]. The light force arises from repeated momentum transfer in a series of many absorption–spontaneous emission cycles. The alkali-like ^{9}Be$^+$ provides a strong dipole-allowed transition line in the near-UV spectral range (D_2 line $2^2S_{1/2} \to 2^2P_{3/2}$ at a restframe wavelength of 313.13 nm). As shown in Fig. 1, the laser beam is merged copropagating with the ion beam in one of the straight sections of the storage ring over a length of about 5 m (corresponding roughly to 1/10 of the ring circumference $C = 55.4$ m). In the laboratory frame, the resonance

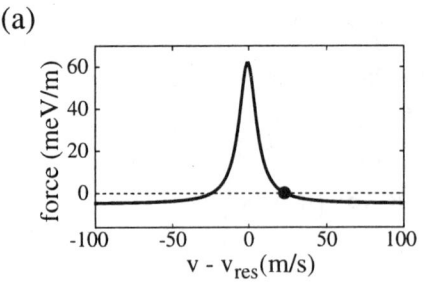 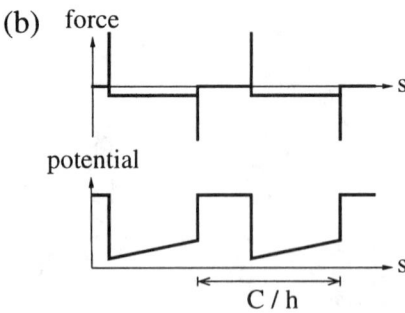

FIGURE 2. (a) Light-pressure force plus velocity-independent counterforce as a function of the velocity v relative to the resonant velocity v_{res}. The dot indicates the stable force equilibrium at the velocity v_*. (b) Bunching force and resulting pseudopotential as a function of the longitudinal coordinate in the comoving frame s for a square-well bucket providing a constant counterforce. C denotes the ring circumference ($C = 55.4\,\text{m}$) and h is the bunch harmonics.

wavelength is Doppler-shifted to 300.35 nm at the chosen ion beam energy corresponding to a velocity of $\beta = 0.0417$. This particular choice of the velocity allows us to use powerful fixed-frequency argon ion lasers. Due to the hyperfine structure of the D_2 line, two laser frequencies separated by 1.30 GHz are required to realize a closed excitation cycle. The total power of the UV laser beam in the interaction region is typically 60 mW focussed into a diameter of about 2 mm yielding a considerable saturation of the transition. The resulting maximum radiation-pressure force amounts to approx. 60 meV/m averaged over the storage ring round trip.

The Doppler effect leads to a very sharp dependence of the radiation-pressure force on the ion velocity. The resonance width corresponding to typically 20 m/s determines the capture range of the light force. To obtain stationary conditions over long time intervals, the accelerating laser force has to be counteracted by an additional, oppositely directed force (counterforce). The combination of these two forces as depicted in Fig. 2 (a) provides friction along the longitudinal degree of freedom necessary for ion beam cooling, i.e. damping of the momentum fluctuations in the frame moving at the mean velocity. Different schemes have been demonstrated to create the counterforce [7,17–19]. Here, we will concentrate on laser cooling of bunched ion beams [17,19], where the counterforce is provided by a longitudinal radio-frequency field tuned to a harmonics of the ion's revolution frequency (225 kHz under our conditions). The RF field applied to the non-resonant kicker device (see Fig. 1) confines the longitudinal ion motion in so-called RF buckets. In Fig. 2 (b), a square-well bucket with counterforce is shown as a specific example which is will be discussed in Sect. III A.

Intrabeam Coulomb scattering (IBS) leads to bimodal longitudinal velocity distributions of the laser-cooled ions: a very cold ($T \lesssim 10\,\text{K}$) fraction of ions experiencing the strong friction force, and a hot ($T \simeq 1000\,\text{K}$) background of ions that have undergone a large longitudinal momentum change by a close Coulomb collision [20]. These momentum changes are much larger than the capture range of the light force. Besides providing a counterforce, longitudinal confinement in RF buckets has the additional advantage to recycle ions into the laser-cooling process after they have undergone a close Coulomb collision. The most effective longitudinal cooling scheme to date is a combination of RF bunching with recently developed extensions of the capture range through broadband laser excitation [20,21]. Attainable damping times in our experiments, referring to a $1/e$ reduction of the longitudinal mean kinetic energy in the comoving frame, are below 10 ms, i.e. two orders of magnitude faster than achievable damping times with electron cooling.

In the experiments presented here, the capture range extension is based on the ion excitation through optical rapid adiabatic passage accompanied by the corresponding momentum transfer. This process offers a much broader dependence on the ion velocity than resonant excitation [21]. Rapid adiabatic passage is realized by applying appropriate voltages to the HV tubes indicated in Fig. 1. The induced local change in the ion velocity is equivalent to fast switching of the laser frequency in the ion's rest frame which leads to the desired adiabatic excitation of the ion.

B Transverse laser cooling

The momentum transfer between the merged ion beam and laser beam is restricted to the longitudinal degree of freedom. Direct transverse laser cooling is known to work very efficiently for slow atomic beams [16] and has been proposed for fast ion beams by several authors [22–24]. In the reality of a storage ring, however, implementation of the proposed schemes for direct transverse cooling appears to be extremely difficult, mainly due to two effects related to the high ion velocities: First, the interaction times for transverse laser irradiation are very short, and second, first-order Doppler shifts make the atomic resonance extremely sensitive to slightest angular misalignments between ion beam and laser beam.

In contrast to direct transverse cooling, *indirect* methods for transverse laser cooling have successfully been demonstrated [18,19,14]. To extend the longitudinal light force to also damp the tranvserse degrees of freedom, one has to find appropriate mechanisms to couple longitudinal and transverse motion. One coupling mechanism is provided by a dense ion beam itself, namely longitudinal-transverse thermal relaxation through intrabeam Coulomb scattering as described in Refs. [18] and [19]. However, this *collective* transverse cooling relies on the same coupling mechanism, namely Coulomb collisions, responsible for beam heating [11] which poses a fundamental limitation to the achievable transverse emittances. In addition, for dilute ion beams, IBS coupling, and thus indirect transverse cooling, becomes inefficient. In order to overcome these limitations, especially in view of Coulomb ordering, we have recently realized transverse laser cooling based on *single-particle* interaction of the ions with the laser light [14]. This cooling scheme, first suggested by A. Wolf [25], exploits longitudinal-horizontal coupling arising from storage ring dispersion in combination with the transverse gradient of the longitudinal light force arising from the Gaussian laser beam profile. Thus, damping of the longitudinal momentum fluctuations is transferred to the horizontal degree of freedom. True 3D cooling is achieved by additionally mixing horizontal and vertical motion through betatron coupling. Although interacting only longitudinally with the ion beam, laser light can therefore efficiently cool a stored ion beam in all three dimensions to very high phase-space densities.

C Diagnostics at the TSR

The longitudinal density distribution of the ion bunches can be probed by an electrostatic pickup system (see Fig. 1). The system measures the voltage induced by the ion current on a capacitor consisting of two metal plates. Any temporal change in the induced voltage is proportional to the change of the ion current. Given the revolution frequency of the ions one can therefore directly deduce the longitudinal ion distribution from the pickup signal. The longitudinal velocity distribution is determined by ramping a bias voltage applied to drift tubes in the laser-cooling section and detecting the intensity of the fluorescence light from the cooling laser. Since the local velocity of the ions inside the drift tube is proportional to the applied voltage, the fluorescence spectrum as a function of the bias voltage can be interpreted as the Doppler spectrum of the velocity distribution [9].

When no voltage is applied to the drift tubes, the fluorescence signal is proportional to the number of atoms in a small velocity interval ($\simeq 50\,\mathrm{m/s}$) around the mean velocity because of the resonant character of the laser-ion interaction. The drift tubes can also be used to implement the extension of the light-force capture range as described above [21]. In this case, the fluorescence signal yields the amount of ions undergoing the rapid adiabatic passage, i.e. ions with large deviations ($\lesssim 1000\,\mathrm{m/s}$) from the mean velocity. Since such large deviation are caused by close Coulomb collisions, the fluorescence signal from the capture range extension provides a measure of the probability of these collisions.

Transverse emittances are measured by a beam profile monitor (BPM) schematically indicated in Fig. 1. The BPM spatially resolves ions produced by collisions of the ion beam with residual gas. From the Gaussian spatial distribution of the ion beam at the BPM position one can deduce the horizontal and vertical emittances from the known lattice functions of the storage ring [9].

III NOVEL BUNCH FORMS AND SPACE-CHARGE EFFECTS

The standard technique for creating bunched beams consists in applying a longitudinal electric field oscillating at some harmonics h of the revolution frequency. This can be accomplished by, e.g., feeding an RF voltage of some tens of volts amplitude (approx. 30 dBm power) to the longitudinal kicker plates shown in Fig. 1 or to a resonator. The longitudinal pseudopotential experienced by the ions is proportional to the applied voltage leading to sinusoidal bunch potentials in the comoving frame. The longitudinal ion density distribution in such a sinusoidal potential is strongly inhomogeneous. For some applications, in particular the formation of Coulomb-ordered ion beams, a spatially homogeneous distribution of the laser-cooled beam would be preferable.

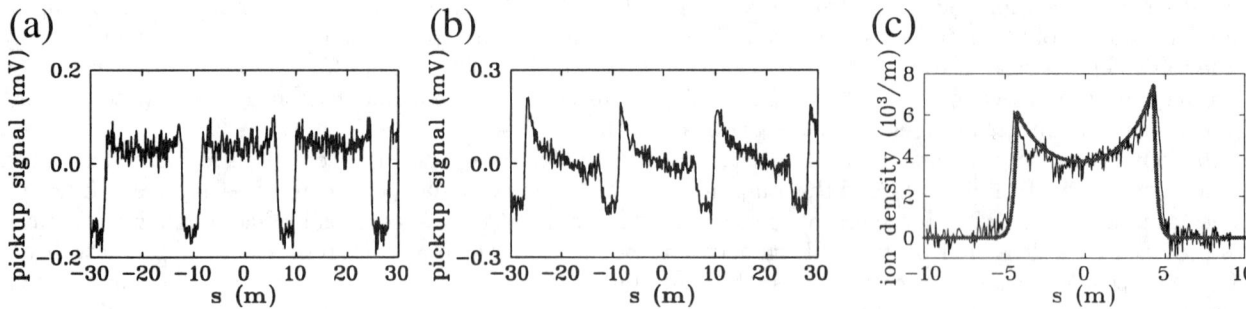

FIGURE 3. (a) Pickup signal of a square-well bunch versus the longitudinal coordinate in the comoving frame. The pickup signal reflects the density distribution of ions. The beam is bunched at the third harmonics ($\nu_b = 676\,\text{kHz}$) of the revolution frequency with a duty-cycle of 80%. Stable velocity and synchroneous velocity are prefectly matched. In the thermal regime, the rectangular shape of the distribution directly reflects the potential form. (b) Density distribution under the same conditions as in (a) but with the bunch frequency decreased by only 2 Hz. (c) Longitudinal density distribution of an ion ensemble in the space-charge regime at a duty cycle of 50%. The distribution exhibits strong effects of Coulomb repulsion. The result of a model calculation is shown by the solid line on the experimental data.

We have therefore developed a method which combines homogeneous ion distributions as in a coasting beam with the longitudinal ion confinement in buckets for the suppression of IBS cooling losses.

A Square-well bunch potentials

For homogeneous ion distributions inside the bunch, a square-well potential is required. Since the bunch potential in the comoving frame is proportional to the applied voltage, this can readily be realized by applying a voltage with rectangular waveform. A particularly nice feature of rectangular bunch potentials is the possiblity to externally adjust the length of the bunch, and thus the longitudinal density, by simply changing the on/off-ratio (duty cycle) of the applied rectangular wave. To exclude distortions of the potential due to the limited bandwidth of our RF amplifier, we have used the third harmonics ($h = 3$) of the revolution frequency for our experiments ($\nu_b = 676\,\text{kHz}$). The counterforce to the light-pressure force is generated by a sawtooth wave at the same frequency which is added phase-coherently to the voltage. In the comoving frame of the ions, one thus creates a potential well with a bottom of constant slope [b] as schematically depicted in Fig. 2 (b). The resulting counterforce is constant and independent of the ion velocity. The sum of light-pressure force and counterforce can be linearly expanded as $F \approx -\alpha(v - v_*)$ with α denoting the friction coefficient and v_* the velocity at stable force equilibrium (see Fig. 2 (a)). Ions are cooled towards the stable velocity, which depends on the counterforce and the laser intensity as does the friction coefficient.

To create constant ion density in the bucket, one has to achieve force equilibrium at the synchoneous velocity $v_s = C\nu_b/h$ which is determined by the bunch frequency ν_b (C = ring circumference). Therefore, the stable velocity v_* has to be exactly matched with the synchronous velocity. An experimental realization of a constant longitudinal density distribution is presented in Fig. 3 (a). If the two velocities are unequal, the ions experience a constant net force $F \approx -\alpha(v_* - v_b)$. This force will push them towards one end of the potential well yielding density distributions as shown in Fig. 3 (b).

B Space-charge effects

At large longitudinal temperatures [c] and low densities, one can neglect the energy associated with the mutual interaction of the ions due to Coulomb repulsion (*thermal regime*). The longitudinal density distribution is

---

[b] Typical ramp slopes are $10\,\text{V}/\mu\text{s}$ corresponding to ring-averaged counterforces around $3\,\text{meV/m}$.

[c] Due to the bimodal character of the longitudinal velocity distribution, it is, strictly speaking, not appropriate to ascribe a temperature to the ensemble. In the context here, temperature is meant as a measure of the mean kinetic energy of the ions in the frame moving at the mean ion velocity.

determined by a Boltzmann distribution in the pseudopotential which includes the combined action of bunching fields, light-pressure force, and a term associated with a possible mismatch between equilibrium velocity and synchronous velocity. Density distributions observed in the thermal regime are shown in Fig. 3 (a) and (b).

The bimodal character of the velocity distribution can be accounted for by assuming two different temperatures for the two subensembles. Since v_* represents the velocity at force equilibrium between light-pressure force and counterforce, the cold partition of the ions interacts with a pure square-well potential. The hot fraction of the ion velocity distribution is far detuned from resonance with the laser light. The counterforce is not balanced by the light pressure force leading to an asymmetric ion distribution in the bucket. However, because the large temperature of the hot partition, the energy gain related to the counterforce is generally much smaller than the thermal energy. The asymmetry of the distribution is therefore not very pronounced as can be inferred from Fig. 3 (a).

As the longitudinal temperature gets lower and the density increases, the influence of the ionic space-charge on the density distribution becomes important (*space-charge regime*). Qualitatively, the repulsive electrostatic force between the ions will push them to the edges of the potential well which leads to density distributions as shown in Fig. 3 (c). Space-charge effects have been observed before in sinusoidal buckets [17,26], and models have been introduced to describe these space-charge dominated bunches [26–28]. The models are based on a mean-field Coulomb potential proportional to the *local* ion density. From the condition that, in equilibrium, the mean field energy has to compensate the external potential, these models predict a constant ion density for square-well buckets, in contradiction to the experimental data. In square-well potentials, the local-density approximation for the Coulomb potential has to be given up, and the potential has to be calculated *self-consistently* taking into account the external confining forces, the Coulomb interaction and the thermal motion at finite temperatures.

To adequately describe our experimental findings, we have developed a novel approach [29] based on a thermodynamic mean-field model first introduced by Debye and Hückel for ionic solutions. The mean longitudinal Coulomb potential in this model is derived self-consistently from a Poisson-Boltzmann equation [30]. The density distribution is calculated by the second derivative of the Coulomb potential with respect to the longitudinal coordinate. A comparison between a measured ion distribution and the model calculations is presented in Fig. 3 (c). The model has no free parameters, since all relevant parameters, like beam current and temperatures of the two subensembles, are determined experimentally. From the comparison between model and experiment, we can infer that only the cold partition of the velocity distribution contributes to the density enhancement at the potential edges. The apparant asymetry of the density distribution stems from the interaction of the the cold partition with the slightly asymetric distribution of the hot ions experiencing the counterforce. As mentioned above, the asymmetry of the hot distribution is small, yet it creates a measurable effect through the Coulomb interaction with the cold partition.

IV ANOMALOUS BEAM BEHAVIOUR

When the thermal energy per ion becomes smaller than the Coulomb potential energy between two neighboring ions ($\Gamma \gtrsim 1$), the ions start to show ordering phenomena. The ion configuration representing the energetic ground state sensitively depends on the local ion density. Based on Molecular Dynamics for cylindrically confined ion plasmas, Hasse and Schiffer have derived stability ranges for particular ion structures as a function of the linear particle density [31]. At low enough densities and $\Gamma \simeq 1$, the ions are ordered in a linear Coulomb string ("1D crystal"). As the density is increased and $\Gamma \gg 1$, the ions evade into the transverse directions to form 2D or 3D crystalline structures. The transition to different ordered structures as a function of the linear density has been observed in linear ion traps nicely confirming the predictions of Hasse and Schiffer [2,3].

In storage rings like the TSR, a 1D Coulomb string and possibly a vertical zig-zag band are predicted to represent the only stable configurations [32]. Three-dimensional structures are excluded due to shear and other destructive effects exerted by the ring lattice. The simulations suggest that the longitudinal laser cooling rates achievable with bunched beams should suffice to generate a linear Coulomb string. In addition, the Hasse-Schiffer stability analysis tells one, that the mean distance between adjacent ions must become larger than a critical value ($d_{\text{crit}} \gtrsim 30\,\mu\text{m}$) to guarantee that the linear string represents the ground state. At higher linear densities, stable Coulomb structures are excluded in the TSR. The homogeneous ion distribution achieved in square-well buckets ensures that the Hasse-Schiffer condition of minimum ion distance is fulfilled over the whole bunch length. This is why square-well buckets are more advantageous than sinusoidal ones. Single-particle transverse cooling as described in Sect. II B has to be applied since, at the required low densities, collective

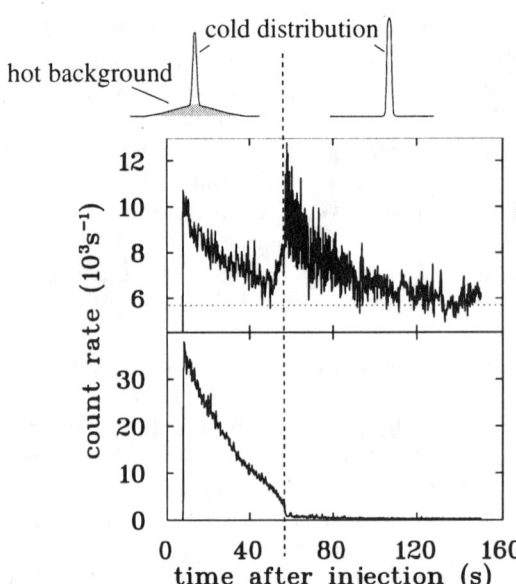

FIGURE 4. Temporal behaviour of the ion fluorescence during laser cooling. Longitudinal velocity distributions are schematically depicted on top of the figure. Upper trace: signal from the cold ion distribution (the dotted line indicates the stray light background). Lower trace: signal from the hot ion background. The two photomultiplier tubes measuring the fluorescence signals have different sensitivity. Laser cooling starts at $t = 8\,\mathrm{s}$. At $t \approx 60\,\mathrm{s}$ the hot background suddenly disappears, as seen from the lower trace, and the fluorescence emitted by the cold fraction of ion increases by nearly an order of magnitude.

transverse cooling through IBS becomes inefficient.

A Recent obervations

In a very recent series of experiments, we have investigated laser cooling in square-well buckets in combination with single-particle transverse cooling. The bunch frequency is adjusted such that the stable velocity is exactly matched with the synchronous velocity to achieve a homogeneous density distribution in the bucket. The duty cycle is set to 50%, and the capture range extension of the laser force (see Sect. II A) is applied to one pair of the HV drift tubes. As explained in Sect. II C, we can record different fluorescence signals providing information on the longitudinal velocity distribution of the ions, the fraction of cold ions, and the rate of intrabeam collisions. We measure these fluorescence signals as a function of time during the decay of the number of stored ions through rest gas collisions.

A typical example of a single time scan is presented in Fig. 4. Quite surprisingly, at about 60 s after ion injection intrabeam collisions suddenly vanish, and the fraction of cold atoms drastically increases. At the same time, the hot background of ions disappears, and the longitudinal temperature of the cold partition drops from some Kelvin to $T \approx 1\,K$. Thus, the ion distribution changes into a state where all ions gather in a narrow velocity interval around the stable velocity without further undergoing hard Coulomb collisions.

Such behaviour can be perfectly reproduced in subsequent injections, but depends extremely critical on the bunch frequency, i.e. the matching between synchronous velocity and stable velocity [d]. When, after the transition has occured, the bunch frequency is switched to a value differing by some Hz from the matched frequency, collisions and the hot background reappear. They dissappear again when the bunch frequency is switched back to its original value after some seconds. We have also shuttered the laser beam for several seconds after the disappearance of collisions. During this time interval we can naturally observe no fluorescence light. But when the laser is turned on again, again no indication for collisions or a background of hot ions is found.

---

[d] The dependence on the bunch frequency, and thus on the exact revolution frequency of the ions, is so critical (on the level of 10^{-6}) that we are limited by the stability of the storage ring magnets.

B Interpretation

Our observations show strong similarities with the anomalous longitudinal temperature reduction in dilute beams of electron-cooled, highly-charged ions at the GSI as reported by M. Steck *et al.* [33]. As an important difference, we observe the sudden disappearance of IBS heating at mean distances between the particles which are 3 orders of magnitude smaller than in the GSI experiments. The signatures we have observed so far are consistent with the formation of a Coulomb ordered structure. The transition occurs typically after 2.5 beam lifetimes when about 10^6 ions are left in the ring. The fact, that no transition occurs at larger particle numbers, i.e. smaller average particle spacings, could be interpreted in terms of the stability criterion by which the ion spacing has to be above the critical value d_{crit} to exclude unstable higher-dimensional Coulomb-ordered structures (see discussion above). The mean distance of some tens of μm, at which the transition happens, is consistent with predictions for the formation of a Coulomb string [31] as calculated with the actual TSR parameters [32]. The same holds for the temperature: At the transition we estimate from our data a longitudinal plasma parameter $\Gamma_\| \equiv e^2/(4\pi\varepsilon_0 a k_B T_\|) \approx 1$ ($2a$ = average longitudinal particle distance, $T_\|$ = longitudinal temperature). Once ordering is achieved, hard Coulomb collisions are suppressed [11,13] which would explain the lower trace in Fig. 4. Due to the reduction of this heating mechanism, the mean kinetic energy of the ion ensemble in the comoving frame is strongly decreased. Furthermore, the stability of an ion crystal should very critically depend on the bunch frequency due to the additional forces arising from a velocity mismatch (see Sect. III A), in agreement with the observed behaviour.

However, all our observations refer only to the longitudinal dynamics. Due to the limited resolution of our BPM (beam emittances $\gtrsim 10^{-3}\,\pi$ mm mrad) and the poor statistics at the low particle numbers, we can make no definite statement on the transverse beam dynamics. Strong heating of the transverse motion would lead to the disappearance of hard Coulomb collisions which would then result in signals similar to the ones shown in Fig. 4. A blowup of the transverse motion, however, would not be reversible after changes in the bunch frequency, as we know from experiments on transverse heating on coupling through ring dispersion [14]. In some preliminary measurements on the transverse degree of freedom we could also find no evidence for a dramatic transverse beam blowup. The observed disappearance of hard Coulomb collisions might alternatively indicate transverse beam instabilites. Possible sources of such transverse instabilities could be space-charge induced tune shifts [34] or transverse action of the light pressure force via higher-order longitudinal-transverse coupling processes through, e.g., the ring chromaticity. Such instabilites competing with laser cooling might possibly lead to an increase of the transverse phase-space, and thus to the observed disappearance of IBS, without necessarily resulting in a dramatic blowup of the beam. At the present stage of our analysis neither of the possibilities, beam crystallization or transverse instabilities, can be definitely proven or ruled out.

V CONCLUSIONS

The quest for crystalline beams of fast ions has entered a decisive phase. Recent achievements in laser cooling at storage rings, in particular the realization of a single-particle transverse cooling mechanism [14] and the introduction of quasi-coasting bunched beams confined in square-well buckets, have opened an intriguing new regime for the investigation of cold and dense ion beams. We have found clear evidence for a sudden disappearance of hard Coulomb collisions at longitudinal temperatures around 1 K. The transition occured when the mean distance between the ions became larger than some tens of μm. Our observations are consistent with expected features of Coulomb ordering, but they might possibly also be explained by longitudinal-transverse decoupling through transverse beam instabilities. Our limited knowledge on the dynamics of the transverse degree of motion does, at present, not allow us to experimentally distinguish between these two processes. We need further experiments, with improved diagnostics on the transverse degree of freedom, to unambiguously interpret our observations. The next laser cooling beam times at the TSR are awaited with tension and will certainly lead to further clarification.

ACKNOWLEDGEMENTS

We thank T. Schätz and C. Podlech for their support during beam times. Stimulating discussions with A. Wolf and H.-J. Miesner are greatfully acknowledged, as well as invaluable technical contributions by H. Krieger.

REFERENCES

1. Diedrich, F., et al., *Phys. Rev. Lett.* **59**, 2931 (1987); Wineland, D.J., et al., *ibid.*, 2935 (1987).
2. Birkl, G., Kassner, S., Walther, H., *Nature* **357**, 310 (1992).
3. Drewsen, M., et al., *Phys. Rev. Lett.* **81**, 2878 (1998).
4. For an overview on crystalline one-component plasmas, see contribution by J.J. Bollinger et al. in this volume.
5. Habs, D., and Grimm, R., *Annu. Rev. Nucl. Part. Sci.* **45**, 391 (1995), and references therein.
6. For a recent overview on crystalline beams see Maletic, D.M., and Ruggiero, A.G., (eds.) *Crystalline Beams and Related Issues*, Singapore: World Scientific, 1996.
7. Schröder, S., et al., *Phys. Rev. Lett.* **64**, 2901 (1990).
8. Hangst, J.S., et al., *Phys. Rev. Lett.* **67**, 1238 (1991).
9. Petrich, W., et al., *Phys. Rev. A* **48**, 2127 (1993).
10. Bosser, J., (ed.), *Proc. of the Workshop on Beam Cooling and Related Topics*, CERN Report 94-03, 1994.
11. Seurer, M., Reinhard, P.-G., Toepffer, C., *Nucl. Instr. Meth. Phys. Res. A* **351**, 286 (1994); Spreiter, Q., Seurer, M., Toepffer, C., *ibid.* **364**, 239 (1995); Seurer, M., Spreiter, Q., Toepffer, C., in Ref. [6], p. 311;
12. Schiffer, J.P., in Ref. [6], p. 217, and references therein.
13. Wei, J., Okamoto, H., Sessler, A.M., *Phys. Rev. Lett.* **80**, 2606 (1998); see also Wei, J., et al., in Ref. [6], p. 229.
14. Lauer, I., et al., *Phys. Rev. Lett.* **81**, 2052 (1998).
15. A discussion of both methods is presented by Grimm, R., et al., in *Proc. of the Workshop on Quantum Aspects of Beam Physics*, Monterey, 1998, to be published.
16. Metcalf, H., and van der Straten, P., *Phys. Rep.* **244**, 203 (1994).
17. Hangst, J.S., et al., *Phys. Rev. Lett.* **74**, 4432 (1995).
18. Miesner, H.-J., et al., *Phys. Rev. Lett.* **77**, 623 (1996)
19. Miesner, H.-J., et al., *Nucl. Instrum. Meth. Phys. Res. A* **383**, 634 (1996).
20. Atutov, S.N., et al., *Phys. Rev. Lett.* **80**, 2129 (1998).
21. Wanner, B., et al., *Phys. Rev. A* **58**, 2242 (1998).
22. Channel, P.J., *J. Appl. Phys.* **52**, 3791 (1981).
23. De Salvo, L., Bonifacio, R., Barletta, W., *Opt. Comm.* **116**, 374 (1995).
24. Calabrese, R., et al., *Opt. Comm.* **123**, 530 (1996).
25. Wolf, A., "Transverse cooling of a stored ion beam in a collinear laser beam", presented at the 2nd Workshop on Diagnostics of Laser Cooled Beams, Sandbjerk, Denmark, 1991.
26. Ellison, T.J.P., et al., *Phys. Rev. Lett.* **70**, 790 (1993).
27. Reiser, M., and Brown, N., *Phys. Rev. Lett.* **71**, 2911 (1993).
28. Nagaitsev, S.S., et al., in Ref. [10], p. 405.
29. A detailed description of the model and the comparison with experiments is going to published elsewhere: Eisenbarth, U., et al., in preparation.
30. Cambel, A.B., and Shapiro, A.H., *Real Gases*, New York: Academic Press, 1963.
31. Hasse, R.W., and Schiffer, J.P., *Ann. Phys. (N.Y.)* **203**, 419 (1990).
32. Wei, J., Li, X.P., Sessler, A.M., in *Proc. of the Particle Accelerator Conf. and Int. Conf. on High-Energy Accelerators*, 1995, p. 2946.
33. Steck, M., et al., *Phys. Rev. Lett.* **77**, 3803 (1996).
34. Bryant, P.J., and Johnson, K., *The Principles of Circular Accelerators and Storage Rings*, Cambridge: Cambridge University Press, 1993.

Storage of keV Ion Beams

Daniel Zajfman[*], Oded Heber[†], Michael Rappaport[†], Kris G. Bhushan[*]

[*]*Department of Particle Physics, Weizmann Institute of Science, Rehovot, 76100, Israel*
[†]*Physics Services, Weizmann Institute of Science, Rehovot, 76100, Israel*

Abstract. A new ion trap for storing fast (keV) ion beams is presented. The trap, which is electrostatic, stores the ions between two electrostatic mirrors. Two different examples of utilization of the trap are given. The first one required the extraction of the trapped particles after storage, in order to study their collision with an external target, while the second example measured the lifetime of the mestastable He$^-$ levels. The advantage of storage using pure electrostatic fields is discussed.

I INTRODUCTION

The use and development of ion trapping techniques, which started more than 50 years ago [1], have lead to a broad range of discoveries and new experiments in physics and chemistry. In particular, one can cite high precision spectroscopy, mass measurements, particle dynamics, nuclear and atomic processes and the measurement of fundamental constants [2]. In most of these ion traps, the ions are confined in a small region of space using a combination of electrostatic and magnetostatic or time-dependent fields. The trajectories of the ions are usually complex functions of these fields, although the motion is, in general, well understood.

Another technique used for ion trapping is the heavy-ion storage ring method. In these large-scale devices, ions are stored with high kinetic energy (MeV's) using a combination of magnetic steering and focusing fields. The use of this technique has yielded important breakthroughs in atomic and molecular physics, as well as in the understanding of beam dynamics [2,3]. The storage ring technique requires the use of an external ion source together with either an accelerator which allows direct injection at energies of several MeV's [3], or a high-voltage platform together with synchrotron acceleration capabilities in the ring itself [4,5].

In the following, we present a new type of ion trap which is located somewhere between the world of storage rings ("fast" beams with well-defined direction in space) and ion traps (small "table-top" devices). The storage time is limited only by the residual gas and is on the order of several seconds for heavy, singly charged ions at a background pressure of 1×10^{-10} Torr. Two examples of experiments which can be carried out with such a device are presented.

II ION TRAP CONFIGURATION

It is well known that the motion of charged particles in an electrostatic field is similar to the propagation of light in a medium where the index of refraction is proportional to the square root of the electric field. Based on such an analogy, we have developed an electrostatic ion storage device which is based on the same principle as the equivalent "photon storage" device which is known as an optical resonator. A detailed account of the trap operation and its characteristics has already been published [6,7].

Figure 1 shows a general schematic view of the electrostatic bottle ion trap. The ions are created in an external ion source (not shown), accelerated up to an energy of a few keV, and mass selected. After focusing and collimation, the beam is directed into the ion trap along its axis. The trap is made of two cylindrically symmetric "electrostatic mirrors", each of which is made up of a stack of cylindrical electrodes which both trap the beam in the longitudinal direction, and focus the beam in the lateral direction. Upon injection, the entrance set of electrodes is grounded so that the beam can reach the exit set of electrodes. The potentials of these electrodes are set so that the beam is stopped, reflected and focused. When the reflected beam reaches

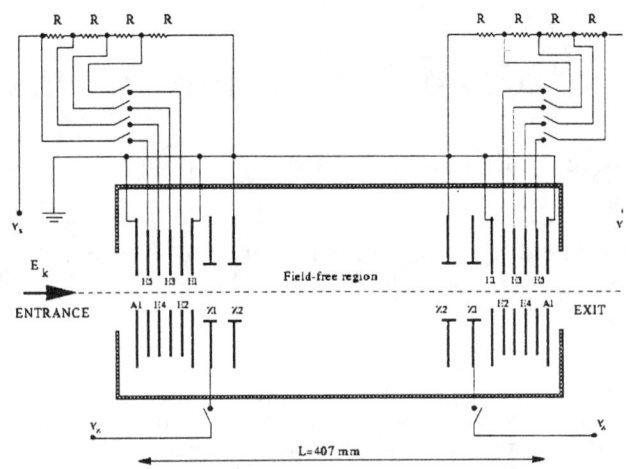

FIGURE 1. Schematic view of the ion trap.

the entrance electrodes, their potentials are rapidly switched on (about 100 ns rise time) to the same values as for the exit electrodes. For a proper choice of voltages [7], the ions are trapped between the two mirrors, and they bounce back and forth. It has already been proved, both theoretically and experimentally, that stable operation requires the focal length of the mirrors f to obey the following inequality:

$$\frac{L}{4} \le f \le \infty, \qquad (1)$$

where L is the length of the trap.

Such a condition is easy to satisfy, and depends only on the geometrical configuration of the electrodes and on the field strength. Once Eq. 1 is satisfied, the lifetime of the ion beam in the trap is mainly limited by collisional effects with the residual gas in the trap. As an example, for a 4.2 keV Xe$^+$ beam and a base pressure

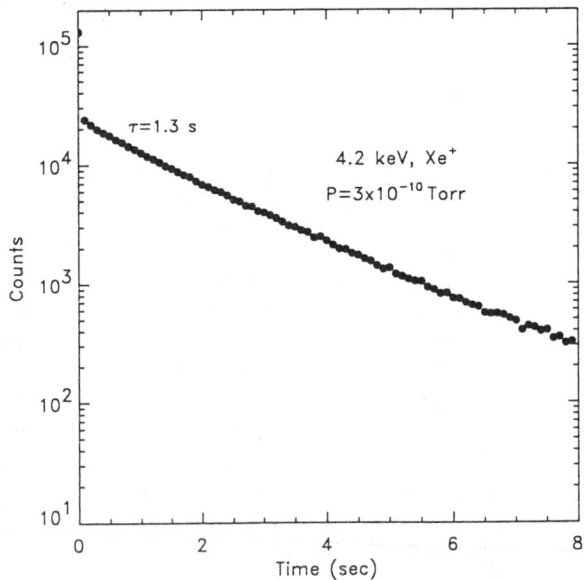

FIGURE 2. Lifetime of 4.2 keV Xe$^+$.

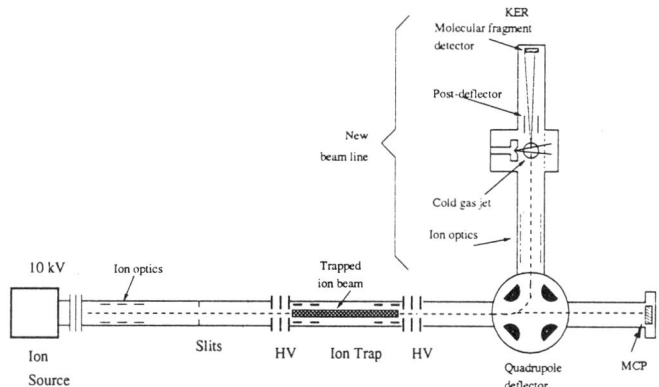

FIGURE 3. Experimental setup used for the extraction of the beam from the ion trap and collision induced dissociation with a gas jet.

of 3×10^{-10} Torr, a lifetime of about 1.3 s is obtained (see Fig. 2). The geometrical configuration of the trap, as seen in Fig. 1, allows for easy injection and extraction of the beam, by switching the electrodes on and off. Fig. 3 shows an overview of the whole setup, including an extraction beam line which allows to direct the extracted beam toward an external target.

III FIRST EXPERIMENTAL RESULTS

A large variety of new experiments can be carried out with the system described above, and in the following we are presenting two examples. The first is related to the vibrational cooling of molecular ions in the trap, and their dissociation after electron capture, while the second deals with lifetime of metastable states of negative ions.

A Capture Induced Dissociation of Vibrationally Cold HeH$^+$

Molecular ions are known to be highly reactive species and are of special importance in astrophysical and laboratory plasmas. The various dissociation processes due to collisions with neutral atoms are of great importance for understanding the behavior of these media. Among these processes, the dissociative charge exchange (DCE) where an electron is transferred from the atomic target to the molecular ion projectile plays an important role as it produces neutral atomic fragments with relatively large kinetic energies. In a typical DCE process, a molecular ion collides with an atomic target, and captures one of its electrons, leaving the target ionized. The electron can be captured either in a bound or dissociative state, which then leads to the production of atomic fragments. For the molecular ion HeH$^+$ colliding with Ar, this process can be depicted as

$$HeH^+(\nu) + Ar \rightarrow He(n_{He}) + H(n_H) + Ar^+, \qquad (2)$$

where ν is the initial vibrational state of the molecular ion, and n_{He} and n_H are the final quantum states of the He and H, respectively. A schematic description of the process using the potential curves of HeH and HeH$^+$ is shown in Fig. 4. It is clear that the final kinetic energy release is a direct function of the initial vibrational state of HeH$^+$. Fig. 4 shows the expected kinetic energy release distribution for a ground state HeH$^+$, assuming that the electron capture is independent of the internuclear distance (the "reflection" model).

One of the main problems in the experimental investigation of the DCE process is related to the initial vibrational state ν of the molecular ion beam. It is well known that standard ion sources produce molecular ion beams with a wide distribution of initial vibrational states, which makes it difficult to obtain results for the DCE process under controlled conditions.

Using the setup shown in Fig. 3, we have performed the first DCE experiment with HeH$^+$ in the ground vibrational state. A 4.2 keV HeH$^+$ beam was produced from a standard electron impact ion source, injected

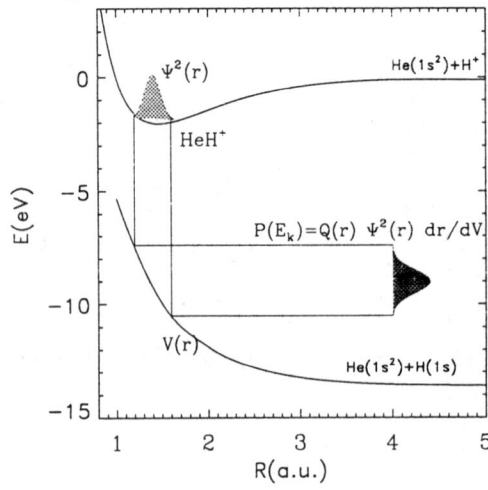

FIGURE 4. Potential curves of HeH and HeH$^+$ involved in the DCE process. The kinetic energy release plotted on the right side is calculated assuming that the electron capture is independent of the internuclear distance

in the ion trap, and stored for a time which is long enough to achieve complete vibrational relaxation through natural ro-vibrational transitions. The beam was then extracted from the trap, by lowering the voltages of the exit electrodes, and was directed with the help of a quadrupole electrostatic deflector toward a free gas jet of argon. About 10^5 particles were extracted in a bunch width of about $2\mu s$. The Ar gas jet density was set so that less than one DCE event was produced per extraction.

The kinetic energy release of the He and H fragments in the center-of-mass frame of reference was then analyzed using a three-dimensional position and time detector [8]. Fig. 5 shows the results for two different trapping times, where the effects of vibrational cooling can be clearly seen. After 10 ms, no more changes in the kinetic energy distribution were observed, in agreement with theoretical calculations which predict that ro-vibrational transitions in HeH$^+$ are of the order of few ms [9]. Complete analysis of the kinetic energy distribution, based on the initial wave function of the vibrational ground state of HeH$^+$, is underway.

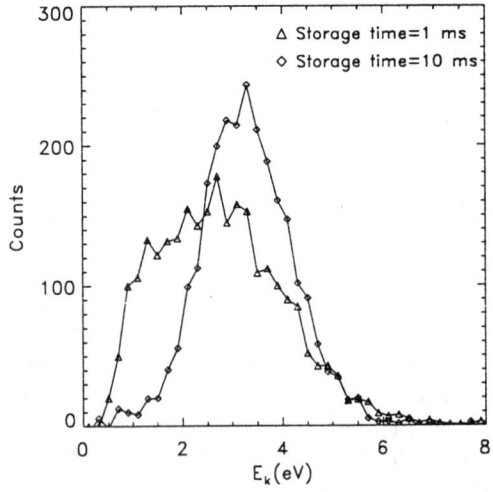

FIGURE 5. Kinetic energy release in the for the dissociative charge exchange reaction of HeH$^+$ with Ar after two different storage time in the trap.

B Lifetime of metastable He⁻

The physics of negative ions has attracted extensive experimental and theoretical attention during the last decades. The experimental progress has been very much related to the introduction of new techniques which allow for detailed study of the negative ion structure and lifetime. During the last decade, storage rings have been an important tool for such studies [10] as they have made possible the long time storage of heavy-ion beams stored at energies between tens of keV to few MeV. For stable negative ions, the limit on the storage time is due to neutralization through collisions with the residual-gas, setting an upper limit of a few seconds. For weakly bound systems (tens of meV), the decay induced by blackbody radiation represents another major restriction on the storage time, which is very much dependent on the value of the binding energy, but can be as short as a few hundreds microseconds [11]. Such storage enables the study of the lifetime of metastable negative ions in the range of 10μs-100 ms. However, one of the main drawbacks of the heavy-ion storage ring technique is the presence of magnetic fields which can mix the magnetic substates from the different, but close-lying, fine-structure components with the same magnetic quantum number. One of the simplest negative metastable negative ions is He⁻, which is known to be formed in the $1s2s2p\ ^4P$ state, and is bound by 77 meV relative to the first excited state $1s2s\ ^3S$ of Helium. This ion has received a great deal of attention, both theoretically [12–18] and experimentally [19–24]. The He⁻ is known to be metastable and the decay of the three fine structure components ($^4P_{5/2}$, $^4P_{3/2}$, and $^4P_{1/2}$) is due to spin-orbit or spin-spin coupling [19]. Calculations and experiments have shown that the $^4P_{3/2}$ and $^4P_{1/2}$ have much shorter lifetime than the $^4P_{5/2}$ as the decay mode of the latter is induced by spin-spin interaction only.

On the experimental side, the most accurate measurement of the $^4P_{5/2}$ was conducted by Andersen et al. [24], using the heavy-ion storage ring ASTRID. As pointed out above, the ring is equipped with a number of dipole and quadrupole magnets to store the beam. In order to correct for magnetic field effect, Andersen [24], have measured the lifetime of He⁻ at different beam energies, thus sampling different values of the magnetic fields. The data were then extrapolated to zero magnetic field using a two-parameter fit based on a theoretical function which takes into account the Zeeman mixing in the dipole magnetic field. Although the presence of magnetic field makes the direct measurement of the $^4P_{5/2}$ lifetime difficult, it has, as pointed out by Andersen [24], the advantage of providing information on the lifetime of the short lived $^4P_{3/2}$ state through the fitting procedure.

In the present experiment, we have measured the lifetime of He⁻ using the electrostatic ion trap [6,7], thus avoiding all together the presence of magnetic fields. A He⁺ beam is produced by an electron impact ionization source, accelerated to an energy of 4.2 keV, selected by a Wien filter, and subsequently passed through a windowless target cell filled with cesium vapor, produced by a small oven. It is well known that He⁻ can be efficiently produced from He⁺ by double charge exchange with cesium atoms at keV energy [25]. In the present case, about 0.25% of the He⁺ was transformed to He⁻, resulting in a beam of ≈ 0.4 nA.

Injection and trapping are performed at a repetition rate of 30 Hz. For each injection, the rate of neutral particles hitting an MCP downstream, after the quadrupole, is measured as a function of storage time. Fig. 6 shows this time dependence for a total about 50,000 injections. The spectrum can clearly be divided into three different components: two exponential decays, and a constant. The spectrum was fitted with such a function, with a total of five free parameters, and the resulting fit is shown as the solid line in Fig. 6. We have assigned the fast decay to the lifetime of the average of the $^4P_{3/2}$ and $^4P_{1/2}$ states of He⁻, and the slow decay to the $^4P_{5/2}$ state. The flat background at times greater than 1.5 ms is due to the noise in the MCP detection system. The lifetimes obtained with the fitting procedure as described above are $\tau_{<3/2,1/2>} = 8.8 \pm 0.1 \mu$s for the mean value of the $^4P_{3/2}$ and $^4P_{1/2}$ states and $\tau_{5/2} = 290 \pm 2 \mu$s for the $^4P_{5/2}$ level.

Numerical integration of the decay induced by blackbody-radiation has already been performed by Andersen [24], and yielded a decay rate of 0.534 ms^{-1} at room temperature. Subtracting this decay rate from the measured values, the lifetime of the $^4P_{5/2}$ increases to $\tau_{5/2} = 343 \pm 10 \mu$s, where the error bar is mainly due to the uncertainty in the blackbody-radiation cross section. The mean lifetime for the $^4P_{3/2}$ and $^4P_{1/2}$ states changed by only 0.1 μs: $\tau_{<3/2,1/2>} = 8.9 \pm 0.2 \mu$s. The $^4P_{5/2}$ lifetime is in very good agreement both with the storage ring experiment carried out by Andersen [24] and with the last theoretical value calculated by Miecznik, Brage, Froese Fischer [18].

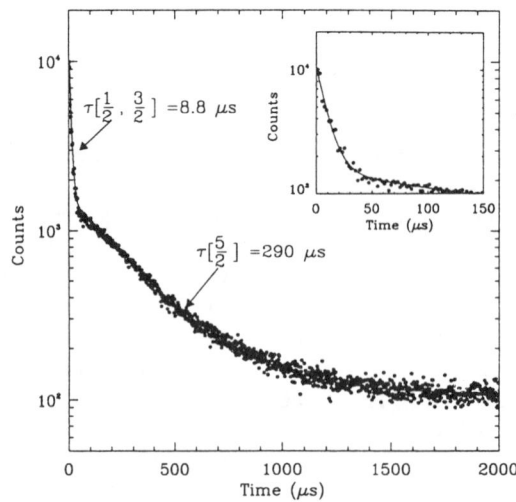

FIGURE 6. Neutral He signal from the channel plate detector as a function of time. The solid line is the fit to the data as described in the text. The two lifetimes $\tau_{<1/2,3/2>}$ and $\tau_{5/2}$ are not corrected for blackbody radiation induced decay. Inset: expanded scale for short times.

IV CONCLUSIONS

The above results are only preliminary and are presented here only to demonstrate the power of our electrostatic ion trap. Because the trap is based on electrostatic fields only, there are no limits to the mass of the ions which could be stored, so that very heavy clusters can be trapped with such a system. Injection and extraction of beams in the trap is very easy, and more complex trap geometries are possible, including that of a "bend" trap where an electrostatic deflector is located at the center of the trap. Bunching of the ions has been tried, using a small additional time-dependent voltage on one of the electrodes. This allows to increase the local density of ions, a technique which is useful for crossed-beam experiments. The advantageous configuration of this trap will open new and exciting possibilities in other fields, and it is expected that new types of experiments will be suggested and carried out in the near future.

This work was supported by the Minerva Foundation and by the Federal Ministry of Education, Science, Research and Technology (BMBF) within the framework of the German-Israeli Project Cooperation in Future-Oriented Topics (DIP).

REFERENCES

1. W. Paul, and H. Steinwedel, Z. Naturfoschung, **A8**, 448 (1953).
2. For a recent review on ion trapping, see Phys. Scr. **T59** (1995).
3. D. Habs et al., Nucl. Instrum. Methods Phys. Res. B **43**, 390 (1989).
4. R. Stensgaard, Physica Scripta, **T22**, 315 (1988).
5. K. Abrahamsson et al., Nucl. Instr. Meth. **B79**, 269 (1993).
6. D. Zajfman, O. Heber, L. Vejby-Christensen, I. Ben-Itzhak, M. Rappaport, R. Fishman, and M. Dahan, Phys. Rev. A **55**, 1577 (1997).
7. M. Dahan, R. Fishman, O. Heber, M. Rappaport, N. Altstein, D. Zajfman, and W. J. van der Zande, Rev. Sci. Instrum. **69**, 76 (1998).
8. Z. Amitay and D. Zajfman, Rev. Sci. Instrum. **68**, 1 (1997).
9. S. Datz and M. Larsson, Phys. Scripta **46**, 343 (1992)
10. L. H. Andersen, T. Andersen, and P. Hvelplund, Adv. Atom. Mol. Opt. Phys. **38**, 155 (1997).
11. H. K. Haugen, L. H. Andersen, T. Andersen, P. Balling, N. Hertel, P. Hvelplund, and S. P. Möller, Phys. Rev. A **46**, 1 (1992).
12. G. N. Estberg and R. W. LaBahn, Phys. Lett. **28A**, 420 (1968).
13. C. Laughlin and A. L. Steward, J. Phys. B **1**, 151 (1968)
14. G. Miecznik, T. Brage, and C. Froese Fischer, (unpublished).
15. T. Brage, and C. Froese Fischer, Phys. Rev. A **44**, 71 (1991).
16. G. N. Estberg and R. W. LaBahn, Phys. Rev. Lett. **24**, 1265 (1970).
17. B. F. Davis, and K. T. Chung, Phys. Rev. A **36**, 1948 (1987).
18. G. Miecznik, T. Brage, and C. Froese Fischer, Phys. Rev. A **47**, 3718 (1993).
19. L. M. Blau, R. Novick, Phys. Rev. Lett. **24**, 1268 (1970).
20. R. Novick and D. Weinflash, in *Proceedings of the International Conference on Precision and Fundamental Constants*, Natl. Bur. Stand. (US) Spec. Publ. No. 343, edited by D. N. Langenberg and N. N. Taylor (U.S. GPO, Washington, D.C., 1970), p. 403.
21. D. J. Nicholas, C. W. Trowbridge, and W. D. Allen, Phys. Rev. **167**, 38 (1968).
22. F. R. Simpson, R. Browning, and H. B. Gilbody, J. Phys. B **4**, 106 (1971).
23. G. D. Alton, R. N. Compton, and D. J. Pegg, Phys. Rev. A **28**, 1405 (1983).
24. T. Andersen, L. H. Andersen, P. Balling, H. K. Haugen, P. Hvelplund, W. W. Smith, and K. Taulbjerg, Phys. Rev. A **47**, 890 (1993).
25. B. L. Donnally and G. Thoeming, Phys. Rev. **159**, 87 (1967).

Storage Rings at RIKEN RI Beam Factory

M. Wakasugi[*], Y. Batygin[*], N. Inabe[*], T. Katayama[†], K. Maruyama[†], K. Ohtomo[*], T. Ohkawa[*], M. Takanaka[*], T. Tanabe[*], I. Tanihata[**], S. Watanabe[†], Y. Yano[*], and K. Yoshida[**]

[*]*RI Beam Factory Project Office, RIKEN, Wako, Saitama 351-0198, Japan*
[**]*Linac Laboratory, RIKEN, Wako, Saitama 351-0198, Japan*
[†]*Center of Nuclear Study, Univ. of Tokyo, Tanashi, Tokyo 188-0002, Japan*

Abstract. We will construct two different types of storage rings at RIKEN Radioactive Isotope Beam Factory (RIBF). Accelerator complex including these storage rings is called MUSES (Multi-USe Experimental Storage rings). One is an accumulator cooler ring (ACR). An electron cooling device and a stochastic cooling device will be installed in the ACR. The RI beams quickly cooled by the combination of the two cooling methods are not only used for experiments at the ACR but also transported to double storage rings (DSR). The DSR is a new type of storage ring. One ring of the DSR is for RI (heavy ion) beams, and another ring accepts not only heavy ion beams but also an electron beam. In this paper, we present outline of the RIBF project and two kinds of unique experiments planned at the DSR i.e. the RI-electron collision experiment and the RI-X-ray collision experiment.

OUTLINE OF RIBF PROJECT

The Radioactive Isotope Beam Factory (RIBF) is an expansion of the existing heavy ion accelerators facility at RIKEN[1]. The construction of the RIBF is separated into two phases. The first phase consisting of an intermediate ring cyclotron (IRC), a superconducting ring cyclotron (SRC), two RI beam separators (Big RIPS) and some experimental halls has been started from this year. The second phase is named MUSES (Multi-USe Experimental Storage rings) project[2]. The MUSES is an accelerators complex consisting of an accumulator cooler ring[3] (ACR), a booster synchrotron ring[4] (BSR), a 300-MeV electron linac[5] (e-linac) and double storage rings[6] (DSR). Figure 1 shows a plan view of the RIBF. Heavy ion beams from the RRC (Riken Ring Cyclotron K=540) are boosted up to 400A MeV for light ions and more than 100A MeV for heavy ions by the IRC (K=950) and the SRC (K=2500). With this beam energy, we can produce RI beams for all elements using projectile-fragmentation process. Details of the SRC and the IRC are described elsewhere[7,8].

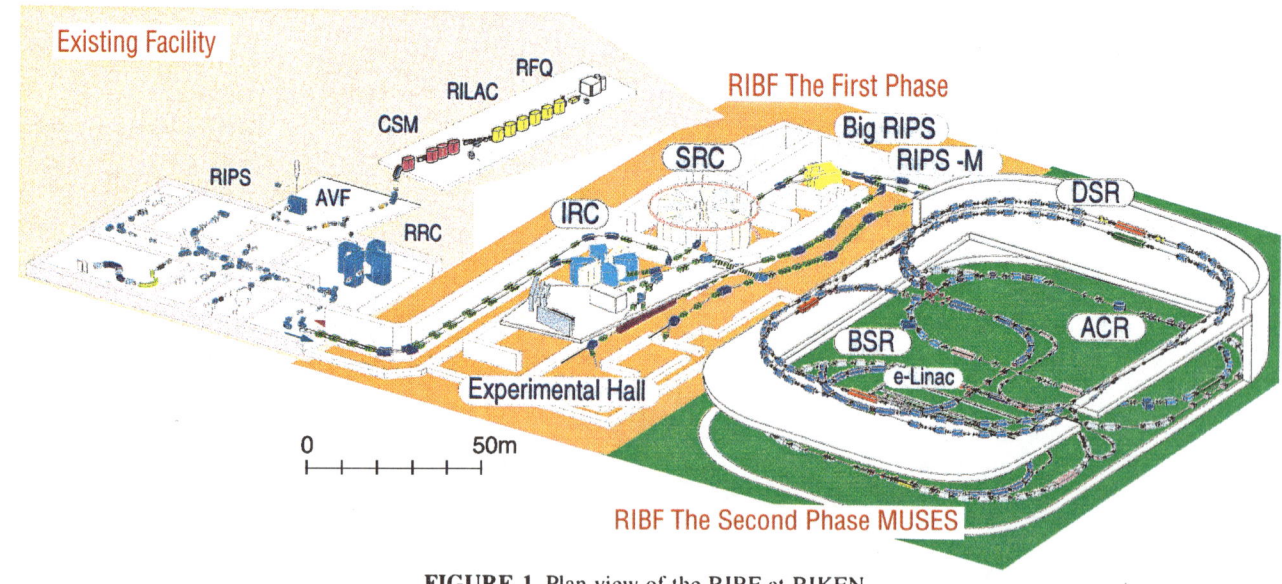

FIGURE 1. Plan view of the RIBF at RIKEN.

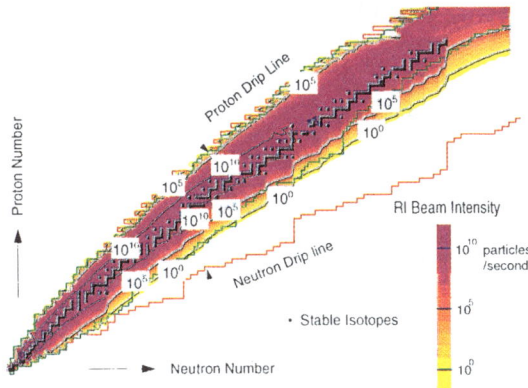

FIGURE 2. Estimated RI beam intensity from the Big RIPS. The intensity of the primary beam is assumed to be 1 pμA. The primary beam energy and target thickness are optimized in this calculation.

RI Beam Separators

At the downstream of the SRC, we construct three RI beam separators. Two separators (Big RIPS) provide RI beams for the experimental halls, and another (RIPS-M) for the MUSES system. The primary beam is supplied for three separators with time sharing technique (see Fig. 4). A pulse beam with the beam intensity of 100 particles μA (pμA) is supplied for RIPS-M, and the maximum duty factor is 10^{-3} (the beam duration of 30 μsec and the interval of 30 msec). DC beams with the intensity of 1 pμA are supplied for the Big RIPS. RI beam intensity from the Big RIPS estimated using the computer code INTENSITY2[9] is shown in Fig. 2. About 3000 radioactive isotopes including about 1000 new isotopes can be used for experiments. The RIPS-M used for the MUSES has a momentum acceptance of ±2.5 % and an angular acceptance of ±10 mrad. The momentum spread of the RI beams from the RIPS-M is expected to be ±0.5 %. Since this value is too large from the cooling time in the ACR point of view, we place debunchers at about 80-m downstream of the RIPS-M. The momentum spread is reduced to ±0.15 % by the debunchers. The maximum RF voltage required here is totally 4.23 MV[10].

Accumulator Cooler Ring (ACR)

Figure 3 shows schematic view of the MUSES system. The RI beams are injected into the ACR by means of a multi-turns injection method (about 30 turns per one injection). Injected RI beam is stacked by controlling the supplied RF voltage and the frequency in the ACR. During RF stacking process, the beam is cooled down in both the transverse and the longitudinal directions by combination of a stochastic cooling[11] and an electron cooling methods[12]. A cycle of the injection, i.e. the multi-turns injection, the RF stacking and the cooling, is repeated until that the number of stored particles reaches to the equilibrium number which depends on the lifetime of the RI and the space charge limit. This cycle is shown in Fig. 4(b). The combination of the stochastic cooling and the electron cooling makes the cooling time shorter than that for the case of only the electron cooling. The cooling time is, roughly speaking, less than 1 sec for all RI beam in our estimation. If we do not need the cooling, only the stacking process takes about 30 msec. This is why the maximum duty factor of primary beam for the RIPS-M is 10^{-3}. Details of the ACR, the injection system and the cooling devices installed in the ACR are described

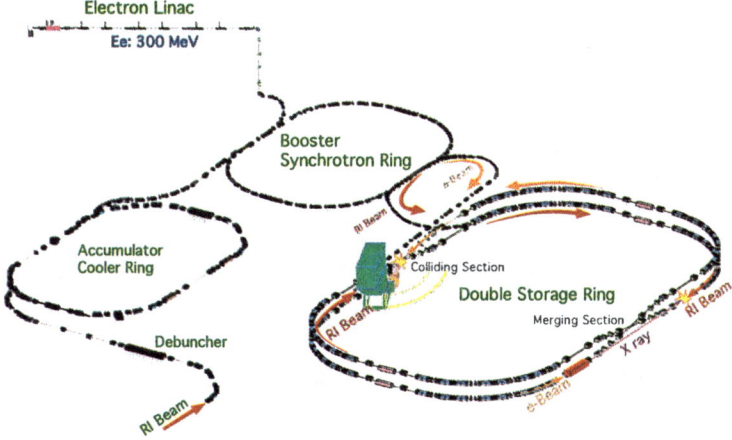

FIGURE 3. Schematic view of the MUSES system.

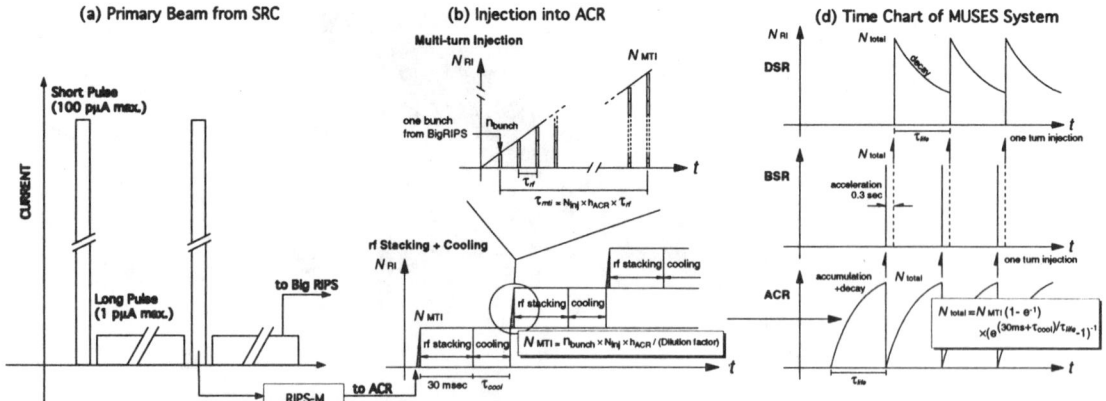

FIGURE 4. Time charts of ion beam in the MUSES system. (a) indicates the time sharing of the primary beam from the SRC, (b) the injection cycles, and (c) the time charts of the ACR-BSR-DSR complex.

in Refs. 3, 10, 11 and 12. The ACR itself is not only a cooling device but also an experimental device. We provide an electron cooler, schottky devices, four dispersive positions in arcs section (the maximum dispersion is 4.52 m), two achromatic straight sections where internal targets can be placed, etc. in the ACR so that the ACR is responsible to various experiments. Some experiments have been already proposed at the ACR, and they are presented elsewhere.

Booster Synchrotron Ring (BSR)

The cooled RI beam is extracted from the ACR and injected into the BSR to boost up to the required energy, and the beam is immediately transported to the DSR as shown in the time chart of Fig. 4(c). The BSR has a circumference of 179.7 m, the maximum magnetic rigidity of 14.6 Tm, a repetition rate of 1 Hz and the acceleration time of 0.3 sec. The ion beams can be accelerated up to 1.4 GeV for proton and 0.8A GeV for uranium. Relatively wide range of RF frequency of 25-53 MHz is required to boost up to the maximum energy[13]. Two kind of extraction methods are provided[14], which are a fast (one turn) extraction and a slow extraction using 1/3 resonance technique. The fast extraction is for transporting the beams to the DSR, and the slow extraction is used for experiments at the experimental halls. The BSR can accept not only ion beams coming from the ACR but also an electron beam from the e-linac. The electron beam can be accelerated from 300 MeV up to required energy, the maximum energy is 4.8 GeV. Depending on the use of the electron beam at the DSR, either the single bunch or the full bunch operation mode is chosen in the BSR. Corresponding to that, the operation mode of the e-linac is also changed to the short pulse mode (1-nsec pulse length, 1-A peak current) or the long pule mode (5 μsec, 100 mA).

Double Storage Rings (DSR)

As shown in Fig. 3, the DSR is a new type of experimental storage ring that consists of vertically stacked two rings which are called e-ring and I-ring, respectively. It has a circumference of 269.5 m and two colliding points in long straight sections which are called the colliding section and the merging section, respectively. The colliding section is for nearly head-on colliding experiments and the crossing angle is 20 mrad. The RI-electron collision experiment, which is described later, is planned at the colliding section. The betatron function of the RI beams and the electron beam at the colliding point are designed to be 10 cm and 2 cm, respectively, and the collision length is 10 cm. On the other hand, the merging section with

TABLE I. Specifications of the ion and the electron beams in the presently designed DSR.

	e-Ring		I-Ring	
	Large emittance mode	Small emittance mode	Colliding mode	Merging mode
Harmonic No.	450	450	48	48
Momentum compaction	0.042	0.0014	0.039	0.039
Betatron tune v_x/v_y	6.754/8.164	16.046/9.106	6.235/5.018	5.637/5.732
Emittance $\varepsilon_x/\varepsilon_y$ (πμmrad)	0.97/0.01 at 1 GeV	0.0016/0.0047 at 1 GeV	1.0/1.0 typical	1.0/1.0 typical
Betatron function β_x/β_y (m)	0.02/0.02 [a]		0.1/0.1 [a]	0.6/0.6 [b]
Natural chromaticity ξ_x/ξ_y	-37.7/-90.7	-29.7/-34.7	-62.7/-47.6	-11.4/-10.3
Momentum spread ($10^4 \Delta E/E$)	2.64 at 1 GeV	2.74 at 1 GeV		
Radiation loss (keV/turn)	10.6 at 1 GeV	10.6 at 1 GeV		

[a] at the colliding point.
[b] at the merging point.

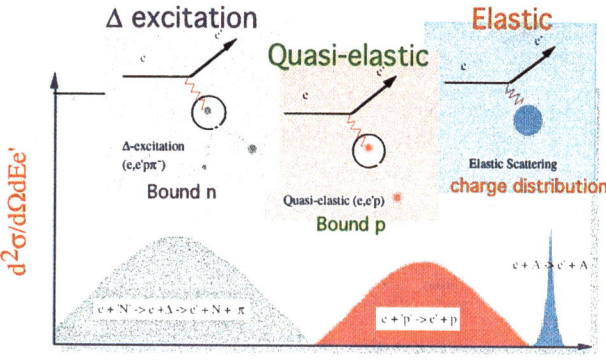

FIGURE 5. Schematic drawing of spectrum of scattered electron in RI-electron collision.

the crossing angle of 175 mrad is for the ion-ion merging experiments. In this section, we can make low energy collision experiments such as a fusion reaction. In this straight section, the RI-X-ray colliding section is also provided. An undulator is installed as a source of high-brilliant X ray in this section. Details of this experiment is described later. The DSR has different operation modes corresponding to different types of collision experiments. There are the colliding mode and the merging mode for the RI (ion) beams. For the electron beam, we have the small emittance operation mode required to produce high-brilliant X ray, and the large emittance operation mode is also required to get larger luminosity for the RI-electron collision experiment. According to requirements for the small emittance mode, a double bend achromatic (DBA) lattice is adopted in the arc sections, and the emittance of order of 10^{-9} mrad is presently designed. On the other hand, the emittance for the large emittance mode is designed to be about 10^{-6} mrad. Specifications of the ion and the electron beams for each operation mode are summarized in Table I, and details of design of the DSR is described in Refs 6 and 15.

RI-ELECTRON COLLISION EXPERIMENT

One of unique experiments at the DSR is the RI-electron collision[16]. Figure 5 shows schematic view of typical spectrum observed in the electron scattering experiment. The sharp peak at the highest energy indicates the elastic scattering (e,e') from which we can determine the nuclear charge distribution of RI's. Physical interests are the proton skin structure, the neutron skin/halo structure, difference in collective structure between protons and neutrons etc. in RI's that have unbalanced numbers of protons and neutrons. This experiment allow us to make systematic study on these problems. This kind of study that has so far been performed only for light elements can be extended to heavier elements. The second peak in Fig. 5 shows the quasi-elastic scattering (e,e'p). This experiment is useful to determine the momentum distribution and the wavefunction of protons existing around nuclear surface. This experiment for RI's can be performed at only the DSR. The third peak is the Δ-excitation reaction (e,e'pπ). Because the Δ-excitation is the M1 transition, we can find many information for not only protons but also neutrons in RI's. In this chapter, we concentrate and describe about the elastic scattering experiment.

Nuclear Charge Distribution

As well known, measured differential cross sections of the elastic scattering are represented using the cross section for the point charge (Motto scattering) and the form factor as

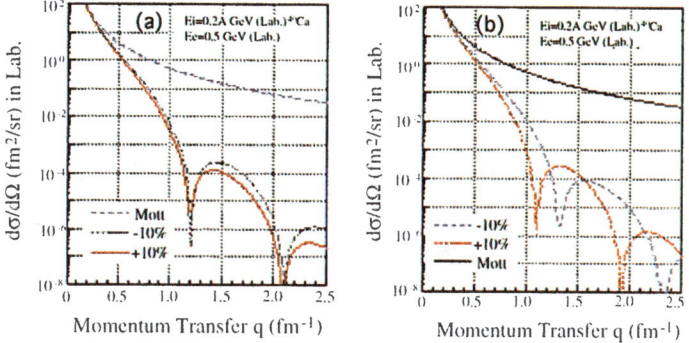

FIGURE 6. The sensitivity of the cross section against the change in the diffuseness (a) and the nuclear radius (b). In this example, the ^{40}Ca nuclei with the energy of 200A MeV and the 500-MeV electron beam are used. The Woods-Saxon type distribution is assumed. Two curves of the cross sections are for ±10-% different parameters.

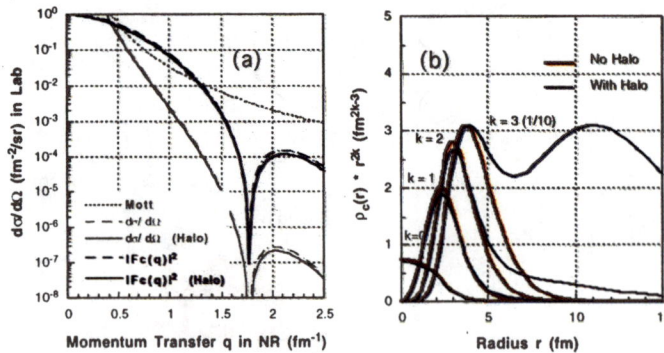

FIGURE 7. The sensitivity against the halo component for ^{8}B case. (a) and (b) shows the cross sections and k-th order of $\rho(r)r^{2k}$ values, that the k-th-order moments are give by integrating this function, with and without the halo component, respectively. Parameters are c=2.3 fm, z=0.57 fm, ρ_0=0.05569, x=1.83 and ρ_0'=0.00283.

$$\frac{d\sigma}{d\Omega} = \left(\frac{d\sigma}{d\Omega}\right)_{point} |F(q)|^2 \quad , \tag{1}$$

where q is the momentum transfer. The nuclear charge distribution is expressed by Fourier transform of the form factor

$$\rho(r) = \frac{Ze}{2\pi^2} \int_0^\infty F(q) e^{-iqr} q^2 \, dq \quad . \tag{2}$$

In our estimation, required range of q value is 0.5 - 2 fm^{-1} to determine the charge distribution around the nuclear surface. This corresponds to the scattering angle from 10 deg. to 60 deg. in the laboratory frame for the case of electron beam energy of less than 1 GeV and the RI beam energy of less than 1A GeV.

First, we have checked the sensitivity of the expected cross section against the nuclear charge distribution and the halo component. Cross sections for the case of ±10-% changes in the surface diffuseness (a) and the nuclear radius (b) are shown in Fig. 6, where we used the Woods-Saxon type distribution. It is found that the cross section is insensitive against the change in the diffuseness but quite sensitive against the change in the nuclear radius. This means that less than 0.1-fm sensitivity with respect to the nuclear charge radius can be expected. The sensitivity against the halo component is shown in Fig. 7. The charge distribution with the halo component is assumed as

$$\rho(r) = \frac{\rho_0}{1+\exp\left(\frac{r-c}{z}\right)} + \rho_0' \exp\left(-\frac{r}{x}\right) \quad , \tag{3}$$

where the second term is the halo component. The halo component gives small change in the cross section but about 16-%

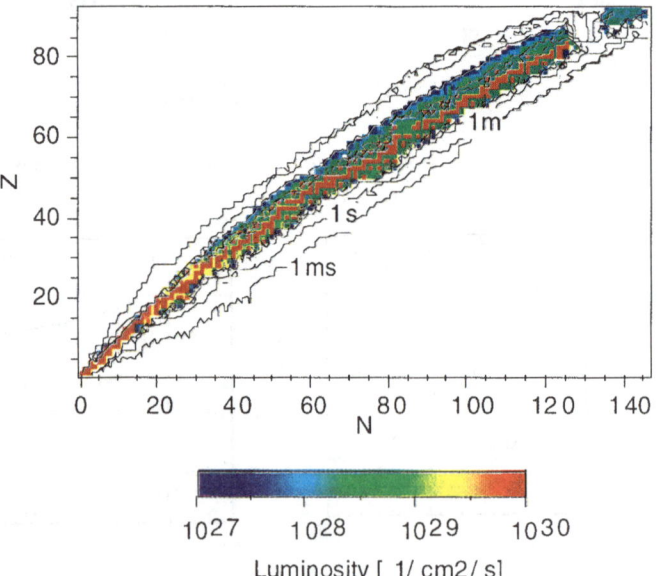

FIGURE 8. Luminosity of the RI-electron collision at the DSR. Bunch length is assumed to be 50 cm for RI beam and 2 cm for electron beam.

change in the second-order moment $<r^2>^{1/2}$ and 42-% change in the third-order moment $<r^4>^{1/4}$. This means that we need sufficient yield to derive some information about the halo structure from the measured cross section and the measurement of moments gives us clear evidence of existing the halo component.

Luminosity and Yield Estimations

The essential point of this experiment is that how big luminosity can we have at the DSR. The luminosity of the head-on collision of beams with Gaussian distribution in space is calculated as

$$L = n_1 n_2 c N_c E(\beta_1,\beta_2,\varphi) \frac{1}{(2\pi)^3} \int \frac{dx\,dy\,dz\,dt}{\sigma_{x1}\sigma_{y1}\sigma_{z1}\sigma_{x2}\sigma_{y2}\sigma_{z2}} exp\left[-\frac{1}{2}\left\{\frac{x_1^2}{\sigma_{x1}^2} + \frac{x_2^2}{\sigma_{x2}^2} + \frac{y_1^2}{\sigma_{y1}^2} + \frac{y_2^2}{\sigma_{y2}^2} + \frac{(z_1-c\beta_1 t)^2}{\sigma_{z1}^2} + \frac{(z_2-c\beta_2 t)^2}{\sigma_{z2}^2}\right\}\right] \quad (4)$$

where n_i (i=1,2) is the number of particle in a bunch, N_c the number of collisions in unit time, σ_x, σ_y and σ_z bunch size in x, y, and z directions, respectively, φ the collision angle, suffix i=1, 2 represent the electron and the RI beams, and $E(\beta_1,\beta_2,\varphi)$ expressed as

$$E(\beta_1, \beta_2, \varphi) = \sqrt{\beta_1^2 + \beta_2^2 + 2\beta_1\beta_2 \cos\varphi - \beta_1^2\beta_2^2 \sin^2\varphi} \quad . \quad (5)$$

The calculated luminosity is shown in Fig. 8, where we took into account the RI production rate, the lifetime of RI and the cooling time at the ACR, and the electron beam current is assumed to be 500 mA. We have more than 10^{27} cm^{-2}sec^{-1} for RI's having the lifetime of longer than 1 min.

From the calculated luminosity, we estimated the yield of the scattered electrons, and some examples are sown in Fig. 9. It has been found that the minimum luminosity to determine the nuclear charge distribution is 10^{27} cm^{-2}sec^{-1}, and for the case of 10^{23}-10^{27} cm^{-2}sec^{-1}, we can determine the root mean square charge radius $<r^2>^{1/2}$. Our estimation concludes that the nuclear charge distribution can be determined for RI's having the lifetime of longer than 1 min, and the root mean square charge radii can be determined for RI's with the lifetime of longer than 1 sec.

Beam-Beam Effect

We have to consider the instability problems for both the ion amd the electron beams in the DSR, because the instability strongly limits the luminosity. The beam-beam interaction, which is nonlinear phenomenon, results in an change of the betatron tune of the ion beam, because the electron beam is much stronger than the ion beam. The strength of the beam-beam interaction is defined by the linear part of the betatron tune shift ξ represented as

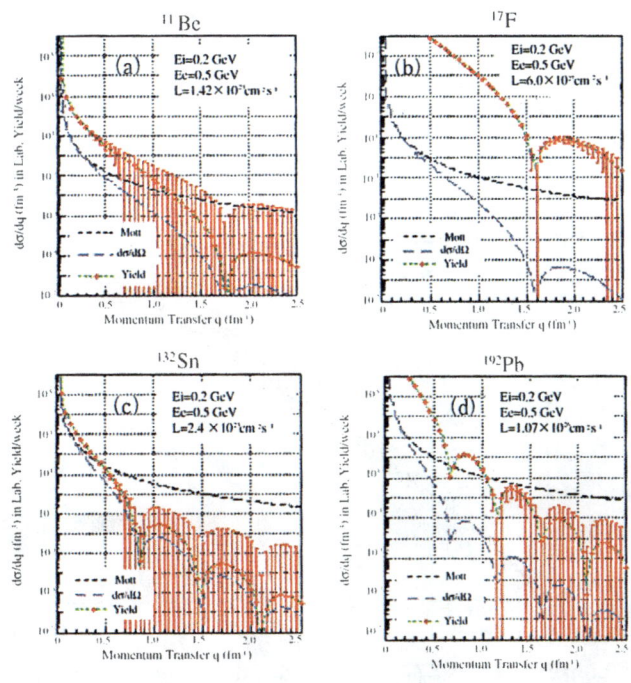

FIGURE 9. Examples of estimated yield per week.

$$\xi = \frac{r_p \, \beta_x \, (Z/A) \, N_e \, (1+\beta_e\beta_i)}{4 \, \pi \, \gamma_i \, \beta_i^2 \, \sigma_e^2} \quad , \tag{6}$$

where N_e is the number of electrons per bunch, β_x the betatron function at the colliding point, β_e, β_i velocity of electrons and ions, respectively, γ_i a reduced ion energy, σ_e the half-size of the electron beam envelope and r_p the classical radius of a proton. In our simulation study[17], the phase space trajectory of particles are ellipses for the ξ of smaller than 0.01. In the case of $\xi > 0.01$, the 7th, the 14th and the higher order characteristics overlap each other, and behavior of particles are chaotic (unstable). Taking into account the influence of random noise which fluctuates electron beam size, the maximum tune shift is found to be $\xi=0.005$ to maintain the stable beam trajectory[18]. Finally, the limitation of the luminosity is found to be

$$L \leq 2 \times 10^{18} \, N_i^{total} \quad cm^{-2} \, sec^{-1} \quad , \tag{7}$$

where N_i^{total} is the total number of ions in the DSR. To compensate of the instability, we will install an electron cooler in the DSR.

RI-X-RAY COLLISION EXPERIMENT

The purpose of this experiment is to determine the mean square nuclear charge radii $<r^2>$ and the electromagnetic moment by means of isotope shift measurements in the 2S-2P atomic transitions of the Li-like RI ions[19,16]. We provide an undulator and an X-ray spectrometer as a monochromatic X-ray source in the e-ring of the DSR. The advantage of this experiment is that this method can be applied to small number of RI's stored in the DSR because the atomic transition has much bigger cross section compared with nuclear reactions.

According to the multi-configuration Dirac-Fock calculation[20], the excitation energy of $2S-2P_{1/2}$ (D1) transition has less than 300 eV for all element. X ray in this energy region is easy to produced at the DSR. The isotope shift in this transition is expected to be larger than the other charge states. Because of simple electronic structure, calculations for nuclear effects in the isotope shift is possible with high accuracy. These are why we choose the Li-like charge state as a target of this experiment. The problem is that we have to make RI ions having three electrons. Figure 10 shows the production rate of the Li-like RI beam from the RIPS-M. Because of required ion beam energy for the projectile-fragment reaction, the Li-like ions can be produced for elements of Z>36. So this experiment is mainly applied to heavier elements.

Requirements for RI beam and X ray

We wish to derive the $<r^2>$ values with the error of less than 5 % from the isotope shifts. The requirement for RI beams is that the momentum spread $\Delta P/P$ should be 10^{-4} - 10^{-3} for Z=40 - 92. This beam is provided in two ways: one is the use of the RI beam cooled by the ACR (storage mode), and another way is that the RI beam is directly transported from the RIPS-M to the DSR after momentum selection by slits at the RIPS-M (direct mode). Both ways have advantages and disadvantages. Former way is useful for RI's having relatively long lifetime because of the cooling time at the ACR, and latter way is useful for RI's having shorter lifetime and large production rate. Depending on the lifetime and the production rate, we can choose either way. On the other hand, requirements for the X ray are follows. The X-ray energy is 30 - 800 eV to excite the D1 transition of Z>40 elements, and the energy resolution is about $\Delta E_x/E_x=10^{-4}$. The X ray intensity should be

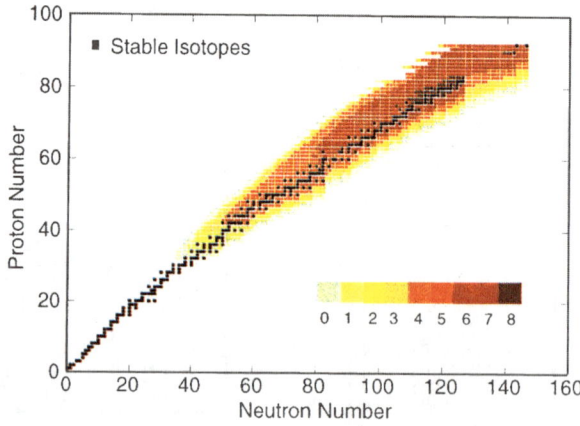

FIGURE 10. Production rate of Li-like RI beams. Conditions such as the primary beam energy, target thickness, etc. are optimized to produce Li-like RI ions. The primary beam intensity is assumed to be 1 pµA.

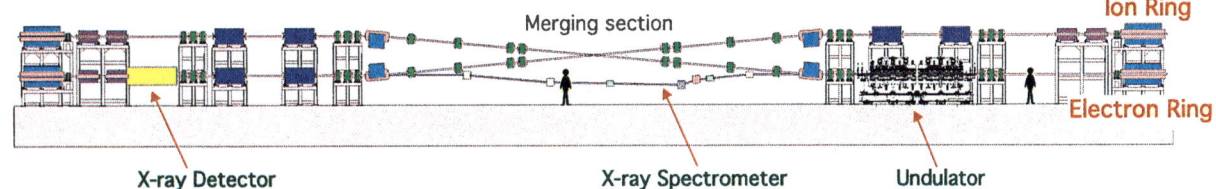

FIGURE 11. Experimental setup of the RI-X-ray collision experiment at the DSR.

at least 10^{12} photons/sec/0.01%b.w. at the RI-X-ray colliding section.

Experimental setup

Figure 11 shows experimental setup installed near merging section in the DSR. This consists of an undulator, an X-ray spectrometer and a fluorescence X-ray detector. High-brilliant X ray is produced by the undulator, its energy resolution is reduced to 10^{-4} by the X-ray spectrometer, and the X ray injected again into the DSR collides with RI ions at the detector position (RI-X-ray colliding section). The resonance energy of the D1 transition is shifted due to the Doppler effect. Calibration of the resonance energy is performed in two ways. One is the velocity measurement using schottky signals in the ring. Another way is that the measurement is repeated for RI beams propagating in both parallel and antiparallel directions to the X ray beam. From measured two Doppler shifted resonance energy of E_p and E_{ap}, we can derive true resonance energy E_0 as

$$E_0 = \sqrt{E_p E_{ap}} , \qquad (8)$$

independently of the ion beam velocity.

The Undulator and the Photon Flux

Presently designed undulator is the Apple-II type of undulator[21] so that it is possible to scan not only the energy but also the polarization. The length of a period is 3 cm, which consists of 16 permanent magnets of Nd-Fe-B, and the total length is 4.8 m including 160 periods. The tunable range of gap width is 20 - 27 mm, corresponding to the bending factor K of 0.712 - 0.35. Required X ray energy range is covered by changing both the gap width and the electron beam energy from 0.3 GeV to 1.7 GeV. Figure 12 shows calculated photon flux from the undulator at the electron beam current of 500 mA. Assuming the transmission efficiency of the X-ray spectrometer $\varepsilon_{spt}=10^{-3}$, the intensity of X-ray at the colliding section exceeds the minimum intensity required from the experiment. In this case, quality of the electron beam in the DSR is important. As shown in Table I, the electron beam emittance in the small emittance mode is order of 10^{-9} mrad and the beam size at the undulator section is about 100 μm. This can produce high brilliant X ray as shown in Fig. 12. The specifications for the small emittance mode is the same like the third generation synchrotron light source. In such machine, instability of the electron beam is always big problem especially at lower energy. The instability is caused by the ring broadband impedance and the narrowband impedance at the high-Q cavities. We are now investigating the instability[22] and designing the vacuum tube and cavities of the DSR.

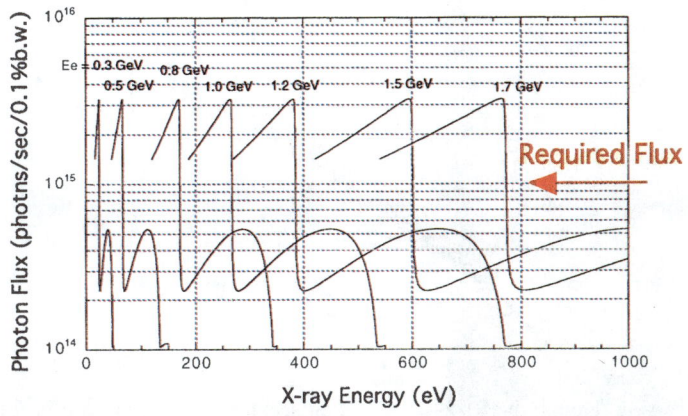

FIGURE 12. Calculated X-ray flux from the presently designed undulator.

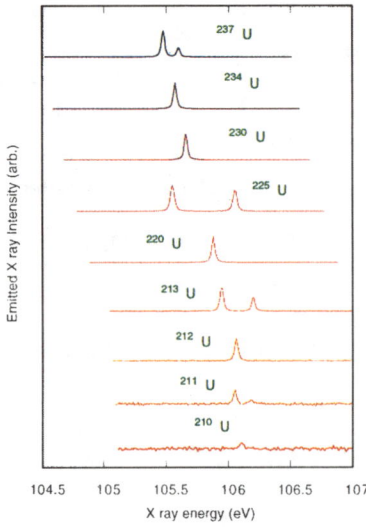

FIGURE 13. Expected spectrum for uranium isotopes for the storage mode. The following conditions are assumed: the ion beam energy of 500A MeV, the momentum spread of 0.1 %, the beam size of 2 mm in diameter, the X ray energy resolution of 10^{-4}, the photon flux of 10^{12} photons/sec/0.01%b.w., and the cross section 2×10^{-13} cm^2.

Simulations

Taking into account expected conditions of the RI beams and the X ray mentioned above, we have simulated the expected fluorescence spectrum and results for uranium isotopes are shown in Fig. 13. The bottom spectrum is for ^{210}U which is close to the proton dripline, and about 80 ions of ^{210}U are stored in the DSR. In this simulation study, we tried to decrease the number of stored particles in the DSR, and it is found that only one ion stored in the DSR is enough to make isotope shift measurements. Assuming the maximum measurement time of 1 month and averaged counting rate of noise of 0.01 counts per second, we tried to find the lower limit of the number of particles for this experiment. This simulation results that the required minimum number of ions passing through the RI-X-ray colliding section in unit time is 10^4 ions/sec for this experiment. Figure 14 shows the number of ions passing through the RI-X-ray colliding section in unit time for both the

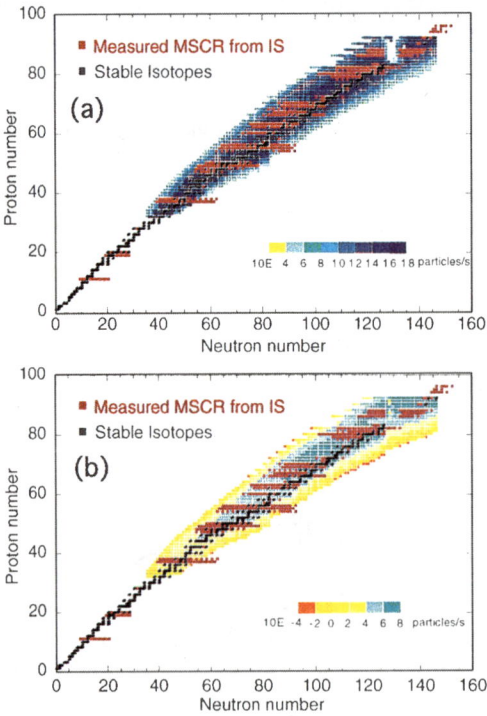

FIGURE 14. Number of ions passing through the RI-X-ray colliding section in unit time. (a) is for the storage mode and (b) is for the direct mode.

storage mode and the direct mode described above. Especially, the direct mode is useful for isotopes located at unstable valley around N=130, because these isotopes have short lifetime and large production rate. The total number of isotopes having more than 10^4 ions/sec is about 1900 isotopes.

CONCLUDING REMARKS

In the MUSES project, we will construct two storage rings the ACR and the DSR. The DSR is a unique storage ring because not only the RI beams but also the electron beams can be stored at the same time. This allow us to perform new type of collision experiments such as the RI-electron collision and the RI-X-ray collision. According to many requirements of various experiments, the DSR has been designed to be a highly potential machine.

The luminosity of the RI-electron collision is obtained to be mode than 10^{27} cm^{-2}ces^{-1} for RI's having the lifetime of longer than 1 min. The nuclear charge distribution of these RI's are possible to be measured, and the root mean square nuclear charge radius $<r^2>^{1/2}$ is measured for RI's having the lifetime of longer than 1 sec. To get such higher luminosity, the RF stacking-cooling process at the ACR is quite important.

With respect to measurement of the mean square nuclear charge radii, the RI-X-ray collision experiment further expands the boundary of the limit of the measurement. The measurements for RI's having the lifetime of order of 1 μsec is possible using the direct mode. Totally, we can make measurements for about 1900 isotopes with Z>36. The key of this experiment is the production of the intense monochromatic X ray i.e. providing a large-current small-emittance electron beam in the DSR. The details of design of the accelerator system and the detector system are in progress.

REFERENCES

1. Yano Y., et al., Proc. of PAC97, 930 (1998).
2. Katayama T., Nucl. Phys. A626, 545c (1997).
3. Ohtomo K., et al., Proc. of PAC97, 1072 (1998).
4. Ohkawa T., et al., Proc. of PAC97, 1024 (1998).
5. Kamino Y., et al., Proc. of EPAC98, 722 (1998).
6. Inabe N., et al., Proc. of PAC97, 1400 (1998).
7. Goto A., et al., Proc. of 16th Cyclotron and Their Applications, to be published.
8. Kawaguchi T., et al., Proc. of PAC97, 3419 (1998).
9. Winger J.A., et al., Nucl. Instrum. Meth., B70, 380 (1992).
10. Ohtomo K., et al., Proc. of EPAC98, 2126 (1998).
11. Inabe N., et al., Proc. of EPAC98, 1037 (1998).
12. Watanabe I., et al., Proc. of EPAC98, 2258 (1998).
13. Watanabe I., et al., Proc. of EPAC98, 1823 (1998).
14. Ohkawa T., et al., Proc. of EPAC98, 2123 (1998).
15. Inabe N., et al., Proc. of EPAC98, 897 (1998).
16. Tanihata I., Nucl. Phys. A588, 253c (1995).
17. Batygin Y., et al., RIKEN Accel. Prog. Rep., 29, 254 (1995).
18. Batygin Y., et al., Proc. of EPAC96, 1170 (1997).
19. Wakasugi M., et al., Proc. of EPAC96, 611 (1997).
20. Cheng K.T., et al., Atom. Data Nucl. Data Tables 24, 111 (1979).
21. Wakasugi M., et al., Proc. of PAC97, 3521 (1998).
22. Wakasugi M., et al., Proc. of EPAC98, 1017 (1998).

Clusters in Storage Rings

P. Hvelplund, J.U. Andersen and K. Hansen

Institute of Physics and Astronomy
University of Aarhus, DK – 8000 Aarhus C, Denmark

Abstract. Anions of fullerenes and small metal clusters have been stored in the storage rings ASTRID and ELISA. Decays on a millisecond time scale are due to electron emission from metastable excited states. For the fullerenes the decay curves have been interpreted in terms of thermionic emission quenched by radiative cooling. The stored clusters were heated by a Nd:YAG laser resulting in increased emission rates. With an OPO laser this effect was used to study the wavelength dependence of the absorption of light in hot C_{60}^- ion molecules.

INTRODUCTION

The present paper describes studies of lifetimes of cold and hot clusters in ASTRID (the Aarhus STorage RIng Denmark). The first lifetime studies of clusters in the new electrostatic storage ring ELISA (ELectrostatic Ion Storage Ring Aarhus) are also included. ASTRID (1) has a perimeter of 40 m, two bending magnets in each of the four corners, 16 quadrupoles and 16 correction dipoles (cf. Fig. 1). The negative clusters were produced in a plasma source or in a sputter source and were accelerated and mass selected before injection into the ring at an energy around 50 keV. The rate of neutral particles detected behind one of the dipole magnets was then recorded as a function of time after injection and a so-called lifetime spectrum was obtained. The clusters are normally "born" in the ion source with a broad distribution in internal excitation ('temperature') and the decay of hot clusters by electron emission is observed. The injected clusters can also be heated in the ring in one of the straight sections with the beam from a Nd:YAG laser or an OPO pumped by this laser.

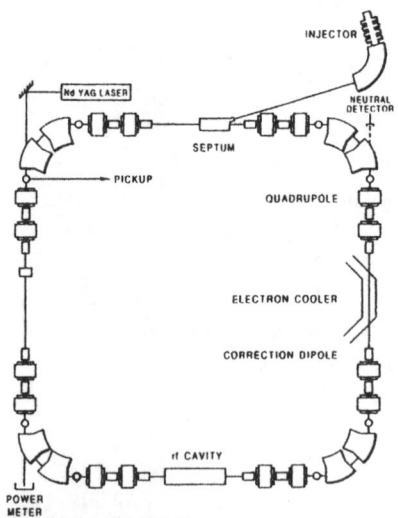

Figure 1. Layout of the ASTRID storage ring with injector and a laser for beam heating.

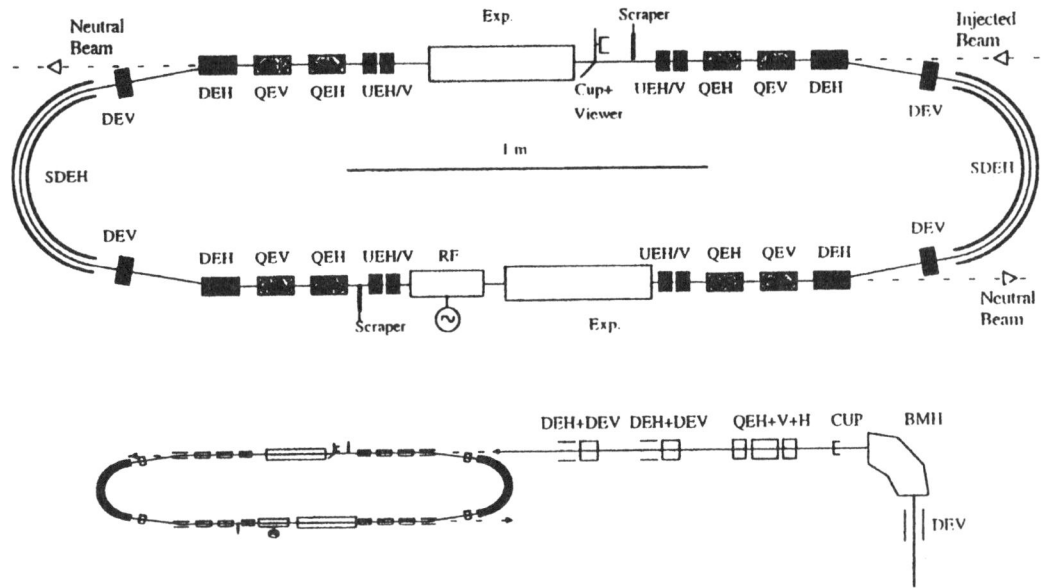

Figure 2. Layout of the ELISA storage ring (top) with injector (bottom). SDEH designates the spherical 160° horizontal deflectors, DEH and DEV horizontal and vertical electrostatic deflectors, QEH and QEV horizontally and vertically focusing electrostatic quadrupoles, UEH and UEV horizontal and vertical pick-up electrodes, RF the drift-tube RF-system, and BMH the separator magnet in the injector

The electrostatic storage ring ELISA (2) has a race-track shape as shown in Fig. 2. The injection system with a separator is also shown. ELISA has a perimeter of 7.62 m with two 160° spherical electrostatic deflectors, four 10° parallel plate deflectors and four pairs of electrostatic quadrupoles in two straight sections. The revolution time of 25 keV C_{60} ions is about 100 µs as compared to about 500 µs for C_{60} ions at the same energy in ASTRID. At present, the first test experiments have been performed on ELISA with C_{60}^- and improved short time (~ 1 ms) information has been obtained as compared to similar measurements at ASTRID. We are currently testing an electro-spray ion source (3) for production of multiply charged biomolecules and plan in the near future to install this ion source on ELISA in order to study free protein molecules. Both the spontaneous decay of hot, metastable molecules and decay induced by laser excitation can be measured, in analogy to the cluster studies.

LIFETIME STUDIES OF FULLERENE ANIONS

A typical lifetime spectrum for C_{60}^- is shown in Fig. 3. The storage lifetime is limited by collisional destruction in rest gas interactions and the decay has been observed to follow an exponential law. The lifetime is ~ 10 s at a rest gas pressure of ~ $2 \cdot 10^{-11}$ mbar. The high point at $t \approx 0$ shows that a major part of the C_{60}^- ions decay at very short times, by auto-detachment from hot

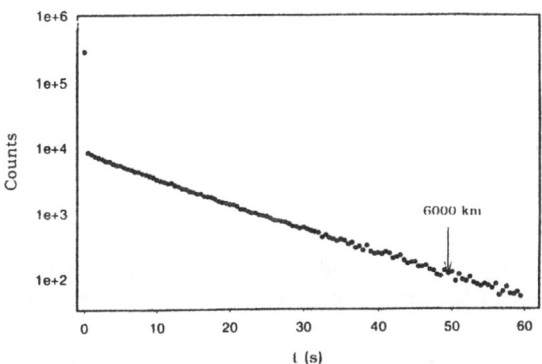

Figure 3. The yield of neutral particles at the detector as a function of time after injection of 50 keV C_{60}^-. The average presure in the ring was 2×10^{-11} mbar, and the collisions-induced lifetime is found to be ~7 s. The arrow points to the time where the fullerene ions have traversed a distance of 6000 km.

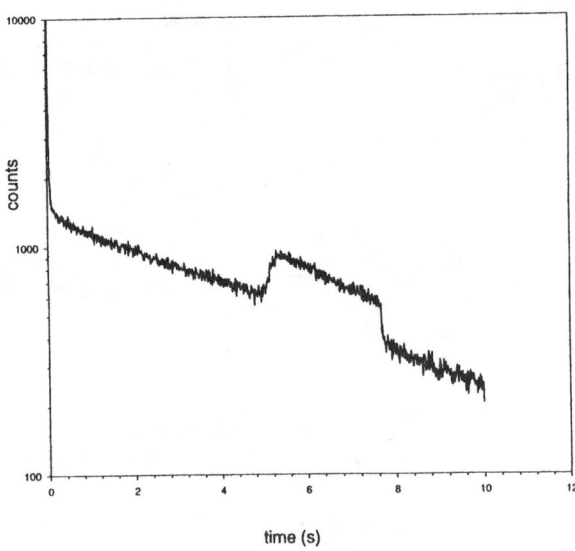

Figure 4. A spectrum like the one shown in Fig. 3 but with a 4.5 W Nd:YAG laser beam merged with the ion beam in a 2.6 s time interval starting 5 s after ion injection in the ring.

molecules (4, 5, 6). We shall return to this point later. In Fig. 4 is shown the effect of laser heating of the beam. A 4.5 W Nd:YAG laser is turned on 5 sec after ring injection and turned off 2.6 s later. In this time interval the detachment rate is increased by almost a factor of 2 due to heating and autodetachment of the circulating C_{60} anions.

LIFETIME STUDIES OF SMALL NEGATIVE METAL CLUSTERS

An example of lifetime measurements involving small metal clusters is shown in Fig. 5 (see next page). The decay curves for anions Al_n^- ($2 \leq n \leq 7$), which are "hot" when they leave the sputter source, are observed to be similar to the one observed for fullerene anions (7). The rate is seen to drop two orders of magnitude within the first few milliseconds. For these metal clusters the electron affinity and the binding energy per atom are similar (1-2 eV) (8) and the hot anions may decay via either electron emission or unimolecular fragmentation. Radiative cooling of the hot metal clusters is also believed to be active but has not yet been modelled. The stored ions were heated by a CW YAG laser with a power of ~ 25 W from 10 ms to 140 ms after injection. Note that the intensity of the detected neutrals drops for Al_2^- when the laser is on whereas it increases for the larger clusters. This observation indicates that destruction of Al_2^- is a "prompt" process whereas destruction of the larger clusters is delayed by at least half the revolution time in the ring, ~ 0.1 ms. This is consistent with the fact that the electron binding for Al_2^- is smaller than the photon energy, 1.17 eV. It should also be noted that the decay rate within the first few ms changes dramatically as a function of cluster size. The variation is probably related to differences in the rate of radiative cooling.

STATISTICAL DESCRIPTION OF THE DECAY AND COOLING OF STORED IONS

Only if the injected cluster ions are stable against all forms of spontaneous particle decay will the evolution of the number of stores particles be described by a simple exponential time dependence due to destruction in collisions with the rest gas. If clusters possess internal energy, decay channels such as unimolecular fragmentation and thermionic emission open up (5). Figure 6 shows the decay of stored C_{60}^- ions in the millisecond range after subtraction of the nearly constant contribution from rest-gas collisions, illustrated in Fig. 3. The rapid decay on the millisecond time scale is caused by electron emission, since the electron affinity of C_{60} (2.7 eV) is much smaller (9, 10) than the activation energy for unimolecular fragmentation ~ 10 eV (11).

The functional form of the decay rate at short times carries information about the internal state of the clusters and the development of that state over time. In the atomic case, when all or a fraction of the ions are in a well-defined metastable state, the decay is described by a single exponential. If several metastable states are populated, the decay function will be more complicated, containing exponentials with different lifetimes. However, in the limit, where many states are populated, with a broad distribution of lifetimes, the decay function again becomes simple and the decay is described approximately by a t^{-1} law (5, 12).

This description may be expected to apply to negative fullerene ions stored in ASTRID or ELISA. In the ion source, the

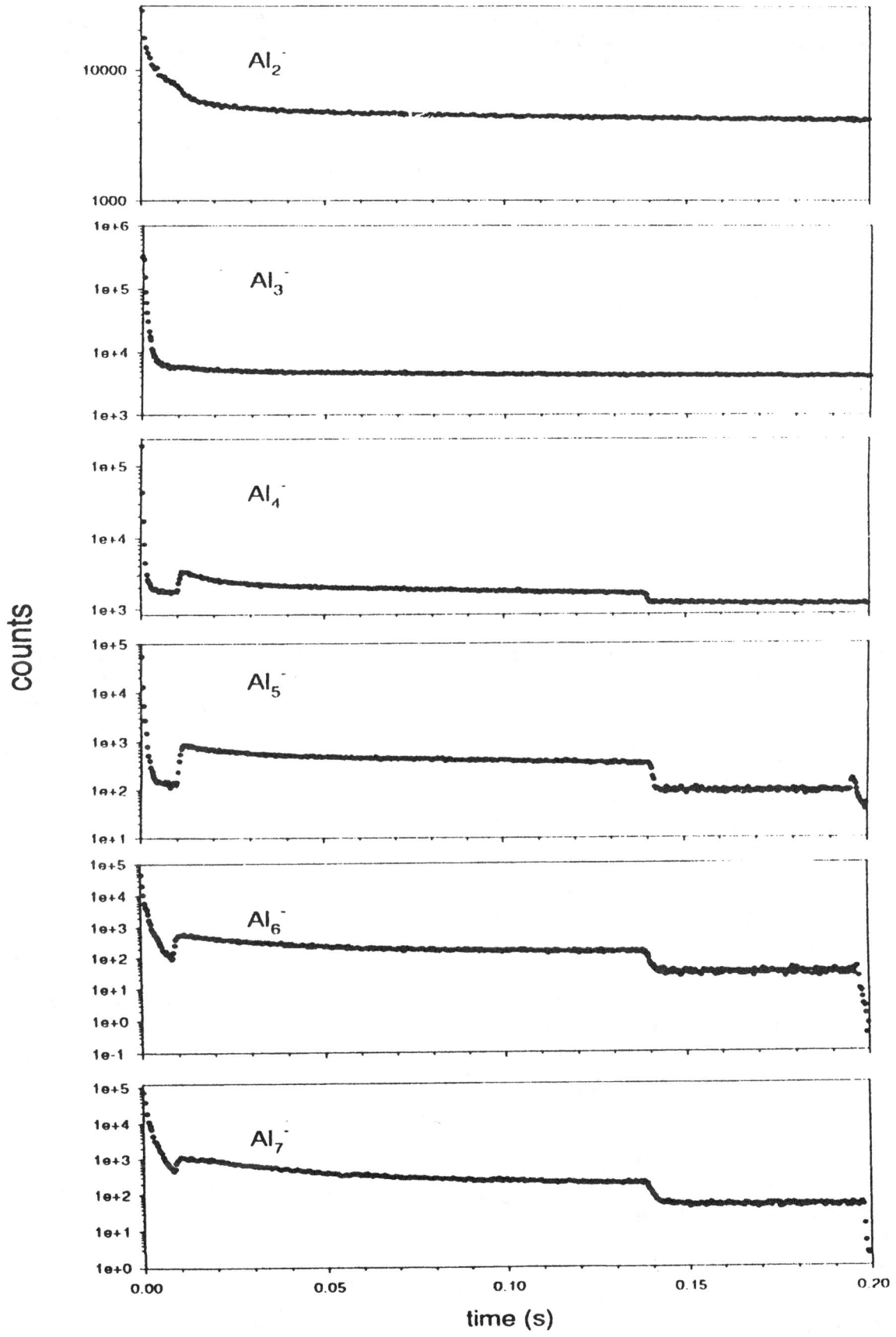

Figure 5. Lifetime spectra of Al_n^- ($2 \leq n \leq 7$). The beam from a 25 W Nd:YAG laser has been merged with the ion beam in the time interval from 10 to 140 ms.

clusters are bombarded with electrons, and the extracted negative ions have a broad distribution in excitation energy. Furthermore, even at moderate excitation energies, the vibrational level density is enormous. With the new storage ring ELISA we have extended our earlier measurements at ASRID to shorter times, as seen in Fig. 6a. In the range 0.2 – 1 ms, the expected t^{-1} law is seen to describe the decay quite well, while the signal decreases faster at longer times. The results from ASTRID in Fig. 6b illustrate the nearly exponential decrease in the millisecond range. We have interpreted this as evidence for radiative cooling of the hot clusters (5). Just as for the negative atomic ions with very small binding energy, the interaction with the radiation field can have a strong influence on the decay rate (13), but here it is a quenching of the decay due to *emission* of radiation.

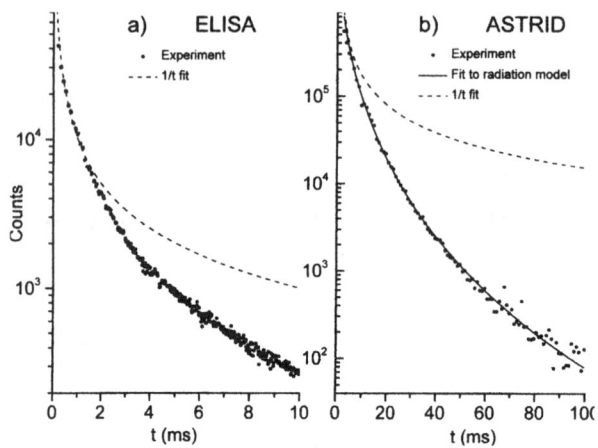

Figure 6. Rate of decay by thermionic emission of a stored C_{60}^- beam. A contribution from collisions with the rest gas has been subtracted.

As discussed in Ref. 5, there are strong theoretical arguments and experimental evidence for the description of electron detachment from excited C_{60}^- ions as a thermally activated process analogous to thermionic emission from a hot filament. Without cooling, the distribution is depleted from the high-energy side by electron emission. Cooling by radiation will quench the electron emission when the increase of the lifetime becomes significant on the time scale given by the decay rate.

The curve through the data points in Fig.6b is a fit calculated from a statistical model of the competition between electron emission and cooling. The radiation intensity is about 190 eV/s at an internal temperature of 1500 K and is approximately proportional to T^7. The major contribution to the radiative cooling comes from a thermally stimulated transition at 1.16 eV of the electron attached to C_{60} to form the anion (14, 15). In the following, we describe a spectroscopic study of absorption of radiation by this transition in hot C_{60}^- molecules.

THERMIONIC EMISSION LASER SPECTROSCOPY OF C_{60}^-

Thermionic emission of electrons from clusters is, as we have seen, enhanced by absorption of photons. The process can therefore be used to monitor the wavelength dependence of the photo-absorption cross sections of hot molecules and clusters. A stored C_{60}^- beam can at a preselected time be irradiated with a pulse from a tunable, Nd:YAG pumped OPO laser. In Fig. 7 is shown a neutrals spectrum, where the laser was fired 7.1 ms after C_{60}^- injection in the ring. The resulting enhanced electron

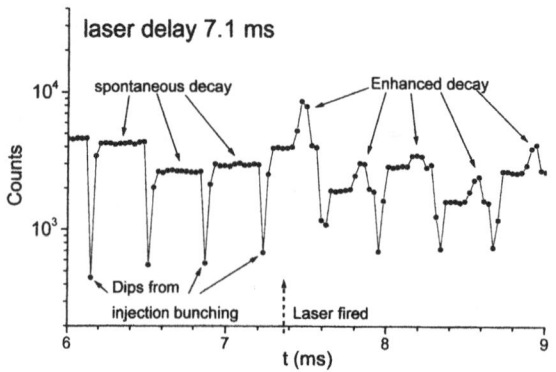

Figure 7. "Enhancement" spectrum for a laser firing time of 7.1 ms.

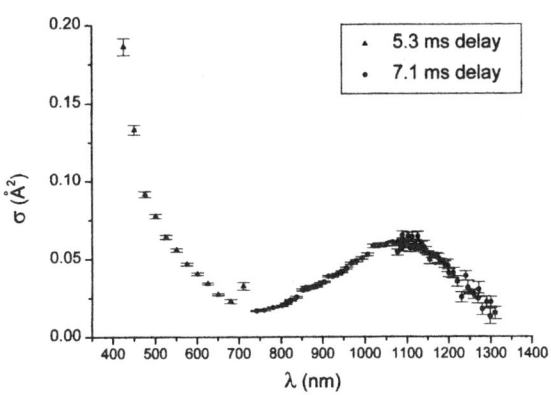

Figure 8. Photon absorption cross section as a function of laser light wavelength at a laser firing time around 5–7 ms.

emission, measured with a ~ 200 µs delay, reflects the photon absorption cross section. The absorption strength decreases with photon wavelength up to 700 nm, followed by a broad absorption peak around 1070 nm (Fig. 8). This absorption peak has been studied extensively for C_{60}^- ions at lower temperature, for example for C_{60}^- in solution (14, 15) and the data in Fig. 8 have been normalized to give the same peak area as obtained in those experiments. The peak is ascribed to the lowest transition ($t_{1u} \rightarrow t_{1g}$) of the additional electron in the anion. Strong sidebands at shorter wavelengths have been observed, corresponding to vibrational excitation. At the high temperatures of the stored ions (~ 1400 K according to the analysis illustrated in Fig. 6b), such excitations should be stronger because dipole matrix elements increase with excitation of an oscillator, $|<n+1|x|n>|^2 \propto n+1$, and also hot bands with vibrational deexcitation should be observed. These expectations are consistent with the observed, very broad absorption peak. We had hoped to be able to deduce a temperature from the asymmetry between the sidebands corresponding to vibrational excitation and deexcitation, respectively, but since the sidebands are not resolved there is a large uncertainty in the analysis from the broadening and a possible shift of the central electronic transition.

ACKNOWLEDGEMENT

This work was supported by the Danish National Research Foundation through the Aarhus Center for Atomic Physics (ACAP). Collaboration with the staff of ISA (Institute for Storage Ring Facilities Aarhus) is greatly appreciated.

REFERENCES

1. Møller, S. P., *Conference Record of the 1991 IEEE Particle Accelerator Conference*, San Francisco, ed.: K. Berkner, p. 2811 (1991).
2. Møller, S. P., *Nucl. Instr. & Meth. A* **394**, 281 (1997); Møller, S.P., in *Proceedings of the 6th European Particle Accelerator Conference*, Institute of Physics Publishing, Stockholm 1998.
3. Fenn, J. B., Mann, M., Meng, C. K., Wong, S. F. and Whitehouse, C.M., *Mass. Spec. Rev.* **9**, 37 (1990).
4. Hvelplund, P., *Phys. Scripta* **T59**, 244 (1995)
5. Andersen, J. U., Brink, C., Hvelplund, P., Larsson, M. O., Bech Nielsen, B., and Shen, H., *Phys. Rev. Lett.* **77**, 3991 (1996).
6. Andersen, J.U., Brink, C., Hvelplund, P., Larsson, M. O., and Shen, H.., *Z. Phys. D* **40**, 365 (1997).
7. Larsson, M.O., Gottrup, C., Hvelplund, P., and Andersen, J.U., to be published.
8. de Heer, W. A., *Rev. Mod. Phys.* **65**, 611 (1993) and Calaminici, P., Russo, N., and Toscano, M., *Z. Phys. D* **33**, 281 (1995).
9. Wang, L.-S., Conceicao, J., Jin, C., and Smalley, R. E., *Chem. Phys. Lett.* **182**, 5 (1991).
10. Brink, C. Andersen, L. H., Hvelplund, P., Mathur, D., and Voldstad, J. D., *Chem. Phys. Lett.* **233**, 52 (1995).
11. Hansen, K. and Echt, O., *Phys. Rev. Lett.* **78**, 2337 (1997).
12. Hansen, K. and Campbell, E. E. B., *J. Chem. Phys.* **104**, 5012 (1996).
13. Andersen, T., Andersen, L. H., Balling, P., Haugen, H. K., Hvelplund, P., Smith W. W., and Taulbjerg, K., *Phys. Rev. A* **47**, 890 (1993).
14. Greaney, M. A. and Gorun, S. M., *J. Phys. Chem.* **95**, 7142 (1991).
15. Lawson, D. R., Feldheim, D. L., Foss, C. A., Dorhout, P. K., Elliott, C. M., Martin, C. R., and Parkinson, B., *J.*

Electrochem. Soc. **139**, L68 (1992).

16. Hansen, K., Andersen, J. U., Cederquist, H., Gottrup, C., Hvelplund, P., Larsson, M. O., Petrunin, V. V. and Schmidt, H. T.: Thermionic emission laser spectroscopy of stored C_{60}^-. To be published.

Negative Ion Spectroscopy with Stored H⁻ Ions

T. Andersen, H. H. Andersen, P. Balling, and V. V. Petrunin

Institute of Physics and Astronomy, University of Aarhus, DK-8000 Aarhus C, Denmark

Abstract. This paper reviews the results obtained in recent years from spectroscopic studies of the negative hydrogen ion at the ASTRID storage ring. The two lowest-lying members of the $^1P^0$ dipole series of autodetaching resonances in H⁻ located just below the H(n=2) threshold have been observed and characterised, using Doppler-tuned collinear laser spectroscopy. The resonance positions have been determined for both H⁻ and D⁻, allowing also a critical test of the predicted isotope effects. Further studies are in progress based on electron cooling of the negative hydrogen ion beam.

INTRODUCTION

The negative hydrogen ion, H⁻, belongs to the simplest three-body two-electron atomic systems and among these is in many ways the most interesting because the electron correlations are especially strong. The independent particle approximation, which is well suited to describe neutral atoms and positive ions, is grossly inadequate in connection with modelling the H⁻ ion. The correlation between the interacting atomic electrons can be intricate, making calculations difficult. At the ASTRID storage ring, University of Aarhus, Denmark the study of the structure, dynamics and collisions of the negative hydrogen ion is part of the ongoing negative ion research program (for recent reviews see Ref. 1-3), but this paper only deals with the spectroscopy of H⁻ summarizing the results obtained so far[4,5].

The structure of atomic negative ions is usually associated with the characteristics of short-range potentials[6]. The consequence is that negative ions only possess one or a few bound states, but no infinite series of such states as known for neutral atoms. However, negative ions formed by attaching an electron to an excited state of the hydrogen atom ($n \geq 2$) provide a remarkable exception: The mixing of degenerate angular-momentum states of a given n level by the electric field of a distant electron results in an asymptotic long-range dipole tail of the potential experienced by the latter electron. As a consequence, infinite series of bound states may exist below each threshold of hydrogen[7].

In order to study the $^1P^0$ resonances located in the vicinity of the n=2 threshold of hydrogen we had to rely on the ASTRID storage ring and apply Doppler-tuned spectroscopy, using an H⁻ beam stored in the heavy-ion storage ring collinearly overlapped with a vacuum ultraviolet laser beam at 118 nm. Spectroscopy of the H⁻ ion is related to the investigation of transient states that reveal themselves as resonances in the photodetachment cross section. The resonances correspond to doubly excited states of the negative ion, where both electrons are promoted to spacially extended orbits, and they are particular sensitive to the interelectron repulsion. The spectroscopic studies of H⁻ in the region near the n=2 threshold of the neutral atom are complicated by the need for photon energies around 11 eV. For this reason we have utilized fixed frequency laser light and varied the effective photon energy by adjusting the velocity of the H⁻ ions stored in ASTRID.

The region under investigation is dominated by two types of pronounced structures: below the n=2 threshold one or more narrow resonances, which are usually called Feshbach resonances, and above the threshold a broad structure, a shape resonance. When we initiated the study of the H⁻ ion a few years ago, only Bryant and coworkers[8-12] had previously performed experimental studies in the region of interest. In a series of experiments performed at LAMPF using an 800 MeV H⁻ beam, overlapped under a variable angle by visible or ultraviolet laser light, they had discovered one of the Feshbach resonances and the shape resonance. Their experiments had covered an impressive range of photon energies from below the n=2 detachment threshold to above the limit for two-electron ejection. The resolution had, however, been Doppler limited to approximately 8 meV. This is an important constraint for studies of the narrow Feshbach resonances, since recent theoretical studies[13-18] have predicted that the negative hydrogen ion may possess more than one of this type of resonances below n=2, which should be rather narrow, far below the limits available at Los Alamos. The aims of our first experiments have been: to observe some of the narrow resonances located below the H(n=2) limit, to determine their positions, to test the predicted widths of the resonances, and to investigate possible isotope effects by studying H⁻ as well as D⁻. Not all of these goals have been reached by now.

CLASSIFICATION OF RESONANCES

The doubly excited states in the negative hydrogen ion may be classified according to the $_n\{v\}_m^A \, ^{2S+1}L^\pi$ scheme which is developed from the approximate symmetries of the two-electron Hamilton[19,20]. Here n denotes the hydrogenic threshold to which the doubly excited series is converging, while v and A are related to the angular and radial correlations of the two electrons, respectively. Combined with the total spin (S), angular momentum (L) and parity(π), a particular choice of n, v, and A will define a channel which may bind a series of states of similar symmetry. The m label denotes the degree of excitation within a particular channel, but it should not be considered a quantum number of the same type as n. Two qualitatively very different types of series may exist.

In the A= + states, the electrons can be considered to move radially in phase, meaning that the electrons approach or leave the nucleus simultaneously. In these states, the electrons spend considerable time at equal and small distances from the nucleus, resulting in a large interaction between the electrons. This will lead to a short lifetime for such states, and consequently result in relatively broad resonance structures. In a series of experiments, Bryant and coworkers[8,11,12] have observed part of such $^1P^0$ series with A= + below the n=3-8 thresholds, the observation of high-lying members being prohibited by the experimental resolution (8 meV). The corresponding n=2 channel, $_2\{0\}^+ \, ^1P^0$, is not sufficiently attractive to support a series of bound states, but it is able to temporarily trap an electron behind a repulsive potential barrier (the so-called shape-resonance above the n=2 threshold). In the A= − states, the electrons move radially out of phase; when one of the electrons is close to the nucleus, the other is at a great distance. This leads to a significant suppression of the photoexcitation strength compared to the more localized A= + states. In addition, the interelectron interaction is much weaker, and thus considerably narrower structures are expected. Due to the limited resolution of the LAMPF experiments it was only possible for Bryant and coworkers to observe a single A=− resonance, the socalled Feshbach resonance below n= 2 (in the present notation called $_2\{0\}_3^-$). It was, however, also difficult at the LAMPF experiment to determine the position of this resonance with a good accuracy. The experimental value reported was more than three standard deviations from the theoretical values available now, which could indicate either an incorrect experimental value or a significant deviation between theory and experiment.

EXPERIMENTAL APPROACH

A schematic of the experimental setup is shown in Fig. 1. Negative hydrogen ions are extracted from a duoplasmatron ion source and accelerated to 150 keV, typically producing currents of 10 μA. Following mass analysis the H⁻ beam is injected into the ASTRID storage ring and accelerated in a radio-frequency (RF) cavity to approximately 1 MeV for H⁻ and 2 MeV for D⁻. The RF field is turned off and the beam is left circulating at a fixed energy. Along one of the straight sections of the storage ring (approx. 8m) the ion beam is overlapped with a vacuum ultraviolet (VUV) laser beam. The VUV beam is introduced counterpropagating with respect to the H⁻ beam under a very small angle of 0.18° to allow the detection of neutral atoms generated along the straight section of the ring(see Fig. 1). When the ion beam is circulating at fixed energy the neutral atoms formed in a short time gate during each laser shot are counted on a detector when they pass straight through the bending magnet.The atoms counted consist of those stemming from the interaction with the VUV light and those originating from the collisional detachment in the residual gas ; the latter contribution is eliminated by counting separately the collisional background. The relative photodetachment cross section is obtained by normalizing the light-induced signal to the exponentially decreasing current. The current right after the RF acceleration is typically a few μA, as can be measured by a current transformer when the RF field is kept on after acceleration.

The VUV light is of fixed wavelength: the outout from a pulsed (10 Hz, pulse length approx. 10 ns) injection seeded Nd:YAG laser with a linewidth of less than 0.4 μeV is frequency doubled and sum-frequency mixed in non-linear optical crystals. Pulses of the generated light (355 nm) are focussed into a gas cell containing 5 mB of Xe providing approx. 10^{10} photons per laser shot of the ninth harmonic (photon energy 10.48299(2) eV) of the fundamental Nd:YAG frequency. The photon energy seen by the moving ions is then controlled by tuning their velocity. If θ denotes the angle of incidence on the ion beam with θ = 0° being head on collision, v the velocity of the ion beam, and $h\nu_0$ the laboratory photon energy, then the effective photon energy is

$$h\nu = h\nu_0 \gamma [\,1 + (v/c)\cos\theta\,] \qquad [1]$$

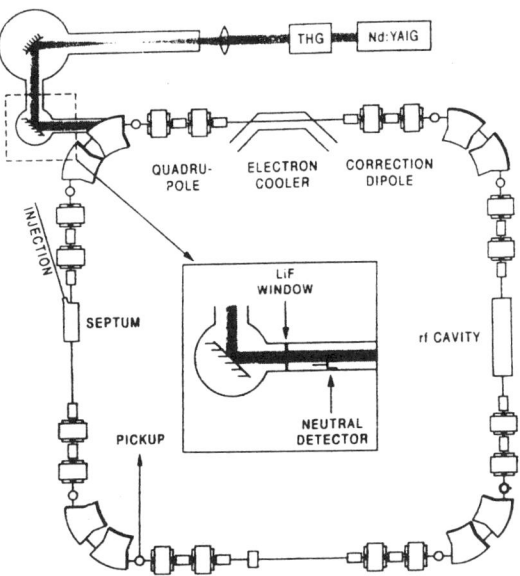

Figure 1. A schematic of the experimental setup showing the VUV laser system and the ASTRID storage ring with injector, acceleration section and detection system.

where $\gamma = (1-v^2/c^2)^{-1/2}$ is the Lorentz factor. The velocity of the ions is measured by determining their revolution frequency in the storage ring (measured on a Schottky pickup) and using the known circumference of the ring, 39.99(2) m. This leads to an uncertainty in the absolute calibration of 0.2 meV, which constitutes the main uncertainty in all the measured energy positions. The circumference may be determined with a factor of 5-10 better accuracy indicating that it may be possible to reduce the quoted error bars. In general, the resolution in the effective photon energy is limited by three factors: the bandwidth of the laser, the angular spread of the two beams, and the velocity spread of the ion beam. The contribution from the laser bandwidth is negligible compared with other error sources, and the broadning due to divergence of the two beams is strongly reduced by applying parallel beams. In the collinear geometry the resolution is normally limited by the momentum spread of the ion beam.

RESULTS

Structural properties of the A = − resonances

Figure 2 shows the measured relative photodetachment cross section for D⁻ and similar measurements were performed for H⁻ with the same structures being observed in the two measurements. The two narrow resonances correspond to the first two members of the A= − series converging to the n= 2 threshold, while the broad resonance above the threshold is the only member of the A= + series, the shape resonance. The cross section in the vicinity of the Feshbach resonances is expected to be well described by a Fano parametrization[13]

$$\sigma(h\nu) = \sigma_o (q + \epsilon)^2/(1+\epsilon^2) \qquad [2]$$

where $\epsilon = 2(h\nu - E_R)/\Gamma$. Here E_R, Γ, and q represent the energy, width and the asymmetry parameters of the resonance, respectively. As can be seen from the expanded views of the $_2\{0\}_3^-$ and $_2\{0\}_4^-$ resonances presented in Fig. 2b and 2c the measured widths of the resonances are significantly larger than 30-65 μeV predicted by theory for the former, and the approx. 2 μeV for the latter of the two resonances. This reflects the resolution of the the experiments performed so far: The velocity spread of the stored ions leads to a broadening of the the resonance structures. Whereas the shape of the $_2\{0\}_4^-$ resonance(see Fig.2c) is completely dominated by this Doppler broadening, and hence a fit to a Gaussain form is justified, the $_2\{0\}_3^-$ resonance (Fig. 2b) is clearly asymmetric as can be seen in the tails of the observed profile. The actual width of this resonance is approximately a factor of three larger than the theoretically predicted width, so even a moderate reduction in the Doppler broadening (obtained by means of electron cooling of the stored negative hydrogen beam) may allow testing of the width of this resonance in the near future.

The analysis of the $_2\{0\}_3^-$ resonance has been performed utilizing the cross section profile (Eq.2) convolved with the distribution of effective photon energies stemming from the velocity spread (cf. Eq.1). The resonance positions of 10.9243(2) eV for H⁻ and

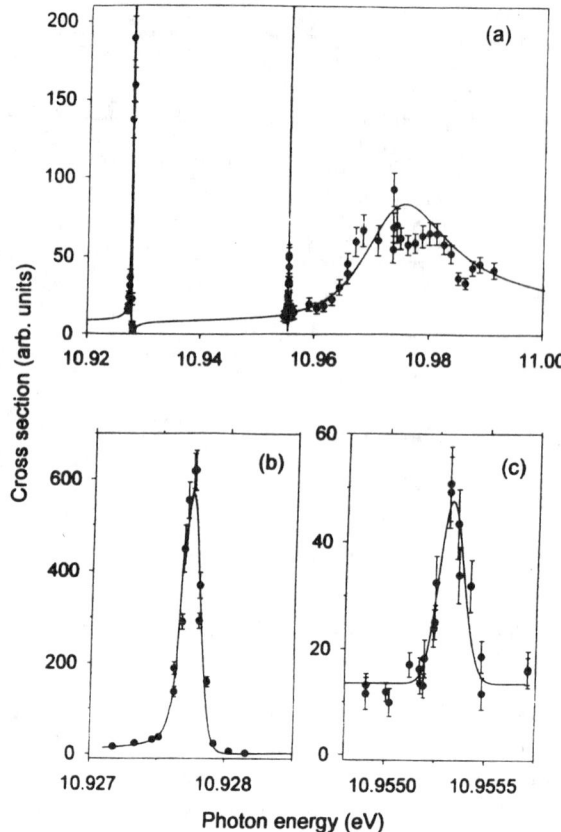

Figure 2. Photodetachment cross section of D$^-$. (a) The measured cross section has been normalized to the theoretical results of Lindroth[13] (solid curve). The data for the $_2\{0\}_3^-$ resonance at 10.9277 eV exceeds the vertical scale of the plot. (b) Blowup of the region near the $_2\{0\}_3^-$ resonance. The solid curve shows a fit to a Fano profile convolved with the the photon energy resolution. (c) Blowup of the region near the $_2\{0\}_4^-$ resonance at 10.9553 eV. Here the solid curve is a fit to a Gaussian profile.

10.9277(2) eV for D$^-$ are within these uncertainties independent of the other parameters of the fit. The measured cross section also constrains the Γ and q parameters (values outside the range presented in Table I lead to a systematic deviation from the data). The results obtained so far[4,5] are summarized in Table I and compared to previous experimental and recent theoretical calculations. The position of the $_2\{0\}_3^-$ resonance was first obtained in 1996[4] and redetermined under improved experimental conditions in 1997[5]. The positions obtained in the two experiments agreed very well, but clearly deviated from the result obtained at LAMPF[12]. The recent calculations of resonance positions are all in good agreement with the new experimental values, but the numerical accuracy is now challenged. The data for the shape resonance did not allow for an accurate evaluation of its resonance parameters. The signal fluctuations in Fig. 2a may be attributed to variation in the overlap of the laser and ion beam over the large range of ion storage energies required to cover the shape resonance, but the figure indicates that the measured cross section is consistent with the calculation of Lindroth[13].

Since the overlap between the ion and the laser beams may vary when changing between resonances it has not been possible to determine the absolute photodetachment cross sections in the experiments performed so far. Thus direct comparison between the strengths of the resonances, as given as the cross section integrated over the resonance, cannot be made. Normalization of the data has therefore been performed by adjusting the flat tails of the measured resonances to the calculated levels obtained by Lindroth[13]. This allows us to determine the relative strength between the two narrow resonances to be approx. 19, with $_2\{0\}_3^-$ being the resonance with the largest strength of these, a value which is in rather good agreement with the calculated ratio[13] of 23.

Specific Mass Shift

The kinetic energy in the center-of-mass frame for the three-body system (H$^-$ or D$^-$) can be expressed as:

$$(\mathbf{p}_1^2 + \mathbf{p}_2^2)/2m + (\mathbf{p}_1^2 + \mathbf{p}_2^2)/2M + \mathbf{p}_1\mathbf{p}_2/M \qquad [3]$$

Table I

Resonance	Energy (H⁻)/ eV	Energy (D⁻)/eV	Width (Γ)/μeV	Asymmetry parameter (q)
$_2\{0\}_3^-$ $^1P^o$				
Experiment				
ASTRID[5]	10.9243(2)	10.9277(2)	20 < Γ < 60	-30 < q < -10
LAMPF exp.[12]	10.9264(6)			
Theory				
Lindroth et al.[16]	10.9245	10.9277	37.2	-16.5
Chen[15]	10.92452	10.92774	35.6	
Gien[18]	10.9245	10.9279	37.7	
$_2\{0\}_4^-$ $^1P^o$				
Experiment				
ASTRID[5]	10.9519(2)	10.9553(2)		
Theory				
Lindroth et al[16]	10.9521	10.9553		
Chen[15]	10.95212	10.95535		
Gien[18]	10.9520	10.9554		

where p_1 and p_2 are the momenta of the two electrons, m the electron mass, and M the mass of the nucleus. The first two terms combined simply require rescaling of the energies according to the appropriate reduced electron mass. The last term, often called the specific mass shift, depends directly on the momentum correlation of the two electrons and scales according to the mass of the nucleus. The narrow A= - resonances facilitates an investigation of the specific mass shift. It has been predicted[21] to be quite large (486 μeV) for the ^1S ground state of H⁻, but much smaller for the $_2\{0\}_3^-$ resonance, only approximately 10μeV[13]. The data in Table I show that a shift in resonance energy of 3.4(2) meV is observed for both the two A=- resonances between H⁻ and D⁻. This is consistent with the 3.2 meV found by scaling the transition energies according to reduced electron mass and assuming that the specific mass shift for D⁻ can be scaled from that of H⁻ by the nuclear mass ratio. Thus, the specific mass shift for the excited A=- states is small. The present data are consistent with the theoretical predictions. It should be noted, however, that Rislove et al[22] have measured the specific mass shift for the lowest lying autodetaching ^1D resonance, obtained by two-photon absorption, and reported a specific mass shift of 2.4(1.1) meV, considerably larger than the approx. 0.1 meV predicted from theoretical calculations for this resonance.

The relative small size of the specific mass shift for the excited states of A=- symmetry may be understood from a qualitative analysis of the kinetic energy, Eq. 3. Since the A=- states are characterized by the out-of-phase motion of electrons, this means that while one electron is close to the nucleus (has large momentum) the other is far from the nucleus (has small momentum). While the sum of the terms p_1^2 or p_2^2 always will collect a large contribution to the isotope shift, the last term, $p_1 p_2$ will be significantly suppressed by the smaller of the two momenta.

DISCUSSION

Since the position of resonances in a dipole potential is characterized by the exponential convergence towards the threshold, the observation of two successive resonances can be used to predict the behavior of the entire series, taking into consideration that the deviation from a dipole potential at small distances from the nucleus is expected only to have a weak influence on the spatially extended A=- resonances. Since the constant ratio between the binding energy with respect to the threshold of successive members in the series can be determined to be 26(5), the third member of this series should be located only about 40μeV below threshold. Recent theoretical studies[16,17] have predicted that the lifting of the degeneracy of the H(n=2) levels by relativistic and QED effects will truncate the series after its third member. The experimental data obtained so far did not reveal the third member of the series.

More accurate experimental studies of the width of the $_2\{0\}_3^-$ resonance, observation of the predicted third member of the dipole series, $_2\{0\}_m^-$ $^1P^0$, or studies of the predicted narrow resonances below the H(n=3) threhold will all demand improvement in the resolution from the present one, 180μeV, obtained with an uncooled ion beam produced in a duoplasmatron ion source at rather low energy (150 keV acceleration voltage). With an improved resolution it may also be possible to identify the n=2 threshold in the cross section.This would provide an inherent calibration of the effective photon energy and allow a significant reduction of the uncertainties of the resonance positions.

Electron cooling of positive ions, particularly fully stripped ions as H^+ or D^+, is relatively easy to obtain due to the long lifetimes of these ions in the storage ring. Negative ions with rather loosely bound electrons have a much shorter lifetime in the storage ring due to collisional destruction by restgas atoms or molecules, limiting the lifetime to approx. 2-4 s under the experimental conditions used so far. Experimental problems related to position stability of the electron cooled negative ion beam have so far prohibited new experiments, but these problems seem now to be understood, so there is a good prospect for further studies of the negative hydrogen ion within the coming years.

ACKNOWLEDGEMENTS

The support of the ASTRID staff, particularly S.Pape Møller, has been of great importance for the negative hydrogen ion project. Discussions with or inputs to the project from experimental or theoretical colleagues have been very valuable, particularly the role played by H.K.Haugen, McMaster University, Canada during the early phase of this project, the exchange of information and ideas with H. C. Bryant, LAMPF, and the discussions with E.Lindroth, University of Stockholm, who also communicated results prior to publications. The project described is part of the research program of the ACAP center, which is funded by the Danish National Research Foundation.

REFERENCES

1. Andersen, L. H., Andersen, T., and Hvelplund,P., *Adv. At. Mol. Opt. Phys.* **38**, 155- 191 (1997)
2. Andersen, T., *Physica Scripta* **T59**, 230 -35 (1995)
3. Andersen, T., in *Photonic, Electronic and Atomic Collisions,* Singapore: World Scientific, 1998, pp 401-19
4. Balling, P. et al., *Phys. Rev. Lett.* **77**, 2905-08 (1996)
5. Andersen, H. H. et al., *Phys. Rev. Lett.* **79**, 4770-73 (1997)
6. Buckman, S. J. and Clark, C. W. *Rev. Mod. Phys.* **66**, 539- 655 (1994)
7. Gailitis, M. and Damburg, R., *Proc. Phys. Soc.* **82**, 192- 200 (1963)
8. Bryant, H. C. and Halka, M., in *Coulomb Interactions in Nuclear and Atomic Few-Body Collisions*, New.York 1996, Plenum Press, ch.4, pp. 221-80 and references therein.
9. Harris, P. G. et al., *Phys. Rev. A.* **42**, 6443- 65 (1990)
10. Halka, M. et al., *Phys. Rev. A.* **44**, 6127 - 29 (1991)
11. Bryant, H. C. et al., *Phys. Rev. Lett.* **38**, 228-31 (1977)
12. MacArthur, D. W. et al., *Phys. Rev. A.* **32**, 1921 -23 (1985)
13. Lindroth, E., *Phys. Rev. A.* **52**, 2737- 49 (1995)
14. Tang, J. et al., *Phys. Rev. A.* **49**, 1021-28 (1994)
15. Chen, M. K., *J. Phys. B.* **30**, 1669-76 (1997)
16. Lindroth, E., Burgers, A. and Brandefelt, N., *Phys.Rev. A.* **57**, R685-88 (1998)
17. Purr, T., Friedrich, H., and Stelbovics, A.T., *Phys.Rev. A.* **57**, 308-11 (1998)
18. Gien, T. T., private information about electron-hydrogen scattering calculation (1998), to be published.
19. Watanabe, S. and Lin, C. D., *Phys. Rev. A.* **34**, 823-37 (1986)
20. Sadeghpour, H. R. and Greene, C. H., *Phys. Rev. Lett.* **65**, 313-16 (1990)
21. Drake, G. W. F., *Nucl. Instrum. Methods Phys. Res., Sect. B* **31**, 7- 13 (1988)
22. Rislove, D. C. et al., *Phys. Rev. A. in press* (September 1998)

SECTION 6

STUDIES OF LOW ENERGY TRAPPED IONS

RETRAP: An Ion Trap for Laser Spectroscopy of Highly-Charged Ions

D. A. Church[a], J. Steiger[b], B. R. Beck[b], L. Gruber[b], J. P. Holder[a], J. McDonald[b], and D. Schneider[b]

[a]Physics Department, Texas A&M University, College Station, TX 77843-4242 USA
[b]Lawrence Livermore National Laboratory, P. O. Box 808, Livermore, CA 94551 USA

Abstract. The possibility of highly-charged ions, captured and stored in an ion trap, and cooled by elastic collisions with confined Be ions, has been achieved in RETRAP; a cryogenic Penning trap system coupled to the Electron Beam Ion Trap (EBIT), which was used as a source of highly-charged ions. Be^+ cooling in RETRAP has been carried out with a combination of resistive damping of the axial motion of the ions by a tuned circuit, and laser cooling. Spectroscopic goals of the research include metastable level lifetime measurements and precision laser spectroscopy on magnetic dipole transitions of selected highly-charged ions. Potential measurements of the hyperfine structure splitting, level lifetime, and bound state g-factor of a high-Z hydrogen-like ion $^{165}Ho^{66+}$ already studied by emission spectroscopy in Super-EBIT, are discussed with relation to current progress.

INTRODUCTION

A collaboration between the EBIT group at Lawrence Livermore National Laboratory (LLNL) and the group of D. Church of Texas A&M University was initiated to capture, cool, and perform a variety of measurements on highly-charged high-Z ions in RETRAP, a cryogenic Penning ion trap (1,2). The current status and some goals of the laser spectroscopy portion of the program are described here. Precision laser spectroscopy on ground-state singly-charged atomic ions is concentrated on those few ions with transitions that fall in the tunable laser range, roughly 200 – 800 nm if frequency-doubling is used. A pertinent example is Be^+, with an S ↔ P transition near 313 nm. As ion charge increases, most electric dipole transition wavelengths are rapidly shifted outside the laser range, but laser-induced magnetic dipole transitions between certain fine structure levels become feasible, and for highly-stripped hydrogenic ions, magnetic dipole transitions between ground term hyperfine structure levels can be excited (3). Although magnetic dipole transition rates are 10^{-5} the rates of electric dipole transitions, fluorescence from laser excitation of magnetic dipole transitions of ions in a Penning trap should be detectable. Potential precision laser measurements on the ground state hyperfine structure splitting and level lifetime, and the bound state g-factor of hydrogen-like $^{165}Ho^{66+}$, are discussed. Laser measurements on other hydrogen-like ion ground states, and on the n = 2 triplet fine-structure of certain Be-like ions with Z near 22, are also planned.

CURRENT STATUS

Once formed in EBIT, the highly-charged ions are extracted at low voltage (4), analyzed on a charge-to-momentum basis, and transported electrostatically in a short (5 – 10 μs) pulse to RETRAP (1,5). The ion pulse is abruptly decelerated in the fringe magnetic field of the RETRAP cryogenic coils as the ions enter a deceleration tube biased near the EBIT extraction potential. This potential is rapidly (≈10 ns) decreased, permitting those ions inside the tube to emerge at low energy. These ions enter the Penning trap through an upper electrode, are reflected by a potential at the lower end, and are captured by rapidly raising the potential of the upper electrode.

A cylindrical Penning trap (1,6) was employed for the inaugural measurements.. The advantages of this trap for this research included a large aperture for ion capture, harmonic axial ion motion near the trap center, and efficient coupling to an axial tuned circuit using the "compensation" electrodes. The end cylinder electrodes can be used to capture and release the ions. Electron capture cross sections of low-energy stored ions with charges ranging from 11+ to 80+ colliding with H_2

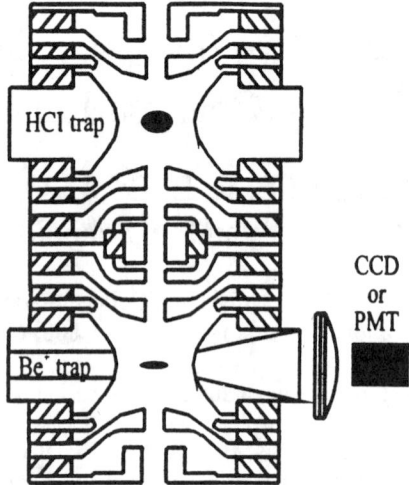

FIGURE 1. Two hyperbolic Penning traps in series. The lower trap is for laser-precooling Be+ ions, which serve to collisionally cool the highly-charged ions. This is the spectroscopic trap. The upper trap is intended to pre-cool the highly-charged ions, before merging them with the Be ions.

were measured (5). It was verified that the rate coefficient k_q, and the total electron capture cross section $\sigma_q \equiv k_q/v_{rms}$ in terms of the root mean squared ion speed, increased approximately linearly with charge state q (5,7) in agreement with the predictions of the absorbing sphere model (8). For a residual density of H_2 molecules $n(H_2)$ near 8×10^4 cm^{-3} in the cryogenic trap volume, the ion storage time constant $\tau_q = [n(H_2)k_q]^{-1}$ was measured to be near 50 s for Th^{80+}. This time is already sufficiently long for spectroscopy, but more recent measurements under improved conditions show that this time constant can be extended by at least a factor of ten (9).

Following these measurements RETRAP was rebuilt to include two hyperbolic-electrode Penning traps in series (see Fig. 1). Two additional non-harmonic traps (not shown) were also included at the ends, but these will be ignored henceforth. The plan for these traps was to initially laser-cool Be$^+$ ions (10) in the lower trap, next cool highly-charged ions by interaction with a tuned circuit in the upper trap, and then transfer the (hotter) highly-charged ions to the lower trap, to be further cooled by elastic collisions with the laser pre-cooled Be$^+$ ions. Once the highly-charged ions were cooled and crystalized, laser spectroscopy on the highly-charged ions would be initiated.

Both Penning traps were constructed identically, with the exception that apertures were placed in the ring electrode of the lower trap to permit the introduction of laser beams and to collect fluorescence. Each trap had end-cap separations $2z_0 = 1$ cm, and central ring diameter $2\rho_0 = 1.16$ cm (11), so that $d^2 = (z_0^2 + \rho_0^2/2)/2 = 0.209$ cm^2. The trap electrodes were constructed from gold-plated OFHC copper, and were separated by alumina spacers. The trap structure was clamped to the liquid helium dewar. The ring electrode was divided into four quadrants to permit cyclotron excitation and detection of the ions. Compensation electrodes were installed to control trap harmonicity. They were also split for possible use in electronic centering of the ions. In the lower trap, six 3.2 mm diameter apertures in the ring electrode, oriented toward corresponding tubes through the helium and nitrogen dewars and shields, provided for laser excitation and fluorescence light detection through fused quartz windows in the vacuum shell. Each tube contained a quartz window to block gas influx to the low temperature region from other portions of the vacuum, and mechanical shutters at the top and bottom of the trap structure could similarly isolate the ion beam path. Beside one aperture, a quartz lens was mounted to the trap structure to efficiently collect fluorescence and focus it outside the chamber. The trap dc bias U was applied to the ring electrode, and an external inductor was connected between the end caps to form a tuned circuit. The circuit for the upper trap was tuned to 2.26 MHz, and for the lower trap to 2.51 MHz. Quality factors Q were usually near 500. The external magnetic field B produced by the cryogenic coils was operated near 4.5 T for the lower trap, but was closer to 4.1 T in the center of the upper trap. The traps were separated by spacer electrodes for shielding, and for use to capture and release the ions. Small coils were also connected between ring electrode quadrants to aid cyclotron excitation and detection.

Ions with charge (qe) and mass m in a Penning trap have an axial oscillation frequency $\omega_z = (qeU/md^2)^{1/2}$ and radial oscillation frequencies at $\omega_+ = \omega_c - \omega_-$ and $\omega_- \approx \omega_z^2/2\omega_c$. Here $\omega_c = (qe)B/m$ is the true cyclotron frequency. For Xe^{44+}, $\omega_c/2\pi \approx 23.2$ MHz near 4.5 T, and $\omega_-/2\pi \approx 140$ kHz, if $\omega_z/2\pi = 2.51$ MHz. The axial frequency resonance with the tuned circuit occurred near U = 170 V.

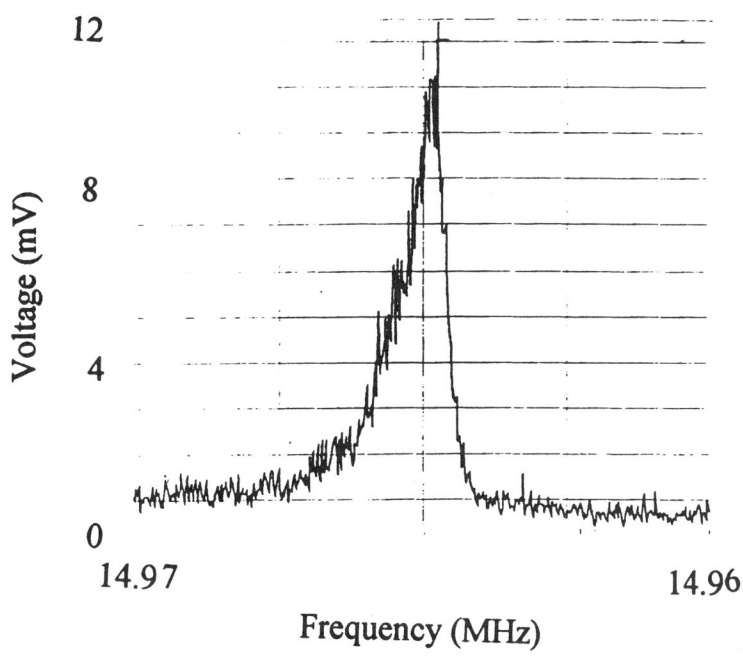

FIGURE 2. Cyclotron resonance signal for Be^{2+} ions, measured by the heating of the axial motion as observed using the tuned circuit. The exciting frequency was swept from 14.97 to 14.96 MHz over 20 seconds. The ion-ion collisional coupling had a time constant estimated to be about 2 s. The signal asymmetry is thought to arise from frequency shifts for the hot ions.

Using the split ring electrodes, radial resonant excitation of the cyclotron motion of the ions heated the axial motion of the ions, which was coupled by ion-ion elastic collisions. This heating could be detected on the tuned circuit, using the induced currents from the fluctuating ion motion at resonance, which produced a signal voltage proportional to $(NT)^{1/2}$, where N is the stored ion number and T is the ion temperature. Fig. 2 shows such a signal for Be^{2+} ions when ω_c was excited by the non-uniform radial electric fields between ring electrode quadrants. Excitation at the ion frequency ω_+ was separately observed. The excitation voltage amplitude was 5 mV during a single 20 s sweep of the excitation frequency from 14.97 to 14.96 MHz. The asymmetry in the signal shape is thought to be due to a frequency shift of the hot ions. Measurement precision for the cyclotron frequency is about 3×10^{-5} in this crude result, but with care and improved technique, the precision of the frequency measurement, and consequently the magnetic field determination, should be improved by more than a factor of 100. Other techniques, including direct detection of the excited motion of the ions using the split ring, will be implemented.

A mixture of Be$^+$ and Be^{2+} produced in a short pulse by a metal vapor vacuum arc (MeVVA) source mounted above RETRAP, can be captured into the lower trap. The ions are initially quite hot, but can be slowly cooled by tuning the axial frequency of either charge state to resonance with the tuned circuit. A frequency-doubled laser beam, tuned initially to a wavelength $\lambda = 313.155$ nm for the 4.5 T field of RETRAP, is used to cool the Be$^+$ ions, and hence the Be^{2+} ions through ion-ion collisions, by exciting the 2s ^{2}S ($m_J = -1/2$, $m_I = -3/2$) – 2p ^{2}P ($m_J = -3/2$, $m_I = -3/2$) cycling transition a few GHz below resonance (10).

Figure 3 schematically shows the laser cooling and detection geometry. The fluorescence rate resulting from Be$^+$ excitation was measured through one port using a photomultiplier tube (PMT), while the fluorescence was also imaged using a CCD camera at another port. The laser beam position was monitored after passing through the trap. Finally, the axial oscillation frequency of either Be$^+$ or Be^{2+} was adjusted to resonance with the tuned circuit to monitor the voltage signal proportional to $(NT)^{1/2}$ directly. Since cooled ion numbers change only slowly, the tuned circuit signal followed temperature changes. Fig. 4(a) shows the decrease with time of the Be^{2+} tuned circuit signal due to laser cooling of Be$^+$ using a fixed laser wavelength. Fig. 4(b) shows heating following cooling, when the laser wavelength is retuned near 313.152 nm, on the short wavelength side of the Be$^+$ excitation resonance.

After cooling the ions, the laser frequency was linearly swept across the Be$^+$ excitation resonance, while monitoring the photon count rate detected by the PMT. The full width at half maximum of the resonance was 0.3 GHz. Taking this to be the Doppler width of the cold ion cloud, the corresponding ion temperature was near 1 K.

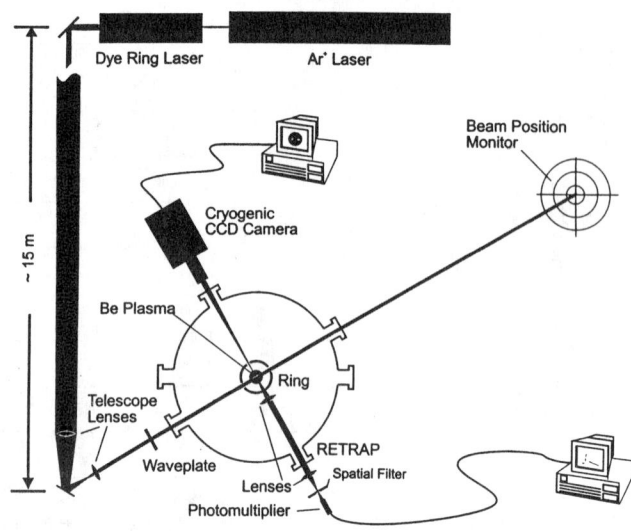

FIGURE 3. Schematic diagram showing the laser excitation system and the two modes of detection of light emitted by the ions.

With continued cooling, it was observed using the imaging detector that the light emitted by the Be$^+$ ions appeared to form a ring with a dark center (12), as shown in Fig. 5. This was interpreted as a possible centrifugal separation of the Be^{2+} and Be$^+$ ions, due to their factor-of-two different mass-to-charge ratios, with the smaller mass-to-charge ratio at the trap center. These measurements, calculations, and a discussion of data are presented in the contribution of J. Steiger et al. (13). The significance of this result to spectroscopy is that the cold, highly-charged ions can be expected to be located near the trap center, so that the exciting laser light can be tightly focussed. There will also be little interference from fluorescence associated with the laser cooling of Be$^+$, although the cooling beam can be chopped in any event. Overall, these measurements clearly demonstrate that multiply-charged ions stored with Be$^+$ ions can be cooled via collisions while using a laser to cool the Be ions.

Manipulation of the ions, including ion transfer between traps, has also been studied. It was found that previously captured ions could be transferred into an initially empty trap, but when hot or cold highly-charged ions were to be transferred from the upper trap, to merge with laser-cooled Be ions in the lower trap, then unacceptable ion losses occurred. This was the case when the highly-charged ions were transferred in a pulse, or by adiabatically decreasing the depth of the well in which the ions resided, so that individual ions could be transferred at low energy into the Be ion cloud where collisions would capture them. On the other hand, it was found that a pulse of ions from EBIT could be successfully captured into a trap containing Be ions, which then were observed to collisionally cool the highly-charged ions (13). Further study of collisional cooling of highly-charged ions is planned.

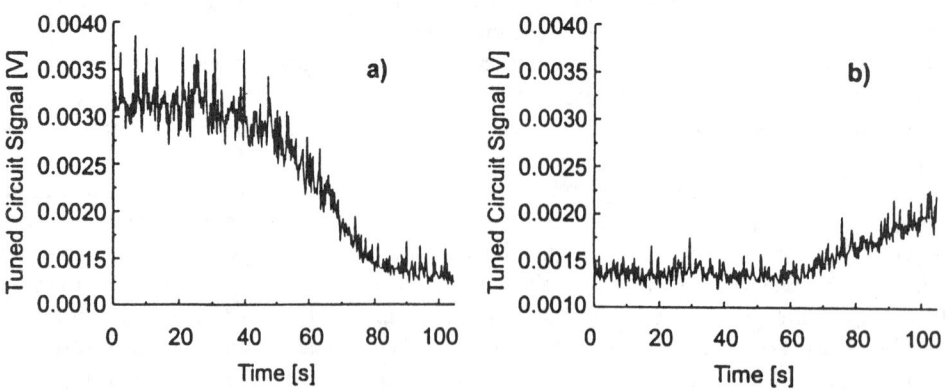

FIGURE 4. Tuned circuit voltage signals, proportional to the square root of ion temperature, plotted vs. time. In (a), Be+ ions are cooled with the laser wavelength tuned above resonance, while in (b) the laser wavelength is retuned below resonance to heat the ions.

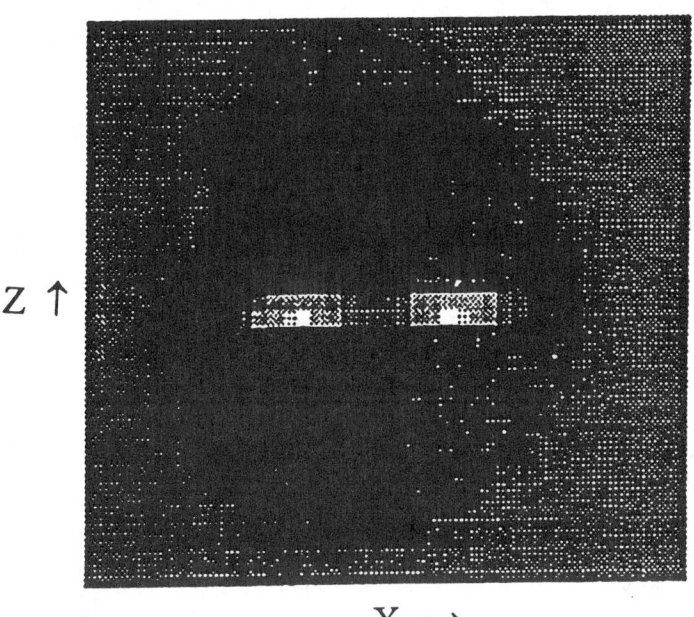

FIGURE 5. Redistribution of very cold Be$^+$ ions, when stored and cooled with Be^{2+} ions present, into an annulus about the trap center. as imaged using the CCD detector (11) mounted as shown in Figure 3.

As shown in Fig. 6, hydrogen-like Xe^{53+} ions, among other charge states, have been successfully extracted from Super-EBIT, which has the electron energy to fully strip any element (14). Presently the beam line is being extended from Super-EBIT to RETRAP and the extraction improved, so that hydrogen-like ions can be studied in RETRAP.

ION SPECTROSCOPY

Hydrogen-like ^{165}Ho^{66+} has a nuclear spin I = 7/2, resulting in the two hyperfine levels F = 4 and F' = 3. The nuclear moment is μ_I = 4.132(5)μ_n in terms of the nuclear magneton μ_n. The ground state hyperfine structure splitting (hfs) scales as Z^3, and the magnetic dipole hfs transition has been observed in emission in the spectrum of H-like Ho from Super-EBIT at a

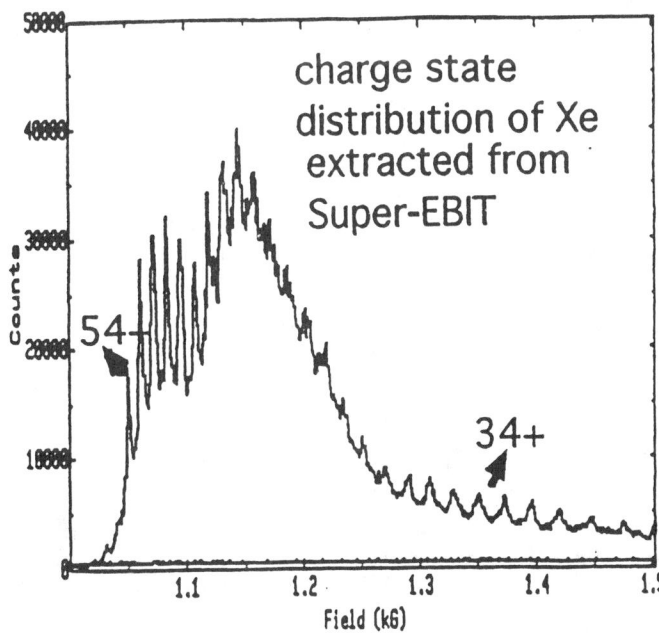

FIGURE 6. Plot of the extracted ion current for Xe^{q+} ions from Super-EBIT, after charge-to-momentum separation. Fully-stripped Xe is indicated.

wavelength of 572.79(15) nm (15). The wavelength uncertainty is caused primarily by the Doppler broadening of the source ions in Super-EBIT. The 0.15 nm uncertainty in the measurement corresponds to about 135 GHz, only a few 30 GHz scans of the dye laser, so the transition should not be difficult to find. Estimates of the equilibrium number of confined Ho^{66+} ions in Super-EBIT, based on the observed photon count rate and other considerations, together with the extraction efficiency, indicate that sufficient ions for a spectroscopy measurement should be available in a single pulse for capture into RETRAP.

In RETRAP, the magnetic field B splits the magnetic sublevels by the "weak field" Zeeman effect, since the hfs is near 5×10^{14} Hz. To a good approximation, the magnetic sublevel energies E_F for hyperfine level F are described by $E_F \approx g_F \mu_B B m_F$ where μ_B is the Bohr magneton $e\hbar/2m_e$, g_F is the electron g-factor for the coupled nuclear spin I and electronic spin J = 1/2, and m_F is the magnetic sublevel quantum number. The Zeeman energy-level separation within a given F level is then $\Delta E_F = g_F \mu_B B$, which is numerically ≈ 15 GHz for B = 4.5 T (see below). The g-factor $g_F = g_J[\{F(F+1) + J(J+1) - I(I+1)\}/2F(F+1)] - g_I'[\{F(F+1) + I(I+1) - J(J+1)\}/2F(F+1)]$. Here, $g_I' = g_I(m_e/m_p)$ in terms of the electron-proton mass ratio m_e/m_p and the nuclear g-factor g_I. For the F = 4 level of Ho^{66+}, $g_4 = g_J/8 - 5.627(6) \times 10^{-4}$ and for the F = 3 level, $g_3 = -g_J/8 - 7.2338(8) \times 10^{-4}$. Full diagonalization of the Hamiltonian generates the exact dependence of the sublevel energies with B, which becomes important near the 10^{-6} level.

The g-factor of an electron bound in the ground state of a hydrogen-like ion with atomic number Z is modified from the free-space value g_e by relativistic and QED effects. Several terms have been calculated. There is a relativistic correction $(2/3)[2(1 - (Z\alpha)^2)^{1/2} + 1]$, often called the Breit correction, which produces the largest deviation from g_e. The first order radiative corrections calculated by Grotch and Hegstrom are $+\alpha/\pi + \alpha(Z\alpha)^2/6\pi$ (16). The all-order one loop radiative corrections calculated by Blundell et al. are $+7\alpha(Z\alpha)^4/\pi$ (17). First and second order recoil corrections calculated by Grotch and Hegstrom are $+(Z\alpha)^2(m_e/M)(1 + \alpha/\pi)$ and $-(Z\alpha)^2(m_e/M)^2(1 + Z + 5(Z\alpha)/12\pi)$ (16). In the preceding expressions, the fine-structure constant is α, and M is the ion mass. Figure 7(a) and (b) show calculations of the relativistic and radiative corrections vs. Z. For H-like Ho, from the calculation $g_J(67) = 1.833021$ including these terms. Of course, better experimental precision would begin to test the recoil corrections as well. A measurement on Ho^{66+} with fractional precision of 5×10^{-7} would test the radiative corrections to $\approx 2 \times 10^{-4}$, and first-order recoil within a factor of two. Using the laser linewidth of 300 MHz already measured on cooled Be^+ (see above), the fractional precision of a laser wavelength measurement on Ho^{66+} could be at least 3×10^{-7} without interpreting the lineshape.

Assuming that the matrix element for an angular momentum component in an H-like system is $\hbar$, the magnetic dipole transition rate $A_{ab}(M1) = (2\pi\alpha\hbar^2/3m_e^2c)\lambda^{-3} \approx 144$ s^{-1} for Ho^{66+}. Using the relations $A_{ab} = (\hbar\omega^3/2\pi^3c^3)B_{ab}$ and $B_{ba} = (g_b/g_a)B_{ab}$ where in this equation g denotes a statistical weight, then $B_{ba} = 13 \times 10^{15}$ m^3/js^2. A laser spot diameter of 300 μm should include at least 10^3 cold highly-charged ions near the trap center, so a laser power of 10 mW will produce a spectral energy density $\rho(\omega) = 1.3 \times 10^{-9}$ js/m^3. Taking into account the fraction of the laser linewidth within the Doppler width of the ions, and the fraction of the transition linewidth within the laser linewidth, the excitation rate per ion should be 65 s^{-1} at

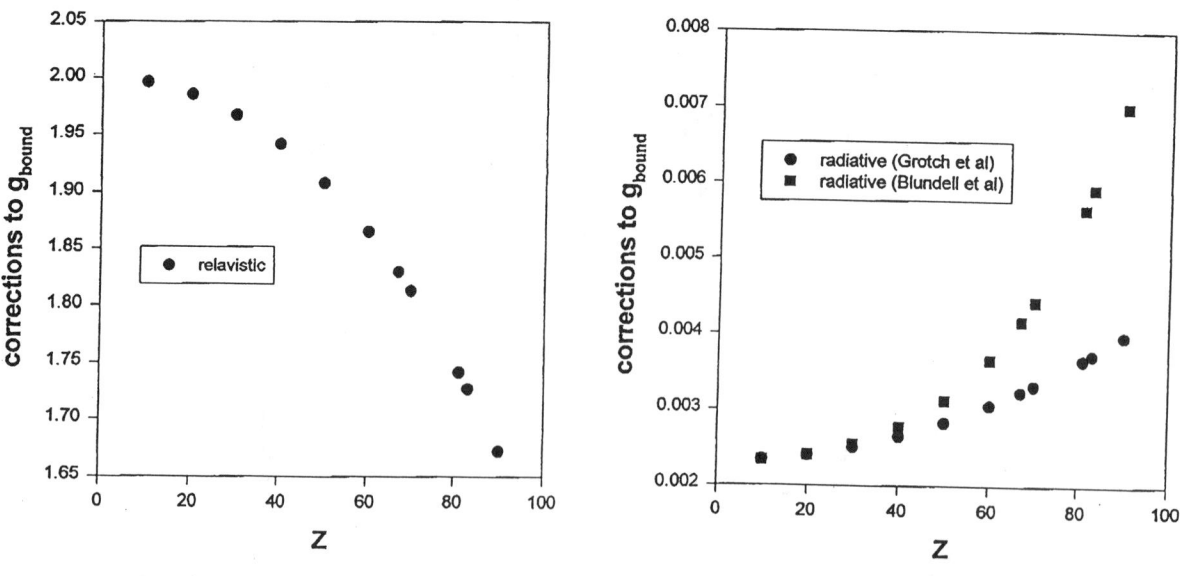

FIGURE 7 (a) Relativistic corrections to the bound state g-factor of the electron in a hydrogen-like ion, plotted vs. atomic number Z. (b) Radiative corrections to the bound state g-factor, plotted vs. Z. The equations plotted are from ref. 16 and 17.

resonance for ions at 1 K. The fractional photon detection solid angle (with the lens) is 4×10^{-2}. With a PMT efficiency of 0.3, and 50% light loss at optical surfaces, the expected detected count rate from 25 cold highly-charged ions is 10 s^{-1}. This number will be increased if the ion temperature is further reduced, or other improvements made. Since the level lifetime is several milliseconds, the cooling and exciting lasers can be chopped, so the fluorescence can be detected free of scattered light background. The fluorescence decay yields the level lifetime. Averaged photon detection using a CCD camera is also feasible.

Optical pumping between sublevels of the lower F = 3 level will occur, unless excitation of a cycling transition such as (3,-3) ↔ (4,-4) is used. The excitation rate will be reduced by the level multiplicity under these conditions. Microwave optical double resonance measurements, using a phase-locked loop operating near 15 GHz as a source, should be feasible, since the microwave wavelength $\lambda_m \leq 2d$. Transitions induced between magnetic sublevels would be detected as changes in the intensity of the scattered laser light. This type of measurement can potentially increase measurement precison considerably, since the first order Doppler shift would be suppressed. Single-ion measurement techniques may also be employed eventually, if they are found advantageous.

CONCLUSION

Data and calculations have been presented to support the conclusion that precision spectroscopic measurements using lasers can successfully be performed on the magnetic dipole transitions of the hyperfine structure of cold, high-Z hydrogen-like ions such as Ho^{66+} stored in RETRAP. Measurements on other electronic configurations of ions should similarly be feasible, if the transitions can be induced by a tunable laser. The ions are captured into the trap from the extracted beam of an external source, and are cooled using tuned circuit technology, and collisionally-coupled laser cooling methods. The next experimental steps are to observe fluorescence light from magnetic dipole transitions induced in multiply-charged ions, and to improve the cyclotron resonance measurements.

ACKNOWLEDGMENTS

This research is supported by the U. S. Department of Energy under Contract No. W-7405-ENG-48 with LLNL, and by the Texas Advanced Research Program through Texas A&M University. We thank Ed Magee and Dan Nelson for technical support of this research.

REFERENCES

1. Schneider, D., Church, D. A., Weinberg, G., Steiger, J., Beck, B., McDonald, J., Magee, E., and Knapp, D., *Rev. Sci. Instrum.* **65**, 3472 (1994).
2. Church, D. A.., *Nuc. Instrum. Meth. Phys. Res.* B**53**, 504 (1991)..
3. Klaft, I., Borneis, S., Engel, T., Fricke, B., Grieser, R., Huber, G., Kühl, T., Marx, D., Neumann, R., Schröder, S., Seelig, P., and Völker, L., *Phys. Rev. Lett.* **73**, 2425 (1994).
4. Schneider, D., DeWitt, D., Clark, M. W., Schuch, R., Cocke, C. L., Schmieder, R., Reed, K. J., Chen, M. H., Marrs, R. E., Levine, M., and Fortner, R., *Phys. Rev. A***42**, 3889 (1990).
5. Beck, B. R, Steiger, J., Weinberg, G., Church, D. A., McDonald, J., and Schneider, D., *Phys. Rev. Lett.* **77**, 1735 (1996).
6. Gabrielse, G., Haarsma, L.,.and Rolston, S. L., *Int. J. Mass Spectrom. Ion Proc.* **88**, 319 (1989).
7. Weinberg, G., Beck, B. R., Steiger, J., Church, D. A., McDonald, J. and Schneider, D., *J. Phys. Rev. A***57**, 4452 (1998).
8. Olson R. E., and Salop, A., *Phys.Rev. A***14**, 579 (1976).
9. RETRAP collaboration, unpublished (1998).
10. Brewer, L. R., Prestage, J. D., Bollinger, J. J., Itano, W. M., Larson, D. J., and Wineland, D. J., *Phys. Rev. A***38** 859 (1988).
11. Gabrielse, G., *Phys. Rev. A***29**, 462 (1984).
12. Gruber, L. et al., "Separation of laser cooled Be$^+$ and Be^{2+} in RETRAP", LLNL EBIT Annual Report 1996 - 7 (1998), (in press)
13. Steiger, J., et al., Proceedings of this conference.
14. Marrs, R. E., Beiersdorfer, P., Elliott, S. R., Knapp, D. A., and Stoehlker, Th., *Physica Scripta* T**59**, 183 (1995).
15. Crespo Lopez-Urrutia, J. R., Beiersdorfer, P., Savin, D. W., and Widmann, K., *Phys. Rev. Lett.* **77**, 826 (1996).
16. Grotch, H. and Hegstrom, R. A., *Phys. Rev. A***4**, 59 (1971).
17. Blundell, S. A , Cheng, K. T., and Sapirstein, J., *Phys. Rev. A***55**, 1857 (1997).

A Quantum Mechanical Model of Rabi Oscillations Between Two Interacting Harmonic Oscillator Modes and the Interconversion of Modes in a Penning Trap

Martin Kretzschmar

Institut für Physik,
Johannes-Gutenberg-Universität,
55099 Mainz, Germany

Abstract. When a Penning trap is operated with an additional quadrupole driving field with a frequency that equals a suitable combination (sum or difference) of the frequencies of the fundamental modes of motion (modified cyclotron, magnetron and axial frequency), then a periodic conversion of the participating modes into each other is observed, strongly resembling the Rabi oscillations in a 2-level atom driven by a laser field tuned to the transition frequency. This investigation attempts to understand on a fundamental level how and why the motion of a classical particle in a macroscopic apparatus can be truely analogous to the oscillations of states of quantum mechanical 2-level systems (2-level atom or magnetic resonance).

Ion motion in a Penning trap with an additional quadrupole driving field is described in a quantum mechanical frame work. The Heisenberg equations of motion for the creation and annihilation operators of the interacting oscillators have been explicitly solved, the time development operator of the Schrödinger picture has been determined. The driving field provides for two types of intermode interaction: Type I preserves the total number of excitation quanta present in the two interacting modes, the system oscillates between the modes with a frequency corresponding to the Rabi frequency in two-level systems. Type II preserves the difference of the numbers of excitation quanta present in the two interacting modes, it causes the ion motion to become unbounded. The two types of interaction are associated in a natural way with a $SU(2)$ and a $SU(1,1)$ Lie algebra. The three generators of these algebras form a vector operator that we denote as the Bloch vector operator. The Hilbert space decomposes in a natural way into invariant subspaces, finite dimensional in the case of type I interaction ($SU(2)$-algebra) and infinite dimensional in the case of type II interaction ($SU(1,1)$-algebra).

The physics of the 2-level atom in the laser field can be described in the 2-dimensional (the lowest nontrivial) sector of the Hilbert space associated with the type I ($SU(2)$-algebra) interaction. The Bloch vector, well known from quantum optics, is the expectation value of our Bloch operator. On the other hand, the description of ion motion in the Penning trap requires the whole infinite dimensional Hilbert space of our model. Classical ion trajectories are obtained by calculating for the observables corresponding to position and momentum the expectation values with respect to minimum uncertainty coherent oscillator states.

I INTRODUCTION

Some years ago Bollen, Moore, Savard and Stolzenberg [1–3] developed a new technique for ion cyclotron resonance (ICR) precision mass spectroscopy, which has found important practical applications. They used a hyperbolic Penning trap with a ring electrode that was divided into four sectors. This modification made it possible to introduce a driving azimuthal electric quadrupole field into the trap. Choosing for the driving frequency the cyclotron frequency of the trapped ions, they demonstrated that magnetron motion could be converted into cyclotron motion and vice versa. The phenomenon was in fact a periodic one, strongly resembling the Rabi oscillations of an atomic 2-level system that is exposed to a strong laser beam tuned to the transition frequency of the atom. It could also be seen as being an analogue to magnetic resonance phenomena observed on spin-$\frac{1}{2}$ particles with a magnetic moment. In subsequent developments the new technique was adapted to cubic traps by Schweikhard, Guan and Marshall [4,5], it was further extended by Guan, Xiang and Marshall [6] to permit conversion of axial motion into cyclotron motion or magnetron motion and vice versa. For a recent review of the present state of development of techniques and applications see references [7,8].

Theoretical insight was initially based on numerical and approximate analytical integration of the classical equations of motion of the trapped ion [1,2]. Later on Guan and Marshall [9] pointed out that after taking into account a conservation theorem the equations of motion for the remaining variables could be transformed into a form of the Bloch equations known from nuclear magnetic resonance [10]. Further more extensive studies of the classical equations of motion were carried out by Schweikhard and Marshall [5] and by König et al. [11,12].

The present paper was motivated by the desire to understand on a more fundamental level the apparent analogy between the classical motion of a point particle on the one hand and quantum mechanical 2-level systems on the other hand. How can a classical physical system, whose quantized version requires an infinite dimensional Hilbert space, be similar to a 2-level system? It is natural for this investigation to adopt the quantum mechanical point of view right from the beginning and only later to consider the classical limit.

As a starting point the theory of ion motion in an ideal Penning trap is presented in quantized form using the Heisenberg picture. An additional quadrupole driving field introduces two types of intermode interaction: Type I interactions preserve the total number of excitation quanta present in the motion and the Hilbert space decomposes into finite dimensional subspaces that are invariant under this type of interaction. Because of the equal spacing of oscillator levels it is then possible to introduce a triplet of operators, with commutation relations corresponding to a $SU(2)$ Lie algebra, that act in these invariant subspaces as ladder and number operators in a way familiar from the theory of angular momentum. This operator triplet has been named here the Bloch vector operator, its time dependent expectation values can be visualized as an ordinary vector in a formal 3-space (the Bloch vector) that is precessing in a prescribed fashion. Type II interactions preserve the difference between the numbers of excitation quanta present in the two interacting modes, the Hilbert space correspondingly decomposes into infinite dimensional invariant subspaces, and the components of the Bloch vector operator generate a $SU(1,1)$ Lie algebra. Due to the linearity of the Heisenberg equations of motion a complete solution can be worked out for the time development of all Heisenberg operators of interest and for the unitary operator $U(t)$ describing the time development of the model in the Schrödinger picture.

Trajectories of classical point particles moving in the Penning trap are calculated as expectation values of the observables representing position and momentum of the particle, using minimum uncertainty coherent oscillator states. The results reproduce those obtained by using the equations of classical mechanics. Beyond the classical realm our model covers all situations where nonrelativistic quantum phenomena become important. For some general considerations on the relevance of quantum effects for trapped particles see the review article by Brown and Gabrielse [13]. In connection with experiments using ultracold trapped particles the topic of "nonclassical states of motion" has recently found increasing interest [14].

One observes that the formalism required for the theory of the 2-level atom or of magnetic resonance of spin-$\frac{1}{2}$ particles [15] resides in the lowest nontrivial ($n=1$) sector of our model, while the application to the conversion of modes in a Penning trap requires the full infinite dimensional Hilbert space.

II ION MOTION IN AN IDEAL PENNING TRAP

In the canonical formalism (for details see ref. [16]) the motion of a trapped ion of mass m and electric charge Ze is governed by the hamiltonian

$$H = \omega_+ \cdot \tfrac{1}{2}\left(q_+^2 + p_+^2\right) - \omega_- \cdot \tfrac{1}{2}\left(q_-^2 + p_-^2\right) + \omega_z \cdot \tfrac{1}{2}\left(q_3^2 + p_3^2\right). \tag{1}$$

The q_k and p_k are the canonical coordinates of the ion, $\omega_c = ZeB/(mc)$ is the cyclotron frequency, ω_z the axial frequency, $\omega_+ = \tfrac{1}{2}(\omega_c + \omega_1)$ the modified cyclotron frequency, $\omega_- = \tfrac{1}{2}(\omega_c - \omega_1)$ the magnetron frequency, and ω_1 is defined as $\omega_1 = \sqrt{\omega_c^2 - 2\omega_z^2}$. In order to quantize this hamiltonian we use the Heisenberg picture and postulate the equal time commutation relations $[q_j, p_k] = i\hbar \delta_{jk} \cdot \mathbb{1}$. Since the hamiltonian eq. (1) is obviously composed of three 1-dimensional harmonic oscillator hamiltonians, one can also use the annihilation and creation operators known from harmonic oscillator theory [13],

$$A_k = \frac{1}{\sqrt{2\hbar}}\left(q_k + ip_k\right), \qquad A_k^\dagger = \frac{1}{\sqrt{2\hbar}}\left(q_k - ip_k\right), \qquad (k = +, -, 3). \tag{2}$$

For these operators the equal time commutation rules $\left[A_j(t), A_k^\dagger(t)\right] = \delta_{jk} \cdot \mathbb{1}$ hold for $j, k = +, -, 3$ and the hamiltonian becomes

$$H = \hbar\omega_+ \left(A_+^\dagger(t)A_+(t) + \tfrac{1}{2}\cdot \mathbb{1}\right) - \hbar\omega_- \left(A_-^\dagger(t)A_-(t) + \tfrac{1}{2}\cdot \mathbb{1}\right) + \hbar\omega_z \left(A_3^\dagger(t)A_3(t) + \tfrac{1}{2}\cdot \mathbb{1}\right). \tag{3}$$

In terms of these annihilation and creation operators the original cartesian coordinates are

$$x(t) = \frac{1}{2}\sqrt{\frac{2\hbar}{m\omega_1}} \left(A_+(t) + A_+^\dagger(t) + A_-(t) + A_-^\dagger(t)\right), \tag{4}$$

$$y(t) = \frac{1}{2i}\sqrt{\frac{2\hbar}{m\omega_1}} \left(A_+(t) - A_+^\dagger(t) - A_-(t) + A_-^\dagger(t)\right), \tag{5}$$

$$z(t) = \frac{1}{2}\sqrt{\frac{2\hbar}{m\omega_z}} \left(A_3(t) + A_3^\dagger(t)\right). \tag{6}$$

III   QUADRUPOLE EXCITATIONS IN A PENNING TRAP

The three modes of motion of the ion can be partially or totally converted into each other, by suitable time dependent electric quadrupole fields acting on the ion. For all experimental and technical details we refer the reader to the original papers [1–6] and the recent review articles [7,8]. For the following we assume that a general time dependent (driving) electric quadrupole field has been added to the configuration of the ideal Penning trap. The additional potential $\Phi_d(x,y,z,t)$ can be written as a superposition of spherical harmonics $Y_{2,m}$ [16]. Because Φ_d is real, and because with a suitable choice of coordinates it may be assumed to be symmetric under reflections by the xz-plane [1,6], the most general form reduces to

$$\Phi_d(x,y,z,t) = C_0(t)\cdot\left(z^2 - \frac{x^2+y^2}{2}\right) + C_1(t)\cdot xz + C_2(t)\cdot(x^2-y^2) \tag{7}$$

Assuming a driving field characterized by a single driving frequency ω_d and a phase δ_d, we write

$$C_m(t) = C_m \cos\phi_d(t) \qquad \text{with} \qquad \phi_d(t) = \omega_d t + \delta_d. \tag{8}$$

The real coefficients C_m are proportional to the applied voltage difference, but they also reflect geometric details of the specific trap design.

The $m=0$ term in eq. (7) is usually associated with parametric excitation of the various modes of motion, in this paper it is of no further interest. The $m=1$ term in eq. (7) is the one with the simplest structure. After quantization its contribution to the hamiltonian is rewritten as follows in terms of creation and annihilation operators, using eqs. (4, 5, 6),

$$\hbar g_1 \cdot \left(e^{i\phi_d(t)} + e^{-i\phi_d(t)}\right)\left(A_3(t)+A_3^\dagger(t)\right)\left(A_+(t)+A_+^\dagger(t)+A_-(t)+A_-^\dagger(t)\right) \tag{9}$$

with the coupling constant $g_1 = (Ze\,C_1)/(4m\sqrt{\omega_1\omega_z})$. In absence of the driving potential $\Phi_d(x,y,z,t)$ the time dependence of the annihilation operators is

$$A_+(t) = e^{-i\omega_+ t}A_+(0), \qquad A_-(t) = e^{+i\omega_- t}A_-(0), \qquad A_3(t) = e^{-i\omega_z t}A_3(0), \tag{10}$$

with corresponding adjoint equations for the creation operators. In the presence of the driving potential the relations still hold approximately. The 16 terms obtained by expanding the product in expression (9) can now be grouped into two classes: (i) Some terms can be made nearly stationary with an appropriate choice of the driving frequency ω_d (for example $A_3^\dagger(t)A_-(t)e^{-i\phi_d(t)}$ is slowly varying when $\omega_d \approx \omega_z + \omega_-$). They become dominant and will be retained, (ii) The other terms are rapidly varying with any choice of ω_d (for example the expression $e^{+i\phi_d(t)}A_3^\dagger(t)A_-(t)$ varies as $\exp[i(\omega_z + \omega_- + \omega_d)t]$). They give only very small contributions and are assumed to be negligible. The procedure of dropping the rapidly varying terms is known as the rotating wave approximation. The expression (9) then takes the following form:

$$\hbar g_1 \cdot \left(e^{-i\phi_d(t)}A_+^\dagger(t)A_3(t) + e^{+i\phi_d(t)}A_3^\dagger(t)A_+(t)\right) \tag{11}$$

$$+ \hbar g_1 \cdot \left(e^{-i\phi_d(t)}A_3^\dagger(t)A_-(t) + e^{+i\phi_d(t)}A_-^\dagger(t)A_3(t)\right) \tag{12}$$

$$+ \hbar g_1 \cdot \left(e^{-i\phi_d(t)}A_+^\dagger(t)A_3^\dagger(t) + e^{+i\phi_d(t)}A_3(t)A_+(t)\right) \tag{13}$$

$$+ \hbar g_1 \cdot \left(e^{-i\phi_d(t)}A_-^\dagger(t)A_3^\dagger(t) + e^{+i\phi_d(t)}A_3(t)A_-(t)\right) \tag{14}$$

The four terms can be made slowly varying by choosing the driving frequency to be $\omega_d \approx \omega_+ - \omega_z$, $\omega_z + \omega_-$, $\omega_+ + \omega_z$, or $\omega_z - \omega_-$ respectively. All these terms represent interactions between different modes of motion in the Penning trap. As we shall show in the following sections, the first two expressions (11), (12) give rise to the interconversion between the axial and the cyclotron modes and between the axial and the magnetron modes respectively, while the remaining two interactions (13) and (14) lead to unstable motions in the trap.

The last term in eq. (7) can be discussed in an analogous fashion. It contains the expression

$$\hbar g_2 \cdot \left(e^{-i\phi_d(t)} A_+^\dagger(t) A_-(t) + e^{+i\phi_d(t)} A_-^\dagger(t) A_+(t) \right) \tag{15}$$

which is responsible for the interconversion between the cyclotron and magnetron modes of motion, when the driving frequency is chosen to be $\omega_d \approx \omega_+ + \omega_- = \omega_c$.

To summarize, a driving quadrupole potential provides two basically different types of intermode interaction,

$$\text{type I:} \qquad H_1^I = \hbar g \cdot \left(e^{-i\phi_d(t)} A_b^\dagger(t) A_a(t) + e^{+i\phi_d(t)} A_a^\dagger(t) A_b(t) \right), \tag{16}$$

$$\text{type II:} \qquad H_1^{II} = \hbar g \cdot \left(e^{-i\phi_d(t)} A_a^\dagger(t) A_b^\dagger(t) + e^{+i\phi_d(t)} A_b(t) A_a(t) \right) \tag{17}$$

where a, b denote any two of the three modes of motion $(+, -, z)$. The first type of interaction preserves the total number of excitation quanta present in the system and results, as we shall see below, in Rabi like oscillations between the modes. It becomes dominant when the driving frequency is chosen $\omega_d \approx \omega_a - \epsilon \omega_b$, with $\epsilon_+ = \epsilon_z = 1$ and $\epsilon_- = -1$. The second type of interaction causes the number of excitation quanta to increase without bound, resulting in exponentially increasing amplitudes of the motion. It becomes dominant when the driving frequency is chosen $\omega_d \approx \omega_a + \epsilon \omega_b$. A short qualitative discussion of the two types of interaction can be found already in the review article by Brown and Gabrielse [13].

IV QUANTUM THEORY OF TWO HARMONIC OSCILLATORS INTERACTING WITH AN EXTERNAL DRIVING FIELD

The results of the preceding section suggest the investigation of a quantum mechanical model consisting of two 1-dimensional harmonic oscillators that interact by emitting energy into or absorbing energy from a classical external (electromagnetic) driving field via the interactions (16) or (17). The third oscillator plays the role of a spectator.

A The hamiltonian and the Heisenberg equations of motion

The model hamiltonian in the Heisenberg picture[1] reads in the case of a type I interaction

$$H^I(t) = \hbar \omega_a \left(A_a^\dagger(t) A_a(t) + \tfrac{1}{2} \cdot \mathbb{1} \right) + \epsilon \hbar \omega_b \left(A_b^\dagger(t) A_b(t) + \tfrac{1}{2} \cdot \mathbb{1} \right)$$
$$+ \hbar g \left(A_a^\dagger(t) A_b(t) e^{-i\phi_d(t)} + A_b^\dagger(t) A_a(t) e^{+i\phi_d(t)} \right) \tag{18}$$

and in the case of a type II interaction

$$H^{II}(t) = \hbar \omega_a \left(A_a^\dagger(t) A_a(t) + \tfrac{1}{2} \cdot \mathbb{1} \right) + \epsilon \hbar \omega_b \left(A_b^\dagger(t) A_b(t) + \tfrac{1}{2} \cdot \mathbb{1} \right)$$
$$+ \hbar g \left(A_a^\dagger(t) A_b^\dagger(t) e^{-i\phi_d(t)} + A_b(t) A_a(t) e^{+i\phi_d(t)} \right) \tag{19}$$

with $\phi_d(t) = \omega_d t + \delta_d$. The constant g measures the strength of the coupling of the two oscillators to the external driving field with (circular) frequency ω_d, while δ_d is an arbitrary phase. Without restriction of generality we assume $\omega_a \geq \omega_b$. Since for all practical purposes the magnetron frequency ω_- is the smallest of the three frequencies, the magnetron mode shall always be associated with oscillator b. The oscillator of the magnetron mode is "inverted", we take this into account by choosing $\epsilon = -1$ when b represents the magnetron mode and

---

[1] Operators $O(t)$ and states $|\psi\rangle$ shall always be understood as being given in the Heisenberg picture.

$\epsilon = +1$ otherwise. The annihilation and creation operators of the two oscillators obey the usual equal times commutation relations.

The Heisenberg equations of motion for the annihilation and creation operators are:

For type I interactions:

$$i\hbar \frac{d}{dt} A_a(t) = \left[A_a(t), H^I(t)\right] = \hbar\omega_a \cdot A_a(t) + \hbar g\, e^{-i\phi_d(t)} \cdot A_b(t) \tag{20}$$

$$i\hbar \frac{d}{dt} A_b(t) = \left[A_b(t), H^I(t)\right] = \hbar g\, e^{+i\phi_d(t)} \cdot A_a(t) + \epsilon\hbar\omega_b \cdot A_b(t) \tag{21}$$

For type II interactions:

$$i\hbar \frac{d}{dt} A_a(t) = \left[A_a(t), H^{II}(t)\right] = \hbar\omega_a \cdot A_a(t) + \hbar g\, e^{-i\phi_d(t)} \cdot A_b^\dagger(t) \tag{22}$$

$$i\hbar \frac{d}{dt} A_b^\dagger(t) = \left[A_b^\dagger(t), H^{II}(t)\right] = -\hbar g\, e^{+i\phi_d(t)} \cdot A_a(t) - \epsilon\hbar\omega_b \cdot A_b^\dagger(t) \tag{23}$$

Because of the linearity of these equations exact solutions can be obtained. Omitting all details we state the final result.

For type I interactions:

$$A_a(t) = e^{-i(\omega_a + \delta/2)t} \cdot \left[\left(\cos(\omega_R t/2) + i\frac{\delta}{\omega_R}\sin(\omega_R t/2)\right) \cdot A_a(0) - i\frac{2g}{\omega_R}\sin(\omega_R t/2)e^{-i\delta_d} \cdot A_b(0)\right] \tag{24}$$

$$A_b(t) = e^{-i(\epsilon\omega_b - \delta/2)t} \cdot \left[-i\frac{2g}{\omega_R}\sin(\omega_R t/2)e^{+i\delta_d} \cdot A_a(0) + \left(\cos(\omega_R t/2) - i\frac{\delta}{\omega_R}\sin(\omega_R t/2)\right) \cdot A_b(0)\right] \tag{25}$$

where $\delta = \omega_d - (\omega_a - \epsilon\omega_b)$ denotes the detuning of the driving frequency and $\omega_R = \sqrt{4g^2 + \delta^2}$ is the Rabi frequency.

For type II interactions:

$$A_a(t) = e^{-i(\omega_a + \delta/2)t} \cdot \left[\left(\cos(\omega_R t/2) + i\frac{\delta}{\omega_R}\sin(\omega_R t/2)\right) \cdot A_a(0) - i\frac{2g}{\omega_R}\sin(\omega_R t/2)e^{-i\delta_d} \cdot A_b^\dagger(0)\right] \tag{26}$$

$$A_b^\dagger(t) = e^{i(\epsilon\omega_b + \delta/2)t} \cdot \left[i\frac{2g}{\omega_R}\sin(\omega_R t/2)e^{+i\delta_d} \cdot A_a(0) + \left(\cos(\omega_R t/2) - i\frac{\delta}{\omega_R}\sin(\omega_R t/2)\right) \cdot A_b^\dagger(0)\right] \tag{27}$$

with detuning $\delta = \omega_d - (\omega_a + \epsilon\omega_b)$ and Rabi frequency $\omega_R = \sqrt{-4g^2 + \delta^2}$. For small detuning δ (i.e. $\delta^2 < 4g^2$) the Rabi frequency ω_R becomes imaginary, $\omega_R = i|\omega_R|$. The trigonometric functions must then be replaced by the corresponding hyperbolic functions.

B The structure of the Hilbert space

A convenient basis of the Hilbert space of our model is obtained by acting with the creation operators $A_a^\dagger(0)$, $A_b^\dagger(0)$ at time $t = 0$ on the zero excitation state[2] $|0, 0\rangle$ of the model,

$$|n_a, n_b\rangle = \frac{1}{\sqrt{n_a! n_b!}} \left(A_a^\dagger(0)\right)^{n_a} \left(A_b^\dagger(0)\right)^{n_b} |0, 0\rangle, \quad (n_a, n_b = 0, 1, 2, \ldots). \tag{28}$$

For type I interactions the total number of excitation quanta of the two oscillators is conserved, or expressed differently, the "total number" operator $N_{\text{tot}} = 2T_0^I - \mathbb{1}$ with

$$T_0^I = \tfrac{1}{2}\left(A_a^\dagger(t) A_a(t) + A_b^\dagger(t) A_b(t) + \mathbb{1}\right). \tag{29}$$

---

[2] In the case $\epsilon = -1$ (inverted oscillator) there exists no ground state in the sense of a state of lowest energy.

commutes with $H(t)$ and is therefore a time independent operator.

The total Hilbert space therefore splits into an infinite direct sum of invariant subspaces corresponding to the eigenspaces of T_0^I, $\mathcal{H} = \sum_{n=0}^{\infty} \oplus \mathcal{H}_n^I$. The subspace $\mathcal{H}_n^I$ is spanned by the set of all states with exactly n oscillator excitation quanta, it thus belongs to the eigenvalue $n/2$ of T_0^I and is $(n+1)$-dimensional. For each subspace $\mathcal{H}_n^I$ we have a basis $|k, n-k\rangle$ such that the k-th basis vector is the state describing $n_a = k$ excitation quanta for oscillator a and $n_b = n - k$ excitation quanta for oscillator b (with $k = 0, 1, 2, \ldots, n$). Furthermore, the states $|n_a, n_b\rangle = |k, n-k\rangle$ are eigenstates with eigenvalue $\tau_3 = (n_a - n_b)/2 = k - \frac{1}{2}n$ of the operator $T_3^I(0)$, defined in eq. (33) below.

For type II interactions the difference between the numbers of excitation quanta in the two oscillator modes is conserved. The operator

$$T_0^{II} = \tfrac{1}{2}\left(A_a^\dagger(t)A_a(t) - A_b^\dagger(t)A_b(t)\right). \tag{30}$$

commutes with $H(t)$ and is therefore a time independent operator.

The total Hilbert space splits into an infinite direct sum of invariant subspaces corresponding to the eigenspaces of T_0^{II}, $\mathcal{H} = \sum_{n=-\infty}^{\infty} \oplus \mathcal{H}_n^{II}$. The subspace $\mathcal{H}_n^{II}$ is spanned by the set of all states $|n_a, n_b\rangle$ with $n_a - n_b = n$, where $n = \ldots -2, -1, 0, +1, +2, \ldots$. All subspaces $\mathcal{H}_n^{II}$ are obviously of infinite dimension.

C The Bloch vector operator

For each of the two types of interaction we shall now define a set of operators that generate a characteristic Lie algebra and play an important role in the description of the physical phenomena.

For type I interactions let us consider the set of operators

$$T_1^I(t) = \frac{1}{2}\left(A_a^\dagger(t)A_b(t) + A_b^\dagger(t)A_a(t)\right) \tag{31}$$

$$T_2^I(t) = \frac{1}{2i}\left(A_a^\dagger(t)A_b(t) - A_b^\dagger(t)A_a(t)\right) \tag{32}$$

$$T_3^I(t) = \frac{1}{2}\left(A_a^\dagger(t)A_a(t) - A_b^\dagger(t)A_b(t)\right) \tag{33}$$

They obey at equal times the commutation rules

$$[T_1^I(t), T_2^I(t)] = i\,T_3^I(t), \qquad [T_2^I(t), T_3^I(t)] = i\,T_1^I(t), \qquad [T_3^I(t), T_1^I(t)] = i\,T_2^I(t), \tag{34}$$

and are, therefore, the generators of a $SU(2)$ Lie algebra. From general principles all irreducible representations of this Lie algebra must be finite dimensional, and in fact our subspaces $\mathcal{H}_n^I$ are the invariant subspaces of the Lie algebra (i.e. invariant under the action of $T_1^I(t)$, $T_2^I(t)$, $T_3^I(t)$). The operator triplet thus behaves like a familiar angular momentum operator. It plays an important role for the following developments and shall henceforth be referred to as the *Bloch vector operator*.

Straightforward algebra and the commutation relations eq.(34) yield the result

$$(T_1^I(t))^2 + (T_2^I(t))^2 + (T_3^I(t))^2 = (T_0^I)^2 - \tfrac{1}{4}\mathbb{1}. \tag{35}$$

The operators $T_+^I(t) = T_1^I(t) + iT_2^I(t) = A_a^\dagger(t)A_b(t)$ and $T_-^I(t) = T_1^I(t) - iT_2^I(t) = A_b^\dagger(t)A_a(t)$ can conveniently be used as ladder operators connecting the basis vectors in $\mathcal{H}_n^I$, similar as with angular momentum operators. The model hamiltonian eq. (18) can now be expressed in terms of the components of the Bloch vector operator as follows

$$H^I(t) = \hbar(\omega_a + \epsilon\omega_b)T_0^I + \hbar(\omega_a - \epsilon\omega_b)T_3^I(t) + \hbar g\left(T_+^I(t)e^{-i\phi_d(t)} + T_-^I(t)e^{+i\phi_d(t)}\right). \tag{36}$$

For type II interactions we consider the operators

$$T_1^{II}(t) = \frac{1}{2}\left(A_a^\dagger(t)A_b^\dagger(t) + A_b(t)A_a(t)\right) \tag{37}$$

$$T_2^{II}(t) = \frac{1}{2i}\left(A_a^\dagger(t)A_b^\dagger(t) - A_b(t)A_a(t)\right) \tag{38}$$

$$T_3^{II}(t) = \frac{1}{2}\left(A_a^\dagger(t)A_a(t) + A_b^\dagger(t)A_b(t) + \mathbb{1}\right) \tag{39}$$

They obey at equal times the commutation rules

$$[T_1^{II}(t), T_2^{II}(t)] = -iT_3^{II}(t), \quad [T_2^{II}(t), T_3^{II}(t)] = iT_1^{II}(t), \quad [T_3^{II}(t), T_1^{II}(t)] = iT_2^{II}(t), \tag{40}$$

and are, therefore, the generators of a $SU(1,1)$ Lie algebra. The irreducible representations of this Lie algebra are of infinite dimension, our subspaces $\mathcal{H}_n^{II}$ are the invariant subspaces of this Lie algebra (i.e. invariant under the action of the $T_i^{II}(t)$).

Straightforward algebra and the commutation relations eq.(40) yield the result

$$-(T_1^{II}(t))^2 - (T_2^{II}(t))^2 + (T_3^{II}(t))^2 = (T_0^{II})^2 - \tfrac{1}{4}\mathbb{1}. \tag{41}$$

Again the operators $T_+^{II}(t) = T_1^{II}(t) + iT_2^{II}(t) = A_a^\dagger(t)A_b^\dagger(t)$ and $T_-^{II}(t) = T_1^{II}(t) - iT_2^{II}(t) = A_b(t)A_a(t)$ can be used as ladder operators connecting the basis vectors in $\mathcal{H}_n^{II}$.

The model hamiltonian eq. (19) can now be expressed in terms of the components of the Bloch vector operator as follows

$$H^{II}(t) = \hbar(\omega_a - \epsilon\omega_b)T_0^{II} + \hbar(\omega_a + \epsilon\omega_b)T_3^{II}(t) + \hbar g\left(T_+^{II}(t)e^{-i\phi_d(t)} + T_-^{II}(t)e^{+i\phi_d(t)}\right). \tag{42}$$

V OBSERVABLES AND THEIR EXPECTATION VALUES

Contact to experiment is made by constructing hermitean observables as polynomials in the fundamental operators $A_a(t)$, $A_b(t')$, $A_a^\dagger(t'')$, $A_a^\dagger(t''')$ and by calculating expectation values for these observables[3]. Examples for observables linear in the fundamental operators are the coordinates $x(t)$, $y(t)$, $z(t)$ of an ion moving in the Penning trap, examples for bilinear observables are the components of the Bloch vector operator $T_i(t)$ or the variances of the linear observables.

The obvious method for the calculation of expectation values uses the density operator ρ (time independent in the Heisenberg picture) to encode the state of our knowledge about the physical system [18]. The expectation value of a Heisenberg operator $O(t)$ is the given by

$$\langle O(t)\rangle = \text{Tr}\{O(t)\rho\} = \sum_{n_a,n_b}\sum_{n_a',n_b'} \langle n_a', n_b'|O(t)|n_a, n_b\rangle \rho_{n_a n_b, n_a' n_b'}. \tag{43}$$

In the last step matrix representations with respect to the Fock basis (28) were used. The matrix elements of the observable $O(t)$ at time t can be reduced to the matrix elements $O(0)$ at time $t = 0$, using our solutions eqs. (24, 25, 26, 27) of the Heisenberg equations of motion for the operators $A_a(t)$, $A_b(t)$.

As an example consider the Bloch vector operator for type I interactions. One finds the expectation values

$$\langle T_1^I(0)\rangle = \frac{1}{2}\sum_{n_a n_b}\left(\sqrt{(n_a+1)n_b}\,\rho_{n_a+1\,n_b-1,n_a n_b} + \sqrt{n_a(n_b+1)}\,\rho_{n_a-1\,n_b+1,n_a n_b}\right), \tag{44}$$

$$\langle T_2^I(0)\rangle = \frac{1}{2i}\sum_{n_a n_b}\left(\sqrt{(n_a+1)n_b}\,\rho_{n_a+1\,n_b-1,n_a n_b} - \sqrt{n_a(n_b+1)}\,\rho_{n_a-1\,n_b+1,n_a n_b}\right), \tag{45}$$

$$\langle T_3^I(0)\rangle = \frac{1}{2}\sum_{n_a n_b}(n_a - n_b)\,\rho_{n_a n_b, n_a n_b}. \tag{46}$$

These three real numbers may be taken as the components of an ordinary 3-dimensional vector that may be called the Bloch vector at time $t = 0$. Using the solution of the Heisenberg equations of motion for the fundamental creation and annihilation operators eqs. (24,25) a time dependent Bloch vector is obtained that can be visualized as precessing in space in a prescribed manner.

Let us look at the motion of this vector in the simplest possible case $\langle T_1^I(0)\rangle = \langle T_2^I(0)\rangle = 0$, $\langle T_3^I(0)\rangle \neq 0$. The components of the precessing Bloch vector at time t are obtained as

---

[3] For simplification the spectator oscillator, which does not participate in the interaction, is ignored in this section.

$$\langle T_1^I(t)\rangle = \left[\cos\phi_d(t)\left(-\frac{2g\delta}{\omega_R^2} + \frac{2g\delta}{\omega_R^2}\cos\omega_R t\right) + \frac{2g}{\omega_R}\sin\phi_d(t)\,\sin\omega_R t\right]\langle T_3^I(0)\rangle, \tag{47}$$

$$\langle T_2^I(t)\rangle = \left[\sin\phi_d(t)\left(-\frac{2g\delta}{\omega_R^2} + \frac{2g\delta}{\omega_R^2}\cos\omega_R t\right) - \frac{2g}{\omega_R}\cos\phi_d(t)\,\sin\omega_R t\right]\langle T_3^I(0)\rangle, \tag{48}$$

$$\langle T_3^I(t)\rangle = \left[\frac{\delta^2}{\omega_R^2} + \frac{4g^2}{\omega_R^2}\cos\omega_R t\right]\langle T_3^I(0)\rangle. \tag{49}$$

The third component that describes the difference of the number of excitation quanta present in the oscillator modes a and b is seen to oscillate between the values $\langle T_3^I(0)\rangle$ and $-(1 - 2\delta^2/\omega_R^2)\langle T_3^I(0)\rangle$. In a reference frame that rotates with the driving frequency ω_d the Bloch vector is seen to be composed of a constant vector whose length $(|\delta|/\omega_R)\langle T_3^I(0)\rangle$ is determined by the detuning and another vector rotating with the Rabi frequency about the axis with direction $(2g/\omega_R, 0, -\delta/\omega_R)$ and of length $\sqrt{1 - (\delta/\omega_R)^2}\langle T_3^I(0)\rangle$.

Each invariant subspace $\mathcal{H}_n^I$ contributes its share to the expectation values $\langle T_i^I(0)\rangle$. We note here that with $n = n_a + n_b = 1$ the contribution from the 2-dimensional subspace $\mathcal{H}_1^I$ is

$$\langle T_1^I(0)\rangle_{n=1} = \frac{1}{2}(\rho_{10,01} + \rho_{01,10}) = \frac{1}{2}(\rho_{10,01} + \rho_{10,01}^*) = \mathrm{Re}(\rho_{10,01}) \tag{50}$$

$$\langle T_2^I(0)\rangle_{n=1} = \frac{1}{2i}(\rho_{10,01} - \rho_{01,10}) = \frac{1}{2i}(\rho_{10,01} - \rho_{10,01}^*) = \mathrm{Im}(\rho_{10,01}) \tag{51}$$

$$\langle T_3^I(0)\rangle_{n=1} = \frac{1}{2}(\rho_{10,10} - \rho_{01,01}) \tag{52}$$

This is the result familiar from the theory of 2-level atoms, where the Bloch vector is usually introduced as a "vector model of the density matrix" [15,17].

VI CLASSICAL ION TRAJECTORIES

The classical ion trajectories are calculated by computing the expectation values of the time dependent observables $x(t), y(t), z(t)$ with respect to a minimum uncertainty coherent state $|\alpha, \beta, \gamma\rangle$. Coherent states are those quantum states that correspond most closely to a classical state of the system. They can be introduced through their expansion in terms of Fock states, for other definitions and more on the properties of these states see [15,17]. For any three complex numbers α, β, γ one defines

$$|\alpha, \beta, \gamma\rangle = \exp\left[-\tfrac{1}{2}(|\alpha|^2 + |\beta|^2 + |\gamma|^2)\right] \sum_{n_a=0}^{\infty}\sum_{n_b=0}^{\infty}\sum_{n_c=0}^{\infty} \frac{\alpha^{n_a}\beta^{n_b}\gamma^{n_c}}{\sqrt{n_a!\,n_b!\,n_c!}}\,|n_a, n_b, n_c\rangle. \tag{53}$$

The complex numbers α, β, γ characterizing the coherent state are related to the initial position and velocity of the ion at time $t = 0$ through the eqs. (4), (5), (6) and

$$\mathrm{Re}(\alpha) = \frac{1}{\sqrt{2\hbar}}\langle\alpha|q_+(0)|\alpha\rangle, \qquad \mathrm{Re}(\beta) = \frac{1}{\sqrt{2\hbar}}\langle\beta|q_-(0)|\beta\rangle, \qquad \mathrm{Re}(\gamma) = \frac{1}{\sqrt{2\hbar}}\langle\gamma|q_3(0)|\gamma\rangle, \tag{54}$$

$$\mathrm{Im}(\alpha) = \frac{1}{\sqrt{2\hbar}}\langle\alpha|p_+(0)|\alpha\rangle, \qquad \mathrm{Im}(\beta) = \frac{1}{\sqrt{2\hbar}}\langle\beta|p_-(0)|\beta\rangle, \qquad \mathrm{Im}(\gamma) = \frac{1}{\sqrt{2\hbar}}\langle\gamma|p_3(0)|\gamma\rangle. \tag{55}$$

As a simple example let us consider the trajectory of an ion in an ideal Penning trap. For this case the time dependence of the annihilation and creation operators is given by eq. (10). The trajectory is calculated as

$$\langle\alpha, \beta, \gamma|x(t)|\alpha, \beta, \gamma\rangle = \sqrt{\frac{2\hbar}{m\omega_1}}\left(|\alpha|\cos(\omega_+ t - \arg\alpha) + |\beta|\cos(\omega_- t + \arg\beta)\right), \tag{56}$$

$$\langle\alpha, \beta, \gamma|y(t)|\alpha, \beta, \gamma\rangle = -\sqrt{\frac{2\hbar}{m\omega_1}}\left(|\alpha|\sin(\omega_+ t - \arg\alpha) + |\beta|\sin(\omega_- t + \arg\beta)\right), \tag{57}$$

$$\langle\alpha, \beta, \gamma|z(t)|\alpha, \beta, \gamma\rangle = \sqrt{\frac{2\hbar}{m\omega_z}}|\gamma|\cos(\omega_z t - \arg\gamma). \tag{58}$$

For the discussion of the interconversion of modes a driving quadrupole interaction of type I (see eq.(16)) has to be added to the hamiltonian of the ideal Penning trap, eq. (3). One of the three oscillators does not participate in the interconversion process, it acts, so to say, as a spectator (denoted by index c). The explicit solutions of the equations of motion for the operators $A_a(t)$, $A_a^\dagger(t)$, $A_b(t)$, and $A_b^\dagger(t)$ have been obtained in eqs.(24, 25), for the spectator we have ($\epsilon_c = -1$ for the magnetron mode, $= +1$ otherwise)

$$A_c(t) = e^{-i\epsilon_c \omega_c t} A_c(0). \tag{59}$$

These solutions are used to calculate the expected position of the particle at time t, given the initial values at time $t = 0$. It is obvious from eqs.(24, 25) that with type I interactions an oscillatory motion results, determined by the frequencies ω_a, ω_b, ω_c, and the Rabi frequency ω_R. On the other hand, with type II interactions the solutions of the equations of motion, eqs.(26, 27), contain hyperbolic functions, so that the particle trajectories run away to infinity. This result agrees with the conclusions that Guan and Marshall [9] have reached within their classical approach.

Interconversion of cyclotron and magnetron motion: Including the interaction term (15) at a driving frequency $\omega_d \approx \omega_+ + \omega_- = \omega_c$, oscillator a is identified with cyclotron motion, oscillator b is inverted ($\epsilon = -1$) and identified with magnetron motion, while axial motion assumes the spectator role.

$$\langle \alpha, \beta, \gamma | x(t) | \alpha, \beta, \gamma \rangle = \frac{1}{2}\sqrt{\frac{2\hbar}{m\omega_1}} \langle \alpha, \beta, \gamma | \left(A_a(t) + A_a^\dagger(t) + A_b(t) + A_b^\dagger(t)\right) | \alpha, \beta, \gamma \rangle, \tag{60}$$

$$\langle \alpha, \beta, \gamma | y(t) | \alpha, \beta, \gamma \rangle = \frac{1}{2i}\sqrt{\frac{2\hbar}{m\omega_1}} \langle \alpha, \beta, \gamma | \left(A_a(t) - A_a^\dagger(t) - A_b(t) + A_b^\dagger(t)\right) | \alpha, \beta, \gamma \rangle, \tag{61}$$

$$\langle \alpha, \beta, \gamma | z(t) | \alpha, \beta, \gamma \rangle = \sqrt{\frac{\hbar}{2m\omega_z}} \langle \alpha, \beta, \gamma | \left(A_c(t) + A_c^\dagger(t)\right) | \alpha, \beta, \gamma \rangle. \tag{62}$$

Here the solutions (24), (25), (59) have to be inserted with the initial values

$$\langle \alpha, \beta, \gamma | A_a(0) | \alpha, \beta, \gamma \rangle = \alpha, \quad \langle \alpha, \beta, \gamma | A_b(0) | \alpha, \beta, \gamma \rangle = \beta, \quad \langle \alpha, \beta, \gamma | A_c(0) | \alpha, \beta, \gamma \rangle = \gamma. \tag{63}$$

Interconversion of axial and magnetron motion: Including the interaction term (12) at a driving frequency $\omega_d \approx \omega_z + \omega_-$, oscillator a is identified with axial motion, oscillator b is inverted ($\epsilon = -1$) and identified with magnetron motion, while cyclotron motion assumes the spectator role.

$$\langle \alpha, \beta, \gamma | x(t) | \alpha, \beta, \gamma \rangle = \frac{1}{2}\sqrt{\frac{2\hbar}{m\omega_1}} \langle \alpha, \beta, \gamma | \left(A_c(t) + A_c^\dagger(t) + A_b(t) + A_b^\dagger(t)\right) | \alpha, \beta, \gamma \rangle, \tag{64}$$

$$\langle \alpha, \beta, \gamma | y(t) | \alpha, \beta, \gamma \rangle = \frac{1}{2i}\sqrt{\frac{2\hbar}{m\omega_1}} \langle \alpha, \beta, \gamma | \left(A_c(t) - A_c^\dagger(t) - A_b(t) + A_b^\dagger(t)\right) | \alpha, \beta, \gamma \rangle, \tag{65}$$

$$\langle \alpha, \beta, \gamma | z(t) | \alpha, \beta, \gamma \rangle = \sqrt{\frac{\hbar}{2m\omega_z}} \langle \alpha, \beta, \gamma | \left(A_a(t) + A_a^\dagger(t)\right) | \alpha, \beta, \gamma \rangle. \tag{66}$$

Here the solutions (24), (25), (59) have to be inserted with the initial values

$$\langle \alpha, \beta, \gamma | A_a(0) | \alpha, \beta, \gamma \rangle = \gamma, \quad \langle \alpha, \beta, \gamma | A_b(0) | \alpha, \beta, \gamma \rangle = \beta, \quad \langle \alpha, \beta, \gamma | A_c(0) | \alpha, \beta, \gamma \rangle = \alpha. \tag{67}$$

Interconversion of axial and cyclotron motion: Including the interaction term (11) at a driving frequency $\omega_d \approx \omega_+ - \omega_z$, oscillator a is identified with cyclotron motion, oscillator b is identified with axial motion ($\epsilon = +1$), while magnetron motion assumes the spectator role.

$$\langle \alpha, \beta, \gamma | x(t) | \alpha, \beta, \gamma \rangle = \frac{1}{2}\sqrt{\frac{2\hbar}{m\omega_1}} \langle \alpha, \beta, \gamma | \left(A_a(t) + A_a^\dagger(t) + A_c(t) + A_c^\dagger(t)\right) | \alpha, \beta, \gamma \rangle, \tag{68}$$

$$\langle \alpha, \beta, \gamma | y(t) | \alpha, \beta, \gamma \rangle = \frac{1}{2i}\sqrt{\frac{2\hbar}{m\omega_1}} \langle \alpha, \beta, \gamma | \left(A_a(t) - A_a^\dagger(t) - A_c(t) + A_c^\dagger(t)\right) | \alpha, \beta, \gamma \rangle, \tag{69}$$

$$\langle \alpha, \beta, \gamma | z(t) | \alpha, \beta, \gamma \rangle = \sqrt{\frac{\hbar}{2m\omega_z}} \langle \alpha, \beta, \gamma | \left(A_b(t) + A_b^\dagger(t)\right) | \alpha, \beta, \gamma \rangle. \tag{70}$$

Here the solutions (24), (25), (59) have to be inserted with the initial values

$$\langle \alpha, \beta, \gamma | A_a(0) | \alpha, \beta, \gamma \rangle = \alpha, \quad \langle \alpha, \beta, \gamma | A_b(0) | \alpha, \beta, \gamma \rangle = \gamma, \quad \langle \alpha, \beta, \gamma | A_c(0) | \alpha, \beta, \gamma \rangle = \beta. \tag{71}$$

VII CONCLUDING REMARKS

Ion motion in a Penning trap has been discussed in a quantum mechanical framework. Aside from being of interest in itself, this approach has served to treat seemingly disparate topics under one unifying view point, on the one hand the 2-level atom driven by a laser field tuned to its transition frequency or the magnetic resonance of spin-$\frac{1}{2}$ particles, on the other hand the interconversion of modes of ion motion in a Penning trap under the influence of a driving electric quadrupole field. The concepts of Rabi frequency and Bloch vector arise here in the development of the underlying general framework, they therefore must apply to all these phenomena in the same fashion.

A final remark concerns our model hamiltonians expressed in terms of the Bloch operator, eqs. (36,42). When the external driving field is also quantized and the annihilation and creation operators of the field quanta are denoted by a and $a^\dagger$, then both model hamiltonians appear in the form

$$H(t) = \hbar(\omega_a + \eta\epsilon\omega_b)T_0 + \hbar(\omega_a - \eta\epsilon\omega_b)T_3(t) + \hbar g\left[T_+(t)a(t) + T_-(t)a^\dagger(t)\right]. \tag{72}$$

The operators $T_3(t)$, $T_+(t)$, $T_-(t)$ generate a reducible or irreducible representation of a $SU(2)$ or $SU(1,1)$ Lie algebra, depending on whether we have $\eta = +1$ (type I interactions) or $\eta = -1$ (type II interactions). The operator T_0 is the Casimir operator of the Lie algebra.

The hamiltonian eq. (72) has a similar structure as the Jaynes-Cummings hamiltonian, which is well known from quantum optics. The hamiltonian in fact generalizes the Jaynes-Cummings hamiltonian from the 2-dimensional $SU(2)$ representation to an arbitrary finite or infinite dimensional representation of the $SU(2)$ or $SU(1,1)$ Lie algebra. It can be shown that many features carry over to the generalized theory, however, the Heisenberg equations of motion of the fundamental operators are linear only in the case of the 2-dimensional $SU(2)$ representation, in all other situations nonlinear terms appear. A more detailed investigation of the generalized Jaynes-Cummings hamiltonian eq. (72) shall be published elsewhere.

Acknowledgements: The author gratefully acknowledges the support and the hospitality extended to him by the Department of Physics of Florida State University, Tallahassee (Florida), where part of this work was performed.

REFERENCES

1. Bollen G., Moore R. B., Savard G., and Stolzenberg H., *J. Appl. Phys.* **68** (1990), 4355 - 4374
2. Bollen G., *Erste Massenmessung an instabilen Isotopen mit Hilfe einer Penningfalle*, Dissertation (1989), Johannes-Gutenberg-Universität Mainz (Germany)
3. Savard G., Becker S., Bollen G., Kluge H.-J., Moore R. B., Schweikhard L., Stolzenberg H., and Wiess U., *Phys. Lett.* **A 158** (1991), 247 - 252
4. Schweikhard L., Guan Shenheng, and Marshall A. G., *Int. J. Mass Spectrom. Ion Proc.* **120** (1992), 71 - 83
5. Schweikhard L., and Marshall A. G., *J. Amer. Soc. Mass Spectrom.* **4** (1993), 433 - 452
6. Guan Shenheng, Xiang Xinzhen, and Marshall A. G., *Int. J. Mass Spectrom. Ion Proc.* **124** (1993), 53 - 67
7. Marshall A. G., and Guan Shenheng, *Observation, manipulation and uses for magnetron motion in ion cyclotron mass spectrometry*, in the Proceedings of the Nobel Symposion 91 (Lysekil (Sweden), August 19 - 26, 1994) on "Trapped Charged Particles and Related Fundamental Physics", (Eds.: Bergström I., Carlberg C., and Schuch R.), *Physica Scripta* **T59** (1995), 155 - 164
8. Guan Shenheng, and Marshall A. G., *Int. J. Mass Spectrom. Ion Proc.* **146/147** (1995), 261 - 296
9. Guan Shenheng, and Marshall A. G., *J. Chem. Phys.* **98** (1993), 4486 - 4493
10. Guan Shenheng, *J. Chem. Phys.* **96** (1992), 7959 - 7964
11. König M., Bollen G., Kluge H.-J., Otto T., and Szerypo J., *Int. J. Mass Spectr. and Ion Proc.* **142** (1995) 95 - 116
12. König M., *Präzisionsmassenbestimmung instabiler Cäsium- und Bariumisotope in einer Penningfalle und Untersuchung der Ionenbewegung bei azimuthaler Quadrupolanregung*, Dissertation (1995), Johannes-Gutenberg-Universität Mainz (Germany)
13. Brown L. S., and Gabrielse G., *Rev. Mod. Phys.* **58** (1986), 233 - 311
14. Blatt R., Cirac J. I., Parkins A. S., and Zoller P., *Quantum Motion of Trapped Ions*, in the Proceedings of the Nobel Symposion 91 (Lysekil (Sweden), August 19 - 26, 1994) on "Trapped Charged Particles and Related Fundamental Physics", (Eds.: Bergström I., Carlberg C., and Schuch R.), *Physica Scripta* **T59** (1995), 294 - 302
15. Meystre P., and Sargent M. III, *Elements of Quantum Optics* (2nd Ed.), Springer-Verlag Berlin 1990
16. Kretzschmar M., *Physica Scripta* **46** (1992) 544 - 554
17. Mandel L., and Wolf E. *Optical Coherence and Quantum Optics*, Cambridge Univ. Press, Cambridge (UK) 1995
18. Cohen-Tannoudji C., *Quantum Mechanics*, Vol. 1 and 2, J. Wiley & Sons, New York 1977

Quantum Measurement and Nonclassical Vibration of an Ion in a Trap

R. Huesmann, Ch. Balzer, B. Appasamy, Y. Stalgies and P. E. Toschek

Institut für Laser-Physik, Jungiusstraße 9, 20355 Hamburg

Abstract. The microwave-induced dynamics of the nuclear spin and the laser-induced dynamics of the centre-of-mass motion of individual ions have been studied: (i) The phase shift of the hyperfine Larmor precession of a ground-state $^{171}Yb^+$ ion upon pulsed variation of the ambient magnetic field has been measured by microwave-optical double resonance interpreted in terms of Mach-Zehnder interferometry and quantum ergodicity. Even a single measurement yields (partial) phase information. At the extremes of the fringes, the results of measurements are *deterministic* where ion probing is compatible with ion preparation, as demonstrated by laser-exciting an ion on an $E2$ line. (ii) Excitation and de-excitation spectra of the $E2$ line $S_{1/2} - D_{5/2}$ of an individual trapped $^{138}Ba^+$ ion show different first-order sidebands of radial and axial ion vibration in the trap, since the vibrational distribution in the excited $D_{5/2}$ level is modified by the ion being reduced to this level in null observations of resonance light. Delayed sideband deexcitation eventually leaves the ion in the Fock state $|n = 1>$ ("stochastic cooling"). Sideband modulation is identified as stroboscopic detection of the light-induced nutation.

INTRODUCTION

A localized individual cold ion may interact — by way of its various degrees of freedom — with radiation in various wavelength domains, and this interaction brings to the fore the specific quantum phenomena to the largest extent. This situation is highlighted by experiments of microwave-optical double resonance (MODR, [1,2]) and opto-optical double resonance (OODR) on an ion, as well as by the coupling of electronic degrees of freedom with the ion's center-of-mass vibration in the trap. Measurements on a single ion, on the other hand, benefit from the ion's quantized dynamics: pulsed interaction with radiation that prepares the ion in a particular state is succeeded by probing the state of the ion when attempting to detect many quanta of laser-excited resonance fluorescence. This scheme of "quantum amplification" [3,4] allows one to detect small numbers of ions, or even individual ones, and to implement measurements of their variables with efficiency close to unity. For the measurement of expectation values of variables, the average extends over the ensemble of measurements rather than the ensemble of quantum systems. Sensitive phase or frequency measurements on the radiation-induced moments make use of two temporally separated pulses for the preparatory interaction [2]. Varying the frequency gives rise to the recording of Ramsey fringes in the time domain upon averaging over many measurements [5]. This procedure may be interpreted as Mach-Zehnder interferometry, in configuration space, of the ionic wave function [6-8]. With a single particle, it represents a selective measurement since the state of this system with respect to the particular variable is completely determined.

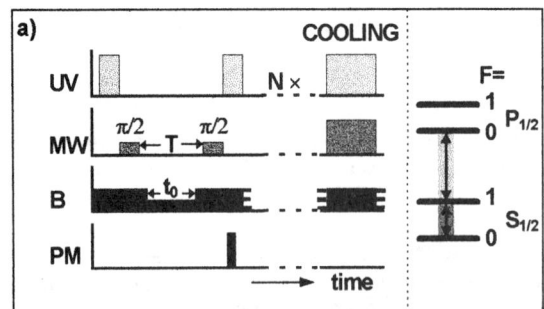 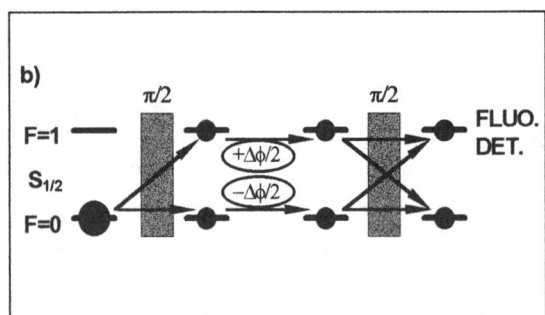

FIGURE 1. (a): Gating of UV ion preparation and probe, microwave, magnetic field (B), and detector (PM). (b): Atom-wave Mach-Zehnder interferometer in configuration space. The beam splitters are replaced by $\pi/2$ pulses.

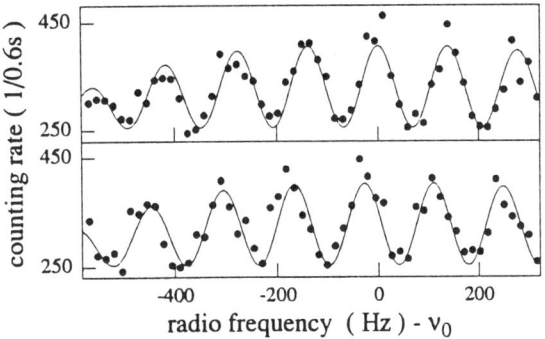

FIGURE 2. Photo counting rate of single-ion fluorescence *vs* microwave detuning. Each data point shows accumulated results of 300 attempts; dc magnetic field B_o constant (top), and set to $B_o - \Delta B$, during t_o, yielding the phase shift of fringes $\langle \Delta \phi \rangle_{+1/2} = 1.28$ rad (bottom).

ATOM INTERFEROMETRY BY MODR

The excitation of free atoms by light transfers recoil momentum onto the atoms such that the successfully excited component of the wave function spatially separates from the unexcited rest, and separated spatial trajectories emerge as in Mach-Zehnder interferometry [9-12]. In contrast, excitation by microwaves varies, in the semi-classical picture, the angle of inclination of the atomic angular momentum versus the direction of the ambient magnetic field [8]. We have MODR applied to an individual localized atomic ion and show that this interaction represents Mach-Zehnder interferometry in configuration space.

The present experiment is made up of a 2-mm-sized Paul-type electrodynamic trap that contains up to 50 ions. Alternatively, an individual ion is localized in the trap's electric centre. Tunable frequency-doubled cw laser light at 369nm and of 80-kHz bandwidth excites the ion(s) on the $|S_{1/2}, F = 1\rangle \rightarrow |P_{1/2}, F = 0\rangle$ resonance line, and the scattered light is photo-counted over a preselected time interval, typically 2ms. The light is down-tuned by some 20MHz (150MHz) in order to laser-cool the ion(s). Optically pumping the ion(s) into the metastable $^2D_{3/2}$ level is undone by illumination with 609-nm light that retrieves the ion(s) to the ground state via the $|D_{3/2}, F=1\rangle \rightarrow |[1/2]_{1/2}, F = 0\rangle$ excitation. The ion is also irradiated, from a horn antenna, by 12.6 GHz microwave radiation that drives the $|S_{1/2}, F = 0, m_F = 0\rangle \rightarrow |S_{1/2}, F = 1, m_F = 0\rangle$ hyperfine transition shifted by quadratic Zeeman interaction. Optical pumping into the $|F = 1, m_F = \pm 1\rangle$ levels is avoided when the E vector of the linearly polarized light subtends 45° with the direction of the dc magnetic field. A cycle of measurement (Fig. 1a) includes (*i*) preparing the ion in the $|F = 0, m_F = 0\rangle$ state by a uv laser pulse that spuriously and nonresonantly excites the $F = 1 \rightarrow 1$ transition, (*ii*) the first microwave $\pi/2$ pulse, (*iii*) a time interval T of free evolution of the magnetic dipole, (*iv*) the second $\pi/2$ pulse, and (*v*) probing the fluorescence by another uv pulse which simultaneously prepares the ion in the $F = 0$ state for the next cycle. After N-fold repetition, the microwave is applied simultaneously with the uv light for recooling the ion. Subsequent sequences of N cycles involve the microwave frequency stepwise scanned across the hyperfine resonance. Fig. 1b shows the interpretation of the procedure as Mach-Zehnder interferometry, with the $\pi/2$ pulses acting as beam splitters for the wave functions of upper and lower state in configuration space. The state of the ion(s), after the second $\pi/2$ pulse, results from interference of two pathways of evolution. Such a Ramsey interference pattern of a *single* ion is shown in Fig. 2a, where each data point represents averaging over 300 individual observations. When in the two interferometric "arms" – i.e., in the component states – the phases are shifted differently between the two $\pi/2$ pulses, e.g. by pulsed variation of the ambient magnetic field, the interference pattern although *not* its envelope shifts in frequency (Fig. 2b). The solid line represents the calculated signal of the interference fringes, $P(F = 1, \theta, \phi)$, where

$$\sqrt{P(1,\theta,\phi)} = \cos\chi \sin\theta \cos\frac{\phi}{2} - \sin 2\chi \sin^2\frac{\theta}{2}\sin\frac{\phi}{2}, \quad (1)$$

$\tan\chi = \Delta/\Omega$, the azimuthal angle of Larmor precession is $\phi = -(1/\hbar)\int \mu B dt$, and $\theta = \sqrt{\Omega^2 + \Delta^2}\tau$ is the polar or "nutation" angle in the SU2 configuration space, Ω, ν_o are the Rabi and resonance frequencies of the ground-state hyperfine transition, respectively, and $\nu = \nu_o + \Delta$ is the microwave frequency, τ the microwave pulse length.

Variation of the magnetic field by $\Delta B = B - B_o$ makes shift the Larmor phase by $\Delta\phi = \phi - \phi_o$, and also the phase of the fringes, $\langle\Delta\phi\rangle$. The latter relationship is shown in Fig. 3 where 300 superimposed results of detections on a single ion are combined in each of the dots, and individual results on an ensemble of some 30 ions are shown by crosses. Note that the two averages of the phase shift agree. This agreement in fact amounts to an experimental proof of what has been named

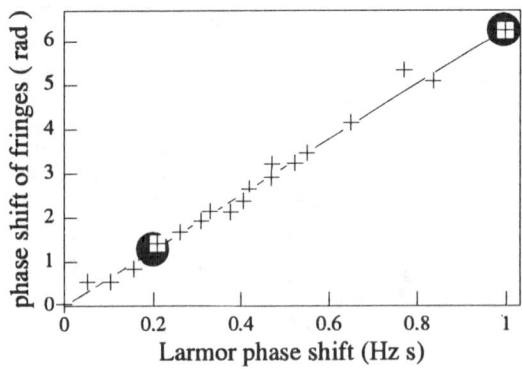

FIGURE 3. Phase shift of fringes $\langle \Delta \phi \rangle_{+1/2}$, vs Larmor phase shift $\Delta \phi$, for ensemble of some 30 ions (+), and for 300 detections on *individual* ion, as in Fig. 2 (●). Variation of $\Delta \phi$ by $0, 5ms \leq t_o \leq 5ms$.

"quantum ergodicity": Matrix elements, calculated with almost all eigenfunctions of the set that characterizes the quantum system, approach the average over the microcanonical ensemble in the semiclassical limit [13].
The expectation value for the phase shift $\Delta \phi$ under the condition of observing the ion in state $F = 0, 1$,

$$\langle \Delta \phi \rangle_F = P(F, \theta, \phi) \cdot \frac{\phi}{2} - P(F, \theta, \phi_o) \cdot \frac{\phi_o}{2}, \qquad (2)$$

is shown, as the solid line in Fig. 3, in the perturbative approximation $P(\phi) \simeq P(\phi_o)$. On resonance, $\Delta = 0$,

$$P(F, \theta, \phi) = 1 - F + (2F - 1) \sin^2 \theta \cos^2 \frac{\phi}{2}, \qquad (3)$$

and $P(F, \phi) d\phi$ is the probability of finding the phase of precession between ϕ and $\phi + d\phi$ in the measurement. The variation of P on the phase ϕ indicates that even a *single* measurement on an individual quantum system provides information – although incomplete one – since all phase values except one are either excluded, or their probability to occur is reduced. Each consecutive result of a subsequent measurement modifies this probability, and for a trajectory of n measurements at signal frequency $\nu \simeq \nu_o$, of which r yield the fluorescence signal, the probability per unit phase is

$$P_{n,r}(1, \theta(\nu), \phi) = \binom{n}{r} \sin^{2r} \theta(\nu) \cos^{2r}(\phi/2) \times [1 - \sin^2 \theta(\nu) \cos^2(\phi/2)]^{n-r}. \qquad (4)$$

This evolution of the knowledge on the phase is shown in Fig. 4a where $P_{n,n}(\nu_o)$, the probability for the appearance of homogeneous sequences of results on resonance, is plotted for various values of n. Note that with $n \to \infty$ a fringe spectrum emerges that features a well-defined phase, lacks projection noise and agrees with that of a measurement on a large ensemble.

The probability $P(1, \theta(\nu))$ of detecting light at the end of a cycle, versus the detuning of the microwave frequency – i.e., the fringe pattern – is shown in Fig. 4b. Superimposed is its standard deviation δP, where $(\delta P)^2 = P(F, \theta, \phi)(1 - P(F, \theta, \phi))$ [14], with an individual measurement, and with the results of n measurements averaged. This spectrally modulated variance represents the projection noise that arises from the detection of the ion in one of its energy eigenstates being incompatible with the ion state that evolved from the pulsed microwave interaction and free precession *except* at frequency values that correspond to the peaks and dips of the fringes where the ion has in fact evolved into an energy eigenstate again. This spectral distribution of the projection noise is particularly undesirable when the slope of a fringe is supposed to serve as the discriminant for the frequency control of a laser that represents a frequency standard, since for this purpose the signals at both wings of a fringe are compared in order to derive feedback for frequency reset. The distribution may be shifted by another rotation in configuration space such that the minimum variance appears at the halfway frequencies ("spin squeezing", [15]). Unfortunately, the excitation of fluorescence on the line $|S_{1/2}, F = 1, m_F = 0\rangle \to |P_{1/2}, F = 0, m_F = 0\rangle$ goes along with spurious scattering in the wing of the line $|S_{1/2}, F = 1\rangle \to |P_{1/2}, F = 1\rangle$ that optically repumps the ion into the $|S_{1/2}, F = 0\rangle$ level and spoils the efficiency of the scheme of quantum amplification. Thus, averaging over at least some twenty measurements per frequency setting are required in order to overcome the noise. This drawback is absent in the excitation of an *optical* signal transition, i.e. in OODR.

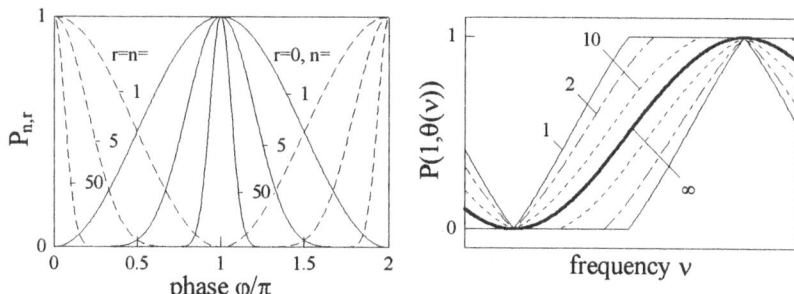

FIGURE 4. (a): Probability $P_{n,r}(\phi)$ of finding r equal results (out of n accumulated attempts) showing resonance scattering (dashed line), or no resonance scattering (solid line). The phase $\langle \phi \rangle_{F=0,1}$ of Ramsey fringes is the mean weighted by $P_{n,r}(\phi)$. (b): Probability $P_{n,r}(1,\theta(\nu))$ of "on" signal, vs microwave frequency ν. Superimposed is the standard deviation ("projection noise") for averaging over 1, 2, or 10 measurements.

QUANTUM JUMPS AND RABI NUTATION IN OODR

The excitation of the ion to its $D_{5/2}$ state, on the $E2$ line at 411nm, lacks competing channels. Consequently, the concomitant excitation of resonance fluorescence, upon probing the resonance line, becomes efficiently suppressed, and "quantum jumps" to the "dark" state of quenched fluorescence and back to light scattering [3,4] are observed. With alternating excitation on the $E2$ and resonance lines, one may take pairs of measurements with the results "on-off" as digitized quantum jumps of the ion to the dark state (absorption), pairs with the results "off-on" as reversed jumps to the bright state (re-emission). The rate of on-off jumps varies as a function of the detuning of the blue signal light and displays the spectrum of an absorption line. With the light being sufficiently coherent, the corresponding spectral probability is represented by the Rabi solution

$$R(\theta(\nu,\tau)) = \cos^2 \chi \sin^2 \theta(\nu,\tau), \tag{5}$$

where ν is now the optical signal frequency. This solution displays modulation by the optical nutation of the ionic quadrupole, whose period depends on the light frequency. Such a spectrum, excited on a $^{172}Yb^+$ ion by the frequency-doubled light of a stepwise detuned and well-controlled diode laser at 822nm, is shown in Fig. 5. A light pulse induces a certain number of nutation periods that varies upon frequency tuning. A single fringe corresponds to a few tuning steps each of which comprises a varying shift of θ by many nutation periods. Therefore, the fringe pattern is established as beat note of the frequencies of nutation (θ/τ) and measurement ($1/2\tau$), and it amounts to a stroboscopic recording. Note that a sideband appears that indicates Doppler phase modulation generated by the center-of-mass vibration of the ion in the trap.

It is remarkable, that qualitatively quite different temporal trajectories of data correspond to particular points of the spectrum: At the positions of the peaks (dips), the results are predetermined to be "on" ("off") as a consequence of the deterministic nature of the ion's interaction with the blue radiation field, and these results are *certain* – save for technical im-

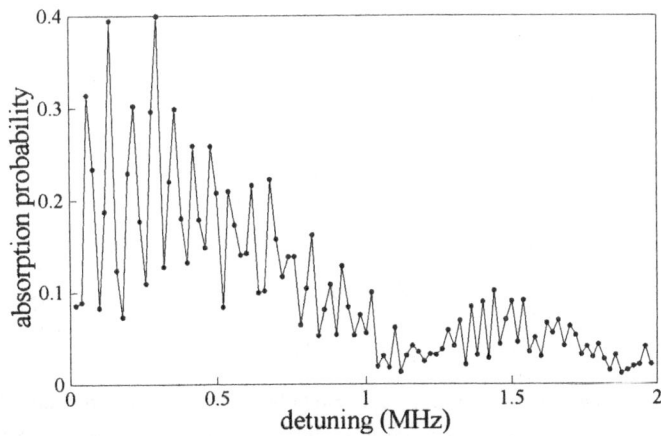

FIGURE 5. Absorption probability, after pulse duration τ, on $E2$ line $Yb^+(S_{1/2} - D_{5/2})$ vs detuning of 411-nm light. Sideband of secular vibration at 1.5MHz.

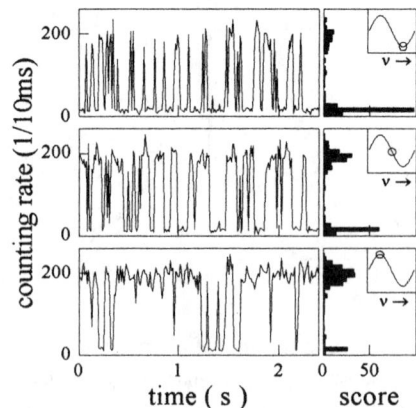

FIGURE 6. Photo counting rate of $^{172}Yb^+$ single-ion resonance scattering after attempted laser excitation on $S_{1/2} - D_{5/2}$ E2 line, at laser detuning to dip (top), mid-slope (centre), and peak of fringe (bottom) shown in the insets. Distributions of counting rates (right-hand graphs) correspond to deterministic (top, bottom) and stochastic (centre) measurements. Spurious counts in the upper score and "no counts" in the lower one are caused by imperfections in the frequency setting.

perfections – at nutation angles per pulse that are even (odd) multiples of π (see Fig. 6). At midslope of the fringes, i.e. at nutation angles that are odd multiples of $\pi/2$, the results are approximately equipartitioned over "on" and "off", and they display projection noise. So far, projection noise has been observed with ensembles of trapped ions [14].

VIBRONIC DYNAMICS

The center-of-mass motion of the ion in the trap imposes a time-varying phase shift – via the Doppler effect – upon the ion's internal moments. Thus, the ion vibration within its confinement couples to the internal degrees of freedom. In an rf trap the normal modes of ion vibration, in addition to the rf-driven micro-motion, are three orthogonal modes of secular vibration, whose excitation shows up as vibronic sidebands to absorption lines narrow enough to warrant spectral resolution of those sidebands.

We have trapped and laser-cooled also a single $^{138}Ba^+$ ion and laser-excited it, to its metastable $D_{5/2}$ level, on the E2 line $S_{1/2} - D_{5/2}$ at 1,76μm, this excitation having been alternated with the excitation of fluorescence on the $S_{1/2} - P_{1/2}$ resonance line at 493nm [16]. Optical pumping into the level $D_{3/2}$ is avoided by repumping the ion to the ground state by light at 650nm. Upon successful excitation of the $D_{5/2}$ level, resonance light does not appear when the ion is probed for it. Again, the number

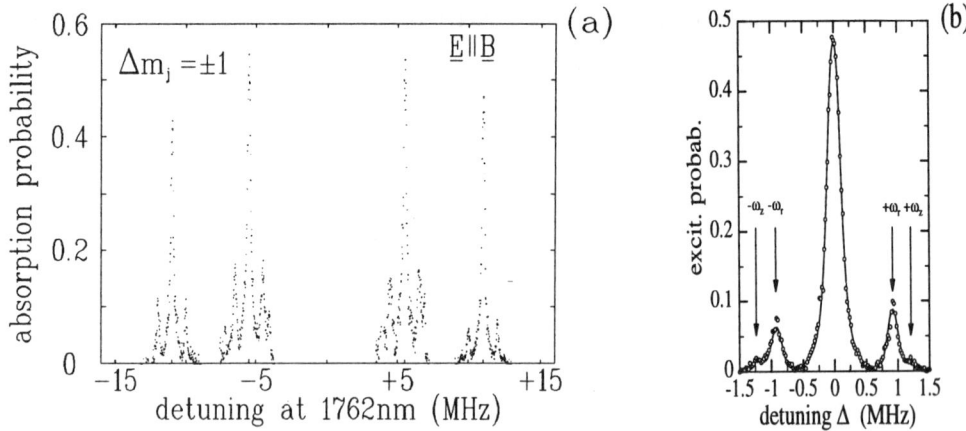

FIGURE 7. (a): Zeeman-split conponents of E2 absorption line $Ba^+(S_{1/2} - D_{5/2})$ that are allowed when E field of light parallel with ambient B field. Resonances accompanied by vibronic sidebands. (b): Component $m_S = +1/2 \rightarrow m_D = +3/2$. Different strengths of upper and lower sidebands yield the mean quantum number $\langle n_r \rangle = 2.5 \pm 0.5$ of the radial, and $\langle n_z \rangle = 0.9 \pm 1.7$ of the axial vibration.

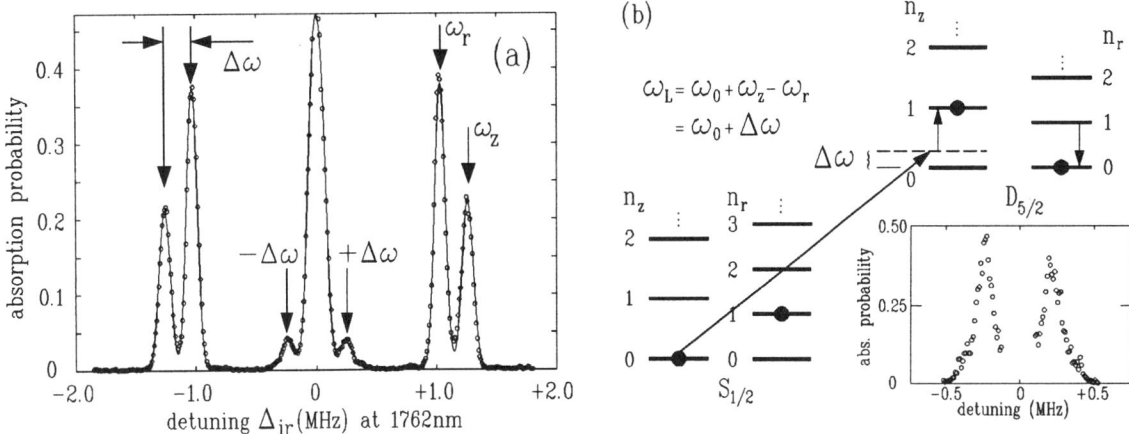

FIGURE 8. (a): Excitation spectrum with r and z vibrational sidebands resolved, and r sideband and carrier lines strongly saturated. Mean quantum number of radial vibration $\langle n_r \rangle = 85 \pm 12$, of axial vibration $\langle n_z \rangle = 14 \pm 2$. The carrier is accompanied by 2nd-order difference sidebands that mark light-induced interchange of r and z vibrational quanta ("inverse Raman effect"). Bandwidth 15kHz. (b): Exchange of radial and axial vibronic quanta, mediated by light absorption. Inset: Strongly excited difference sidebands.

of pairs of measurements with resonance scattering "on-off", normalized by the total number of measurements, represents the probability of excitation on the $E2$ line. A set of such measurements recorded upon stepwise scanning the frequency ν of the ir laser yields a spectrum of absorption (Fig. 7a). This spectrum shows those four of the ten Zeeman components of the $E2$ line that do not vanish under excitation by light whose polarization vector is parallel with the ambient magnetic field. The carrier of each of these lines is less than 60kHz wide; this width is attributed mostly to power broadening even at the low level of irradiation, since the bandwidth of the laser emission is much smaller. The upper and lower first-order vibronic sidebands vary as $(\langle n \rangle + 1)\eta^2\Omega^2$ and $\langle n \rangle \eta^2 \Omega^2$, respectively, where η is the Lamb-Dicke parameter, Ω the $E2$ Rabi frequency and $\langle n \rangle$ the mean vibronic quantum number of the particular vibronic mode (Fig. 7b) [17]. The sidebands depend on Ω equally when the distribution over the vibronic energy eigenstates is initially thermal. From the two first-order sidebands, one derives $\langle n \rangle$. Sub-unity values of $\langle n \rangle$ have been observed, and we attribute the low kinetic energy to the combined action of Doppler and Raman cooling [18-21]. With better resolution, two pairs of vibronic first-order sidebands are resolved which pertain to the axial secular frequency ω_z and the radial secular frequency ω_r that are nondegenerate in the spheroidal pseudo-potential of the trap (Fig. 8a). Carrier and even first-order sidebands appear saturated. There are second-order unsaturated sidebands close to the carrier; they are generated by the simultaneous exchange of a radial and an axial vibronic quantum $\hbar\omega_r$ and $\hbar\omega_z$ respectively. The energy defect of these *vibronic* quanta is made up by the offset of the light from resonance (Fig. 8b). This situation is just opposite to the one encountered in stimulated Raman excitation, where the energy difference of *electronic* pump and Stokes quanta is made up by the *vibronic* excitation. Consequently, this type of fundamental atom-light interaction qualifies as an *inverse Raman* process. With the combined excitation of an internal resonance and a vibrational degree of freedom both serving as qubits for the operation of a quantum-logic gate, the second vibrational degree of freedom provides the ion with an extra memory qubit. Excitation of those second-order sidebands would allow one to shift the ion from a particular state of the r vibration to the corresponding one of the z vibration or *vice versa* without *a priori* knowledge of the setting [22]. The enlarged recording of the inverse Raman sidebands shown in the inset of Fig. 8b displays asymmetry opposite to that of the first-order sidebands. Since $\omega_z > \omega_r$ and $\langle n_r \rangle > \langle n_z \rangle$, the strength of the lower difference sideband exceeds that of the upper one:

$$\sigma_r^+ \sigma_z^- = \Omega^4 \eta^4 (\langle n_r \rangle + 1)\langle n_z \rangle > \Omega^4 \eta^4 (\langle n_z \rangle + 1)\langle n_r \rangle = \sigma_z^+ \sigma_r^-, \qquad (6)$$

where σ^+, σ^- designate first-order matrix elements of excitation and deexcitation, respectively.

After the ion having been excited, on a sideband, to the $D_{5/2}$ level, its vibrational distribution is non-thermally bunched around a value of $\langle n \rangle$ that is higher than before, in the $S_{1/2}$ ground state [23]. Triples of measurements with the result of probing – the resonance light – found "on-off-on" represent events of prompt stimulated *deexcitation*. The spectra derived from these events show both upper and lower first-order sidebands having the size $(2\langle n \rangle + 1)\eta^2\Omega^2$ (Fig. 9a). Moreover, there are events of delayed deexcitation of the "dark" electronic state $D_{5/2}$, with more than one "off" result sandwiched between initial and final "on" observations. Sequences of such null detections have been shown to modify the ion's distribution over its vibronic levels into a state of lower $\langle n \rangle$, i.e. to stochastically cool the ion [24]. In fact, $\langle n \rangle$ is supposed to decrease monotonously upon

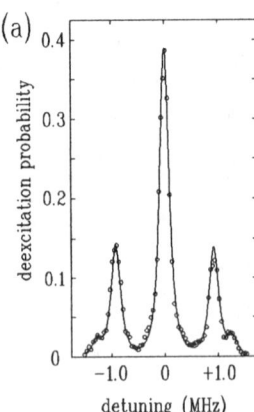 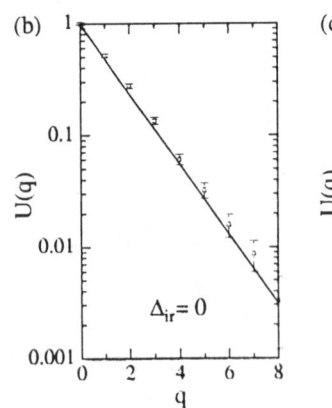 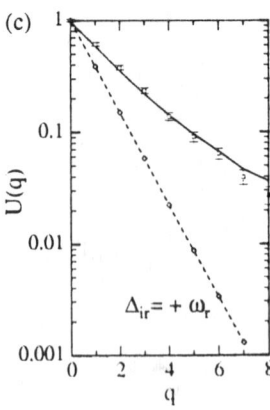

FIGURE 9. (a): *Reemission* line (same Zeeman component as in Fig. 7b). Vibronic sidebands are equal and twice as large as with absorption. Corresponding absorption spectrum reveals $\langle n_r \rangle + \langle n_z \rangle = 7$.   (b): Probability $U(q)$ for q sequential null detections of resonance light upon excitation at the carrier line (vibration unaffected).   (c): Same, upon excitaton at the upper sideband ω_r where vibration and probability of deexcitation decrease such that $U(q)$ exceeds corresponding value calculated with modification of vibrational distribution neglected (dashed line).

increasing number q of intermediate "off" results of consecutive measurements. This number represents the length of time the metastable ion undergoes intermittant interaction with the driving light but fails to get actually deexcited to its ground state.

We have recorded deexcitation spectra by plotting, versus light detuning, the number $W(q)$ of "on" measurements, with detection of fluorescence, *after* q consecutive detections of darkness, normalized to the number of measurements, which is

$$W(q) = -dU(q)/dq. \tag{7}$$

From these data, one derives the probability

$$U(q) = \prod_{p=1}^{q} \rho_{11}(\tau, p) \tag{8}$$

of q sequential "off" results. Here, $\rho_{11}(\tau, p) = \sum_n \rho_{11}^n(\tau, p)$ is the ion's probability of being found in the electronic "dark" state $D_{5/2}$, after the p-th measurement, and τ is the duration of its interaction with the driving light [23].

In Fig. 9b, $U(q)$ has been plotted as derived from the carrier line, and as calculated from the initial ground-state thermal distribution in turn derived from an absorption spectrum. The events of reemission that contribute to the carrier line do not modify the vibronic state of the ion, and the slope of U represents just the scarcity of sequences of q "off" results increasing with q. In contrast, the plot of Fig. 9c shows $U(q)$ derived from the upper first-order radial sideband. Here, frustrated attempts to deexcite the ion narrow its vibrational distribution, and the probability for q consecutive "off" results exceeds by far a result computed neglecting renormalization of ρ_{11} after each individual measurement (dashed line). This excess, growing with q, demonstrates the dramatic decrease of the probability of deexcitation on the sideband upon decreasing $\langle n \rangle$. This stochastic cooling process generates a predetermined vibronic energy when the preparation-detection cycles are repeated up to the appearance of the required sequence of $q_{\min}$ "off" results. Since that probability vanishes at $\langle n \rangle \to 0$, a long enough sequence will prepare the ion in its vibronic Fock state $n = 1$. When starting a trajectory of measurements with a first excitation of the ion on the lower sideband or the carrier, and then proceeding on the upper sideband, eventually the ion will become prepared in its vibronic vacuum state, which is a trapped state since it cannot be deexcited any more by radiation on the upper sideband.

The *axial* vibronic sidebands are less strongly saturated in the recorded spectra, and thus they are more susceptible to variations. Their probability of deexcitation, at the q-th attempt, $P_{deex}(q) = W/U$, also decreases, but shows superimposed modulation whose contrast increases with q (Fig. 10a). The nonthermal velocity distribution of the ion in the dark state $D_{5/2}$ peaks approximately at $\langle n \rangle$, and for deexcitation on the upper and lower sidebands, the chance of the ion to show up in the ground state is [25]

$$P_{\pm}(S_{1/2}, \langle n \rangle) = P_{deex}(q) = \sin^2 q\tau \Omega_{\langle n \rangle, \langle n \rangle \pm 1}, \tag{9}$$

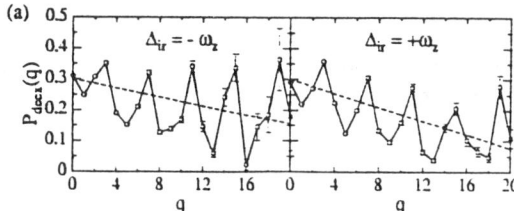 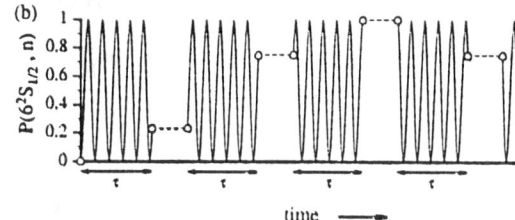

FIGURE 10. (a): Peak values on sidebands are modulated, *vs* q or time, by coherent vibronic ion dynamics during uninterrupted null detections. (b): Laser-driven coherent vibronic ion dynamics adds, during each half cycle τ of interaction, another increment $\Delta\theta = \tau \cdot \Omega_{\langle n \rangle,\langle n \rangle+1}$ to the nutation phase $\theta = (q-1) \cdot \Delta\theta$ which stays constant during half cycle of probing. θ is detected via $P_{deex}(q)$.

where $\Omega_{n,n\pm 1} = \Omega\eta\sqrt{n+1/2 \pm 1/2}$ are the Rabi nutation frequencies on the sidebands. The driving light drives both the induced quadrupole *and* the nutation dynamics between vibronic states $|n>$ and $|n\pm 1>$ during the first half of each cycle of measurement. During the second half – the probing – the vibronic superposition state and its phase remains unaffected by the reduction to the electronic $D_{5/2}$ state going along with the detection of no fluorescence for sake of the degeneracy of the vibronic quanta of the harmonic oscillator (Fig. 10b). The phase of nutation varies stepwise and periodically with the number p of cycles in any sequence of length q, until the sequence is terminated by the appearance of fluorescence in cycle $q+1$. The probability of this particular sequence to appear depends on the phase finally acquired in the q-th cycle, and it varies periodically with q. Thus, the interrupted nutation is detected *stroboscopically*. The stroboscopic period is infinite if the duration τ of the driving-light irradiation is an integer multiple m of the nutation period $\tau_N = 2\pi/\Omega_{\langle n \rangle,\langle n \rangle+1}$; in general it varies as $|\tau - m \cdot \tau_N|^{-1}$, and the stroboscopic frequency is

$$\nu_s = (1/\tau_N)\,(\tau - m \cdot \tau_N)/\tau. \tag{10}$$

The data in Fig. 10a are compatible with $\nu_s \approx \tau_N^{-1}/50$ and $m=5$. Note that the contrast of modulation increases with q, since the ion *dynamically* approaches a Fock state of its vibration.

SUMMARY

- We have applied microwave-optical double resonance on a single $^{171}Yb^+$ ion and demonstrated single-atom interferometry.
- The phase shift of hyperfine Larmor precession upon variation ΔB_0 of the ambient magnetic field has been measured by adding up results of many single-ion measurements.
- Comparison with the results of single measurements on 30 ions prove quantum ergodicity.
- At peaks and dips of the interferogram, results of single-ion measurements are deterministic, but elsewhere stochastic.
- This feature has been demonstrated by excitation of the ion on an $E2$ line.
- Excitation spectra ("absorption") of a $^{138}Ba^+$ show different motional sidebands characterized by the mean vibronic excitation $\langle n_S \rangle$ in the ground electronic state; $\langle n_S \rangle_{min} < 1$.
- Laser-induced coupling of vibronic r and z modes has been shown.
- Deexcitation spectra ("emission") show equal vibronic sidebands; they are characterized by $\langle n_D \rangle$ in the metastable electronic state.
- The distribution of population over the vibronic states is modified by sideband excitation and deexcitation.
- Frustrated attempts of deexcitation *reduce* $\langle n_D \rangle$ and stochastically cool the ion.
- The vibronic distribution of population in the dark metastable electronic state, P_D, oscillates with the number q of frustrated attempts to deexcite the ion.
- The ion vibration approaches the Fock state $|n=1\rangle$.

These findings are important prerequisites for the application of trapped ions for quantum computing.

ACKNOWLEDGEMENTS

Klaus Abich helped with the preparation of the Figures.– This work was supported by the Körber-Stiftung, Hamburg, by the Hamburgische Wissenschaftliche Stiftung, by the ZEIT-Stiftung, Hamburg and in part by the Deutsche Forschungsgemeinschaft, Bonn.

REFERENCES

1. A. Kastler and J. Brossel, C.R. Acad.Sci. **229**, 1213 (1949).
2. N.F. Ramsey, Phys.Rev. **76**, 996 (1949).
3. W. Nagourney, J. Sandberg, and H. Dehmelt, Phys.Rev.Lett. **56**, 2797 (1986).
4. Th. Sauter, W. Neuhauser, R. Blatt, and P.E. Toschek, Phys.Rev.Lett. **57**, 1696 (1986).
5. R. Huesmann, Ch. Balzer, Ph. Courteille, W. Neuhauser, and P.E. Toschek, submitted for publication.
6. K. Sangster, E.A. Hinds, S.M. Barnett, and E. Riis, Phys.Rev.Lett. **71**, 3641 (1993).
7. S. Nic Chormaic, Ch. Miniatura, O. Gorceix, B. Viaris de Lesegno, J. Robert, S. Feron, V. Lorent, J. Reinhardt, J. Baudon, and K. Rubin, Phys.Rev.Lett. **72**, 1 (1994).
8. A. Görlitz, B. Schuh, and A. Weis, Phys.Rev. **A 51**, R4305 (1995).
9. F. Riehle, A. Witte, T. Kisters, J. Helmcke, Appl.Phys. **B 54**, 333 (1992).
10. C.J. Bordé, Phys.Rev. **A 140**, 10 (1989).
11. K. Zeiske, G. Zinner, F. Riehle, and J. Helmcke, Appl.Phys. B **60**, 205 (1995).
12. J.H. Müller, D. Bettermann, V. Rieger, K. Sengstock, U. Sterr, and W. Ertmer, Appl.Phys. **B 60**, 199 (1995).
13. M. Feingold and A. Peres, Phys.Rev. **A 34**, 591 (1986).
14. W.M. Itano, J.C. Bergquist, J.J. Bollinger, J.M. Gilligans, D.J. Heinzen, F.L. Moore, M.G. Raizen, and D.J. Wineland, Phys.Rev. **A 47**, 3554 (1993).
15. D. Wineland, J.J. Bollinger, W.M. Itano, and D.J. Heinzen, Phys.Rev. **A 50**, 67 (1994).
16. B. Appasamy, Y. Stalgies, and P.E. Toschek, Phys.Rev.Lett. **13**, 2805 (1998).
17. D.J. Wineland and W.M. Itano, Phys.Rev.Lett. **A 20**, 1521 (1979).
18. W. Neuhauser, M. Hohenstatt, P.E. Toschek, and H.G. Dehmelt, Phys.Rev.Lett. **41**, 233 (1978); App.Phys. **17**, 123 (1978).
19. D.J. Wineland, R.E. Drullinger, and F.L. Walls, Phys.Rev.Lett. **40**, 1639 (1978).
20. M. Lindberg and J. Javanainen, J.Opt.Soc.Am. B **3**, 1008 (1986).
21. H. Gilhaus, Th. Sauter, W. Neuhauser, R. Blatt, and P.E. Toschek, Optics Communic. **69**, 25 (1988).
22. J. Steinbach, J. Twamley, and P.L. Knight, Phys.Rev. **A 56**, 4815 (1997).
23. J. Eschner, B. Appasamy, and P.E. Toschek, Optics Communic. **118**, 123 (1995).
24. J. Eschner, B. Appasamy, and P.E. Toschek, Phys.Rev.Lett. **74**, 2435 (1995).
25. D.M. Meekhof, C. Monroe, B.E. King, W.M. Itano, and D.J. Wineland, Phys.Rev.Lett. **76**, 1796 (1996).

Enhanced-Micromotion Reduction and Elimination

Nan Yu* and Hans Dehmelt

Department of Physics, University of Washington, Seattle, WA 98199
*Jet Propulsion Laboratory, California Institute of Technology, 4800 Oak Grove Drive, Pasadena, CA 91109

Abstract. Micromotion is the driven motion of trapped ions in rf traps. In practical traps, the micromotion is often greatly enhanced due to "dirty effects" such as stray dc fields from surface patch effects and/or rf phase differences between the endcaps of a quadrupole trap. Large micromotions can have adverse effects on precision measurements and quantum experiments that the trapped ions are best suited for. We present here a new approach to reduce and eliminate the enhanced micromotion. The approach uses a novel Paul-Straubel-Kingdon trap in which ions are loaded indirectly to avoid contamination of the trap surfaces.

INTRODUCTION

Radio-frequency(rf) ion traps have been widely used to confine charged atomic particles in experiments of high resolution spectroscopy, metrology and quantum measurements. The success owns largely to the high degree of localization and the perturbation-free environment of single trapped ions. A conventional rf quadrupole trap, also known as Paul trap, consists of a ring and two endcaps. An ideal quadrupole trap with the ring radius R_0 and the endcap separation $2Z_0 = \sqrt{2}R_0$ has the electric potential distribution $\Phi = \frac{V}{2R_0^2}(x^2 + y^2 - 2z^2)$, where $V = V_{rf}\sin(\Omega t)$ is the rf voltage applied across the ring and the endcaps. A charged particle of mass M moving in this field can have stable closed orbits and its center of mass moves as if in a pseudo harmonic potential well $\bar{\Phi} = \frac{1}{2}M(\omega_x^2 \bar{x}^2 + \omega_y^2 \bar{y}^2 + \omega_z^2 \bar{z}^2)$, where $\omega_z = 2\omega_x = 2\omega_y$ are referred as secular frequencies.

The motion of trapped ions in the above potential can be characterized by two oscillatory motions. The first is the secular motion of the ions in the pseudo potential well. It is the averaged center-of-mass motion. A single ion laser-cooled to a temperature T has the secular motion orbit size of $\sqrt{2kT/M\omega^2}$. A Ba$^+$ ion, for example, at the Doppler cooling limit of $kT = \frac{1}{2}\hbar\Gamma$ has an orbital size on the order of 40 nm for a typical $\omega \approx 1 MHz$. The second oscillatory motion is the fast synchronous motion driven by the trapping rf field, known as the "micromotion". It is this averaged micromotion in the trapping field gradient which gives rise to the ac trapping force [1]. In an ideal quadrupole trap, single cold ions occupy only a small region at the center of the trap where the rf field is close to zero. Therefore, the "intrinsic" micromotion is small, smaller than its secular motion by a factor of $\sqrt{2}\omega/\Omega$(typically about 0.1 to 0.2). In the above Ba$^+$ example, the intrinsic micromotion would be about 6 nm, a small amplitude indeed.

There are however several mechanisms in practical devices that enhance the micromotions. One such mechanism is stray dc fields which push the trapped ions away from the center of the trap to a point where the trapping rf field is larger. Common stray fields are generated by non-uniform surface potentials("patch effects") on the trap and nearby electrodes. Insulating layers, when charged up, result in large stray fields as well. Another cause of the enhanced micromotion is the possible phase difference of the applied rf voltages on the endcaps. It has been explicitly analyzed by Wineland et al. [2] Briefly, when a phase difference θ exists between the two endcaps, there is an effective rf voltage $V_{rf}\theta\cos\Omega t$ between the two endcaps. An ion, even at the center of the trap, will see a rf field of amplitude $E_\theta \approx V_{rf}\theta/2Z_0$.

Let's again use Ba$^+$ in a typical trap as an example to illustrate the micromotion effects quantitatively. At the presence of the stray dc field E_s, a single trapped ion is displaced from the center to z_0 where the trap restoring force $M\omega_z^2 r_0$ equals to the static electrical force eE_s. If E_s is caused by a 0.5 eV potential difference across a 1 mm trap ring, the ion will be displaced away from the trap center by $r_0 \approx 9$ μm. The rf field at this point is given by $E_{rf} = \sqrt{2}(\Omega/\omega)E_s$, or about 70 V/cm. The corresponding micromotion amplitude $r_\mu = \sqrt{2}(\omega/\Omega)r_0 \approx 1.3\ \mu m$, which is much greater than its secular motion size.

What will be the effects of such a micromotion amplitude? The larger oscillatory motion amplitude induces larger motional sidebands in the ion spectrum and the spectral lines are broadened. The micromotion sidebands can also cause ion heating even at large frequency detunings [3]. Therefore, the Doppler cooling limit may not be reached. The second-order Doppler shift due to the micromotion is increased regardless of the ion temperature. The second-order Doppler shift is given by $\Delta\nu/\nu = \Omega^2 r_\mu^2/2c^2$, or about 4×10^{-14} in the above Ba$^+$ example. Furthermore, the ac Stark shift by the trapping field is also increased to $\delta_s = \frac{1}{2}\sigma_s E_{rf}^2$. This added Stark shift to Ba$^+$ clock transition would be 30 Hz(1.5×10^{-13}) [4]. Admittedly, the example is about the worst case one may have. But it shows the significance of the micromotion problem in precision measurements and metrology experiments using single trapped ions. Other adverse effects of the large micromotion include higher rf heating rates during collisions, compromised tight spatial localization, and possibly causing decoherence of the ion motional quantum state.

The usual approach to reduce the enhanced micromotion due to stray fields has been using compensation fields to null out the stray fields [5-7]. This scheme works well when combined with a sensitive micromotion monitoring scheme such as spectroscopic phase-sensitive detection. However, there are limitations to this scheme. Since one can only detect the modulation spectroscopically in the laser direction, The compensation can be made only in that direction. Lasers in three orthogonal directions have to be arranged to reduce the overall micromotions. In addition, patch effects change over time, even between ion loadings. The phase imbalance on the endcaps can also be nulled out by injecting "compensating" rf voltage to one of the endcaps. Again, spectroscopic monitoring of the micromotion amplitudes is needed.

In this paper, we describe our efforts to reduce and eliminate the overall micromotions. We will first describe spectroscopic observations and detections of micromotion. Then, we will discuss our approaches of micromotion reduction by the Paul-Straubel(ring only) trap [8] and the new Paul-Straubel-Kingdon(PSK) trap with the indirect ion loading scheme [10]. The preliminary results of the PSK trap will be presented together with discussions of improvements and other possible applications of the new scheme.

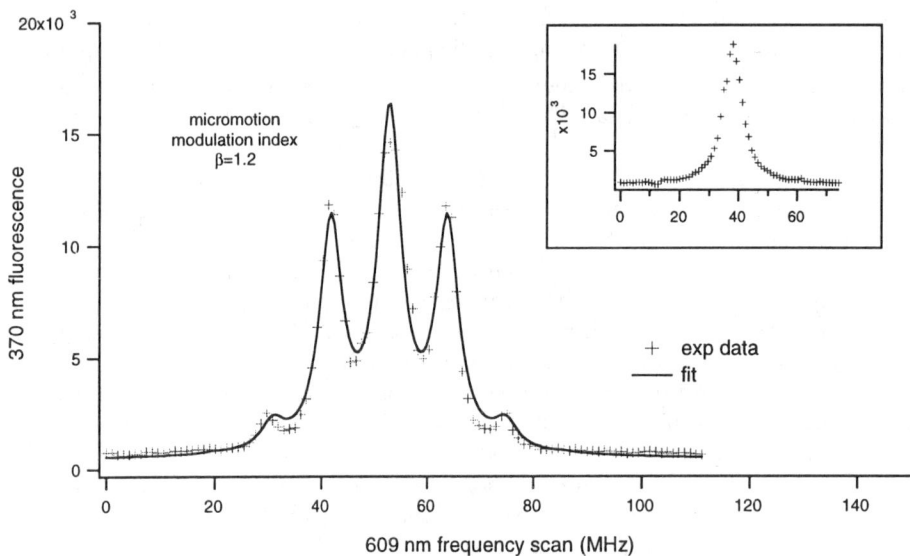

Fig. 1. Yb$^+$ spectrum with resolved micromorion sidebands is obtained by scaning the 609 nm clearing laser. The insert shows the vanished sidebands after compensation.

II. SPECTROSCOPIC EFFECTS OF MICROMOTION

As mentioned in the introduction, enhanced micromotions can grossly affect the spectrum of trapped ions. In fact, the spectrum of a single trapped ion with significant micromotions looks much like that of a small ion cloud. It is often convenient to analyze the spectroscopic effects of micromotion in terms of phase modulation.

Considering the modulation only at the micromotion frequency Ω, the resulting phase modulation is given by $\sin(2\pi\nu_0 + \beta\cos\Omega t)$ with the modulation index $\beta = 2\pi r_\mu/\lambda$. The ion spectrum consists of the carrier and the modulation sidebands at $\pm n\Omega$ with the intensities $J_n^2(\beta)$.

Indeed, resolved micromotion sideband spectra have been observed mostly in the narrow clock transitions of trapped single ions. Fig. 1 shows a spectrum of a single Yb$^+$ ion done JPL [9]. It was obtained by scanning the 609 nm clearing transition from $4f^{14}5d^2D_{3/2}$ to $4f^{13}5d6s^3D[1/2]_{1/2}$. The intrinsic transition width of about 4 MHz resolves the $\Omega = 11$ MHz sidebands. A theoretical function fit gives a modulation index of 1.2, corresponding to 0.12 μm micromotion amplitude in the direction of the laser.

The ability to obtain quickly a spectrum like Fig. 1 provides one with a convenient micromotion monitoring scheme. This is useful for reducing the micromotion through compensation(at least in the laser direction). After applying some compensation field, the micromotion sidebands vanished as shown in the insert of Fig. 1. With Ba$^+$(and many ion candidates used in trapping), however, such a revolved spectrum can not be easily obtained. The coupled energy levels and broader widths of the transitions often result in a broadened spectrum which offers little information about the micromotion.

To measure the micromotion in Ba$^+$ then, one will have to use a different method. The method of detecting the amplitude modulation in the ion fluorescence is commonly used [5-7]. This modulation is easily understood when the photon scattering rate is much large than the modulation frequency. As the ion oscillates back and forth, the first-order Doppler shift tunes the effective laser frequency in and out of the resonance at the ion's motional frequency. So the ion fluorescence will have the intensity modulation at the frequency. In the simplest case of a two-level system with a Lorentzian line profile, the instantaneous photon scattering rate is $S = S_0\Gamma^2/((\Delta\nu - v_\mu/c\nu)^2 + \Gamma^2)$. Setting $\Delta\nu = \Gamma$ and further assuming small micromotion amplitude ($r_\mu < \lambda$), one obtains the modulation amplitude in fluorescence intensity $\delta S = 1/2(\Omega/\Gamma)(r_\mu/\lambda)S_0$.

This modulation signal can be picked up by various phase-sensitive-detection(PSD) arrangements. We have used a simple digital PSD setup which consists of two fast switching gates for the photon pulses and two separate counters [5]. The gates are switched on and off 180° out of phase so that the two counters record the photon counts in two half rf periods. The difference of the two counts is taken as the modulation signal. The detection system was checked and calibrated with an intensity-modulated LED at the micromotion frequency. The correct reference phase was set with the actual ion PSD signal.

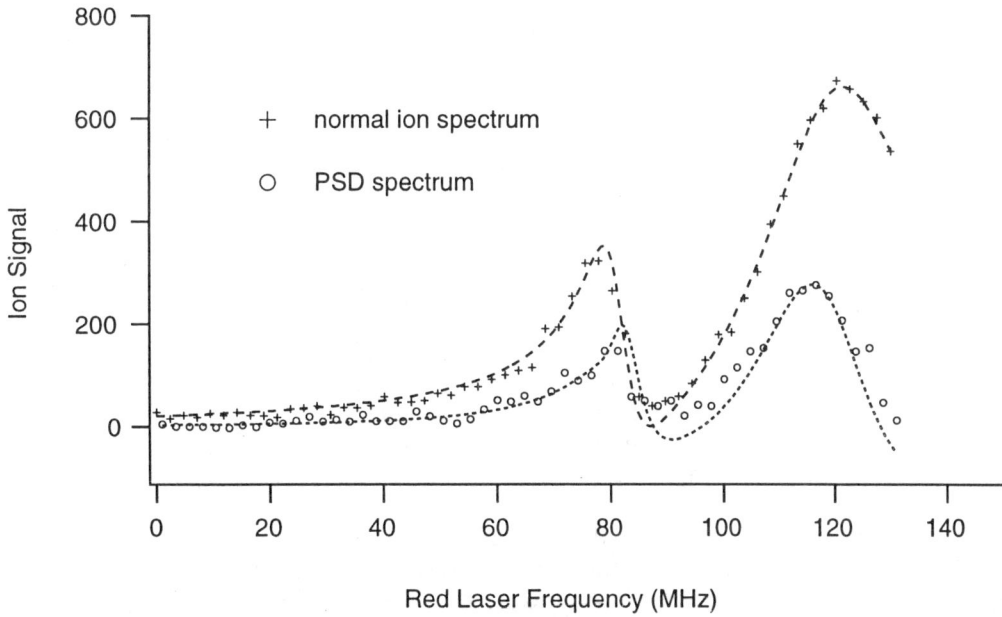

Fig. 2. Spectra of a single trapped Ba$^+$ ion. The PSD spectrum is obtained at a lower rf voltage to increase the micromotion modulation.

The expression of the intensity modulation signal for a coupled three-level energy structure of Ba$^+$ is more complicated but can be obtained in a similar fashion using the photon scattering rate given in reference [5]. Fig. 2 shows one of the observed micromotion PSD signal from a single Ba$^+$. The dotted line is the least-square fit to the expected theoretical function with the ion's micromotion size r_μ being the only free fitting parameter. Most experimental operating parameters were obtained first from the normal ion fluorescence spectrum also shown in Fig. 2. The detection sensitivity to the micromotion amplitude depends on the detunings of the cooling laser. The two-photon feature to the left can be very sensitive but tends to be very sharp. The right single photon feature is preferred if a longer integration time is necessary. The actual sensitivity is slightly reduced in this case because the 11 MHz micromotion frequency is comparable to the 20 MHz transition linewidth. It should be pointed out here that the data was taken with co-propagating laser beams. The sensitivity to the micromotion modulation can be increased by choosing counter-propagating laser beams.

TRAPS FOR ELIMINATING ENHANCED MICROMOTIONS

As we have seen, the enhanced micromotions can be reduced by externally applying compensation fields for the stray dc fields and carefully balancing the rf voltages on the endcaps, as long as a sensitive micromotion monitoring scheme is available. However, the complete elimination of micromotions in all directions at all times seems experimentally difficult. It is therefore desirable to have a rf trap completely free of enhanced micromotions. To this end, we first designed and successfully operated the Paul-Straubel traps [8], a kind of rf trap significantly different in structure than the conventional quadrupole traps. The trap consists mainly of a ring with the supporting leads. The endcaps are reduced to a pair of distant compensation plates. The trap structure is more open to laser beams and allows resistive heating to a high temperature to eliminate any surface deposits which may generate stray dc fields. The advantages of the surface cleaning by heating were strikingly demonstrated [8] some time ago by a considerable reduction in the width of the allowed transition in Ba$^+$ after two successive heatings to ≈1200°C while retaining the same ion in the trap! There is no micromotion due to endcap phase mis-match for the "ring-only" traps.

It turns out that the mere Paul-Straudel trap arrangement could not completely eliminate the surface patch effects and therefore the enhanced micromotion. Each ion loading process adds deposits. Maintaining ions in the trap while heating could not be achieved consistently. Furthermore, the refractory material(W or Ta) of the trap electrode tends to recrystalize after many high temperature heating cycles, creating permanent surface potential difference from different crystal faces. In addition, the patches on the compensation plates can not be always ignored and final compensations are still needed.

To remedy these shortcomings and to completely avoid trap electrode surface contamination during the ion loading, we have investigated a new approach for complete elimination of enhanced micromotions. In this approach, we first prepare all trap electrodes for extreme uniformity of surface potential. Then a novel ion loading scheme in PSK traps is used to avoid contamination.

Clean electrode surfaces can be prepared by coating all the electrodes of the trap and nearby conducting surfaces with fine-gained graphite such as Aerodag. The electric potential uniformity of the surfaces with graphite and gold coatings have been studied carefully [11]. Briefly, the electric field caused by the coated surface at a point away from the surface is determined by the *rms* value of the surface potential fluctuations and the distance from the surface. The residual stray field at the center of a ring trap is given by $E_s = \delta\phi_{rms}\epsilon/R_0^2$, where ϵ is the grain size of the coating and $\delta\phi_{rms}$ is the *rms* value of the surface potential fluctuations. Taking the surface potential variations of 0.1 eV, the nominal trap size of 1 mm, and the gain size of 1 μm, the electric field at the center of the trap is estimated to be on the order of 4 mV/cm. Again, using the usual trapping parameters, this corresponding to an enhanced micromotion orbit of 1 nm, less than the intrinsic micromotion size.

The PSK trap consists of a previously mentioned Paul-Straubel trap with straight wire leads enclosed in a kingdon trap housing, as shown in Fig. 3. When the wire (and the trap) is biased with a negative dc voltage, the straight-wire and the housing form a large dynamic Kingdon trap for the positive ions. The dc attractive force is dominant at larger distance away from the wire and the ac repulsive force prevails close to the wire. An effective potential trough parallel to the wire is therefore formed. Unlike the classical dc Kingdon trap, trapped ions will not fall into the center filament(wire) when they are damped. Furthermore, the two ends are sealed off because of the larger rf fields at the end openings. The dynamic kingdon trap has been studied theoretically by Flümel *et al.* and can be stable for ion confinement [12].

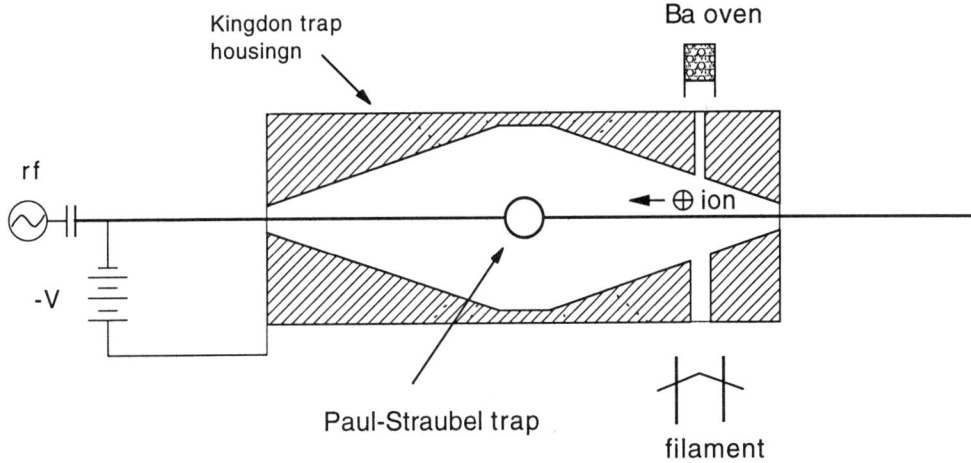

Fig. 3. Sketch of the Paul-Straubel-Kingdon trap.

In the above described setup, the ions are created near one end away from the center Paul-Straubel trap. These ions should be retained in the Kingdon trap at least for a while in the absence of damping. Collisions among the trapped ions may provide some damping by evaporation. Some ions in the larger Kingdon trap can fall into the much smaller Paul-Straubel trap at the center and subsequently laser-cooled for permanent storage. To see this indeed would work, one is reminded that the effective potential at the center must be the lowest since the repulsive pseudo force vanishes there. Of course, the negative bias on the ring can not be too large as to completely remove the trapping potential well in the radial direction of the ring trap. On the other hand, there will be a potential barrier in the axial direction surrounding the ring trap if the negative bias is not enough. Clearly there is an appropriate bias voltage window to be found for the proper operation of the trap. To further facilitate the capture of the ions at the center, the inner wall is made into a tapered cone structure, making the effective potential trough deeper towards the center.

A full analysis of the potential distribution in the PSK trap would be useful to optimize the operation, but requires rather involved numerical calculations. Short of that, one can gain some very rough quantitative understanding of the potential distribution by approximating the trap with a combination of simple structures and adjusting the parameters to fit the boundary conditions [10].

To do this, let's consider a Kingdon trap of the infinite length. The potential distribution with the line charge density λ can be written out right away as $\Phi(r) = V_{ring} \ln(r/r_2)/\ln(r_1/r_2)$ with r_1 and r_2 being the radii of the wire and the housing. To approximate the ring, we first cut out a line section of length $2R_0$ and replace it with the trap ring of charge density $\alpha\lambda$. We know that the potential at the trap center in the plane of the ring(xy plane) is close to being symmetric. To compensate this, two point charges of $\beta\lambda$ are added on the ring in the direction perpendicular to the long wire. Again, the potentials of the cut-out section, the ring, and the point charges are readily spelled out. The parameters α and β (≈ 1) are to be determined by fitting the potential points at some specific boundary points.

By combining all the potentials from various pieces into a total potential for a given applied voltage on the ring, we can proceed to write out both the dc and the rf pseudo potentials for a set of given applied dc and rf voltages on the ring. Details of the calculations are somewhat lengthy. We refer interested readers to reference [10]. But we will show some results here. Fig. 4(a) plots the various potentials along the radial direction of the Kingdon trap away from the ring. The dotted line shows the log distribution of dc potential($\propto \ln(r)$) which pulls the ion towards the wire. The broken line shows the ac pseudo potential($\propto r^{-2}$) which pushes ion away from the center wire. A combination of the two forms the total dynamic Kingdon trapping potential shown in the solid line. It can be readily shown that the trough potential is deeper and the minimal point r_M moves farther away from the wire as the housing radius r_2 becomes larger. This is why the housing is made of a hallow cone shape as mentioned earlier. One should avoid making r_M too small at the ion loading point. Otherwise, the large driven rf micromotion of the ions near the loading point could destroy the ions into the wire. Furthermore, the initial ion energy is determined by the micromotion energy. An excessively large initial energy may make the ions harder to load into the center ring trap.

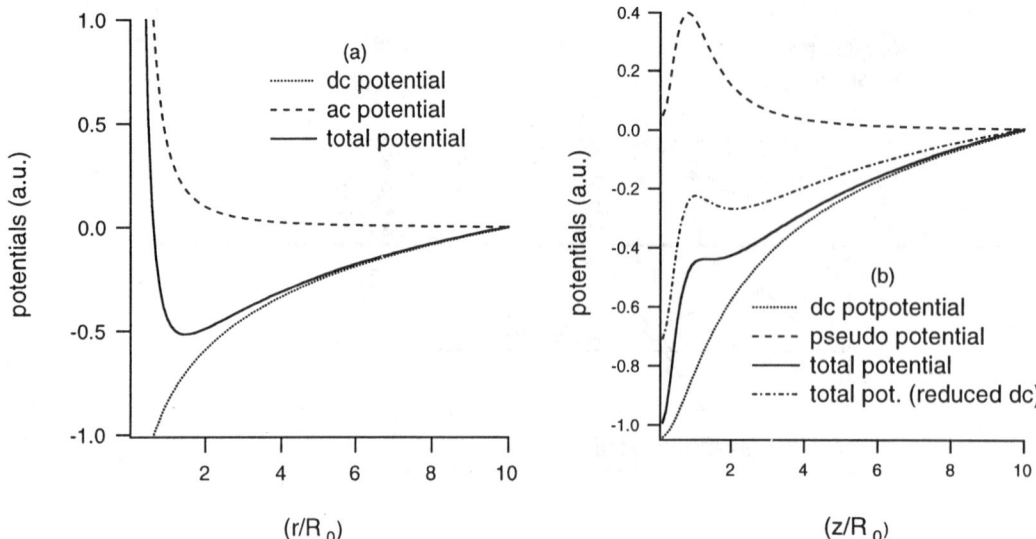

Fig. 4. (a) The potentials in the radial direction of the Kingdon trap far away from the center trap. (b) The potentials along the ring trap z axis.

To see that the ions can smoothly fall into the ring trap at the center from the bigger Kingdon trap, we plot the effective potential long the z axis(the trap axial direction) at some negative dc biases in Fig. 4(b). At a proper dc bias(which reduces the radial trapping depth by 55% here), a smooth potential fall from Kingdon trap to the ring trap is formed(the solid curve). This would be an ideal situation for the PSK trap. Ions are created at the loading region, trapped in the Kingdon trap trough with a leak passage along the z axis into the center Paul-Straubel trap, and then laser cooled and permanently trapped. If one reduce the dc bias, say, by a factor of $\sqrt{2}$ from that of the solid curve, a barrier results between the center ring trap and the outside Kingdon trap reservoir, preventing evaporate-cooled ions falling into the center trap.

The above semi-quantitative analysis clearly confirms the validity of the idea of combining the Paul-Straubel trap with a dynamic Kingdon trap. It also shows that the trapping parameters are critical for the successful operation. Another important aspect not touched here is the stability of Kingdon trap for ions, especially near the loading point where the rf field is high. It is well known in a conventional rf trap that ion instability results when one of the secular frequencies approaches half of the trapping rf frequency. The instability can be understood by the parametric excitation of the trapped ions [13]. A simple minded calculation of the equivalent secular oscillation frequency of the ions inside the Kingdon trap confirms that the parameters used are within the stable region.

RESULTS AND DISCUSSION

The actual PSK construction has a twisted wire ring trap of 1 mm in diameter. The wire diameter is 125 μm. The Kingdon housing is a cone shaped cylinder of 20 mm in length similar to the sketch in Fig. 3. The cone has a 6 mm diameter at the middle with 1 mm at the closing ends. Four apertures are cut out at the side for the laser access and photon collecting. The ion and electron orifices are about 2 mm from one end. The oven orifice also serves as the barium beam collimation. The entire trap assembly is coated with Aerodag for surface potential uniformity.

In order to find the operation window for efficient ion loading, the trapping parameters have been varied extensively. In particular, the trapping rf voltage was used between 100 to 700 volts peak. The dc bias voltage was varied accordingly between 0 to over -30 volts. It was found that the ions were loaded more readily at lower rf voltages. This is probably due to the high initial micromotion ion energy when the trapping rf is high. For instance, when the rf voltage is 700 volts, the ions at r_M in the loading region is over 30 eV. In the absence of effective damping, these ions are hard to be captured at the center ring trap.

The result on the overall efficiency of ion loading has unfortunately turned out disappointing. It was much less efficient than normal Paul-Straubel traps where the oven material is aimed right at the trap. More importantly, the initial trapping of the ions in Kingdon trap could not be consistently confirmed so far. To do this, we compared the loading efficiency with and without the negative dc bias on the ring electrode. With two versions of the PSK trap tested, only initial results showed higher loading efficiency with the negative bias. But the difference diminished over time. Therefore, this seems to suggest that the deposits in the loading area have some detrimental effects on the operation of the Kingdon trap. It is possible that either the contact potential or the charge-up effect may create a leak for the Kingdon trap and therefore the ions could not be effectively contained in the Kingdon trap. Further study is needed and some modification would have to be made for the PSK trap to have its full potential.

Despite of the incomplete results on the trapping efficiency, the test results showed the significantly lower overall micromotions of the trapped ions compared to the previous traps used. To measure the residual micromotion of trapped single Ba^+ ions, the digital PSD method discussed in the previous section was used. The micromotion amplitude is now too small to detect directly at the normal trap well depth. It is easy to show however that the micromotion amplitude is inversely proportional to the trapping rf field. To purposely increase the micromotion to a detectable level, the applied rf voltage could be lowered by as much as a factor of 8 without significant ion signal loss due to the increased micromotion. This corresponds to the trap well depth reduction of a factor over 60. The very fact that the ions were not lost at the shallow trap well depth also indicates little residual micromotion. In the previous traps, this was not possible.

Before the micromotion PSD measurement, the ion spectrum was first obtained at normal rf level from which the laser intensity and detuning parameters were obtained. With the laser setting left unchanged, the trapping rf level was lowered to 160 volts from 700 volts and the PSD spectrum was taken. The obtained spectra similar to that in Fig. 2 were fitted, yielding a micromotion amplitude of 28 nm. Scaling to the actual micromotion size at the rf level of 700 volts and taking into account the response time of the ion, we obtain the residual micromotion amplitude of 7 nm along the laser direction. Since the stray dc field is most likely in the plane of the ring trap and the laser is approximately along the body diagonal of the trap, it is reasonable to assume that the overall micromotion amplitude of the ion is $\sqrt{3}$ times that of the measured value, $i.\ e.$ 12 nm. This is already smaller than the secular motion amplitude but is twice as large as that of the intrinsic micromotion. The corresponding second order Doppler shift in the clock transition of Ba^+ is then about 3×10^{-18}, and the corresponding ac Stark shift 2 mHz(or 1.3×10^{-17}). It should be also emphasized that micromotion amplitude depends on trap operation parameters. Therefore, it is useful to look at the reduction of the actual stray field in general. From the previous discussion, we found $E_s = 23\ mV/cm$, which is still much larger than the estimated value.

In conclusion, we have presented a new approach to reduce the overall enhanced micromotions. With the PSK trap, the patch effect can be eliminated by a combination of uniform surface coating and indirect ion loading. The rf phase imbalance is completely avoided with the ring-only Paul-Straubel trap. Preliminary results have already shown the much reduced micromotion without any compensations. But the PSK trap has not prove to be working as designed. A detailed modeling of the trap potential may help to understand the trap better and to narrow down the more appropriate operating parameter window.

Finally, we would like to suggest other possible applications with the techniques discussed. First, an enhanced-micromotion-free rf trap allows one to have an extreme shallow well depth, or the secular motions at low radio frequencies. Such mono-ion oscillator has extremely low noise and high Q and could be used as a very sensitive radio frequency detector/amplifier. Second, the indirect loading method channels a few ions to the trapping region, avoiding spreading atoms and charged particles at the trap center. It may be useful in such ion trap application as cavity QED experiments with trapped ions. Finally, when the PSK trap loading efficiency is made high, one can use it for rare isotope ion loading where only minimal oven charges for the rare isotopes would be necessary. The ions may even be captured in a minuscule amount of residual gases. In view of all these, one may wish to use the indirect loading scheme with endcap-only traps [14] as well.

The Ba^+ work at the University of Washington is supported by NSF. The Yb^+ work is carried out at JPL under a contract from NASA. N. Yu would like to thank L. Maleki for reading the manuscript and offering comments.

REFERENCES

1. For example, H. Dehmelt, *Advances in Atomic and Molecular Physics*, **3**, 53 (1967).
2. D. J. Wineland, W. M. Itano, J. C. Bergquist, and R. G. Hulet, *Phys. Rev. A* **36**, 2220 (1987).

3. R. G. Devoe, J. Hoffnagle, and R. G. Brewer, *Phys. Rev. A* **39**, 4362 (1989); J. I. Cirac, L. J. Garay, R. Blatt, A. S. Parkins, and P. Zoller, *Phys. Rev. A* **49**, 421 (1994).
4. N. Yu, X. Zhao, H. Dehmelt, and W. Nagourney, *Phys. Rev. A* **50**, 2738 (1994).
5. G. Janik, Ph.D. thesis, University of Washington, (1984).
6. J. T. Hoffges, H. W. Baldauf, W. Lange, and H. Walther, *J. Mod. Opt. (UK)* **44**, 1999 (1997).
7. D. J. Berkeland, J. D. Miller, J. C. Bergquist, W. M. Itano, and D. J. Wineland, *J. Appl. Phys.* **83**, 5025 (1998).
8. N. Yu, W. Nagourney, and H. Dehmelt, *J. Appl. Phys.* **69**, 2779 (1990).
9. N. Yu, W. Wo, and L. Maleki, to be published.
10. H. Dehmelt and N. Yu, *Proc. Natl. Acad. Sci. USA* **94**, 10031 (1997).
11. J. B. Camp, T. W. Darling, and R. E. Brown, *J. Appl. Phys.* **71**, 783 (1992).
12. R. Blumel, *Phys. Rev. A* *51*, R30 (1995).
13. R. J. Cook, D. G. Shanland, and A. L. Wells, *Phys. Rev. A* **31**, 564 (1984).
14. C. A. Schrama, E. Peik, W. W. Smith, and H. Walther, *Opt. Commun.* **101**, 32 (1993).

Towards crystalline ion beams - the PALLAS[1] ring trap

T. Schätz, D. Habs, C. Podlech, J. Wei* and U. Schramm

Ludwig-Maximilians-Universität München, Sektion Physik, 85748 Garching, Germany
* Brookhaven National Laboratory, Upton, New York 11973

Abstract. To experimentally elucidate fundamental issues of crystalline ion beams at low velocities we presently set up PALLAS[1], a table top circular RF quadrupole storage ring for acceleration and laser cooling of, e.g., ^{24}Mg$^+$ ions. Applying the smooth approximation to PALLAS we compare its beam dynamics to heavy ion synchrotrons like TSR Heidelberg and thereby demonstrate the necessity of the highly symmetric lattice for the attainment of crystalline structures. Furthermore, dedicated molecular dynamics simulations are presented, affirming the feasibility of beam crystallization in PALLAS.

INTRODUCTION

The fascination of crystalline ion beams [1], representing the ultimate form of space charge dominated beams in accelerator physics and reaching far beyond standard limitations of emittance dominated beams, has strongly driven the improvement of storage ring cooling techniques like electron and laser cooling throughout the last decade [2,3]. However, since the first discussion of crystalline ion beams following experiments at the NAP-M proton storage ring [4], no such beam has been definitely proven up to now. Only evidence for the formation of a chain like structure of highly charged ions electron cooled at the ESR [5] as well as a preliminary hint for beam ordering of a longitudinally and dispersively [6] laser cooled ^{9}Be$^+$ ion beam at the TSR [7] have been reported. On the other hand, early theoretical predictions concerning the structure of ion crystals under storage ring like focusing conditions [8] were confirmed by elaborate studies of ion crystals at rest in traps (see e.g. [9,10]). Regarding especially higher order structures like helices, which are nicely observable only in trap experiments up to now, this fact is believed to be due to the too low symmetry and periodicity of the lattice functions of existing storage rings [2,3]. Nevertheless, overcoming shear forces in the bending sections by a well adapted gradient laser cooling and applying direct transverse cooling, low order crystalline structures were proposed to be reachable in the present machines [11]. Here dedicated molecular dynamics (MD) simulations [12] were performed, which, besides full inter-particle Coulomb interaction take into account the individual lattice parameters of any given storage ring.

SMOOTH APPROXIMATION

As depicted in fig. 1, beam heating (or to be more precise emittance growth in the transverse and broadening of the momentum distribution in the longitudinal degree of freedom) in present day heavy ion storage rings originates from the unavoidable polygon-like arrangement of the ring focusing and bending magnets. This feature on the one hand effects orbital variations of the focusing strength, expressed by the β-function, which cause beam envelope oscillations representing via intra beam scattering (IBS) the major heating mechanism. On the other hand shear is introduced predominantly in the bending regions, expressed by the dispersion function D, leading to less important heating in the emittance dominated regime. Since clearly bending but

---

[1] PAuL Laser cooling Acceleration System

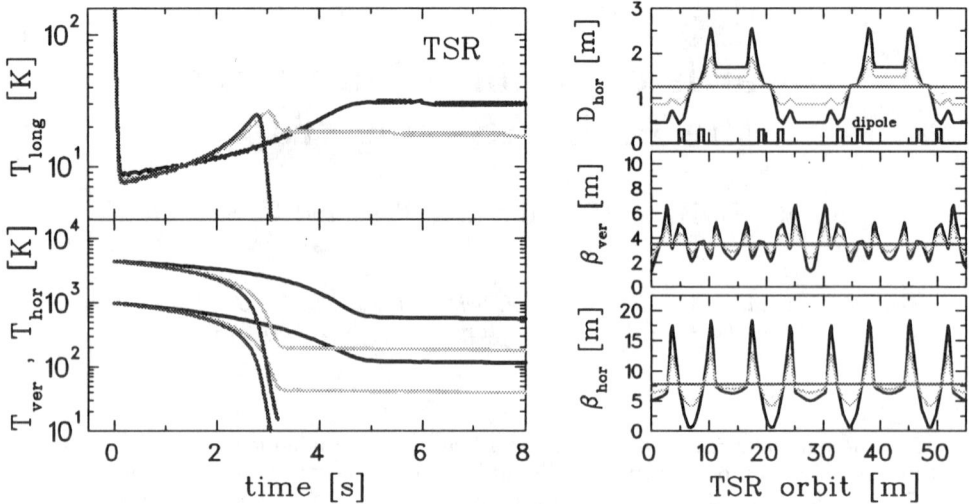

FIGURE 1. Simulation (Code INTRABSC [13]) of the temporal development of the longitudinal (upper figure) and transverse (lower figure) beam temperatures of an uncooled ^{9}Be$^+$ beam ($7 \cdot 10^5$ ions) at the TSR Heidelberg, subject to longitudinal laser cooling at a typical rate of 0.04 1/s. Regarding realistic conditions (black line) laser cooling leads to a fast reduction of the longitudinal temperature followed by a pronounced indirect transverse cooling [6] mediated by intra beam Coulomb scattering (IBS). The equilibrium temperatures now strongly depend on the heating mechanisms inherent in heavy ion storage rings: The orbital variation (right figures) of the focusing strength (β-function) and therefore of the mean beam radius causes beam heating via IBS as also does (less important) shear (dispersion function D) in the bending regions (the position of the bending dipoles is marked in the upper right figure). Artificially reducing β_{hor}, β_{ver} and D to half amplitude (light grey curves) already illustrates reduced heating. For the smooth approximation (dark grey curves), eliminating local variations of the ring lattice functions, IBS beam heating almost vanishes in the simulation, when the phase space density of the beam is reduced sufficiently.

also focusing primarily acts in the horizontal plane (see right part of fig. 1), longitudinal laser cooling in storage rings generally leads to a strongly anisotropic velocity distribution at rather high equilibrium temperatures [6].

We now apply the smooth approximation to, e.g., the TSR Heidelberg by artificially using the mean values of the corresponding lattice β-functions β_{hor}, β_{ver} and the dispersion function and therefore eliminate the influence of envelope oscillations (dark grey curves in fig. 1). This results in a collapse of the transverse and longitudinal temperatures after a sufficient reduction of the transverse phase space density and thus of the remaining heating rate (due to the constant dispersion). Despite the fact that the temperature region below the latter cannot be reasonably simulated by the employed code INTRABSC [13] (only employing frictional cooling and neglecting static inter-particle Coulomb forces), we want to emphasize the obvious advantage of realizing a *smooth* ring like the RF quadrupole ring PALLAS [14], details described below, for the scope of beam crystallization. A more rigid definition of the validity of the mentioned smooth approximation is given in the following.

The smooth approximation (SA) is a universal approximation method for integrating Hill's equation or generally differential equations with periodic coefficient like

$$\ddot{y} + K(s)y = 0, \quad K(s+L) = K(s). \tag{1}$$

As a general result of this method one, e.g., obtains the wavelength λ of the solution $y(s)$ implying λ to be large compared to the length of the periodic focusing structure L. For an accuracy within a few percent the relation

$$\lambda \gtrsim 2.6 \cdot L \tag{2}$$

has to be satisfied [15]. The position dependent restoring force $-K(s)y$ may then be replaced by an equivalent average restoring force $-\overline{K}y$. It is interesting to see, that the upper condition 2, only stating the validity of the SA, coincides with the well-known necessary condition for the stability of crystalline beams, namely to avoid the excitation of crystal phonon modes by the periodic focusing structure. In this case the superperiodicity

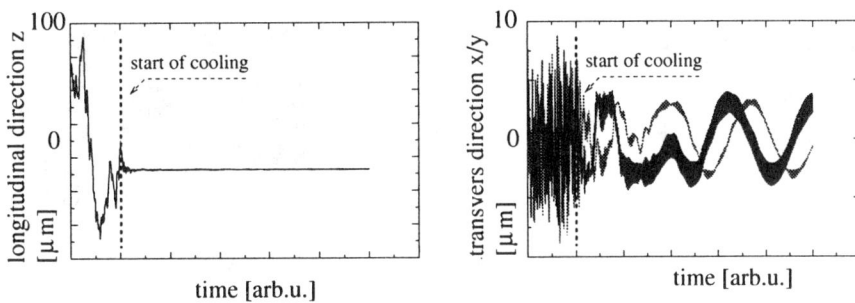

FIGURE 2. Longitudinal respectively transverse position of a test particle in the MD simulation (one out of 10^5 in the whole ring) versus time. While the longitudinal (beam orbit) position coincides with its equilibrium value (co-moving frame) immediately after the beginning of longitudinal cooling, the behavior in both transverse degrees of freedom is different: requiring time for the ordering process, the transverse motion is finally described by two phase shifted oscillations corresponding to a twisted helix-like trajectory.

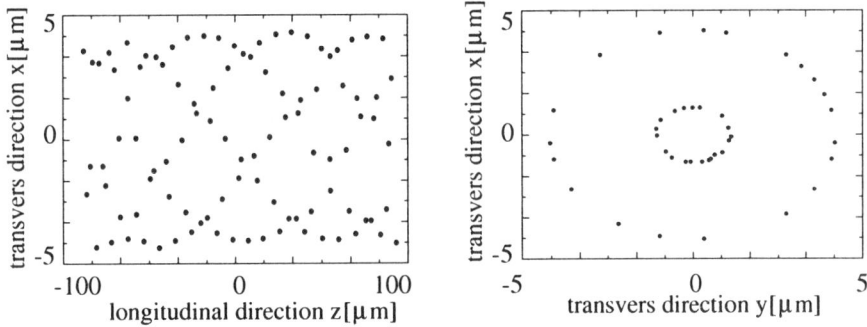

FIGURE 3. Snapshot of the ion distribution in the beam after the phase transition demonstrated in fig. 2 has occurred.

(C/L) of the ring lattice (circumference C) has to be larger than $2\sqrt{2}$ times the maximum tune $Q = C/\lambda$ (see, e.g., [16] and [1]), or in other words $\lambda \gtrsim 2\sqrt{2} \cdot L$.

To compare the upper condition with the relevant (compared to an equivalent strong focusing synchrotron) design parameters of the RF quadrupole storage ring PALLAS we introduce [15]: the tune Q_x, the average beta function $\beta_{x,y} = \beta_o$, the average dispersion function of the ring D_o, the momentum compaction factor α and the transition energy E_t as

$$\begin{aligned} Q_x &= 2\pi R_o \nu_\beta / v_o &= 100 \\ \beta_o &= R_o / Q_x &= 5 \cdot 10^{-4}\,\text{m} \\ D_o &= R_o / Q_x^2 &= 5 \cdot 10^{-6}\,\text{m} \\ \alpha &= 1/Q_x^2 &= 1 \cdot 10^{-4} \\ E_t &= Q_x \cdot Mc^2 &= 2.4\,\text{TeV} \end{aligned} \qquad (3)$$

Here $R_o = 57.5$ mm corresponds to the radius of PALLAS, ν_β to the secular frequency in a RF quadrupole trap, respectively to the the betatron frequency in a synchrotron, of 1 MHz (at about 5 MHz driving RFQ frequency), Mc^2 to the ^{24}Mg$^+$ ion mass and $v_o = 3000$ m/s to the average ion velocity, respectively the kinetic energy of 1 eV. Consequently the centrifugal force will not cause a significant change (of the order of 10 μm for a typical secular potential of 10 eV) of the orbital radius of PALLAS. Due to the highly symmetric setup of PALLAS the periodicity length L of the focusing parameter K is less than 1 mm for the given velocity, while the wavelength λ of the betatron oscillation is of the order of 3 mm. Therefore, PALLAS fulfills the requirement given in eq. 2 (in contrast to TSR, where $\lambda \gtrsim L$) and will thus be treated according to the SA in dedicated MD simulation of laser cooling of ion beams in PALLAS, presented below.

Fig. 2 qualitatively shows the behavior of one particle out of the sum of 40 in the MD simulation cell, representing 10^5 ions in the storage ring respectively. The longitudinal (beam direction) position of test particle reaches its equilibrium position (in the co-moving frame) rapidly after the beginning of (up to now unrealisticly strong) longitudinal cooling (left figure). The transverse motion shows a lengthy transition phase

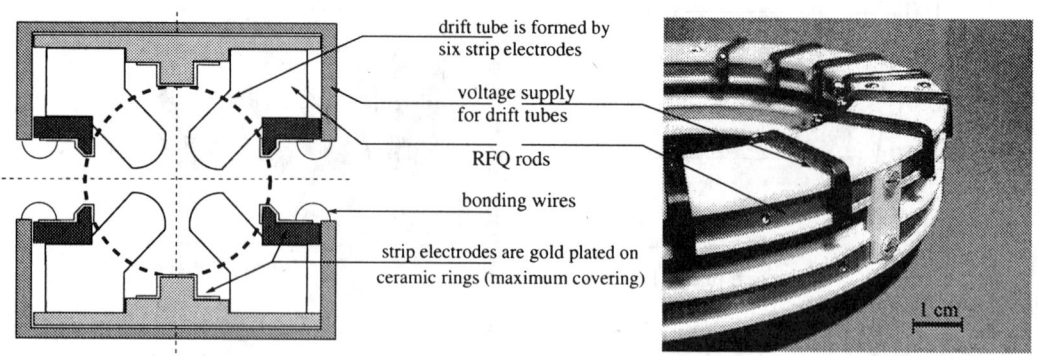

FIGURE 4. Schematic cross section and side view of the present status of the partially assembled PALLAS storage ring: It principally consists out of the four circular RFQ rods providing the storage field on axis and the drift tube arrangement for ion acceleration. The latter are segmented into six stripe electrodes metalized in sandwich layers onto precision Al_2O_3 rings which also guarantee the alignment of the main rods to better than 2/100 mm. The yokes provide electric contacting of the segments via bonding wires.

to the ordered regime, where finally the particles trajectory is described by two phase shifted oscillations in the transverse degrees of freedom (right figure), corresponding to a rotating helix-like structure. The turning is believed to be caused by the non-vanishing angular momentum of the initial velocities in the ion cloud. It remains an interesting experimental task, to verify, whether this rotation survives in PALLAS.

In fig. 3 a snapshot of the simulated ion cloud after the phase transition is presented. The left part of the figure shows the longitudinal dimension (for simplicity only those ions belonging to the outer helix are shown). The right part shows the transverse cross section of the resulting crystalline beam structure (both helices).

Even though not every possible ion position of the helices comes out to be occupied, the crystalline beam structure remains stable in time. The inter particle distance of about $10\,\mu m$ is comparable to the one observed for corresponding static ion crystals. Varying the ion density, the MD simulation results reasonably follow the crystalline structures, predicted earlier [1].

To now extract the dynamics of beam crystallization and melting (respectively heating and required cooling rates) from ongoing simulations, we have to introduce realistic laser cooling rates and possible external sources for beam heating like patch potentials caused by Mg atoms on the rods, cross talk of disturbing potentials from outside PALLAS and dark ions (e.g. $Mg^{25/26}$ respectively N_2^{28}) in the beam.

Summarizing the upper results, we demonstrated the possibility of reaching crystalline beams in PALLAS, employing an approved MD code [12], which was initially developed for the discussion of the stability of crystalline beams in synchrotrons. Obviously one major reason for the predicted stability of higher order crystalline structures is the assumed *smoothness* of the PALLAS lattice, as can be concluded also from the above IBS simulations, where the realistic TSR lattice was compared to an artificially smoothened lattice.

EXPERIMENTAL SETUP

The idea for the PALLAS storage ring has developed starting from the successfully operating RF quadrupole ring trap at the MPQ Munich, where the first beam-like ion crystals of laser cooled $^{24}Mg^+$ ions were reported [10]. PALLAS now stands for a circular RF quadrupole setup, dedicated especially to the acceleration of stored ions while preserving the established environmental conditions of the ring trap. To visualize the underlying idea, let us assume to start with a 3D crystalline structure at rest. Now forming a crystalline beam were equivalent to rotating the ring trap without affecting the ion crystal due to the high symmetry of the ring, centrifugal forces being negligible in PALLAS, as described in the preceding section. Despite the simplicity of this argument, the realization of crystalline beams in existing RFQ ring traps [10] emerged to be technically impossible, since the light pressure of the cooling lasers was by far not sufficient to overcome potential wells along the beam axis due to, e.g., mechanical or chemical imperfections.

We therefore invented an external acceleration mechanism, based on 16 individually biasable drift tubes evenly distributed along the ring circumference. For practical reasons, each tube consists out of six stripe electrode elements which are metalized onto Al_2O_3 rings, as sketched in the cross section of fig. 4. The efficiency of the proposed acceleration scheme was tested in MD simulations, reported in [14], as also more details of the construction of PALLAS. In addition to this description we want to emphasize further improvements concerning the planned operation of the ion source. To avoid any known heating source, we prepare to use a Mg oven with isotope enriched ^{24}Mg (from naturally 79 % to 99.9 %) thus reducing the quota of dark ions in the crystal. Furthermore we introduce focusing elements to the electron gun, used for ionizing Mg atoms inside the ring, to optimize the electron beam quality and prepare a pulsed gun mode, triggered by the zero-crossing of the ion storage field, to avoid a deviation of the electron beam.

With the stored ions then being accelerated, their velocity distribution will be reduced by standard longitudinal laser cooling, employing the closed optical transition between the $3^2S_{1/2} - 3^2P_{3/2}$ levels of 279.6 nm. The detection of the aimed beam crystallization should be possible via the well-known [9] hysteresis behavior of the monitored fluorescence signal within the cycle of beam crystallization and melting.

CONCLUSION

The introduction of the smooth approximation for the description of PALLAS in terms of the synchrotron terminology obviously enables the transfer of theoretical models and methods between the well established ion trap and storage ring communities. With the experimental realization of PALLAS fully satisfying the conditions for the smooth approximation, we predict the existence of stable 3D crystalline ion beam configurations for PALLAS employing a dedicated MD simulation code.

Once obtaining a crystalline ion beam of laser cooled ^{24}Mg$^+$ ions in PALLAS, we may continuously introduce disturbances with the acceleration electrodes to investigate the limit of smooth lattice deviations. Comparing and scaling anticipated results from PALLAS (approx. 0.4 m circumference) with simulations based on the smooth approximation should enable the detailed study of the properties of beam crystallization in typical ion storage rings (approx. 50 m circumference).

REFERENCES

1. D. Habs and R. Grimm, Ann. Rev. Nucl. Part. Sci. 45 (1995) 391.
2. Proc. 31st INFN Eloisatron Workshop, Erice, Italy, 1995, edts. D.M. Maletic and A.G. Ruggiero, World Scientific (1996).
3. Proc. Euroconf. on "Atomic Physics with Highly Charged Ions" I, Hyp. Int. 99 (1996) and II, Hyp. Int. 108 (1997) and III, Hyp. Int. in press (1998).
4. V.V. Parkhomchuk et al., Proc. "Ecool" Karlsruhe 1984, edt. H. Poth, KfK Report No. 3846 (1984) 71 and in [2].
5. M. Steck et al., Phys. Rev. Lett. 77 (1996) 3803.
6. I. Lauer et al., Phys. Rev. Lett. 81 (1998) 2052 and references therein.
7. M.Weidemüller et al., Max-Planck-Institut für Kernphysik (MPI-K), Heidelberg, contribution to this conference.
8. A. Rahman and J.P. Schiffer, Phys. Rev. Lett. 57 (1986) 1133 and R.W. Hasse and J.P. Schiffer, Annals of Phys. 203 (1990) 419.
9. H. Walther, Adv. At. Mol. Opt. Phys. 31 (1993) 137 and 32 (1994) 379.
10. G. Birkl, S. Kassner and H. Walther, Nature 357 (1992) 310.
11. J. Wei, H. Okamoto, A.M. Sessler, Phys. Rev. Lett. 80 (1998) 2606.
12. J. Wei, X.-P. Li, A. Sessler, Phys. Rev. Lett. 73 (1994) 3089 and in Ref. [2] (1996) 229.
13. R. Giannini and D. Möhl, private communication, following M. Martini, CERN Report PS/84-9 AA (1984) and M. Conte and M. Martini, Part. Accel. 17 (1985) 1.
14. T.Schätz, U.Schramm, D.Habs, Proc. 3rd Euroconf. on "Atomic Physics with Stored Highly Charged Ions", Ferrara, Italy, Sept. 22.-26. 1997, Hyp. Int. in press (1998).
15. H. Bruck, *Circular Particle Accelerators*, La-TR-72-10 Rev., Los Alamos, reprint, MPI-K, Heidelberg, 1986.
16. J. Schiffer in ref. [2] p. 217 and J. Wei et al., in ref. [2] p. 229.

SECTION 7

PLASMA AND COLLECTIVE BEHAVIOR

Steady-State Confinement of Electron Plasmas Using Trivelpiece-Gould Modes Excited by a "Rotating Wall"

F. Anderegg, E.M. Hollmann, and C.F. Driscoll

Department of Physics
and
Institute for Pure and Applied Physical Sciences
University of California at San Diego, la Jolla, CA 92093-0319 USA

Abstract.
A "rotating wall" electric field can give steady-state confinement of more than 10^9 charges in a Penning-Malmberg trap at 4 Tesla. For both pure ion plasmas and pure electron plasmas, the torque exerted on the plasma by the rotating wall exhibits peaks at the frequencies of $k_z \neq 0$ Trivelpiece-Gould modes. As expected, modes with $\omega > \omega_R$ (i.e. propagating faster than the plasma rotation) give positive torque and cause plasma compression; and modes with $\omega < \omega_R$ give adverse torque and cause plasma expansion. By increasing the frequency of the rotating wall, we observed a plasma central density compression of about a factor of 20. These techniques may be useful for a variety of trapping experiments.

Nonneutral electron or ion plasmas confined in Penning-Malmberg traps have inherent confinement times which are long, but finite. In practice, background neutral gas and small confinement field asymmetries exert a drag on the rotating plasma, causing slow radial expansion and eventual particle loss. Previous work [1] on small ion plasmas has demonstrated radial compression and steady-state confinement using laser techniques to apply a torque which counteracts the drag on the plasma. However, there is considerable interest in containment of elementary particles, including antimatter [2], where laser techniques are not applicable.

Previously, modest density and angular momentum changes of electron plasmas were reported [3,4] when applied dipolar electric fields excited a plasma mode, but strong heating and background gas ionization made the technique impractical at low magnetic fields ($B \leq 400G$). In other experiments, this background gas ionization is used to maintain a steady state electron target for high energy beams [5]. Recently, these "rotating wall" electric fields applied to the end of a column of 10^9 Mg$^+$ ions have been shown to give steady-state confinement and compression up to 20% of the Brillouin density limit [6].

In this paper, we describe electron plasmas confined by rotating dipole ($m_\theta = 1$) and quadrupole ($m_\theta = 2$) electric fields applied at one end of the plasma column. We show that the rotating wall fields apply a torque which can be used to compress or expand the plasma, and the torque is shown to arise from Trivelpiece-Gould plasma modes. The rotating wall fields also cause plasma heating: for electron plasmas the cyclotron radiation cooling at $B = 4T$ keeps the plasma temperature low; for ion plasmas, collisions with neutral gas or laser cooling keeps the ion temperature low. The IV apparatus [7] used here normally contains Mg$^+$ ions which are continuously diagnosed by laser-induced fluorescence. When containing electrons, the apparatus operates in a standard inject/hold-and-manipulate/dump-and-measure cycle [8].

Figure 1 shows the Penning-Malmberg trap consisting of cylindrical electrodes of radius $R_w = 2.86$ cm in

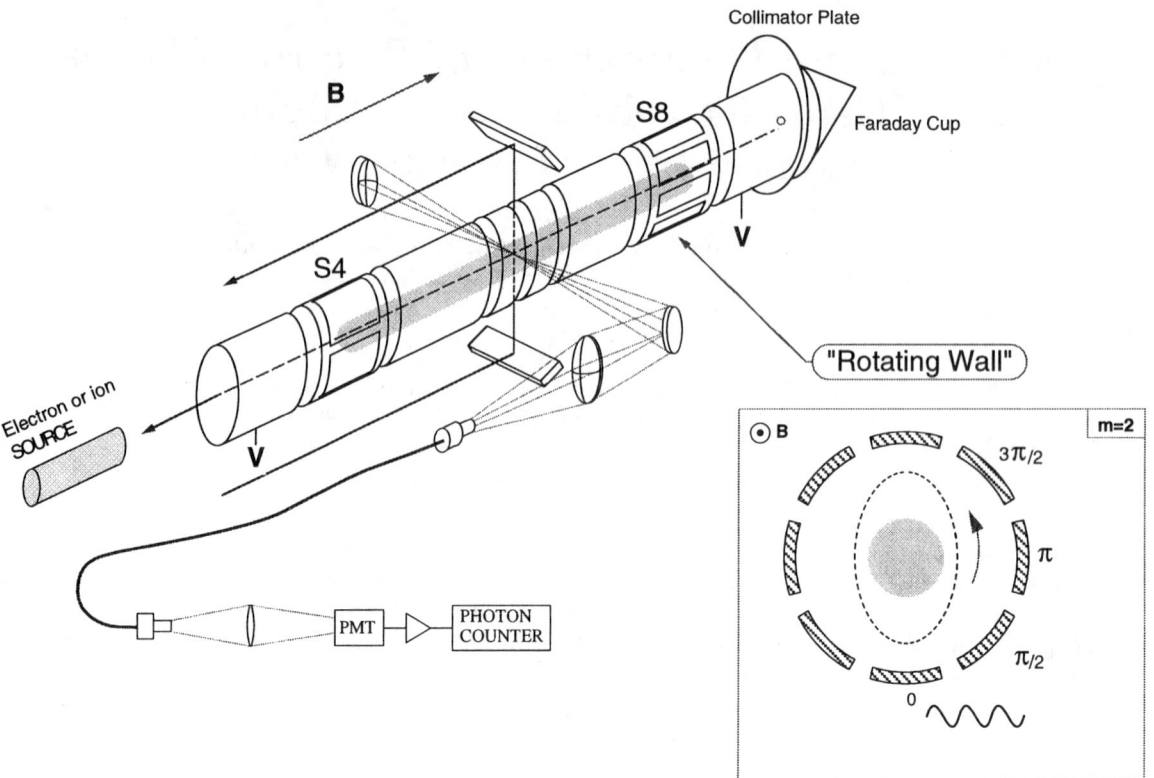

Fig. 1. Schematic diagram of the cylindrical trap, with inset representing the rotating wall drive on sectored cylinder S8.

ultra-high vacuum ($P \approx 3 \times 10^{-9}$ Torr, 97% H_2), in a uniform axial magnetic field ($B = 4T\hat{z}$). Electron injection from a thoriated tungsten filament gives $N_{\text{tot}} \approx 3 \times 10^9$ electrons in a column of length $L_p \approx 35$ cm and radius $R_p \approx 0.27$ cm, with central density $n_0 \approx 4 \times 10^8$ cm^{-3}.

The electron plasma density profile $n(r)$ and an estimate of the thermal energy T are obtained by dumping the plasma axially and measuring the charge passing through a hole in a (rotatable) collimator plate [9]. Both measurements require shot-to-shot reproducibility of the injected plasma, and we typically obtain variability $\delta n/n \lesssim 1\%$.

The radial expansion or compression of the plasma is determined by changes in the total angular momentum $P_\theta \equiv \sum_j [mv_{\theta_j} r_j - eBr_j^2/2c] \approx -(eB/2c)\sum_j r_j^2$, with the sum over the N_{tot} particles. At low temperature and low density, the angular momentum in the electromagnetic field dominates, so conservation of angular momentum implies conservation of the mean-square radius $\sum_j r_j^2$ of the plasma.

In practice, inherent "background" asymmetries in the magnetic or electric confinement fields [10] exert a weak drag on the rotating plasma, causing a decrease in P_θ and a bulk expansion of the plasma. Measurements show that this "mobility" expansion rate scales roughly as $\tau_m^{-1} \equiv -(\dot{n}_0/n_0)_{\text{bkg}} \approx (6 \times 10^{-4}$ sec$^{-1})(n_0/10^8$ cm$^{-3})^2$ for the electron columns described here ($L_p = 35$ cm, $B = 4$T). To maintain or compress the plasma, the rotating wall drive must supply a positive torque as large or larger than this drag; alternately, a reverse-rotating drive can substantially increase the background expansion rate.

The rotating wall drive consists of sinusoidal voltages $\Phi_{wj} = A_w \cos(m_\theta \theta_j - 2\pi f_s t)$ applied to the eight sectors at $\theta_j = 2\pi j/8$. Here, f_s is the signal generator frequency, and the wall perturbation effectively rotates at $f_w = f_s/m_\theta$.

We find that the applied drive couples to the plasma through discrete $k_z \neq 0$ Trivelpiece-Gould (T-G) plasma mode resonances [11]. Figure 2 shows the measured peaks in the compression rate versus drive frequency when

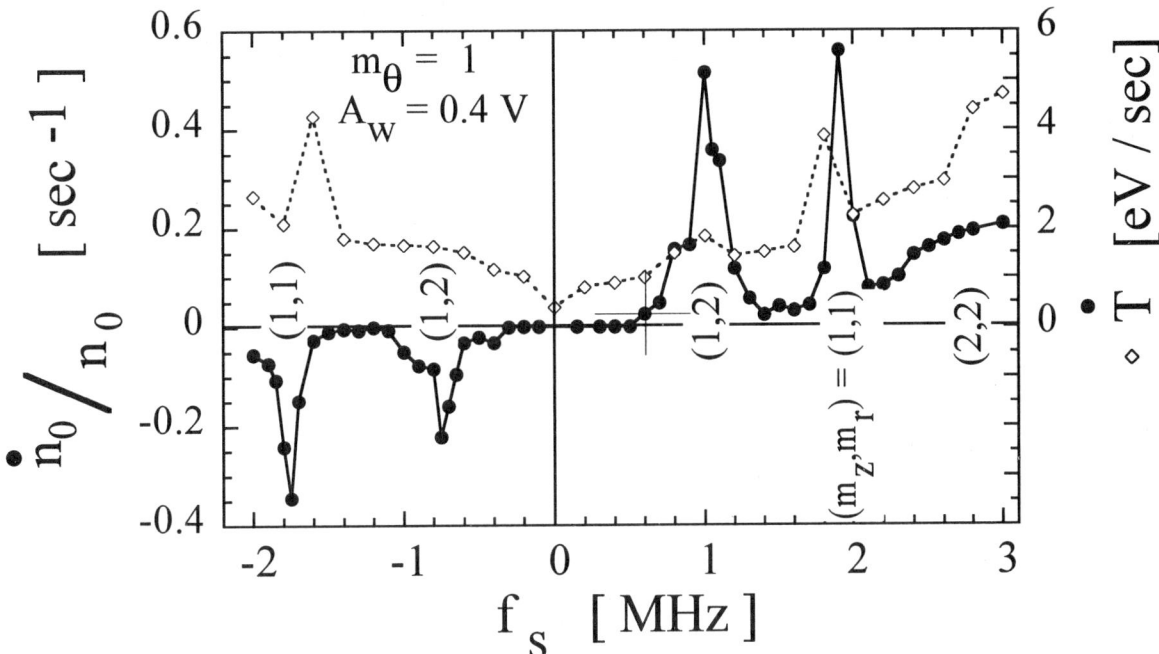

Fig. 2. Density compression rate for *strong* drive and plasma heating rate for $m_\theta = 1$ rotating drive. The compression peaks are associated with shifted (m_z, m_r) modes.

a *strong* drive of amplitude $A_w = 0.4$V is applied to the injected plasma profile. Here, an $m_\theta = 1$ rotating drive at a chosen frequency is applied to the sectored electrode S8 for 5 sec, and the initial compression (or expansion) rate $\dot{n}_0/n_0$ is measured. The measured background expansion rate of $(\dot{n}_0/n_0)_{bkg} = -4 \times 10^{-3}$ sec^{-1} (somewhat less than expected from the n^2 scaling) has been subtracted from the data, so the plot indicates torque from the rotating drive alone. Two strong compression peaks and one broader compression region are observed; and two negative torque peaks are clearly visible in the reverse drive direction. Figure 2 also shows the rate of temperature change $\dot{T}$, suggesting that the drive causes general heating as well as heating directly associated with T-G mode resonances. These temperature changes shift and broaden the T-G modes, making precise comparison with theory difficult.

For comparison to linear mode theory, we apply a *weak* $m_\theta = 1$ rotating wall, with $A_w = 0.025$V. The resulting compression peaks are shown in Fig. 3. This small amplitude does not measurably heat the plasma, so the temperature remains low, with $T \approx 0.1 - 0.2$ eV. We observe many narrow T-G compression peaks, and these correspond closely with wave transmission peaks, i.e. 10–30dB enhancement in the wave signal received at S4. The observed wave transmission peaks correspond closely with numerical drift-kinetic predictions for T-G plasma modes varying as $h(r, m_r) \exp(im_\theta \theta + im_z z\pi/L_p)$ where $h(r, m_r)$ represents the radial eigenfunction with m_r zeros in the radial eigenfunction (counting the one at $r = 0$). The six observed wave transmission and plasma compression peaks agree quantitatively with the (m_z, m_r) mode frequencies calculated numerically using two "fit" parameters of $N_{tot} = 2.7 \times 10^9$ and $T = 0.1$eV. These parameters are consistent with the measured $N_{tot} = (3 \pm 0.6) \times 10^9$ and $T = 0.1 - 0.2$eV. This correspondence has been further verified by varying the plasma length and by tailoring the antenna configuration to distinguish even and odd m_z.

The T-G modes for long columns within a cylindrical wall are predicted to have a rotationally-shifted "acoustic" dispersion relation, given approximately by

$$f - m_\theta f_R \approx \pm g(m_r, T) \frac{\omega_p}{2\pi} R_p \frac{\pi m_z}{L_p} . \qquad (1)$$

The left hand side of Eq. (1) represents the frequency of the mode in the plasma rotating frame f_R, which

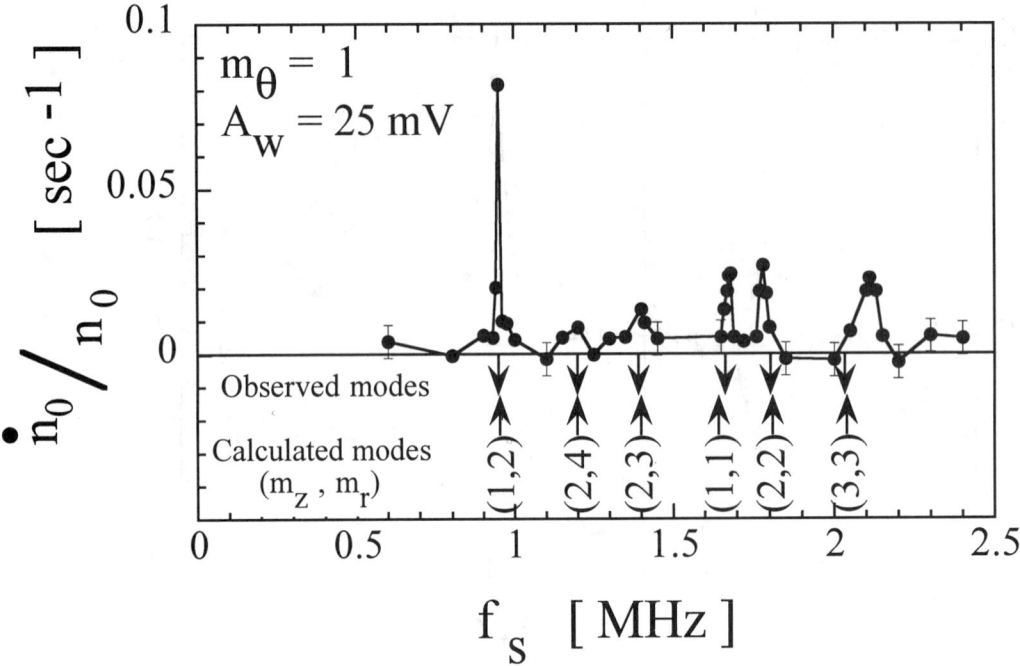

Fig. 3. Density compression rate for *weak* drive $m_\theta = 1$, compared to the observed and calculated Trivelpiece-Gould (T-G) plasma mode frequencies for various (m_z, m_r).

can be approximated by $f_R \approx f_E$ when the diamagnetic and centrifugal drift terms are small. The shifted frequencies are proportional to $N_{\text{tot}}^{1/2}$ through $\omega_p \equiv [4\pi n e^2/m]^{1/2}$ and R_p, are proportional to $k_z \equiv \pi m_z/L_p$, and depend functionally on T and m_r. In contrast, the radial density profile $n(r)$ and absolute column size R_p have little effect on the mode frequencies except through f_R.

The observed compression peaks and wave transmission peaks are also consistent with Eq. (1) when the column length is varied. Changing L_p from 35 to 17.5 cm moved the first (1,2) compression peak of Fig. 2 from 1.0 to 1.8 MHz, and moved the (1,2) transmission peak from 0.97 to 1.75 MHz. Separate $m_\theta = 1$ wave transmission experiments agree reasonably well with Eq. (1) for $L_p = 17.5$ to 40.9 cm and $m_z = 1, 2$, $m_r = 1, 2$.

The rotating wall technique enables practical plasma manipulation; for example, Fig. 4 demonstrates plasma compression (solid dots) by slowly ramping the drive frequency from 0.5 to 2.13 MHz in 1000 seconds. From 0.5 MHz to 0.65 MHz, the central density slowly decreases, indicating that there is no significant torque from the rotating wall drive and that no torque-balanced equilibrium is reached. From 0.65 MHz to 1.95 MHz, the torque provided by the rotating wall coupling through the (1,2) mode exactly balances the background drags, and the plasma is in equilibrium. Above 1.95 MHz, the background drags are larger than the rotating wall torque, and the plasma expands rapidly before reaching a new equilibrium with torque coupled through the (2,2) mode.

The range of torque-balanced equilibria obtained in Fig. 4 is quantitatively explained by the compression peaks of Fig. 2 modified by the temperature- and density-induced shifts in mode frequencies. That is, the equilibria of Fig. 4 with $0.65 < f_s < 1.95$ and $4.5 < n_0 < 14$ represent a torque-balanced equilibrium "riding up" the left side of the (1,2) peak of Fig. 2; here, the increasing background drag $((\dot{n}_0/n_0)_{\text{bkg}} \propto n_0^2)$ is balanced by the (increasing) drive torque as f_s approaches the resonant peak at $f_{m_z,m_r}^{m_\theta}$. The "crash" at $f_s = 2$ occurs because the drive at peak supplies insufficient torque, so the plasma expands until a new equilibrium is obtained on the left side of the next peak.

To see this quantitatively, we numerically calculate the shifted mode frequencies $f_{m_z,m_r}^{m_\theta}(n_0, T)$ appropriate to the measured n_0 and T during the compression ramp of Fig. 4. For example, a drive at $f_s = 0.95$ gives

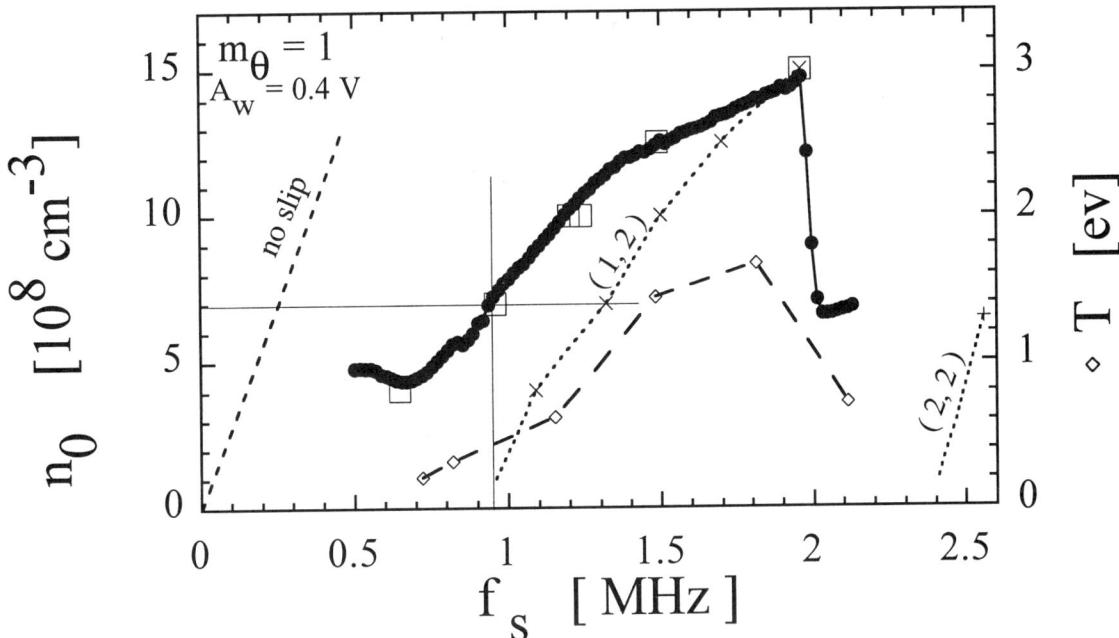

Fig. 4. Central density versus ramped rotating drive frequency f_s for $m_\theta = 1$ and measured temperature during the ramp.

a plasma density $n_0 = 7$ and temperature $T \approx 0.4$, and we calculate the resonant mode frequency to be $f^{(1)}_{(1,2)}(7,.4) = 1.32$ as shown by an X. The background drag on this plasma, $(\dot{n}_0/n_0)_{bkg} \simeq 3 \times 10^{-2} \text{sec}^{-1}$, approximately equals the torque represented by the first non-zero dot on Fig. 2, which is to the left of the (1,2) peak by an amount $\delta f \approx -0.4$. For simplicity, we presume that the (1,2) compression peak shifts in frequency so as to remain centered at $f^{(1)}_{(1,2)}(n_0, T)$; and that the peak maintains the same height and width. We then conclude that this (small) torque should be obtained at a drive frequency $f_s^{tb} = f^{(1)}_{(1,2)} + \delta f = 0.95$, shown by the square in Fig. 4. The torque-balanced drive frequencies f_s^{tb} predicted by this construction closely agree with the actual ramp frequencies f_s over the range of equilibria shown. At the maximum density obtained($n_0 = 15$), δf is approximately zero, indicating that any further increase in f_s would move the equilibrium to the right-hand side of the (1,2) compression peak; but this side of the peak is unstable [12], leading to the crash. The plasma then expands and locks to a new equilibrium on the left side of the (2,2) mode.

The nonlinear nature of the coupling to the (1,1) mode is shown in Fig. 5. The measured compression rate (dots) scales as $\dot{n}/n_0 \propto A_w^{1.1}$ for the experimentally accessible range of $A_w \geq 0.025$V. To understand this result, we measured the amplitude A_{rec} of the received signal in a transmission experiment, and obtained scalings of $A_{rec} \propto A_w^{1.1}$ for $A_w < 0.02$V and $A_{rec} \propto A_w^0$ for $A_w > 0.03$V. Simple perturbation theory suggests that the compression should scale as $(\dot{n}/n_0) \propto \delta n \cdot \delta \psi \cdot \cos(\phi)$, where δn is the plasma density perturbation (with $\delta n \propto A_{rec}$), $\delta \psi$ is the applied potential perturbation (with $\delta \psi \propto A_w$), and ϕ is the phase shift between δn and $\delta \phi$ (with measurements showing $\phi \approx$ const). Since the density perturbation δn is observed to be saturated for $A_w > 0.03$V, this theory perspective "predicts" that $\dot{n}_0/n_0 \propto A_w^1$, as observed experimentally.

We have interpreted the rotating wall coupling as a collective effect, in contrast with "side band cooling" which is interpreted as a single particle effect, i.e. the energy of a single particle transferred from the magnetron motion into damped axial or cyclotron motion.

Quadrupole rotating perturbations are also observed to couple to electron plasmas through $m_\theta = 2$, (m_r, m_z) Trivelpiece-Gould modes. Furthermore, recent experiments with pure ion plasma columns indicate that $k_z \neq 0$ Trivelpiece-Gould mode resonances are the dominant torque coupling mechanism.

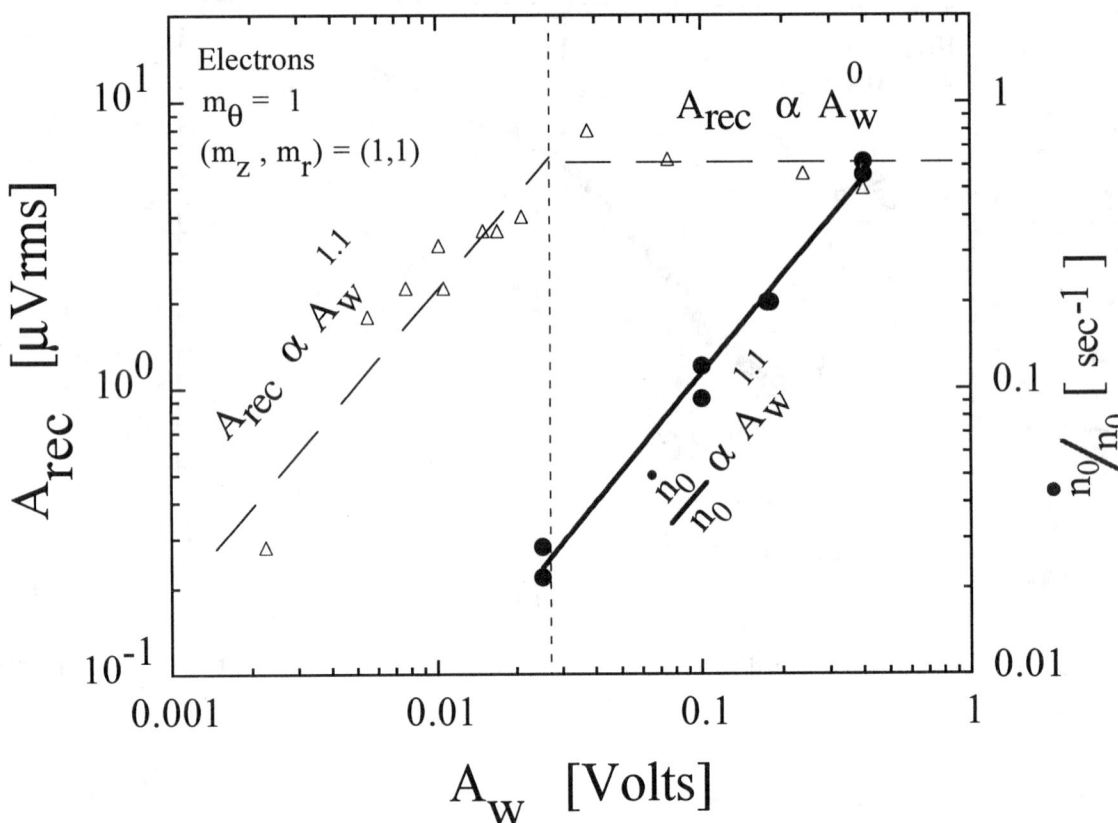

Fig. 5. Peak density compression rate and amplitude of the received signal on S4 for the $m_\theta = 1$, (1,1) mode driven by S8.

A rotating wall technique has also been applied to spheroidal ion crystals [13] using an axially uniform rotating electric field. Here, the torque is applied to a *solid* object, and in this "rotating brick" case, the crystal rotation was generally observed to be phase locked with the rotating field [13]. For an electron plasma, finite slip is required to apply a torque on the *fluid*; the T-G modes rotate faster (or slower) than the plasma, and the angular momentum carried by the wave is transferred to the particles. However, further experiments will be needed to clarify the distinction between the finite-slip $k_z \neq 0$ couplings described here and the zero slip $k_z = 0$ coupling obtained with spheroidal coulomb ion crystals [13]. Further, the wave-particle interaction which generates the torque is not understood theoretically: if the interaction is essentially Landau damping, the measurements imply that this damping is not in the linear regime. Further experiments may clarify this issue.

We thank Drs. Thomas M. O'Neil, Daniel H.E. Dubin, X.-P. Huang, Travis Mitchell, John J. Bollinger, Robert E. Pollock, and Mr. James Danielson for stimulating discussions, Dr. Ross Spencer for use of his drift kinetic computer code; and Mr. Robert Bongard for construction of a custom 8 channel digital function generator. This work is supported by Office of Naval Research Grant No. N00014-96-1-0239 and National Science Foundation Grant PHY94-21318.

REFERENCES

1. D.J. Heinzen, J.J. Bollinger, F.L. Moore, W.M. Itano, and D.J. Wineland, Phys. Rev. Lett. **66**, 2080 (1991).

2. D.S. Hall and G. Gabrielse, Phys. Rev. Lett. **77**, 1962 (1996); ATHENA Collaboration, Hyperfine Interactions **109**, 1 (1997).
3. D.L. Eggleston, T.M. O'Neil, and J.H. Malmberg, Phys. Rev. Lett. **53**, 982 (1984).
4. T.B. Mitchell, Ph.D. Thesis, UCSD (1993); see also T.B. Mitchell et al. in this book.
5. R.E. Pollock and F. Anderegg, in *Nonneutral Plasma Physics II*, AIP Conf. Proc. **31**, 139 (1995); R.E. Pollock, D. Stoller, A. Sarrazine, H. Gerberich, and T. Sloan, Bull. Am. Phys. Soc. **41**, 1603 (1996).
6. X.-P. Huang, F. Anderegg, E.M. Hollmann, C.F. Driscoll, and T.M. O'Neil, Phys. Rev. Lett. **78**, 875 (1997). Recent results on pure electron plasmas can be found in: F. Anderegg, E.M. Hollmann, and C.F. Driscoll, "Rotating Field Confinement of Pure electron Plasma Using Trivelpiece-Gould Mode," submitted to Phys. Rev. Lett.
7. F. Anderegg, X.-P. Huang, E. Sarid, and C.F. Driscoll, Rev. Sci. Instrum. **68**, 2367 (1997).
8. J.S. deGrassie and J.H. Malmberg, Phys. Fluids **23**, 63 (1980).
9. B.R. Beck, J. Fajans, and J.H. Malmberg, Phys. Plas. **3**, 1250 (1996); D.L. Eggleston, C.F. Driscoll, B.R. Beck, A.W. Hyatt, and J.H. Malmberg, Phys. Fluids B **4**, 2432 (1992).
10. C.F. Driscoll, K.S. Fine, and J.H. Malmberg, Phys. Fluids **29**, 2015 (1986).
11. S.A. Prasad and T.M. O'Neil, Phys. Fluids **26**, 665 (1983); A.W. Trivelpiece and R.W. Gould, J. Appl. Phys. **30**, 1784 (1959).
12. T.M. O'Neil and D.H.E. Dubin, Phys. Plasmas **5**, 2163 (1998).
13. X.-P. Huang, J.J. Bollinger, T.B. Mitchell, and W.M. Itano, Phys. Rev. Lett. **80**, 73 (1998).

Coulomb Clusters in RETRAP

J. Steiger*, B. R. Beck*, L. Gruber*, D. A. Church[†], J. P. Holder[†], D. Schneider*

Lawrence Livermore National Laboratory, P.O. Box 808, Livermore, CA 94550, USA
[†] *Texas A&M University, College Station, TX 77843-4242, USA*

Abstract. Storage rings and Penning traps are being used to study ions in their highest charge states. Both devices must have the capability for ion cooling in order to perform high precision measurements such as mass spectrometry and laser spectroscopy. This is accomplished in storage rings in a merged beam arrangement where a cold electron beam moves at the speed of the ions. In RETRAP, a Penning trap located at Lawrence Livermore National Laboratory, a sympathetic laser/ion cooling scheme has been implemented. In a first step, singly charged beryllium ions are cooled electronically by a tuned circuit and optically by a laser. Then hot, highly charged ions are merged into the cold Be plasma. By collisions, their kinetic energy is reduced to the temperature of the Be plasma. First experiments indicate that the highly charged ions form a strongly coupled plasma with a Coulomb coupling parameter exceeding 1000.

INTRODUCTION

The development of efficient cooling schemes for ions confined in storage rings and ion traps enabled precision measurements which greatly improved our understanding of basic atomic physics, collisions and charge exchange processes as well as the behavior of strongly coupled, one component plasmas. Multi component, strongly coupled plasmas on the other hand, which are of considerable interest in an astrophysical context [1–4], have not been studied experimentally at all. The main reasons for this lack of investigations lie in the experimental and technical difficulties to produce these kind of plasmas. For example, in order to create a two component, strongly coupled plasma in a Penning trap, the mass to charge ratios of the different ion species have to match, otherwise a centrifugal separation of the plasma will take place [5–7] before the plasma becomes strongly coupled. Only a very limited number of low charge state elements fulfill this criterion ($^9Be^+$ which can be laser cooled effectively and $^{27}Al^{3+}$ are possible candidates); whereas, this difficulty can be avoided by using highly charged ions. A large variety of heavier ions with a mass to charge ratio matching $^9Be^+$ can be found. Prominent candidates include $^{45}Sc^{5+}$, $^{63}Cu^{7+}$, $^{81}Br^{9+}$, $^{180}Hf^{20+}$, $^{198}Hg^{22+}$ and $^{207}Pb^{23+}$. This short list of elements is by no means complete, many other combinations are feasible. Heavy elements in these charge states are easily produced in an electron beam ion trap (EBIT) and $^9Be^+$ can be delivered by a metal vapor vacuum arc (MEVVA) ion source.

Unfortunately the most effective cooling technique, laser cooling, cannot be applied to high charge state ions due to a lack of suitable transitions accessible with lasers. Therefore, a sympathetic cooling scheme has been implemented at RETRAP. In brief: First $^9Be^+$ is caught and confined in the trap. Then the ions are electronically cooled with a tuned circuit and optically by a laser. Highly charged ions are merged into the cold Be plasma, energy is exchanged by collisions, reducing the kinetic energy of the highly charged ions to the Be plasma temperature.

EXPERIMENTAL SETUP

The experimental setup used for the cooling experiments is shown in Fig. 1. Singly and doubly charged Be ions are produced in a MEVVA ion source. After momentum analysis in a 90° bending magnet, Be^+ is decelerated and caught in one of RETRAP's [8] Penning traps (for a description of the trap geometry see the article by D. A. Church et al. in these proceedings). A high impedance tuned circuit, consisting of the

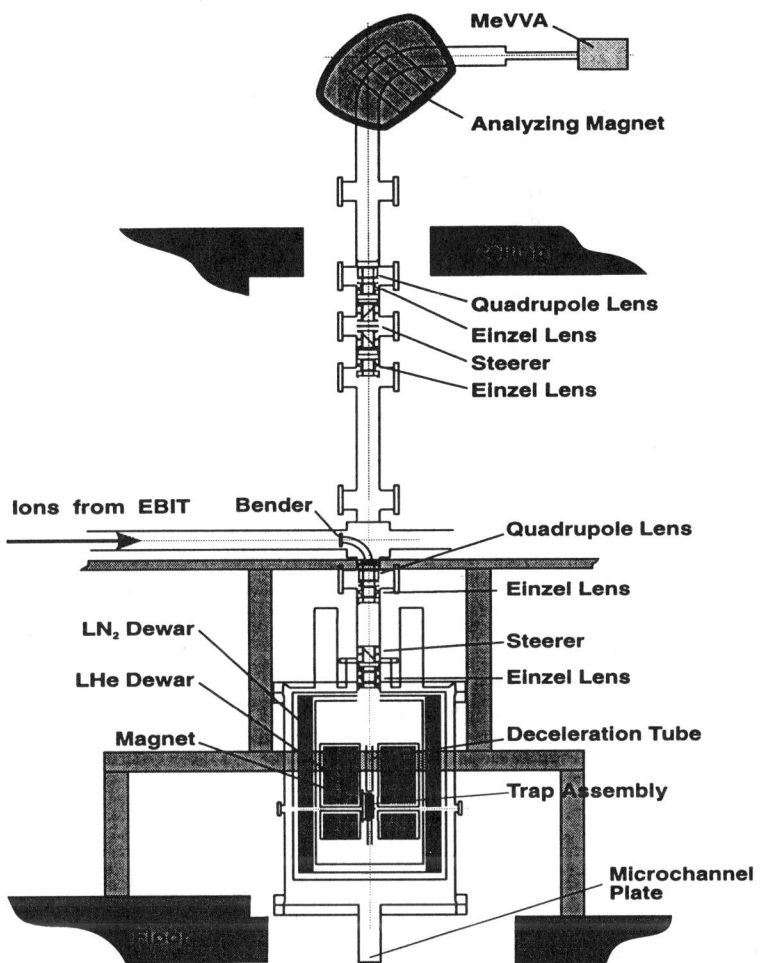

FIGURE 1. Experimental setup used for the cooling experiment. Two ion sources are being used. A Metal Vapor Vacuum Arc provides for singly and doubly charged beryllium while the highly charged ions are extracted from EBIT and transported to RETRAP. Both ion beams are momentum analyzed to select a certain charge state. Only the analyzing magnet for the Be beam is shown here. The ion energy is typically 6 $keV \cdot q$, too high for direct catching. A deceleration tube situated above the trap is used to reduce the kinetic energy of the ions to about 50 $V \cdot q$.

endcap electrode capacitance and an inductor connecting these electrodes, precools the ions with a cooling time constant of about 80 s for Be^+. A laser beam, entering the trap through holes in the ring electrode and tuned to the red side of the Be^+ $2s\,^2S_{\frac{1}{2}}(m_j = -\frac{1}{2}, m_I = -\frac{3}{2}) - 2p\,^2P_{\frac{3}{2}}(m_j = -\frac{3}{2}, m_I = -\frac{3}{2})$ transition reduces the temperature of the ion cloud below the tuned circuit temperature. Highly charged ions are produced in EBIT, extracted and a certain charge state is selected for transport to RETRAP. Here they are decelerated and caught into the trap which already confines the Be ions. Eventually, the highly charged ion-Be plasma reaches thermal equilibrium, both ion species will assume the same temperature, which is, in principle, limited only by the laser cooling force. To gain information about the number and radial distribution of trapped ions, they can be released to a microchannel plate-phosphor screen combination below RETRAP. Also, two additional holes in the ring electrode allow the detection of scattered photons. A lens, mounted in front of one of the holes, collects light emitted from the trapped ions. For light detection, a photo multiplier tube and a cryogenically cooled CCD camera are placed in the radial plane outside the vacuum vessel. The photo multiplier tube provides the time development of the photon yield, whereas the CCD camera images the plasma cloud by integration over a certain time interval (typically in the order of 10 s). In addition, the tuned circuit can be used to determine the plasma constituents. By changing the electrical field inside the trap, ions with different mass to charge ratios are tuned onto the tuned circuit resonance, thereby increasing the noise at that frequency. After background subtraction the relative amount of ions with different mass to charge ratios can be determined.

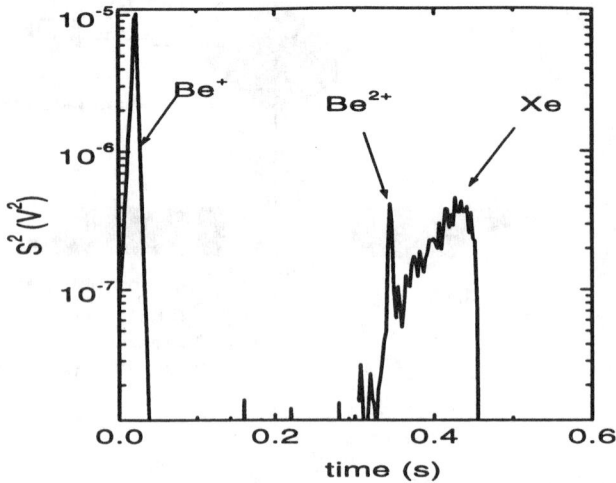

FIGURE 2. Tuned circuit signal (S^2) as a function of time, while the ring electrode potential is lowered in order to tune different ion species onto resonance. Assuming that this system has reached thermal equilibrium the relative number of ions can be determined by integrating the area under each peak. In this case the plasma consists of 98.4% Be^+, 1.2% Be^{2+} and 0.4% Xe^{44+}. The broad peak for the Xe ions is either due to a large energy spread indicating that thermal equilibrium has not been reached yet or a broad distribution of Xe charge states. In both cases a larger electrical field is necessary to tune the Xe ions onto resonance with the tuned circuit.

RESULTS

A typical tuned circuit signal is shown in Fig. 2. After the capture of Be and Xe ions the ring electrode potential is ramped in order to tune the axial oscillation frequency of different ion species to the tuned circuit resonance. After amplification a spectrum analyzer is used to measure the noise at this frequency.

Assuming that the system has reached thermal equilibrium, the tuned circuit signal can be used, after subtracting the Johnson noise, to determine the plasma composition. The measured voltage squared is given by $S^2 = \frac{NkTR}{2\tau_z}$ where N is the number of ions on resonance with the tuned circuit, $\frac{1}{2}kT$ is the thermal energy in the axial degree of freedom, R is the resistance of the tuned circuit and τ_z is the electronic cooling time constant given by $\tau_z = \frac{4mz_0^2}{\kappa^2 q^2 R}$ [9]. Here m and q are the mass and charge of the ion respectively, z_0 is the half length of the trap and κ is a constant depending on the trap geometry. Since the voltage is changed slowly compared to the characteristic frequencies of the ions the ion motion changes adiabatically. Under these conditions the action integral $\oint p_z dz = \frac{4\pi}{\omega_z}\frac{1}{2}kT$ is a constant of motion, with ω_z being the ion frequency along the magnetic field lines. Therefore the change in temperature has to be taken into account when the concentration of different ion species present in the plasma is determined. For the spectrum shown in Fig. 2, the concentrations are: 98.4% Be^+, 1.2% Be^{2+} and 0.4% Xe^{44+}. The reader should note that only Be^+ and Xe^{44+} had been injected into the trap. The doubly charged Be seems to built up during the first few seconds after the catch, possibly due to collisions with residual gas molecules. The exact process, however, is unknown. The broad Xe^{44+} signal is either due to a wide energy distribution or the presence of lower charge state Xe. Ions with larger energies can leave the harmonic region of the trap which results in a slightly lower oscillation frequency; Xe ions with $q \leq 44$ would oscillate with a lower frequency as well. A different potential has to be applied to the ring electrode in order to tune these ions onto the tuned circuit resonance.

As mentioned above, a CCD-camera is mounted onto one of RETRAP's radial ports to gain information about the spatial distribution of the Be^+ ions. The camera is placed at a 90° angle with respect to the cooling laser beam and collects a small fraction of the emitted photons. A typical image is shown in Fig. 3. At low temperatures the cloud is expected to have an ellipsoidal shape where the ratio of the two axes ($\frac{a}{b}$) is defined by the constant density of the cloud [10–13]:

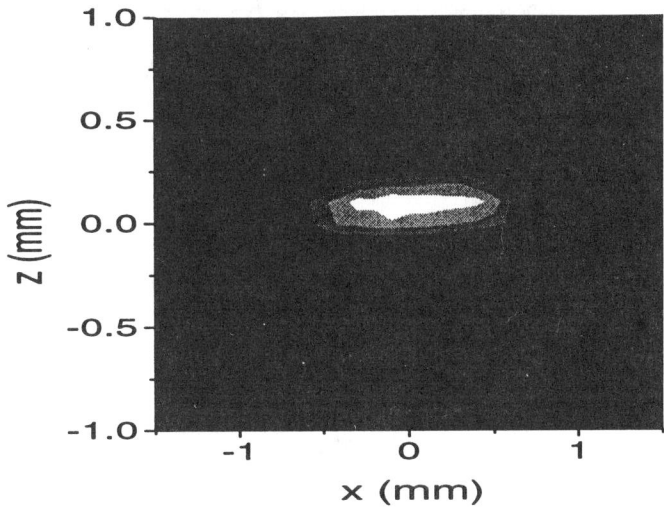

FIGURE 3. A cryogenic CCD camera is used for gaining information about the spatial distribution of the trapped Be^+ ions. The image shown here is a side view of the plasma cloud. The elliptical shape can be used to determine the Be^+ density which is constant over the full extend of the cloud. The total number can be determined by measuring the size of the plasma.

$$n = \frac{3m\omega_z^2}{4\pi q^2 \epsilon(\frac{a}{b})}. \tag{1}$$

Here $\epsilon(\frac{a}{b})$ is a function of the ratio of the axes given in [13]. In RETRAP the so obtained density for Be^+ is typically between $1 \cdot 10^9 \ cm^{-3}$ and $2 \cdot 10^9 \ cm^{-3}$. The total number of Be^+ lies between $1 \cdot 10^5$ and $4 \cdot 10^5$ ions. Combining the relative concentration obtained with the tuned circuit and the absolute number for Be^+ ions the number of Be^{2+} and Xe^{44+} can be calculated as well. The plasma shown in Fig. 2 contains at least 400 Xe^{44+} ions, which is in good agreement with the number of highly charged ions detected by the microchannel plate detector underneath the trap when the ions are released.

In addition, images of the plasma can be used to show the location of the highly charged ions within the Be^+ cloud. At low temperatures a centrifugal separation of ions with different mass to charge ratios is expected to occur. After laser-cooling and before the introduction of Xe^{44+} ions, the CCD image shows the appearance of a dark center in the middle of the trap. Figure 4 shows a projection of the central, radial plane. The drop of the photon yield to background level in the vicinity of $x = 0 \ mm$ is clearly visible even before highly charged ions are present in the trap (solid line in Fig. 4). Composition measurements at higher temperatures revealed the presence of trapped singly and doubly charged Be ions. Be^{2+} is not visible in the laser light. At low temperatures it accumulates in the center of the trap forcing Be^+ to form an annulus around the Be^{2+} cloud. After catching and cooling Xe^{44+} the dark center increases in diameter (Fig. 4, dotted line) which can be explained by an accumulation of the highly charged ions in the very center of the trap; the two Be charge state plasmas form rings around the Xe^{44+} ions. The increase of the dark center of the trap is not observed when no Xe^{44+} is present in the trap.

The temperature of the plasma has been estimated in three different ways:

1.) When the cooling laser beam frequency is swept over the Be^+ resonance a width of 0.3 GHz has been measured indicating a temperature below 1.7 K.

2.) The observed centrifugal separation of Be^+ and Be^{2+} can be used to establish an upper temperature limit as well. O'Neil introduced a set of scaling lengths [5] l_{ij} where the l_{ij} are given by $l_{ij}^{-1} = \frac{d}{dr}[q_i|\frac{m_i}{q_i} - \frac{m_j}{q_j}|\frac{\omega^2 r^2}{2kT}]$. Here m_i, m_j, q_i, q_j are the mass and charge of the ion species i and j, ω is the density dependent rotation frequency of the plasma and r the radial extend of the plasma cloud. According to O'Neil's model complete separation takes place when the conditions $l_{ij} < \lambda_D$ are fulfilled, where λ_D is the Debye length of the plasma. The temperature dependence of λ_D and l_{ij} then gives an upper bound for the temperature of the plasma.

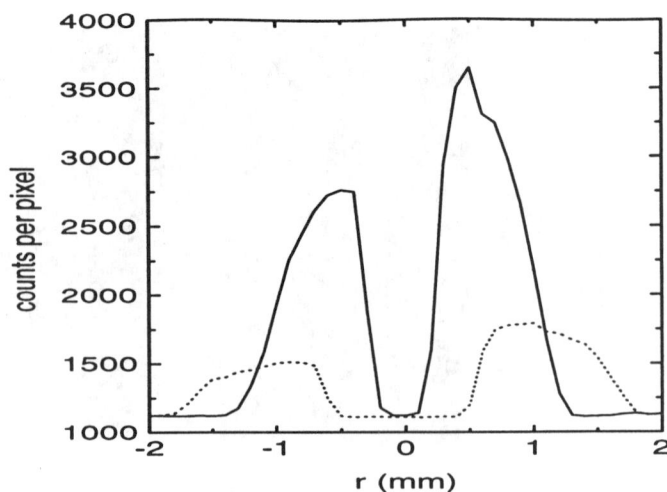

FIGURE 4. Projection of the scattered photons from the central, radial plane of the trap. Even before highly charged ions are introduced into the trap a dark center appears due to the presence of Be^{2+} (solid line). After catching and cooling highly charged ions the radius of the dark center increases (dotted line) indicating the accumulation of Xe^{44+} in the trap center. The left-right asymmetry is due to a slight misalignment of the laser beam with respect to the axis of the trap.

For typical parameters in RETRAP the temperature of the plasma should be below 0.5 K, in agreement with observed Doppler width of the cooling transition.

3.) Numerical simulations of a small number (128) of Be^+ and Be^{2+} ions show a separation only when the temperature is below 1 K [14].

Therefore we assume 1.7 K as a conservative upper temperature limit for the plasma in the following discussion.

In order to determine the Coulomb coupling parameter, $\Gamma = \frac{q^2}{akT}$ with $\frac{4\pi a^3 n}{3} = 1$, for the highly charged ion plasma the density has to be known. Unfortunately, with the present setup no direct measurement of the density is possible in RETRAP. Nevertheless, an estimate of the lower limit can be achieved with the help of equation 1, assuming that the fluid model can be applied for the highly charged ion plasma. The function $\epsilon(\frac{a}{b})$ has a maximum value of three, leading to a minimum density of $3 \cdot 10^7$ cm^{-3}, well below the Brillouin density of $3.8 \cdot 10^8$ cm^{-3} for Xe^{44+}. This lower density limit is consistent with the observed increase of the size of the dark trap center (Fig. 4). Using a density of $3 \cdot 10^7$ cm^{-3} and a temperature of 1.7 K, a Coulomb coupling parameter $\Gamma = 1370$ is calculated, indicating the formation of a strongly coupled, highly charged ion plasma and possibly the existence of an ordered structure.

CONCLUSION

The sympathetic cooling of highly charged ions with laser cooled Be^+ ions has been demonstrated for the first time. The measured Doppler broadened line width of the cooling transition, the observed centrifugal separation and numerical simulation established an upper temperature limit of 1.7 K. A rough estimate of the density leads to $n \geq 3 \cdot 10^7$ cm^{-3}. These parameters indicate the formation of a strongly coupled plasma consisting of highly charged ions with a Coulomb coupling parameter exceeding 1000. Several hundred Xe^{44+} ions could be confined and cooled in a single cycle. This number can possibly be increased by applying stacking techniques and/or improvements of the trap geometry and a more efficient deceleration scheme. A direct proof of the existence of an ordered structure was not possible at the time. An UHV compatible CCD camera is being developed in order to implement an ion projection scheme for measuring the spatial distribution of the highly charged ions.

ACKNOWLEDGEMENTS

This work was performed under the auspices of the Department of Energy by the Lawrence Livermore National Laboratory under contract No. W-7405-ENG-48 and supported in part by the Division of Chemical Physics in the Office of Basic Energy Sciences of the Department of Energy and by the Texas Advanced Research Program.

REFERENCES

1. Brush S. G. Sahlin H. L., Teller E., *J. Chem. Phys.* **45**, 2102 (1966).
2. Hansen J. P., *Phys. Rev. A***8**, 3096 (1973).
3. Stringfellow G. S., DeWitt H. E., Slattery W. L., *Phys. Rev. A* **41**, 1105 (1990).
4. Segretain L., Chabrier G., Hernanz M., García-Berro E., Iserin J., *Astrophys. J.* **434**, 641 (1994).
5. O'Neil T. M., *Phys. Fluids* **24**, 1447 (1981).
6. Larson D. J., Bergquist J. C., Bollinger J. J., Itano W. M., Wineland D. J., *Phys. Rev. Lett.* **57**, 70 (1986).
7. Imajo H., Hayasaka K., Ohmukai R. Tanaka U., Watanabe M., Urabe S., *Phys. Rev. A***55**, 1276 (1997).
8. Schneider D., Church D. A., Weinberg G., Steiger J., Beck B., McDonald J., Magee E., Knapp D., *Rev. Sci. Instrum.* **65**, 3472 (1994).
9. Dehmelt H. G., Walls F. L., *Phys. Rev. Lett.* **21**, 127 (1968).
10. Malmberg J. H., O'Neil T. M., *Phys. Rev. Lett.* **39**, 1333 (1977).
11. Prasad S. A., O'Neil T. M., *Phys. Fluids* **22**, 278 (1979).
12. Wineland D. J., Bollinger J. J., Itano W. M., Prestage J. D., *J. Opt. Soc. Am. B***2**, 1721 (1985).
13. Brewer L. R., Prestage J. D., Bollinger J. J., Itano W. M., Larson D. J., Wineland D. J., *Phys. Rev. A***38**, 859 (1988).
14. DeWitt H. E., Pollock E. L., private communication.

Chaos and order in ion traps and storage rings

Reinhold Blümel

Fakultät für Physik, Albert-Ludwigs-Universität,
Hermann-Herder-Str. 3, D-79104 Freiburg, Germany

Abstract. Chaos and order play a major role in the physics of Coulomb crystallization in traps and storage rings. This fact is illustrated with the help of the following four topics: (i) fractional frequency parametric resonances in a Paul trap, (ii) the chaos scenario in the dynamic Kingdon trap, (iii) beam crystallization criteria and (iv) suppression of synchrotron radiation by crystallized beams.

I. PARAMETRIC RESONANCE IN A PAUL TRAP

The equations of motion of a single charged particle in a Paul trap can be reduced to three decoupled Mathieu equations of the type (see, e.g., [1])

$$\ddot{x} + [a - 2q\cos(2t)]x = 0, \qquad (1)$$

where a and q are control parameters. The parameter space $\mathcal{P}$ of (1) is two dimensional. For a given point $(q, a) \in \mathcal{P}$ the solution $x(t)$ of (1) may either be stable or unstable. Thus we can decompose $\mathcal{P}$ into regions that correspond to stable and unstable solutions of (1), respectively. For the range $0 \leq q \leq 1$, $0 \leq a \leq 10$, the decomposition of $\mathcal{P}$ is shown in Fig. 1. The white regions correspond to stable solutions of (1), the black regions to unstable solutions. The unstable regions form tongues that touch the a axis at $a = n^2$, $n = 1, 2, \ldots$. The black regions in Fig. 1 are called regions of parametric instability.

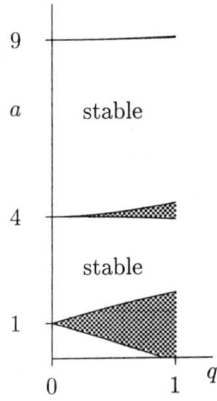

FIGURE 1. Stable (white) and unstable (black) regions of the parameter space $\mathcal{P} = \{(a, q)\}$ of the Mathieu equation.

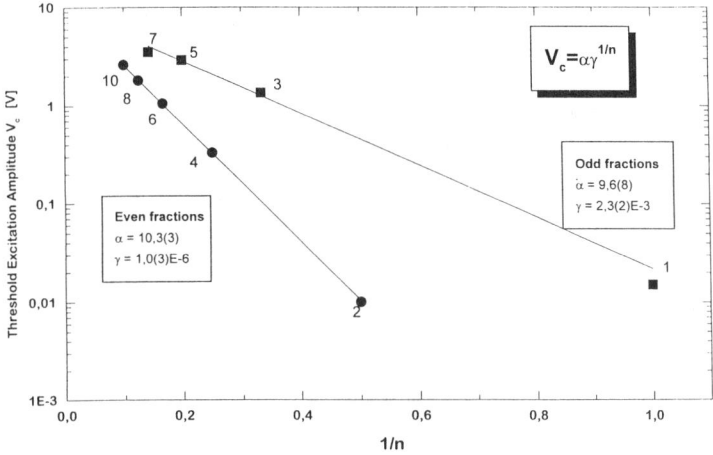

FIGURE 2. Measured critical voltages vs. $1/n$ show odd-even staggering.

Knowledge of the stable and unstable regions of $\mathcal{P}$ is useful for analyzing a recent experiment [2] on the parametric excitation of a large uncooled cloud of N_2^+ ions stored in a Paul trap. In this experiment a weak excitation voltage (on the order of 100 mV to a few Volts, frequency ω) is applied to the end caps of the trap in addition to the trap voltage (on the order of 100 V, frequency $\Omega \approx 2\pi \times 3$ MHz). Denoting by ω_z the axial secular frequeny of the center-of-mass motion of the ion cloud, the experiment detected very efficient excitation of the stored ions accompanied by a large particle loss at excitation frequencies $\omega_n \approx 2\omega_z/n$, $n = 1, 2, \ldots$.

The simplest model for describing the axial motion of the ion cloud retains only its center-of-mass coordinate Z and uses the pseudopotential approximation [3]. The equation of motion reads

$$\ddot{Z} + \omega_z^2 Z = F\cos(\omega t)Z, \tag{2}$$

where F is the strength of the additionally applied excitation field. The substitution $\tau = \omega t/2$ turns (2) into a Mathieu equation of the type (1) with $a = (2\omega_z/\omega)^2$. Since, as mentioned above, the Mathieu equation is unstable in the vicinity of $a_n = n^2$, the solutions of (2) are unstable for $\omega_n = 2\omega_z/n$ as observed in the experiments.

Since the instability tongues of Fig. 1 touch the a axis, the simple equation (2) predicts that for any nonzero value of F (F not too large) there is an ω in the vicinity of ω_n, such that 100% particle loss occurs. But this is not what is observed in the experiments. Experimentally there is a measurable particle loss only if F exceeds a critical excitation strength F_n. In order to understand this, we add a damping term to (2). This is because it is known from the theory of the damped Mathieu equation that in the presence of damping the instability tongues shown in Fig. 1 move away from the a axis, resulting in a contiguous region in the vicinity of the a axis where the fixed point $Z = 0$ of (2) is stable. Thus damping is a possible mechanism for explaining the existence of a critical excitation strength. Adding a damping term to (2) followed by appropriate scaling results in [2]

$$\ddot{Z} + \gamma \dot{Z} + Z = f\cos(\nu\tau)Z. \tag{3}$$

Here γ is the damping constant, f is the scaled excitation strength, ν is the scaled excitation frequency, τ is the scaled time and the dot refers to differentiation with respect to τ. According to a formula stated without derivation in Landau and Lifshitz [4] the critical excitation strength f_n at $\nu_n = 2/n$ is given by

$$f_n = \alpha_n(\gamma)\,\gamma^{1/n}. \tag{4}$$

We checked in various representative cases that $\alpha_n(\gamma)$ depends only weakly on n and γ. Thus plotting the logarithms of the measured critical excitation voltages versus $1/n$ should result in a collection of data points that essentially fall onto a single straight line with a shift that is associated with α and a slope given by $\log(\gamma)$. Figure 2 shows the experimental result. Instead of the expected single line the data points fall onto two different straight lines corresponding to even and odd n, respectively. This effect was called "odd-even staggering" in [2]. The reason for this effect is currently not known.

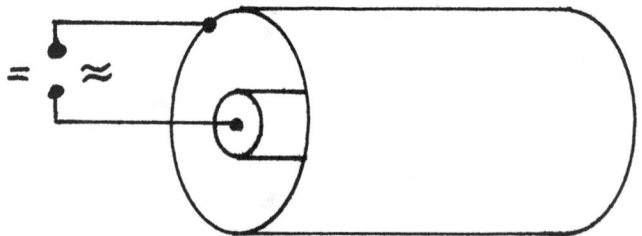

FIGURE 3. Sketch of the cylindrical dynamic Kingdon trap.

II. CHAOS SCENARIO IN THE DYNAMIC KINGDON TRAP

The cylindrical version of the dynamic Kingdon trap is shown in Fig. 3. It consists of a cylinder capacitor with a superposition of an ac voltage V_{ac} and a dc voltage V_{dc} applied between the inner and the outer cylinder of the trap. This trap was developed by E. Teloy at Freiburg University in the 1960s. It was subsequently investigated analytically, numerically and experimentally in the framework of two student theses [5,6]. In contrast to the Paul trap which relies on oscillating quadrupole fields, the dynamic Kingdon trap relies on a superposition of static and dynamic monopole fields. There are various versions of this trap. The early Freiburg trap used an axially symmetric design [5,6]. A radially symmetric version of the dynamic Kingdon trap was recently investigated experimentally [7].

In contrast to the single particle Paul trap which is completely integrable, the dynamic Kingdon trap exhibits chaos and a period doubling scenario even in the case of a single trapped particle [8–12]. It can be shown [12] that in the presence of damping the equation of motion of the radial coordinate x of a charged particle stored in a dynamic Kingdon trap can be cast into the form of a nonlinear damped Mathieu equation

$$\ddot{x} + \gamma \dot{x} + [1 - 2\eta \cos(2t)] x^\alpha = 0, \tag{5}$$

where γ is the damping constant, $\eta \sim V_{ac}/V_{dc}$ is the control parameter and α is the nonlinearity exponent. For the cylindrical trap $\alpha = -1$; for the radially symmetric trap $\alpha = -2$. Both types of traps exhibit a mixed phase space with regular and chaotic regions. In the following we focus on the cylindrical trap. For $\gamma = 0$ (the Hamiltonian case) a phase-space portrait of the cylindrical trap is shown in Fig. 4 (a) for $\eta = 4$. The fixed point at $(x, \dot{x}) \approx (2.2, 0)$

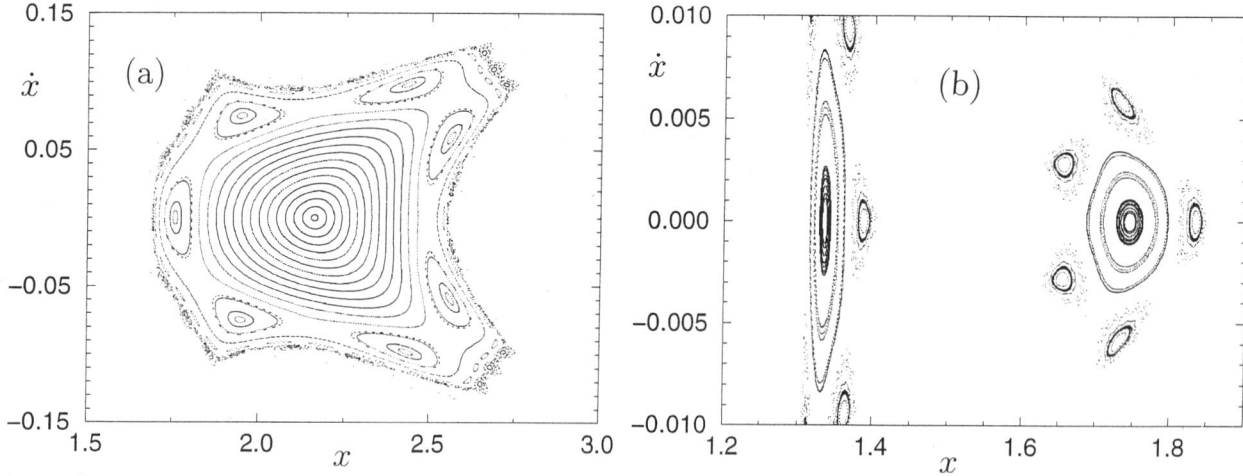

FIGURE 4. Phase-space portrait of the cylindrical dynamic Kingdon trap in the vicinity of the primary fixed point. (a) $\eta = 4$, (b) $\eta = 3.05$.

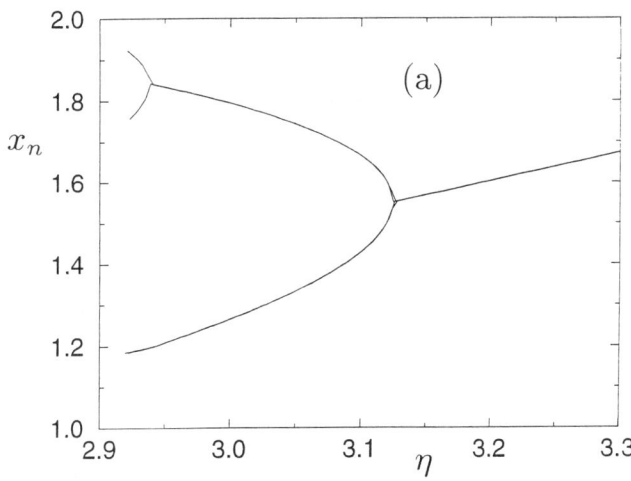

FIGURE 5. Bifurcation diagram of the cylindrical dynamic Kingdon trap.

corresponds to the main trapping region of the dynamic Kingdon trap. It is surrounded by island chains and chaotic regions. Decreasing the value of the control parameter η, the main fixed point in Fig. 4 (a) exhibits a series of bifurcations at well defined values η_n, $n = 1, 2, ...$ of the control parameter η. The first few η_n values are given by $\eta_1 \approx 3.125$, $\eta_2 \approx 2.938$, $\eta_3 \approx 2.917$ and $\eta_4 \approx 2.915$. At $\eta = \eta_1$ the central island in Fig. 4 (a) splits into two. The resultig situation immediately after the bifurcation is shown in Fig. 4 (b). Each of the two resulting islands splits again at $\eta = \eta_2$.

The bifurcation scenario of the main fixed point of the dynamic Kingdon trap is shown in Fig. 5. It reminds of the bifurcation diagram of the logistic mapping [10]. The bifurcation diagram shows the position of the stable islands of the dynamic Kingdon trap as a function of η. With the help of a Fourier analysis an approximate analytical value for the position η_1 of the first bifurcation was recently computed [12]. It is given by $\eta_1 \approx (15 + \sqrt{485})/12 \approx 3.085$ and is in good agreement with the numerical value.

III. DIAGNOSTIC METHOD FOR BEAM CRYSTALLIZATION

Crystallization of a large number of ions in a circular model storage ring was recently demonstrated [13]. Because of their near-relativistic speeds, however, crystallization of heavy ions in large-scale storage rings is much harder to achieve. Several laboratories are currently working on this problem [14]. But even if a crystallized beam is actually present in a storage ring, there remains the question of how to prove it. For the relatively slow ions in traps and model storage rings this problem has been solved more than a decade ago using direct optical imaging [15,16]. In the case of fast beams, however, this method does not work and one has to resort to more indirect diagnostic methods. One of these methods is to pick up the Schottky noise of the beam. But this method, as implemented in current storage rings, is mainly sensitive to the longitudinal characteristics of the ion beam. Indeed, longitudinal ordering has already been inferred this way [17]. Another indirect method was proposed recently [18,19]: Absence of heating of a crystalline beam. This method is based on the mixed nature of phase space with its regular and chaotic regions (see, e.g., Fig. 4). The crystalline phase of a beam in a storage ring corresponds to a fixed point of phase space. In the case of a simple geometric structure of the beam, such as e.g. a linear chain of ions, this fixed point corresponds to a stable island. It is surrounded by a chaotic sea. If an ion beam in a storage ring is just cold, but not yet crystallized, its phase-space trajectory explores the chaotic sea. Without a cooling mechanism the beam's thermal energy increases due to chaotic diffusion. But if the beam is cold enough, it enters the stability region of the fixed point and no heating occurs even in the absence of a cooling mechanism. Thus, in order to prove that the beam in a storage ring is crystallized, the following three steps may be followed:

1) The cooling devices are switched on. The beam reaches a final cold state whose nature (crystallized or not) is to be determined.
2) All the cooling devices are switched off.
3) Evaluation: (i) If the beam heats up immediately following the shut-down of the cooling devices, the beam

was cold, but not crystallized. (ii) If the beam's temperature does not change following the shut-down of the cooling devices, the beam was crystallized.

The validity of this criterion was established with the help of detailed molecular dynamics simulations [19].

IV. SYNCHROTRON RADIATION OF CRYSTALLIZED BEAMS

All the early atomic models based on classical mechanics suffered from a serious shortcoming: the circulating charges, undergoing accelerated motion, continuously radiated energy eventually resulting in the collapse of the atom. In this respect Thomson's raisin cake model [20] was no exception. But Thomson hit upon a brilliant idea to save his model: suppression of electromagnetic radiation due to Coulomb crystallization [21]. For the six electrons of the carbon atom, e.g., arranged equi-spaced on a ring, he computed a suppression factor of $\approx 10^{-17}$ [21]. Thus Thomson may have been the first to dicuss suppression of radiation by geometrically ordered charges.

Nowadays geometrically ordered charges are of topical interest again in connection with crystallized beams. Here Thomson's idea of suppression of electromagnetic radiation of ordered accelerated charges leads to the suppression of synchrotron radiation emitted by a crystallized beam [22]. This effect may be used in two different ways: (i) for the development of new electron accelerator schemes and (ii) as a further indirect diagnostic criterion for beam crystallization. Concerning (i), the idea is to cool the electron beam in a cyclic electron accelerator sufficiently such that the energy loss from synchrotron radiation is no longer a serious limitation of the maximal reachable energy of the accelerator. Concerning (ii), a sharp drop in the power of the emitted synchrotron radiation indicates beam crystallization.

ACKNOWLEDGMENTS

The author gratefully acknowledges financial support by the Deutsche Forschungsgemeinschaft (SFB 276).

REFERENCES

1. Ghosh, P. K., *Ion Traps*, Oxford: Clarendon Press, 1995.
2. Razvi, M. A. N., Chu, X. Z., Alheit, R., Werth, G., and Blümel, R., Phys. Rev. A **58**, R34–R37 (1998).
3. Dehmelt, H., Adv. At. Mol. Phys. **3**, 53–72 (1967).
4. Landau, L. D., and Lifshitz, E. M., *Mechanics*, Oxford: Pergamon Press, 1960.
5. Bahr, R. E., *Untersuchung der Eigenschaften eines Hochrequenz-Ionen-Speichers*, Freiburg: Diplom thesis, 1969.
6. Behre, E., *Bau und Erprobung einer massenselektiven Speicherionenquelle*, Freiburg: Zulassungsarbeit, 1972.
7. Peik, E., and Fletcher, J., J. Appl. Phys. **82**, 5283–5286 (1997).
8. Blümel, R., Appl. Phys. B **60**, 119–122 (1995).
9. Blümel, R., Phys. Rev. A **51**, R30–R33 (1995).
10. Blümel, R., Physica Scripta T **59**, 126–130 (1995).
11. Blümel, R., Physica Scripta T **59**, 369–379 (1995).
12. Blümel, R., Bonneville, E., and Carmichael, A., Phys. Rev. E **57**, 1511–1518 (1998).
13. Birkl, G., Kassner, S., and Walther, H., Europhys. News **23**, 143–145 (1992).
14. Habs, D., and Grimm, R., Ann. Rev. Nucl. Part. Sci. **45**, 391–428 (1995).
15. Diedrich, F., Peik, E., Chen, J. M., Quint, W., and Walther, H., Phys. Rev. Lett. **59**, 2931–2934 (1987).
16. Wineland, D. J., Bergquist, J. C., Itano, W. M., Bollinger, J. J., and Manney, C. H., Phys. Rev. Lett. **59**, 2935–2938 (1987).
17. Steck, M., Beckert, K., Eickhoff, H., Franzke, B., Nolden, F., Reich, H., Schlitt, B., and Winkler, T., Phys. Rev. Lett. **77**, 3803–3806 (1996).
18. Blümel, R., Phys. Rev. A **51**, 620–624 (1995).
19. Primack, H., and Blümel, R., Phys. Rev. E, in press.
20. Thomson, J. J., *Die Korpuskulartheorie der Materie*, Braunschweig: Vieweg, 1908.
21. Thomson, J. J., *Elektrizität und Materie*, Braunschweig: Vieweg, 1909.
22. Primack, H., and Blümel, R., "Suppression of synchrotron radiation due to beam crystallization", Eur. Phys. J. A, in press.

Crystalline Order in Strongly Coupled Plasmas*

J. J. Bollinger, T. B. Mitchell, X.-P. Huang, W. M. Itano, J. N. Tan,[†] B. M. Jelenković,[‡] and D. J. Wineland

Time and Frequency Division, National Institute of Standards and Technology, Boulder, CO 80303

Abstract. Laser-cooled trapped ions can be strongly coupled and form crystalline states. We describe experimental studies which measure the spatial correlations of Be$^+$ ion crystals formed in Penning traps. Both Bragg scattering of the cooling-laser light and spatial imaging of the laser-induced ion fluorescence are used to measure these correlations. In spherical plasmas with more than 2×10^5 ions, body-centered-cubic (bcc) crystals, the predicted bulk structure, are the only type of crystals observed. We are able to phase-lock the orientation of the ion crystals to a rotating electric-field perturbation. With this "rotating wall" technique and stroboscopic detection, images of individual ions in a Penning trap are obtained. The rotating wall technique also provides a precise control of the time-dilation shift due to the plasma rotation, which is important for Penning trap frequency standards.

INTRODUCTION

This manuscript summarizes recent progress on the study of strongly coupled ion OCP's in Penning traps. Trapped ions are a good example of a one-component plasma (OCP). An OCP consists of a single charged species immersed in a neutralizing background [1]. In an ion trap, the trapping fields provide the neutralizing background [2]. Examples of OCPs include such diverse systems as the outer crust of neutron stars [3] and electrons on the surface of liquid helium [4]. The thermodynamic properties of the classical OCP of infinite spatial extent are determined by its Coulomb coupling constant [1]

$$\Gamma \equiv \frac{1}{4\pi\epsilon_o} \frac{e^2}{a_{WS} k_B T}, \qquad (1)$$

which is a measure of the ratio of the Coulomb potential energy of nearest neighbor ions to the kinetic energy per ion. Here, ϵ_o is the permittivity of the vacuum, e is the charge of an ion, k_B is Boltzmann's constant, T is the temperature, and a_{WS} is the Wigner-Seitz radius, defined by $4\pi(a_{WS})^3/3 = 1/n_o$ where n_o is the ion density. For low temperature ions in a trap, n_o equals the equivalent neutralizing background density provided by the trapping fields. Plasmas with $\Gamma > 1$ are called strongly coupled. The onset of fluid-like behavior is predicted at $\Gamma \approx 2$ [1], and a phase transition to a body-centered cubic (bcc) lattice is predicted at $\Gamma \approx 170$ [1,5]. From a theoretical perspective, the strongly coupled OCP has been used as a paradigm for condensed matter for decades. However, only recently has it been realized in the laboratory [6].

Experimentally, freezing of small numbers ($N < 50$) of laser-cooled atomic ions into Coulomb clusters was first observed in Paul traps [7-9]. With larger numbers of trapped ions, concentric shell structures were observed directly in Penning [10] and linear Paul [11,12] traps. The linear Paul traps provided strong confinement in the two dimensions perpendicular to the trap axis and very weak confinement along the trap axis. This resulted in cylindrically shaped plasmas whose axial lengths are large compared to their cylindrical diameters. Cylindrical-shell crystals which are periodic with distance along the trap axis were observed. The diameter of these crystals was limited to $\sim 10\ a_{WS}$ in Ref. [11] and $\sim 30\ a_{WS}$ in Ref. [12], presumably due to rf heating [13] which is produced by the time-dependent trapping fields and increases with the plasma diameter. These

---

*) Work of the U.S. Government. Not subject to U.S. copyright
†) Present address: Physics Dept., Harvard Univ., Cambridge, MA 02138
‡) On leave from the Institute of Physics, University of Belgrade, Belgrade, Yugoslavia

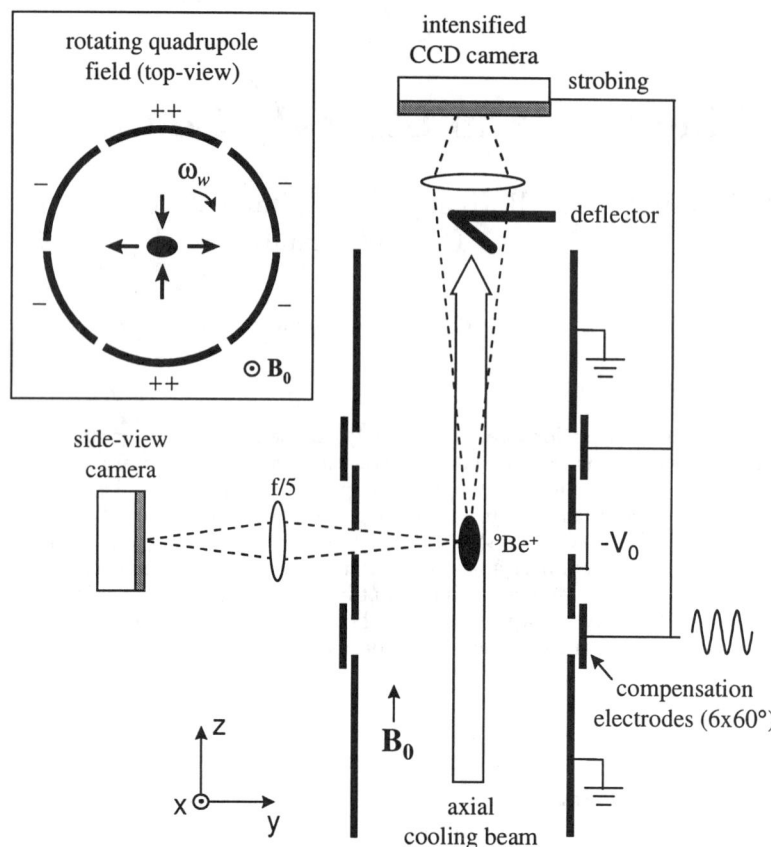

FIGURE 1. Schematic view of the cylindrical trap with real space imaging optics for the side-view camera and Bragg diffraction detection system for the axial cooling beam. The size of the plasma is exaggerated. Cross section of the rotating quadrupole field (in the x-y plane) is shown in the insert. From Ref. [19].

plasma diameters appear to be too small to observe the 3-D periodic crystals predicted for the infinite, strongly coupled OCP. Strong coupling and crystallization have also been observed with particles interacting through a screened Coulomb potential. Examples include dusty plasma crystals [14] and colloidal suspensions [15,16].

Because Penning traps use static fields to confine charged particles, there is no rf heating. This has enabled ion plasmas which are large in all three dimensions to be laser-cooled. For example, we have laser-cooled $\sim 10^6$ Be$^+$ ions in an approximately spherical plasma with diameter $\sim 200 a_{WS}$. With these large ion plasmas we have used Bragg scattering of the cooling laser light to detect the formation of bcc crystals [17,18], the predicted state for a bulk OCP with $\Gamma > 170$. In addition, we have studied the spatial correlations in planar, lens-shaped plasmas with axial thickness $\lesssim 10 a_{WS}$. These plasmas consist of extended, two dimensionally periodic lattice planes. The importance of the plasma boundary in this case results in different crystalline structures depending on the details of the plasma shape.

A potential drawback of the Penning trap versus the rf trap is that the ions rotate about the trap magnetic field, and this has previously prevented the imaging of the ion crystals as done in Paul traps. This is because the rotation, created by the $\mathbf{E} \times \mathbf{B}$ drift due to the radial electric and the trap magnetic fields is, in general, not stable. For example, fluctuations in the plasma density or shape produce fluctuations in the ion space charge fields which change the plasma rotation. However, we are able to phase-lock the rotation of the laser-cooled ion crystals to a rotating electric field perturbation [19,20]. The success of this "rotating wall" technique enables us to strobe the cameras recording the ion fluorescence synchronously with the plasma rotation and obtain images of individual ions in the plasma crystals [21].

Figure 1 is a schematic of the cylindrical Penning trap we use to confine ^{9}Be$^+$ ions. The trap consists of a 127 mm long vertical stack of cylindrical electrodes with an inner diameter of 40.6 mm, enclosed in a room temperature, 10^{-8} Pa vacuum chamber. The uniform magnetic field $\mathbf{B_o} = 4.46$ T is aligned parallel to the trap axis within 0.01° and produces a ^{9}Be$^+$ cyclotron frequency $\Omega = 2\pi \times 7.61$ MHz. A quadratic, axially symmetric potential $(m\omega_z^2/2e)[z^2 - r^2/2]$ is generated near the trap center by biasing the central electrodes to a negative voltage $-V_o$. At $V_o = 1$ kV, the single-particle axial frequency $\omega_z = 2\pi \times 799$ kHz and the magnetron $\mathbf{E} \times \mathbf{B}$ drift frequency $\omega_m = 2\pi \times 42.2$ kHz. The trapped Be$^+$ ions are Doppler laser-cooled by two 313 nm laser beams. The principal cooling beam (waist diameter ~ 0.5 mm, power ~ 50 μW) is directed parallel to $\mathbf{B_o}$. A second, typically weaker cooling beam with a much smaller waist (~ 0.08 mm) is directed perpendicularly to $\mathbf{B_o}$ (not shown in Fig. 1). This beam can also be used to vary the plasma rotation frequency by applying a torque with radiation pressure. With this configuration, ion temperatures close to the 0.5 mK Doppler laser-cooling limit are presumably achieved. However, experimentally we have only placed a rough 10 mK upper bound on the ion temperature [22]. For a typical value of $n_o = 4 \times 10^8$ cm^{-3}, this implies $\Gamma > 200$.

Two types of imaging detectors were used. One is a charge-coupled device (CCD) camera coupled to an electronically gateable image intensifier. The other is an imaging photomultiplier tube based on a microchannel-plate electron multiplier and a multielectrode resistive anode for position sensing. For each detected photon, the position coordinates are derived from the current pulses collected by the different electrodes attached to the resistive anode. This camera therefore provides the position and time of each detected photon. However, in order to avoid saturation, we placed up to 20 dB of attenuation in front of this camera to lower the detected photon counting rate to less than ~ 300 kHz.

In thermal equilibrium, the trapped ion plasma rotates without shear at a frequency ω_r where $\omega_m < \omega_r < \Omega - \omega_m$ [23,24]. For the low temperature work described here, the ion density is constant and given by $n_o = 2\epsilon_o m\omega_r(\Omega - \omega_r)/e^2$. With a quadratic trapping potential the plasma has the simple shape of a spheroid, $z^2/z_o^2 + r^2/r_o^2 = 1$, where the aspect ratio $\alpha \equiv z_o/r_o$ depends on ω_r [22,24]. This is because the radial binding force of the trap is determined by the Lorentz force due to the plasma's rotation through the magnetic field. Thus low ω_r results in a lenticular plasma (an oblate spheroid) with large radius. As ω_r increases, r_o shrinks and z_o grows, resulting in an increasing α. However, large ω_r ($\omega_r > \Omega/2$) produces a large centrifugal acceleration which opposes the Lorentz force and lenticular plasmas are once again obtained for $\omega_r \sim \Omega - \omega_m$. In our work, torques from a laser or a rotating electric field are used to control ω_r and therefore the plasma density and shape. The plasma shape is observed by imaging the ion fluorescence scattered perpendicularly to $\mathbf{B_o}$ with an f/5 objective. (See Fig. 1.) All possible values of ω_r from ω_m to $\Omega - \omega_m$ have been accessed using both methods of applying a torque [20,25,26]. Azimuthally segmented compensation electrodes located between the main trap electrodes are used to apply the rotating electric-field perturbation. Both a rotating quadrupole (see inset in Fig. 1) and rotating dipole field (not shown in Fig. 1) have been used to control ω_r. Below we explain how the rotating quadrupole field provides precise control of ω_r.

BRAGG SCATTERING

BCC Crystals

An infinite OCP with $\Gamma \gtrsim 170$ is predicted to form a bcc lattice. However, the bulk energies per ion of the face-centered-cubic (fcc) and hexagonal-close-packed (hcp) lattices differ very little from bcc ($< 10^{-4}$) [27]. Because some of the fcc and hcp planes have lower surface energies than any of the bcc planes, a boundary can have a strong effect on the preferred lattice structure. One calculation [27] estimates that the plasma may need to be $\gtrsim 100 a_{WS}$ across its smallest dimension to exhibit bulk behavior. For a spherical plasma this corresponds to $\sim 10^5$ ions.

We used Bragg scattering to measure the spatial correlations of approximately spherical plasmas with $N > 2 \times 10^5$ trapped Be$^+$ ions [17,18]. The cooling-laser beam directed along the trap axis was used for Bragg scattering as indicated in Fig. 1. First the plasma shape was set to be approximately spherical. (In early experiments this was done with the perpendicular laser beam; more recent experiments used the rotating wall.) The parallel laser beam was then tuned approximately half a linewidth below resonance, and a Bragg scattering pattern recorded ($\sim$1–30 s integration). The plasma was then heated and recooled, and another Bragg scattering pattern was recorded. Because the 313 nm wavelength of the cooling laser is small compared to the inter-ion separation ($\sim$10–20 μm), Bragg scattering occurs in the forward (few degree) scattering direction. In order for a diffracted beam to form, the incident and scattered wave vectors $\mathbf{k}_i$ and $\mathbf{k}_s$ must differ by a reciprocal lattice vector (Laue condition) [28]. In a typical x-ray crystal diffraction case, satisfying the Laue condition for many reciprocal lattice vectors requires that the incident radiation have a continuous range of wavelengths. Here the

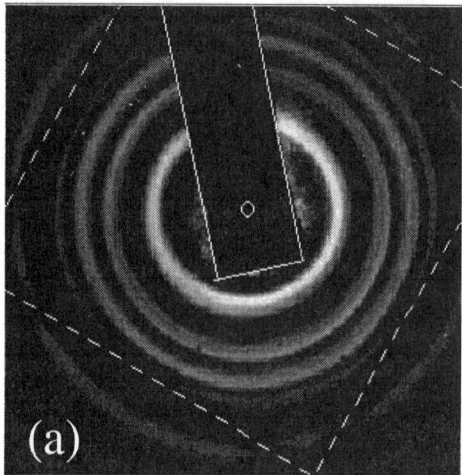 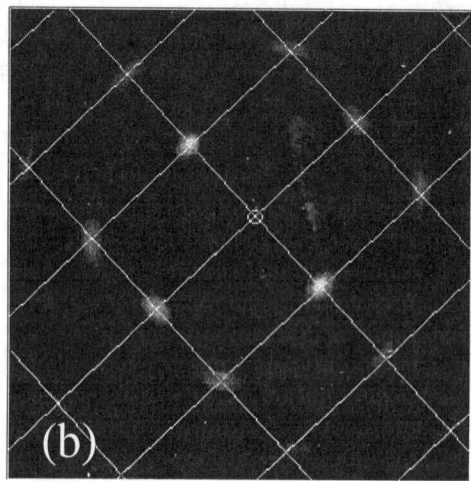

FIGURE 2. Bragg diffraction patterns from a plasma phase locked to a rotating quadrupole field ($\omega_r = 2\pi \times 140$ kHz, $n_o \approx 4.26 \times 10^8$ cm^{-3}, $\alpha \approx 1.1$). (a) 1 s time-averaged pattern. The long rectangular shadow (highlighted by solid lines) is from the deflector for the incident beam; four line shadows (highlighted by dashed lines) that form a square are due to a wire mesh at the exit window of the vacuum chamber. The small open circle near the center of the figure marks the position of the undeflected laser beam. (b) Time-resolved pattern obtained nearly simultaneously with (a) by strobing the camera with the rotating field (integration time ≈ 5 s). A spot is predicted at each intersection of the rectangular grid lines for a bcc crystal with a [110] axis aligned with the laser beam. The grid spacings were determined from the n_o calculated from ω_r and are not fitted. From Ref. [20].

Laue condition is relaxed because of the small size of the crystal, so a crystalline Bragg diffraction pattern is frequently obtained even with monochromatic radiation.

Figure 2(a) shows a time-averaged diffraction pattern obtained on a spherical plasma with $N \sim 7.5 \times 10^5$. The multiple concentric rings are due to Bragg scattering off different planes of a crystal. A concentric ring rather than a dot pattern is observed because the crystal was rotating about the laser beam. In general, many different patterns were observed, corresponding to Bragg scattering off crystals with different orientations. Figure 3 summarizes the analysis of approximately 30 time-averaged patterns obtained on two different spherical plasmas with $N > 2 \times 10^5$. It shows the number of Bragg peaks as a function of the momentum transfer $q = |\mathbf{k}_s - \mathbf{k}_i| = 2k\sin(\theta_{scatt}/2)$ ($\simeq k\theta_{scatt}$ for $\theta_{scatt} \ll 1$), where $k = 2\pi/\lambda$ is the laser wave number and θ_{scatt} is the scattering angle. The density dependence of the Bragg peak positions is removed by multiplying q by a_{WS}, which was determined from ω_r. The positions of the peaks agree with those calculated for a bcc lattice, within the 2.5% uncertainty of the angular calibration. They disagree by about 10% with the values calculated for an fcc lattice. The ratios of the peak positions of the first five peaks agree within about 1% with the calculated ratios for a bcc lattice. This provides strong evidence for the formation of bcc crystals in spherical plasmas with $N > 2 \times 10^5$ ions. This result is significant because it is the first evidence for bulk behavior in a strongly coupled OCP in the laboratory.

Rotating Wall

By strobing the camera recording the Bragg scattering pattern synchronously with the plasma rotation, we should be able to recover a dot pattern from the time-averaged concentric ring pattern in Fig. 2(a). Initially we used the time dependence of the Bragg scattered light to sense the phase of the plasma rotation [18,29]. More recently we used a rotating electric field perturbation to phase-lock the ion plasma rotation [19,20].

Consider the rotating quadrupolar perturbation shown in the inset of Fig. 1. This z-independent perturbation produces a small distortion in the shape of the spheroidal plasma. In particular, the plasma acquires a small elliptical cross section normal to the z-axis. (In our work the distortion created by the rotating quadrupole field was typically less than 1% of the plasma diameter.) The elliptical boundary rotates at the applied rotating wall frequency ω_w. An ion near the plasma boundary experiences a torque due to this rotating bound-

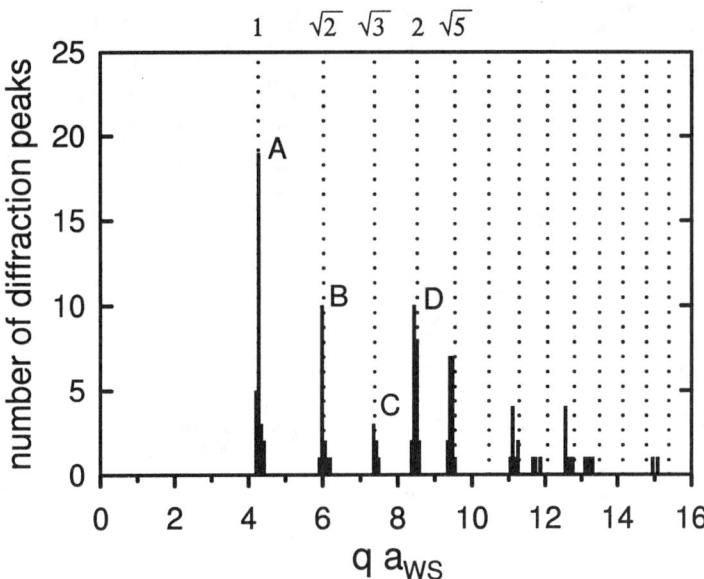

FIGURE 3. Histogram showing the numbers of peaks (not intensities) observed as a function of $q \cdot a_{WS}$ (defined in the text) for 30 time-averaged Bragg scattering patterns obtained on two different spherical plasmas with $N > 2 \times 10^5$. The dotted lines show the expected peak positions for a bcc crystal, normalized to the center of gravity of the peak at A (corresponding to Bragg reflections off {110} planes). From Ref. [18].

ary. If the ion is rotating slower than ω_w, the torque will speed it up. If it is rotating faster than ω_w, the torque will slow it down. Through viscous effects, this torque is transmitted to the plasma interior. Therefore, if other external torques are small, the rotating wall perturbation will make ω_r equal ω_w. Crystallized plasmas behave more like a solid than a liquid or gas. Because the viscosity is high, the whole plasma will tend to rotate rigidly with its boundary. In particular, the orientation of the ion crystals can phase-lock to the rotating quadrupolar perturbation if the frequency difference between ω_r and ω_w is small.

To check for phase-locked control of ω_r, we strobed the camera recording the Bragg scattering pattern in Fig. 2(a) with the synthesizer used to generate the rotating wall signal. Specifically, once each $2\pi/\omega_w$ period, the rotating wall signal gated the camera on for a period $\lesssim 0.02(2\pi/\omega_w)$. The resulting Laue dot pattern in Fig. 2(b) shows that the plasma rotation was phase-locked to the rotating electric-field perturbation. The dot pattern provides detailed information on the number and orientation of the crystals which contributed to the Bragg scattering signal. For example, the pattern in Fig. 2(b) was due to a single bcc crystal with a [110] axis aligned along the laser beam. For phase-locked operation of the rotating wall, other external torques must be small. For example, a misalignment of the trap magnetic field with the trap electrode symmetry axis of $> 0.01°$ prevented phase-locked control of the plasma rotation. In our work, alignment to $\lesssim 0.003°$ was obtained by minimizing the excitation of zero-frequency plasma modes [25,26].

In addition to the rotating quadrupole perturbation, phase-locked control was also achieved with a uniform rotating electric field (a "dipole" field). In fact under many circumstances a uniform oscillating field worked equally well. In these cases the co-rotating component of the oscillating field controlled the plasma rotation while the perturbing effects due to the counter-rotating component were minimal. The simplicity of the oscillating dipole field makes it a convenient tool for controlling ω_r. However, in a quadratic trap, control of ω_r with a uniform rotating or oscillating electric field requires an effect which breaks the separation of center-of-mass and internal degrees of freedom of the plasma. In our work this is done by impurity ions which experience a different centrifugal potential than the $^9Be^+$ ions [20].

REAL-SPACE IMAGES

Bragg scattering measures the Fourier transform of the spatial correlations of the trapped ions. It provides a picture of these correlations in reciprocal-lattice space. With phase-locked control of ω_r, real-space imaging

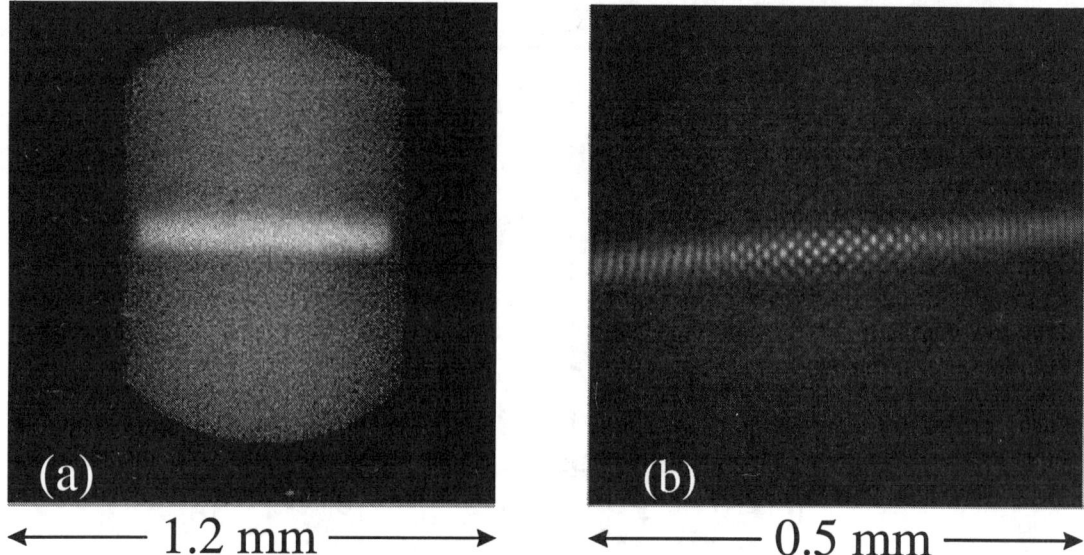

FIGURE 4. Real-space images of an $N \sim 1.8 \times 10^5$ ion plasma phase-locked with an oscillating dipole field at $\omega_r = 2\pi \times 120$ kHz. (a) Time-averaged side-view image showing the overall plasma shape. The bright line of fluorescence through the plasma center is due to a laser beam directed perpendicularly to $\mathbf{B_o}$. The plasma shape is approximately spherical. The presence of heavier-mass ions, which centrifugally separate from the ^{9}Be$^+$ ions, produces the straight vertical boundaries in the image. (b) Strobed top-view image, obtained simultaneously with (a), showing the presence of a bcc crystal in the plasma center. The distance scales in (a) and (b) are different as noted.

of individual ions in a Penning trap becomes possible. To obtain real-space images with high resolution, we replaced the Bragg scattering optics (see Fig. 1) with imaging optics, starting with an f/2 objective, which formed a real, top-view image of the ion plasma. The combined resolution limit of the optics and camera was less than 5 μm near the optimal object plane of the f/2 objective. This is less than the $\sim$10 μm resolution limit required to resolve individual ions. However, the depth of field of an f/2 objective for 10 μm resolution is $\sim$80 μm. For lenticular plasmas with $2z_o \lesssim 80$ μm, all of the ions within the plasma were resolvable. For plasmas with $2z_o > 80$ μm, the cooling-laser beam directed perpendicularly to $\mathbf{B_o}$ was used to illuminate a section of the plasma within the depth of field.

Figure 4 shows side-view and top-view images of an approximately spherical plasma with $N \sim 1.8 \times 10^5$. The fluorescence from the perpendicular laser beam used to highlight a small region of the plasma is clearly visible. In the top-view image a square grid of dots is observed near the plasma center. The measured spacing between nearest neighbor dots is 12.8 ± 0.3 μm, in good agreement with the 12.5 μm spacing expected for viewing along a [100] axis of a bcc crystal with density determined by the ω_r set by the rotating field. Real-space imaging provides direct information on the location and size of the crystals. In Fig. 4 the crystal was located in the radial center of the plasma and was at least 230 μm across, or at least 1/4 of the plasma diameter.

For lenticular plasmas with $2z_o \lesssim 80$ μm, all of the ions within the plasma are resolved without the use of the perpendicular laser beam. Lenticular plasmas are obtained with ω_r slightly greater than ω_m. For small plasmas ($N \lesssim 2000$ ions) we were able to use the rotating-dipole electric field to lower ω_r and obtain a single plane while maintaining long-range order in the top-view images. Figure 5(a) shows a top- and side-view image of such a plasma. Near the plasma center a 2-D hexagonal lattice is observed, the preferred lattice for a 2-D system. Here each dot is the image of an individual ion.

Starting with a single plane like that shown in Fig. 5(a), we studied the structural phase transitions that occur as ω_r is increased [21]. With increasing ω_r, the radial confining force of the Penning trap increases, which decreases r_o. At a particular point, there is a structural phase transition near the plasma center from a single, hexagonal lattice plane to two lattice planes where the ions form a square grid in each plane, as shown in Fig. 5(b). Further increases in ω_r increase the number of ions per unit area of each plane as well as the spacing between the planes. During this process the square lattice planes smoothly change into rhombic lattice planes and eventually there is a sudden transition to hexagonal lattice planes. Further increases in ω_r eventually

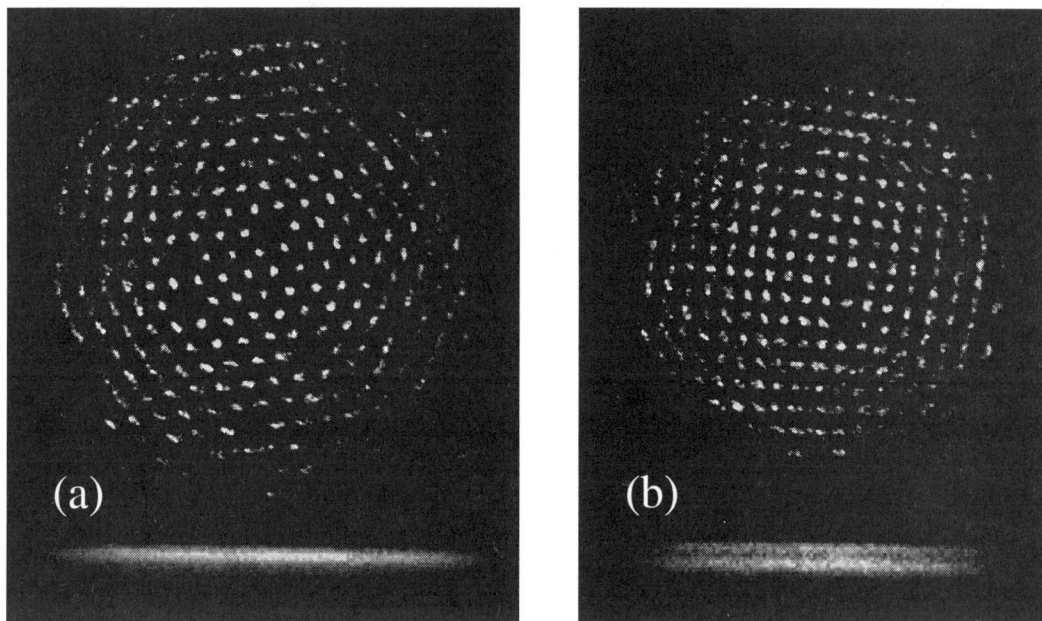

FIGURE 5. Strobed top-view images of a small ($N \sim 300$ Be$^+$) ion plasma phase-locked with a rotating dipole field at (a) $\omega_r = 2\pi \times 65.7$ kHz and (b) 66.5 kHz. Below are unstrobed side-views showing the axial lattice planes. Heavier-mass ions are located outside the ^{9}Be$^+$ ions.

produce a structural transition to three square lattice planes, and the basic pattern repeats.

The structure of the crystallized ions depends sensitively on the projected areal density σ of the plasma. The side- and top-view images were analyzed to characterize the phase structure. Within a layer, the structural order is characterized by the primitive vectors $\mathbf{a_1}$ and $\mathbf{a_2}$ (which are observed to be equal in magnitude) and the angle θ ($\leq 90°$) between them. The interlayer order is characterized by the axial positions z_n of the n lattice planes (measured by the side-view camera) and the interlayer displacement vector $\mathbf{c_n}$ between layers 1 and n. Hence, the equilibrium positions in the (x,y) plane of ions in axial planes 1 and n are given by $\mathbf{R_1} = i\mathbf{a_1} + j\mathbf{a_2}$ and $\mathbf{R_n} = i\mathbf{a_1} + j\mathbf{a_2} + \mathbf{c_n}$, where i,j are integers. Three different types of intralayer ordering are observed: hexagonal ($\theta = 60°$), square ($\theta = 90°$) and rhombic ($90° > \theta \geq 65°$). The observations were compared to the results from Dubin [21], who performed an analytic calculation of the energies of lattice planes which are infinite and homogeneous in the (x,y) direction but are confined in the axial direction by a harmonic external electrostatic confinement potential $\phi_e = 1/2(m/e)\omega_z^2 z^2$. Since this potential is identical to the confinement potential of a Penning trap as seen in the rotating frame in the $\alpha \to 0$ planar limit, the minimum energy phase structures predicted by the theory should match the structures observed in the central regions of the oblate plasmas of the experiments.

Figure 6 displays the agreement between theory and experiment for the interlayer quantities, with measurements taken on different plasmas with $N < 10^4$. Lengths have been normalized by $a_{ws2D} = (3e^2/4\pi\epsilon_0 m\omega_z^2)^{1/3} = 10.7$ μm, which is the Wigner-Seitz radius in the planar limit. As the central areal density is increased the lattice planes move further apart axially in order to match their average density to the neutralizing background. Eventually it becomes energetically favorable to form an additional lattice plane. The symbols indicate whether the lattices had an interlattice displacement vector $\mathbf{c_2}$ characteristic of the hexagonal phases (triangles) or the square and rhombic phases (squares).

Figure 7 displays the agreement between experiment and theory for the dependence of the angle θ (between the primitive vectors) on central areal charge density σ. The trend is that when a new lattice plane is formed, θ changes discontinuously from $\approx 60°$ to a higher value. As the central areal density of the crystal is further increased, θ smoothly decreases to $\approx 65°$ until there is a second discontinuous transition to a hexagonal structure. This latter transition has been predicted [30] to become continuous in liquid ($\Gamma < 80$) bilayer systems. The lines indicate the minimum energy structures predicted by the 2D theory.

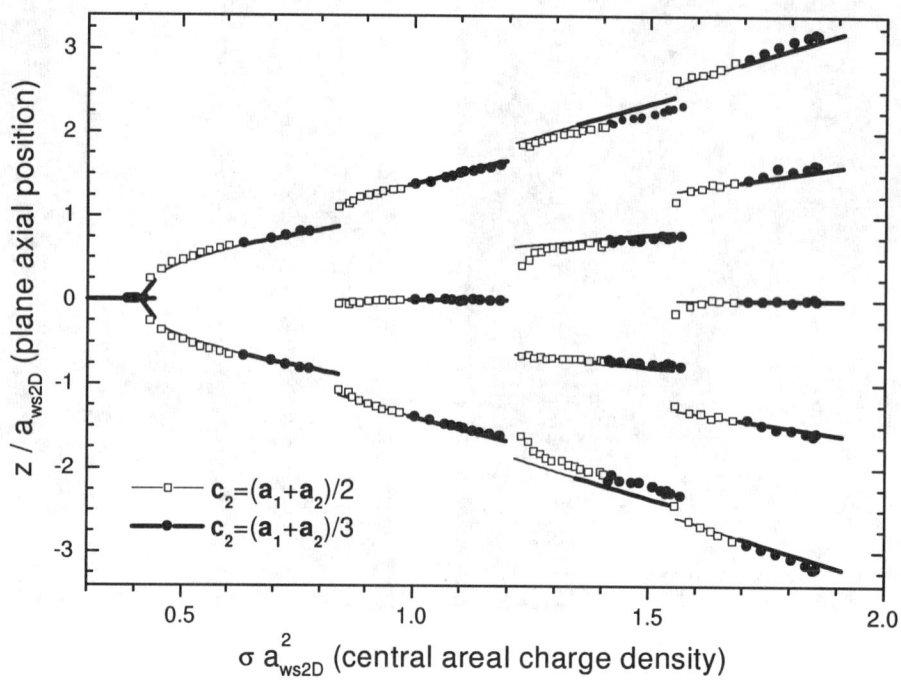

FIGURE 6. Interlayer structure (plane axial positions and displacement vectors) as a function of normalized areal charge density. The lines are the predictions of theory, and the symbols are experimental measurements.

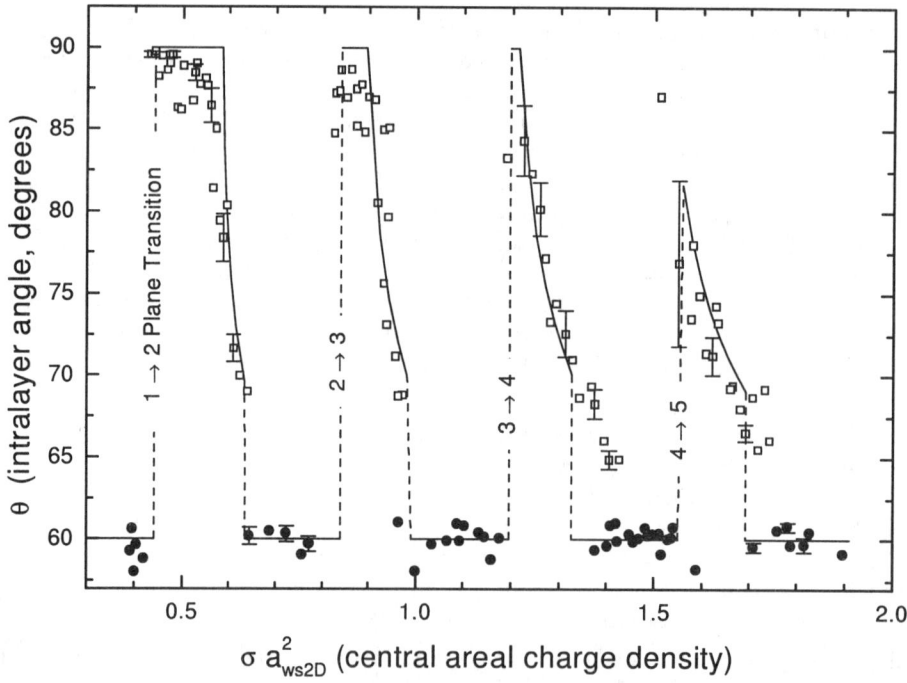

FIGURE 7. Intralayer angle θ structure as a function of normalized areal charge density. The lines are the predictions of theory, and the symbols are experimental measurements. Representative error bars are included with some of the measurements.

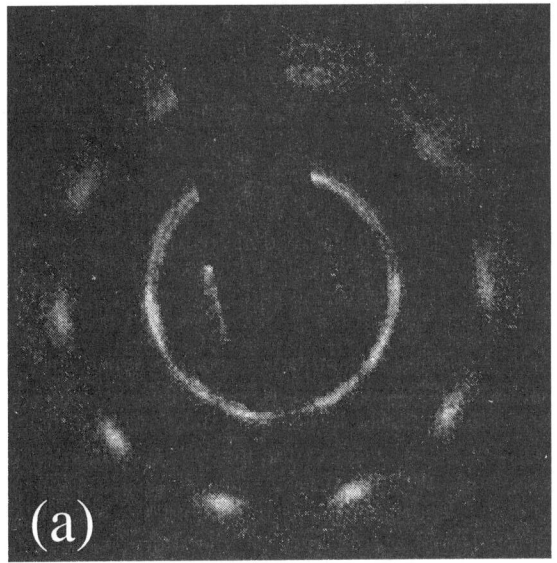 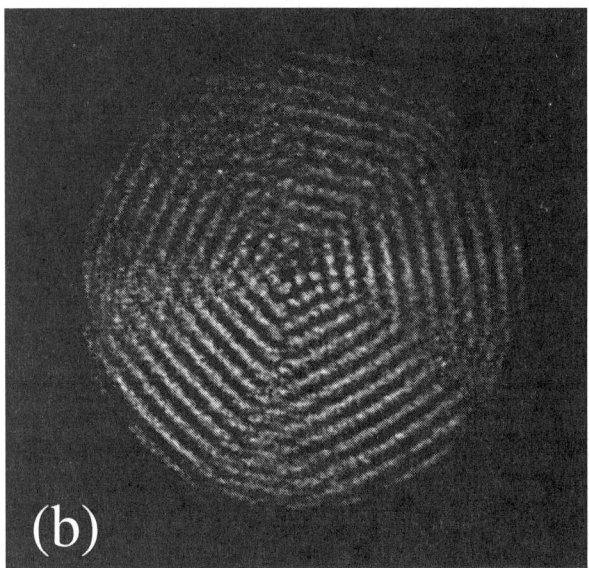

FIGURE 8. Fivefold Bragg scattering and real-space patterns obtained by strobing the intensified CCD camera synchronously with the rotating electric field perturbation. (a) Bragg scattering pattern obtained on an $N \sim 1.2 \times 10^5$ ion plasma phase-locked with a rotating dipole field at $\omega_r = 2\pi \times 166.84$ kHz. Here $V_o = 500$ V and $\alpha = 2.6$. (b) Real-space image of a lenticular plasma consisting of 4 horizontal planes in the plasma center. The rotating dipole field was used to set $\omega_r = 2\pi \times 74.35$ kHz.

DISCUSSION

With Bragg scattering and spatial imaging, we have measured the correlations in both highly oblate and spherical strongly coupled ^{9}Be$^+$ ion plasmas. The planar geometry permits a detailed comparison with theoretical calculations. We have measured the preferred lattice structures for up to five lattice planes in lenticular plasmas and obtain good agreement with theory. By increasing the number of planes (by adding more ions to the plasma), the transition from surface-dominated to bulk behavior in the planar geometry can be studied. Ions in a trap have been proposed as a register for a quantum computer [31]. Work in this area has focussed on a string of a few ions in a linear Paul trap [32]. A single lattice plane of ions as in Fig. 5 could provide a 2-D geometry of trapped ions for studies of quantum computing or entangled quantum states.

In spherical plasmas with more than 2×10^5 ions, we have observed the formation of bcc crystals, the predicted state for the infinite strongly coupled OCP. The crystals occupied the inner quarter of the plasma diameter. Outside the crystal there was a complicated transition to shell structure. In this system we have not observed the thermodynamic liquid-solid phase transition predicted for the bulk OCP. Our measurements have concentrated on the correlations obtained at the coldest temperatures (therefore maximum Γ) where the ion fluorescence is maximum. The phase transition may take place in the present system, but we have experimentally missed detecting it, or possibly larger crystals (for example, where the number of ions in the crystal is large compared to the number of ions in the shells) may be required in order for a sharp phase transition to be exhibited.

We have observed structures for which we do not have any current theoretical understanding. Figure 8(a) shows an approximate fivefold Bragg scattering pattern that was observed a number of times under different experimental circumstances. A fivefold Bragg scattering pattern is characteristic of a quasi-crystal. However, more sets of dots would be present in a true quasi-crystalline Bragg scattering pattern. We now think that the fivefold Bragg scattering pattern of Fig. 8(a) is due to a structure like that shown in Fig. 8(b). Figure 8(b) is a top-view image of a lenticular plasma which consisted of four horizontal planes. Even though it is difficult to distinguish individual ions in this figure, it is possible to see that there are five distinct regions where the ions resided in vertical planes. The planes from these different regions form a five-sided structure that would produce a Bragg scattering pattern like Fig. 8(a). (With the small crystals and forward Bragg scattering angles of this work, each set of vertical planes produces two Bragg peaks.) Once formed, this fivefold structure was stable and persisted for reasons which we do not understand.

In addition to enhancing studies of Coulomb crystals, the phase-locked control of ω_r has improved the prospects of a microwave frequency standard based on a hyperfine-Zeeman transition of ions stored in a Penning trap. This is because the time-dilation shift due to the plasma rotation is one of the largest known systematic shifts in such a standard. Reference [33] discusses the potential frequency stability and accuracy of a microwave frequency standard based on 10^6 trapped ions. For ions such as ^{67}Zn$^+$ and ^{201}Hg$^+$, fractional frequency stabilities $\lesssim 10^{-14}/\tau^{1/2}$ with time-dilation shifts due to the plasma rotation of $\sim$few$\times 10^{-15}$ are possible. Here τ is the measurement time in seconds. With phase-locked operation of the rotating wall, we think it should be possible to stabilize and evaluate the rotational time-dilation shift within 1%. Therefore the inaccuracy due to this shift would contribute a few parts in 10^{-17}.

ACKNOWLEDGEMENTS

We gratefully acknowledge the support of the Office of Naval Research. We thank S. L. Gilbert and R. J. Rafac for their comments and careful reading of the manuscript.

REFERENCES

1. Ichimaru, S., Iyetomi, H., and Tanaka, S., *Phys. Rep.* **149**, 91–205 (1987).
2. Malmberg, J. H., and O'Neil, T. M., *Phys. Rev. Lett.* **39**, 1333–1336 (1977).
3. Horn, H. M. V., *Science* **252**, 384–389 (1991).
4. Grimes, C. C., and Adams, G., *Phys. Rev. Lett.* **42**, 795–798 (1979).
5. E. L. Pollock and J. P. Hansen, *Phys. Rev. A* **8**, 3110–3122 (1973); W. L. Slattery, G. D. Doolen, and H. E. DeWitt, *ibid.* **21**, 2087–2095 (1980); W. L. Slattery, G. D. Doolen, and H. E. DeWitt, *ibid.* **26**, 2255–2258 (1982); S. Ogata and S. Ichimaru, *ibid.* **36** 5451–5454 (1987); G. S. Stringfellow and H. E. DeWitt, *ibid.* **41**, 1105–1111 (1990); D. H. E. Dubin, *ibid.* **42**, 4972–4982 (1990).
6. Schiffer, J. P., *Science* **279**, 675 (1998).
7. Diedrich, F., et al., *Phys. Rev. Lett.* **59**, 2931–2934 (1987).
8. Wineland, D. J., et al., *Phys. Rev. Lett.* **59**, 2935–2938 (1987).
9. Strongly coupled clusters of highly charged, micrometer-sized aluminum particles were previously observed in Paul traps. See R. F. Wuerker, H. Shelton, and R. V. Langmuir, *J. Appl. Phys.* **30**, 342–349 (1959).
10. Gilbert, S. L., Bollinger, J. J., and Wineland, D. J., *Phys. Rev. Lett.* 2022-2025 (1988).
11. Birkl, G., Kassner, S., and Walther, H., *Nature* **357**, 310–313 (1992).
12. Drewsen, M., et al., *Phys. Rev. Lett.* (1998), in press.
13. Walther, H., *Adv. At. Opt. Phys.* **31**, 137–182 (1993).
14. Melzer, A., Homann, A., and Piel, A., *Phys. Rev. E* **53**, 2757 (1996).
15. Murray, C. A., and Grier, D. G., *American Scientist* **83**, 238–245 (1995).
16. Vos, W. L., Mehens, M., van Kats, C. M., and Bösecke, P., *Langmuir* **13**, 6004–6008 (1997).
17. Tan, J. N., Bollinger, J. J., Jelenković, B., and Wineland, D. J., *Phys. Rev. Lett.* **75**, 4198-4201 (1995).
18. Itano, W. M., et al., *Science* **279**, 686–689 (1998).
19. Huang, X.-P., Bollinger, J. J., Mitchell, T. B., and Itano, W. M., *Phys. Rev. Lett.* **80**, 73–76 (1998).
20. Huang, X.-P., Bollinger, J. J., Mitchell, T. B., and Itano, W. M., *Phys. Plasmas* **5**, 1656–1663 (1998).
21. Mitchell, T. B., et al., in preparation.
22. Brewer, L. R., et al., *Phys. Rev. A* **38**, 859–873 (1988).
23. Davidson, R. C., *Physics of Nonneutral Plasmas*, New York: Addison-Wesley Publishing, 1990, pp. 39–75.
24. O'Neil, T. M., and Dubin, D. H. E., *Phys. Plasmas* **5**, 2163–2193 (1998).
25. Heinzen, D. J., et al., *Phys. Rev. Lett.* **66**, 2080–2083 (1991).
26. Bollinger, J. J., et al., *Phys. Rev. A* **48**, 525–545 (1993).
27. Dubin, D. H. E., *Phys. Rev. A* **40**, 1140–1143 (1989).
28. Ashcroft, N. W., and Mermin, N. D., *Solid State Physics*, Philadelphia: Saunders College, 1976, pp. 95–110.
29. Tan, J. N., et al., in *Proceedings of the International Conference on Physics of Strongly Coupled Plasmas*, Kraeft, W. D., and Schlanges, M., ed. World Scientific: 1996, pp. 387–396.
30. Valtchinov, V. I., Kalman, G., and Blagoev, K. B., *Phys. Rev. E* **56**, 4351–4355 (1997).
31. Cirac, J. I., and Zoller, P., *Phys. Rev. Lett.* **74**, 4091–4094 (1995).
32. Wineland, D. J., et al., *J. Res. Natl. Inst. Stand. Technol.* **103**, 259–328 (1998).
33. Tan, J. N., Bollinger, J. J., and Wineland, D. J., *IEEE Trans. Instrum. Meas.* **44**, 144–147 (1995).

Sympathetic Cooling and Crystallization of Ions in a Linear Paul Trap

M. Drewsen*, P. Bowe*, L. Hornekær*, C. Brodersen*, J. P. Schiffer[†] and J. S. Hangst*

*Institute of Physics and Astronomy, Aarhus University, DK-8000 Aarhus C, Denmark
[†]Argonne National Laboratory, Argonne, IL 60439, U.S.A., and University of Chicago, Chicago, IL 60637, U.S.A.

Abstract. Coulomb crystals, containing up to a few hundred ions of which more than 50 % were cooled sympathetically by the Coulomb interaction with laser cooled Mg^+ ions, have been produced in a linear Paul trap. By controlling the balance of the radiation pressure from the two cooling lasers, the Coulomb crystals could be segregated according to ion species. Previuos studies of ion crystals and molecular dynamics simulations suggest that the temperture may be around 10 mK or lower. The obtained results indicate that a wide range of atomic and molecular ions, which due to their internal structures are not amenable to direct laser cooling, can be effectively cooled and localized (crystalized) in linear Paul traps. For high resolution spectroscopy of such ions this may turn out to be very useful.

INTRODUCTION

Trapped ions, when cooled sufficiently, form spatially ordered structures (Coulomb crystals). For smaller crystals, where surface effects play a role, shell and string like structures are equilibrium states [1–3], while molecular dynamics (MD) simulations of infinite single component plasmas, predict a body centred cubic (BCC) structure [4] as the equilibrium state. In both Paul and Penning traps string and shell structures have been produced by applying laser cooling [5–11], and recently BCC structures at the centre of very large ion crystals in Penning traps have been reported [12,13]. Since only ions with simple level schemes accesible to lasers can easily be laser cooled, most atomic ion species and all molecular ions, due to their complex vibrational and rotational structure, are excluded from this type of cooling. Hence, to date only very few singly charged atomic ion species have been laser cooled and crystallized. The equation of motion in vaious traps allow, however, ions in several different charge states and with a wide range of masses to be trapped simultaneously, which make sympathetic cooling though the Coulomb interaction possible.

Previously, several authors have investigated sympathetic cooling, where directly laser cooled ions were used to cool ions of different species through mutual Coulomb interaction [14–17]. In most of these experiments, the typically achieved temperatures of the sympathetic cooled ions were some 100 mK, which did not lead to ordering of the whole plasmas. However, in a few cases a few dark sites in crystals consisting of laser cooled $^{24}Mg^+$ were attributed to indirectly cooled impurity ions [18,19]. Recently, a crystal consisting entirely of Ca^+ was observed to stay crystallised when some of the constituent ions were decoupled from the cooling laser by being optically pumped into a metastable dark state [9].

In this proceeding, Coulomb crystals in a linear Paul trap consisting of upto a few hundred ions where the fraction of sympathetically cooled ions is greater than 50 % are presented. In one case particular intersting for spectroscopy, 14 sympathetically cooled ions were maintained in an ordered string structure by only one directly cooled $^{24}Mg^+$ ion. We have not measured the temperature of the sympatetically cooled ions in the crystals, but since ordering is observed it must be below temperatures previously reported in experiments with comparable large fractions of sympathetically cooled ions [15–17]. Recent studies of the formation of $^{24}Mg^+$ ion crystals [10] and molecular dynamics (MD) simulations of infinitely long cylindrical plasmas [20] indicate that the temperature may well be around 10 mK or lower. The radiation pressure of the cooling lasers has furthermore made it possible to segregate the ions according to species. Besides the potential interst of such mixed Coulomb crystals within plasma physics, the very cold and well-localized sympathetically cooled ions are very interesting in connection with high resulution spectroscopy, where both transit time broadening, Doppler shift/broadening as well as laser induced light shifts can be problematic.

EXPERIMENTAL SETUP

The experimental setup is practically identical to the one described in the contribution by J. S. Hangst *et al.* in this proceeding and is described in details in ref. [10]. In the experiments, the ions are confined in a linear Paul trap [21] operating at a RF frequency of $\Omega = 2\pi \times 4.2$ MHz and a RF amplitude of $U_{RF} = 15 - 100$ V, which give values of the stability parameter q within the range [0.07,0.5] for ^{24}Mg$^+$ ions. Since loading of the trap is done by electron bombartment of an atomic magnesium beam sent through the trap center, ions from the background gas are produced as well and will be trapped if their charge to mass ratio gives rise to values of q below the stability limit of 0.9. The fraction of trapped background gas ions can be controlled to some extent by varying the atomic beam flux. The ^{24}Mg$^+$, and in some cases also ^{26}Mg$^+$, ions are laser cooled axially by Doppler cooling on the $3s^2 S_{1/2} - 3p^2 P_{3/2}$ transition. Sympathetic cooling of the transverse degrees of freedom of the laser cooled ions as well as all degrees of freedom of the other trapped ions is accomplised through the Coulomb interaction between the ions. When cooling a single species two counter propagating laser beams at the same frequency and with adjustable intensity balance are sent into the trap region, while when cooling two isotopes an extra laser beam at a separately tunable frequency is overlapped. The positions of the ions are monitored by imaging the fluorescence light onto an image-intensified video camera, the time resolution of which is 20 ms. The camera system views the trap perpendicular to the trap axis.

EXPERIMENTAL RESULTS

We have observed sympathetic crystallization of crystals containing of up to some 200 ions where 60% of the ions were cooled indirectly. Fig.1 shows a sequence where ions are loaded into the trap using two counter propagating lasers resonant with ^{24}Mg$^+$ and ^{26}Mg$^+$, respectively. Due to the net radiation pressure force exerted on the cooled ions in the direction of the laser beams, the crystal is spatially segregated according to species. Since this forces can at most displace the ions some 50 μm along the axial trap potential, we conclude that the dark regions between the two magnesium isotopes in Fig. 1 must be filled with other ionic species. The visible magnesium ions are clearly crystallized at the boundary of the dark region, indicating that the dark ions in the dark region are also crystallized, since the Coulomb interaction is long range and the otherwise chaotic motion at the centre of the crystal would lead to heating of the entire ion plasma. By using two counter propagating laser beams resonant with one of the magnesium isotopes and temporarily changing their power balance, we can cause the visible ions to move back and forth though the centre. The movement takes the form of discrete hops from one well-defined site to another, in agreement with a full ordered plasma.

Previous studies of larger ^{24}Mg$^+$ crystals of similar shapes Ref. [10] show that the onset of spatial ordering happens at a plasma coupling parameter comparable to the value predicted for infinitely long cylindrical plasmas [20]. The plasma coupling parameter is defined as:

$$\Gamma = \frac{E_{Coul}}{k_B T},$$

where T is the temperature and E_{Coul} is the nearest neighbour Coulomb energy. From the simulation presented in Ref. [20] fully ordered shell structures are expected when $\Gamma \geq 100$. With the trap parameters used in the present experiment this lower limit of Γ sets an upper temperature limit of about 10 mK. Due to the strong coherent transverse micro-motion of the ions, the maximum transverse kinetic energy of the outer shell ions in the last image of Fig. 1 corresponds to a temperature of a few Kelvin. Since the coupling of transverse micro-motion into the axial motion is found to be extremely small for very prolate crystals, this leads to small additional first order Doppler shifts along the trap axis, while the second order Doppler shifts will be larger than implied by the estimated temperature for crystallization.

In Fig. 4a. in the contribution by Hangst *et al.* is shown a video frame of a single ^{24}Mg$^+$ ion in a string with 14 non-fluorescing ions, which are either ions from the background gas or other magnesium isotope ions. The evidence for the string comes about by monitoring the various positions the single ^{24}Mg$^+$ ion visits due to diffusion in a 2 min. video sequence. A composite picture of the positions visited by the visible ion fits perfect with simulated positions of a string of 15 singly charged ions at zero-temperature (See fig 4b. in contribution by

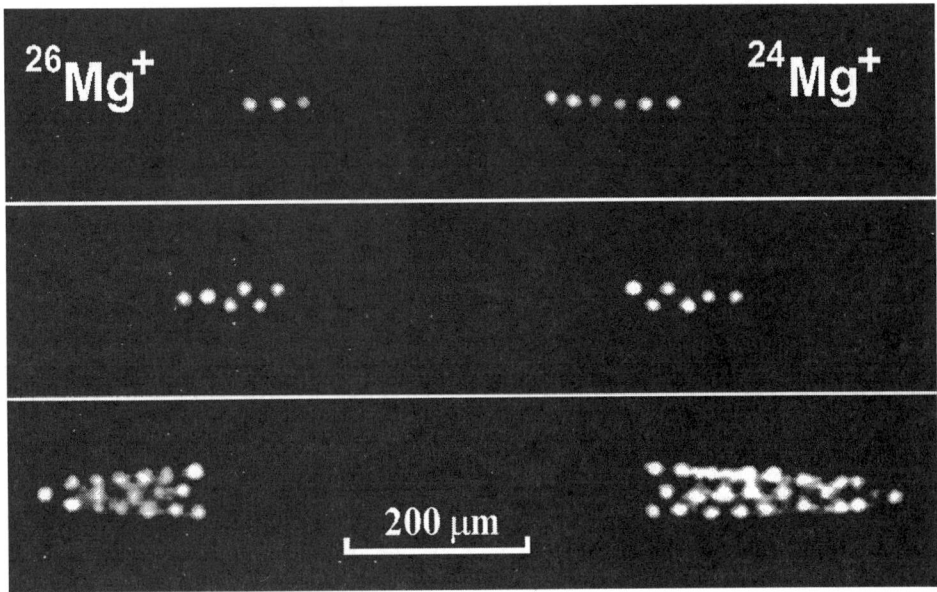

FIGURE 1. Three successive images taken during a loading sequence. The ^{24}Mg$^+$ ions are pushed to the right and ^{26}Mg$^+$ ions to the left by their respective near resonant cooling lasers. The central region contains sympathetically cooled ions.

J. S. Hangst et al.). Since each video frame always shows the visible ion in a single position, the rearrangement time of the string after an detectable disturbance must be faster than the frame integration time of 20 ms. This is in accordance with typical reordering times (∼1ms) found in molecular dynamics (MD) simulations. For the observed positional diffusion of the visible ion with a jump rate of 0.36 per second there are two probable processes: Occasional elastic collisions with background gas atoms/molecules and random walk of the ions induced by the photon scattering events through laser cooling. For the experimental background pressure of approx. 10^{-10} Torr, we estimate 0.03 elastic collisions per ion per second [22]. If each such collision leads to a positional jump this can explain the observed jump rate. However, future experiments conducted at various pressures as well as MD simulations should reveal the importance of the two mechanism for positional jumps. We have initiated MD simulations where the RF field of the trap, collisions with the background gas and the random momentum kicks given to the ions in the photon scattering processes are included.

CONCLUSION

In conclusion, we have demonstrated that in a linear Paul trap it is possible to make Coulomb crystals consisting of up to at least a few hundred ions where more than 50% of the ions are cooled sympathetically. We have proven it is possible to create and keep a string of 15 cold ions where only a single ion is directly cooled. Earlier studies of the formation of ^{24}Mg$^+$ ion crystals and MD simulations indicate that the temperature may well be around 10 mK or lower. Our findings indicate that a wide range of atomic and molecular ions, can in a substantial amount be effectively cooled and localized by the use of linear Paul traps. This points towards improvements in spectroscopic studies of such ions.

REFERENCES

1. A.Rahman and J.P.Schiffer, Phys. Rev. Lett. **57**, 1133 (1986).
2. R.W.Hasse and J.P.Schiffer, Ann. Phys. **203**, 419 (1990).

3. D.H.E.Dubin and T.M.O'Neil, Phys. Rev. Lett. **60**, 511 (1988).
4. D.H.E.Dubin, Phys. Rev A **42**, 4972 (1990).
5. F. Diedrich, E.Peik, J.M.Chen, W.Quint and H.Walther, Phys. Rev. Lett. **59**, 2931 (1987).
6. D.J.Wineland, J.C.Bergquist, W.M.Itano, J.J. Bollinger and C.H.Manney, Phys. Rev. Lett. **59**, 2935 (1987).
7. S.L.Gilbert, J.J.Bollinger and D.J.Wineland, Phys. Rev. Lett. **60**, 2022 (1988).
8. G.Birkl, S.Kassner and H.Walther, Nature **357**, 310 (1992).
9. W.Alt, M.Block, P.Seibert and G.Werth, Phys. Rev. A. **58**, R23 (1998).
10. M.Drewsen, C.Brodersen, L.Hornekaer, J.P.Schiffer and J.S.Hangst, to appear in Phys. Rev. Lett.
11. T. B. Mitchell, J. J. Bollinger, . H. E. Dubin, X.-P. Huang, W. M. Itano, and R. H. Baughman, to appear in Nature
12. J.N.Tan, J.J.Bollinger, B.Jelenkovic and D.J.Wineland, Phys. Rev. Lett. **75**, 4198 (1995).
13. W.M.Itano, J.J. Bollinger, J.N.Tan, B.Jelenković and D.J.Wineland, Science **279**, 686 (1998).
14. R.E.Drullinger, D.J.Wineland and J.C.Bergquist, Appl. Phys. **22**, 365 (1980).
15. D.J.Larson, J.C.Bergquist, J.J.Bollinger, W.M.Itano and D.J.Wineland, Phys. Rev. Lett. **57**, 70 (1986).
16. H.Imajo, K.Hayasaka, R.Ohmukai, U.Tanaka, M.Watanabe and S.Urabe, Phys. Rev. A, **53**, 122 (1996).
17. T.Baba and I.Waki, Jpn. J. Appl. Phys. **35**,1134 (1996).
18. I.Waki, S.Kassner, G,Birkl and H.Walther, Phys. Rev. Lett. **68**, 2007 (1992).
19. M. G. Raizen, J. M. Gillian, J. C. Bergquist, W. M. Itano, and D. J. Wineland, J. Mod. Opt. **39** , 233 (1992)
20. J.P.Schiffer in *Proceedings of the Workshop on Crystalline ion beams*, Wertheim, Gemany, Oct. 1988, Ed. R.W.Hasse, I.Hofmann and D.Liesen, GSI-89-10 Report, April 1989.
21. J.D.Prestage, G.J.Dick and L.Maleki, J. Appl. Phys. **66**, 1013 (1989).
22. L.D.Landau amd E.M.Lifshitz, *Quantum Mechanics - Non-relativistic Theory* (1977).
München, München 1993.

Mode and Transport Studies of Laser-Cooled Ion Plasmas in a Penning Trap*

T. B. Mitchell, J. J. Bollinger, X.-P. Huang and W. M. Itano

Time and Frequency Division, National Institute of Standards and Technology, Boulder, CO 80303

Abstract. We describe a technique and present results for imaging the modes of a laser-cooled plasma of ^{9}Be$^+$ ions in a Penning trap. The modes are excited by sinusoidally time-varying potentials applied to the trap electrodes, or by static field errors. They are imaged by changes in the ion resonance fluorescence produced by Doppler shifts from the coherent ion velocities of the mode. For the geometry and conditions of this experiment, the mode frequencies and eigenfunctions have been calculated analytically. A comparison between theory and experiment for some of the azimuthally symmetric modes shows good agreement. Enhanced radial transport is observed where modes are resonant with static external perturbations, such as those caused by misaligning the trap with respect to the magnetic field. Similarly, the plasma angular momentum can be changed through the deliberate excitation of azimuthally asymmetric modes. The resultant torque can be much greater than that from the "rotating wall" perturbation, which is not mode-resonant.

INTRODUCTION

Non-neutral plasmas consisting exclusively of particles of a single sign of charge have been used to study many basic processes in plasma physics [1], partly because non-neutral (as opposed to neutral or quasi-neutral) plasmas can be confined by static electric and magnetic fields and also be in a state of global thermal equilibrium [2,3]. A particularly simple confinement geometry for non-neutral plasmas is the quadratic Penning trap, which uses a strong uniform magnetic field $\mathbf{B_0} = B_0\hat{\mathbf{z}}$ superimposed on a quadratic electrostatic potential

$$\phi_T(r,z) = \frac{m\omega_z^2}{2q}\left(z^2 - \frac{r^2}{2}\right). \qquad (1)$$

Here m and q are the mass and charge of a trapped ion, and ω_z is the axial frequency of a single ion in the trap. The global thermal equilibrium state for a single charged species in a quadratic Penning trap has been well studied [3,4]. For sufficiently low temperatures, the plasma takes on the simple shape of a uniform density spheroid. An interesting result is that all of the electrostatic modes of a magnetized, uniform density spheroidal plasma can be calculated analytically [5,6]. This is the only finite length geometry for which exact plasma mode frequencies and eigenfunctions have been calculated for a realistic thermal equilibrium state.

In this manuscript we describe a technique for measuring these frequencies and eigenfunctions, and compare theory predictions and experimental results for some of the magnetized plasma modes. We also discuss several potential applications for the modes in Penning trap experiments. In general, the mode frequencies depend on the density and shape of the plasma spheroid. Therefore measurement of a mode frequency provides a nondestructive method for obtaining basic diagnostic information about the plasma. This is especially important in anti-matter plasmas [7,8], where conventional techniques for obtaining information about these plasmas involve ejecting the plasma from the trap. Measurement of the damping of the modes can provide information on the plasma's viscosity [9,10]. Other applications arise from the fact that the modes can strongly influence the dynamical behavior of trapped plasmas. For example, certain azimuthally asymmetric modes can have zero frequency in the laboratory frame and be excited by a static field error of the trap. These zero-frequency modes can strongly limit the achievable density in a Penning trap [11]. Similarly, the plasma angular momentum can

---

*) Work of the U.S. Government. Not subject to U.S. copyright.

be changed through the deliberate excitation of azimuthally asymmetric modes [12,13], and the applied torque can be much greater than that from the "rotating wall" perturbation [14,15], which is not mode-resonant.

Previous experimental mode studies on spheroidal plasmas have been limited to frequency measurements on a small class of modes. With laser-cooled Be$^+$ ion plasmas, some quadrupole mode frequencies have been measured and agree well with theory [6,11]. Mode frequencies have also been measured on spheroidal cryogenic electron plasmas [16], 0.025–0.5 eV electron and positron plasmas [17], and room temperature Ar$^+$ ion plasmas [18]. In these cases qualitative agreement with theory was observed and the modes provided some basic diagnostic information. However, deviations from the model of a constant density spheroid in a quadratic trap limited the comparison with the ideal linear theory. Here, in addition to measuring mode frequencies, we also measure the mode eigenfunctions. The eigenfunctions permit direct identification of the modes. In addition, they contain much more information than the frequencies and therefore may be useful for observing nonlinear effects such as mode couplings. Mode eigenfunctions have been measured for low frequency, z-independent (diocotron) modes on cylindrical electron columns [19]. In that work, the mode measurements were important in identifying two coexisting modes.

EXPERIMENTAL APPARATUS

Figure 1 shows a schematic of the apparatus [20,21] used for the mode measurements. The trap consists of a 127 mm long stack of cylindrical electrodes at room temperature with an inner diameter of 40.6 mm, enclosed in a 10^{-8} Pa vacuum chamber. A uniform magnetic field $B_0 = 4.465$ T is aligned parallel to the trap axis within $0.01°$, and results in a ^{9}Be$^+$ cyclotron frequency $\Omega = qB_0/m = 2\pi \times 7.608$ MHz. The magnetic field is aligned by minimizing the excitation of zero-frequency modes produced by a tilt of the magnetic field with respect to the trap electrode symmetry axis [6,11]. Positive ions are confined in this trap by biasing the central ring electrode to a negative voltage $-V_0$ with respect to the endcaps. Because the dimensions of the Be$^+$ plasmas ($\lesssim 2$ mm) are small compared to the diameter of the trap electrodes, the quadratic potential of Eq. (1) is a good approximation for the trap potential. For most of the work reported here, V_0 was set at 2.00 kV which results in $\omega_z = 2\pi \times 1.13$ MHz and a single-particle magnetron frequency $\omega_m = [\Omega - (\Omega^2 - 2\omega_z^2)^{\frac{1}{2}}]/2 = 2\pi \times 84.9$ kHz.

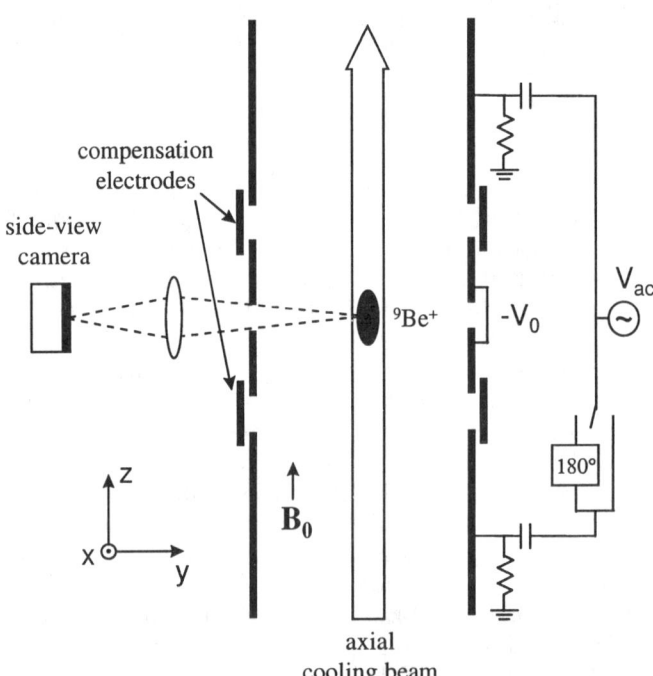

FIGURE 1. Schematic of the experimental apparatus. Azimuthally symmetric $m = 0$ modes were excited by applying in-phase or $180°$ out-of-phase sinusoidal potentials to the trap endcaps.

We create a Be$^+$ plasma by ionizing neutral Be atoms in a separate trap (not shown) and then transferring the ions to the main trap. For the work discussed here, the number of ions was typically 6×10^4. While the total charge in the trap is conserved after loading, the relative abundance of contaminant, heavier-mass ions increases, presumably due to reactions between Be$^+$ ions and background neutral molecules. Because we analyze our experimental results using an existing theory [5] for the electrostatic modes of a single-species plasma, we took data for mode studies only with relatively clean clouds (<3% impurity ions). The plasmas were cleaned approximately every 30 minutes by transferring the ions to the load trap where, with a shallow 3 V deep well, contaminant ions were driven out of the trap by exciting their axial frequencies. Cleaning therefore results in a decrease in the number of trapped ions. Over a 12–14 hour period, the number of ions is reduced by a factor of 2. Because the mode frequencies and eigenfunctions in a quadratic trap are independent of the number of ions, the mode measurements described here are not affected.

The trapped Be$^+$ ions are Doppler cooled by two laser beams at wavelength $\lambda \approx 313.11$ nm. The main cooling beam is directed parallel to $\mathbf{B_0}$ as shown in Fig. 1, and a second cooling beam propagating perpendicular to $\mathbf{B_0}$ (not shown and turned off during measurements) is also used to compress the plasma by applying a radiation pressure torque [3,11]. For mode eigenfunction measurements the axial cooling-laser frequency is fixed about one natural linewidth (~ 20 MHz) below the transition frequency. Ions which, due to excitation of a mode, have an axial velocity $v_z < 0$ therefore fluoresce more strongly than ions with $v_z > 0$. The ion temperature was not measured; however, based on previous work [3], we expect $T \lesssim 20$ mK.

An $f/5$ imaging system detects the Be$^+$ resonance fluorescence scattered perpendicularly from the axial cooling beam (waist ≈ 0.5 mm, power ≈ 50 μW) to produce a side-view image of the Be$^+$ ions. The side-view image is obtained with a photon-counting camera system which records the spatial and temporal coordinates of the detected photons. This data is processed to obtain the mode eigenfunctions by constructing side-view images as a function of the phase of the external drive used to excite the modes.

ELECTROSTATIC MODES OF A SPHEROIDAL PLASMA

A constant-density, spheroidal plasma model is a good approximation for our work. In thermal equilibrium, a Penning trap plasma rotates as a rigid body at frequency ω_r, where $\omega_m < \omega_r < \Omega - \omega_m$, about the trap's $\hat{\mathbf{z}}$ axis [2,4]. In this work the rotation frequency was precisely set by a rotating dipole electric field [14,15]. As the ions rotate through the magnetic field they experience a Lorentz force which provides the radial confining force of the trap. This ω_r-dependent confinement results in an ω_r-dependent ion density and plasma shape. At the low temperatures of this work, the plasma density is uniform over distances large compared to the interparticle spacing (~ 10 μm) and is given by $n_0 = \epsilon_0 m \omega_p^2 / q^2$ where $\omega_p = [2\omega_r(\Omega - \omega_r)]^{\frac{1}{2}}$ is the plasma frequency. With the confining potential of Eq. (1), the plasma is spheroidal with boundary $z^2/z_0^2 + x^2/r_0^2 + y^2/r_0^2 = 1$. The spheroid aspect ratio $\alpha \equiv z_0/r_0$ is determined by ω_r [3,4]. We have neglected the effect of image charges, because the plasma dimensions are small compared to the trap dimensions.

The modes of these spheroidal plasmas can be classified by integers (l, m), where $l \geq 1$ and $0 \leq m \leq l$ [5,6]. For an (l, m) mode with frequency ω_{lm} [22] the perturbed potential of the mode inside the plasma is given by a symmetric product of Legendre functions,

$$\Psi^{lm} \propto P_l^m(\bar{\xi}_1/\bar{d}) P_l^m(\bar{\xi}_2) e^{i(m\phi - \omega_{lm} t)}. \tag{2}$$

Here $\bar{\xi}_1$ and $\bar{\xi}_2$, discussed in Ref. [5], are scaled spheroidal coordinates where the scaling factor depends on the frequencies ω_r, Ω, and ω_{lm}, and $\bar{d}$ is a shape-dependent parameter which also depends on these frequencies. In general, for a given (l, m) there are many different modes. In this paper we report measurements of the mode frequencies and eigenfunctions of several magnetized plasma modes, which are defined as those modes with frequencies $|\omega_{lm}| < |\Omega - 2\omega_r|$ [5,6]. For $\omega_r \ll \Omega/2$, these modes principally consist of oscillations parallel to the magnetic field at a frequency on the order of ω_z. In the experiment we detect the axial velocity of a mode. In the linear theory, this is proportional to $\partial \Psi^{lm}/\partial z$.

We excite azimuthally symmetric ($m = 0$) plasma modes by applying sinusoidally time-varying potentials to the trap electrodes. Even-l $(l, 0)$ modes are excited by applying in-phase potentials to the endcaps (even drive), while odd-l $(l, 0)$ modes are excited by applying 180° out-of-phase potentials to the endcaps (odd drive). Azimuthally asymmetric ($m \neq 0$) modes can be excited by applying potentials to the compensation electrodes, which have 6-fold azimuthal symmetry. In Refs. [6,11] quadrupole ($l = 2$) mode frequencies were measured by observing the change in the total ion fluorescence from the plasma, averaged over the phase of the drive, which occurred when the drive frequency equaled the mode frequency. However, in order to observe such a change, the mode excitation must be large enough so that either the fluorescence from an ion nonlinearly depends on its velocity or there is some heating of the plasma by the mode. The large amplitude drive required by this technique decreases the precision of the mode measurements.

The technique described here entails reducing the drive amplitude until the change in the phase-averaged ion fluorescence is negligible, and detecting the mode's coherent ion velocities by recording side-view images as a function of the phase and frequency of the external drive [23]. These Doppler images provide direct measurements of the mode's axial-velocity eigenfunction [24]. In addition, an accurate measurement of the mode's frequency (both real and imaginary parts) can be obtained from measurements of the mode amplitude as a function of drive frequency. High order modes have been excited and detected with this technique, such as the $(11,0)$ and $(12,1)$ modes. Imaging is not required for the $(1,0)$ and $(1,1)$ modes because there is no spatial variation in their eigenfunction. The driven mode amplitude and phase of these center-of-mass modes can therefore be obtained by coherently detecting the spatially integrated fluorescence as a function of the phase of the external drive [25].

EXPERIMENTAL RESULTS

Mode Frequency And Eigenfunction Measurements

In Fig. 2 we plot measured mode frequencies, along with the theoretical predictions, for several azimuthally symmetric magnetized plasma modes as a function of ω_r for $\omega_z/2\pi = 1.13$ MHz and $\Omega/2\pi = 7.608$ MHz. Many different mode frequencies at various values of ω_z have been measured with the Doppler imaging technique, and on very clean clouds agreement between the observed and predicted mode frequencies is typically better than 1%. However, as the percentage of impurity ions increases, the shift between the measured frequency and the value predicted by the single-species theory also increases. Both positive and negative frequency shifts have been observed. We think that these frequency shifts are caused by changes in the cloud shape which perturb the spheroidal geometry of the single-species cloud, arising because impurity ions centrifugally separate from the Be^+ [26].

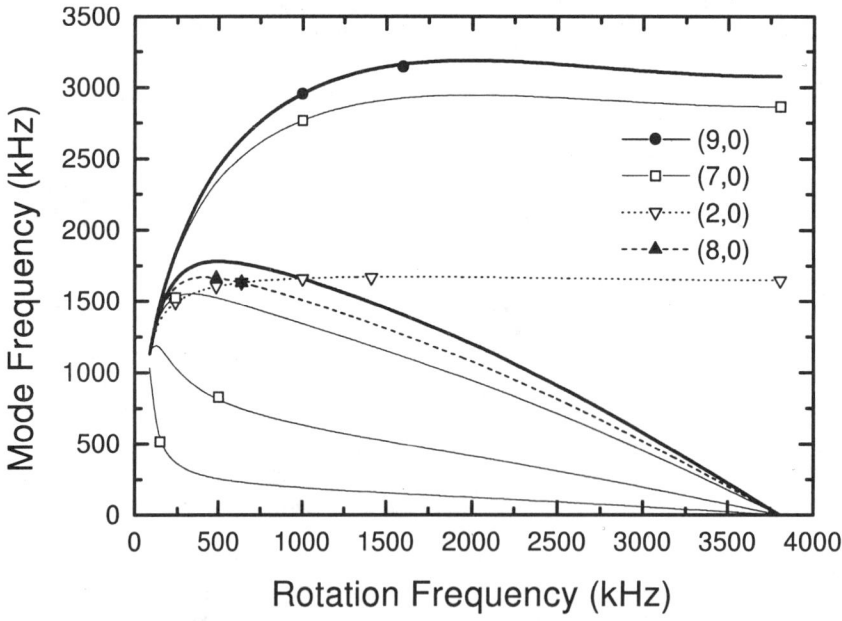

FIGURE 2. Plots of the frequencies of several $m = 0$ magnetized plasma modes as a function of rotation frequency for $\Omega/2\pi = 7.608$ MHz and $\omega_z/2\pi = 1.13$ MHz. The solid lines are the theoretical predictions and the symbols are experimental measurements. Only the highest frequency $(9,0)$ plasma mode and the second highest frequency $(8,0)$ plasma mode are plotted.

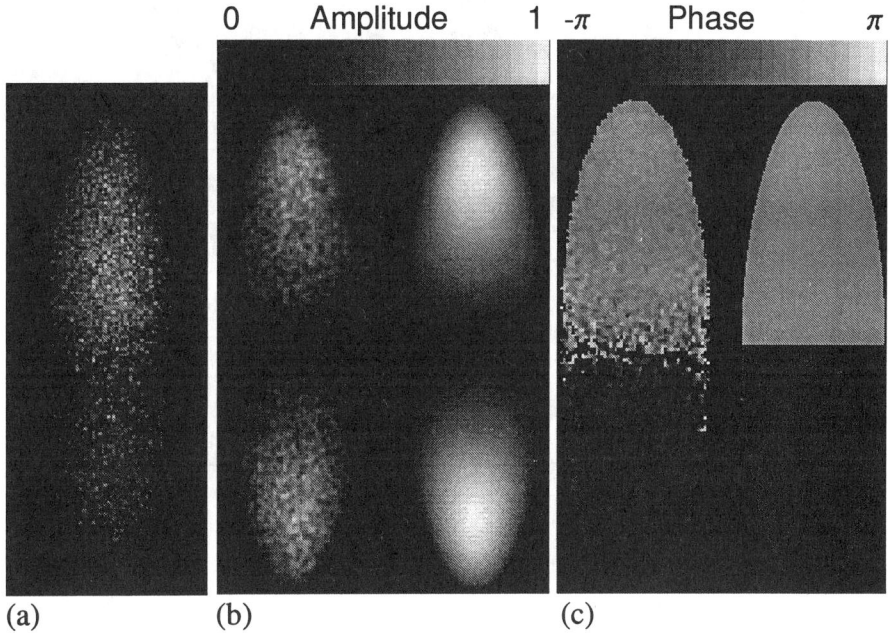

FIGURE 3. (a) Phase-coherent sideview image data obtained on a plasma with $\omega_r/2\pi = 1$ MHz while driving a $(2,0)$ mode at $\omega_{2,0}/2\pi = 1.656$ MHz. The magnetic field and axial laser beam point up. The ion cloud dimensions are $2z_0 = 0.76$ mm and $2r_0 = 0.24$ mm, and the density $n_0 = 2.70 \times 10^9$ cm^{-3}. Comparison of the amplitude (b) and phase (c) extracted from the $(2,0)$ mode in (a) with the predictions of linear theory. The theory predictions for (b) and (c) are on the right. From Ref. [23].

Figure 3 illustrates phase-coherent detection and Doppler imaging of the $(2,0)$ mode. This is one of the simplest modes that is not merely a center-of-mass oscillation of the plasma. In this mode the plasma stays spheroidal but the aspect ratio (and density) oscillate at $\omega_{2,0}$. For $\omega_r \ll \Omega/2$, the oscillation in r_0 is very small, so the mode principally consists of oscillations in z_0 at $\omega_{2,0}$. Ions above the $z = 0$ mid plane oscillate 180° out of phase with ions below $z = 0$.

Figure 3(a) shows one of a sequence of 18 side-view images taken as a function of the phase of the mode drive at $\omega_{2,0}/2\pi = 1.656$ MHz. A movie of the entire sequence is included in Ref. [23]. The plasma's rotation frequency was set to $\omega_r/2\pi = 1$ MHz and the $m = 0$ even drive rms amplitude was 7.07 mV. In the images, the magnetic field and the axial laser beam point up. As expected for the $(2,0)$ mode, the detected fluorescence in the upper half of the plasma is bright when the lower half is dark and vice versa. We analyze the data of Fig. 3(a) by performing a least-squares fit of the intensity at each point to $A_0 + A_{2,0}\cos(\omega_{2,0}t + \varphi_{2,0})$. Figures 3(b) and 3(c) show the resultant images of the measured mode amplitude $A_{2,0}(x,z)$ and phase $\varphi_{2,0}(x,z)$. These are compared with the theoretically predicted values of these quantities. Because the plasma is optically thin, the theoretical predictions were obtained by integrating $\partial \Psi^{lm}/\partial z$ over y. The amplitude of the theoretical prediction is scaled to match the experiment, and both amplitudes are normalized to 1.

From the fitted values of $A_{2,0}$ and A_0 we can estimate the coherent-ion mode velocities if the dependence of the ion fluorescence on velocity (through Doppler shifts) is known. For the low temperatures of this experiment a good approximation is to assume a Lorentzian profile with a full width at half maximum of 19 MHz due to the natural linewidth of the optical cooling transition. With the 20 MHz detuning used in this measurement, we estimate for the data of Fig. 3 that the maximum coherent mode velocity, which occurs at $z = \pm z_0$, is ~ 1.5 m/s. The spatial and density changes in the plasma spheroid for this excitation are too small to be resolved ($\Delta z/z_0, \Delta n/n_0 < 10^{-3}$). Therefore the observed variation in the fluorescence intensity is entirely due to Doppler shifts induced by the coherent ion velocities of the mode.

We have measured the mode eigenfunctions of a number of different azimuthally symmetric modes including the $l = 2, 3, 4, 5, 7$, and 9 modes. Like the data of Fig. 3, good agreement with the predicted eigenfunction amplitude and phase distribution is obtained in the limit of low laser power and drive amplitude. Surprisingly high-order odd modes could be excited with the odd drive on the trap endcaps. Figure 4(a) shows one

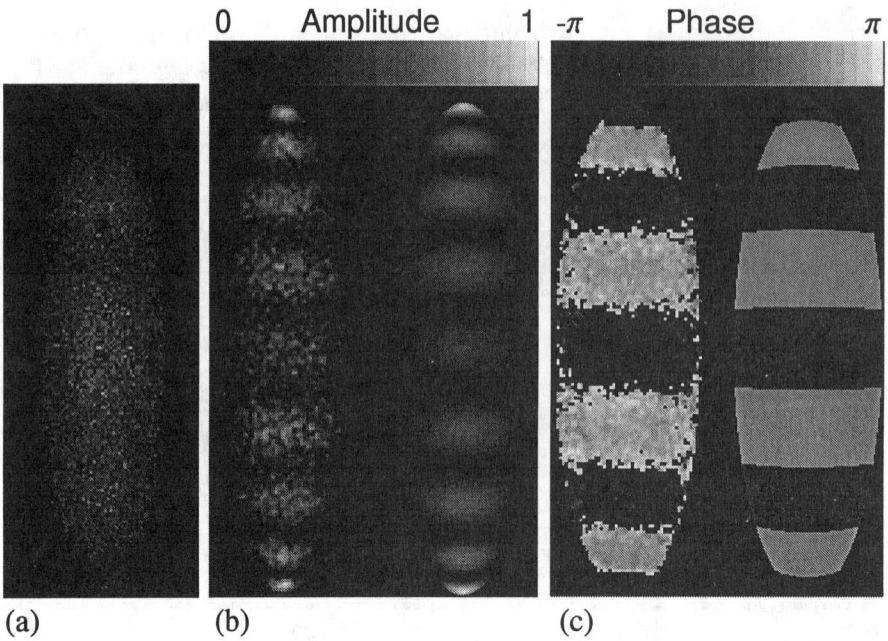

FIGURE 4. (a) Phase-coherent sideview image data obtained on the plasma of Fig. 3 with $\omega_r/2\pi = 1.00$ MHz while driving a (9,0) mode at $\omega_{9,0}/2\pi = 2.952$ MHz. Comparison of the amplitude (b) and phase (c) extracted from the (9,0) mode in (a) with the predictions of linear theory. The theory predictions are on the right. From Ref. [23].

of a sequence of 18 sideview images obtained with the highest frequency $(9,0)$ mode excited by a drive at $\omega_{9,0}/2\pi = 2.952$ MHz. For a given $(l,0)$, the highest frequency magnetized plasma mode does not have any radial nodes. Figures 4(b) and 4(c) show the fitted amplitude and phase from the sequence, along with the predictions from theory. Similar high-order even $(l,0)$ modes are more difficult to excite. The mode eigenfunctions of some of the azimuthally asymmetric ($m=1$ and $m=2$) modes, such as the (1,1), (2,1), (3,1) (4,1), (6,1), (8,1) and (3,2) modes, have also been imaged. In general, the qualitative agreement with the predictions of theory is good.

Figure 5 shows images from a plasma with $\omega_r/2\pi = 638$ kHz driven by an even drive at 1.619 MHz. This case demonstrates the utility of the Doppler imaging diagnostic. These data were initially taken during a survey of the $(2,0)$ mode eigenfunction as a function of the plasma's rotation frequency. Analysis of the phase-coherent data revealed additional, higher-order structure. An examination of the predictions for the mode frequencies revealed that at this particular rotation frequency, as shown in Fig. 2, both the $(2,0)$ mode and an $(8,0)$ mode with a radial node have similar frequencies. Characteristics of both modes are seen in the data. However, subsequent measurements of the $(2,0)$ mode frequency near this crossing indicated that any frequency shifts due to a nonlinear coupling with the $(8,0)$ mode are less than a few kilohertz. The $(2,0)$ mode driven in Fig. 3 occurs near a crossing with a $(9,0)$ mode (see Fig. 2). In this case no evidence for an excitation of a $(9,0)$ is observed, presumably because it is an odd mode which does not couple to even drives, and because there is little or no mode coupling between the $(2,0)$ and $(9,0)$.

Doppler imaging also provides a technique for measuring the damping of plasma modes. This is done by sweeping the frequency ω_{pert} of the sinusoidally time-varying perturbation through a mode frequency, while measuring the mode's resultant amplitude and phase. If the perturbation amplitude is kept low to avoid large amplitude effects the system can be modeled as a damped harmonic oscillator driven by a periodic external force, which has a characteristic lineshape for its amplitude response and a phase difference of π above and below resonance [27].

Figure 6 shows a measurement of the $(2,0)$ mode amplitude and phase response. The axial laser intensity was reduced in an attempt to make mode damping from viscous dissipation dominant over that from laser cooling, and the $z > 0$ upper half of the plasma was blocked off to permit phase-coherent detection without spatial discrimination. At each perturbation drive frequency ω_{pert} the fluorescence intensity was fitted to

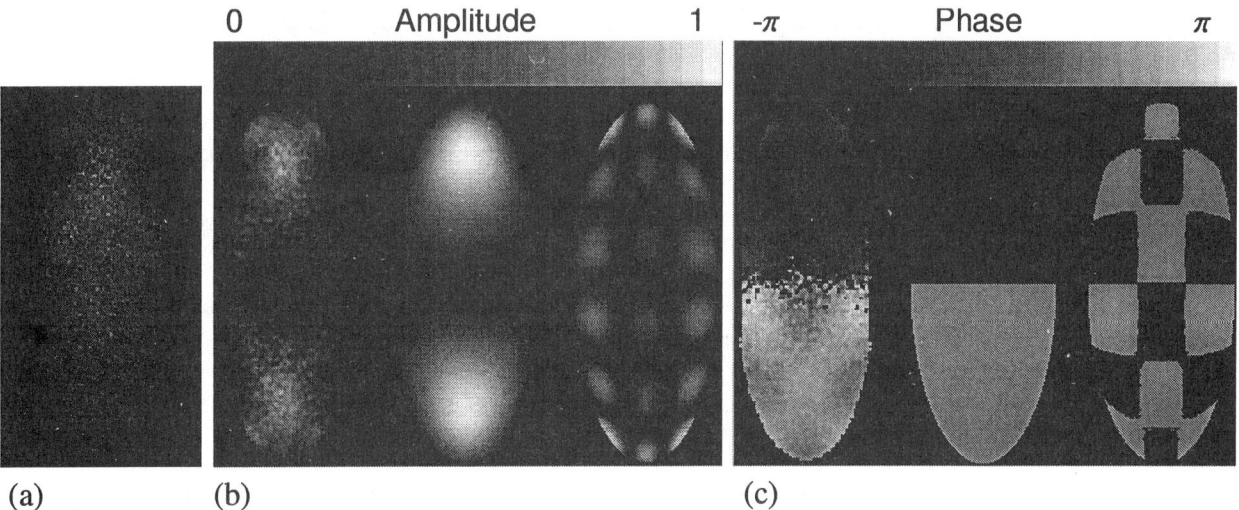

FIGURE 5. (a) Phase-coherent sideview image data obtained on a plasma with $\omega_r/2\pi = 638$ kHz while driving with an even drive at 1.619 MHz. At this rotation frequency there is a crossing of the $(2,0)$ mode and an $(8,0)$ mode with a radial node. Comparison of the amplitude (b) and phase (c) extracted from the data in (a) with the predictions of linear theory. The predictions of both the $(2,0)$ and $(8,0)$ modes are given. For this plasma $2z_0 = 0.70$ mm and $2r_0 = 0.29$ mm. From Ref. [23].

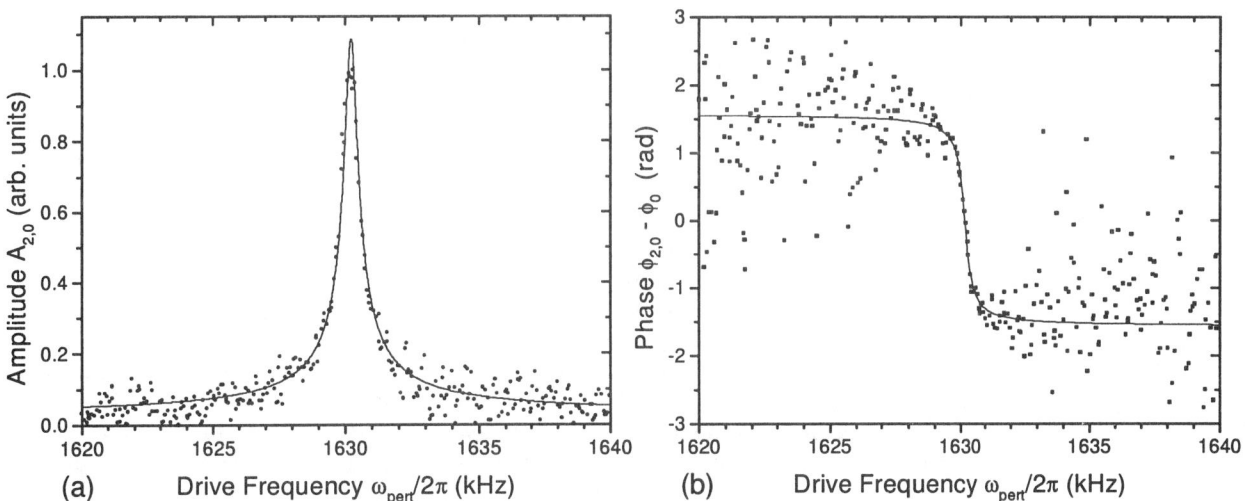

FIGURE 6. Measurement of the $(2,0)$ mode amplitude (a) and phase (b) as a function of perturbation drive frequency. The lines are fits to theory curves for the velocity of a driven, damped harmonic oscillator [27].

$A_0 + A_{2,0}\cos(\omega_{pert}t + \varphi_{2,0})$. Figure 6(a) shows the measured amplitude $A_{2,0}$ along with a 4-parameter fit to the damped harmonic oscillator lineshape $a_0 + a_1\omega_{pert}/(\omega_{2,0}\sqrt{(\omega_{pert} - \omega_{2,0})^2 + \lambda^2})$, where λ is the damping coefficient. Figure 6(b) shows the measured phase $\varphi_{2,0}$ along with a 3-parameter fit to $\varphi_0 - \arctan((\omega_{pert} - \omega_{2,0})/\lambda)$. The two fits give damping coefficient values of $\lambda = 1405$ and $1409\ s^{-1}$. These are consistent with the rates of viscous damping seen in simulations [9].

Angular Momentum Transport From Resonant Modes

In principle, the confinement time of non-neutral plasmas in Penning traps is infinite because angular momentum conservation in an ideal, cylindrically symmetric trap places a constraint on the radial transport of the plasma [28]. In practice radial transport of the plasma always occurs, and at rates which, with ultrahigh vacuum, are greater than can be explained by collisions of the plasma with the neutral background gas. Because the rates of this "ambient" transport increase with increasing static field errors in the trap, it is thought to be caused by couplings between the confining field asymmetries and the plasma [29].

Although ambient transport is at present poorly understood, progress has been made on the related but simpler mechanism of mode-resonant transport. Here, torques are imparted because azimuthally asymmetric plasma modes can have zero frequency in the laboratory frame and hence be excited by the static field errors [6,11,30–32]. Because these modes need to have negative (backward) frequencies in the rotating frame of the plasma to come into resonance with a static field error, any torque they exert will slow the plasma down and hence increase transport. An analysis based on the second law of thermodynamics yields the same result [28].

When the trap walls are well away from the plasma and the ambient field errors are small, as in our experiment, it is particularly easy to study mode-resonant transport. Reference [11] demonstrated that torque and heating of the plasma occur when one of the (2,1) plasma modes is resonant with a static field error produced by a tilt between the trap symmetry axis and the magnetic field. The presence of additional heating resonances at lower rotation frequencies was also noted. We have used Doppler imaging to identify these resonances and find that they arise from $(l,1)$ modes which come into resonance with the tilted-field error. We have also established that they exert a torque when they are resonant. Experimentally this tilt can be applied either mechanically by tilting the trap electrodes, or electrically with $m = 1$ perturbations applied to the compensation electrodes; we find no difference in the transport caused by the two methods.

Figure 7 is a plot of rotation frequency (as determined by side-view images [3]) versus time for a plasma when the trap has been electrically tilted from its aligned value by an amount equivalent to $\sim 5 \times 10^{-4}$ radians of mechanical tilt. Radial transport, which is measured here by decreases in rotation frequency (corresponding in this experimental regime to increases in the plasma radius), is enhanced by roughly a factor of 10 as compared with the aligned case. The rotation frequencies where transport is especially rapid can be identified with the mode resonances indicated on the plot. The lines show the predictions from theory for where the indicated mode has $\omega_{lm} = -\omega_r$, and Doppler imaging was used to verify the identity of these resonant zero-frequency modes. We find that the tilted-field error couples to modes with $m = 1$ and odd axial symmetry. Since with a single-species cloud only the (2,1) is predicted in linear theory to couple with a tilted-field error, the transport displayed in Fig. 7 might require the presence of impurity ions. In ion traps these are usually present at some level, and comprised $\sim 20\%$ of the cloud of Fig. 7.

Because small-amplitude static field errors can be so effective in causing outward transport, it is not surprising that the process can be usefully inverted by actively driving modes which travel faster than the cloud's rotation. With the laser-cooled Be$^+$ plasmas, we have demonstrated mode-resonant inward transport with the (1,1), (2,1), (3,1) and (2,2) modes. The (1,1) is particularly easy to excite, as only an axially uniform rotating dipole field is required, and is useful for driving clouds into the regime where ω_r approaches Ω_c. The mode-resonant technique can do this in a few seconds, while doing the same thing with laser torque takes many seconds, and with the rotating wall perturbation at least several minutes. We note that the (1,1) mode requires an effect to break the separation between the center-of-mass and the internal degrees of freedom of the plasma. In our work, a small number of impurity ions could do this.

In comparison with the rotating wall technique for controlling an ion cloud's rotation frequency [14,15], the mode-resonant technique is less precise. The mechanism by which the rotating wall is believed to work with strongly correlated plasmas is that the plasma comes into equilibrium with rotating distortions of its surface which are imposed by the perturbation. Hence, the torque from a perturbation applied at frequency f_{RW} goes to 0 when $f_{RW} = f_r$, and changes sign about this point. In contrast, the torque imparted by a driven (l,m) mode usually has only one sign and is experimentally observed to depend sensitively upon such parameters

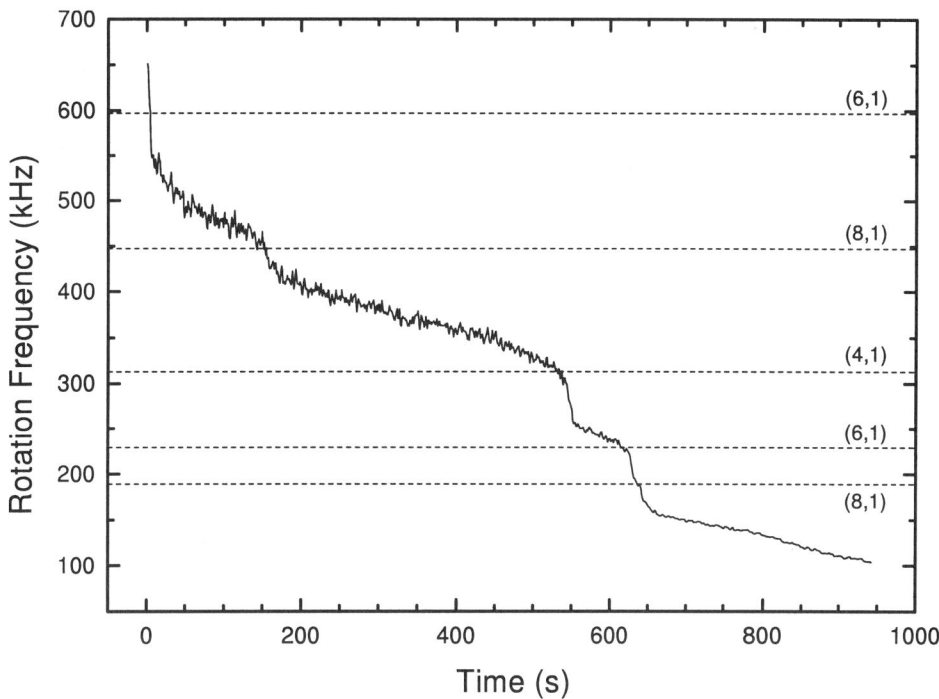

FIGURE 7. Measured plasma rotation frequency vs. time obtained when the trap was electrically tilted ~5 ×10⁻⁴ radians from its aligned value. The lines are predictions from theory for where the indicated modes have zero frequency in the lab frame.

as the temperature of the plasma. As a consequence the rotation frequency at which the cloud comes into equilibrium with the mode, which is determined by a balance between the inward mode torque and the outward ambient torque, is difficult to calculate in advance and is experimentally observed to change with variations in the cooling laser power or frequency.

However, an important advantage of mode-resonant coupling is that it can be used to transfer angular momentum to hot (uncorrelated: $\Gamma \ll 1$) plasmas. The phase-locked rotating wall control [14,15] described above has only been demonstrated with laser-cooled plasmas. In contrast, mode-resonant coupling has been used to increase angular momentum in non-neutral plasmas with temperatures up to 5 eV (which is where ionization of neutrals begins to change the density profile). Reference [12] demonstrated inward radial transport of a hot (T=0.9 eV) spheroidal electron cloud through the use of $(l, 1)$ modes. This transport was accompanied by a heating of the electrons, since there were no cooling processes in the experiment. At higher magnetic fields the heating can be balanced by cyclotron radiation cooling; steady-state confinement of uncorrelated electron plasmas in a 4 T field through the application of azimuthally asymmetric modes has recently been demonstrated [33].

SUMMARY AND FUTURE DIRECTIONS

We have described a technique, Doppler imaging, for studying the mode properties of laser-cooled ion plasmas. In general, for the magnetized plasma modes of spheroidal plasmas discussed here, good agreement is obtained between linear theory and the experimental measurements we have made to date. In the future the technique should be a useful tool for studying deviations from the linear theory such as large amplitude frequency shifts, non-linear corrections to the mode eigenfunction, and mode coupling. Because the width of the resonant lineshape of the mode amplitude as a function of the drive frequency provides a measurement of the mode damping, lineshape measurements may be able to provide information on the collisional viscosity of the strongly correlated plasma, about which little is currently known. Enhanced radial transport is observed when modes are resonant with static external perturbations, and future work may permit a quantitative comparison to be made between experiment and theory for this basic transport process. Finally, we described how the plasma angular momentum can be usefully changed through the deliberate excitation of azimuthally asymmetric modes.

ACKNOWLEDGMENTS

We thank D. H. E. Dubin, C. F. Driscoll, E. M. Hollmann and D. J. Wineland for useful discussions, and L. B. King and B. M. Jelenković for their comments and careful reading of the manuscript. This work is supported by the Office of Naval Research.

REFERENCES

1. Fajans, J., and Dubin, D. H. E., *Non-Neutral Plasma Physics II*, New York: AIP, 1995.
2. Malmberg, J. H., and O'Neil, T. M., *Phys. Rev. Lett.* **39**, 1333–1336 (1977).
3. Brewer, L. R., et al., *Phys. Rev. A* **38**, 859–873 (1988).
4. Bollinger, J. J., Wineland, D. J., , and Dubin, D. H. E., *Phys. Plasmas* **1**, 1403–1414 (1994).
5. Dubin, D. H. E., **66**, 2076–2079 (1991).
6. Bollinger, J. J., et al., *Phys. Rev. A* **48**, 525–545 (1993).
7. Greaves, R. G., and Surko, C. M., *Phys. Plasmas* **4**, 1528–1543 (1997).
8. Holzscheiter, M. H., et al., *Phys. Lett. A* **214**, 279–284 (1996).
9. Dubin, D. H. E., and Schiffer, J. P., *Phys. Rev. E* **53**, 5249–5267 (1996).
10. Dubin, D. H. E., *Phys. Rev. E* **53**, 5268–5290 (1996).
11. Heinzen, D. J., et al., *Phys. Rev. Lett.* **66**, 2080–2083 (1991).
12. Mitchell, T. B., Ph.D. thesis, University of California at San Diego, 1993, pp. 58–61.
13. Huang, X.-P., et al., *Phys. Rev. Lett.* **78**, 875–878 (1997).
14. Huang, X.-P., Bollinger, J. J., Mitchell, T. B., and Itano, W. M., *Phys. Rev. Lett.* **80**, 73–76 (1998).
15. Huang, X.-P., Bollinger, J. J., Mitchell, T. B., and Itano, W. M., *Phys. Plasmas* **5**, 1656–1663 (1998).
16. Weimer, C. S., Bollinger, J. J., Moore, F. L., and Wineland, D. J., *Phys. Rev. A* **49**, 3842–3853 (1994).
17. Tinkle, M. D., Greaves, R. G., and Surko, C. M., *Phys. Plasmas* **2**, 2880–2894 (1995).
18. Greaves, R. G., Tinkle, M. D., and Surko, C. M., *Phys. Rev. Lett.* **74**, 90–93 (1995).
19. Driscoll, C. F., *Phys. Rev. Lett.* **64**, 1528–1543 (1990).
20. Tan, J. N., Bollinger, J. J., Jelenković, B., and Wineland, D. J., *Phys. Rev. Lett.* **75**, 4198-4201 (1995).
21. Itano, W. M., et al., *Science* **279**, 686–689 (1998).
22. Here ω_{lm} is the mode frequency in a frame rotating with the plasma. For the $m = 0$ modes discussed here this distinction is not necessary because their frequency is the same in either the laboratory or rotating frame.
23. Mitchell, T. B., Bollinger, J. J., Huang, X.-P., and Itano, W. M., *Opt. Exp.* **2**, 314–322 (1998).
24. Information on the mode eigenfunction can be obtained from the side-view images even when there is a change in the phase-averaged ion fluorescence. However, the images may no longer provide a linear measure of the mode axial velocity.
25. Thompson, R. C., et al., *Phys. Scripta* 24–33 (1997).
26. Larson, D. J., et al., *Phys. Rev. Lett.* **57**, 70–73 (1986).
27. Landau, L. D., and Lifshitz, E. M., *Mechanics and Electrodynamics*, Oxford: Pergamon Press, 1972, pp. 68–70.
28. O'Neil, T. M., and Dubin, D. H. E., *Phys. Plasmas* **5**, 2163–2193 (1998).
29. Driscoll, C. F., Fine, K. S., and Malmberg, J. H., *Phys. Fluids* **29**, 2015–2017 (1986).
30. Driscoll, C. F., and Malmberg, J. H., *Phys. Rev. Lett.* **50**, 167–170 (1983).
31. Keinigs, R., *Phys. Fluids* **27**, 1427–1433 (1984).
32. Eggleston, D. L., O'Neill, T. M., and Malmberg, J. H., *Phys. Rev. Lett.* **53**, 982–984 (1984).
33. Anderegg, F., Hollmann, E. M., and Driscoll, C. F., "Rotating Field Confinement of Pure Electron Plasmas Using Trivelpiece-Gould Modes," submitted to *Phys. Rev. Lett.*

Measurement of Cross-Field Heat Transport in a Nonneutral Plasma

E. M. Hollmann, F. Anderegg, and C. F. Driscoll

*Institute for Pure and Applied Physical Sciences,
University of California at San Diego, La Jolla, CA 92093-0319, USA*

Abstract. The cross-magnetic field heat transport in a Mg$^+$ plasma confined in a Penning-Malmberg trap is found to be dominated by novel long-range collisions. The measurement uses two lasers: a strong cooling beam creates a controlled temperature gradient in the plasma and a weak probe beam measures the ion temperature evolution. The temperature gradient drives a heat flux which dominates the measured temperature evolution, with corrections due to small external heating terms. Classical theory of collisional heat conductivity considers only ion-ion collisions with collisionality ν and impact parameter $\rho < r_c$, giving thermal diffusivity $\chi_c \propto \nu r_c^2$. Here, long-range collisions with impact parameter $r_c < \rho < \lambda_D$ can dominate the heat transport, giving $\chi_L \simeq \frac{1}{2}\nu\lambda_D^2$. Initial measurements taken for a plasma with temperature T $\simeq$ .02 eV and density $n \simeq 5 \times 10^7 \text{cm}^{-3}$ at B = 4 T show enhanced heat transport consistent with thermal diffusion dominated by these long-range collisions.

I INTRODUCTION

The study of heat transport in plasmas is an area of extensive research in astrophysics [1], laser physics [2], plasma processing [3], and especially in magnetic fusion research [4]. Quiescent plasmas with a single sign of charge (nonneutral plasmas) are, because of their superior stability and confinement properties, important for basic plasma physics and atomic physics research [5,6]. These properties also make nonneutral plasmas ideal systems for the study of collisional heat transport.

Heat can be transported across the magnetic field from both short-range and long-range collisions. The standard "classical" theory of heat flow across a magnetic field was derived for neutral plasmas in the regime $r_c > \lambda_D$, where $\lambda_D \equiv \sqrt{\frac{kT}{4\pi n e^2}}$ is the Debye shielding length and $r_c \equiv \frac{\bar{v}}{\Omega_c} = \frac{\sqrt{kT/m}}{eB/mc}$ is the cyclotron radius. The resulting thermal diffusivity is given by $\chi_c = \frac{16}{15}\sqrt{\pi}\ln(r_c/b)\nu r_c^2$, where $\nu \equiv n\bar{v}b^2$ is the collisionality and $b \equiv \frac{e^2}{kT}$ is the classical distance of closest approach [7]. Here, the basic transport mechanism is velocity-scattering collisions with impact parameter $\rho \leq r_c$, so that heat is transported across the magnetic field by particles taking cyclotron radius-sized steps. Nonneutral plasma experiments typically operate in the regime $\lambda_D \gg r_c$ because of the Brillouin density limit [8]. In this regime, calculations predict that the thermal diffusivity is dominated by "long-range" collisions, giving thermal diffusivity $\chi_L \simeq 0.49\nu\lambda_D^2$ [9]. Here, the particles feel each others' electric fields on the scale of λ_D and the basic transport mechanism is the tranfer of parallel velocities over distances of order λ_D. The perpendicular velocities of the particles remain fairly constant during these long-range collisions, and the interacting particles remain well-separated in space. There is a slight cross-field drift due to these collisions, which is negligible for the purposes of heat transport calculations, but is important for particle and angular momentum transport [10].

Here, we present heat transport measurements which support the presence of long-range collisional transport. Data is presented for a nonneutral Mg$^+$ plasma with B = 4 T, $n \simeq 5 \times 10^7 \text{cm}^{-3}$, and T $\simeq$ .02 eV. The corresponding equipartition rate, $\nu_{\perp\parallel} \equiv \frac{8}{15}\sqrt{\pi}\ln(r_c/b)\nu \simeq 400/\text{s}$, is large, so that the temperatures parallel and perpendicular to the magnetic field are well-equilibrated, i.e. $T_\parallel = T_\perp = T$. For diffusive heat transport, the local heat flux Γ_q is proportional to the local plasma density n, thermal diffusivity χ, and temperature gradient ∇T:

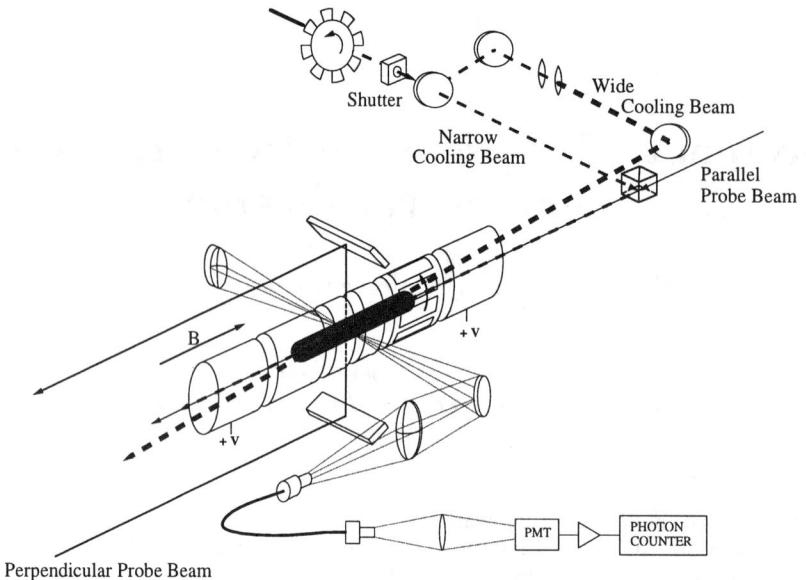

FIGURE 1. Ion trap schematic showing cooling beam and probe beam geometries.

$$\Gamma_q = -\frac{5}{2} n \chi \nabla \mathrm{T} \ . \tag{1}$$

The local energy density, $q \equiv \frac{3}{2} n \mathrm{T}$, changes at a rate related to the heat flux by conservation of energy:

$$\dot{q} = -\nabla \cdot \Gamma_q + \dot{q}_{ext} \ . \tag{2}$$

Here, $\dot{q}_{ext}$ represents the local sum of external heat sinks and/or sources acting on the plasma. We believe $\dot{q}_{ext}$ to be dominated by collisions with the background gas and by Joule heating due to the slow radial expansion of the ion cloud. We measure $\dot{q}_{ext}$ by creating an initial condition with $\nabla \mathrm{T} \simeq 0$, so that the measured heating rate at each point in space is dominated by the external heating term, i.e. $\dot{q} \simeq \dot{q}_{ext}$. For the data presented here, the density n evolves on a time scale which is slow (~ 1000 sec) compared to the time scale on which the temperature T evolves (~ 1 sec), so that $\dot{q} \simeq \frac{3}{2} n \dot{\mathrm{T}}$.

II EXPERIMENTAL SETUP

Figure 1 shows the experimental setup used. The magnesium ions are created with a metal vacuum vapor arc (MEVVA) [11] and are trapped in a Penning-Malmberg trap with uniform axial magnetic field B = 4 Tesla and end-confinement potentials of 200 V. For the data presented here, 5×10^8 Mg$^+$ ions are confined in a plasma cylinder with length $L_p \simeq 14$ cm, radius $R_p \simeq 0.5$ cm, and density $n \simeq 5 \times 10^7$ cm^{-3}. The radial electric field from the ion space charge causes the plasma column to $\mathbf{E} \times \mathbf{B}$ rotate at a frequency of $f_E \simeq 18$ kHz. This rotation is rapid compared to the transport time scales discussed here, so the radial, cross-field transport measurements are azimuthally-averaged. The conducting electrodes used in the trap have a wall radius $R_w = 2.86$ cm. The trap operates with a neutral background pressure of $P_N \simeq 4 \times 10^{-9}$ Torr of H_2.

The ion plasma is held in a steady, near-thermal equilibrium state by a "rotating wall" field applied using a sectored ring at one end of the plasma. The rotating wall torques on the plasma and balances the background drags due to neutrals and trap asymmetries, thus allowing steady-state plasma confinement [12]. Here, the applied field is turned off during the heat transport experiment; however, we find that the results obtained for the thermal diffusivity in this plasma are the same with or without the rotating wall applied.

Heat transport experiments are performed using two continuous 280 nm laser beams: a weak ($\simeq 10$ μW) probe beam is used to nonperturbatively measure the plasma density and temperature, while a stronger ($\simeq 1$ mW) cooling beam is used to cool the plasma. Normally, the plasma relaxes to an equilibrium temperature T $\simeq$ .05 eV, determined by the balance between cooling from neutral collisions and heating due to the

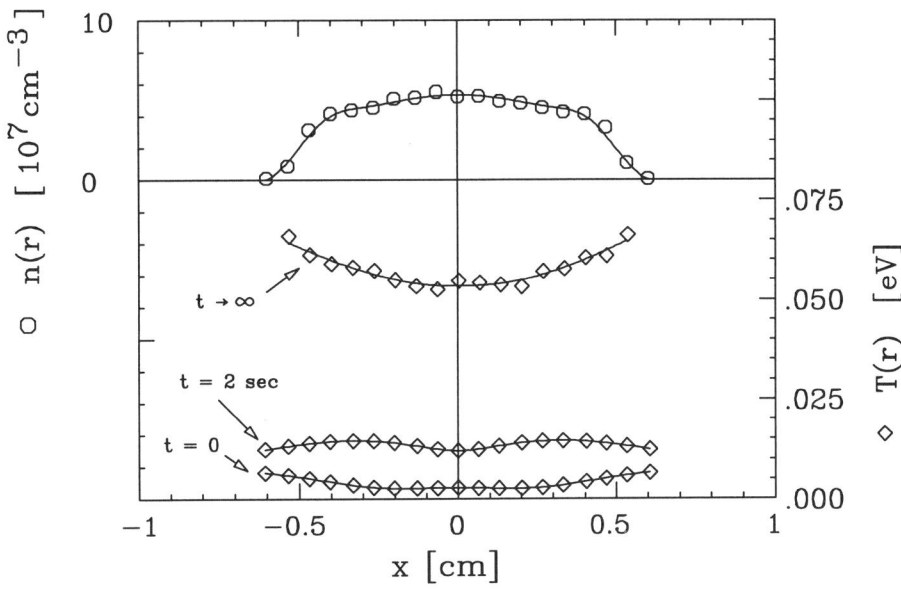

FIGURE 2. Measured plasma heating starting from uniformly cooled initial condition.

slow plasma expansion or due to the rotating wall drive. By detuning the cooling beam slightly off-resonance (- 2 GHz) from the cyclic $3^2 S_{1/2} \rightarrow 3^2 P_{3/2}$ Mg$^+$ transition, the plasma equilibrium temperature is cooled to about 3×10^{-3} eV. This laser cooling can be localized to the center of the plasma cylinder ($r \simeq 0$) by shining the cooling beam along the plasma axis, or made uniform over the whole plasma by widening the beam and shining it at an angle through the column.

The plasma density and temperature evolution, $n(r,t)$ and T(r,t), are measured using the probe beam. The probe beam is scanned in frequency through the cyclic transition at each radial position, and the resulting Doppler-broadened scattered photon signal, $f(v, r, t = 0)$, is fit to three shifted Maxwellians (one for each Mg isotope) to obtain $n(r, t = 0)$, T$(r, t = 0)$ [11]. The probe beam can be used to measure the temperatures both parallel and perpendicular to the magnetic field: for the data presented here, we find the temperatures to be well-equilibrated. The cooling beam is chopped at 50 Hz and the photon counting is performed synchronously with the chopper, so that only scattered light from the probe beam is measured. At t = 0, the cooling beam is blocked with a shutter so that the plasma temperature profile relaxes toward its normal equilibrium value, T$(r, t \rightarrow \infty) \simeq .05$ eV. Here, since the temperature evolution is much faster than the density evolution, i.e. $n(r, t) \simeq n(r, t = 0)$, we can measure the rapid temperature evolution by recording the peak of the scattered signal, $f(v = 0, r, t)$. Repeating experiment at different radii, r, gives the temperature evolution of the entire radial profile, T(r, t).

III MEASUREMENT OF BACKGROUND HEATING TERMS

Measurement of the background heating terms, $\dot{q}_{ext}$, is performed by using the wide cooling beam so that the plasma is cooled uniformly. After the plasma has reached a uniformly-cooled equilibrium, the cooling beam is turned off. The subsequent measured time evolution of the plasma, T(r,t), gives $\dot{q}_{ext}$ through Equations (1) and (2) since ∇T $\simeq 0$. Figure 2 shows the result of this experiment: at $t = 0$, the plasma is cooled uniformly to T$(r, t = 0) \simeq 3 \times 10^{-3}$ eV. When the cooling beam is turned off, the plasma is observed to heat toward T$(r, t \rightarrow \infty) \simeq .05$ eV. For clarity, only T$(r,t = 0)$ and T$(r,t = 2$ sec$)$ are shown; actually 100 time steps ($t = 0$ to $t = 4$ sec) are measured at each radial position.

From the data shown in Figure 2, we obtain an external heating rate which agrees reasonably well with the calculated values for this plasma assuming Joule heating from the plasma expansion plus heating due to collisions with the room-temperature background gas. The Joule heating contribution is estimated by performing an independent experiment to measure the slow density evolution of this plasma, $n(r, t)$, when the rotating wall is turned off. This gives the bulk radial particle flux, Γ_b, from particle conservation: $\partial n/\partial t + \nabla \cdot \Gamma_b = 0$. We find $\Gamma_b \simeq 10^4$ cm^{-2}s$^{-1}(n/10^7$cm$^{-3})(r/1$cm$)$, which gives an estimated Joule heating near

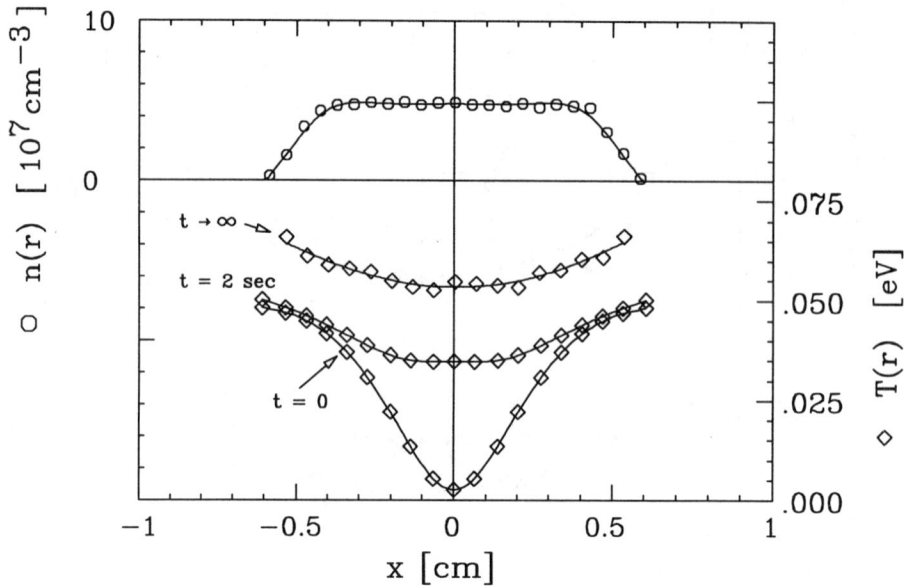

FIGURE 3. Measured thermal diffusion starting from locally (r=0) cooled initial condition.

$r = 0$ of $\dot{q}_J = env_r E_r = \frac{eB}{c}\omega_E r\Gamma_b \simeq 1.8 \times 10^5 \frac{\text{eV}}{\text{cm}^3\text{s}} \rho^2$, with $\rho \equiv (r/0.3\,\text{cm})$. To estimate the heating rate due to the collisions with the neutral gas, we use $\dot{q}_N = \frac{3}{2}n\dot{T} \simeq -\frac{3}{2}n\sigma n_N u(T - T_N)$, where u is the average relative particle velocity and T_N is the effective neutral gas temperature as seen by the rotating plasma: $T_N \simeq .024$ eV $[1 + .4\rho]$. To estimate the collision cross-section, σ, we use the Langevin cross-section [13], $\sigma = 2\pi \frac{e}{u}\sqrt{\alpha/\mu}$, where $\alpha \simeq 5\,a_0^3$ is the polarizability of H_2 and μ is the reduced mass. This gives a neutral heating rate $\dot{q}_N \simeq 2.8 \times 10^5 \frac{\text{eV}}{\text{cm}^3\text{s}} [1 + .4\rho]$ and a total predicted heating rate $\dot{q}_{ext} = \dot{q}_J + \dot{q}_N \simeq 2.8 \times 10^5 \frac{\text{eV}}{\text{cm}^3\text{s}} [1 + .4\rho + .6\rho^2]$. This agrees reasonably well with the measured heating of Figure 2 between $t = 0$ and $t = 2$ seconds: $\dot{q}_{ext} \simeq 3.7 \times 10^5 \frac{\text{eV}}{\text{cm}^3\text{s}} [1 + .1\rho]$.

IV MEASUREMENT OF THE HEAT FLUX

The heat flux experiment is shown in Figure 3. Here, the plasma is cooled locally at $r = 0$, creating an initial condition with a strong temperature gradient with a central temperature $T(r = 0, t = 0) \simeq 3 \times 10^{-3}$ eV. When the cooling beam is turned off, the plasma temperature is observed to rise rapidly toward the equilibrium temperature of $T(r, t \to \infty) \simeq .05$ eV. As in Figure 2, we show only time steps $t = 0$ and $t = 2$ sec for clarity. It can be seen, however, that the central temperature rises more rapidly than is observed in the experiment with no temperature gradient (Figure 2). At early times ($t = .1$ sec), the measured rate of change of the central temperature is found to be about $10\times$ larger than for the data with no temperature gradient. Thus, we have a clear signal of heat flowing radially inward from the edges of the plasma.

The radial heat flux, Γ_q, is obtained from the measured $T(r,t)$ using Equation (2) and the measured $\dot{q}_{ext}(r)$. The best signal-to-noise for Γ_q was obtained at radii $r \simeq 0.1 \to 0.2$ cm. Figure 4 shows the heat flux measured at radii $r = 0.1 \to 0.2$ cm and at times $t = .1 \to 1.9$ sec. Over this range of r and t, the density remains fairly constant, $n \simeq 5 \times 10^7$ cm^{-3}, but the temperature varies by about a factor of four: $T \simeq .01 \to .04$ eV. Because of this, we divide the displayed Γ_q by the expected temperature scaling of the thermal diffusivity, $\chi \propto T^{-1/2}$.

The data in Figure 4 indicates that the heat flux is diffusive in nature and is in reasonable agreement with the predictions of the long-range collisional theory. From the straight (solid) line fit through the data, it can be seen that our model $\Gamma_q \propto T^{-1/2}\nabla T$ describes the data well. Both classical and long-range collisions predict a heat flux which satisfies this scaling. However, as can be seen from the theory predictions for this plasma (dashed lines), the magnitude of the measured heat flux is about $8\times$ too large to be described by classical, short-range collisions alone, but is within reasonable agreement (about 80%) of the predictions of short-range plus long-range collisions.

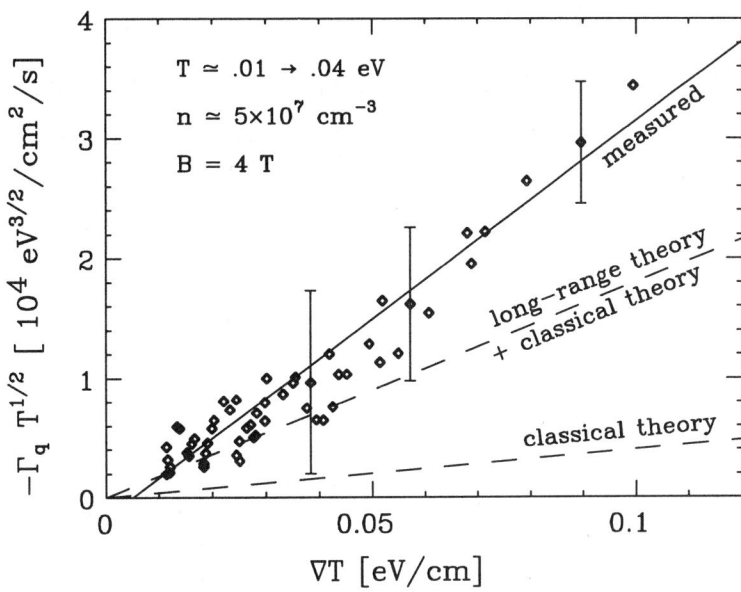

FIGURE 4. Measured radial heat flux vs. temperature gradient. The heat flux is multiplied by the measured $T^{1/2}$ at each point to remove the expected temperature scaling of the thermal diffusivity. Solid line fit passes through data and origin, indicating diffusive heat flux. Fit is within 80% of long-range theory + classical theory, but 8× larger than classical theory alone.

This preliminary measurement provides evidence that the thermal diffusivity in nonneutral plasmas is dominated by long-range collisions with impact parameter of order of a Debye length; the standard thermal diffusivity based on collisions with impact parameter of order the cyclotron radius can therefore usually be neglected for these plasmas. For very large and/or strongly correlated nonneutral plasmas ($R_p > 100\,\lambda_D$ and/or $\Gamma > 1$), the heat conduction is predicted to be dominated by wave (nondiffusive) transfer of energy. Future experiments will attempt to test these predictions.

We gratefully acknowledge the support of the Office of Naval Research (ONR N00014-96-1-0239) and the National Science Foundation (NSF PHY94-21318).

REFERENCES

1. S. M. Ichimaru, H. Iyetomi, and S. Tanaka, Phys. Rep. **149**, 93 (1987); S. Pistinner, A. Levinson, and E. Eichler, Astrophys. J. **467**, 162 (1996).
2. T. Ditmire, E. T. Gumbrell, R. A. Smith, A. Djaoui, and M. H. R. Hutchinson, Phys. Rev. Lett. **80**, 720 (1998).
3. M. Tuszewski, Phys. Plasmas **5**, 1198 (1998).
4. F. Wagner and U. Stroth, Plasma Phys. and Contr. Fusion **35**, 1321 (1993).
5. T. M. O'Neil, in *Non-neutral Plasma Physics*, edited by G. M. Bunce, AIP Conf. Proc. No. 175 (AIP, New York, 1988), 1.
6. J. N. Tan, J. J. Bollinger, and D. J. Wineland, IEEE Trans. Instrum. Meas. **44**, 144 (1995).
7. M. N. Rosenbluth and A. N. Kaufmann, Phys. Rev. **109**, 1 (1958).
8. L. Brillouin, Phys. Rev. **67**, 260 (1945); R. C. Davidson, *Physics of Nonneutral Plasmas* (Addison-Wesley, Redwood City, 1989) p. 42.
9. D. H. E. Dubin and T. M. O'Neil, Phys. Rev. Lett. **78**, 3868 (1997).
10. D. H. Dubin, Phys. Plasmas **5**, 1688 (1998).
11. F. Anderegg, X.-P. Huang, E. Sarid, and C. F. Driscoll, Rev. Sci. Instrum. **68**, 2367 (1997).
12. X.-P. Huang, F. Anderegg, E. M. Hollmann, C. F. Driscoll, and T. M. O'Neil, Phys. Rev. Lett. **78**, 875 (1997); "Rotating Field Confinement of Pure Electron Plasmas Using Trivelpiece-Gould Modes", F. Anderegg, E. M. Hollmann, and C. F. Driscoll, submitted to Phys. Rev. Lett.
13. P. Langevin, Ann. Chim. Phys. **8**, 245 (1905).

The decay instability of Langmuir waves in the non-neutral electron plasma column

H. Higaki

Institute of Physics, Graduate School of Arts and Sciences,
University of Tokyo
Komaba, Meguro-ku, Tokyo 153-8902 Japan

Abstract. The behaviour of large amplitude Langmuir waves is experimentally observed in a nonneutral electron plasma column. It is found that a Langmuir wave whose amplitude is above a threshold decays into lower frequency Langmuir waves. Since the dispersion relation is different from that of the unbounded neutral plasma, the three wave decay process of Langmuir waves becomes possible. In these processes, sidebands due to a large amplitude Langmuir wave satisfy the energy and momentum conservation relation. The threshold of this decay instability depends on the plasma temperature. It becomes lower as the temperature becomes higher.

The theory of an unbounded neutral plasma explains that three wave processes $\omega_k \to \omega_{k'} + \omega_{k''}$ leads to the parallelogram on $\omega - k$ plane. A remarkable feature in this phenomenon is that it has a certain power threshold for initial pump waves. By using the damping rate $\Gamma_{k'}$ and $\Gamma_{k''}$ of the waves, it is shown that the threshold for this process is proportional to $4\omega_{k'}\omega_{k''}\Gamma_{k'}\Gamma_{k''}$ [1,2]. The process in which a Langmuir wave decays into two Langmuir waves are forbidden in the plasma. Because the energy conservation law

$$\omega_k = \omega_{k'} + \omega_{k''} \qquad (1)$$

is not satisfied [3]. However, the dispersion relation of Langmuir waves are dominated by the boundary condition and the density distribution [4-7]. Equation (1) might be satisfied in a non-neutral electron plasma column through a nonlinear process.

A nonneutral electron plasma is confined in the multi-ring electrode trap shown in Fig.1. The uniform axial magnetic field $B = 280$ G for the radial confinement and the electrostatic potential for the axial confinement are supplied. Eleven ring electrodes of 3 cm inner radius are aligned along the axis and mesh electrodes are set on each endside of the trap. These ring electrodes ($No.2 \sim 12$) are grounded through resistors and also used as probes. Mesh electrodes ($No.1, 13$) are negatively biased to $V = -23$ V to form the electrostatic potential for trapping. Each ring electrode numbered by 4, 7 and 10 is azimuthally divided into four sectors for observing axially asymmetric motion of the plasma. A cathode of a spiral tungsten wire is set outside the confinement region [6] and also a set of collectors are installed on the other side. The vacuum chamber is evacuated to 8×10^{-8} Torr. Total axial length available for a confinement is 28 cm.

The experimental procedure is as follows. At first, the potential applied to $No.1$ mesh electrode is slightly changed so that electrons can flow into the confinement region. After injection of electrons, the potential of the electrode is returned to $V = -23$ V and electrons are confined. The plasma temperature becomes less than 0.1 eV through collisions with background neutrals. The electron cyclotron heating of about 784 MHz is

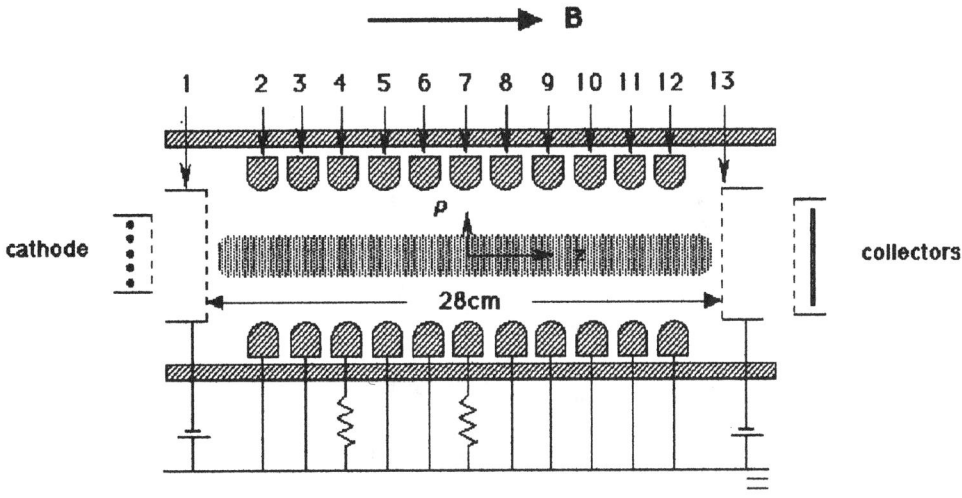

FIGURE 1. Multi-Ring Electrode Trap

used for controlling the plasma temperature up to 1.0 eV. Then at a constant temperature, RF perturbation is applied to excite an axially symmetric Langmuir wave. Oscillations of the plasma are detected with other ring electrodes after the RF excitation is ceased. Immediately after the oscillation measurement, the electrode $No.13$ is grounded to measure N and the radial profile. A plasma temperature is measured by changing the potential of the electrode $No.13$ [8].

The measured linear dispersion relation of Langmuir waves in a nonneutral electron plasma column is shown in Fig.2. In this case, the plasma density on the axis $n \sim 6.9 \times 10^6 cm^{-3}$ ($N \sim 4 \times 10^8$) and the plasma temperature $T \sim 0.2$ eV give the Debye length $\lambda_D \sim 0.13$cm and the electron plasma frequency $\omega_p \sim 148 \times 10^6$ rad/s (23.5MHz). The number for each measured point denotes its mode number ℓ. The solid line in Fig.2 is calculated from the Trivelpiece-Gould dispersion relation with the effective plasma density $n_{eff} \sim 5.0 \times$

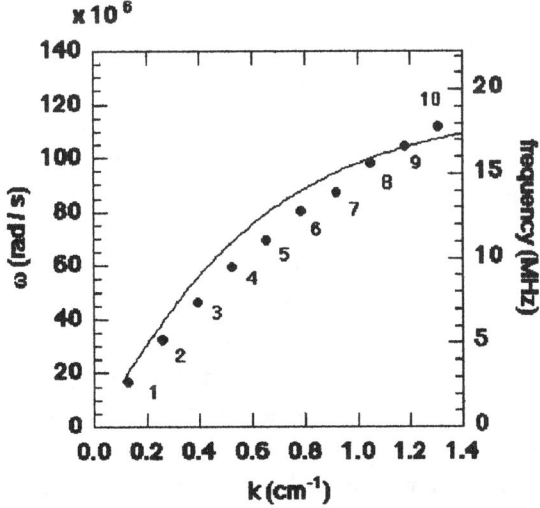

FIGURE 2. The dispersion relation in the electron plasma column.

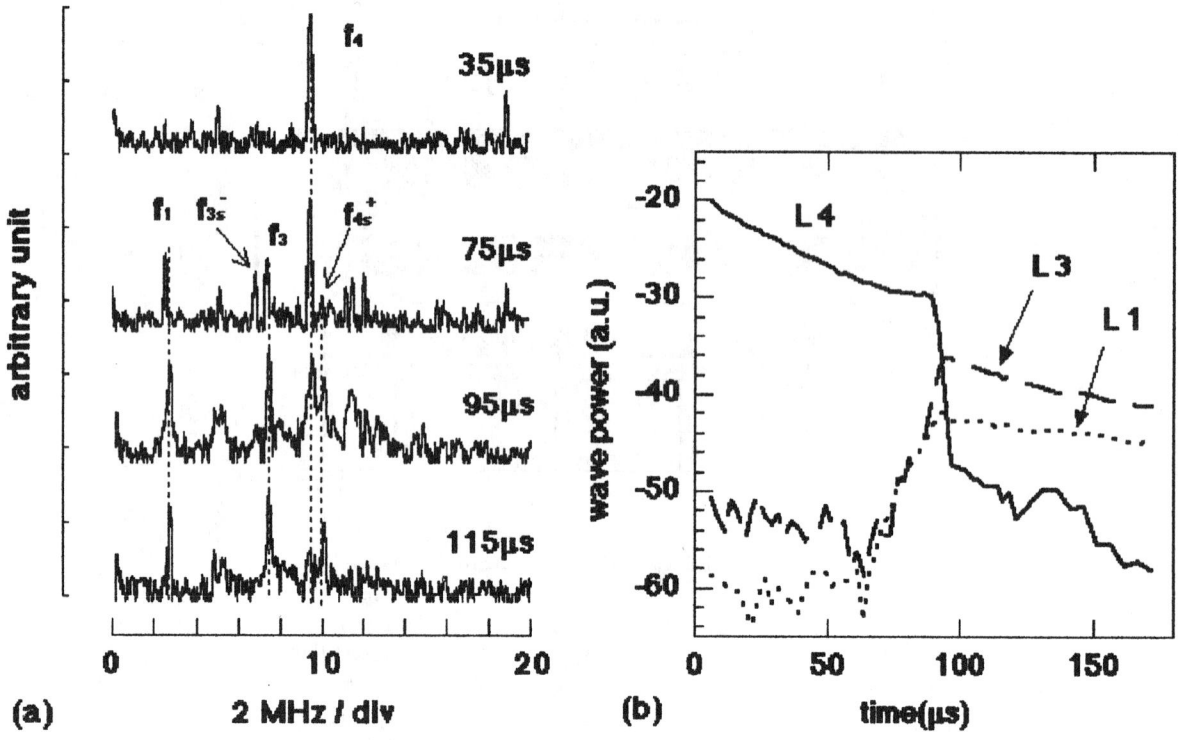

FIGURE 3. The nonlinear decay instability in the electron plasma column.

$10^6 cm^{-3}$. The deviation from the measurement is mainly due to the density gradient in the radial direction. Observed Langmuir waves are standing waves and resonance frequencies become discrete because the plasma is axially finite [4,6,7]. The electrostatic potential of these standing waves has antinodes at the ends of the plasma and the wave number k_ℓ of the mode ℓ is approximately given by $k_\ell = \pi \ell/L$ where $L \sim 24$ cm is the axial length of the plasma [9]. The mode of oscillations can be identified by measuring the phase differences of detected signals. Odd number modes have a node at the center electrode and they are neither excited nor detected with it. On the other hand, even number modes have an antinode at the center. In the case of $\ell = 2$, signals at $No.2 \sim 4$ and $10 \sim 12$ have the same phase and those at $No.5 \sim 9$ have antiphase.

When a large amplitude Langmuir wave is excited in the plasma, nonlinear processes occur. A typical behavior of a large amplitude Langmuir wave whose initial power is beyond the threshold is shown in Fig.3. Hereafter, the Langmuir wave of mode ℓ is represented by L_ℓ. The large amplitude L_4 is excited with the center electrode by applying the RF perturbation of 9.5 MHz during $4\mu s$. The plasma parameters are the same as those in Fig.2. The signal detected at the electrode $No.4$ is fast fourier transformed in Fig.3 (a) and the power of the main three peaks ($\ell = 1, 3, 4$) are plotted as functions of time in Fig.3 (b). It is seen that only the excited L_4 dominates the oscillation of the plasma and it decays exponentially until $70\mu s$. Then L_1 and L_3 begin to grow exponentially from the noise level until $90\mu s$. On the other hand, the decay rate of L_4 changes suddenly at $90\mu s$ and L_4 decays drastically until $100\mu s$. As a result, L_4 decays and L_1 and L_3 are created through this process. In this example, it is thought that the processes $\omega_4 = 2\omega_2$ and $2\omega_4 = \omega_3 + \omega_5$ occur at the same time. These are the decay instability among Langmuir waves, because no other oscillations such as diocotron oscillations and electron cyclotron oscillations can be observed during this process and there

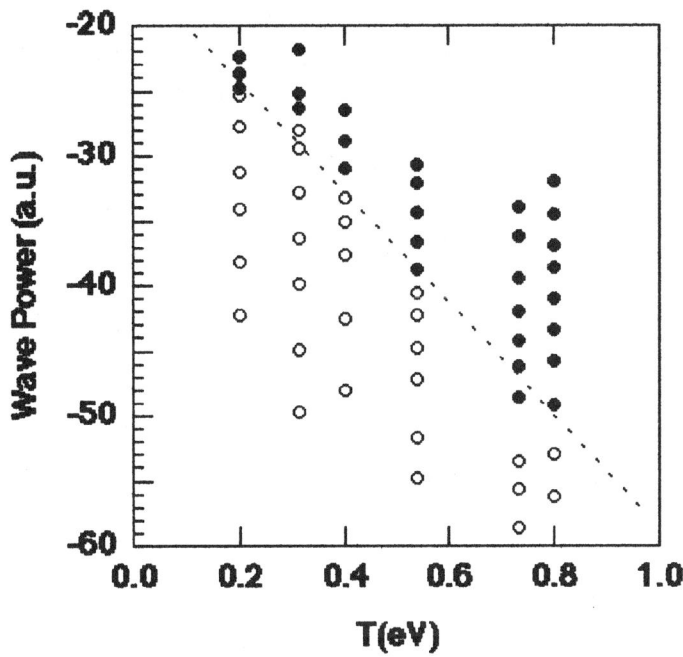

FIGURE 4. The threshold depends on the plasma temperature.

is no ions which lead to low frequency ion sound waves. When the initially excited Langmuir wave is below the threshold, it just decays away exponentially. This is because the mode frequencies of $f_1 = 2.6$ MHz, $f_3 = 7.4$ MHz and $f_4 = 9.5$ MHz do not satisfy the condition $\omega_4 = \omega_3 + \omega_1$. In the case of Fig.3, the excited large amplitude Langmuir wave is accompanied with sidebands of $f_{\ell s}{}^{\pm} = f_\ell \pm 0.5$ MHz and these sidebands satisfy

$$\omega_4 = \omega_{3s}{}^{-} + \omega_1 \quad and \quad \omega_{4s}{}^{+} = \omega_3 + \omega_1. \tag{2}$$

In fact, amplitude oscillation of Langmuir waves which causes sideband instability is sometimes detected by a ring electrode [10,11]. The frequency of this amplitude oscillation is about 0.5 MHz. Therefore, the sideband instability associated with the large amplitude Langmuir wave is fundamental to these decay processes. In addition to the energy conservation relation, the decay process must satisfy the momentum conservation $k_\ell = k_{\ell'} + k_{\ell''}$. In the case of Fig.3, $k_4 \sim k_3 + k_1 (k_4 \sim 0.52, k_3 \sim 0.39, k_1 \sim 0.13)$ is satisfied. This implies that this process is basically a three wave process.

The process shown in Fig.3 is an example and does not mean that L_4 always decays into L_1 and L_3. What kind of process occur in a plasma depends on parameters such as a plasma density, initial amplitude and so on. If the plasma density is lower than that in Fig.3, it is observed that L_4 decays into L_2 and L_1. When the initial amplitude is larger than that of Fig.3, it is observed that L_4 decays into many other modes $L_1 \sim L_5$ (mainly into L_1). The similar process is also observed when the larger amplitude L_6 is excited. It is difficult to figure out these processes only with the three wave or four wave interaction. The fact is that sidebands make the frequency differences of neighbouring modes almost equal and wave energy is transferred to lower modes.

These decay processes have thresholds for the power of the initially excited wave. It is found that the threshold depends on the plasma temperature. Shown in Fig.4 is the case in which L_4 is initially excited. Open circles mean that an initially excited L_4 does not decay into other waves and filled circles mean that L_4 decays into other waves. The obtained results show that the threshold becomes lower as the plasma temperature becomes higher. The temperature dependence of the threshold fits the exponential function $\exp(-\alpha T)$ with

$\alpha \sim 4$, which is represented by the dotted line. When T becomes higher, ω_ℓ and Γ_ℓ become larger. Thus the obtained result is contrary to the theory that threshold is proportional to $4\omega_{k'}\omega_{k''}\Gamma_{k'}\Gamma_{k''}$. It is thought that the proportional coefficient in this system depends on temperature or that the broadening of frequency spectra due to temperature causes this temperature dependence. However, the details are still unclear for the moment.

In summary, it was shown experimentally that the decay instability among Langmuire waves occurs in nonneutral electron plasma column when an excited Langmuire wave exceeds a certain threshold. Not only four wave processes but also a three wave process is possible in the plasma. The threshold for the decay instabilities depends on the plasma temperature and it becomes lower as the plasma temperature becomes higher.

ACKNOWLEDGEMANT

This paper is based on an original which appeared in *Plasma Physics and Controlled Fusion*, **39** (1997) 1793, published by Institute of Physics Publishing Ltd.

REFERENCES

1. Nishikawa, K., and Liu, C. S., *Advances in Plasma Physics vol.6* New York: John Wiley & Sons, 1967
2. Chen, F. F., *Intrduction to Plasma Physics* New York: Plenum, 1974
3. Tsytovich, V. N., *Lectures on Non-linear Plasma Kinetics* Heidelberg: Springer-Verlag, 1995
4. Malmberg, J. H., and deGrassie, J. S., *Phys. Rev. Lett.* **35** 577 (1975).
5. Trivelpiece, A. W., and Gould, R. W., *J. Appl. Phys.* **30** 1784 (1959).
6. Malmberg, J. H., and Wharton, C. B., *Phys. Rev. Lett.* **17** 175 (1966).
7. Dimonte, G., *Phys. Rev. Lett.* **46** 26 (1981).
8. Hyatt, A. W., Driscoll, C. F., and Malmberg, J. H., *Phys. Rev. Lett.* **59** 2975 (1987).
9. Prasad, S. A., and T. M. O'Neil, T. M., *Phys. Fluids* **26** 665 (1983).
10. Wharton, C. B., Malmberg, J. H., and O'Neil, T. M., *Phys. Fluids* **11** 1761 (1968).
11. Starke T. P., and Malmberg, J. H., *Phys. Rev. Lett.* **37** 505 (1976).

Fractional frequency parametric resonances in a Paul trap

M. A. N. Razvi[a,b], X. Z. Chu[a,c], R. Alheit[a], R. Blümel[d] and G. Werth[a]

[a] *Institut für Physik, Johannes-Gutenberg-Universität, D55099 Mainz, Germany*
[b] *Spectroscopy Division, Bhabha Atomic Research Centre, Trombay, Mumbai 400 085, India*
[c] *Department of Electrical and Computer Engineering,
University of Illinois, 1308 West Main Street, Urbana, IL 61801-2307*
[d] *Fakultät für Physik, Albert-Ludwigs-Universität, Hermann-Herder-Str. 3, D-79104 Freiburg, Germany*

Abstract. Excitation of the motional frequencies of clouds of H_2^+ and N_2^+ ions, confined in a Paul trap, by an additionally applied radiofrequency quadrupole field leads to the observation of parametric resonances that are predicted to occur at frequencies $2\omega_z/n$, $n = 1, 2, 3...$, where ω_z is the axial secular frequency. In the case of clouds of N_2^+ ions resonances up to $n = 10$ are detected. They can be explained by parametric instabilities of the center-of-mass motion in the axial direction. The fractional resonances are observable only if the excitation field surpasses a critical strength. We find an odd-even staggering of the thresholds.

I. INTRODUCTION

In addition to its use as a tool in spectroscopy, the Paul trap is an ideal device for the investigation of nonlinear dynamics and complex systems. Many effects initially observed experimentally in Paul traps were eventually explained within the framework of nonlinear dynamics and chaos. Examples are the phenomenon of rf heating [1] and crystallization and melting of Coulomb crystals [2–6]. Another example of a complex system where nonlinear dynamics plays a role are large clouds of H_2^+ or N_2^+ ions stored in a Paul trap under the influence of an additionally applied excitation quadrupole field whose frequency ω is scanned [7,8]. Interesting resonance doublets appear in the vicinity of the discrete excitation frequencies $\omega_n = 2\omega_z/n$, $n = 1, 2, 3, ...$ Our experiments show that one of the resonances in the doublet appears only if the excitation voltage surpasses a threshold value that depends on n. The dependence of the threshold voltage on n is rather peculiar and depends on whether n is even or odd. We call this effect "odd-even staggering" [8]. While we succeeded in explaining most of the features of the resonance doublet at least qualitatively, including the very existence of a threshold voltage, we have not yet succeeded in explaining the odd-even staggering effect.

II. EXPERIMENTAL SETUP AND PROCEDURES

A schematic sketch of our set-up is shown in Fig. 1. Most of the components of our experiment are described in detail in [9,10]. In our current experiments we use a Paul trap with an inner radius of the ring electrode of $r_0 = 2$ cm and end-cap distance of $2z_0 = \sqrt{2}\, r_0$. The trap is operated at a frequency of $\Omega/2\pi = 3$ MHz.

In order to study the nonlinear response of large ion clouds stored in the Paul trap to an additionally applied ac excitation voltage we measure the survival rate of the ions in the presence of the excitation field for well defined excitation frequency, amplitude and interaction time. We refer to this type of experiments as excitation experiments. A typical excitation experiment consists of three stages.

(i) Creation stage
In this stage, of temporal duration T_c, the ions are created inside the trap by electron bombardment of the rest gas which is held at 10^{-9} mbar. We refer to T_c as the creation time. It is typically of the order of 1 s.

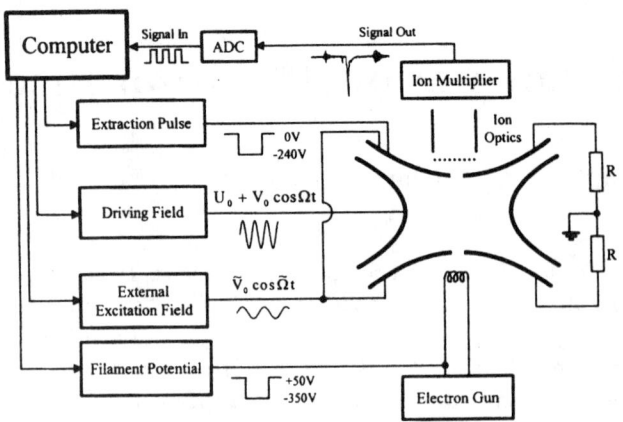

FIGURE 1. Sketch of the experimental set-up.

(ii) Interaction stage
After creating the ions, they are exposed to a superposition of the trap fields and the excitation field during a time T_i, referred to as the interaction time. The interaction time T_i can be changed experimentally from a few ms to arbitrarily long times. In practice, however, an upper limit of T_i is given by the ion storage time which is typically of the order of 8–10 s under our experimental conditions. During the interaction stage both end caps of the trap are electrically connected. This observation is important, since it rules out dipole excitation of the ion clouds.

(iii) Detection stage
Following the interaction stage, we extract the ions with the help of a field pulse through the upper end cap of the trap. The extraction pulse is phase correlated with the trap's ac driving field. The ions arrive at the first dynode of a multiplier tube and create an electron pulse whose total charge is proportional to the ion number. The pulse is amplified and digitized and fed to a PC for further data handling. Different mass ions arrive at the detector at different times. A particular species of ions (e.g. H_2^+ or N_2^+ ions) is selected by setting an amplifier gate at proper timing. The total detection efficiency including ion loss in the time-of-flight region and quantum efficiency of the multiplier is estimated to be 10%.

A motional resonance is detected by a decrease of the ion number arriving at the detector. We should emphasize that for every data point in our experimental observations, the ions are lost from the trap and new ions have to be created for the next point. Thus the ions under investigation do not have a "memory" concerning previous excitations. This is of importance for the shape of the observed resonances.

III. THEORETICAL PRELUDE: WHAT TO EXPECT?

For a first qualitative analysis of a system as complex as an externally excited ion cloud in a Paul trap it is best to resort to the simplest possible model in the hope to capture qualitatively the most prominent effects. Following this recipe we focus only on the center-of-mass coordinates (X, Y, Z) of the ion cloud. Moreover, since in our experiments we investigated mainly axial excitations of the stored ions, we retain only the Z coordinate. In the pseudopotential approximation [1] the equation of motion of the Z coordinate of the cloud perturbed by the additional excitation voltage reads

$$\ddot{Z} + \omega_z^2 Z = F\cos(\omega t) Z, \tag{1}$$

where ω_z is the axial pseudopotential frequency and F is the strength of the additionally applied excitation field. The substitution $\tau = \omega t/2$ turns (1) into the Mathieu equation

$$\ddot{Z} + [a - 2q\cos(2t)]Z = 0, \tag{2}$$

where $a = (2\omega_z/\omega)^2$ and $q = 2F/\omega^2$ are the control parameters of the Mathieu equation. Note that (1) is not the equation of motion of an ordinary driven harmonic oscillator, since Z occurs on the right hand side of the equation. Thus, independently of F, $Z = 0$ is always a solution of (1) and (2). Appreciable amplitudes Z accompanied by appreciable particle loss can be obtained only for a and q combinations which turn the fixed point $Z = 0$ of (2) unstable. It is known from the theory of the Mathieu equation [11] that for small q the Mathieu equation (2) is unstable for $a = n^2$. It follows that (1) is unstable for $\omega_n = 2\omega_z/n$, $n = 1, 2, 3, \ldots$. These instabilities are also called parametric instabilities. Based on this simple theory we obtain a first prediction: Large particle loss at frequencies in the vicinity of the parametric instabilities at ω_n.

So far the theory was Hamiltonian, i.e. no damping was present. In an experiment, however, damping cannot be avoided. Appropriate scaling of the time variable and the addition of a damping term turns (1) into

$$\ddot{Z} + \gamma \dot{Z} + Z = f \cos(\nu \tau) Z. \tag{3}$$

Here γ is the damping constant, f is the scaled excitation strength, ν is the scaled excitation frequency, τ is the scaled time and the dot refers to differentiation with respect to τ. Addition of the damping term changes the qualitative features of the theory. While in the theory without damping 100% particle loss occurs in the parametric instability regions, in the presence of damping the fixed point $Z = 0$ of (3) is stable for small enough, but nonzero f. Thus, in order to induce particle loss at $\omega \approx \omega_n$, f has to exceed a *threshold* f_n. According to a formula stated without derivation in Landau and Lifshitz [12] the critical excitation strength f_n is given by

$$f_n = \alpha_n(\gamma) \gamma^{1/n}. \tag{4}$$

We checked in various representative cases that $\alpha_n(\gamma)$ depends only weakly on n and γ.

A last comment concerns the efficiency of the expected particle loss as a function of the excitation frequency ω. Apparently the particle loss from the cloud is the more efficient the larger the amplitude Z will turn out in the parametric instability regions. In the case of an ideal Paul trap the trapping field is an ideal quadrupole, and in the parametric instability region Z grows exponentially in time without bounds. This holds even in the presence of damping. But in a real Paul trap there are always field imperfections present which add nonlinearities to the trapping field. These nonlinearities limit the maximally achievable amplitude Z at given excitation strength F. Again we refer to Landau and Lifshitz who predict that the form of Z in the regions of parametric instability has a peculiar saw-tooth shape [12], rising sharply, but continuously from $Z = 0$ at the beginning of the instability and falling to zero sharply and discontinuously at the end of the instability. This behavior translates immediately into a prediction for the shape of the particle-loss curve as a function of ω: we should see a saw-tooth shaped particle-loss curve.

In summary, based on the simple theory of parametric instability, we should observe the following three effects: (i) Large particle loss at ω_n, (ii) a saw-tooth shaped particle loss curve and (iii) excitation thresholds that follow the law: $f_n \approx \alpha \gamma^{1/n}$.

IV. EXPERIMENTAL RESULTS

We now compare the theoretical predictions with the results of our excitation experiments described in Section II. For the case of a cloud of H_2^+ ions, Fig. 2 shows the number of surviving ions as a function of the excitation frequency in the vicinity of the $n = 1$ parametric instability. Surprisingly, instead of the single expected instability resonance we obtain a doublet of two resonances, a broad one and a narrow one. A detailed analysis showed that it is the narrow resonance of the doublet that is described by the model of the center-of-mass motion [7]. Thus we called this resonance the "collective resonance" [7]. The broad resonance is explained by the same phenomenon, parametric instability, but now on the level of every single particle in the cloud. In addition to the trap fields, the particles in the cloud also see the space charge of the cloud and the position of the parametric resonance is thus shifted depending on the positions of the individual ions in the cloud. This gives rise to the incoherent part of the doublet and results in a broad resonance feature. Nevertheless, apart from the doublet nature of the parametric resonance, the experiment confirms the predicted instablity at $\omega \approx \omega_1$. Figure 2 also confirms our second prediction, namely the saw-tooth shape of the resonance. This shape is clearly seen in the collective part of the resonance doublet in Fig. 2.

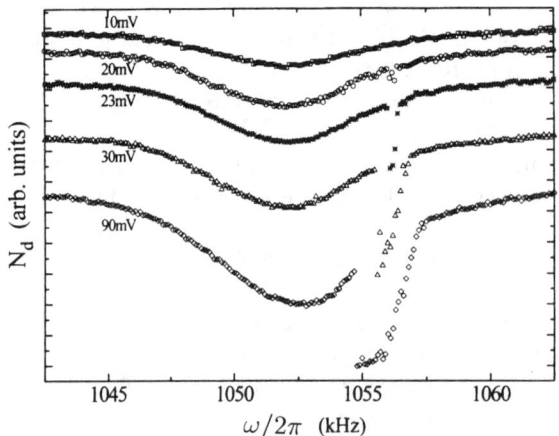

FIGURE 2. Number of detected H_2^+ ions in the vicinity of the $n = 1$ parametric resonance as a function of the excitation frequency for five different excitation voltages. The initial number of ions for the five different curves are the same. We shifted the curves vertically for clarity of presentation.

Recently we also observed the parametric instabilities for $n = 1, 2, ..., 10$ in the case of large clouds of N_2^+ ions. A scan showing these resonances is presented in Fig. 3.

Another prediction of the theory is the existence of excitation thresholds due to the presence of damping. Our experiments confirm the existence of thresholds. This is demonstrated, e.g., in Fig. 2 which shows that in this case efficient excitation starts only for excitation voltages around 20 mV with no excitation at 10 mV and lower. Our experiments with clouds of N_2^+ ions confirm the existence of thresholds also for the higher order parametric resonances ($n = 2, ..., 10$).

We now turn to a check of the threshold law $f_n = \alpha \gamma^{1/n}$. According to this law, plotting the logarithms of the measured critical excitation voltages versus $1/n$ should result in a collection of data points that essentially fall onto a single straight line with a shift that is associated with α and a slope given by $\log(\gamma)$. Figure 4 shows the experimental result. Instead of the expected single line, the data points fall onto two different straight lines depending on whether n is even or odd. This effect was called "odd-even staggering" in [8]. The reason for this effect is currently not known. On the basis of the threshold law $f_n = \alpha \gamma^{1/n}$ the staggering can be interpreted as a manifestation of two vastly different damping constants, $\gamma \approx 2 \times 10^{-3}$ for odd and $\gamma \approx 10^{-6}$ for even n, respectively. The constant of proportionality α, however, appears to be the same for both curves, since they intersect at $1/n \approx 0$.

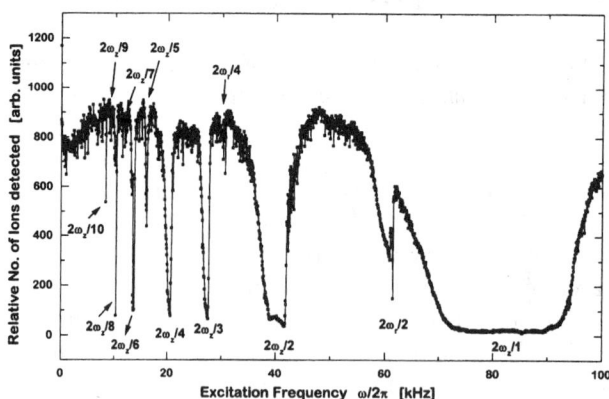

FIGURE 3. Experimental signatures of the first ten fractional parametric resonances for large clouds of N_2^+ ions.

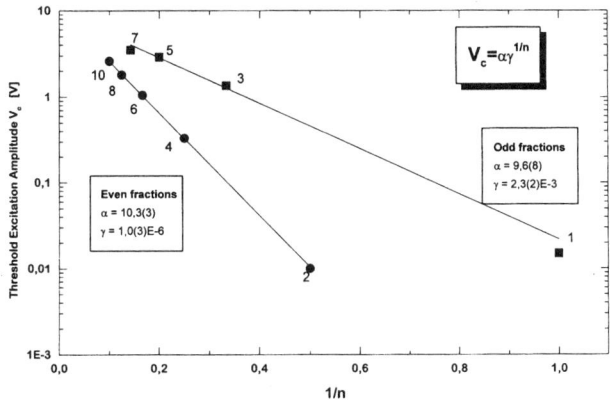

FIGURE 4. The logarithms of the measured critical (peak-to-peak) voltages V_c versus $1/n$ for fractional resonances of order $n = 1, 2, ..., 10$ for N_2^+ ions. The solid line shows a least squares fit of the experimental data points by the function $V_c = \alpha \gamma^{1/n}$. The data point for $n = 9$ is missing due to experimental problems.

ACKNOWLEDGMENTS

The experiment was supported by the Deutsche Forschungsgemeinschaft. M. A. N. R. is grateful for financial support from the DLR International Buro's Indo-German bilateral scientific exchange program. X. Z. C. acknowledges a grant from the Deutscher Akademischer Austauschdienst. R. B. gratefully acknowledges financial support by the Deutsche Forschungsgemeinschaft (SFB 276).

REFERENCES

1. Dehmelt, H., Adv. At. Mol. Phys. **3**, 53–72 (1967).
2. Diedrich, F., Peik, E., Chen, J. M., Quint, W., and Walther, H., Phys. Rev. Lett. **59**, 2931–2934 (1987).
3. Wineland, D. J., Bergquist, J. C., Itano, W. M., Bollinger, J. J., and Manney, C. H., Phys. Rev. Lett. **59**, 2935–2938 (1987).
4. Hoffnagle, J., DeVoe, R. G., Reyna, L., and Brewer, R. G., Phys. Rev. Lett. **61**, 255–258 (1988).
5. Blümel, R., Chen, J. M., Peik, E., Quint, W., Schleich, W., Shen, Y. R., and Walther, H., Nature (London) **334**, 309–313 (1988).
6. Blümel, R., Kappler, C., Quint, W., and Walther, H., Phys. Rev. A **40**, 808–823 (1989).
7. Alheit, R., Chu, X. Z., Hoefer, M., Holzki, M., Werth, G., and Blümel, R., Phys. Rev. A **56**, 4023–4031 (1997).
8. Razvi, M. A. N., Chu, X. Z., Alheit, R., Werth, G., and Blümel, R., Phys. Rev. A **58**, R34–R37 (1998).
9. Alheit, R., Hennig, C., Morgenstern, R., Vedel, F., and Werth, G., Appl. Phys. B **61**, 277–283 (1995).
10. Alheit, R., Gudjons, Th., Kleineidam, S., and Werth, G., Rapid Comm. Mass Spectrom. **10**, 583–590 (1996).
11. Abramowitz, M., and Stegun, I. A., *Handbook of Mathematical Functions*, Washington DC: National Bureau of Standards, 1964.
12. Landau, L. D., and Lifshitz, E. M., *Mechanics*, Oxford: Pergamon Press, 1960.

SECTION 8

FREQUENCY STANDARDS

Lasers for an Optical Frequency Standard using Trapped Hg$^+$ Ions[1]

Brenton C. Young, Flavio C. Cruz,[2] Dana J. Berkeland,[3] Robert J. Rafac, James C. Bergquist, Wayne M. Itano, and David J. Wineland

National Institute of Standards and Technology, Boulder, Colorado 80303

Abstract. We are developing an optical frequency standard based on the narrow 281.5 nm transition of trapped ^{199}Hg$^+$ ions. A major step toward the completion of this standard is the construction of an isolated high-finesse Fabry-Pérot cavity to stabilize the local oscillator. The cavity system that we have assembled has enabled the creation of an optical frequency source with good short-term stability. Eventually, this frequency source will derive long-term stability from a lock to the Hg$^+$ transition. We have recently demonstrated an improved linewidth of 0.6 Hz (40 s averaging time) for a 563 nm dye laser locked to our stable cavity. Additionally, we are developing solid-state laser replacements for gas and dye lasers presently used for driving 194 nm and 281.5 nm Hg$^+$ transitions.

INTRODUCTION

The next major advance for frequency standards probably lies in the development of optical frequency standards. Optical frequency standards are attractive since the potential fractional frequency instability of a quantum system is inversely proportional to the transition frequency. Because optical frequencies are approximately 10^5 times higher than the 9.2 GHz microwave transition used in cesium standards, much higher fractional stability might be achieved in a given measurement time.

In the 1970s, Dehmelt noted that single trapped and laser-cooled ions might be nearly ideal references for optical frequency and time standards [1,2]. High resolution is possible because perturbations can be made small and interrogation times long [1–4]. In addition, laser cooling considerably reduces first- and second-order Doppler shifts [5,6].

Several groups are developing optical frequency standards based on a variety of ions [7–20]. Among proposed standards that use trapped and laser-cooled ions, ^{199}Hg$^+$ ions are attractive because they offer both a suitable microwave and optical transition. Figure 1(a) shows the ^{199}Hg$^+$ electric dipole transitions at 194 nm used for laser cooling, optical pumping, and detection, and the electric quadrupole transition at 281.5 nm that is the reference for the optical frequency standard. Our group has recently demonstrated an accurate microwave frequency standard based on the 40.5 GHz, $F = 0 \to F = 1$, ground-state hyperfine splitting in trapped and cooled ^{199}Hg$^+$ ions [21]. We expect to achieve significant gains in statistical precision, and likely in accuracy, for an optical frequency standard that interrogates the ultraviolet transition, with a frequency over 25 000 times that of the microwave frequency standard.

The optical standard is based on the $^2S_{1/2} \to {}^2D_{5/2}$, 281.5 nm electric-quadrupole transition [20]. An optical oscillator locked to this transition can have a fractional frequency instability approximately equal to 1×10^{-15} at 1 s even for a single laser-cooled ion. However, reaching such low instabilities requires a laser (local oscillator) whose frequency fluctuations are less than approximately 1 Hz during time intervals as long as a few seconds.

One of the main technical difficulties of working with Hg$^+$ is the development of reliable and economical ultraviolet laser sources. Recent advances in solid-state lasers have made possible optical sources at these

---

[1] Work of the U.S. Government, not subject to U.S. copyright.
[2] Present address: Universidade Estadual de Campinas, Campinas, SP, 13083-970, Brazil.
[3] Present address: Los Alamos National Laboratory, Los Alamos, NM 87545.

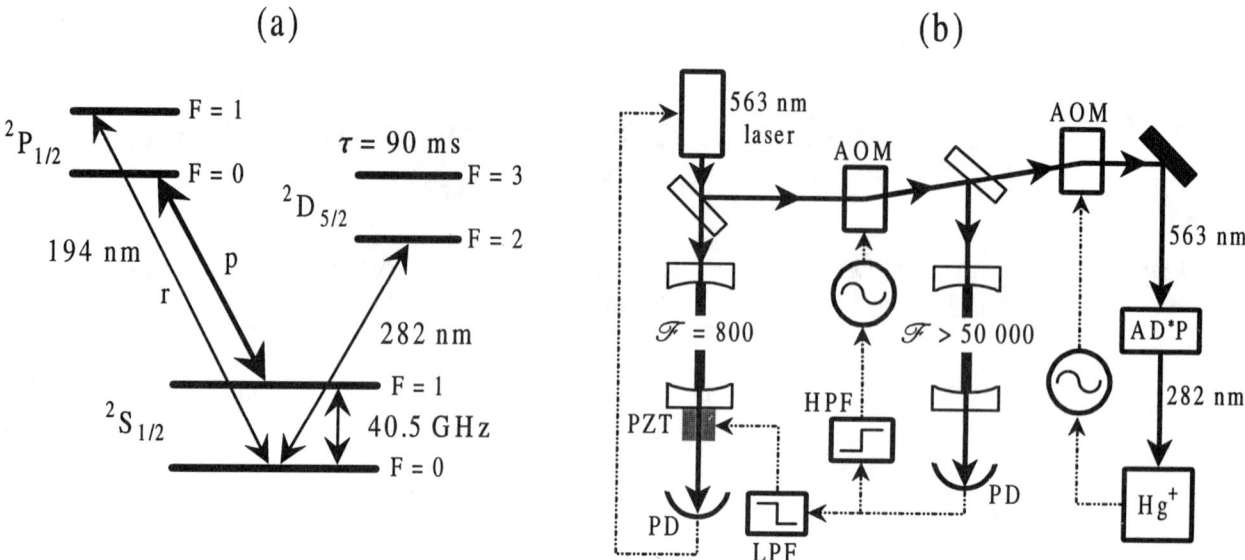

FIGURE 1. (a) Energy level diagram of ^{199}Hg$^+$. We cool the ions using the $^2S_{1/2} \to {}^2P_{1/2}$ transitions at 194 nm. Because the $^2P_{1/2}$, F=0 $\to {}^2S_{1/2}$, F=0 transition is forbidden, transition p is a cycling transition. A second laser on transition r repumps atoms that decayed to $^2S_{1/2}$, F=0 after off-resonant excitation to $^2P_{1/2}$, F=1 by the first laser. (b) Simplified schematic of the proposed optical frequency standard. A dye laser is prestabilized to a Fabry-Pérot cavity ($\mathcal{F} = 800$). Further stabilization to a much higher finesse cavity ($\mathcal{F} > 50\,000$), and eventually to a narrow transition of trapped Hg$^+$ ion(s) should provide a highly stable frequency source. Solid lines denote optical paths and dotted lines represent electrical connections. AD*P, deuterated ammonium dihydrogen phosphate crystal for frequency doubling, AOM, acousto-optic modulator; $\mathcal{F}$, finesse; HPF, high-pass filter; LPF, low-pass filter; PD, photodiode, PZT, piezoelectric transducer.

wavelengths with lower initial costs and operating costs, higher reliability and efficiency, and lower intrinsic noise than for Ar$^+$ and dye lasers. Consequently, we are developing solid-state replacements for our present laser systems.

OVERVIEW OF THE OPTICAL FREQUENCY STANDARD

Figure 1(b) shows a simplified diagram of our proposed optical frequency standard. When interrogating a narrow atomic resonance, the laser must have a frequency width narrower than the transition linewidth to prevent the frequency instability of the laser from limiting the performance of the frequency standard. Consequently, one of the major steps in the development of the optical frequency standard is the construction of an optical local oscillator with sufficient spectral purity. For the 1.7 Hz linewidth ^{199}Hg$^+$ transition, the stability of the standard will not be significantly degraded if the laser linewidth is below 1 Hz for interrogation times as long as a few seconds.

A central component of this system is a high-finesse ($\mathcal{F} > 50\,000$) Fabry-Pérot cavity [20], which is described in detail in a later section. We use a dye laser at 563 nm as the optical source that is locked to this reference cavity. (The light is frequency-doubled to 281.5 nm in a crystal close to the Hg$^+$ trap.) Not shown in Fig. 1(b) is an iodine reference cell that we use to locate the Hg$^+$ transition whenever changes are made to the high-finesse reference cavity.

Rather than locking the laser directly to the high-finesse cavity, we first prestabilize it to a cavity with a finesse of approximately 800 using a Pound-Drever-Hall FM lock [22]. This prestabilization provides several advantages, including an increased locking range, a higher loop bandwidth for the lock, and improved versatility and tunability of the laser. An intracavity electro-optic modulator (EOM) in the dye laser provides high-frequency correction of laser frequency noise. A piezoelectric transducer (PZT) behind one of the dye-laser cavity mirrors eliminates long-term frequency drifts between the dye laser and the cavity. A loop bandwidth of approximately 2 MHz in this prestabilization stage narrows the dye laser short-term ($\tau < 1$ s) linewidth to approximately 1 kHz.

An optical fiber delivers light from the dye-laser table to a vibrationally isolated table that supports the high-finesse cavity. An acousto-optic modulator (AOM) mounted on the isolated table shifts the frequency of the incoming light to match a cavity resonance. Again, we implement the lock using the Pound-Drever-Hall technique. The feedback loop performs corrections at low frequencies by adjusting a PZT on the prestabilization cavity, and at higher frequencies as high as approximately 90 kHz by varying the AOM drive frequency. With the lock enabled, the light entering the cavity has a spectral width less than 1 Hz, as we demonstrate later.

Finally, the frequency-stabilized light is transported to the table holding a cryogenic Hg$^+$ trap. The 563 nm radiation is frequency-doubled to 281.5 nm and is focused onto the trapped ion(s). The final AOM in Fig. 1(b) shifts the frequency of the light to match the ion transition. We plan to interrogate the transition using the Ramsey technique [23], with a Ramsey time of approximately 30 ms. A digital servo loop will adjust the AOM frequency to step between both sides of the central fringe, and will periodically record the values of the center frequency [20].

SOLID-STATE LASERS

The inherent frequency stability of solid-state lasers makes them attractive for metrological applications and precision spectroscopy. In addition, a solid-state laser can be compact, reliable, and long-lived. Reliable, commercial diode lasers do not yet exist in the uv, near the transitions needed for the Hg$^+$ system, but high-power, near-infrared diode lasers are available. Consequently, some groups have frequency-quadrupled the output of near-infrared diode lasers that oscillate at a single frequency and in a single spatial mode to obtain cw, single-frequency uv sources [24,25]. An alternative approach is to frequency-quadruple the output of a cw, solid-state laser that is pumped with high-power multimode diode lasers, as has been done using Nd:YAG and Nd:YVO$_4$ lasers [26,27]. We have taken this latter approach in developing an all-solid-state laser for driving the ^{199}Hg$^+$ $S-D$ transition at 281.5 nm. The solid-state generation of 194 nm radiation for the cooling transition employs a diode-pumped solid-state laser in addition to sum-frequency mixing. These two laser systems are described in the following subsections.

Nd:FAP laser

For the 281.5 nm light source, we plan to replace a dye laser and its multiline Ar$^+$ pump laser with a frequency-doubled Nd^{3+}-doped fluorapatite (Nd:FAP) laser [28]. Nd:FAP has a lasing transition at 1.126 μm that, when frequency-quadrupled, coincides with the Hg$^+$ transition at 281.5 nm. The major difficulty in designing this laser is that Nd:FAP has a much stronger transition nearby at 1.063 μm [29] that must be suppressed by the laser optics. With 680 mW of diode pump light at 808 nm, the Nd:FAP output power is approximately 90 mW at 1.126 μm. Frequency doubling in KNbO$_3$ gives approximately 5 mW at 563 nm.

For the second stage of harmonic generation, we use deuterated ammonium dihydrogen phosphate (AD*P) to frequency-double 563 nm radiation to 281.5 nm [30]. Since less than 1 pW can be enough to saturate the narrow $S-D$ transition [30], we simply frequency-double the radiation at 563 nm in a single-pass configuration. approximately 25 nW is generated at 281.5 nm for 5 mW of input power at 563 nm.

Because the free-running frequency instability of this laser is dominated by low-frequency acoustical and mechanical noise, only a moderate-speed servo system is needed to lock the laser frequency tightly to the resonance of a high-finesse cavity [20,31]. In addition, since the frequency of the Nd:FAP laser is quadrupled to reach the atomic transition, this facilitates the first steps in a frequency chain from the optical to the microwave. We have demonstrated tunability of this laser through the Hg$^+$ transition. Poor stability of the pump-diode output mode, however, has forced us to use the original dye laser source for the experimental work described in the remainder of this paper.

Yb:YAG laser

Currently, the 194 nm cooling light is generated using a single-mode Ar$^+$ laser at 515 nm that is frequency-doubled in β-barium borate (BBO) to 257 nm and is then sum-frequency mixed in BBO with a diode source at

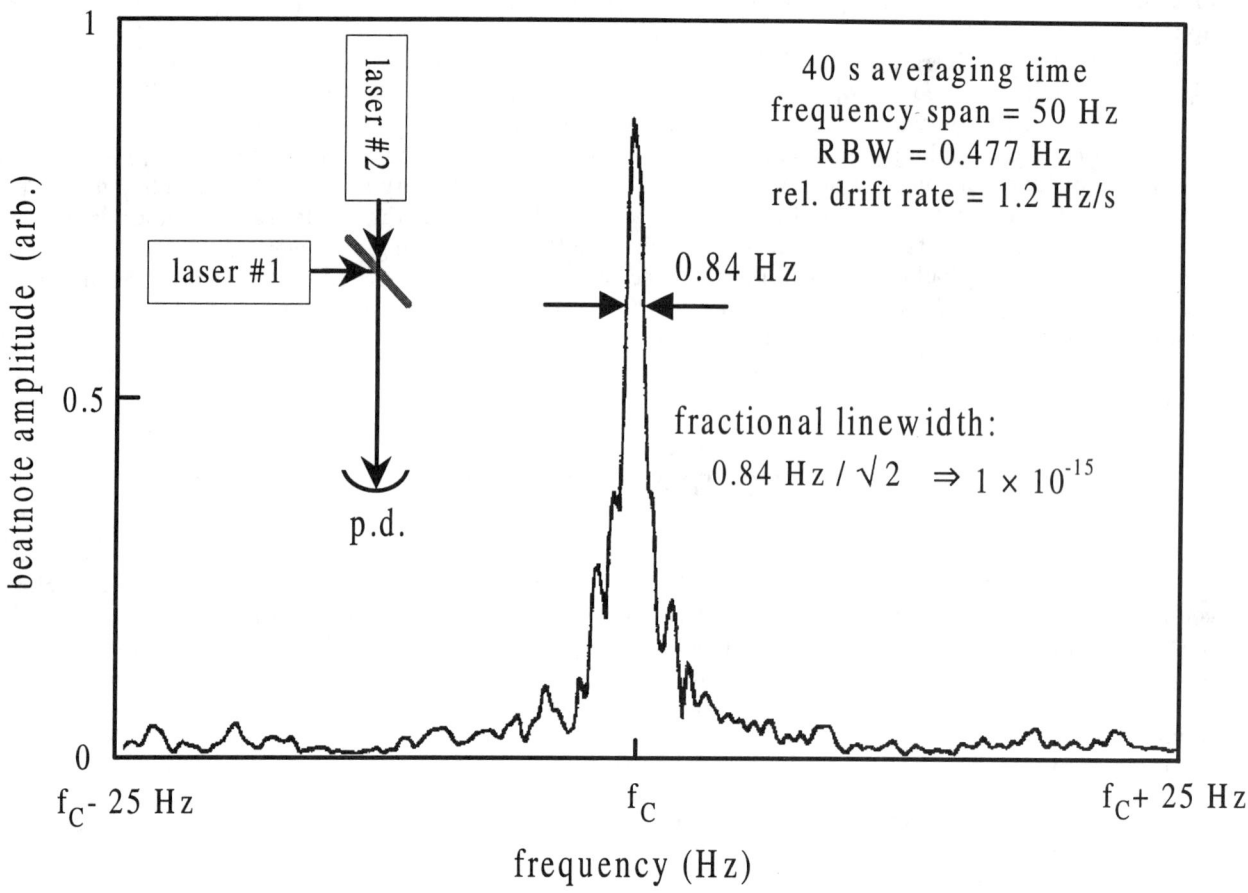

FIGURE 2. Amplitude spectrum of the beat note between two laser beams stabilized to two independent cavities. The dashed line shows the -3 dB level. The averaging time is 40 s. A nearly uniform relative cavity drift of 1.2 Hz/s is suppressed by mixing the beat note with a swept synthesizer.

792 nm to produce light at 194 nm [32]. We plan to replace the Ar$^+$ laser with a Yb:YAG laser at 1.03 μm [33]. With 3 W of diode pump power at 941 nm, the Yb:YAG output power is 1.2 W at 1.03 μm. We have frequency-doubled the Yb:YAG laser output in KNbO$_3$ to obtain over 400 mW at 515 nm, which should be sufficient power to replace the Ar$^+$ laser. At high optical powers, the frequency-doubling conversion efficiency is limited by losses from blue-light-induced infrared absorption [34]. We anticipate achieving a better conversion efficiency by frequency-doubling with lithium triborate (LBO) instead of KNbO$_3$.

HIGH-FINESSE REFERENCE CAVITY

To achieve a laser linewidth of <1 Hz for the source driving the Hg$^+$ ion transition, we start with a high-finesse cavity that has intrinsically low sensitivity to temperature variations, and then take great care to protect it from environmental perturbations. The separation of the cavity mirrors is set by optically contacting the mirrors to the ends of a hollow cylinder made from a low-thermal-expansion material. The mirror substrates are made of the same material as the cylinder. The cavity is supported inside an evacuated chamber by two thin wires. Keeping the cavity under vacuum both avoids pressure shifts of the cavity resonance and thermally insulates it from the environment. The temperature of the vacuum chamber is held near 30 °C, which is the point of zero coefficient of expansion for the cavity material. We protect the cavity from seismic noise by mounting it on a passively isolated optical table. The table is suspended by strands of surgical tubing approximately 3 m long. The fundamental vibrational mode of the suspension has a frequency of ≈0.3 Hz, which provides an isolation from floor noise that exceeds a factor of 50 in noise amplitude already at 3 Hz (some viscous damping is used). To prevent the coupling of acoustic noise into the cavity, we enclose the optical table in a wooden box lined internally with lead foam [35].

The intracavity light heats the mirror coatings, thereby shifting the cavity resonance. To hold this shift at a reasonable value, we couple only approximately 100 μW of 563 nm light into the cavity. Furthermore, controlling the optical power in the cavity stabilizes this power shift. Active control of the rf power driving the AOM stabilizes the output power from the cavity to $\approx$0.1%.

To characterize the cavity's short-term stability performance without referencing to Hg$^+$, we constructed a second cavity and isolated table similar to that described above. Figure 2 shows the spectrum of the beat note between two independent laser beams stabilized to the two cavities. A nearly uniform relative cavity drift of $\approx$1 Hz/s is suppressed by mixing the beat note with a swept synthesizer. The width of the spectrum at its half-power point is 0.8 Hz (40 s averaging time). This implies that at least one of the lasers has a frequency width less than 0.6 Hz at 563 nm, corresponding to a fractional linewidth of only 1×10^{-15}. This is roughly 40 times better than previous results with only one cavity well isolated from vibrations [20], and may represent the smallest fractional linewidth ever recorded in the optical regime.

CONCLUSIONS

We have demonstrated an optical local oscillator suitable for development of a Hg$^+$ optical frequency standard at 281.5 nm. The frequency source has a linewidth of less than 0.6 Hz at 563 nm (40 s averaging time), corresponding to a fractional linewidth of 1×10^{-15}. We have described work on solid-state laser replacements for gas and dye lasers presently used in the trapped Hg$^+$ work. Future work will involve collaboration with other researchers at NIST and JILA to develop a frequency chain for translating the frequency and the stability of our standard into the microwave regime. This work is supported by ONR and NIST.

REFERENCES

1. Dehmelt, H. G., *Bull. Am. Phys. Soc.* **18**, 1521 (1973).
2. Dehmelt, H. G., *IEEE Trans. Instrum. Meas.* **IM-31**, 83–87 (1982).
3. Wineland, D. J., et al., *J. Phys. (Paris)* **42**, C8-307–C8-313 (1981).
4. Fisk, P. T. H., *Rep. Prog. Phys.* **60**, 761–817 (1997).
5. Wineland, D., and Dehmelt, H., *Bull. Am. Phys. Soc.* **20**, 637 (1975).
6. Wineland, D. J., and Itano, W. M., *Phys. Today* **40**, 34–40 (1987).
7. Bergquist, J. C., ed., *Proceedings of the Fifth Symposium on Frequency Standards and Metrology*, Singapore: World Scientific, 1996.
8. Sugiyama, K., Sasaki, K., Wakita, A., and Yoda, J., "Progress toward high-resolution spectroscopy of the $^2S_{1/2}$-$^2D_{5/2}$ transition of laser-cooled trapped Yb$^+$," in *International Workshop on Current Topics of Laser Technology*, 1998, p. 56.
9. Taylor, P., Roberts, M., Barwood, G. P., and Gill, P., *Opt. Lett.* **23**, 298–300 (1998).
10. Engelke, D., and Tamm, C., *Europhys. Lett.* **33**, 347–352 (1996).
11. Barwood, G. P., et al., *Opt. Commun.* **151**, 50–55 (1998).
12. Bernard, J. E., Marmet, L., and Madej, A. A., *Opt. Commun.* **150**, 170–174 (1998).
13. Madej, A. A., et al., "Precision absolute frequency measurements with single atoms of Ba$^+$ and Sr$^+$," in *Proceedings of the Fifth Symposium on Frequency Standards and Metrology*, 1996, pp. 165–170.
14. Urabe, S., et al., *Appl. Phys. B* **67**, 223–227 (1998).
15. Knoop, M., Vedel, M., and Vedel, F., *Phys. Rev. A* **58**, 264–269 (1998).
16. Fermigier, B., et al., *Opt. Commun.* **153**, 73–77 (1998).
17. Peik, E., et al., "Towards an optical clock with a laser-cooled indium ion," in *International Workshop on Current Topics of Laser Technology*, 1998, pp. 23–24.
18. Nagourney, W., Burt, E., and Dehmelt, H. G., "Optical frequency standard using individual indium ions," in *Proceedings of the Fifth Symposium on Frequency Standards and Metrology*, 1996, pp. 341–346.
19. Yu, N., Dehmelt, H., and Nagourney, W., *Proc. Natl. Acad. Sci. U.S.A.* **89**, 7289 (1992).
20. Bergquist, J. C., Itano, W. M., and Wineland, D. J., "Laser stabilization to a single ion," in *Frontiers in Laser Spectroscopy*, 1994, pp. 359–376.
21. Berkeland, D. J., et al., *Phys. Rev. Lett.* **80**, 2089–2092 (1998).
22. Drever, R. W. P., et al., *Appl. Phys. B* **31**, 97–105 (1983).
23. Ramsey, N. F., *Phys. Rev.* **78**, 695–699 (1950).

24. Zimmermann, C., Vuletic, V., Hemmerich, A., and Hänsch, T. W., *Appl. Phys. Lett.* **66**, 2318–2320 (1995).
25. Matsubara, K., *et al.*, *Appl. Phys. B* **67**, 1–4 (1998).
26. Hollemann, G., Peik, E., and Walther, H., *Opt. Lett.* **19**, 192–194 (1994).
27. Kondo, K., *et al.*, *Opt. Lett.* **23**, 195–197 (1998).
28. Cruz, F. C., Young, B. C., and Bergquist, J. C., "Diode-pumped Nd:FAP laser at 1.126 µm: a possible local oscillator for a Hg^+ optical frequency standard," *Appl. Opt.* (to be published).
29. Ohlmann, R. C., Steinbruegge, K. B., and Mazelsky, R., *Appl. Opt.* **7**, 905–914 (1968).
30. Bergquist, J. C., Hulet, R. G., Itano, W. M., and Wineland, D. J., *Phys. Rev. Lett.* **57**, 1699–1702 (1986).
31. Zhu, M., and Hall, J. L., "Frequency stabilization of tunable lasers," in *Atomic, Molecular, and Optical Physics: Electromagnetic Radiation*, 1997, pp. 103–136.
32. Berkeland, D. J., Cruz, F. C., and Bergquist, J. C., *Appl. Opt.* **36**, 4159–4162 (1997).
33. Fan, T. Y., and Ochoa, J., *IEEE Photon. Technol. Lett.* **7**, 1137–1138 (1995).
34. Mabuchi, H., Polzik, E. S., and Kimble, H. J., *J. Opt. Soc. Am. B* **11**, 2023–2029 (1994).
35. Hils, D., Faller, J. E., and Hall, J. L., *Rev. Sci. Instrum.* **57**, 2532–2534 (1986).

Optical Frequency Standard Based Upon Single Laser-cooled Indium Ion

Warren Nagourney, Justin Torgerson and Hans Dehmelt

University of Washington, Department of Physics, Seattle, Washington 98195

Abstract. The current state of our single indium ion work will be presented, with emphasis on the construction of a narrow, solid-state source of tunable UV radiation which will serve as a "clock" laser for the single-ion standard. Indium is shown to be a favorable candidate for a standard with a long term inaccuracy of less than one part in 10^{17}. Its advantages are small systematic errors which are well-understood and relatively straightforward generation of the cooling and "clock" radiation. The stringent requirements on the long term stability of the "flywheel" clock laser and the difficulty in coherently relating the optical clock laser frequency to a convenient microwave source are problems which should soon be overcome.

INTRODUCTION

Despite the current popularity of using a microwave transition in a beam (or fountain) of laser-cooled neutral atoms as the basis for an ultimate atomic clock, a standard based upon an optical transition in a single laser-cooled ion can be shown to have a number of advantages once certain technical problems are solved. The benefits of a single-ion standard are its high (optical) frequency and the relative smallness of its systematic errors, which are also well-understood. The disadvantages are the technical problems in constructing a "flywheel" clock laser source with adequate long term stability and the problem of "counting" this source if a time standard is desired. Indium is probably the best current candidate for a single-ion standard since it has the advantages of a clock transition which is both insensitive to perturbations from external fields and accessible to current solid-state laser sources. Indium also has energy levels which lend themselves to the use of the "shelving" [1] method, which enables one to count every clock transition. The main disadvantage of indium is its relatively weak cooling transition (200 times weaker than that in Hg^+, for example). A standard based upon a single laser-cooled indium ion has a potential inaccuracy of less than one part in 10^{17}.

ADVANTAGES OF INDIUM

The In^+ levels of interest are shown in Fig. 1. Every other element in group IIIa of the periodic table has a similar level system; indium is the current favorite of this group since it has an adequately strong (intercombination) cooling line together with cooling and clock transition frequencies accessible to currently available solid-state frequency-quadrupled laser sources. The lifetime of the upper cooling level is .43 μs [2] and the lifetime of the upper clock level is .14 s [2]. The latter allows an optical frequency standard with 1 Hz jitter in 1 s to be constructed.

As can be seen from the figure, both levels of the clock transition have zero electronic angular momentum. This is the main advantage of group IIIa ions, since it implies that their magnetic field shifts will be due to the nuclear magnetic moment only and will therefore be $\approx$1000 times smaller than the shifts in group II ions (such as barium, calcium, strontium, mercury, etc.). In addition, the small shift from the quadrupole interaction between the trapping field gradient and a possible quadrupole moment of the electronic charge distribution in the ion will be completely absent. Finally, the level structure makes it straightforward to employ the "shelving" double-resonance scheme for detecting clock transitions. Using this method, every clock transition will be observed by a large and easily-detectable change in the number of cooling-laser fluorescent photons.

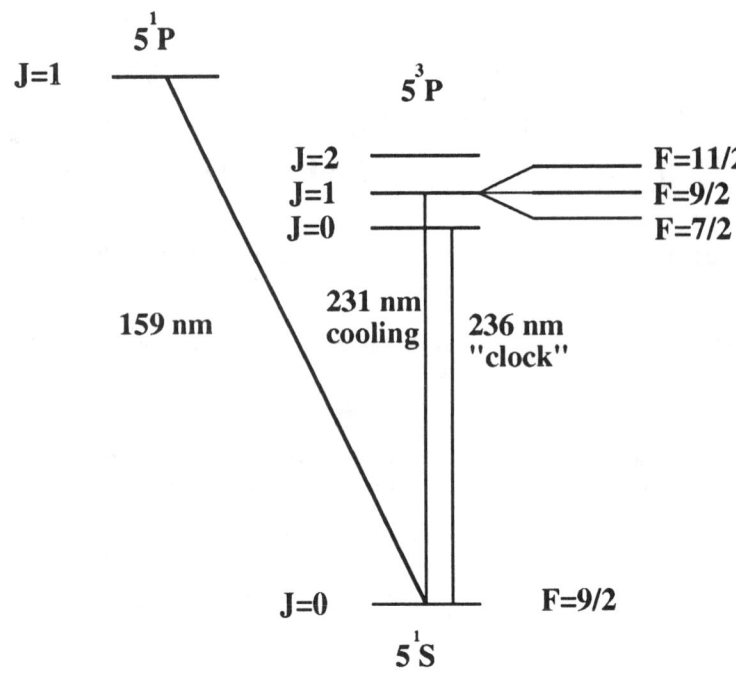

FIGURE 1. Relevant energy levels of In$^+$ showing clock and cooling transitions. The strong singlet to singlet transition is not accessible with current CW laser technology, so the somewhat weaker intercombination line is used for cooling.

This is a form of noise-free amplification, with each (downward) clock transition being promoted into about 300,000 cooling transitions.

It can be shown [3] that all but one of the shifts and broadenings in the indium ion system are less than or equal to 10^{-17} of the clock frequency. The exception is the natural breadth, which is about 1 Hz (10^{-15} of the clock frequency) – this determines the stability of the standard for a given integration time. The first-order Doppler shift is absent since the ion is in the "Lamb-Dicke" [4] regime and the second-order Doppler is less than 10^{-18} of the clock frequency because of the small kinetic energy of the cooled ion. Transit-time broadening is obviously absent as is power broadening with the appropriate detection strategy (infrequent interrogation of the ion and alternation of clock and cooling radiation using the "shelving" scheme). The relative shift due to the external magnetic field is less than 10^{-17} if the field is reduced to the microgauss level and the relative electric field shift is less than 10^{-19} when the ion is cooled to the Doppler limit (about 15 μK). Finally, collisional shifts are less than 10^{-17} of the clock frequency at a room temperature vacuum of 10^{-11} Torr.

EXPERIMENTAL DETAILS

A schematic of the complete standard appears in Fig. 2. The ions are trapped in a small, Paul-Straubel [5] (single ring) radiofrequency trap (ring diameter = 1 mm, trapping frequency = 10 MHz) contained in a quartz enclosure evacuated to a pressure of $\approx 3 \times 10^{-11}$ Torr with an ion pump. Two solid-state frequency-quadrupled lasers supply the cooling and clock radiation and the fluorescence is detected by a photon-counting PMT using a large numerical aperture (NA=.5) Schwartzschild objective, which is well-corrected and intrinsically achromatic (this eases alignment). We are currently using a 462 nm argon-pumped dye laser and BBO-based frequency doubler to generate the cooling radiation, as the solid-state 231 nm source has not yet been built. The figure shows a "tellurium spectrometer" which provides the long-term stability of the cooling laser by locking it to a narrow resonance in saturated Te$_2$; this device employs a dual-pass AO-modulator in the saturating arm which allows the locked laser to be tuned over the cooling transition frequency by tuning the AOM drive frequency. Although the Zerodur cavity to which the clock laser is locked provides long and short term stability for the clock laser, it is possible that a tellurium spectrometer would be also useful here as a "frequency meter" with megahertz resolution.

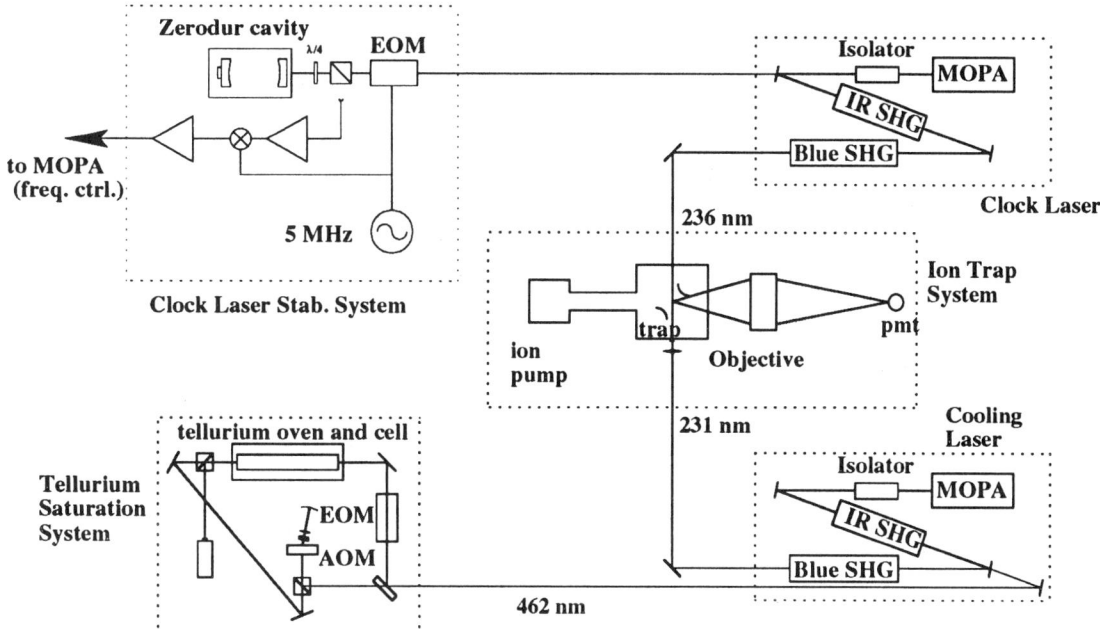

FIGURE 2. Schematic of complete standard showing trap, cooling and clock lasers, clock laser stabilization and tellurium spectrometer used to establish the cooling frequency and provide long-term stability to the cooling laser. The tellurium spectrometer uses an AOM (in dual pass) in the saturating arm to allow one to tune the locked laser by changing the AOM drive frequency

SOLID-STATE CLOCK LASER

Using a commercial Master Oscillator Power Amplifier (MOPA) infrared laser source, we have been able to generate more than 200 mW in the blue and about 25 mW in the UV using two frequency doubling "buildup" cavities. Only about 12 mW of UV is currently available due to reflection losses in a number of optical elements whose surfaces are not anti-reflection-coated.

A schematic of our frequency quadrupled clock laser appears in Fig. 3. The MOPA has a conventional low power external-cavity diode laser "oscillator" which is amplified by a separate tapered amplifier chip, generating about 500 mW of radiation at 946 nm (tunable from 926 nm to 966 nm with little power reduction). Appropriate isolators both inside and outside the MOPA eliminate instabilities due to reflections from the doubling cavities. The separation of the narrow-band oscillator from the amplifier has finally put into practice the wisdom of workers in the analogous radiofrequency domain where it has been known for many years that separate oscillators and amplifiers are necessary to create a stable high-power source. Using a Fabry-Perot spectrum analyzer with 300 MHz free spectral range, the frequency fluctuations of the MOPA were observed using "slope-detection". Most of the frequency jitter is at low audio frequencies (300-600 Hz) with deviations of about 2 MHz; the rest of the noise has somewhat smaller deviations with several kHz modulation frequencies.

The frequency doublers employ build-up cavities to compensate for the intrinsically small nonlinear coefficients of the doubling crystals. The first doubler uses a 3x3x3 mm AR-coated cube of potassium niobate, angle-tuned at 30° to the optical axis. Due to the relatively large doubling efficiency of this material (about 1%/cm-W), the cavity loss is dominated by the depletion of the fundamental as it is converted into the blue. In theory, one should be able to convert most of the 946 nm radiation into the blue; in practice, we only see about 40% conversion, perhaps due to the increased losses at high blue output powers caused by the so-called "blue-light-induced infrared absorption" (BLIIRA). The second doubling cavity uses a 7 mm long crystal of beta-barium-borate (BBO) with Brewster-cut faces angle-tuned at about 60°. Extremely low loss mirrors, coated using the ion-beam process, are used in this cavity to keep the overall losses to a minimum. The result is a power enhancement factor of about 100, which enables us to convert the 200 mW of blue into about 25 mW of UV. The entire system (excluding electronics) is mounted on a 19"x25" aluminum breadboard and the power requirements are quite modest (in contrast to the former argon pumped dye laser system); this should make the entire standard a possible candidate for flight on a future space mission, where the vibrational

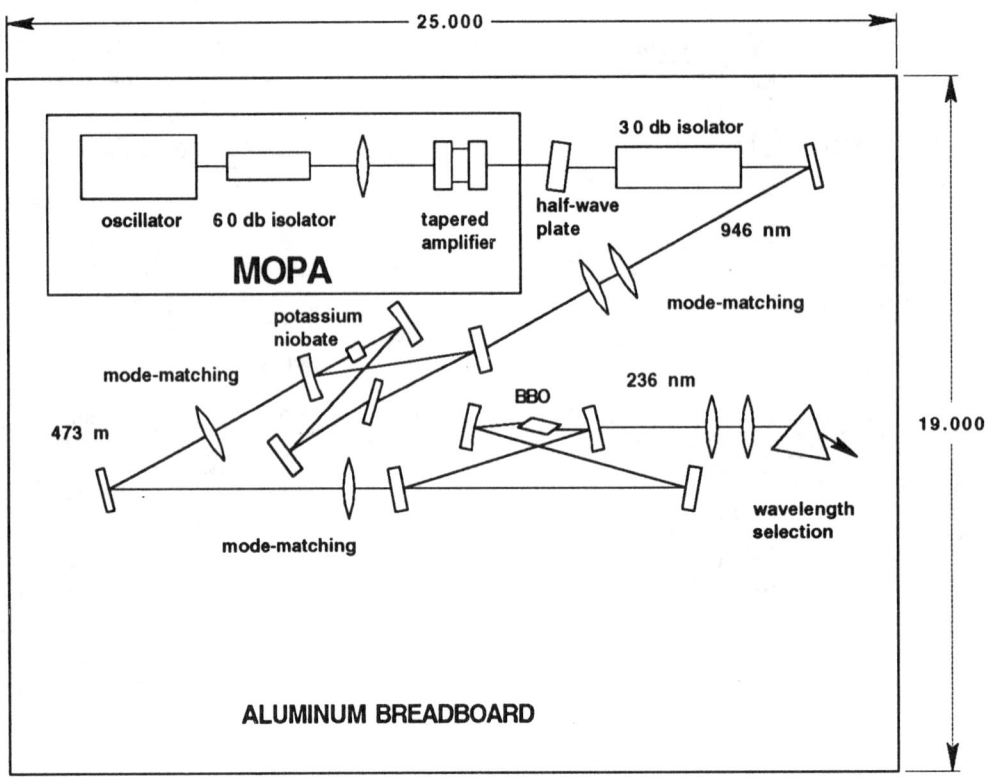

FIGURE 3. Schematic of clock laser using a commercial 500 mW MOPA laser driving a pair of doubling cavities

isolation problem would be more manageable and a number of interesting astrophysical experiments could be performed.

Although the power needed to excite the clock transition is many of orders of magnitude less than 25 mW, high power was sought for two main reasons. First, we are interested in establishing a "proof of principle" that a compact solid-state frequency quadrupled system could be constructed for those experiments needing relatively high power in the UV (for example, a mercury edm experiment). The cooling laser for the In^+ experiment falls into this category, as it is useful to have relatively high power to facilitate initial cooling just after the ion has been loaded. Second, a high power clock laser enormously eases the search for the clock transition frequency, which is known to several hundred MHz at present. Although the clock transition will ultimately be driven with nanowatts of power, searching for the transition with this power level requires a very slow sweep with a very narrow laser and therefore a long search time. With higher power, the search can be done with a broader laser, greatly shortening the search time.

Both cavities are locked to their respective input frequencies using the "polarization" scheme [6], which requires the internal cavity losses to be polarization dependent. The potassium niobate cavity lacks any polarization dependence (the crystal is at normal incidence); installation of a thick quartz plate at Brewster's angle in the nearly collimated arm of the cavity provides the polarization dependence without increasing the cavity loss significantly. This plate also helps compensate for the astigmatism in the cavity due to the non-normal incidence of the curved mirrors. The Brewster-cut BBO crystal provides the polarization dependence and astigmatism correction in the second cavity. The polarization-generated discriminant is integrated and applied to one of the PZT-driven cavity mirrors. By careful adjustment of the loop constants, the cavity tracks the input frequency extraordinarily well. The servo errors were reduced by the use of relatively high resonant-frequency (100 kHz) PZT's and by driving them with high-frequency, high-current "power booster" operational amplifiers. The measured peak amplitude fluctuations of the blue radiation are only .1% of the blue power – part of this is due to the excellent amplitude stability of the MOPA (the tapered amplifier is saturated, which smoothes fluctuations in the oscillator output power) and the rest is due to the high servo gain of the locking system.

The clock laser will be locked to a 30 cm optically-contacted Zerodur cavity placed in an evacuated and temperature stabilized chamber which will be suspended from the laboratory ceiling using bundles of surgical tubing, as has been demonstrated elsewhere [7]. The laser will be locked to the cavity using the Pound-Drever scheme [8] and it will be tuned relative to the (fixed) cavity frequency cavity using an AO-modulator. As of this writing, the cavity chamber is under construction and the clock transition has not yet been seen.

CONCLUSIONS

We are well under way to creating a relatively compact optical frequency standard based upon a single indium ion. The principal remaining problems are the isolation of the clock laser from terrestrial vibrations and the need to "count" the optical frequency if a time standard is desired.

The use of a narrow optical transition of a single ion makes especially stringent demands on the vibrational isolation of the clock laser. The laser needs to drift by less than 1 Hz (clock transition width) between interrogations which occur about once per second (to avoid power broadening). Since this implies that the cavity mirror separation need be constant within 10^{-15} of a meter over this period, extremely good isolation against vibration is absolutely necessary. Either advanced terrestrial vibrational isolation schemes need to be developed or the standard needs to be placed in orbit, where the zero gravity environment makes vibrational isolation much easier, since the cavity suspension need not support the weight of the cavity and can therefore be made extremely weak. The "optical counting" problem is being addressed by chains of lasers which coherently relate an optical frequency to a microwave frequency. One expects that eventual simplification of the latter schemes will make an indium ion clock a possibility in the near future. Before that happens, there are still a number of experiments which would benefit from an ultimate resolution optical *frequency* standard.

REFERENCES

1. H. G. Dehmelt, *Bull. Am. Phys. Soc.* **20**,60 (1975).
2. E. Peik, G. Holleman and H. Walther, *Phys. Rev. A* **49**, 402 (1994).
3. H. G. Dehmelt, *IEEE Trans. Instrum. Meas.* **IM-31**,83 (1982).
4. W. Neuhauser, M. Hohenstatt, P. Toschek and H. Dehmelt, *Phys. Rev. Lett.* **41**, 233 (1978).
5. N. Yu and W. Nagourney, *J. Appl. Phys.* **77**, 8 (1995).
6. T. W. Hänsch and B. Couillaud, *Optics Comm.* **35**, 441 (1980).
7. J. C. Bergquist, W. M. Itano, and D. J. Wineland, in *Frontiers in Laser Spectroscopy,* proc. International School of Physics, Course CXX, ed. by T. W. Hänsch and M. Inguscio, (North Holland, Amsterdam 1994), pp. 359-376.
8. R. W. P. Drever, J. L. Hall, F. W. Kowalski, J. Hough, G. M. Ford, A. G. Manley and H. Wood, *Appl. Phys. B* **31**, 97 (1981).

Frequency Measurement of Visible Light

Fritz Riehle, Harald Schnatz, Burghardt Lipphardt, Götz Zinner, Tilmann Trebst, Jürgen Helmcke

Physikalisch-Technische Bundesanstalt
Bundesallee 100, D-38116 Braunschweig, Germany
E-mail: juergen.helmcke@ptb.de

Abstract. Frequency measurements of the Ca optical frequency standard based on an ensemble of cold atoms are described. They represent the first *phase-coherent* frequency measurement of visible radiation directly related to the primary standard of time and frequency. The optical frequency was provided by the Ca inter-combination line ($^1S_0$-$^3P_1$). Sub-kHz resolution close to the natural linewidth of the optical clock transition was observed by time-separated pulsed excitation. The measured frequency is ν_{Ca} = 455 986 240 494.13 kHz with a total relative uncertainty of $2.5 \cdot 10^{-13}$.

Keywords: Optical frequency standards, optical frequency measurements, laser cooling, precision laser spectroscopy

INTRODUCTION

The development of frequency stabilized lasers based on cold atomic ensembles has opened the door to a new generation of optical frequency standards of unprecedented low uncertainty. Compared to standards operating in the microwave range, the use of optical frequencies is advantageous since a high quality factor $Q = \nu / \delta\nu$ and consequently a high frequency stability can be achieved at short interaction times of the absorbers with the optical radiation. Excellent reference lines are provided, for example, by ions as e.g. Yb$^+$ (1), In$^+$ (2), and Hg$^+$ (3), or by atoms, as e.g. silver (4), hydrogen (5) or alkaline earth atoms (6). The use of such a standard requires the knowledge of its frequency which has to be measured in relation to the primary standard of time and frequency, the Cs atomic clock. The basic difficulty of optical frequency measurements is caused by the fact that the frequency of visible radiation is approximately a factor of 50 000 higher than that of the Cs-clock and no means of direct electronic frequency counting exist for visible radiation.

The first optical frequency measurements were performed in the infrared range (7) by the use of harmonic mixing. With this method, the frequencies of CO_2 lasers and methane-stabilized He-Ne lasers have been determined (8). Extensions of frequency measurements from the IR to the visible were mostly performed by interferometric wavelength comparisons (9, 10), which in some cases were combined with optical frequency mixing (11). Since that time, the accuracy and the range of frequency measurements have been further extended leading to precise frequency measurements of the iodine-stabilized He-Ne laser (12, 13) and of the 1S-2S-transition of hydrogen (5). These measurements were based on frequencies of secondary standards which in turn were linked in separate measurements to the Cs-clock. In these cases, the secondary standards themselves contributed to the uncertainty of the measured frequency value. With the continuous improvement of the reproducibility of optical frequency standards, the demand on the uncertainty of optical frequency measurements was also steadily increasing. Consequently, the frequency of a Ca optical frequency standard at $\lambda \approx 657$ nm has been measured by a frequency chain which allows a phase-coherent link to the Cs atomic clock (14).

This contribution discusses frequency measurements of a Ca optical frequency standard based on cold and trapped atoms. In the next section, we first present the Ca optical frequency standard including the lasers, the source of cold Ca atoms, and the frequency stabilization. The following section then describes the phase-coherent measurement of its frequency and gives an estimation of the frequency uncertainty. The paper is concluded by an outlook briefly presenting novel methods of optical frequency measurements.

CALCIUM OPTICAL FREQUENCY STANDARD

Reference frequencies in the visible range are provided by optical frequency standards. In such a standard, the radiation of a single frequency laser passes through an ensemble of atoms, molecules, or ions providing an absorption line suitable as a reference for the stabilization. If the laser frequency is tuned across the absorption line, a part of the power is transferred from the laser beam to the absorber and an absorption feature is detected versus the laser frequency. The stabilization circuit converts this absorption signal to an error signal which is used to servo-control the laser frequency to the line center.

Among the alkaline earth atoms, Ca is very attractive: The intercombination transition $^3P_1$ - $^1S_0$ is well suited as a reference for the stabilization (clock transition). It has a narrow natural linewidth of $\delta v \leq 400$ Hz corresponding to a high Q factor $v / \delta v > 10^{12}$ and the transition with $\Delta m_J = 0$ has only small quadratic dependencies on electric and magnetic fields. In addition, the $^1P_1$ - $^1S_0$ transition is suitable for laser cooling (15) and trapping (16) to reduce the residual influence of the Doppler effect and to increase the effective interaction time of the light with the atomic samples. The most abundant isotope ^{40}Ca has no hyperfine structure, therefore each atom in the ground state is available for the excitation and can contribute to the signal. As an important technical aspect, the clock transition ($\lambda \approx 657$ nm) and the cooling transition ($\lambda \approx 423$ nm) can be excited by small and efficient diode lasers. We have therefore concentrated our efforts on the development of an optical frequency standard based on cold Ca atoms and on the measurement of its frequency.

Experimental Setup

The development of a Ca frequency standard requires a high resolution laser to probe the clock transition and a laser to cool and trap the atoms. The radiation of the "clock laser" is generated by a laser spectrometer comprising either a dye laser (17) or a diode-laser in an extended cavity configuration (18). In either case, the laser is stabilized to a suitable eigenfrequency of an optical resonator which acts as a "flywheel" to provide sufficient short-term stability. The resonators consist of the low expansion materials Zerodur M and Zerodur for the dye laser and the diode laser, respectively. The mirrors are wrung to the spacer. The resonators are suspended by thin wires inside temperature controlled vacuum tanks to shield them from acoustic and thermal disturbances of the environment. For the stabilization, the method of Pound, Drever, and Hall is applied (19). The spectra of the lasers were investigated by measuring the beat signal between the dye laser and the diode laser. The full width of the beat spectrum at -3 dB was ≤ 0.6 kHz, demonstrating a linewidth well below one kilohertz for each laser. We have observed a constant frequency drift of the laser spectrometer of less than 0,7 Hz/s caused by a slow shrinking of the length of the reference resonator (20). The determination of this drift allowed to predict the frequency of the spectrometer to a few parts in 10^{13} for time intervals of several hours.

The radiation at $\lambda \approx 423$ nm for cooling and trapping was generated either by a dye laser pumped by an UV argon laser or more recently by frequency doubling the radiation of a diode laser. Utilizing $KNbO_3$ as nonlinear crystal placed in an optical ring resonator, we have obtained up to 50 mW blue radiation at a fundamental power of 180 mW with a typical (day to day) second harmonic power in the range of 30 mW (21).

In our experiments, we have used two different setups. In the first apparatus, an effusive Ca beam with a mean velocity of 700 m/s was decelerated by a counter-propagating laser beam and subsequently deflected by a tilted one-dimensional optical molasses (22). The deflected beam was then injected into a magneto-optical trap (MOT). Typically about 10^7 Ca atoms were stored in this trap (23). In a second set-up, the Ca atoms effusing from an oven (≈ 900 K) are directly injected into a second MOT. In this case only the low velocity atoms from the thermal distribution could be trapped resulting in up to 10^6 stored Ca atoms. The root-mean-square velocity of the trapped atoms was less than 1 m/s. Both traps have been used in our optical frequency standards.

Excitation of the Clock Transition

To avoid any light shift and Zeeman shift of the atomic reference frequency, the trapping fields (laser fields and magnetic quadrupole field) were switched off before the clock transition was probed. Consequently, trapping of Ca atoms, probing of the clock transition, and its detection were performed in sequence: First, the trapping fields were switched on for approximately 15 ms and the Ca atoms were trapped. About 0.5 ms after the trapping fields had been switched off and allowed to decay, the clock transition was probed and subsequently, the fluorescence from the excited ($^3P_1$-) atoms decaying into the ground state was detected by a photomultiplier.

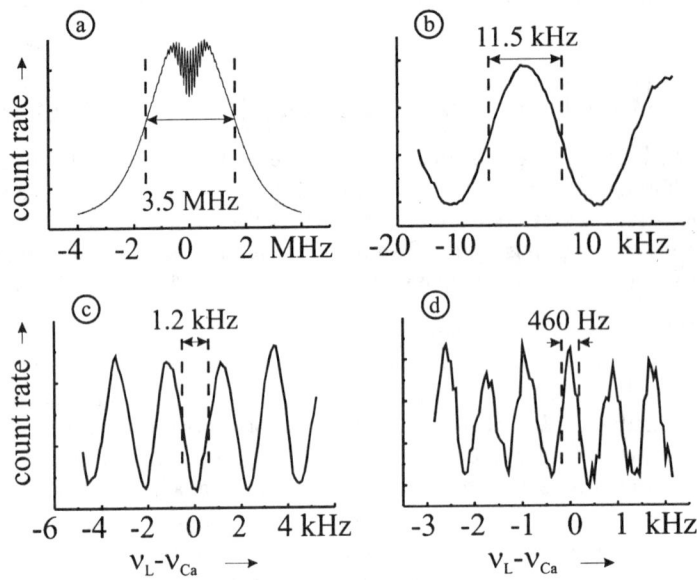

Figure 1. Ca atom interference structures observed by time separated pulsed excitation of the clock transition for various pulse separations T (24): a) The interference structure appears in the center of the Doppler broadened incoherent background of the fluorescence signal. The resolution of the interference structure increases with increasing T. The period of the interference is equal to the recoil splitting $\delta\nu_{rec}$ = 23.1 kHz = $1/2T$ ($T \approx$ 21.6 µs) in curve b), to $\delta\nu_{rec}/10$ in curve c), and to $\delta\nu_{rec}/25$ in curve d).

In order to achieve high spectral resolution combined with a good S/N ratio, we have applied the method of separated field excitation which is well established for the excitation of atomic beams (25, 26). The clock transition was probed either by three pulses of standing laser fields or by a sequence of two counter-propagating pairs of pulsed traveling laser waves. If the laser frequency is tuned close to the resonance, the fluorescence intensity contains a contribution which varies with the cosine of the laser detuning (Fig. 1). It was shown by Bordé that this oscillating behavior in the fluorescence intensity is caused by an atom interference generated by the excitation with separated fields (27). In fact, the signals shown in Fig. 1 represent a superposition of two interference signals caused by the two recoil components which are separated by 23.1 kHz. Since the Doppler broadening of the cold atoms is still orders of magnitude bigger than the natural linewidth, an excitation with short laser pulses strongly increases the interaction time broadening and thereby the number of accessible atoms, resulting in an enhanced fluorescence signal. The necessary high spectral resolution is then achieved by a sufficiently large time separation T between two consecutive pulses (28). If the length of the pulse is small compared to their separation, the width of the interference fringes $\delta\nu \approx 1/(4T)$ is inversely proportional to T. The narrowest fringe width obtained in our standard was 460 Hz which is close to the natural linewidth of the clock transition.

The spectral resolution $\delta\nu = \nu/Q$ and the signal-to-noise ratio (S/N) of the interference structure (depending on the averaging time τ) are important parameters of a frequency standard. The instability of the frequency $\sigma(\tau)$ is determined ultimately by $\sigma(\tau) \approx (S/N \cdot Q)^{-1}$. The observed values of S/N and $S/N \cdot Q$ are shown in Fig. 2 versus the pulse separation T, i.e. versus the spectral resolution. With increasing T, the S/N ratio decreases due to the residual phase noise of the laser and to the natural linewidth of the clock transition. With our present setup, we obtain the maximum value of $S/N \cdot Q \approx 5 \cdot 10^{12}$ at $T \approx 350$ µs and $\tau = 3.2$ s. In most cases, we have performed the stabilization at a fringe width of approximately 1.2 kHz, i.e. at a resolution which is slightly larger than the optimum value for maximum stability.

The error signal for the stabilization is generated from the interference signal by modulating the laser frequency and simultaneously measuring the fluorescence intensity. In the most straightforward way, the frequency is square-wave modulated between two discrete values with the mean frequency tuned close to the center of the central fringe. The difference of the corresponding fluorescence intensities is used as error signal. This method corresponds to a first harmonic detection (1f-method) of a servo-control system using analog electronics and harmonic modulation. The maximum slope is obtained for a total modulation width of $\delta\nu_{mod} = 1/4T$, i.e. if the frequency alternates between the points of maximum slope of the interference signal. After the detection, the error signal is added to the signal controlling the laser frequency leading to a digital integrating servo control. The linear drift of the reference resonator can be determined by the servo-control and compensated by adding a corresponding feed-forward signal to the signal controlling the laser frequency.

The 1f-method can be applied to signals of which the background is symmetric to the center of the absorption line. However, slight asymmetries of the background may occur since the interference structure is superimposed to an incoherent background (see Fig. 1a) caused by the saturation dip and the Doppler broadened spectrum since the Doppler profile is blue shifted by half of the recoil splitting ($\approx 11{,}5$ kHz). In servo control-systems using *analog* electronics, the influence of such asymmetries is reduced by detecting the clock transition at the third harmonic of the modulation frequency (3f-method). We

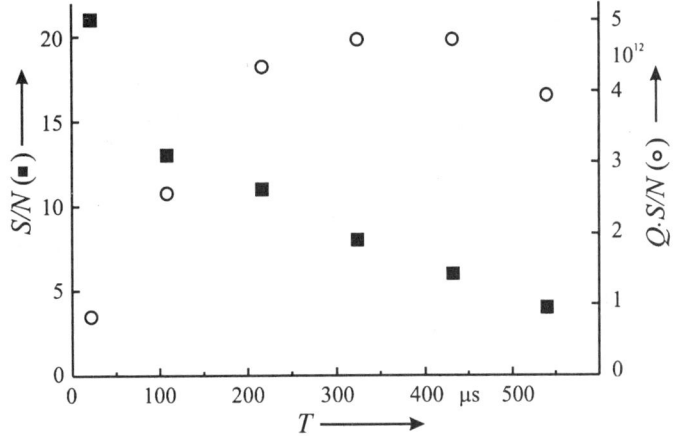

Figure 2. Signal-to-noise ratio (S/N) and $Q \cdot S/N$ versus the pulse separation T (see text).

have approximated this method by measuring the fluorescence signal $S(\nu)$ at the following four frequencies $\nu = \nu_L \pm \delta\nu_{res}/4$ and $\nu = \nu_L \pm 3\delta\nu_{res}/4$ where ν_L is the mean value of the modulated laser frequency and $\delta\nu_{res} = 1/2T$ corresponds to the period of the interference structure. The error signal $E(\nu_L)$ was then calculated as follows:

$$E(\nu_L) = 3 \cdot [S(\nu_L + \delta\nu_{res}/4) - S(\nu_L - \delta\nu_{res}/4)] - [S(\nu_L + 3\delta\nu_{res}/4) - S(\nu_L - 3\delta\nu_{res}/4)].$$

In this equation, $E(\nu_L)$ is not sensitive to a linear and a quadratic frequency dependence of the background. We have checked the suppression of the background by stabilizing the laser frequency alternatingly with the 1f- and the 3f-method. The results obtained at a fringe period of 23.1 kHz show that the shift of the central fringe is in the range of 7 Hz for the 1f-method. The residual shift from the 3f-method was below the detection capability of our setup. A calculation from the measured spectrum resulted in a 50-fold suppression compared to the 1f-method (24).

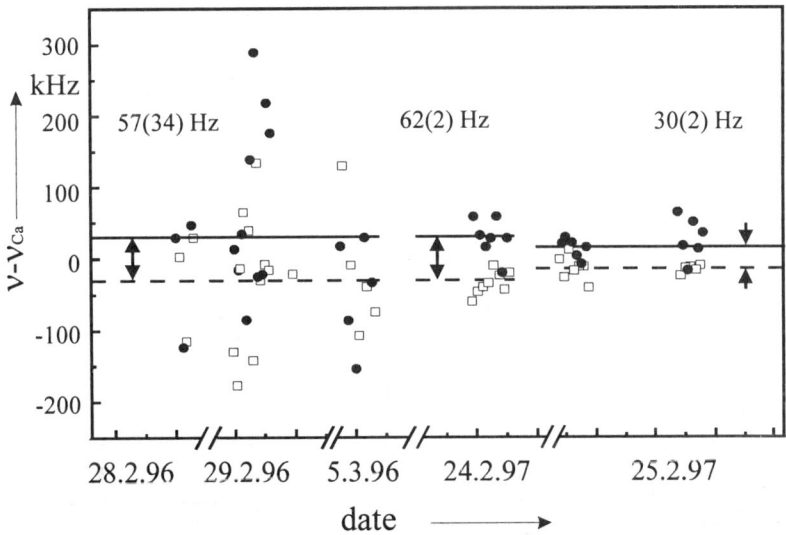

Figure 3. Difference between the transition frequencies realized by two independent ensembles of cold Ca atoms. The reference for the measurements of the first three days was an effusive Ca beam and for later measurements a stable reference resonator. The trap was filled by a laser cooled beam (full dots) or by a thermal beam (open squares, see text).

Comparisons between the transition frequencies realized by the two independent magneto-optical traps have been performed during the past year stabilizing the laser frequency subsequently to the atomic clouds in the two traps. As a reference, we used either an effusive Ca beam or an optical reference resonator. The comparisons show (Fig. 3) that the mean values of the frequencies realized with the two traps differ by less than $10^{-13}\nu_{Ca}$.

OPTICAL FREQUENCY MEASUREMENT

This section describes the frequency measurement of the Ca optical frequency standard. The principle of harmonic mixing is used. The large frequency ratio between the Cs atomic clock (9.2 GHz) and the optical frequency (456 THz) is bridged in several steps leading to a "frequency chain". In each step of the chain, the frequencies of two (or more) oscillators are combined and compared. The lower frequency of the one oscillator is multiplied by a nonlinear device and the corresponding

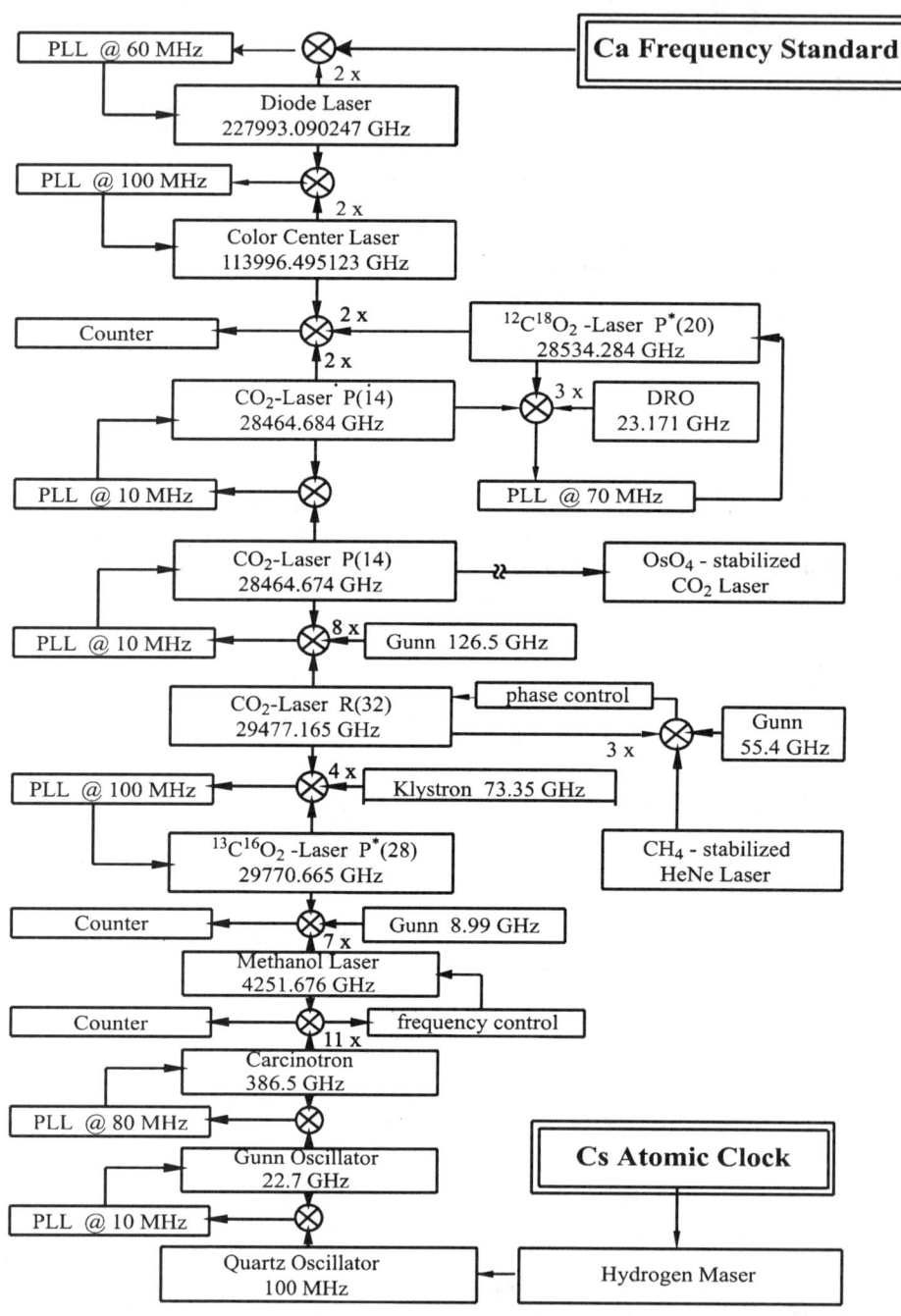

Figure 4. Schematics of the optical frequency chain from the Cs clock to the Ca optical frequency standard.

harmonic is compared to the higher frequency of the second oscillator which is operating close to the harmonic (7). The corresponding beat frequency is detected and one of the two oscillators is servo-controlled by locking the phase of the beat frequency to that of a fixed frequency. Utilizing this method, the frequencies of the two oscillators are phase-coherently related to each other. The nonlinear devices used in the chain are Schottky diodes ($\nu \leq 5$ THz), metal-insulator-metal (MIM) diodes ($\nu \leq 120$ THz), and nonlinear crystals ($\nu \geq 120$ THz).

Fig. 4 shows the scheme of the chain used to measure the frequency of the Ca optical frequency standard. The high frequency part is down locked from the Ca optical frequency standard to a color-center laser (CCL). The lower part is phase-locked from the 100 MHz standard frequency of a hydrogen maser up to the carcinotron. The intermediate part of the chain, consisting of all CO_2-lasers, is locked to a methane stabilized He-Ne laser to improve the frequency stability and to allow the simultaneous frequency measurement of this laser (29). To obtain a value of the Ca transition frequency, we simultaneously counted the beat signals of the methanol laser with the backward wave oscillator and with the CO_2-laser, as well as the beat signal of the two CO_2-lasers with the CCL using totalizing counters. Combining the beat signals yields a frequency ratio independent of fluctuations of the intermediate transfer oscillators. This method allows to track the phase of all intermediate oscillators and therefore leads to a truly phase-coherent measurement.

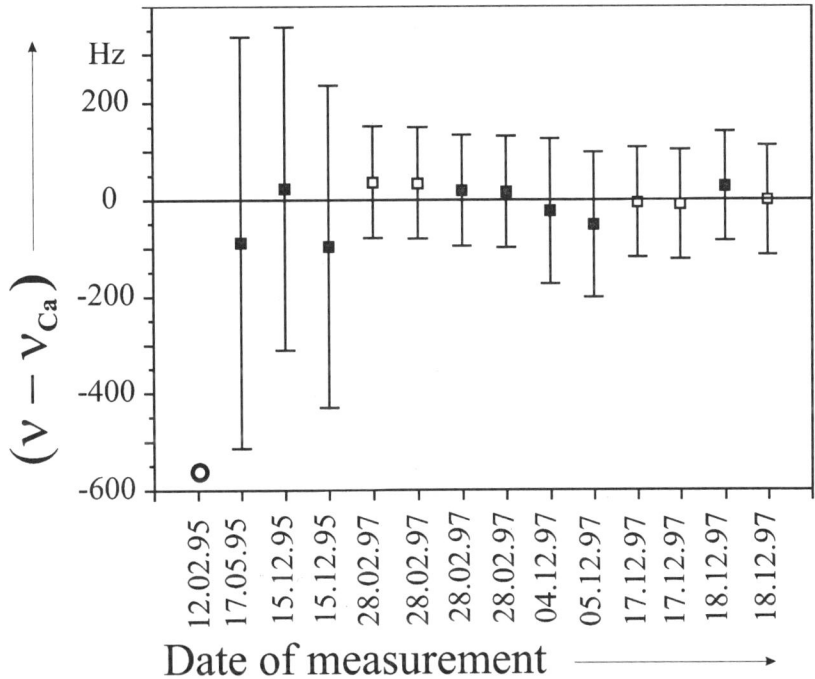

Figure 5. Measured frequency values of the Ca optical frequency standard. The measurements were taken with an effusive beam (circle), a MOT filled by a laser cooled beam (filled squares) and a MOT filled by a thermal beam (empty squares). Different values of the same day correspond to different pulse separations.

Fig. 5 shows the results of the frequency measurements, obtained during a time interval of more than two years. For each measurement, the dye laser was subsequently stabilized to the central fringe of the high- and the low-frequency recoil component. The measurements were performed using both MOTs, different resolutions, and different stabilization schemes (1f and 3f-method). The weighted mean of all frequency measurements is ν_{Ca} = 455 986 240 494.13 (12) kHz.

Fig. 6 shows the Allan standard deviation $\sigma_y(2,\tau)$ versus the integration time of the reference resonator, the Ca standard, the H-maser, and the frequency measurement. The linear dependence of the (reference) resonator represents its linear drift of approximately 0,7 Hz / s. The stability of the Ca standard itself was determined by measuring the frequency difference between the standard and the reference resonator after subtracting its drift. The attack time τ_A of the stabilization to the Ca clock transition is approximately 8.5 s (24). At integration times $\tau > 10$ s, a relative instability below 10^{-14} is reached which is close to the flicker floor of the preset setup. The instability of the hydrogen maser was measured independently by beating two masers. For integration times between $\tau = 1$ s and $\tau = 1000$ s, $\sigma_y(2,\tau)$ decreases with $1/\tau$ and reaches its flicker level at approximately 10^{-15}. For integration times $\tau \leq 100$ s, the standard deviation of the measured optical frequency closely follows that of the hydrogen maser and approaches the flicker noise level of the Ca standard at $\tau = 1000$ s.

Frequency Uncertainty

The frequency uncertainty of the described standard is determined by the uncertainty to realize the line center of the isolated atoms (Cs and Ca) at rest and by that of the frequency measurement. The different contributions to the uncertainty are

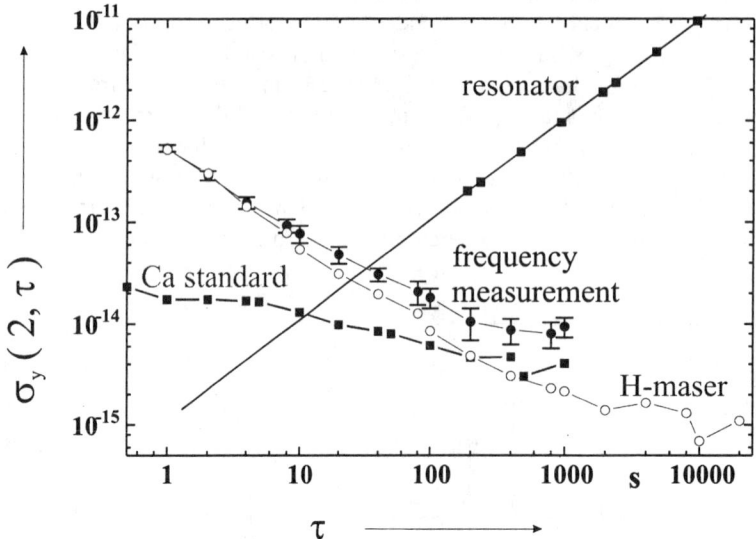

Figure 6. Measured frequency instability $\sigma_y(2,\tau)$ versus integration time τ of the optical frequency measurement, the Ca standard (compared to the drift corrected reference resonator), the reference resonator, and the H-maser.

TABLE 1. Contributions to the standard uncertainty of the Ca optical frequency standard.

Effect	Correction	Contribution to the uncertainty
optical phase		< 6 Hz
influence of Doppler background		< 0.2 Hz
second order Doppler effect		< 0.003 Hz
magnetic field	- 0,6 Hz	0.2 Hz
ac Stark effect		< 5 Hz
quadratic Stark effect ($|E| < 1$ V/cm)		0.06 Hz
black body radiation	+ 5 Hz	< 3.5 Hz
superposition of both recoil components		< 1 Hz
servo electronics (10^{-3} of the linewidth)		< 1 Hz
influence of cold atom collisions		< 10 Hz
reproducibility		< 50 Hz
statistical uncertainty of the line center		4 Hz
counting errors of the frequency measurement		< 100 Hz
Cs clock ($1.5 \cdot 10^{-14}$)		7 Hz
H maser	(Dec. 18, 1997) - 31 Hz	< 5 Hz
total uncertainty $\delta\nu$		**< 113 Hz**
total relative uncertainty $\delta\nu/\nu$		**< $2.5 \cdot 10^{-13}$**

listed in Table 1: The effects contributing to the uncertainty to realize the center of the clock transition of the Ca atom have been investigated experimentally and theoretically (24, 30). Contributions are caused by residual errors in the phase fronts of the laser pulses exciting the clock transition (< 6 Hz), by a potential ac Stark shift caused by scattered light of the cooling laser (< 5 Hz), and collisions of the cold Ca atoms (< 10 Hz). The smaller contributions include influences of the superposition of the two recoil components, the contribution of Doppler background, and the stabilization scheme. Comparisons between the transition frequencies realized by the two independent magneto-optical traps have been performed during the past year stabilizing the laser frequency subsequently to the atomic clouds in the two traps. The origin of this small difference between the two traps of $< 10^{-13}$ is not clear at present. Consequently, we account for this difference by a reproducibility of < 50 Hz.

The described frequency chain allows a phase-coherent optical frequency measurement which does not contribute to the uncertainty of the measured value. However, cycle slips can occur in each multiplication step. The rate of cycle slips critically depends on the S / N ratio of the beat signal (31). We estimated this rate in the most crucial stages of the chain and concluded that they contributed to less than 100 Hz. The total relative uncertainty of the frequency of the Ca stabilized laser of $2.5 \cdot 10^{-13}$ represents the status achieved with our present set-up. Estimations based on these results show that a relative uncertainty of a few parts in 10^{15} can be expected.

CONCLUSION AND OUTLOOK

We have presented a phase-coherent optical frequency chain from the Cs atomic clock to the Ca optical frequency standard. With a present uncertainty of $1.1 \cdot 10^{-13}$ to realize the line center, an Allan standard deviation less than 10^{-14} at $\tau > 10$ s, and an uncertainty of $2.5 \cdot 10^{-13}$ of the measured frequency value, the Ca optical frequency standard belongs to the most advanced standards developed so far. Consequently, the radiation of the Ca optical frequency standard has been recommended by the International Committee of Weights and Measures (CIPM) for the realization of the meter (32). Serving as a reference frequency it supports e.g. precision spectroscopy, atom interferometry and length/wavelength metrology. For example, as a typical application the frequencies of the hyperfine structure lines R(180) 0-16 and R(42) 0-17 in molecular iodine at 815 nm wavelength coinciding with the known frequency difference between the methane stabilized He-Ne laser and the Ca standard (s. Fig. 4) have been determined (33). The dissemination of the Ca frequency to potential other users requires the development of a transportable standard. In a first step, a transportable optical frequency standard based on an effusive Ca beam has been developed at PTB (34) with a relative uncertainty of $1.2 \cdot 10^{-12}$ which is more than an order of magnitude improvement with respect to the iodine stabilized He-Ne laser operating at $\lambda \approx 633$ nm wavelength. Frequency comparisons have recently been performed with a similar stationary setup at NIST (35). Small and efficient light sources for the cooling radiation now allow the development of a transportable standard based on *laser cooled* Ca atoms.

The ultimate goal of optical frequency measurements is to develop a scheme which can synthesize and combine any arbitrary frequency in the optical range, phase-coherently. Several novel concepts of optical frequency synthesis have been proposed and partially developed which may eventually lead to frequency chains more flexible. Such concepts are based on cascaded frequency interval division (5, 36), mixing of subharmonics (37), and optical comb generation (38). A combination of these methods will eventually lead to small and reliable optical frequency synthesizers allowing to measure a variety of different optical frequencies simultaneously and to connect different optical reference frequencies leading to a network of precision reference frequencies in the optical range.

ACKNOWLEDGMENTS

The work was supported by the Deutsche Forschungsgemeinschaft (SFB 407). Contributions of T. Kurosu [1], S. Oshima [1], U. Sterr, and V. Vassiliev [2] are gratefully acknowledged. The $\sigma_y (2, \tau)$ measurements of the hydrogen maser were kindly made available to us by A. Bauch.

REFERENCES

1. Roberts, M., Taylor, P., Barwood, G.P., Gill, P., Klein, H.A., and Rowley, W.R.C., *Phys. Rev. Lett.* **78**, 1876-1879 (1997) and Engelke, D., Tamm, Chr., *Europhys. Lett.* **33**, 347-352 (1996).

---

[1] permanent address: National Research Laboratory of Metrology, Tsukuba-Shi, Japan
[2] permanent address: Lebedev Institute of Physics, Russian Academy of Sciences (FIAN), Moscow, Russia

2. Peik, E., Hollemann, G., Abel, J., von Zanthier, J., Walther, H., "Single-ion spectroscopy of indium: Towards a group III mono-ion oscillator," in *Frequency Standards and Metrology,* (Bergquist, J.C., ed.) pp. 376-379 (1996) World Scientific, ISBN 981-02-2527-X.
3. Young, Y., Cruz, F., Berkeland, D., Itano, W.M., Wineland, D., Bergquist, J.C., "Optical frequency standard using trapped Hg^+ ions" in *Proceedings of the International Conference on Trapped Charged Particle and Fundamental Physics,* Monterey, 1998, pp. XX-XX.
4. Dirscherl, J., Walther, H., "Towards a silver frequency standard ", in *Digest of the 14th International Conference on Atomic Physics (ICAP 94)*, 1994, Boulder.
5. see for example: Ertmer, W., Blatt, R., Hall, J.L., "Some candidate atoms and ions for frequency standards research using laser radiative cooling techniques," in *Laser Cooled and Trapped Atoms* (Phillips, W.D., ed.), U.S. Nat. Bur. of Stand. special publication, 1983, Vol. **653**, pp. 154-161; see also Bergquist, J.C., Barger, R.L., Glaze, D.J., "High resolution spectroscopy of calcium atoms," *in Laser Spectroscopy IV,* (Walther, H., Rothe, K.W., eds.) Springer Series in Optical Sciences **21**, 1979, pp. 120-129.
6. Evenson, K.M., Day, G.W., Wells, J.S., Mullen, L.O., *Appl. Phys. Lett.* **20**, 133-134 (1972).
7. A tabulation of absolute laser-frequency measurements is presented in Knight, D.J.E., *Metrologia* **22**, 251-257 (1986).
8. Bönsch, G., *Appl. Opt.* **22**, 3414-3420 (1983)
9. Layer, H.P., Deslattes, R.D., Schweitzer, W.G., *Appl. Opt.* **15**, 734-743 (1976).
10. Jolliffe, B.W., Rowley, W.R.C., Shotten, K.C., Wallard, A.J., Woods, P.T., *Nature* **251**, No. 5470, 46-47 (1976).
11. Jennings, D.A., Pollock, C.R., Peterson, F.R., Drullinger, R.E., Evenson, K.M., Wells, J.S., Hall, J.L., Layer, H.P., *Opt. Lett.* **8**, 136-138 (1983).
12. Acef, O., Zondy, J.J., Abed, M., Rovera, D.G., Gérard, A.H., Clairon, A., Laurent, Ph., Millerioux, Y., Juncar, P., *Opt. Comm.* **97**, 29-34 (1993).
13. Udem, Th., Huber, A., Gross, B., Reichert, J., Prevedelli, M., Weitz, M., and Hänsch, T.W., *Phys. Rev. Lett.* **79**, 2646 - 2649 (1997).
14. Schnatz, H., Lipphardt, B., Helmcke, J., Riehle, F., Zinner, G., *Phys. Rev. Lett.* **76**, 18-22 (1996).
15. Beverini, N., Giammanco, F., Maccioni, E., Strumia, F., Vassani, G., *Opt. Soc. Am. B* **6**, 2188-2193 (1989).
16. Kurosu, T., Shimizu, F., *Jpn. J. Appl. Phys.* **29**, L2127-L2129 (1990).
17. Helmcke, J., Snyder, J.J., Morinaga, A., Mensing, F., Gläser, M., *Appl. Phys. B* **43**, 85-91 (1987).
18. Vassiliev, V., Velichansky, V., Kersten, P., Trebst, T., Riehle, F., *Opt. Lett.* **23**, 1229-1231 (1998).
19. Drever, R.W.P., Hall, J.L., Kowalski, F.V., Hough, J., Ford, G.M., Munley, A.J., Ward, H., *Appl. Phys. B* **31**, 97-105 (1983).
20. see for example: Riehle, F., *Meas. Sci. Technol.* **9**, 1042-1048 (1998).
21. Kurosu, T., Kisters, Th., Riehle, F., to be published.
22. Witte, A., Kisters, Th., Riehle, F., Helmcke, J., *J. Opt. Soc. Am B.* **9**, 1030-1037 (1992).
23. Kisters, Th., Zeiske, K., Riehle, F., Helmcke, J., *Appl. Phys. B* **59**, 89-98 (1994).
24. Zinner, G., *Ein optisches Frequenznormal auf der Basis lasergekühlter Calciumatome,* Braunschweig: PTB-Bericht, **PTB-Opt-58**, 1998, ISBN 3-89701-131-X.
25. Baklanov, Ye.V., Dubetsky, B.Ya., Chebotayev, V.P., *Appl. Phys.* **9**, 171-173 (1976).
26. Helmcke, J., Zevgolis, D., Yen, B.Ü., *Appl. Phys. B* **28**, 83-84 (1982) and. Bordé, Ch.J, Avrilier, S., van Lerberghe, A., Salomon, Ch., Bassi, D., Scoles, G., in *J. de Physique,* (Audoin, C., ed.), *Coll.* **C8**, suppl. 12, Tome 42, pp. C8-15 - C8-19 (1981).
27. Bordé, Ch.J., *Physics Letters A* **140**, 10-12 (1989).
28. see for example: Sterr, U., Sengstock, K., Ertmer, W., Riehle, F., Helmcke, J., "Atom interferometry based on separated light fields", in *In Atom Interferometry*, Berman, P.R.(ed.) San Diego, Academic Press, 1997, pp. 293-364, ISBN 0-12-092460-9.
29. see for example: Kramer, G., Weiss, C.O., Lipphardt, B., "Coherent frequency measurements of the hfs-resolved methane line", in *Frequency Standards and Metrology*, 1989, pp. 181-186, Springer, Heidelberg.
30. Riehle, F., Schnatz, H., Lipphardt, B., Kersten, P., Zinner, G., Helmcke, J., "Optical calcium frequency standard: Status and prospects" in Helmcke, J., Penselin, S.(eds.), *Frequency Standards Based on Laser-Manipulated Atoms and Ions*, Braunschweig, **PTB-Opt-51**, 1996, pp. 11-20, ISBN 0341-6712
31. Telle, H.R., "Absolute measurement of optical frequencies," in *Frequency control of semiconductor lasers,* 1996, (M. Ohtsu, ed.), pp. 137-172, John Wiley & Sons, Inc, ISBN 0-471-013412.
32. Quinn, T., *Metrologia*, **30**, 523-541, (1994) and *Metrologia*, in preparation.
33. Bodermann, B., Klug, M., Knöckel, H., Tiemann, E., Trebst, T., Telle, H.R., *Appl. Phys. B* **67**, 95-99 (1998).
34. Kersten, P., *Ein transportables optisches Calcium-Frequenznormal*, Braunschweig: PTB-Bericht, **PTB-Opt-59**, 1998, ISBN 3-89701-133-6.
35. Kersten, P., Oates, Chr., Bondu, F., Hollberg, L., to be published.
36. Telle H.R., Meschede, D., Hänsch, T.W., *Opt. Lett.* **15**, 532-534 (1990).
37. Wong, N.C., *Appl. Phys. B* **61**, 143-149 (1995).
38. Kourogi, M., Nakagawa, K., Ohtsu, M., *IEEE J. Quantum Electron.* **29**, 2693-2701 (1993) and Brothers, L.R., Lee, D., Wong, N.C., *Opt. Lett.* **19**, 245-247 (1994).

Hg⁺ Frequency Standards

John D. Prestage, Robert L. Tjoelker and Lute Maleki

California Institute of Technology, Jet Propulsion Laboratory
4800 Oak Grove Drive
Building 298
Pasadena, CA 91104

Abstract. In this paper we review the development of Hg⁺ microwave frequency standards for use in high reliability and continuous operation applications. In recent work we have demonstrated short-term frequency stability of $3 \times 10^{-14}/\sqrt{\tau}$ when a cryogenic oscillator of stability $2-3 \times 10^{-15}$ was used as the local oscillator. The trapped ion frequency standard employs a ^{202}Hg discharge lamp to optically pump the trapped ^{199}Hg⁺ clock ions and a helium buffer gas to cool the ions to near room temperature. We describe a small Hg⁺ ion trap based frequency standard with an extended linear ion trap (LITE) architecture which separates the optical state selection region from the clock resonance region. This separation allows the use of novel trap configurations in the resonance region since no optical pumping is carried out there. A method for measuring the size of an ion cloud inside a linear trap with a 12-rod trap is currently being investigated. At ~10^{-12}, the 2nd order Doppler shift for trapped mercury ion frequency standards is one of the largest frequency offsets and its measurement to the 1-% level would represent an advance in insuring the very long-term stability of these standards to the 10^{-14} or better level. Finally, we describe atomic clock comparison experiments that can probe for a time variation of the fine structure constant, $\alpha = e^2/2\pi\hbar c$, at the level of 10^{-20}/year as predicted in some Grand Unified String Theories.

INTRODUCTION

A small, continuously operating atomic clock with stability of 10^{-15} or better would advance the art and science of spacecraft navigation in deep space and enable space-based tests of general relativity that far exceed the sensitivities of earth-based tests. Similarly, continuously operating earth-based clocks with stability approaching 10^{-16} for averaging intervals between 1000 and 10,000 seconds would enhance the search for low frequency gravitational waves via careful Doppler tracking of the Cassini spacecraft en route to Saturn. These goals have defined and shaped the technologies used for the development of the ultra-stable linear ion trap Hg⁺ at the JPL Frequency Standards Lab. Because ions in a trap undergo very weak hyperfine population relaxation rates, ^{202}Hg rf discharge lamps can substantially optically pump ^{199}Hg⁺ ions into the F=0 ground hyperfine level in a second or so. Pumping into this state proceeds with the scattering of only a few uv photons per ion and thus, a high signal-to-noise ratio in the measured clock resonance can only be achieved with large ion clouds, typically with 10^6 to 10^7 ions. Motivated by these constraints, we first recognized and developed the linear ion trap [1] for storage of ion clouds that were ten or more times larger than could be stored in a conventional Paul trap. Because the linear trap replaces the point node of the Paul trap with a line of nodes, there is no increase in Doppler shift from the excess rf micro-motion caused by space charge repulsion of ions from the vicinity of the node region.

We have built 10 frequency standards based on Hg ions in a linear trap, seven of the style shown in Figure 1 (LITS) [2] and three of the extended type (LITE) [3] as shown in Figure 4. Both the LITS and the LITE architectures have demonstrated frequency stability well into the 10^{-16}-stability range.

RECENT RESULTS WITH LITS/CSO COMBINATION

Figure 1 shows the configuration for the original Linear Ion Trap Standard (LITS). Ions are loaded into the trap with an electron pulse injected along the trap axis that ionizes a vapor of isotopically enriched ^{199}Hg atoms at a pressure of 10^{-9} Torr or less. This low pressure vapor is generated by heating a mercuric oxide (HgO) powder to 200C or higher.

With a background base vacuum system pressure of a few x10^{-10} Torr, ion-trapping times of a few hours can be achieved.

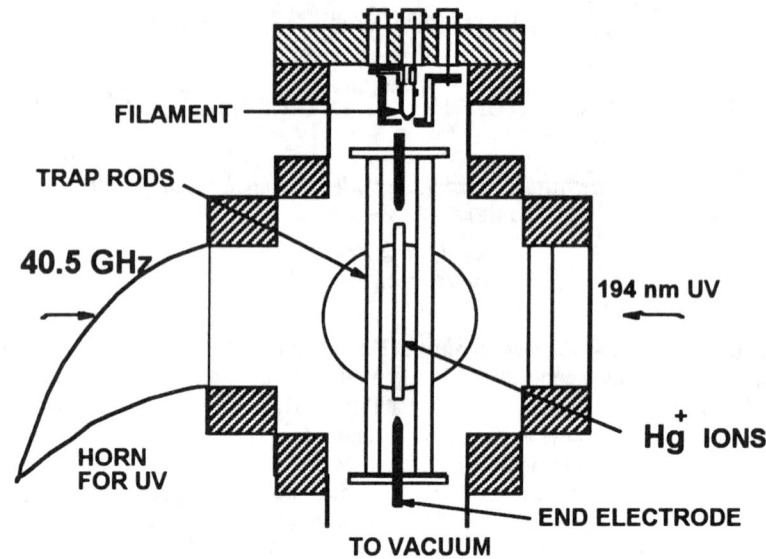

FIGURE 1. The linear ion trap based frequency standard is shown above. The trap is housed in a titanium vacuum cube. The molybdenum trap rods are 5 mm in diameter and are mounted on a 20.3 mm circle. The endcap electrodes are separated by 75 mm. State selection light from a ^{202}Hg discharge lamp enters from the right, is focused onto the central 1/3 of the trapped ions, and is collected in the horn. Fluorescence from the ions is collected in both directions normal to the page.

The LITS has shown excellent signal-to-noise in the measured clock transition and is determined by the size of the ion resonance line and the level of stray light output from the ^{202}Hg lamp. Signal levels as high as 80,000 with a background stray light rate of about 180,000 have been measured in a 1.5-second collection interval with dual fluorescence collection optical modules. These signal levels were measured with an 8 second Ramsey interrogation of the 40.5 GHz Hg$^+$ clock transition and will yield a clock performance $\sigma_y(\tau) = 2 \times 10^{-14}/\sqrt{\tau}$ as calculated from signal-to-noise and line Q. This short term performance exceeds the noise level for all but the very best short term stable oscillators, even the best H-masers available in our lab.

Recently, a cryogenic compensated sapphire oscillator (CSO) [4] has been developed in our lab and has been used to measure the un-degraded LITS performance. This continuously operating cryo-cooled liquid Helium refrigerated sapphire cavity oscillator delivers short-term stability of a few x10^{-15} to a few hundred seconds and is adequate to measure the ion standard short term performance. The LITS standard is compared to the Compensated Sapphire Oscillator [4] as shown in the schematic configuration of Figure 2. During this measurement the LITS calculated stability from signal-to-noise and line-Q was approximately $3 \times 10^{-14}/\sqrt{\tau}$ as was measured. This shows that for short-term performance there are no additional noise sources larger than shot noise of the total collected uv light.

Another measurement shown schematically in Figure 2 is steering the CSO to follow the ion clock resonance and compare that output to a SAO hydrogen maser. This shows that the LITS-CSO combination at $3 \times 10^{-14}/\sqrt{\tau}$, exceeds the hydrogen maser stability for all averaging time intervals over the duration of this measurement. This measurement was carried out with a large ion cloud with a second order Doppler shift of over 10^{-12} and consequently, a potential for noise sources from Doppler instabilities. For this reason and in order to reduce the size of the LITS we have changed the architecture of the LITS clock to exploit the mobility of charged ions and that they can be electrically transported from one end of a linear trap to another.

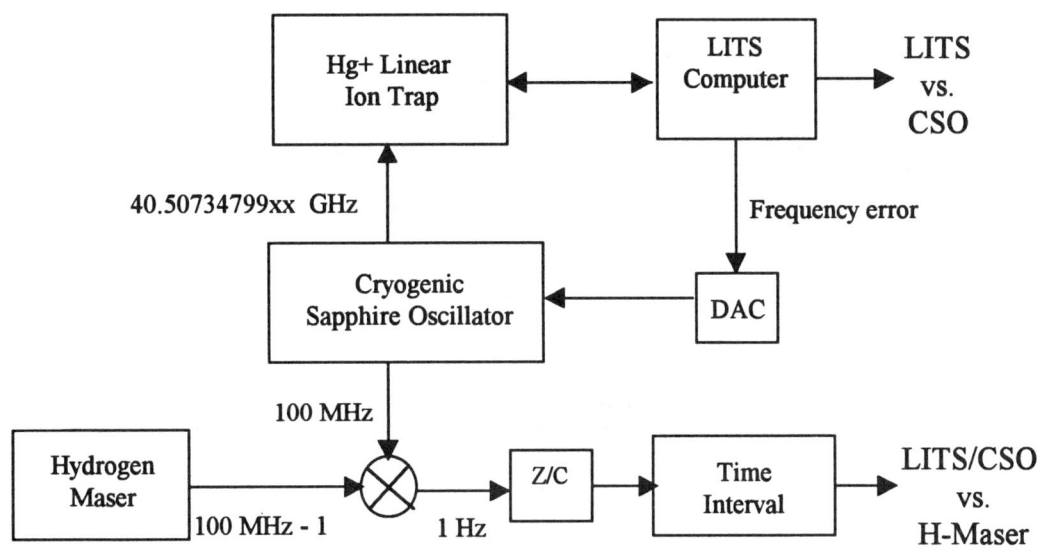

Figure 2. Schematic of the method used to measure the LITS short-term frequency stability against the ultra-stable cryo-cooled sapphire oscillator. Also shown here is the measurement of the frequency stability of the LITS/CSO combination compared to the stability of a hydrogen maser.

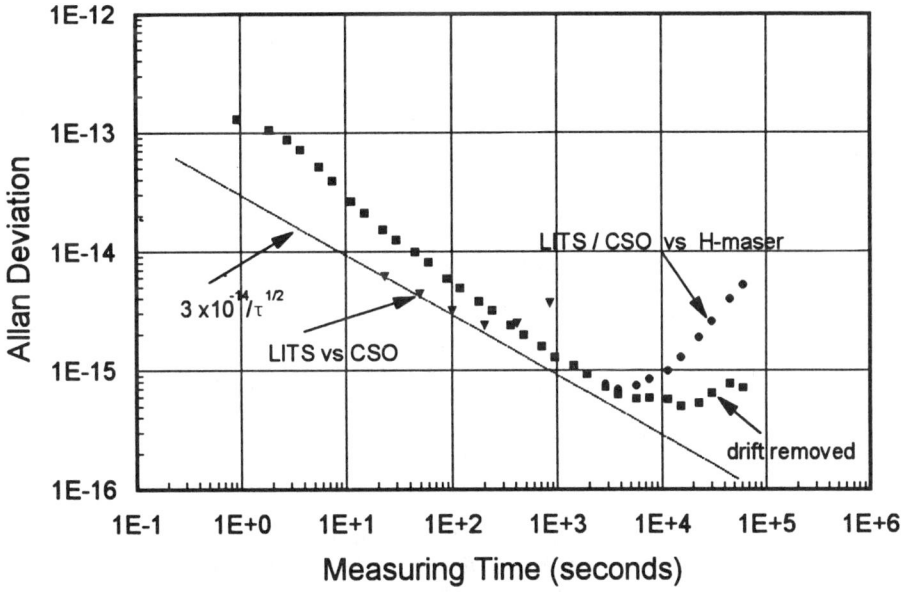

Figure 3. The measured Allan Deviation of the LITS vs CSO (inverted triangles) and the LITS steering the CSO vs a hydrogen maser (circles). The LITS vs CSO data demonstrates that for times less than 200 seconds, where the CSO is better than the LITS, the LITS stability is $3 \times 10^{-14}/\sqrt{\tau}$, better than any passive atomic frequency standard has demonstrated. In the second measurement shown here, the LITS steers the CSO to remove its degrading stability beyond 200 seconds. The LITS/CSO combination exceeds H-maser stability for the duration of the measurement interval.

LINEAR ION TRAP EXTENDED (LITE)

One modification to the LITS architecture that we are investigating is the extended linear ion trap shown in Figure 4. This layout partitions a linear trap into an optical interrogation region, similar to that shown in Figure 1, and adds an extension to carry out the microwave clock transition. The ions are moved from one region to another via a ~5 volt dc voltage bias applied to the trapping rods of that region. There are several advantages to the LITE architecture over the LITS arrangement where the magnetic shields surround the full optical system. In the LITE, the lamp and its rf exciter are outside the magnetic shields which will isolate it both electrically and thermally from the ion resonance region. The lamp exciter dissipates 10-20 Watts at ~170 MHz and can cause temperature variations of the vacuum chamber and stray, unshielded magnetic fields in the ion resonance region. Similarly, ground return currents in the bias feeds to the resistively heated filament, and filament emission grid collection leads, photo-multiplier high voltage supplies, etc. are all inside the multi-layer magnetic shielding of the LITS, unlike in the LITE shown below. In addition, the optical trap region and the resonance trap regions can be designed separately to address each of the two distinct functions. The resonance trap can be designed to allow the ion cloud second order Doppler shift to be measured and that design is described in the next section.

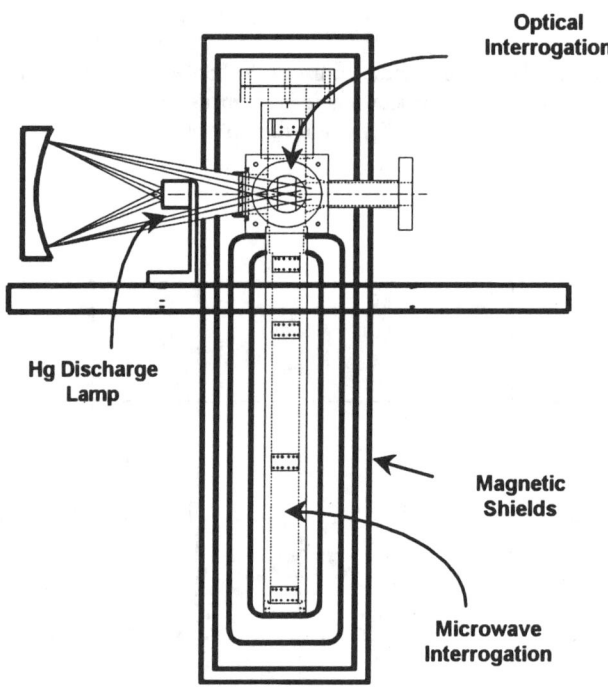

Figure 4. The Linear Ion Trap extended (LITE). The two distinct tasks of optical state selection/preparation and microwave interrogation are carried out in separate regions. Ions are electrically transported from one region to the other by a dc bias on the rods to exclude ions from either region.

HARMONIC LINEAR ION TRAPS

The harmonicity of a traditional four-rod linear ion trap is a function of rod diameter and spacing. Improved harmonicity can be accomplished with variations to this geometry. For example, figure 5 shows a linear ion trap configuration based on a cylinder that has been cut along its length into eight sectors, four at 60° angular width and four at 30° angular width. The quadrupole requirement, $\Phi(\rho,\theta \pm \pi/2) = -\Phi(\rho,\theta)$, leads to the expansion for the potential inside the cylindrical linear trap

$$\Phi(\rho,\theta) = C_0\rho^2\sin(2\theta) + C_1\rho^6\sin(6\theta) + C_2\rho^{10}\sin(10\theta) + ...$$

If the 30° sectors are grounded and the remaining 60° sectors are biased in a quadrupole fashion as shown, the resulting field is very harmonic, i.e., $C_1 = 0$.

An approximate implementation of this 60°/30° arrangement is shown in Figure 5. It consists of 12 circular rods with every 3rd rod grounded with the two intervening rods held at the same potential. This arrangement has the same 60°/30° symmetry of the sectored cylinder.

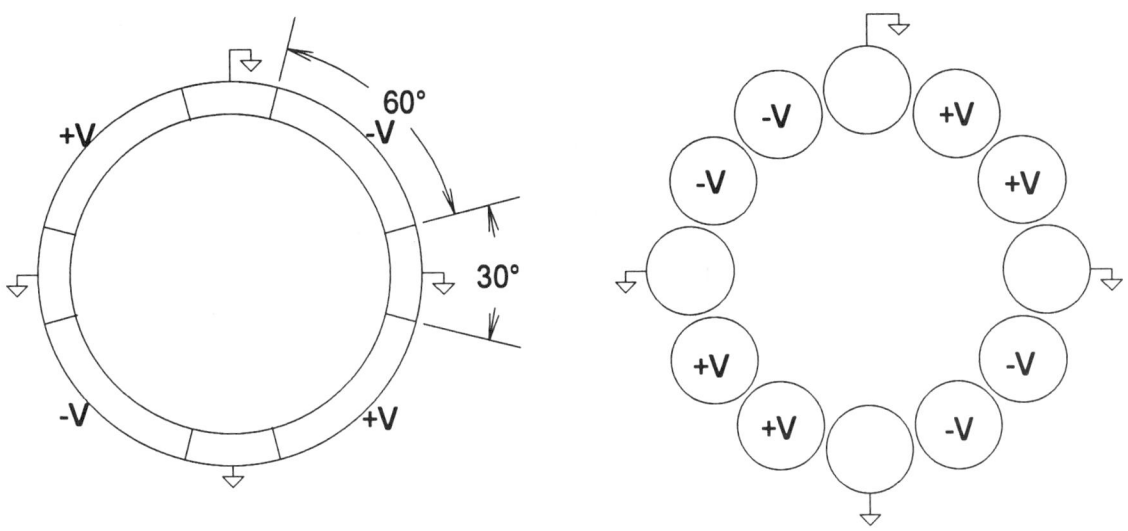

Figure 5. The trap on the left produces a very harmonic trap potential $\Phi(\rho,\theta) = C_0\rho^2\sin(2\theta) + C_2\rho^{10}\sin(10\theta) + ...$. The trap on the right is an approximate implementation of the 60°/30° symmetry constructed with 12 round rods.

ION CLOUD SIZE MEASUREMENT

The four auxiliary grounded rods can be used to generate a quadrupole magnetic field inside the linear trap whose node line coincides with the node line of the rf trapping fields. The shift of the clock transition with applied magnetic field is quadratic, $\nu = \nu_0 + \alpha\,\mathbf{H}^2$ where $\mathbf{H}$ is the total applied field and α is the sensitivity factor; $\alpha = 97$ Hz/Gauss^2 = 97 μHz/mG^2 for the ^{199}Hg$^+$ clock transition. If the field is the sum of a static homogeneous field, $\mathbf{H_0}$, along the trap axis and the transverse quadrupole field, $\mathbf{h_\perp(r)}$, from the four auxiliary trap rods we find $\nu = \nu_0 + \alpha\,(\mathbf{H_0} + \mathbf{h_\perp(r)})^2 = \nu_0 + \alpha\mathbf{H_0}^2 + \alpha\mathbf{h_\perp}^2$. Since $\mathbf{h_\perp(r)} = h'(\hat{x}y + \hat{y}x)$ and therefore, $\mathbf{h_\perp}^2 = (h')^2\rho^2$, the magnetic shift of the clock transition grows quadratically with distance from the node line. The clock frequency is the average of this spatially varying field over the ion cloud distribution, $n(\rho)$. Thus, $\langle\nu\rangle = \nu_0 + \alpha\mathbf{H_0}^2 + \alpha\langle\rho^2\rangle(h')^2$ where the brackets, $\langle\,\rangle$, indicate average over the ion cloud distribution. A measurement of the frequency change of the clock transition when the transverse field $\mathbf{h_\perp(r)}$ is applied can yield a measurement of the ion cloud radius. The quantities which determine $\langle\rho^2\rangle$ are ion number, ion temperature, and the trap rf level and its resulting secular frequency, ω_{sec}. For a fixed secular frequency and buffer gas pressure, ion number and temperature are not independent thus a measurement of $\langle\rho^2\rangle$ could be used to servo the electron emission to hold the ion number (and temperature) constant.

To estimate the size of the shifts suppose $\langle\rho^2\rangle = 1$ mm^2 and h' = 20 mG/mm. The shift when the quadrupole field of this strength is switched on is 97 μHz/mG^2 × 1 mm^2 × 400 mG^2/mm^2 ≈ 40 mHz which corresponds to a 1×10^{-12} shift of the clock transition. With a trap radius R ≈ 5 mm the field gradient produced at the center is h' ≈ $(\mu_0 I)/(\pi R^2)$ ≈ 20 mG/mm at a quadrupole excitation current of I = 125 mA. Even with $\langle\rho^2\rangle$ = 0.1 mm^2, a 400 mA current will produce a 10^{-12} clock shift. A clock with $10^{-13}/\sqrt{\tau}$ short term stability will measure this offset to about 1% in a few minutes of averaging time, τ.

T_2 Relaxation in $h_\perp$ field gradient

One problem that must be avoided with this technique is relaxation of the high Q clock transition in the field gradient of the quadrupole magnetic field, $h_\perp(r)$. This method depends upon the clock transition shifting an amount proportional to $\langle\rho^2\rangle$ with little or no reduction in the signal size and line-Q. Ion motion in the applied field gradient can mix the two Zeeman states F = 1, m_F = ± 1 with the upper clock state F = 1, m_F = 0 and cause a rapid relaxation of the coherence between the two clock levels F =1, m_F =0 and F = 0, m_F = 0. The frequency spacing to these Zeeman levels increases at ≅ 1.4 kHz per mG of applied field H_0, which is typically about 50mG.

The trajectory of an ion in the ρ, θ plane determines the spectrum of time variation of the quadrupole field, $h_\perp(r)$. The coherence in the clock transition will relax rapidly as this spectrum overlaps with the Zeeman states at ω_0 = ± γ H_0, where $\gamma/2\pi$ ≅ 1.4 kHz/mG. Harmonic motion through the trap center leads to a magnetic field variation at the secular frequency of the trap, ω_{sec}. These are the 'free particle' limits. When space-charge repulsion is important (as for a large cloud), these frequencies move down. Thus, it would appear that to avoid relaxation via mixing with the Zeeman states we must run the static field H_0 high enough so that γ H_0 > ω_{sec}. As the static field is increased the clock transition grows more sensitive to field variations and the clock is potentially less stable.

Relaxation rates to the Zeeman states from the upper clock state can be estimated from nmr relaxation rates given in the Redfield theory [5,6]. The time dependent magnetic field seen by an ion moving in the field gradient of $h_\perp(r)$ can have spectral overlap with the frequency splitting to the Zeeman states ω_0 = ± γ H_0 thereby transferring atoms into this state. In this estimate, the rate at which this occurs is assumed to be the same as the rate of coherence loss in the clock transition. The transfer of population occurs at a rate T_1^{-1} ≈ γ^2 (h')^2 ($S_x(\omega_0)$ + $S_y(\omega_0)$)/2 where we have used $|\vec\nabla H_x|^2$ = $|\vec\nabla H_y|^2$ = (h')^2 in the notation of ref [5]. The spectrum of the time variation of the field gradient seen by the moving ion $S_x(\omega)$ = $S_y(\omega)$ ~ $(2k_BT/m\omega_{sec}^2)(t_c(t_c^{-2}+(\omega - \omega_{sec})^2))^{-1}$ is assumed to be a Lorentzian shape centered on the ion secular frequency, ω_{sec}, with width determined by the ion collision rate, t_c^{-1}, which changes the phase of the ion secular motion within the trap. The spectrum is derived from a harmonic oscillator which is randomly re-phased at an average time interval of t_c. These collisions could be with other atoms or ions, or could be with the trap end confinement fields where each turn-around at the end cap field will disrupt the secular frequency phase. The collisions must be of sufficient strength to randomize the phase of the harmonic secular motion. The mean square amplitude of the transverse motion is $2k_BT/m\omega_{sec}^2$ = $\langle\rho^2\rangle$, determined by the secular confinement and the ion temperature, T. We re-write the population rate transfer in terms of the frequency shift when the quadrupole field is applied, δν = $\alpha\langle\rho^2\rangle$(h')^2, as T_1^{-1} ≈ γ^2 (δν/α)$(t_c(t_c^{-2}+(\omega_0 - \omega_{sec})^2))^{-1}$. Taking t_c ~ 1 msec, ω_0-ω_{sec} = 2π 50 kHz, and δν = 40 mHz, we find that T_1^{-1} ≈ 300/sec, a very rapid loss of coherence in the clock transition.

One possible solution to this near resonance relaxation is to apply the quadrupole field at a frequency, Ω, much higher than the secular frequency, ω_{sec}. Since along the path of the ion trajectory, $h_\perp(r(t))$ = h'($\hat x$y + $\hat y$x) = $h_0'\cos\Omega t$ ($y_0\sin\omega_{sec}t\ \hat x$ + $x_0\sin(\omega_{sec}t + \phi)\hat y$), the frequencies of the quadrupole field seen by the moving ion are now up-shifted to Ω ± ω_{sec} which can be 10 or more times higher than γ H_0 to avoid the mixing to the Zeeman states and loss of coherence in the clock transition. In this case the dominant frequency seen by the moving ion is ~ Ω so that T_1^{-1} ≈ γ^2 (δν/α)$(t_c(t_c^{-2}+(\omega_0 - \Omega)^2))^{-1}$. If the quadrupole field is applied at 2 MHz the relaxation rate is, T_1^{-1} ≈ 0.2 /sec, much slower than with the dc quadrupole current and thereby preserving the line Q and signal size.

CLOCK COMPARISONS TO TEST FOR ALPHA VARIATION

A stringent test of modern grand unified theories can be carried out in laboratory clock frequency comparisons. Some theories [ref] predict a slow time variation of the fine structure constant, α = $e^2/2\pi hc$, due to the expansion of the universe. Because α describes the strength of electromagnetism, its time variation would lead to a slow time variation of all atomic energy level spacings. There have been several attempts to find a changing α by comparing clocks or spectral lines of different "electromagnetic composition" that is, atomic transitions which have a different dependence on α. Thus, a microwave superconducting cavity oscillator frequency was compared to a hydrogen maser transition frequency [7], fine structure intervals in Mg were compared to the hydrogen maser interval [8], and more recently, multiple spectral lines from quasars with large cosmological redshifts were compared [9]. In the quasar spectral data, there is evidence that during the early universe α differed from its present day value by $\Delta\alpha/\alpha = -2.64(\pm0.35)\times10^{-5}$. Although among the most precise measurements carried out, atomic clock rate comparisons were thought to be insensitive to any change in α since it was believed that all hyperfine clock transitions scaled with α in the same manner. We recently showed [10] that the relativistic hyperfine interaction strongly violates the uniformity of scaling with α as the atomic number Z of the clock atom increases. Thus a changing fine structure constant would force a fractional frequency change between the hydrogen hyperfine frequency, $\nu_{hydrogen}$, and an alkali atom or ion hyperfine frequency, ν_{Alkali}, according to [10]

$$\frac{\dot{\nu}_{Alkali}}{\nu_{Alkali}} - \frac{\dot{\nu}_{hydrogen}}{\nu_{hydrogen}} = (\alpha Z)^2 \frac{12\lambda^2 - 1}{\lambda^2(4\lambda^2 - 1)} \frac{1}{\alpha}\frac{d\alpha}{dt} \equiv L_d F_{rel}(\alpha Z)\frac{1}{\alpha}\frac{d\alpha}{dt}$$

where $\lambda = [1-(\alpha Z)^2]^{1/2}$. We compared a LITS Hg$^+$ clock frequency to a hydrogen maser frequency for 140 days to limit fractional changes in α to be less than $\sim 4\times10^{-14}$/year. Comparing laser cooled atomic clocks, for example, Rb vs Cs, could measure any linear variation in α larger than $\sim 10^{-16}$/year [11], as good as the best limits placed on a time variation of α [12].

A much more stringent search for an α variation can be carried out by searching for a spatial dependence of α in the strong gravitational potential at four solar radii, where gravitational time dilation slows all clocks by ~ 1 micro-second per second as compared to Earth-based clocks. Because an expanding universe presumably drives a time variation in α, we write

$$\frac{d\alpha}{dt} = \frac{d\alpha}{dU/c^2}\frac{dU/c^2}{dt} = \frac{d\alpha}{dU/c^2}H$$

Where $H \sim 10^{-10}$/year is the Hubble constant. Thus two or more clocks as payload on a four solar radii solar flyby, whose relative rates are compared, could probe for a dependence of α on U_{solar}. Since time dilation effects are $\sim 10^{-6}$ at four solar radii, clocks based on atoms of sufficiently different atomic number Z, with differential stability $\sim 10^{-16}$ could probe for $d\alpha/dU$ at the 10^{-10} level. This would reveal any time variation in α larger than 10^{-20} per year. The design of a spacecraft that can survive and function on a four solar radii flyby, where the solar flux is 4 Mega-Watts per meter^2, has been studied at JPL for many years.

ACKNOWLEDGMENTS

This work was carried out at the Jet Propulsion Laboratory, California Institute of Technology, under a contract with the National Aeronautics and Space Administration.

REFERENCES

1. J. D. Prestage, G. J. Dick, and L. Maleki, *J. Appl. Phys.* **66**, 1013-1017 (1989).
2. R. J. Tjoelker, J. D. Prestage, and L. Maleki, "A Mercury Linear Ion Trap Frequency Standard for the USNO," in *Proceedings of the 1995 IEEE International Frequency Control Symposium*, 1995, pp. 79-81.
3. J. D. Prestage, R. L. Tjoelker, G. J. Dick, and L. Malcki, "Improved Linear Ion Trap Physics Package," *Proceedings of the 1993 IEEE International Frequency Control Symposium*, 1993, pp. 144-147.
4. G. John Dick, Rabi T. Wang, and Robert L. Tjoelker, "Cryo-Cooled Sapphire Oscillator with Ultra-High Stability," *Proceedings of the 1998 IEEE International Frequency Control Symposium*, 1998, pp. 528-533.
5. D. D. McGregor, *Phys. Rev. A* **41**, 2631 (1990).

6. C. P. Slichter, *Principles of Magnetic Resonance* (Harper and Row, New York, 1963), Sec. 5.7.
7. J. P. Turneaure and S. Stein, *Atomic Masses and Fundamental Constants V* (Plenum, London, 1976), pp. 636-642.
8. A. Godone, C. Novero, P. Tavella and K. Rahimullah, *Phys. Rev. Lett.* **71**, pp. 2364-2366 (1993).
9. J. K. Webb et. al., to appear in *Phys. Rev. Lett.*
10. J. D. Prestage, R. L. Tjoelker, and L. Maleki, *Phys. Rev. Lett.* **74**, 3511-3514 (1995).
11. Private Communication, Ch. Salomon and A. Clairon.
12. Th. Damour and F Dyson, *Nuc. Phys. B* **480**, 37-60 (1996).

Probing Ca⁺ ions in a miniature trap

M. Knoop[1], M. Vedel, M. Houssin, T. Schweizer, T. Pawletko, and F. Vedel[2]

Physique des Interactions Ioniques et Moléculaires (UMR 6633 CNRS - UAM1),
Université de Provence, Centre de St-Jérôme, Case C21, F - 13397 MARSEILLE CEDEX 20

Abstract. Different earth-alkaline ions have been proposed to be used as frequency standard in the optical domain. Ca⁺ is a promising candidate that offers a very high line-Q, as well as transitions easy to access. Our medium-term aim is to excite the clock transition of an isolated ion in a miniature trap without broadening by the first-order Doppler effect.

I INTRODUCTION

Among the ions proposed for a frequency standard in the optical domain [1], Ca⁺ takes a special place due to the fact that all the wavelengths necessary for cooling and probing of the ion can in principle be generated by diode lasers. This allows to imagine an ultimate experimental setup which would be simple and compact.

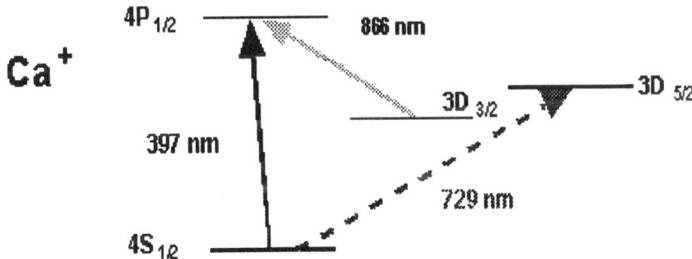

FIGURE 1. Lowest energy levels of the ⁴⁰Ca⁺ ion

The proposed clock transition is the $4S_{1/2}$-$3D_{5/2}$ electric quadrupole transition at 729 nm in the red domain (fig. 1). The metastable $3D_{5/2}$ level has a lifetime of the order of one second [2], thus leading to a natural linewidth inferior to 200 mHz and giving rise to a Q-factor of the transition higher than 10^{15}.

A further advantage of the Ca⁺ ion is the existence of an isotope having odd nuclear spin (⁴³Ca⁺: I=7/2). This may allow to work on transitions which to first order are independent of residual magnetic fields.

We recently investigated the influence of population transferring collisions on the lifetime of the metastable $3D$-doublet of the Ca⁺ ion [3]. We have now started to work on the set-up and the stabilisation of the different laser systems to prepare the probing of a single ion in a miniature trap.

II THE LASER SOURCES

One of the major advantages of working with trapped Ca⁺ is the relative simplicity of the laser sources needed for laser cooling and probing of the clock transition. An all solid-state laser system can be imagined.

---

[1]) mknoop@newsup.univ-mrs.fr
[2]) fern@frmrs12.u-3mrs.fr

A 729 nm

Laser diodes in the wavelength domain 700-750 nm are difficult to fabricate, and besides some applications in metrology (probing of Ca^+, lasercooling of Yb^+...) there is actually no commercial demand for these devices. For this reason we use a low-power, multimode laser diode at room temperature, which has a nominal output of 30 mW at 730 nm and a linewidth of 2 nm in the free-running mode. To improve its spectral properties we have put this diode into an external cavity (Littrow configuration), which allows to reduce the linewidth to about 10-15 MHz. Further stabilisation on a simple reference cavity (FSR 300 MHz) results in a linewidth below 1 MHz, which is necessary for the injection into a high finesse cavity. We are performing this last step using an ULE cavity with a finesse of 15000 and a FSR of 1.5 GHz. As the sum of the different stabilisation procedures consumes quite some laser power, it will be necessary in an ultimate step to inject a slave laser allowing a total laser output at 729 nm of a couple of mW.

B 397 nm

Laser radiation at 397 nm from a diode laser can only be generated by frequency doubling. In a first step we built an external enhancement cavity which allows second harmonic generation by a LBO crystal. The crystal has a length of 7 mm, it is cut under Brewster's angle. The cavity consists of two plane and two spherical mirrors ($r = 10$ cm) arranged in the classical bow-tie geometry [4]. 794 nm light is coupled into the cavity via one of the plane mirrors having a transmission at this wavelength of $T = (5 \pm 1)\%$. The three other mirrors have a coating at 794 nm with a reflection coefficient of 99.8%. The blue light leaves the cavity via one of the spherical mirrors which has a transmission higher than 80% at 397 nm. The described cavity has a quality factor of 33 ± 3 and a measured finesse of 100 ± 5. The obtained resonance lines have a full width at half maximum of 11 MHz. During preliminary measurements this cavity was tested using a TiSa-laser. Figure 2 shows the power of the generated light at 397 nm as a function of the input power. The obtained power output can be increased by improving the impedance matching of our cavity. In fact, the total cavity losses are around $(2.4 \pm 0.3)\%$, whereas our input mirror shows a transmission of 5 %.

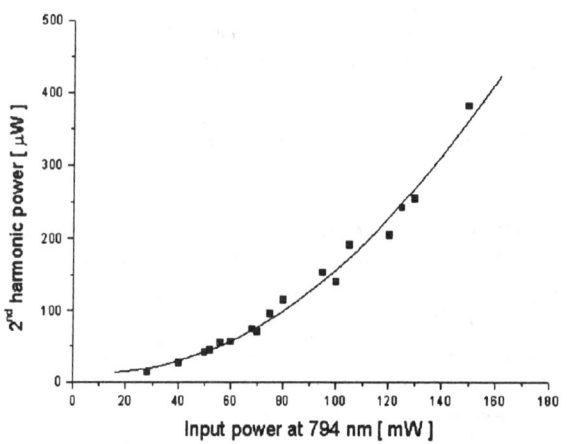

FIGURE 2. Laser output at 397 nm as a function of the pumping TiSa power input

We are now preparing the diode laser system at 794 nm. It is composed by a singlemode master laser with a maximum power of 30 mW which is placed into an external cavity to reduce its linewidth and to be able to tune its wavelength. The output of this diode is injected into a 500 mW multimode diode laser, taking care to geometrically match the dimensions of the lasing junctions. In a preliminary step we obtained 75 mW of singlemode power for an injection power of 1.14 mW. It will be necessary to at least double these performances as the output of the slave laser shows different spatial modes, which should be cleaned by a spatial filter before entering the doubling cavity.

III THE MINIATURE TRAP

High resolution of the ion's clock transition linewidth without broadening by the first order Doppler effect can only be obtained in the Lamb-Dicke regime, where the amplitude of the excursion of the ion is inferior to its emitted wavelength [5]. The access to this regime requires low temperatures as well as high motional frequencies of the observed ion. For technical reasons this necessitates high trapping frequencies and small trap dimensions. Additionnally, a miniature trap geometry has the advantage that the total trapping volume is filled out by the cooling laser beam, thus avoiding residual hot ions in the trap. For our miniature trap design we chose a Paul-Straubel geometry [6] which offers good access for the laser beams and a large solid angle of observation.

The trap consists of a cylindrical ring of molybdene with an inner ring diameter of 1.4 mm and a wall thickness of 300 μm. The total height of the cylinder ($2z_0$) equals 0.85 mm. On top and bottom of the ring compensation electrodes are placed at a distance of 5.5 mm symmetrically from the trap center (see figure 3). They have a diameter of 11 mm and are made from molybdenum mesh with a transmission of 86 %. For compensation of small potential defects in the trap a voltage can be applied between the two mesh electrodes. Additionnally, point electrodes in x and in y axis allow to correct field errors in these directions.

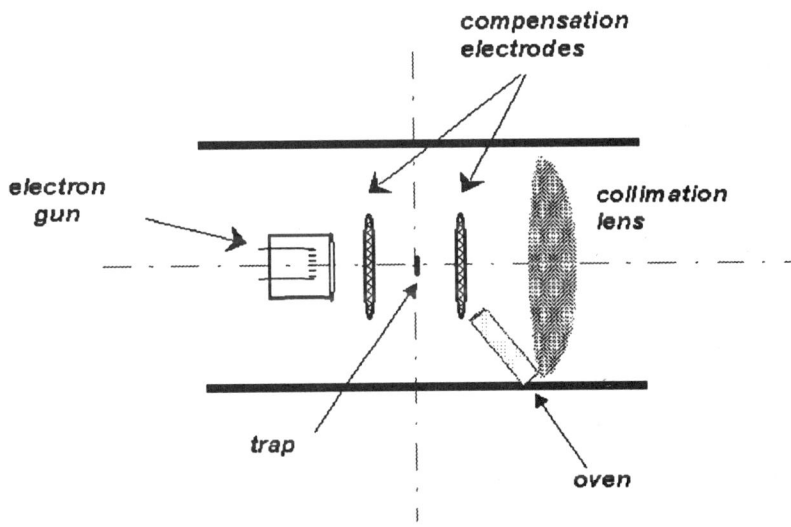

FIGURE 3. Schematic design of the miniature trap; the compensation electrodes in x and y are not shown.

Collimation of the ion's fluorescence is achieved by an aspheric lens at about 16 mm off the trap center. This very open structure provides a solid angle of observation of almost 90°.

Ions are created from electron bombardement of a slow atomic calcium beam. After loading of the trap, the oven and the electron gun are turned off to avoid collisions and an elevated stray light level.

The base pressure is below 7×10^{-10} mbar as measured by a Bayard-Alpert-gauge. A mass-spectrometer allows to determine the composition of partial pressures of the residual gas with a resolution up to 1×10^{-9} mbar.

The trapping parameters are $\Omega/2\pi = 11.6$ MHz with an AC amplitude of V_{AC}=600-900 V_{rms}. This corresponds approximately to $(a_z; q_z) = (0; 0.4-0.75)$. First signals of a cloud of about 50 ions have been obtained using helium buffer gas at around 1×10^{-7} mbar (fig. 4). The temperature derived from the Doppler profile lie between 400 and 700 K, which represents an equilibrium value between the rf heating, the collisions with the infinite heat reservoir of the buffer gas at 300K and the laser-cooling process.

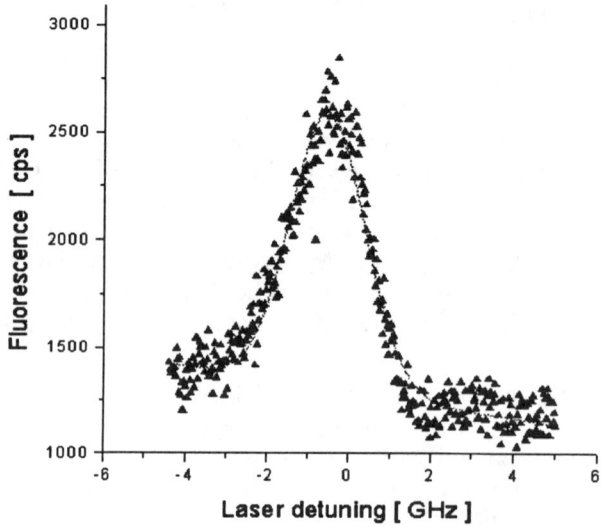

FIGURE 4. Fluorescence of the Ca$^+$ ions in the miniature trap, using a helium buffer gas pressure of 1.7×10^{-7} mbar. The temperature of the ions - fitted from the Doppler width - is $T = 640K \pm 15K$

IV OUTLOOK

We have obtained first signals of small Ca$^+$ ion clouds in a miniature trap. The next step will now be to precisely caracterize and optimize the confinement conditions in the new device. In parallel we are working to improve the spectral performances of the diode lasers needed for laser cooling and probing of the involved transitions.

REFERENCES

1. see i.e. the different contributions in Proceedings of the 5th Symposium on Frequency Standards and Metrology, ed. J.C.Bergquist, World Scientific, Singapore 1996
2. M. Knoop, M.Vedel, F.Vedel, Phys.Rev.**A 52**, 3763-3769 (1995), for a review see E. Biemont and C.J. Zeippen, Comments At. Mol. Phys. **33**, 29 (1996)
3. M. Knoop, M. Vedel, and F. Vedel, Phys.Rev.**A 58**, 264-269 (1998)
4. C.S. Adams, A.I. Ferguson, Opt.Comm.**79**, 4 (1990)
5. R.H. Dicke, Phys.Rev. **89**, 472-473 (1953)
6. C.A. Schrama, E. Peik, W.W. Smith, H. Walther, Opt.Comm.**101**, 32 (1993)

SECTION 9

COHERENT QUANTUM CONTROL

Measurement and Control of Single Atom Motions in the Quantum Regime

Jun Ye, Christina J. Hood, Theresa Lynn, Hideo Mabuchi,
David W. Vernooy, and H. Jeff Kimble

*Norman Bridge Laboratory of Physics 12-33, California Institute of Technology
Pasadena, CA 91125 USA*

Abstract. Using cold atoms strongly coupled to a high finesse optical cavity, we have performed real-time continuous measurement of single atomic trajectories in terms of the interaction energy (E_{int}) with the cavity. Individual transit events reveal a shot-noise limited measurement (fractional) sensitivity of $4 \times 10^{-4}/\sqrt{Hz}$ to variations in $E_{\text{int}}/\hbar$ within a bandwidth of 1-300 kHz. The strong coupling of atom and cavity leads to a maximum interaction energy greater than the kinetic energy of an intracavity laser-cooled atom, even under weak cavity excitation. Evidence of mechanical light forces for intracavity photon number < 1 has been observed. The quantum character of the nonlinear optical response of the atom-cavity system is manifested for the trajectory of a single atom.

I. INTRODUCTION

Optical cavity quantum electrodynamics (QED) in the strong coupling regime [1] plays a significant role in exploring manifestly quantum dynamics where coherent, reversible system evolution dominates over dissipative processes. The large coherent-interaction rate generates a unique capability of monitoring quantum processes in real time, which in turn should lead eventually to the investigation of the strong conditioning of system evolution on measurement results and the realization of quantum feedback control [2]. Another important feature associated with strong coupling is that system dynamics are readily influenced by single quanta. Thus single-atom and single-photon cavity QED provides an ideal stage where the dynamical processes of individual quantum systems can be isolated and manipulated. Such coherently controlled processes are essential to advances of quantum information technologies.

The dipole coupling rate, g_0, between a two-level atom and a small, high-finesse cavity sets the rate of energy exchange between the two constituents. This internal evolution of the composite system is accompanied by two dissipation channels, namely the cavity decay rate κ and the free-space atomic decay $\gamma_\perp$. However, κ can also be regarded as providing a quantum channel through which the information of the open quantum system can be extracted, processed, and fed back with high efficiency. The other relevant time scale, the interaction time T, is associated with the center-of-mass (CM) motion (external degree-of-freedom) of the atom through the spatially-varying cavity eigenmode. The use of laser-cooled atoms in cavity QED [3] effects a dramatic separation of dynamical time scales, with $g_0 > \kappa, \gamma \gg 1/T$. The rate of coherent coupling between the atom and cavity, g_0^2/κ, sets the maximum information bandwidth of the intracavity activity. The optical information per atomic transit is therefore $g_0^2 T/\kappa \gg 1$, eliminating the need to base measurements on averages over an ensemble of atom transits.

In this paper we describe two experiments that explore phenomena uniquely accessible in this regime of true strong coupling. In the first part we investigate the real-time continuous quantum measurement of the interaction energy between the atom and cavity during individual transit events [4]. Within a broader context, we are actively exploring the issue of how the dynamical behavior of a continuously-observed open quantum system is conditioned [5,6] upon the measurement record, which in our case is the broadband photocurrent resulting from cavity leakage photons. In the second part we study the mechanical force on single atoms brought by single intracavity photons via the strong coupling [7]. Laser cooled atoms have sufficiently low kinetic energies that a single quantum (excitation of the composite atom-cavity system) is able to affect profoundly the atomic CM motion.

CP457, *Trapped Charged Particles and Fundamental Physics*
edited by Daniel H. E. Dubin and Dieter Schneider
© 1999 The American Institute of Physics 1-56396-776-6/99/$15.00

II. CONTINUOUS MEASUREMENT OF SINGLE-ATOM DYNAMICS

The cavity we used in this experiment was formed with two mirrors of 1-m radius of curvature. The mean cavity length was 108 μm and its finesse was measured ~ 217,000 at the Cs D2 wavelength (852 nm). The cavity decay rate (HWHM) $\kappa/2\pi \approx 3.2 MHz$ while the dipole decay rate $\gamma_\perp/2\pi \cong 2.6 MHz$ for the Cs $6P_{3/2}$ level. The atom-field coupling coefficient g varies spatially, depending on the position of the atom with respect to the standing-wave structure of the intracavity field. The optimum value $g_0/2\pi$ was ~ 11 MHz for $\sigma_\pm$ transitions ($6S_{1/2}$ (F = 4, $m_F = \pm 4$) → $6P_{3/2}$ (F = 5, $m_F = \pm 5$)) and ~ 6 MHz for π transitions ($6S_{1/2}$ (F = 4, $m_F = 0$) → $6P_{3/2}$ (F = 5, $m_F = 0$)). Cold Cs atoms were provided from a standard magnetooptic trap (MOT), which was located 7mm above the cavity. The trap was typically loaded for about 0.5 s before dropping by quickly turning off the trapping beams and the magnetic field. Individual transit events typically lasted for ~ 250 μs. The trapping beam power was varied to control the number of atoms falling through the cavity. The repumping beam was left on all the time so that falling atoms would be shelved in the F = 4 ground hyperfine level before entering the cavity. No spin polarization was performed on atoms to prepare for a specific Zeeman sublevel.

In hopes of eventually reaching the standard quantum limit for monitoring the position of a free mass as well as observing quantum measurement backaction, we were obliged to use a probe field detuned from the atomic resonance [8,9]. The experimental protocol then involved determination of both the phase and amplitude of the cavity-output field with shot-noise-limited accuracy at a reasonable bandwidth. Figure 1 shows the real-time evolution of the complex field amplitude brought by single atom transits. These quadrature amplitude (QA) signals provide a direct record of the complete time-evolution of the interaction energy ($E_{int} = \hbar g$) between the atom and cavity. For large (> 50 MHz) atom-cavity detunings we seek phase-contrast signals induced by individual atomic transits, corresponding to a regime of strong but dispersive coupling. With this dispersive coupling we minimize not only the random heating generated by on-resonance probing of the atomic motion [10] but also the information loss through the atomic decay to free space. Technically we thus need a probe field void of any frequency noise while still arbitrarily tunable with respect to the atomic resonance. The cavity resonance is then locked tightly on the probe frequency to maintain a stable empty-cavity field and to avoid technical noise in the intrinsically frequency (phase) – discriminating cavity transmission.

The probe field was provided by a Ti:Sapphire laser with its fast linewidth narrowed by a prestabilizing reference resonator. The resonator was locked on the Cs resonance. Tuning of the probe frequency was provided by an in-line electro-optic modulator (EOM). The probe field was too weak to establish a direct frequency lock to the cavity. On resonance, the saturation intracavity photon number for the QED cavity is $m_0 = \gamma_\perp^2 / 2g_0^2 \approx 0.1$, which sets a cavity throughput of $4\pi\kappa m_0 \approx 1 pW$. Tuning off resonance would allow the probe power to be proportionally larger, however, it would still be extremely difficult to obtain a high-quality error signal for cavity locking. We thus used an auxiliary diode laser to stabilize the cavity length on a different longitudinal mode than the probe. The diode laser was detuned 16 nm off the Cs resonance and its power of ~ 40 nW through the cavity incurred an AC Stark shift of only ~ 60 kHz for the Cs resonance. The diode laser itself was stabilized by the same reference resonator shared with the Ti:Sapphire laser. Frequency detuning between the diode laser and the QED cavity was again furnished by a second EOM. (The cavity was locked on an rf sideband of the diode laser.) While this setup allowed us to have arbitrary detunings among the atomic, cavity, and probe frequencies, in the experiment we kept a zero-detuning between the cavity and probe.

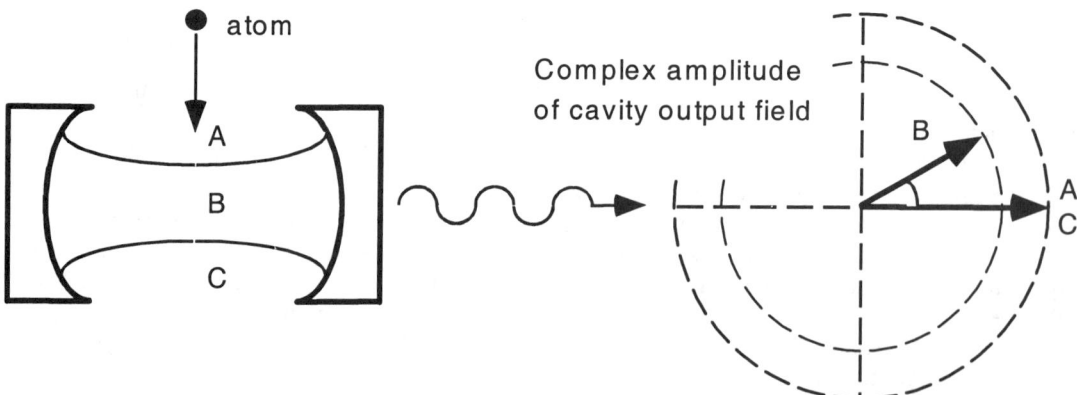

FIGURE 1. Real-time evolution of the complex field amplitude brought by single atom transits. A, B, and C represent atom positions with respect to the cavity mode. Larger detuning from the atomic resonance will elicit more dominant responses in the phase quadrature.

We used a balanced-heterodyne setup in order to achieve high-efficiency, zero-background photodetection of ~ 1 pW level of cavity transmission. The mode-cleaned and intensity stabilized local oscillator was derived from the same laser used for probe with an appropriate frequency offset to avoid electronic noise pickup. The overall heterodyne efficiency (~0.32) was carefully measured in order to deduce the intracavity photon number and make quantitative comparisons with simulations. This relatively low photo-detection efficiency remains a major area for improvement in future experiments. (The difference photocurrent from the balanced heterodyne detectors was 0°-split to two identical copies and they were mixed with the in-phase and quadrature components of an additional rf local oscillator to produce an orthogonal pair of QA signals at baseband. The QA signals were filtered and digitized (with 12-bit resolution) at a 10 MHz sampling rate per channel. Fluctuations of optical phases caused by vibrational disturbances along the beam propagation line were limited to frequencies below ~ 1 kHz and the recorded signal bandwidth typically extended from 1 kHz to 300 kHz. Fortunately, this covered the dominant rates of variation in E_{int}, as the mean duration of individual transit events was 250 μs.

Figure 2 shows the first measurement of the real-time evolution of the complex field amplitude brought by single atom motion. Dynamical variations in both QA (at 300 kHz bandwidth) are displayed vs. time during transit events. Detunings Δ (= atomic resonance - probe frequency) and probe powers (intracavity photon number m when no atoms in cavity) are indicated on the figure. Note we have displaced the amplitude quadrature signals (upper traces) by +400 to prevent overlapping between two traces. Although at times the transit signals seem to show internal structures, we still have yet to devise a clean way of distinguishing between variations caused by atomic motion through the spatial structure (both longitudinal and transverse) and/or optical pumping among atomic internal (Zeeman) states. The transition to dispersive measurement regime is evidenced by the significant phase shift of the probe field and the fact that signals primarily reside in the phase quadrature at large detunings. The full-signal (combining both QAs) to rms-noise ratio (SNR) is estimated at 4.5 at 300 kHz bandwidth, implying a relative sensitivity of ~ $4\times10^{-4}/\sqrt{Hz}$. This SNR lies only a factor of ~ 2.7 (ratio of the excess technical noise over the square root of the detection efficiency) above the fundamental quantum noise level, which is estimated based on an ideal 100% quantum efficiency. Assuming that the largest signals correspond to atoms reaching the maximal coupling strength of $g_0/2\pi$ = 11 MHz, this sets our broadband sensitivity to time-variations in E_{int} to be about

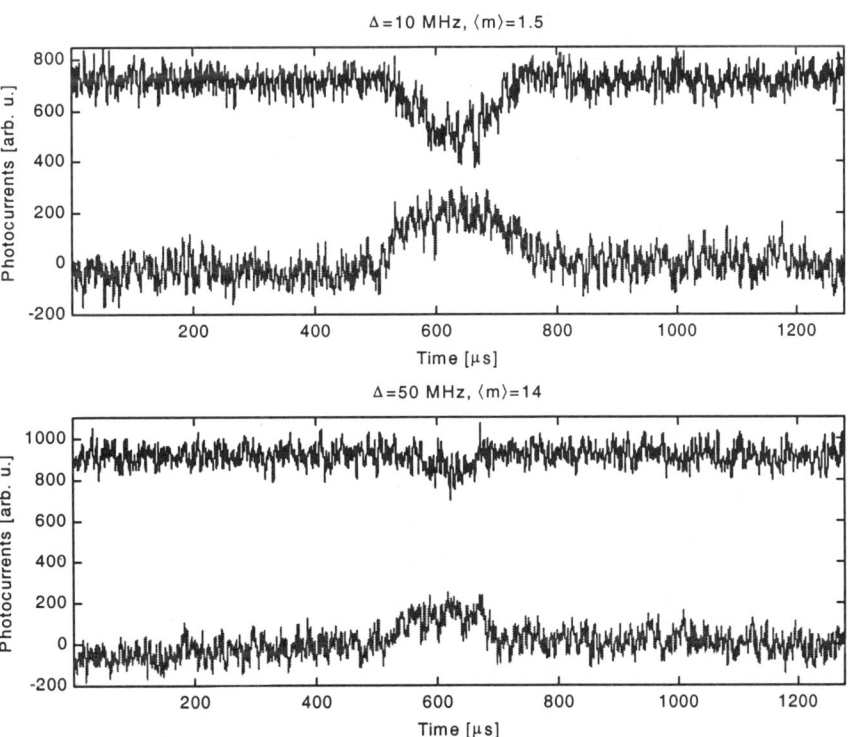

FIGURE 2. Individual atom-transit signals recorded with a 300 kHz bandwidth, with the atom-probe detuning Δ and probe power <m> (intracavity photon numbers in absence of atoms). Upper (lower) traces represent the amplitude (phase) quadrature signals. Upper traces are displaced by an additional + 400 for clarity.

$4.5 kHz/\sqrt{Hz}$. Considering the standing-wave structure of the cavity eigenmode, our data should in principle display a position sensitivity of $1.5\times10^{-10} m/\sqrt{Hz}$ to atomic displacements along the cavity axis. From a large ensemble of atom transit date, we see the shapes of the quadrature amplitude signals are asymmetric in general. Although not much of detailed internal structure is recovered, prominent dips and steps can often be seen in the signal shapes. Simple Monte Carlo simulations of the heterodyne signals effected by atoms falling through the cavity have provided some suggestions for the qualitative interpretation of the experimental data. Atomic motions in the transverse direction, e.g. atoms falling through the cavity mode, provide the overall Gaussian profile for the transit signals. However, it is the atomic motion along the cavity standing wave that inflicts the strongest influences on the dynamical features of the signal shape. The simulation indicates that the longitudinal atomic motion can be initially confined in one standing wave dipole potential. When the atom suddenly escapes from the local confinement and start to pass through many standing waves, the dynamic time scale associated with the variation of the atom-cavity coupling and the subsequent heterodyne signal becomes rather rapid. The detection bandwidth, limited by the shot noise, can provide only the filtered version of the signal shape, resulting in the observed sudden steps and asymmetries. Again, at present we can not rule out the possibility that the signal asymmetries could also arise from the optical pumping process among internal states that may occur during the transit events. We intend to clarify this ambiguity in further experiments.

The atomic-transit signals can also be displayed in parametric plots of amplitude vs. phase to examine the correlation induced between these two quantities by the atom-cavity interaction. This type of parametric plot is dictated by the interaction Hamiltonian for the atom and cavity mode. With fixed values of power (m) and detuning (Δ), the only varying parameter in the plot is the atom-cavity coupling g, in the range of $[0, g_0]$. It is no longer relevant which underlying process (changes of internal states or external motion) controls the change of g. From this perspective the comparison between theory and experiment is thus simplified. Figure 3 (a) shows three transit phasors taken at different probe detunings, and illustrates transition from absorption to dispersive regime. The data in each subplot (in dots, from two individual transit events) overlay theoretical predictions based on quantum and semiclassical theories. The point of no interaction ($g=0$) is marked by a triangle. The $g=g_0$ endpoint (corresponding to the maximum interaction) is marked by a circle (o) for quantum theory and by a cross (×) for semiclassical theory. Figure 3 (b) illustrates the differences between the two theories regarding the nonlinear optical response of the atom-cavity system. The two transit phasors are shown for a fixed detuning but with different probe powers. The data clearly support the quantum theory which shows a discrepancy from the semiclassical theory. We wish to stress here that these nonlinear quantum dynamics are now being explored at single quantum realizations, instead of from ensemble averages [11,12].

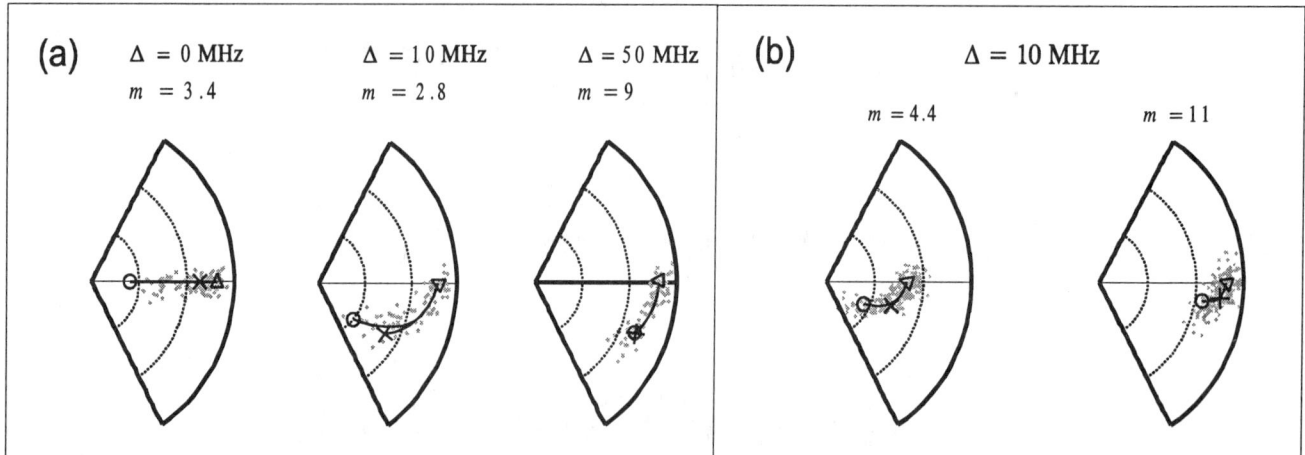

FIGURE 3. (a) Dependence of transit phasor shapes on detuning. Each subplot shows an overlay of two transit data (dots), Master equation-based quantum theory (solid curve ending in circle), and semiclassical theory (solid curve ending in ×). (b) Nonlinear response of the transit phasor, with a fixed detuning (10 MHz) and different probe powers. m is the intracavity photon number in absence of atoms. The triangles in the plot correspond to $g=0$.

III. PHOTON-COVALENT BINDING OF ATOM AND CAVITY

In the second experiment, a very small cavity of length 10.1 μm and finesse 180,000 provided us with the largest coupling g_0 in an optical system to date. The relevant rates were $(g_0, \kappa, \gamma_\perp, T^{-1})/2\pi = (120, 40, 2.6, 0.002) MHz$. These rates correspond to critical photon and atom numbers $(m_0 \equiv \gamma_\perp^2/2g_0^2, N_0 \equiv 2\kappa\gamma_\perp/g_0^2) = (2.3 \times 10^{-4}, 0.015)$ [1]. Cs atoms cooled to ~ 20 μK were again dropped down to the cavity. Here the frequency detuning between the atomic and empty-cavity resonances remained zero ($\nu_{atom} = \nu_{cavity} = \nu_0$). The cavity length was stabilized by a chopped auxiliary beam tuned on Cs resonance. A freely tunable probe beam was used to explore the eigen-spectrum of the atom-cavity system in cavity transmission, again with balanced heterodyne detection.

As an atom falls into the cavity, the increasing $E_{int} = \hbar g(\vec{r})$ causes the otherwise coincident atomic and cavity resonances to split into two "vacuum-Rabi" normal modes, located at $\nu_0 \pm g(\vec{r})/2\pi$, corresponding to two dressed states of the atom-cavity system. We can now watch this mode-splitting process in real time. For a few lucky atoms which actually reach a region of optimal coupling as they fall through the cavity, the vacuum-Rabi sidebands should be swept outward (from the cavity resonance) in frequency to a maximum of $\pm g_0/2\pi = \pm 120 MHz$. Therefore for the probe beam with a fixed frequency detuning at $\Delta = \nu_0 - \nu_{probe} = \pm 120 MHz$, the cavity transmission will show the largest increase. Similarly, for the probe beam of $\Delta = 0$, the cavity transmission shows the largest attenuation. For atoms that do not achieve optimal coupling, we should expect the maximum increase of cavity transmission to occur at some intermediate detunings between 40 to 100 MHz. Figure 4 shows an example of real-time changes in transmission of the atom-cavity system produced by individual atom transits. The transmissions of two probe beams are recorded simultaneously. In the plot the transmissions are normalized against the empty-cavity values. For the probe 1 near resonance, the cavity transmission decreases as an atom falls through the cavity. At the same time, the cavity transmission increases as the lower Rabi sideband for the probe 2 tuned to $\Delta \approx 120 MHz$. We also note that the signal contrasts obtained with two simultaneous probe beams are lower than those in single-probe measurements due to system saturation.

Carrying out repeatedly the transmission measurement, the spectrum of the atom-cavity system can be mapped out with a frequency sweep on a single probe. While it is clear to observe the double-peaked structure of the vacuum-Rabi splitting, with peaks near $\pm g_0/2\pi$, it is equally clear that the spectrum showed some asymmetry between red and blue probe detunings (See [7], Fig. 4). Since the red (blue) Rabi-sideband corresponds to the lower (upper) dressed state, the associated attractive (repulsive) mechanical light force is expected to affect the atom's CM motion. Indeed, weak excitation by a coherent probe tuned to $\Delta_\pm \approx \pm g_0$ gives rise to a pseudopotential (for times $\gg \kappa^{-1} \sim 4$ ns), with depth on the order of $\mp \hbar g_0$ [13]. Since $\hbar g_0/k_B \approx 7 mK$, such light forces can be significant even with a cavity of less than 1 photon, given that

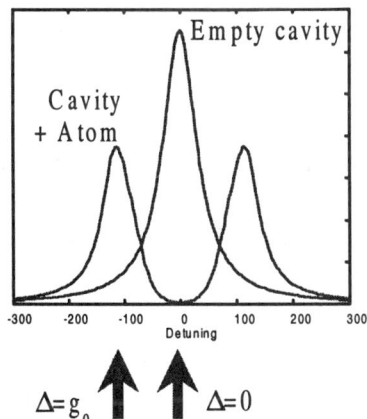
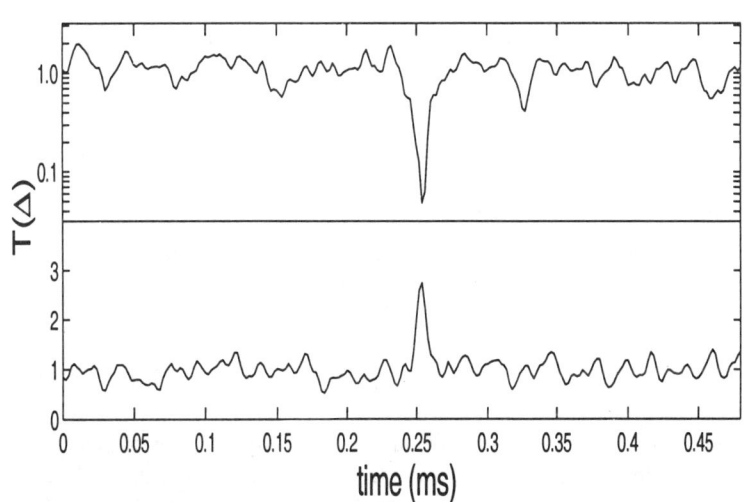

FIGURE 4. Simultaneous recording of two normalized probe transmissions through the atom-cavity system. Upper trace, $\Delta \approx 0$; lower trace, $\Delta \approx 120 MHz$. Bandwidth is 100 kHz.

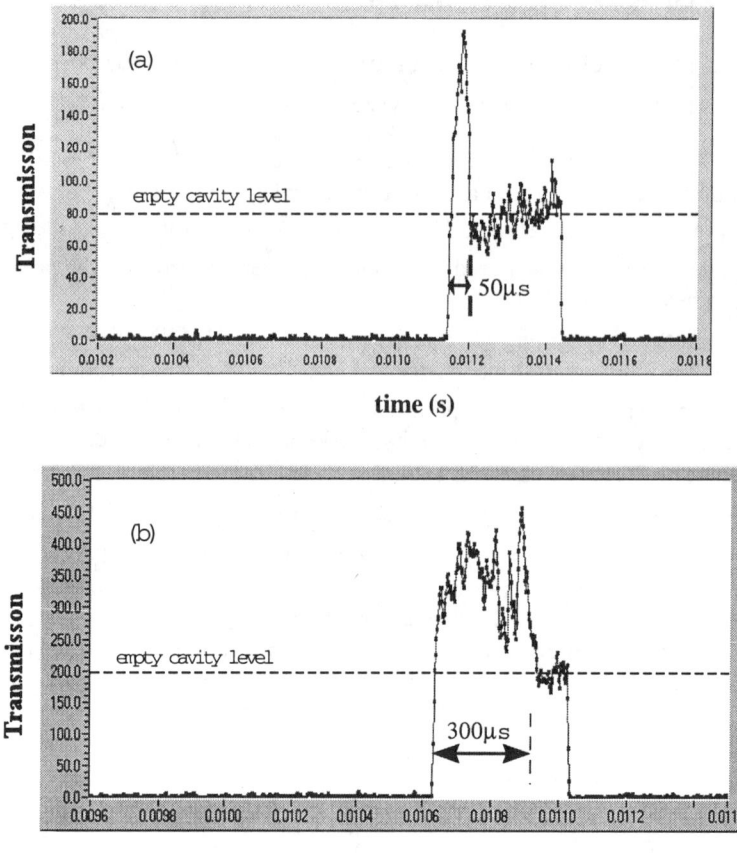

FIGURE 5. Strong-coupling induced mechanical effects on atoms. A probe beam at $\Delta \approx g_0$ is triggered *on* by the entrance of an atom into the cavity. In (b), the atomic transit time is extended to ~ 300 µs, compared with the 100 µs normal duration in the absence of the trapping beam, as in (a).

atoms are pre-cooled and many Rabi-cycles occur during the transit time. Therefore a blue-detuned probe ($\Delta < 0$) creates spatially-varying potential barriers and prevents atoms from reaching areas of optimal coupling. On the other hand, a red-detuned probe ($\Delta > 0$) will coerce atoms to channel through regions of high intensity and strong coupling. This attractive potential can be used as a trap for single atoms. In this context, the quantity $\hbar g_0$ may be interpreted as a covalent atom-cavity binding energy associated with the oscillatory exchange of a single photon. Experimentally, a weak probe beam tuned near the empty-cavity resonance is used to sense the arrival of an atom to the cavity. A drop in transmission of this sensing beam then triggers *off* this resonant field and triggers *on* a second field at $\Delta \approx g_0$ to create the trapping potential. In figure 5 we show an example of an atom which has been held in the cavity field for more than 300 µs (b), whereas the longest possible transit times expected without the strong-coupling-induced mechanical effects are ~ 100 µs (a).

IV. CONCLUSION AND FUTURE OUTLOOK

Combining cold Cs atoms with an optical cavity capable of inducing strong coupling, we have recorded the real-time evolution of the complex field amplitude of cavity transmission with a large bandwidth and nearly quantum noise-limited signal-to-noise ratio. Phasor diagrams built upon one or just a few single-atom transits explore the strong-coupling-induced correlation between two quadratures of the field and reveal the quantum nature of the atom-cavity interaction via the signature of system saturation. Detection in the dispersive regime allows us to map out a single-atom trajectory in the cavity in a non-destructive manner. With a large coherent coupling rate g_0, system dynamics such as vacuum-Rabi splitting are monitored in real time with single atoms falling through the cavity. Mechanical effects on single atom CM motion through atom-cavity strong couplings have also been demonstrated. Trapping of single atoms in the cavity quantum field is being explored.

The demonstrated experimental regime of measurement sensitivity and bandwidth should allow us in the future cavity QED research to observe detailed atomic CM trajectories. With future additional capabilities of manipulating atomic CM in real time, we anticipate progresses in the investigation of feedback control of quantized atomic CM motion. Quantitative experiments on conditional quantum dynamics [14] should also become possible as improvement on the measurement accuracy will shorten the overall observation time before measurement back action becomes manifested. Toward this goal, we are working to trap and localize atoms within the cavity and investigate the interplay between the external degree of freedom (quantized atomic CM motion) and the internal degree of freedom of the composite atom-cavity "molecule" [15].

ACKNOWLEDGMENT

We wish to acknowledge M. S. Champan's vital contribution to the project. This work is supported by the National Science Foundation, by the Office of Naval Research, and by DARPA via the QUIC institute administered by ARO.

REFERENCES

1. H. J. Kimble, "Structure and dynamics in cavity quantum electrodynamics," in *Cavity Quantum Electrodynamics*, P.R. Berman, Ed., San Diego: Academic Press, 1994, pp 203-266.
2. H. M. Wiseman, *Phys. Rev. Lett.*, vol. **70**, p. 548, 1993.
3. H. Mabuchi, Q.A. Turchette, M.S. Chapman, and H.J. Kimble, *Opt. Lett.*, vol. **21**, p. 1393-1395, 1996.
4. H. Mabuchi, J. Ye, and H. J. Kimble, submitted to *Appl. Phys. B*, May, 1998.
5. C. M. Caves and G. J. Milburn, *Phys. Rev. A*, vol. **36**, p. 5543, 1987.
6. H. M. Wiseman, *Quantum Semiclass. Opt.*, vol. **8**, p. 205, 1996.
7. C. J. Hood, M. S. Chapman, T. W. Lynn, and H. J. Kimble, *Phys. Rev. Lett.*, vol. **80**, p. 4157, 1998.
8. P. Storey, T. Sleator, M. Collett, and D. Walls, *Phys. Rev. A*, vol. **49**, p. 2322, 1994.
9. M. A. M. Marte and P. Zoller, *Appl. Phys. B*, vol. **54**, p. 477, 1992.
10. A. C. Doherty, A. S. Parkins, S. M. Tan, and D. F. Walls, *Phys. Rev. A*, vol. **56**, p. 833, 1997.
11. M. Brune, F. Schmidt-Kaler, A. Maali, J. Dreyer, E. Hagley, J. M. Raimond, and S. Haroche, *Phys. Rev. Lett.*, vol. **76**, p. 1800, 1996.
12. R. J. Thompson, Q. A. Turchette, O. Carnal, and H. J. Kimble, *Phys. Rev. A*, vol. **57**, p. 3084, 1998.
13. A. S. Parkins, private communications, 1996.
14. H. Mabuchi, *Phys. Rev. A*, vol. 58, p. 123, 1998.
15. D. W. Vernooy and H. J. Kimble, *Phys. Rev. A*, vol. **56**, p. 4287, 1997.

Quantum Logic with a Few Trapped Ions

C. Monroe, W. M. Itano, D. Kielpinski, B. E. King, D. Leibfried,
C. J. Myatt, Q. A. Turchette, D. J. Wineland, and C. S. Wood

Time and Frequency Division,
National Institute of Standards and Technology
Boulder, CO 80303

Abstract. Small laser-cooled crystals of atomic ions have attracted considerable interest in the last several years for their possible application toward quantum computation. This paper considers quantum logic schemes for small numbers of ions in the context of recent experiments at NIST.

I INTRODUCTION

In its simplest form, a quantum computer is a collection of N two-level quantum systems (quantum bits) which can be prepared in an arbitrary entangled quantum state spanning all 2^N basis states [1,2]. A quantum computer, unlike its classical counterpart, can thus store and simultaneously process superpositions of numbers. Once a measurement is performed on the quantum computer, the superposition collapses to a single number, which in some cases can jointly depend on all of the numbers previously stored. This gives the potential for massive parallelism in particular algorithms [3], most notably an efficient factoring algorithm [2,4] and a fast searching algorithm [5]. Apart from these and other possible applications [6], creating multi-particle entangled states is of great interest in its own right, from the standpoint of quantum measurement theory [7], and for improved signal-to-noise ratio in spectroscopy [8,9].

Unfortunately, very few physical systems are amenable to the task of quantum computation. This is because the quantum bits must (*i*) interact very weakly with the environment to preserve coherence of their superpositions, and (*ii*) interact very strongly with other quantum bits to facilitate the construction of quantum logic gates necessary for computing. In addition to these seemingly conflicting requirements, the quantum bits must be able to be controlled and manipulated coherently and be read out with high efficiency.

In 1995, Cirac and Zoller showed that a collection of trapped and cooled atomic ions can satisfy these requirements and form an attractive quantum computer architecture [10]. In their proposal, each quantum bit is derived from a pair of internal energy levels of an individual atomic ion. The quantum bits are coupled to one another by virtue of the quantized collective motion of the ions in the trap, mediated by the Coulomb interaction. Cirac and Zoller showed that an arbitrary entangled state can be created and permit any quantum computation by applying several laser pulses, each interacting with a single ion at a time. In light of recent experiments at NIST with two ions [11,12], this paper considers alternative quantum logic schemes, where laser pulses simultaneously interact with a few ions and their collective motional modes to produce entangled states.

II BACKGROUND

A Internal Electronic States of Ions as Quantum Bits

Ions can be confined for days in an ultra-high vacuum with minimal perturbations to their internal atomic structure, making particular internal states ideal for representing a quantum bit. Even though the ions interact strongly through their mutual Coulomb interaction, the fact that the ions are localized means that the time-averaged value of the electric field they experience vanishes; therefore electric field perturbations are small.

Although magnetic field perturbations to internal structure are important, the coherence time for superposition states of two internal levels can be made very long by operating at fields where the energy separation between levels is at an extremum with respect to field. For example, a coherence time exceeding 10 minutes between a pair of ^{9}Be$^+$ ground state hyperfine levels has been observed [13]. It is also possible to employ a ground and excited (metastable) electronic state of a trapped ion as a quantum bit. This option seems difficult at present, because the energy splittings are typically in the optical region and thus an extremely high laser frequency stability is required to drive coherent transitions.

Figure 1(a) shows a reduced energy level diagram of a single ^{9}Be$^+$ ion. Although many other ion species would also be suitable for quantum computation, we will concentrate on ^{9}Be$^+$ here for concreteness and to make a connection to the experiments at NIST [11,12,14,15]. We will be interested primarily in two electronic states, the $^2S_{1/2}(F=2, m_F=2)$ and $^2S_{1/2}(F=1, m_F=1)$ hyperfine ground states (denoted by $|\downarrow\rangle$ and $|\uparrow\rangle$ respectively), separated in energy by $\hbar\omega_0$. These long-lived spin states will form the basis for a quantum bit. Standard optical pumping techniques allow the spin to be initialized into either $|\downarrow\rangle$ or $|\uparrow\rangle$. Subsequent detection of the spin states can be accomplished using the technique of quantum jumps [16]. By tuning a circularly polarized laser beam to the $^2P_{3/2}$ transition at $\lambda_{Be} = 313$ nm (Fig. 1(a)), many photons are scattered if the atom is in the $|\downarrow\rangle$ spin state (a "cycling" transition), but essentially no photons are scattered if the atom is in the $|\uparrow\rangle$ spin state. If a modest number of these photons are detected, the efficiency of our ability to discriminate between these two states approaches 100%.

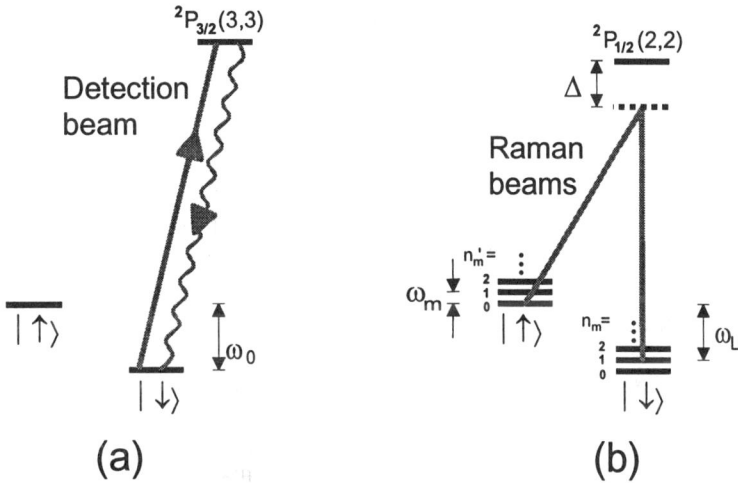

FIGURE 1. (a) Electronic (internal) energy levels (not to scale) of a ^{9}Be$^+$ ion. The $^2S_{1/2}(F=2, m_F=2)$ and $^2S_{1/2}(F=1, m_F=1)$ hyperfine ground states (denoted by $|\downarrow\rangle$ and $|\uparrow\rangle$ respectively), separated in frequency by $\omega_0/2\pi \simeq 1.250$ GHz, form the basis of a quantum bit. Detection of the internal state is accomplished by illuminating the ion with a σ^+-polarized "detection" beam near $\lambda_{Be} \simeq 313$ nm, which drives the cycling $|\downarrow\rangle \rightarrow {}^2P_{3/2}(F=3, m_F=3)$ transition, and observing the scattered fluorescence. The excited P state has radiative linewidth $\gamma/2\pi \simeq 19.4$ MHz. (b) Energy levels of a trapped ^{9}Be$^+$ ion, including the motional states of a single mode m of harmonic motion, depicted by ladders of vibrational states separated in frequency by the mode frequency ω_m. Two Raman beams, both detuned by $\Delta \gg \omega_0, \omega_m$ from the excited $^2P_{1/2}$ state, provide a coherent two-photon coupling between states $|\downarrow\rangle|n_m\rangle$ and $|\uparrow\rangle|n'_m\rangle$ by setting the difference frequency ω_L to match the desired transition frequency. As shown, the Raman beams are tuned to the first red sideband of mode m ($\omega_L = \omega_0 - \omega_m$).

B Collective Motional States of a Linear Crystal

Ions can be confined in several types of electromagnetic traps. Here, we consider the rf (Paul) ion trap, in which an oscillating electric potential is applied to electrodes surrounding the ions. In the standard quadrupole-like rf trap, the potential varies in space and time as

$$\Phi \simeq [U_0 + V_0 cos(\Omega_T t)]\left(\frac{\alpha x^2 + (1-\alpha)y^2 - z^2}{d_T^2}\right), \quad (1)$$

where d_T characterizes the trap electrode dimension and $0 < \alpha \le 1/2$ characterizes the geometrical anisotropy of the trap electrodes ($\alpha = 1/2$ corresponds to a quadrupolar potential). This gives rise to harmonic ponderomotive potentials [17] in all three dimensions with oscillation frequencies

$$\omega_x = \frac{\Omega_T}{2}\sqrt{\alpha a + \frac{\alpha^2 q^2}{2}} \tag{2a}$$

$$\omega_y = \frac{\Omega_T}{2}\sqrt{(1-\alpha)a + \frac{(1-\alpha)^2 q^2}{2}} \tag{2b}$$

$$\omega_z = \frac{\Omega_T}{2}\sqrt{-a + \frac{q^2}{2}}, \tag{2c}$$

where $a = 8eU_0/(m\Omega_T^2 d_T^2)$ and $q = 4eV_0/(m\Omega_T^2 d_T^2)$ and e/M is the charge-to-mass ratio of the ion. In these expressions, it is assumed that $-\alpha q^2/2 < a < q^2/2$ and $q \ll 1$ (or equivalently $\omega_{x,y,z} \ll \Omega_T$)—a condition known as the "pseudopotential approximation" [17]. Motion described by an rf pseudopotential is always accompanied by micromotion at frequencies near Ω_T associated with the rf electric fields. An alternative geometry is the linear trap [18], in which a 2D pseudopotential confines ions radially and an independently applied static potential confines the ions axially. Because the rf fields vanish along the axial node, there is no axial micromotion, and many of the problems associated with micromotion are avoided. For this reason, the linear rf trap is more appropriate when larger numbers of ions are confined and micromotion is not desired.

For a collection of trapped ions, the mutual Coulomb repulsion counteracts the confining potential, and if the ions are sufficiently cold, they will crystallize to an equilibrium configuration. In this case, the three frequencies in Eq. (2) describe center-of-mass (COM) harmonic motion, and a normal mode calculation must be done to solve for the other internal modes of harmonic motion. The simplest crystalline structure is a linear chain, which can be attained with a sufficiently anisotropic trap. Here, we will take the x-axis in the above expressions to represent the weakest (axial) direction of the trap, which can be ensured by setting $a < 0$ and $\alpha < 1/2$. For $N = 2$ ions, the ions will clearly arrange themselves along the x-axis of the trap. For N=3 ions, the ions will form a linear chain along x only if $\omega_x < \sqrt{5/12}\,\omega_{y,z}$. For larger numbers of ions, this anisotropy must be even larger, and has been numerically calculated in Ref [19].

Of the $3N$ normal modes of small oscillation in a linear chain of ions, we are primarily interested in the N collective modes associated with axial motion (assumed to be along the x-direction in the above equations). A remarkable feature of a linear chain of ions is that the axial mode frequencies are nearly independent of

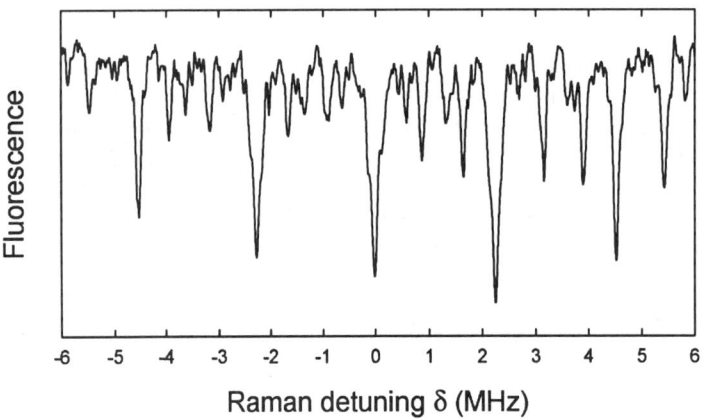

FIGURE 2. Raman absorption spectrum of three trapped ions. The ordinate is the detuning δ of the Raman probe beam difference frequency and the abscissa shows the ion fluorescence, proportional to the number of ions in the state $|\downarrow\rangle$ (the ions are initially prepared in state $|\downarrow\rangle|\downarrow\rangle|\downarrow\rangle$). The carrier appears at $\delta = 0$, and the first sidebands of the three axial normal modes of motion appear at $\delta = \pm 2.25, \pm 3.90,$ and ± 5.43 MHz, in agreement with the theoretical frequency ratios $1 : \sqrt{3} : \sqrt{29/5}$. Several higher order sidebands also appear at sums and differences of harmonics of the normal mode frequencies.

N [10,20,21], offering the possibility that mode interference might be small, even for large numbers of ions. For two ions, the axial normal mode frequencies are ω_x and $\sqrt{3}\omega_x$; for three ions they are $\omega_x, \sqrt{3}\,\omega_x$, and $\sqrt{29/5}\,\omega_x$, as shown in Fig. 2. For $N > 3$ ions, the normal modes must be determined numerically [20,21]. The quantum state of a particular axial mode m of motion at frequency ω_m is represented by the ladder of vibrational eigenstates $|n_m\rangle$ of energy $\hbar\omega_m(n_m + 1/2)$ with vibrational index n_m describing the number of phonons contained in the the mth collective mode of motion. The motional modes can be initialized to the $n_m = 0$ ground state through laser-cooling [11,14]. Whenever possible, operations involving the COM modes should be avoided, as the COM modes are found to lose coherence (through external heating) at an anomalously high rate [14,22] compared to the non-COM modes [11].

C Interaction between Internal and Motional States

We now describe the coupling between the internal electronic levels and the collective axial motion of the ions when a classical radiation field is applied. If the internal levels $|\downarrow\rangle_i$ and $|\uparrow\rangle_i$ of the ith ion in a string are coupled by a dipole moment operator $\boldsymbol{\mu_i}$ (other couplings can be shown to behave analogously), then exposing the ions to traveling-wave electric fields $\boldsymbol{E_i}(\boldsymbol{r}) = \boldsymbol{\tilde{E}_i}cos(\boldsymbol{k}\cdot\boldsymbol{r} - \omega_L t + \phi_L)$ with frequency ω_L, phase ϕ_L, and wavevector $\boldsymbol{k}$ results in the interaction Hamiltonian [22]

$$\mathcal{H} = \sum_{i=1}^{N} -\boldsymbol{\mu_i}\cdot\boldsymbol{E_i}(\boldsymbol{r_i}) = \hbar\sum_{i=1}^{N} g_i\left(S_+^i e^{i(k_x\hat{X}_i - \delta t + \phi_L)} + S_-^i e^{-i(k_x\hat{X}_i - \delta t + \phi_L)}\right) \quad (3)$$

under the rotating wave approximation ($|\delta|, g \ll \omega_0$). In this expression, $g_i = -\langle\uparrow|\boldsymbol{\mu_i}|\downarrow\rangle\cdot\boldsymbol{\tilde{E}_i}/4\hbar$ is the resonant Rabi frequency connecting $|\downarrow\rangle_i$ to $|\uparrow\rangle_i$ in the absence of confinement, $S_\pm^i$ are the internal level raising and lowering operators for the ith ion, $\delta = \omega_L - \omega_0$ is the detuning of the radiation from atomic resonance, and k_x and $\hat{X}_i = \boldsymbol{r_i}\cdot\hat{x}$ are the axial components of the wavevector and position operator of the ith ion, respectively.

In practice, driving direct transitions between $|n_m\rangle|\downarrow\rangle_i$ and $|n'_m\rangle|\uparrow\rangle_i$ with rf or microwave radiation is not feasible, because the coupling between internal and motional states is proportional to powers of k_x from Eq. (3) and would thus be extremely small due to the long wavelength of the radiation. Alternatively, optical fields can be used to drive two-photon stimulated Raman transitions [14]. As depicted in Fig. 1(b), two laser beams detuned by Δ from an excited state of radiative width γ are applied to the jth ion with their difference frequency matched to the desired transition frequency. For sufficient detuning $|\Delta| \gg \gamma$, the excited state may be adiabatically eliminated, and the above coupling of Eq. (3) applies, with g_i replaced by $g_{i1}g_{i2}/\Delta$, where g_{i1} and g_{i2} are the individual Rabi frequencies of the two beams when resonantly coupled to the excited level of ion i. In addition, ω_L (ϕ_L) is replaced by the difference frequency (phase) of the beams, and $\boldsymbol{k}$ is replaced by the difference in wavevectors of the two Raman beams $\boldsymbol{k_1} - \boldsymbol{k_2}$. Since the relevant frequency depends only on the microwave difference between the two laser frequencies, both beams can be generated with a single laser source and a modulator, thereby relaxing the constraints of laser frequency stabilization.

Each ion in a linear array will be displaced from the axial center of the trap due to the other ions by an amount $\bar{X}_i = eF_i/M\omega_x^2$, where F_i is the static axial electric field at the ith ion due to the distribution of charges of the other ions in addition to any externally applied uniform axial field. (The force eF_i is of course balanced by the confining force of the trap.) If the ions are not confined in a linear trap, this displacement will cause each ion to undergo micromotion at the rf drive frequency Ω_T. In the pseudopotential approximation, the axial position operator of the ith ion takes the form [17]

$$\hat{X}_i = \bar{X}_i + \hat{x}_i + \xi_i cos(\Omega_T t). \quad (4)$$

In Eq. (4), the position of ion i is broken into three parts: a static offset term $\bar{X}_i$ which can be combined with ϕ_L in Eq. (3) to give an overall phase $\phi_i = \phi_L + k_x\bar{X}_i$ at ion i, a quantum operator $\hat{x}_i$ associated with small harmonic oscillations about the ion's equilibrium position at the normal mode frequencies, and a classical micromotion oscillation at Ω_T with amplitude $\xi_i = \bar{X}_i\alpha q/2 = eaqF_i/2M\omega_x^2$. A classical treatment of the micromotion is justified because even for unreasonably small fields F_i, the classical amplitude of micromotion is much greater than the zero-point motion ($\xi_i \gg \sqrt{\hbar/M\omega_x}$). The quantum effects of micromotion typically amount to small corrections to the the treatment here [23].

The quantum portion $\hat{x}_i$ of the position can be expressed in terms of the N axial harmonic oscillator normal mode raising and lowering operators $\hat{a}_k^\dagger$ and $\hat{a}_k$:

$$\hat{x}_i = \sum_{k=1}^{N} D_k^i q_{k,0}(\hat{a}_k e^{-i\omega_k t} + \hat{a}_k^\dagger e^{i\omega_k t}). \tag{5}$$

In this expression, D_k^i is the normal mode transformation matrix which relates the ith ion's physical coordinate to the kth collective normal mode coordinate with zero-point characteristic size $q_{k,0} = \sqrt{\hbar/2M\omega_k}$.

When Eqs. (4) and (5) are substituted into the interaction Hamiltonian of Eq. (3), resonances (time independent interaction terms) occur when the detuning is set to sums of harmonics of all normal mode frequencies ω_k and rf drive frequency Ω_T

$$\delta = \sum_{k=1}^{N}(n_k' - n_k)\omega_k + \ell\Omega_T, \tag{6}$$

where n_k', n_k, and ℓ are integers. The resulting Hamiltonian matrix element which couples the quantum states $|\downarrow\rangle_i|\{n\}\rangle$ and $|\uparrow\rangle_i|\{n'\}\rangle$ ($\{n\} = n_1, \ldots, n_N$ and $\{n'\} = n_1', \ldots, n_N'$) is

$$\mathcal{H}'' = \hbar \sum_{i=1}^{N} g_i e^{i\ell\pi/2} J_\ell(k_x \xi_i) \left(S_+^i e^{i\phi_i} \prod_{k=1}^{N} e^{i\eta_{i,k}(\hat{a}_k + \hat{a}_k^\dagger)} + \text{h.c.} \right), \tag{7}$$

where $\eta_{i,k} = k_x D_k^i q_{k,0}$ is the Lamb-Dicke parameter of the ith ion associated with the kth normal mode, and $J_\ell(z)$ is the ℓth Bessel function.

Typically, not all of the motional modes are altered upon application of this interaction, and we will treat two special cases:

(i) **Carrier Interaction.** When the radiation is tuned to the carrier, defined by $\{n'\} = \{n\}$ in Eq. (6), no motional states are altered and there is no entanglement between internal and motional states. Each ion independently evolves between states $|\downarrow\rangle_i|\{n\}\rangle$ and $|\uparrow\rangle_i|\{n\}\rangle$ with coupling

$$\mathcal{H}''_{c_i} = \hbar g_i e^{i\ell\pi/2} J_\ell(k_x \xi_i) \left(\prod_{k=1}^{N} \mathcal{R}_{i,k}(n_k) \right) \left(S_+^i e^{i\phi_i} + \text{h.c.} \right). \tag{8}$$

(ii) **Single-mode Interaction.** Here, the radiation is tuned to $\delta = (n_m' - n_m)\omega_m + \ell\Omega_T$ so that only mode m is altered. This interaction is also called the "$n_m' - n_m$ motional sideband" on mode m and takes the form

$$\mathcal{H}''_1 = \hbar \sum_{i=1}^{N} g_i e^{i\ell\pi/2} J_\ell(k_x \xi_i) \left(\prod_{\substack{k=1 \\ (k \neq m)}}^{N} \mathcal{R}_{i,k}(n_k) \right) \left(S_+^i e^{i\eta_{i,m}(\hat{a}_m + \hat{a}_m^\dagger) + i\phi_i} + \text{h.c.} \right). \tag{9}$$

More-complicated multimode expressions can be similarly derived, allowing the engineering of a large class of interactions by simply tuning the radiation [22].

The factors which appear in Eqs. (8) and (9) are

$$\mathcal{R}_{i,k}(n_k) = \langle n_k | e^{i\eta_{i,k}(\hat{a}_k + \hat{a}_k^\dagger)} | n_k \rangle = e^{-\eta_{i,k}^2/2} \mathcal{L}_{n_k}(\eta_{i,k}^2), \tag{10}$$

describing the effect of motion in spectator motional mode k containing n_k quanta [22], where $\mathcal{L}_n(z)$ is the nth Laguerre polynomial. Both $\mathcal{R}_{i,k}(n_k)$ and $J_\ell(k_x \xi_i)$ factors are ≤ 1 and can be interpreted as a suppression in the laser-ion interaction from the smearing-out of the ion's wavefunction due to spectator normal mode motion and micromotion, respectively. (These factors are known as Debye-Waller factors in many condensed-matter systems [24].)

The central ingredient in the single-mode coupling $\mathcal{H}''_1$ is the coupling of the ith ion's internal states (operated by $S_\pm^i$) to the mth normal mode (operated by $\hat{a}_m$ and $\hat{a}_m^\dagger$) which can allow entangled states to be created. In general, this interaction couples all 2^N internal levels of the N ions in addition to many energy levels of the mth mode of motion. The Schrödinger equation must therefore be intergrated numerically to evaluate the amplitudes of all these energy levels. However, there are special cases which significantly simplify this coupling and can be used for quantum logic.

1 Individual Addressing of Ions by Focussing

In the Cirac-Zoller scheme for quantum logic, the interaction of Eq. (7) is sequentially applied to different ions in the string by focussing laser radiation on the individual ions. For example, if ion j is selected, $g_i = \delta_{i,j} g$ and the sum in Eq. (7) collapses to a single term, resulting in a coupling between the two quantum states $|\downarrow\rangle_j |\{n\}\rangle$ and $|\uparrow\rangle_j |\{n'\}\rangle$ with matrix element

$$\langle\{n'\}|\,{}_j\langle\uparrow|\mathcal{H}''|\downarrow\rangle_j|\{n\}\rangle = \hbar g_j e^{i\ell\pi/2} J_\ell(k_x\xi_j) e^{i\phi_j} \prod_{k=1}^{N} e^{-\eta_{j,k}^2/2} \eta_{j,k}^{|n'_k - n_k|} \sqrt{\frac{n_{k<}!}{n_{k>}!}} L_{n_{k<}}^{|n'_k - n_k|}(\eta_{j,k}^2), \qquad (11)$$

where $n_{k>}$ ($n_{k<}$) is the greater (lesser) of n'_k and n_k, and $L_n^\alpha(z)$ is a generalized Laguerre polynomial.

We will be interested primarily in three types of transitions on ion j involving one mode m of motion, selected by the detuning δ: the carrier, the -1 or "first red" sideband of mode m, and the $+1$ or "first blue" sideband of mode m with Rabi frequencies

$$\Omega_c = g_j J_\ell(k_x \xi_j) \prod_{k=1}^{N} \mathcal{R}_{j,k}(n_k) \qquad (12a)$$

$$\Omega_{-1} = g_j J_\ell(k_x \xi_j) \prod_{\substack{k=1 \\ (k\neq m)}} \mathcal{R}_{j,k}(n_k) \, e^{-\eta_{j,m}^2/2} \eta_{j,m} \frac{L_{n_m-1}^1(\eta_{j,m}^2)}{\sqrt{n_m}} \simeq g_j J_\ell(k_x \xi_j) \eta_{j,m} \sqrt{n_m} \qquad (12b)$$

$$\Omega_{+1} = g_j J_\ell(k_x \xi_j) \prod_{\substack{k=1 \\ (k\neq m)}} \mathcal{R}_{j,k}(n_k) \, e^{-\eta_{j,m}^2/2} \eta_{j,m} \frac{L_{n_m}^1(\eta_{j,m}^2)}{\sqrt{n_m+1}} \simeq g_j J_\ell(k_x \xi_j) \eta_{j,m} \sqrt{n_m+1}, \qquad (12c)$$

where the approximations in Eqs. (12b) and (12c) hold in the Lamb-Dicke regime ($\eta_{j,m}\sqrt{n_m} \ll 1$). The carrier transition, independent of n_m, simply rotates the internal level of ion j and can be used to initialize the quantum bit into the state $(\alpha_j |\downarrow\rangle_j + \beta_j |\uparrow\rangle_j)$, with α_j and β_j arbitrary complex numbers satisfying $|\alpha_j|^2 + |\beta_j|^2 = 1$. The first red sideband interaction is the central ingredient of the Cirac-Zoller scheme. If motional mode m is initially prepared in the $n_m = 0$ zero-point level, then by applying radiation to the jth ion on the first red sideband for a time $\tau = \pi/2\Omega_{-1}$ (a π pulse), the state $(\alpha_j |\downarrow\rangle_j + \beta_j |\uparrow\rangle_j)|0\rangle$ is mapped to $|\downarrow\rangle_j (\alpha_j |0\rangle + \beta_j |1\rangle)$; that is, the quantum bit initially stored in the jth ion is mapped onto the first two states of motion of mode m. This information is shared among all ions which have nonzero $\eta_{j,m}$ and can be subsequently mapped onto another ion j' to produce an arbitrary entangled state between ions j and j'. The operations of Eq. (12) have been realized on a single trapped $^9\text{Be}^+$ ion [14,15].

2 Uniform Coupling

Here, we assume that each ion receives the same coupling from the radiation, or that each term in the sum of Eq. (7) is independent of i, outside of phase factors. This can occur when the applied radiation uniformly illuminates all of the trapped ions ($g_i = g$). Furthermore, we require that (i) $k_x \xi_i$ is independent of i, which holds in a linear trap ($\xi_i = 0$) or in the case of copropagating Raman beams ($k_x = 0$), and (ii) transitions are driven on the carrier or a uniform mode of motion u where the Lamb-Dicke parameter $\eta_{i,u} = \eta_u$ is independent of i and the other (nonuniform) modes are in the Lamb-Dicke regime so that $\mathcal{R}_{i,k} \simeq 1$. Examples of uniform motional modes are the COM mode where $\eta_{i,com} = \eta_{com} = k_x \sqrt{\hbar/2MN\omega_x}$ or the stretch mode for $N = 2$ ions). For transitions on the carrier,

$$\mathcal{H}''_{c_u} = \hbar g \left(J_+ + J_- \right), \qquad (13)$$

and for transitions involving a uniform mode m_u

$$\mathcal{H}''_{1_u} = \hbar g \left(J_+ e^{i\eta_u (\hat{a}_u + \hat{a}_u^\dagger)} + h.c. \right). \qquad (14)$$

In these expressions, $J_\pm \equiv \sum_{i=1}^{N} S_\pm^i e^{\pm i\phi_i}$ are generalized angular momentum raising and lowering operators formed by combining the N spin-$1/2$ ions into an equivalent spin-$J = N/2$ system [8]. If the system is initially in the state $|\downarrow\rangle_1|\downarrow\rangle_2\cdots|\downarrow\rangle_N = |J, m_J = -J\rangle$, then evolution is confined to the $2J+1 = N+1$ coupled eigenstates $|J, m_J\rangle$ with matrix elements $\langle J, m_J \pm 1|J_\pm|J, m_J\rangle = \sqrt{J(J+1) - m_J(m_J \pm 1)}$. (Unlike the usual angular momentum eigenstates, when these states are decomposed to the uncoupled states, there are generally different phase factors in front of each uncoupled state.) In these special cases of uniform couplings, evolution of the quantum state of the N ions is simplified because it is restricted to the subspaces spanned by the $2m_J + 1 \leq N+1$ states of the equivalent spin $J = N/2$ system instead of the complete Hilbert space containing 2^N states.

3 General Treatment of N=2 Ions

For the case of two ions, there are two axial modes of motion: the COM mode at $\omega_{com} = \omega_x$ and the stretch (STR) mode where the ions move with opposite phase at $\omega_{str} = \sqrt{3}\omega_x$. The static fields at the two ion positions (balanced by the trapping fields) are $F_{1,2} = \mp e/4\pi\epsilon_0 s^2 + F_{ext}$, where $s = \sqrt[3]{e^2/(2\pi\epsilon_0 m\omega_x^2)}$ is the spatial separation between the two ions. We can rewrite the general interaction of Eq. (7) for two ions:

$$\begin{aligned}\mathcal{H}''_{2\ ions} &= \hbar g_1 e^{i\ell\pi/2} J_\ell(k_x\xi_1)\left(S^1_+ e^{i\eta_{com}(\hat{a}_{com}+\hat{a}^\dagger_{com})+i\eta_{str}(\hat{a}_{str}+\hat{a}^\dagger_{str})-ik_x s/2} + h.c.\right) \\ &+ \hbar g_2 e^{i\ell\pi/2} J_\ell(k_x\xi_2)\left(S^2_+ e^{i\eta_{com}(\hat{a}_{com}+\hat{a}^\dagger_{com})-i\eta_{str}(\hat{a}_{str}+\hat{a}^\dagger_{str})+ik_x s/2} + h.c.\right),\end{aligned} \quad (15)$$

where $\eta_{com} = \eta_{1,com} = \eta_{2,com}$ and $\eta_{str} = \eta_{1,str} = \eta_{2,str} = \eta_{com}/\sqrt[4]{3}$.

III QUANTUM LOGIC WITH A FEW IONS

A Two Ions

At NIST, we have laser-cooled both axial modes of two $^9Be^+$ ions to the ground state of motion in a trap with $\omega_x/2\pi \simeq 10$ MHz [11]. The spacing between the ions is only $s \simeq 2\mu m$ in such a strong trap, so it is not trivial to focus laser beams to individually address the two ions for quantum logic following Section II C 1. We therefore consider alternate schemes in which both ions are equally illuminated ($g_1 = g_2$).

1 Differential Addressing with Micromotion

It is still possible to differentially address the two ions by tuning the argument of the Bessel functions in Eq. (15) through the externally applied axial field F_{ext}. The micromotion amplitudes of the two ions are

$$\xi_{1,2} = \frac{e\alpha q}{2M\omega_x^2}\left(\mp\frac{e}{4\pi\epsilon_0 s^2} + F_{ext}\right), \quad (16)$$

so the corresponding carrier Rabi frequencies on the two ions can be set to any ratio $\Omega_1/\Omega_2 = J_\ell(k_x\xi_1)/J_\ell(k_x\xi_2)$. In Fig. 3(a), we have measured the Rabi frequencies of the two ions with $\ell = 0$ while varying the externally applied field F_{ext}, showing excellent agreement with theory [12].

In particular, by setting F_{ext} so that $\Omega_1/\Omega_2 = 2$ (vertical arrow in Fig. 3(a)) and initializing the ions in state $|\downarrow\rangle|\downarrow\rangle|0_{str}\rangle$, we can create the state $|\downarrow\rangle|\uparrow\rangle|0_{str}\rangle$ by driving on the carrier for a time $\tau = \pi/\Omega_1$ (vertical arrow in Fig. 3(b)). To make an EPR entangled state [25], we can then switch F_{ext} so that $\Omega_1/\Omega_2 = 1+\sqrt{2}$ and apply radiation tuned to $\delta = -\omega_{str}$ and drive on the STR red-sideband which couples $|\downarrow\rangle|\uparrow\rangle|0_{str}\rangle$, $|\downarrow\rangle|\downarrow\rangle|1_{str}\rangle$, and $|\uparrow\rangle|\downarrow\rangle|0_{str}\rangle$. This three-level system can be solved exactly, and we find that the state evolves to a maximally entangled EPR state

$$|\downarrow\rangle|\uparrow\rangle|0_{str}\rangle \longrightarrow \left(\frac{|\downarrow\rangle|\uparrow\rangle + e^{ik_x s}|\uparrow\rangle|\downarrow\rangle}{\sqrt{2}}\right)|0_{str}\rangle, \quad (17)$$

with the phase controlled by the spacing s of the two ions. In Ref. [12], F_{ext} was fixed for a 2:1 ratio of Rabi frequencies throughout the experiment, which created a similar but slightly less entangled version as Eq. (17).

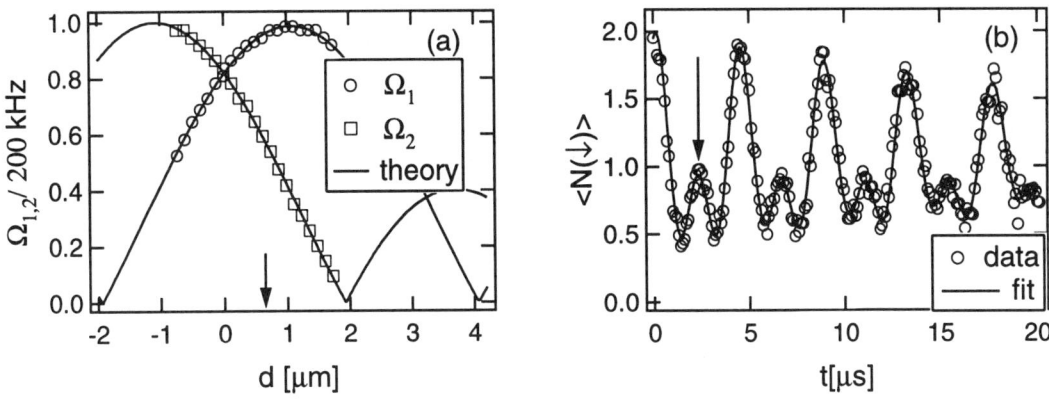

FIGURE 3. (a) Normalized carrier Rabi frequencies of each of two ions as a function of the mean position of the two ions $d = (\bar{X}_1 + \bar{X}_2)/2 = eF_{ext}/m\omega_{com}^2$ set by externally applied uniform field F_{ext}. The solid curves are theoretical Bessel functions $J_0(k_x\xi_1)$ and $J_0(k_x\xi_2)$, where ξ_i is from Eq. (16). (b) Number of ions in state $|\downarrow\rangle$ as a function of the time the carrier is applied, clearly showing two frequency components with $\Omega_1/\Omega_2 = 2$. the arrow in (a) indicates the setting of d for (b).

This technique of tuning each ion's Rabi frequency based on unequal micromotion amplitudes can also be extended to give individual addressing of the ions as in Section II C 1 but without using focussed beams. Here, we set the trap strength and external electric field F_{ext} such that $k_x\xi_1 = 0$ and $k_x\xi_2 = 2.405$. (In the NIST ^{9}Be$^+$ experiments where $k_x = (\mathbf{k}_2 - \mathbf{k}_1) \cdot \hat{x} = \sqrt{2}(2\pi/\lambda_{Be})$ and $\Omega_T/2\pi \simeq 240$ MHz, this condition is met when the axial COM frequency is set to $\omega_x/2\pi \simeq 3.80$ MHz with $U_0 \propto a = 0$ and the external field is set to $F_{ext} = 1.01$ V/cm.) When ion 1 is to be isolated, the radiation is tuned so that $\ell = 0$ in Eq. (6), thus shutting off the coupling to ion 2 since $J_0(2.405) = 0$. When ion 2 is to be isolated, the radiation is tuned so that $|\ell| = 1$, thus shutting off the coupling to ion 1 since $J_1(0) = 0$.

2 EPR State without Micromotion

We describe an alternate scheme for making EPR states where each ion is equally illuminated by the radiation and has the same coupling to the radiation field (in the case of a linear trap or for $F_{ext} = 0$ in Eq. (16) giving $|\xi_1| = |\xi_2|$). Here we can describe the system in terms of an equivalent angular momentum $J = 1$ system described by quantum numbers $|J, m_J\rangle$ and the stretch motional state $|n_{str}\rangle$. If the two ions are initialized in the state $|\downarrow\rangle|\downarrow\rangle|0_{str}\rangle = |1, -1\rangle|0_{str}\rangle$, then we find from Eqs. (11) and (14) that, if we tune to the first blue sideband of the stretch mode ($\delta = \omega_{str}$), the couplings are

$$\langle 1, 0|\langle 1_{str}|\mathcal{H}''|0_{str}\rangle|1, -1\rangle = \hbar g \mathcal{R}_{com}(n_{com})e^{-\eta_{str}^2/2}\eta_{str}\sqrt{2} \tag{18a}$$

$$\langle 1, 1|\langle 2_{str}|\mathcal{H}''|1_{str}\rangle|1, 0\rangle = \hbar g \mathcal{R}_{com}(n_{com})e^{-\eta_{str}^2/2}\eta_{str}(2 - \eta_{str}^2), \tag{18b}$$

where $\mathcal{R}_{com}(n_{com}) = \mathcal{R}_{1,com}(n_{com}) = \mathcal{R}_{2,com}(n_{com})$. Now if the trap strength is set so that $\eta_{str} = \sqrt{2}$, the second coupling (Eq. (18b)) vanishes. The first coupling (Eq. (18a)) can then be driven for a time such that $|1, -1\rangle|0_{str}\rangle$ evolves directly to the EPR entangled state $|1, 0\rangle|1_{str}\rangle = (e^{i\phi_1}|\downarrow\rangle|\uparrow\rangle + e^{i\phi_2}|\uparrow\rangle|\downarrow\rangle)|1_{str}\rangle/\sqrt{2}$.

B Three Ions

For $N = 3$ ions, we are interested in producing the maximally entangled ("GHZ") state [26]:

$$\Psi_{GHZ} = \frac{|\downarrow\rangle|\downarrow\rangle|\downarrow\rangle + e^{i\phi}|\uparrow\rangle|\uparrow\rangle|\uparrow\rangle}{\sqrt{2}}. \tag{19}$$

Such states are of great interest to the measurement of quantum nonlocality and can be employed for rudimentary quantum error correction codes [27]. This state can be produced using the general Cirac-Zoller scheme; here we discuss two schemes for generating (or approximating) Eq. (19) without individual addressing of ions.

1 Approximate GHZ State without Micromotion

We assume that the three ions are equally illuminated by the radiation fields ($g_i = g$), undergo no micromotion, and are initialized in the state $|\downarrow\rangle|\downarrow\rangle|\downarrow\rangle|0_{str}\rangle|0_{eg}\rangle$. Here, $|n_{str}\rangle$ refers to the stretch mode at frequency $\sqrt{3}\,\omega_x$ wherein the middle ion is at rest, and $|n_{eg}\rangle$ refers to the "Egyptian" [28] motional mode at frequency $\omega_{eg} = \sqrt{29/5}\,\omega_x$ wherein the middle ion 2 moves out of phase and with twice the amplitude as the outer ions 1 and 3. We sequentially apply three pulses of radiation to the ions:

(1) The radiation is tuned to the 2nd blue sideband of the Egyptian mode ($n'_{eg} = n_{eg} + 2$ from Eq. (11)). This coupling, proportional to $\eta_{i,eg}^2$, is 4 times larger on the middle ion than on either outer ion. We therefore approximate that only the middle ion is affected. By applying this radiation for an approximate $\pi/2$ pulse duration, we create a superposition of the original state with $|\downarrow\rangle|\uparrow\rangle|\downarrow\rangle|0_{str}\rangle|2_{eg}\rangle$.

(2) The motion in the Egyptian mode is swapped with the motion in the stretch mode. This can be accomplished by tuning the radiation so that $\omega_L = \omega_{eg} - \omega_{str}$. In this case, a different rotating-wave approximation is performed in Section II C, and Eq. (3) is reproduced *without the internal* $S_\pm^i$ *operators*. The three motional states $|0_{str}\rangle|2_{eg}\rangle$, $|1_{str}\rangle|1_{eg}\rangle$, and $|2_{str}\rangle|0_{eg}\rangle$ are coupled without affecting the internal states of the ions, allowing $|0_{str}\rangle|2_{eg}\rangle$ to evolve to $|2_{str}\rangle|0_{eg}\rangle$ through the intermediate $|1_{str}\rangle|1_{eg}\rangle$ state for $\eta_{i,str}, \eta_{i,eq} \ll 1$.

(3) The radiation is tuned to the first red sideband of the stretch mode. This operation does not affect the middle ion ($\eta_{2,str} = 0$) and largely removes the two quanta in the stretch mode while simultaneously flipping the outer two internal states. The three steps roughly give the evolution (ignoring phase factors):

$$|\downarrow\rangle|\downarrow\rangle|\downarrow\rangle|0_{str}\rangle|0_{eg}\rangle \longrightarrow \frac{|\downarrow\rangle|\downarrow\rangle|\downarrow\rangle|0_{str}\rangle|0_{eg}\rangle + |\downarrow\rangle|\uparrow\rangle|\downarrow\rangle|0_{str}\rangle|2_{eg}\rangle}{\sqrt{2}} \tag{20a}$$

$$\longrightarrow \frac{|\downarrow\rangle|\downarrow\rangle|\downarrow\rangle|0_{str}\rangle|0_{eg}\rangle + |\downarrow\rangle|\uparrow\rangle|\downarrow\rangle|2_{str}\rangle|0_{eg}\rangle}{\sqrt{2}} \tag{20b}$$

$$\longrightarrow \frac{|\downarrow\rangle|\downarrow\rangle|\downarrow\rangle|0_{str}\rangle|0_{eg}\rangle + |\uparrow\rangle|\uparrow\rangle|\uparrow\rangle|0_{str}\rangle|0_{eg}\rangle}{\sqrt{2}}. \tag{20c}$$

The actual amplitudes of the states must be solved numerically due to imperfections in the first and third steps. As a consequence, instead of a perfect GHZ state, this scheme generates the state $|\Psi_{MIKKLMTWW}\rangle$ with a fidelity $|\langle\Psi_{MIKKLMTWW}|\Psi_{GHZ}\rangle|^2 \simeq 86\%$.

2 Exact GHZ State with Micromotion

Here, the ions are again equally illuminated, but the three ions have different couplings to the radiation due to different amounts of micromotion in each ion. In particular, we set axial trap strength and external field such that $k_x\xi_{1,3} = 2.405$ and $k_x\xi_2 = 0$. This allows the middle ion 2 to be individually addressed for $\ell = 0$ (since $J_0(2.405) = 0$), and outer ions 1 and 3 to be equally addressed for $\ell = 1$ (since $J_1(0) = 0$). (In the NIST ^{9}Be$^+$ experiments, following the parameters of section III A 1, this condition is met when the axial COM frequency is set to $\omega_x/2\pi \simeq 6.08$ MHz with $U_0 \propto a = 0$ and the external field is set to $F_{ext} = 0$.) We sequentially apply two pulses of radiation to the ions, initially prepared in the state $|\downarrow\rangle|\downarrow\rangle|\downarrow\rangle|0_m\rangle$:

(1) A $\pi/2$-pulse is driven on the first blue sideband of motional mode m with $\ell = 0$, affecting only the middle ion:

$$|\downarrow\rangle|\downarrow\rangle|\downarrow\rangle|0_m\rangle \longrightarrow \frac{|\downarrow\rangle|\downarrow\rangle|\downarrow\rangle|0_m\rangle + e^{i\phi_2}|\downarrow\rangle|\uparrow\rangle|\downarrow\rangle|2_m\rangle}{\sqrt{2}} \tag{21}$$

(2) A pulse is driven on the first red sideband of mode m with $\ell = 1$, affecting the outer two ions identically. If we set the Lamb-Dicke parameter to $\eta_{m,1} = \eta_{m,3} = \sqrt{2-\sqrt{2}}$, we find that the two-ion couplings similar to Eq. (18) are identical, and after a particular time the state evolves to the GHZ state

$$\frac{|\downarrow\rangle|\downarrow\rangle|\downarrow\rangle|0_m\rangle + e^{i\phi_2}|\downarrow\rangle|\uparrow\rangle|\downarrow\rangle|2_m\rangle}{\sqrt{2}} \longrightarrow \frac{|\downarrow\rangle|\downarrow\rangle|\downarrow\rangle|0_m\rangle + ie^{i(\phi_1+\phi_2+\phi_3)}|\uparrow\rangle|\uparrow\rangle|\uparrow\rangle|0_m\rangle}{\sqrt{2}}. \tag{22}$$

Since all axial trap frequencies (proportional to ω_x) are constrained by the micromotion requirements above, one of the nonaxial modes must be used for mode m. Since the Lamb-Dicke parameter is large, this mode should be very weakly bound. For three ions, there are two "zig-zag" modes in which all three ions move along y or z, with the middle ion 180 degrees out of phase with the outer two ions. The oscillation frequency of the z zig-zag mode is $\omega_{zzz} = \sqrt{\omega_z^2 - 2.4\omega_x^2}$, which can be made arbitrarily small while keeping all 8 other modes at relatively high frequencies. For our NIST ^{9}Be$^+$ experiments, we find that $\eta_{1,zzz} = \eta_{3,zzz} = \sqrt{2 - \sqrt{2}}$ implies a zig-zag frequency of $\omega_{zzz}/2\pi \simeq 64$ kHz.

IV OUTLOOK

A scalable scheme for universal quantum logic with trapped ions is the method Cirac and Zoller proposed in which the individual ions in a string are individually addressed with laser radiation (Section II C 1). However, there are many degrees of freedom of up to 3 ions which can allow quantum logic gates and entangled states to be created without individual addressing. Although these schemes do not appear scalable within a single collection of ions, they may prove useful for doing quantum logic on nodes of a few ions which could then be coupled to other nodes of ions with cavity-QED techniques [29,30] or ion accumulators in which an individual ion is physically moved to a nearby separate collection of ions [22].

Work of the U.S. government; not subject to copyright. We gratefully acknowledge support from the National Security Agency, Office of Naval Research, and Army Research Office. We thank Matt Young for critical comments regarding the manuscript.

REFERENCES

1. D. P. DiVincenzo, Science **270**, 255 (1995).
2. A. Ekert and R. Jozsa, Rev. Mod. Phys. **68**, 733 (1996).
3. D. Deutsch, Proc. Roy. Soc. London A **425**, 73 (1989).
4. P. W. Shor, SIAM J. Comp. **26**, 1484 (1997).
5. L. K. Grover, Phys. Rev. Lett. **79**, 325 (1997).
6. S. Lloyd, Science **261**, 1569 (1993).
7. *Quantum Theory and Measurement*, edited by J. A. Wheeler and W. H. Zurek (Princeton University Press, Princeton, N.J., 1983).
8. D. J. Wineland, J. J. Bollinger, W. M. Itano, and D. J. Heinzen, Phys. Rev. A **50**, 67 (1994).
9. J. J. Bollinger, W. M. Itano, D. J. Wineland, and D. J. Heinzen, Phys. Rev. A **54**, R4649 (1996).
10. J. I. Cirac and P. Zoller, Phys. Rev. Lett. **74**, 4091 (1995).
11. B. E. King et al., Phys. Rev. Lett. **81**, 1525 (1998).
12. Q. A. Turchette et al., LANL e-print archive quant-phy/9806012 (1998).
13. J. J. Bollinger et al., IEEE Trans. Instr. Meas. **40**, 126 (1991).
14. C. Monroe et al., Phys. Rev. Lett. **75**, 4011 (1995).
15. C. Monroe et al., Phys. Rev. Lett. **75**, 4714 (1995).
16. R. Blatt and P. Zoller, Eur. J. Phys. **9**, 250 (1988).
17. H. G. Dehmelt, Adv. Atom. Mol. Phys. **3**, 53 (1967).
18. M. G. Raizen et al., J. Mod. Opt. **39**, 233 (1992).
19. J. P. Schiffer, Phys. Rev. Lett. **70**, 818 (1993).
20. A. Steane, Appl. Phys. B **64**, 623 (1997).
21. D. James, Appl. Phys. B **66**, 181 (1997).
22. D. J. Wineland et al., J. Res. Nat. Inst. Stand. Tech. **103**, 259 (1998).
23. G. Schrade et al., Appl. Phys. B **64**, 181 (1997).
24. H. J. Lipkin, *Quantum Mechanics* (North-Holland, New York, 1973).
25. A. Einstein, B. Podolsky, and N. Rosen, Phys. Rev. **47**, 777 (1935).
26. D. M. Greenberger, M. A. Horne, Shimony, and A. Zeilinger, Am. J. Phys. **58**, 1131 (1990).
27. P. W. Shor, Phys. Rev. A **52**, R2493 (1995).
28. W. H. Zurek, private communication.
29. J. I. Cirac, P. Zoller, H. J. Kimble, and H. Mabuchi, Phys. Rev. Lett. **78**, 3221 (1997).
30. T. Pellizzari, Phys. Rev. Lett. **79**, 5242 (1997).

The Quantum Zeno Effect in Trapped Ions

R C Thompson, J-L Hernandez-Pozos*, J Höffges,
D M Segal and J R Vincent

*Blackett Laboratory, Imperial College,
Prince Consort Rd., London SW7 2BZ, UK.
* Present address: School of Physics and Space Science Research,
University of Birmingham, Birmingham, UK.*

Abstract. The quantum Zeno effect is the slowing down of the rate of a quantum mechanical transition by the frequent application of measurements to the system. Following the original suggestion of an experiment to test this in trapped ions, Itano et al performed an experiment in 1990. This was followed by several detailed theoretical treatments of the problem, making new predictions and giving a better understanding of what was happening. Also since then, suggestions for new types of experiment to test the quantum Zeno effect have been made. At Imperial College we are currently setting up an experiment to study the quantum Zeno effect on a single ion, incorporating some of these new ideas. Details of this experiment are discussed in the paper.

INTRODUCTION

Ion traps are well suited to the measurement of long coherent processes in atomic systems because of the long interaction time which is possible, the excellent isolation from the environment, and the stable trapping conditions which can be achieved (see, for example, [1–3] for general reviews). It is possible to use long coherent pulses of radiation to couple two atomic levels together and then to study the evolution of the system under these conditions. It is also possible to isolate single atomic particles, and many significant experiments have been performed on single ions, which have increased our understanding of the way that single particles interact with radiation.

The quantum Zeno effect was first introduced by Misra and Sudarshan in 1977 [4]. These authors pointed out that, in a quantum mechanical system that decays from one state to another, a measurement taking place during the decay collapses the wave function into either the initial state or the decayed state. Now in the very early stages of the decay the probability of such a measurement resulting in the particle ending up in the final (decayed) state rises *quadratically* with time, even though for longer times the decay proceeds exponentially (which can be approximated as a *linear* decay for short times). Since the effect of a measurement is to destroy all the coherences in the system, in effect it restarts the decay process. Now if the decay is continually put back onto the quadratic part of the curve, the end result is that the overall rate of decay is slowed down from the rate expected from the exponential curve.

This was termed the *quantum Zeno effect* by analogy with the classic paradox due to Zeno about an arrow in flight. The point is that if we make many measurements on a decaying system (such that they probe this early quadratic region) we can slow down the decay. In the limit of continuous observation, we can expect the decay to slow to zero. This is a classic example of the measurement of a quantum mechanical system having an effect on the subsequent behaviour of the system.

The trouble with the observation of this phenomenon for spontaneous decay in real atomic systems is that the time period over which the decay is quadratic is extremely short and inaccessible experimentally. The length of this period is related to the bandwidth of the reservoir of states into which the system can decay, which is extremely large for a normal decay of an excited atomic state.

In 1988 Cook suggested that the quantum Zeno effect could in fact be studied experimentally, but by choosing a driven transition instead of a natural decay [5] (see also [6]). The advantage of using a driven transition is that this also has a quadratic phase at the start, but the length of this phase is under experimental control and is much longer than in a natural decay.

Consider a three-level system with a ground state (1) coupled by a microwave transition to a metastable level (2) and by a strong optical transition to an excited atomic state (3) which can only decay back to level 1. Cook suggested taking a π-pulse of microwave radiation between levels 1 and 2 and showed that if this pulse were interrupted by a series of short measurement pulses (a burst of laser radiation coupling level 1 to level 3) then the overall probability of completing the microwave transition, measured at the end of the π-pulse, was reduced dramatically. He suggested that this experiment could be carried out on a single ion in a Paul trap, taking advantage of the ideal conditions offered by the use of trapped ions, in particular the long interaction times possible.

EXPERIMENTAL OBSERVATION

Itano et al [7] performed an experiment in 1990 to observe the quantum Zeno effect. In their experiment they worked on a driven radiofrequency transition between hyperfine Zeeman sublevels in a cloud of Be$^+$ ions in a Penning trap. In this way they avoided the problem of the very small signal which would be available in a single ion experiment. In this case the length of the quadratic period depends on the rate at which the transition is driven, which is under experimental control (they chose a π-pulse with a length of the order of 1 second). Of course the effect is now not quite the same as in the original paper [4], but the principle is very similar.

Since Itano et al used a π-pulse of radiofrequency radiation to drive the transition, in the absence of any measurements the probability of the transition taking place is nearly equal to unity. However, if the coherent transition is interrupted by short measurements, they show that the overall transition probability is reduced dramatically in both theory and experiment [7]. Here the measurement, which consists of a brief laser pulse which is tuned to a transition out of *one* of the two hyperfine Zeeman levels involved, could in principle be used to detect whether the ions were in that state (by the observation of at least one fluorescent photon) or the other one (by the failure to observe any photons). The laser pulse has to be sufficiently long and intense to ensure that an ion in that state has a high probability of being excited by the laser. Since in a large cloud the ions are not inside the laser beam all the time, the length of the pulse must therefore be several times the oscillation period in the trap, in order to ensure that every ion is subjected to the effect of the measurement.

Assume that the ions start in state 1. Then the wave function at time t is given by

$$|\psi(t)\rangle = \cos(\Omega t/2)|1\rangle - i\sin(\Omega t/2)|2\rangle \tag{1}$$

where Ω is the Rabi frequency, and 1 and 2 are the initial and final states. For a π-pulse, the length, T, is given by $T = \pi/\Omega$, and this transfers all the population from state 1 to state 2. If the π-pulse is interrupted at a time $t_1 = \pi/n\Omega$, then we find that the probabilities of the ion being found in the two states are now:

$$P_1(t_1) = \cos^2(\pi/2n) \tag{2}$$

$$P_2(t_1) = \sin^2(\pi/2n). \tag{3}$$

At this point the decay restarts and the calculation can be repeated for each of the $(n-1)$ remaining measurement pulses. In practice the state of the ions is only determined in this experiment at the end of the π-pulse, when the fraction of ions that have made the transition is determined by measuring the amount of fluorescence on a cycling transition which yields a signal of many photons per ion in the first few ms of irradiation. It can be shown by an extension of the arguments above that the fraction of ions in the two states at the end of the π-pulse, after n measurement pulses, is given by

$$P_1(T) = \frac{1}{2}[1 + \cos^n(\pi/n)] \tag{4}$$

$$P_2(T) = \frac{1}{2}[1 - \cos^n(\pi/n)] \tag{5}$$

Itano et al used this simple wavefunction collapse model to predict the fraction of ions to complete the transition from state 1 to state 2 as a function of the number of measurement pulses n (with some small modifications) and the experiment verified these predictions to within the experimental error. They concluded that this wavefunction collapse model gave a good description of the process.

DETAILED THEORETICAL MODELS

Many theorists were unhappy with the approach outlined above, in particular the idea of the collapse of the wavefunction which was employed, and the publication of these experimental results gave rise to much theoretical work which aimed to give a more detailed and more rigorous description of what was going on in this experiment.

One treatment of this problem is given by Frerichs and Schenzle [8] (see also [9]). They treat the system of 3-level ion, radiofrequency radiation and laser radiation as a single system using a Bloch equation approach. The evolution of the complete system can then be calculated both during the periods when just the microwave radiation is present, and also when both radiations are simultaneously present. This gives predictions for how intense the laser pulse has to be for it to count as a measurement. This was something that had to be introduced in an *ad hoc* manner in the treatment of Itano *et al*. The Bloch equation approach shows that the effect of the laser pulse is to destroy the coherences (i.e. the off-diagonal elements of the density matrix) which are built up by the radiofrequency radiation. This is the mechanism for the collapse of the wavefunction in this case. Other treatments of the problem are given, for example, in references [10] and [11].

These treatments show that the measurement pulse can be properly described using standard techniques of quantum optics, so it is not necessary to introduce the idea of the wave function collapse separately. They also show that, for a sufficiently intense measurement pulse, the effect is the same as that obtained by using the wave function collapse model, so this is a good working model under these circumstances.

PROPOSED EXPERIMENT

At Imperial College we are in the process of building an experiment to test the predictions of these theoretical models. We will perform an experiment which is similar to that of Itano *et al* [7] but we need to perform the experiment on a single ion so that we can have a closer control of the experimental conditions. This will also be closer to the original suggestion of Cook [5] (though the experiment will still be performed in a Penning trap rather than a Paul trap). This will clearly reduce the signal level and the experiment will therefore need to run in a stable manner for a long time to collect data. However, it gives us the advantage of being able to make a direct comparison with the predictions of recent formulations of quantum mechanics which calculate single particle trajectories directly (see for example [12,13]).

One point we wish to investigate is the variation of the effect of the measurement pulses with the strength of the pulses in an attempt to verify the theoretical predictions for what constitutes a measurement. This comes naturally out of the Bloch equation treatment of the three level system, including the effects of both radiations. Here the use of a single ion has an advantage because a single ion experiences more of a uniform laser intensity than an ion in a cloud, due to the much reduced amplitude of its motion. There is therefore a more definite relationship between the applied laser intensity and the intensity experienced by the ion.

We will also look at variations in the experiment which should allow us to introduce some decay by coupling to an excited state of the ion in the manner proposed by Plenio *et al* [14]. This will take the experiment closer to the scheme originally proposed by Misra and Sudarshan [4]. In this scheme, the microwave transition takes the system from the ground state (1) to a metastable excited state (2) as before, but this time the metastable state is itself coupled (for example, with a second laser) to a third level (3) which can decay back to the metastable level (2) or down to another excited state (4). The measurements are performed, as before, with a laser pulse coupling state (1) on a strongly allowed transition to an excited state (5). The effect of this is to introduce decay into the system in a controlled manner. The time evolution of the population without the measurement pulses now becomes an exponential at long times, but the early quadratic phase is still there, with its length under experimental control. Both of these schemes suggested by Plenio *et al* can be implemented with trapped ions of Be or Mg using two lasers.

Another possibility is to perform a quasi-continuous measurement, as proposed by Beige and Hegerfeldt [15] for the original three-level system. In this scheme, there is no π-pulse, but the measurement pulses are applied in a continuous stream on a single trapped particle. As the interval between these pulses reduces, the system starts to execute quantum jumps between the ground state and the metastable level. The length of the jumps is proportional to $(\Delta t)^{-2}$, where Δt is the interval between measurements. These quantum jumps are well defined for $\Delta t \approx T/6$ or less [15]. Careful selection of the laser pulse integrated intensity and the values of Ω and Δt should enable a clear quantum jump signal to be obtained. Care has to be taken because in our system

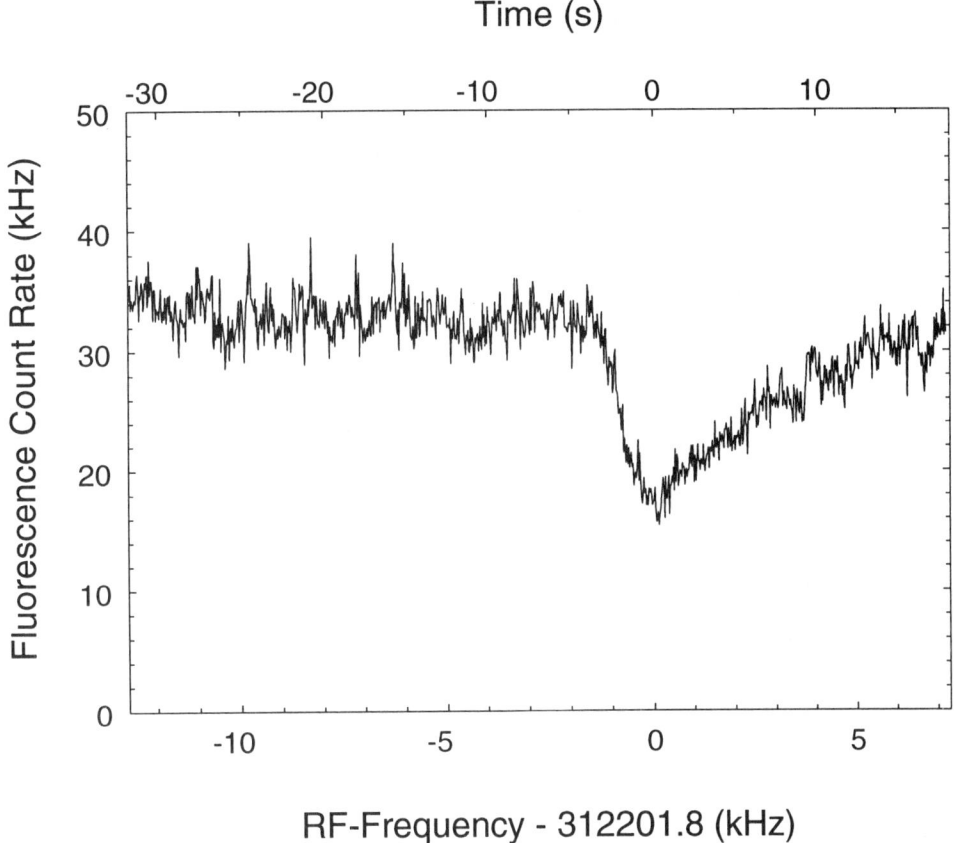

FIGURE 1. A typical RF-optical double resonance in ^{9}Be$^+$. The laser is tuned to the cooling/optical pumping transition, $2S_{1/2}(m_i = -3/2, m_j = -1/2) - 2P_{3/2}(-3/2, -3/2)$, at 313nm. The fluorescence is monitored as the RF frequency is scanned upwards through the resonance. The fluorescence scattering rate decreases to about 50% on resonance. The fluorescence signal returns to its original value with a time constant characteristic of the population repumping [17].

(^{9}Be$^+$ or ^{25}Mg$^+$) the laser pulses can also drive the system into a different atomic state (the state involved in conventional quantum jump measurements in this system: see, for example, [13,16]) so the pulses should not be too intense. On the other hand, the pulses have to be sufficiently intense for us to be able to detect which state the ion is in. This means that we need to detect an average of about 10 photons per pulse for an ion in state (1), leading to a pulse length of the order of 1 ms with our laser.

The trap for this experiment has been constructed and tested. It is a 5-electrode cylindrical Penning trap which is placed in a superconducting magnet with a field of 1.3 T. The experiment is expected to run with Be$^+$ ions, but it is also possible to use Mg$^+$ if required (for example, if sympathetic cooling is required). The laser beam enters the trap at a small angle to the radial plane to give cooling in the axial direction as well as in the radial plane. The ion fluorescence is imaged with a ×25 magnification lens system, specially designed for this trap, onto the photocathode of a photon counting photomultiplier.

We have been able to trap and cool large and small clouds of Be$^+$ ions in this trap, and to drive the microwave resonances between the Zeeman levels in the magnetic field (Figure 1). Quantum jump signals have been seen from single ions, and we expect to make preliminary studies of the Zeno effect in the near future.

ACKNOWLEDGEMENTS

We wish to thank the UK Engineering and Physical Sciences Research Council for their support of our work at Imperial College.

REFERENCES

1. R C Thompson, in *Adv At Mol Opt Phys* **31**, 63-136 (1993).
2. P K Ghosh, *Ion traps*, Oxford University Press (1995).
3. G Zs K Horvath, R C Thompson and P L Knight, *Contemporary Physics* **38**, 25-48 (1997).
4. B Misra and E C G Sudarshan, *J Math Phys* **18**, 756-63 (1977).
5. R J Cook, *Phys Scripta* T **21** 49-51 (1988).
6. R J Cook in *Progress in Optics XXVIII* (E Wolf, ed) Elsevier, pp361-416 (1990).
7. W M Itano, D J Heinzen, J J Bollinger and D J Wineland, *Phys Rev A* **41**, 2295-300 (1990).
8. V Frerichs and A Schenzle, *Phys Rev A* **44**, 1962-8 (1991).
9. A Schenzle, *Contemp Phys* **37**, 303-20 (1996).
10. E Block and P R Berman, *Phys Rev A* **44**, 1466 (1991).
11. L E Ballentine, *Phys Rev A* **43**, 5165 (1991).
12. B M Garraway and P L Knight, *Phys Rev A* **49**, 1266-74 (1994).
13. N Gisin, P L Knight, I C Percival, R C Thompson and D C Wilson, *J Mod Opt* **40**, 1663-71 (1993).
14. M Plenio, P L Knight and R C Thompson, *Opt Comm* **123**, 278-86 (1996).
15. A Beige and G C Hegerfeldt, *J Phys A* **30**, 1323-34 (1997).
16. R G Hulet, D J Wineland, J C Bergquist and W M Itano, *Phys Rev A* **37**, 4544-7 (1988).
17. D J Wineland, J C Bergquist, W M Itano, R E Drullinger, *Optics Letters* **5**, 245-7 (1980)

Spatial separation of atomic states in a laser cooled ion crystal

W. Alt, M. Block, P. Seibert and G. Werth

Institut für Physik, Universität Mainz, D-55099 Mainz, Germany

A laser cooled ion crystal containing several hundred Ca$^+$ ions has been stored in a linear Paul trap. Cooling is provided by a red detund laser at the 4S$_{1/2}$ - 4P$_{1/2}$ resonance transition. A second laser serves for repumping of those ions which decay from the excited 4P$_{1/2}$ level to the metastable 3D$_{3/2}$ state. The ions can be additionally excited by a third laser to a long lived metastable 3D$_{5/2}$ energy level which decouples them from the cooling laser radiation. The light pressure acting upon the laser cooled ions pushes them into the direction of the laser beam. The ions in the metastable 3D$_{5/2}$ state, however, do not experience any light pressure force and diffuse to the crystal side which points towards the cooling laser. Part of the crystal then appears dark. Depending on the number of ions in the crystal and on the fraction of the excited ions, the spatial separation of the ions in the metastable state and the ions in the cooling cycle can be as large as several hundred microns.

When ions, confined in Paul or Penning traps, are sufficiently cooled by laser radiation so that the Coulomb energy between adjacent particles is much stronger than the thermal kinetic energy, they can exhibit phase transitions between a liquid-like behavior and a crystalline. Such crystals have been observed in recent years in different laboratories [1–6]. In these experiments those ions have been investigated, which offer a two-level system and strong cooling transitions such as Mg$^+$ or Be$^+$. More recently Ca$^+$, a heavier ion of the alkaline-earth, has been succesfully crystallized [7,8]. It exhibits a three level system and requires two lasers for a continuous cooling cycle.

We have performed cooling experiments on Ca$^+$ ions suspended in a linear Paul trap. These traps [2,3,6,8] have the advantage to have a field free region along the trap axis. The vanishing micromotion of trapped ions favors the formation of a linear crystalline chain of ions. Fig. 1 shows a series of pictures of linear chains of Ca^+ containing up to 6 ions. Recently the formation of large ion crystals in a linear trap containing several 10^4 ions has been reported [6].

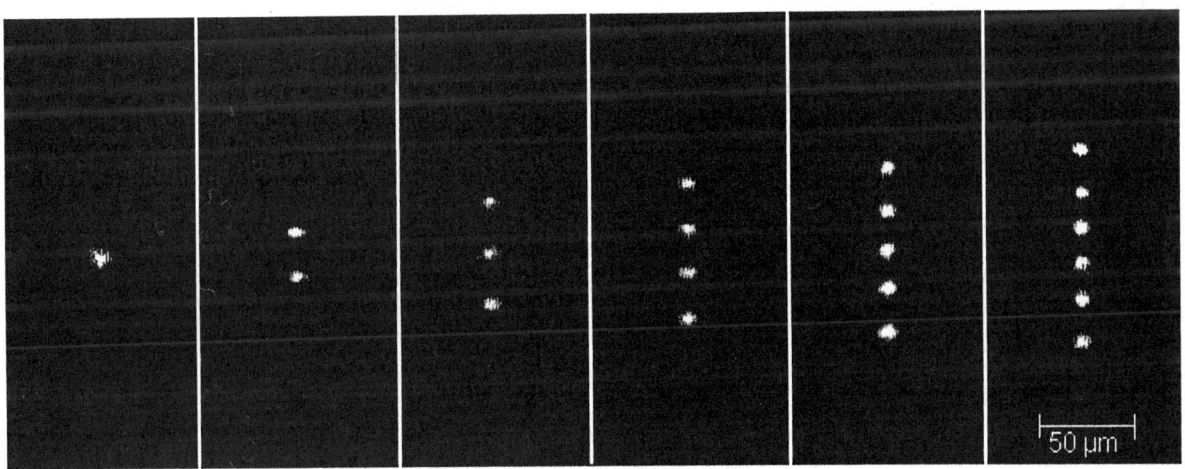

FIG. 1. Linear chains of Ca$^+$ ions containing up to 6 ions directed along the axis of a linear Paul trap

The formation of ionic crystals can be described by molecular dynamics simulations [9] which predicts the existence of parallel shells. This has actually been observed in experiments [3,4], and one of the objectives for investigations on ionic crystals is to test such calculations. The results depend critically on details of the heating process and the numerical accuracy. A second interest in linear ion crystals arises from the fact that they are considered as potential candidates for the experimental realisation of quantum gates [10,11,8] which are based on the entanglement of internal and external degrees of freedom of the trapped ions. Finally ion crystals promise further improvement in the precision of spectroscopic measurements [11,12]. In this contribution we report on a novel effect in an ion crystal. We have observed a spatial separation of an ionic ground and a long lived metastable state. This is caused by light pressure enforced diffusion in an ion crystal of moderate size. It requires the existence of a metastable state in the ion under investigation which is not coupled to the cooling transition. Such levels exist in some alkaline-earth ions such as Ca$^+$, Sr$^+$ or Ba$^+$, and also in Tl$^+$, In$^+$, Hg$^+$ and others.

Our linear Paul trap is made of four parallel cylindrical electrodes. Radial confinement is achieved by a r.f. quadrupole field of 2 MHz frequency and typically 50 V amplitude. The electrodes are segmented into three parts so that a positive d.c. potential can be added to the outer segments to provide axial confinement. The trap electrodes have a diameter of 6 mm and a closest distance from the center of 2.7 mm. Fig. 2 shows a scheme of our experimental setup. Ca$^+$ ions are created by electroionisation of an atomic beam inside the trap. Irradiation by a slightly red detuned laser at the $4S_{1/2}$ - $4P_{1/2}$ resonance transition (397 nm) and a repumping laser at the $3D_{3/2}$ - $4P_{1/2}$ transition (866 nm) cools the ions.

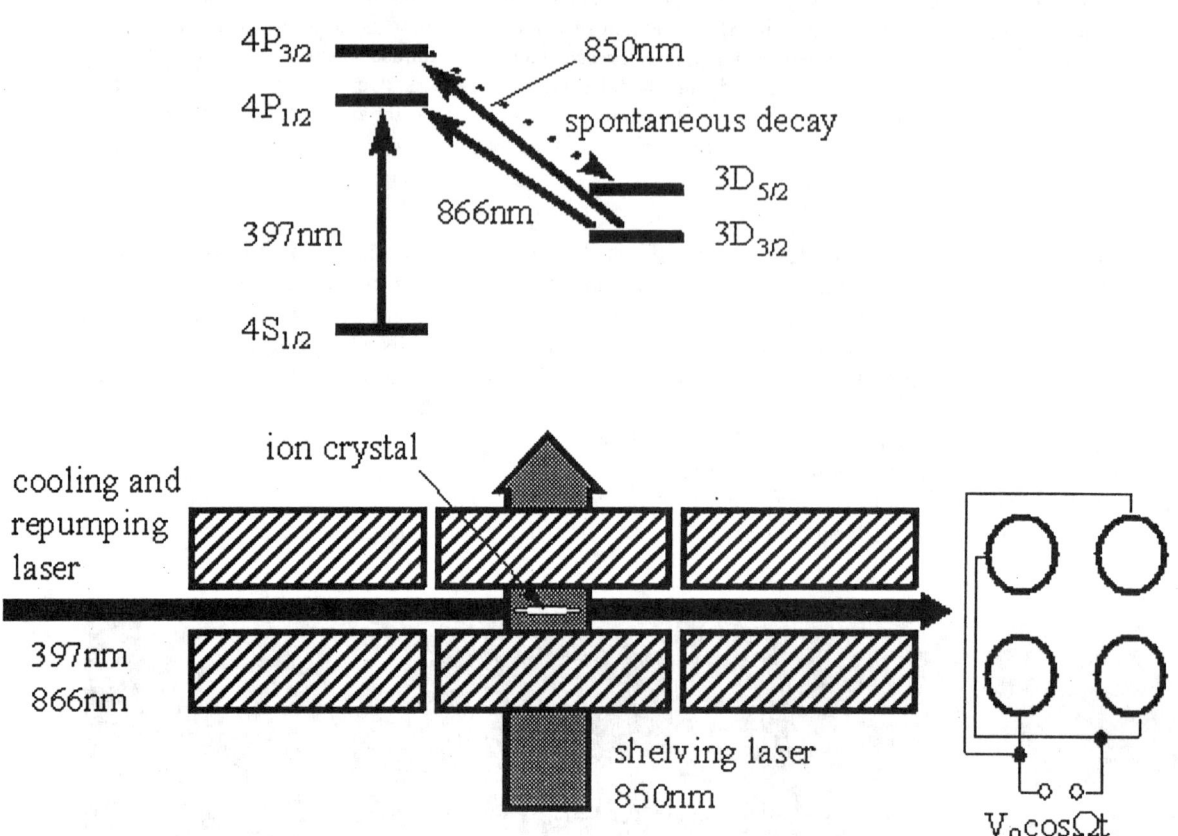

FIG. 2. Partial level scheme of Ca$^+$ and experimental setup showing the segmented trap

At sufficiently high laser power, crystallisation occurs which is detected either by a characteristic kink in the fluorescence intensity or by direct observation by a CCD camera. The spatial resolution of our optical detection system allows the direct observation of individual ions up to a number of about 50, while for higher ion numbers the inter-ionic distance becomes too small to be resolved. The number of ions then can be estimated from the total amount of fluorescence. If we tune a third laser to the $3D_{3/2}$ - $4P_{3/2}$ transition at 850 nm (see level scheme in fig. 2) some ions may spontaneously decay from the $4P_{3/2}$ state into the metastable $3D_{5/2}$ level, whose lifetime has been determined to be about 1 second [14]. During the time which the ions spend in the $3D_{5/2}$ state they are not subjected to any laser radiation and do not emit fluorescence light. They remain, however, in the crystal and are cooled by Coulomb interaction with the neighboring ions. Consequently they would appear as dark spots inside the crystal. This is in fact the case in a small linear crystal which forms along the trap axis. Diffusion of the dark spots inside the crystalline chain can be observed by the change of position to a neighboring ion. This takes place at a typical rate of 1/4 sec-1 at a background pressure of 2.10^{-10} mbar. At higher pressure the jumping rate increases due to ion-neutral collisions.

If the crystal contains many ions, a shell like structure appears as observed previously [6]. In this case the diffusion rate of the dark ions is much faster than in a linear chain due to the higher temperature in larger crystals and a lower two-ion exchange energy barrier. The diffusion is enforced by the light pressure acting on those ions, which are subjected to the continuous cooling laser radiation. This force pushes the ions in the direction of the laser beam. The ions excited to the metastable $4D_{5/2}$ level, however, do not experience any light force and consequently aggregate at the side of the crystal which is directed against the laser. They remain, however crystallized through the Coulomb interaction to the laser cooled ions (sympathetic cooling). The spatial separation of the ions in the $4D_{5/2}$ state from those in the cooling cycle takes place in a time which is faster than the time resolution of our CCD camera (20 msec). We observe a sharp boundary between the excited and the metastable part of the crystal. The number of ions which are driven to the $4D_{5/2}$ state can be varied by changing the power of the 850 nm laser. Fig. 3 illustrates this in a series of pictures of a crystal, containing about 400 ions, at different powers of the 850 nm laser. The direction of the laser beam is perpendicular to the trap axis and the cooling laser direction. Similar results have been obtained when all the lasers are directed colinear along the axis. With increasing power, the dark part of the crystal increases. If the number of cooled ions in the crystal becomes too small, the crystal melts, because then the sympathetic cooling is no longer strong enough to overcome heating effects which act on all the ions in the crystal.

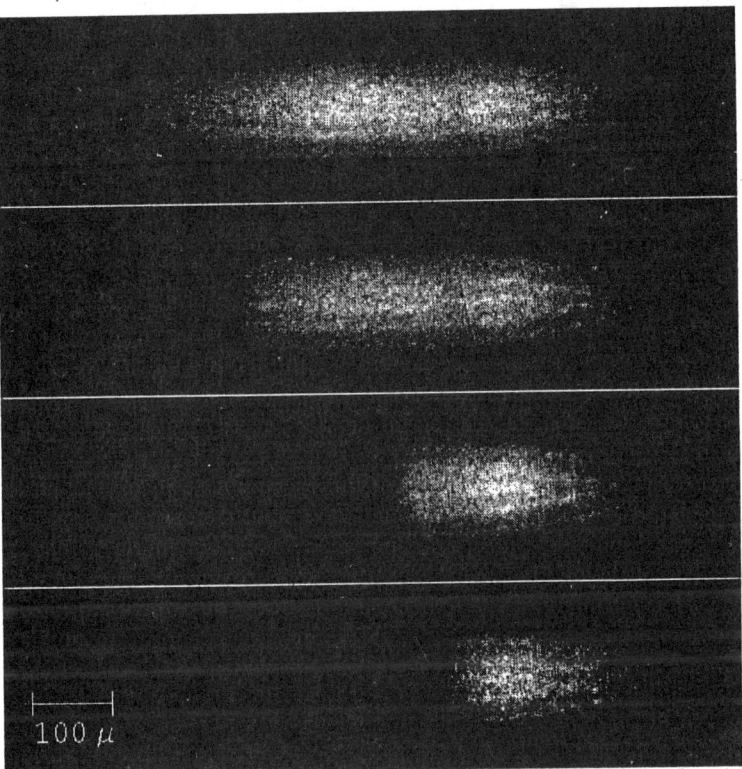

FIG. 3. Moderate size ion crystal containing about 400 ions. In the picture at the top, only the cooling and repumping lasers are present. In the following pictures an additional shelving laser excites a fraction of the ions in the crystal to a long lived metastable state. Those ions aggregate at the left side of the crystal which points in the opposite direction to the cooling and repumping lasers. From top to bottom the shelving laser power is increased (0.4, 0.8, and 1.3 μW, respectively)

Heating is caused by collisions with background gas. Decreasing the 850 nm laser power reestablishes the crystal. The shift in the position of the boundary between the two parts is completely reversible. It should be mentioned, however, that in the case, where individual ions cannot be spatially resolved by our detection optics, we can still unambiguously distinguish between an ionic crystal and a non crystallized ion cloud by the very sharp spatial boundary of the fluorescence in a crystal compared to the diffuse shape of a cloud.

Since the separation between neighboring ions in the crystal is of the order of 10 μm, the separation between the dark and bright part of the crystal is as large as several hundred μm. This allows easy visualization of these parts by the detection optics. We have complete population inversion in the dark part of the crystal which could give rise to lasing, if the trap is placed between high reflecting mirrors. It may also allow precision spectroscopy on the metastable state without perturbation from scattered light background.

A similar effect as described above appears if ionic species other than the ion under inverstigation is produced, confined and sympathetically cooled. Such an impurity ion also would appear dark since is is not excited by any laser beam. This is in particular the case if we store different isotopes of the element. In an experiment on a mixture of $^{40}Ca^+$ and $^{43}Ca^+$ we cooled the even isotope by laser radiation while the odd one does not experience cooling radiation due to the isotpoe shift of the resonance lines [15]. The sympathetically cooled $^{43}Ca^+$ ions thea aggregate on one side of the crystal as above. Due to the mass difference of the two isotopes, however, the radial potential is shallower for the heavier ion. The boundary between the two ion species then is not a sharp line perpendicular to the laser beam direction but the $^{43}Ca^+$ forms a layer around the left part of the $^{40}Ca^+$ crystal. Fig. 4 shows this effect, which may be use to separate spatially different ion species which are simultaneously confined in the trap.

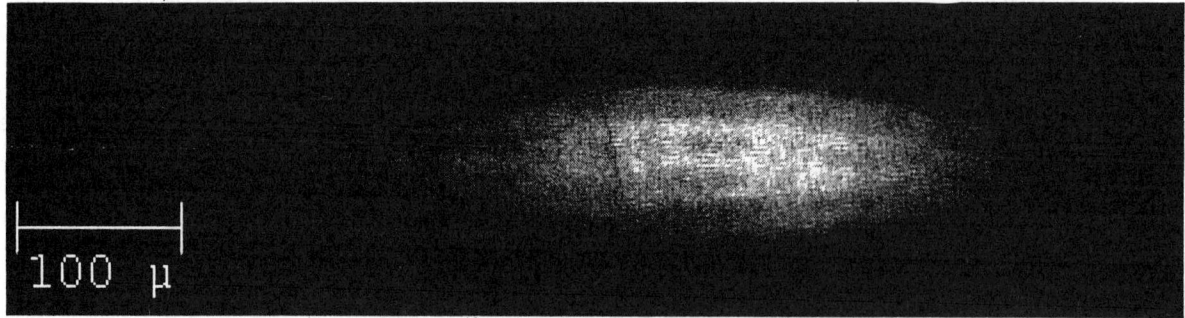

FIG. 4. Ion crystal containing laser cooled $^{40}Ca^+$ ions and sympathetically cooled $^{44}Ca^+$. Due to the absence of radiation pressure and the slightly different radial potential depth the heavier ions form a layer around the left part of the $^{40}Ca^+$ crystal

Our experiments are supported by the Deutsche Forschungsgemeinschaft. We are indebted to T. Nakamura (Tokyo) for support with the trap setup, to M. Holzscheiter (Los Alamos) for placing a CCD camera at our disposal and M. Immel for help with the data handling.

[1] F. Diedrich and H. Walther, Phys. Rev. Lett. **58**, 203 (1987).
[2] D.J. Wineland, J.C. Bergquist, W.M. Itano, J.J. Bollinger and C.H. Manney, Phys. Rev. Lett. **59**, 2935 (1987).
[3] G. Birkl, S. Kassner and H. Walther, Nature **357**, 310 (1992).
[4] S.L. Gilbert, J.J. Bollinger, and D.J. Wineland, Phys. Rev. Lett. **60**, 2022 (1988).
[5] J.N. Tan, J.J. Bollinger, B. Jelenkovic, and D.J. Wineland, Phys. Rev. Lett. **75**, 4198 (1995).
[6] M. Drewsen, C. Brodersen, L. Hornekaer, J.S. Hangst, and J.P. Schiffer (submitted to Phys. Rev. Lett.).
[7] T. Gudjons, F. Arbes, M. Benzing, F. Kurth, and G. Werth, Physica Scripta **T59**, 396, (1995).
[8] H.C. Nägerl, W. Bechter, J. Eschner, F. Schmidt-Kahler, and R. Blatt, Appl. Phys. B (accepted).
[9] A. Rahman and J.P. Schiffer, Phys. Rev. Lett. **57**, 1133 (1986).
[10] J.I. Cirac and P. Zoller, Phys. Rev. Lett. **74**, 4091 (1995).
[11] D.M. Meekhoff, C. Monroe, B.E. King, W.M. Itano, and D.J. Wineland, Phys. Rev Lett. **76**, 1796 (1996).
[12] W.M. Itano, J.C. Bergquist, J.J. Bollinger, J.M. Gilligan, D.J. Heinzen, F.L. Moore, M.G. Raizen, and D.J. Wineland, Phys. Rev. A **47**, 3554 (1993).
[13] J.J. Bollinger, W.M. Itano, D.J. Wineland, and D.J. Heinzen, Phys. Rev. A **54**, R4649 (1996).
[14] T. Gudjons, B. Hilbert, P. Seibert, and G. Werth. Europhys. Lett. **33**, 595 (1996).
[15] W. Alt, M. Block, V. Schmidt, T. Nakamura, P. Seibert, X. Chu and G. Werth. J. Phys. B **30**, L677 (1997).

APPENDIX

List of Participants

Charles Alcock
LLNL, L-413

Gerald D. Alton
Oak Ridge National Laboratory
Bldg. 6000, MS-6368
Oak Ridge, TN 37831

Francois A. Anderegg
University of California San Diego
Physics Department, 0319
9500 Gilman Drive
La Jolla, CA 92093-0319

Jens Ulrik Andersen
Inst. of Physics & Astronomy
University of Aarhus
DK-8000 Aarhus C
DENMARK

Torkild Andersen
Inst. of Physics & Astronomy
University of Aarhus
DK-8000 Aarhus C
DENMARK

Björn O. Asen
Göteborg University
Department of Physics
SE-412 96 Göteborg
SWEDEN

Faouzi Attallah
Planckstr 1
64291 Darmstatdt
GERMANY

James F. Babb
Harvard-Smithsonian Ctr. for Astro.
MS-14, 60 Garden St.
Cambridge, MA 02138

Rudolf Bauer
LLNL, L-056

Bret Beck
LLNL, L-414

Dietrich H. Beck
Institute for Kern-en Stralingsfysica
Celestijnenlaan 200 D
B-3001 Leuven
BELGIUM

John Becker
LLNL, L-414

John Behr
TRIUMF
4004 Wesbrook Mall
Vancouver, B.C.
CANADA V6T 2A3

Peter Beiersdorfer
LLNL, L-414

Lars Ingmar Bergström
Manne Siegbhan Laboratory
Stockholm University
10405 Stockholm, SWEDEN

Jérome Bernard
University of Lyon
19 Avenue, du Vercors
Meylan 38240, FRANCE

Robert Bluhm
Colby College
Physics Department
Waterville, ME 04901

Reinhold Blümel
Fakultät Für Physik
Albert-Ludwigs-Universität
Hermann-Herder-Str.3
D-79104 Freiburg, GERMANY

Georg Bollen
CERN
EP-ISOLDE
CH-1211 Geneva 23
SWITZERLAND

John J. Bollinger
U.S. Dept. of Commerce
NIST
325 Broadway
Boulder, CO 80303-3328

Hans J. Briegel
Institute for Theoretical Physics
Technikerstrasse 25
University of Innsbruck
A-6020 Innsbruck
AUSTRIA

Gregory Brown
LLNL, L-414

Dmitry Budker
UC Berkeley
Department of Physics
Berkeley, CA 94720-7300

James Byrne
University of Sussex
School of CPES
Brighton Sussex, BN1 9QH
UNITED KINGDOM

Conny Carlberg
Stockholm University
Dept. of Atomic Physics
Frescativägen 24
S-104 05 Stockholm, SWEDEN

Marielle Chartier
NSCL/Michigan State University
East Lansing, MI 48824-1321

Mau H. Chen
LLNL, L-041

Kwok-Tsang Cheng
LLNL, L-015

Xinzhao Chu
University of Illinois, Urbana
310 CSRL
1308 West Main St.
Urbana, Illinois 61801-2307

David Church
Texas A & M University
Physics Dept.
College Station, TX 77843-4242

Michel de Saint Simon
CSNSM, bat 104-108
F-91405 ORSAY-CAMPUS
FRANCE

Jens Dilling
GSI
Planckstr. 1
64291 Darmstadt
GERMANY

Michael Drewsen
University of Aarhus
Ny Munkegade
DK-8000, Aarhus C
DENMARK

Fred Driscoll
University of California San Diego
Physics Department, 0319
La Jolla, CA 92093-0319

Dan Dubin
University of California San Diego
Physics Department, 0319
La Jolla, CA 92093-0319

Udo Eisenbarth
Max-Planck, Institut für Kernphysik
Postfach 103980
69029 Heidelberg
GERMANY

Richard Fortner
LLNL, L-051

Stuart Freedman
Lawrence Berkeley Laboratory
1 Cyclotron Road
Berkeley, CA 94720

Stephan Friedrich
LLNL, L-418

Tomas Fritioff
Frescativ. 24
104 05 Stockholm
SWEDEN

Gerald Gabrielse
Harvard Physics Department
Cambridge, MA 02138

Harvey Gould
Lawrence Berkeley National Laboratory
MS 71-259
1 Cyclotron Road
Berkeley, CA 94720

Alex P. Grossmann
Blumenstr. 10
76593 Gernsbach
GERMANY

Lukas Gruber
LLNL, L-421

Martin G. H. Gustavsson
Department of Physics
Götenborg Univ. and Chalmers Univ. of Tech.
Fysik och teknisk fysik
SE-412 96 Göteborg, SWEDEN

Gerald Gwinner
Max-Planck-Insitute for Nuclear Physics
Saupfercheckweg 1
69117 Heidelberg,
GERMANY

Chris Hagmann
LLNL, L-414

Alex Hamza
LLNL, L-414

Jeffrey S. Hangst
Institute of Physics & Astronomy
Aarhus University
DK-8000 Aarhus C
DENMARK

Theodor Hänsch
Max-Planck-Inst. for Quantenoptik
Hans-Kopfermann Str. 1
D-85748 Garching
GERMANY

Andy Hazi
LLNL, L-051

Jurgen Helmcke
PTB, Department 4.3
Physikalisch-Technische Bundesanstalt
Bundesallee 100
D-38116 Braunschweig, GERMANY

Nikolaus Hermanspahn
Staudingerweg 7
55099 Mainz
Rheinland-Pfalz
GERMANY

Hiroyuki Higaki
University of Tokyo
Institute of Physics
Graduate School of Arts & Sci.
3-8-1 Meguro, Tokyo
JAPAN 153-8902

Joe Holder
LLNL, L-421

Eric M. Hollmann
UCSD, Physics Dept. 0350
9500 Gilman Dr.
La Jolla, CA 92093-0319

John F. Holzrichter
LLNL, L-003

Michael Holzscheiter
Los Alamos National Laboratory
Physis Div/P-23
Mail Stop H803
Los Alamos, NM 87545

Richard Hughes
Los Alamos National Laboratory
Phyics Div., H803
Los Alamos, NM 87545

Preben Hvelplund
Institute of Physics & Astronomy
University of Aarhus
DK-8000 Aarhus C
DENMARK

Ulrich D. Jentschura
NIST
Bldg. 225, Rm. B50 (c/o P. Mohr)
Gaithersburg, MD 20878

Walter R. Johnson
Notre Dame University
Physics, Rm. 225
Nieuwland Science Hall
Notre Dame, Indiana 46556

Brian E. King
NIST
Div. 847.10
325 Broadway
Boulder, CO 80303

Hugh Klein
National Physical Laboratory
Teddington
Middlesex TW 11 0LW
UNITED KINGDOM

Heinz-Jürgen Kluge
GSI-Darmstadt
Atomic Physics Group
Planckstr. 1
D-64291 Darmstadt, GERMANY

Alan Kostelecky
Department of Physics
Indiana University
Bloomington, IN 47405

Andreas Krämer
GSI-Darmstadt
PLANCKSTR. 1
D-64291, Darmstadt
GERMANY

Martin Kretzschmar
University of Mainz
Institute of Physics
D-55099 Mainz
GERMANY

Thomas Kühl
GSI/Mainz University
Plancksr. 1
D-64291 Darmstadt
GERMANY

Mats Larsson
Department of Physics
Stockholm University, Box 6730
S-113 85 Stockholm, SWEDEN

Jacob (Yasha) Levin
Max-Planck Institut für Kernphysik
69118 Heidelberg, GERMANY

Eva Lindroth
Stockholm University
Atomic Physics
Frescavr. 24
104 05 Stockhold,
SWEDEN

David Lunney
CSNSM-CNRS
Bat. 108, F-91405
Orsay, FRANCE

Guillaume A. Machicoane
LLNL, L-421

Ross E. Marrs
LLNL, L-050

Gerrit H. Marx
An der kirche 14
56357 Obertiefenbach
GERMANY

Francois Mauger
6, Boulevard du Marechal Juin
CAEN (14050 CAEN cedex)
FRANCE

Joseph W. McDonald
LLNL, L-414

Hans Miesner
Massachusetts Inst. of Technology
Dept. of Physics, Room 26-251
77 Massachusetts Ave.
Cambridge, MA 02139

Travis Mitchell
NIST, MS 847.11
325 Broadway
Boulder, CO 80303

Richard Mittleman
University of Washington
Physics Dept.
Seattle, WA 98195

Peter Mohr
U. S. Department of Commerce
NIST
Building 225, Room B161
Gaithersburg, MD 20899

Luigi Moi
University of Siena
Dept. of Physics
Via Banchi Di Sotto 55
53100 Siena, ITALY

Christopher Monroe
NIST
Time and Frequency Div., 847
325 Broadway
Boulder, CO 80303

Warren Nagourney
University of Washington
Physics Department
Box 351560
Seattle, WA 98195

Michael W. Newman
LLNL, L-414

Thomas R. Niedermayr
LLNL, L-421

Thomas M. O'Neil
University of California, San Diego
Physics Dept., UCSD
La Jolla, CA 92093-0319

Luis A. Orozco
SUNYSB
Physics and Ast. SUNYSB
Stony Brook, NY 11794-3800

John D. Prestage
1141 Heather Sq.
Pasadena, CA 91109

David E. Pritchard
Massachusetts Institute of Technology
Department of Physics
Room 26-239
Cambridge, MA 02139

Wolfgang Quint
GSI
Planck-Str. 1
D-64291 Darmstadt
GERMANY

Simon Rainville
MIT
ICR Lab, Room 26-142
77 Massachusetts Ave.
Cambridge, MA 02139

Matthew Roberts
NPL
Queens Road
Teddington, Middlesex
TWII OLW, ENGLAND

Gary Rouleau
CERN ppe-div
Bat. 23 1-019
CH-1211 Geneva 23
SWITZERLAND

Neil Russell
Swain Hall West 117, Physics Dept.
Indiana University
Bloomington, IN 47405

Jonathan Sapirstein
University of Notre Dame
Physics Dept.
Notre Dame, IN 46556

Michael H. Schacht
University of Washington
Physics Deptartment
Box 351560
Seattle, WA 98195

Tobias Schaetz
LMU-Muenchen
Isarweg 4
Muenchen, GERMANY

Martin Schauer
Los Alamos National Laboratory
MS H803
Los Alamos, NM 87545

Stephan Schlemmer
09107 Chemnitz
Saxony
GERMANY

Jens H. Schmid
3549 W. 32nd Avenue
Vancouver B.C.
CANADA, V6S 1Z1

Dieter Schneider
LLNL, L-414

Reinhold Schuch
Stockholm University
Dept. of Atomic Physics
Frescativägen 24
S-10405 Stockholm, SWEDEN

Paul Schuurmans
Celestijnenlaan 200 D
B-3000 Leuven
BELGIUM

Lutz Schweikhard
Johannes-Gutenberg University
Institut für Physik
D-55099 Mainz, GERMANY

Vladimir M. Shabaev
Oulinovskaya 1, Petrodvorets
St. Petersburg 198904
RUSSIA

Howard A. Shugart
University of California, Berkeley
Department of Physics
Berkeley, CA 94720-7300

Joshua D. Silver
University of Oxford
Clarendon Laboratory
Parks Road
Oxford OX1 3PU ENGLAND

Gerhard E. Soff
TU Dresden
Inst. for Theor. Physik
D-01062 Dresden
Mommsenstr. 13, GERMANY

Markus Steck
GSI-Darmstadt
Planckstr. 1
D-64291, Darmstadt, GERMANY

Joachim Steiger
LLNL, L-414

Thomas Stöhlker
GSI-Darmstadt
Planckstr. 1
D-64291, Darmstadt, GERMANY

Clifford Surko
University of California San Diego
Physics Dept., 0319
La Jolla, CA 92093-0319

Richard C. Thompson
Blackett Laboratory, Imperial College
Prince Consort Rd.
London SW7 2BZ, UNITED KINGDOM

Robert L. Tjoelker
4800 Oak Grove Drive
Pasadena, CA 91109

Peter E. Toschek
University of Hamburg
Institut für Laser-Physik
Jungiusstr. 9
D-20355 Hamburg, GERMANY

Karl R. Umstadter
115 Chamisa
Los Alamos, NM 87545

Steven B. Utter
LLNL, L-414

Robert S. Van Dyck, Jr.
University of Washington
Dept. of Physics,
Box 351560
Seattle, WA 98195-1560

Fernande Vedel
Universite de Provence
PIIM, UMR 6633 CNRS-UAM1
Centre de Saint-Jerome Case C21
F1337 Marseille Cedex 20 FRANCE

David J. Vieira
Los Alamos National Laboratory
CST-11, Mail Stop J514
Los Alamos, NM 87545

Manasori Wakasugi
RIKEN
Hirosawa 2-1, Wako-shi,
Saitama, 351-01
JAPAN

Herbert Walther
Max-Planck-Institut für Quantenoptik
Hans-Kopfermann-Str. 1
D-85748 Garching
GERMANY

Matthias Weidemüller
Max-Planck-Institut für Kernphysik
P.O. Box 103980
D-69029 Heidelberg
GERMANY

Günter Werth
Universität Mainz
Institut für Physik
D-55099 Mainz
GERMANY

David J. Wineland
National Institute of Standard and Technology
Division 847.10
325 Broadway
Boulder, CO 80303

Andreas Wolf
Max-Planck-Institut für Kernphysik
P.O. Box 103 980
D69029 Heidelberg
GERMANY

Chris S. Wood
NIST
Time and Frequency Division
325 Broadway, MS 847.10
Boulder, CO 80303

Valeriy V. Yashchuk
Department of Physics, UCB
Berkeley, CA 94720

Jun Ye
California Institute of Technology
Room 12-33
Pasadena, CA 91125

Brenton C. Young
NIST
325 Broadway
Boulder, CO 80303

Nan Yu
Jet Propulsion Laboratory
MS 298-105C
4800 Oak Grove Dr.
Pasadena, CA 91109

Steven Zafonte
University of Washington
Physics Dept.
Box 351560
Seattle, WA 98195-1560

Daniel Zajfman
Weizmann Institute of Science
Dept. of Particle Physics
Rehovot, 76100,
ISRAEL

Boris N. Zakhariev
Laboratory of Theoretical Physics
Joint Institute for Nuclear Research
Dubna, Moscow District, 141980
RUSSIA

Author Index

A

Alford, W. P., 148
Alheit, R., 329
Alt, W., 393
Ames, F., 111
Anderegg, F., 277, 319
Andersen, H. H., 227
Andersen, J. U., 220
Andersen, T., 227
Appasamy, B., 252
Artemyev, A. N., 22
Asgeirsson, D., 148
Audi, G., 95, 111

B

Ball, G., 148
Balling, P., 227
Balzer, C., 252
Batygin, Y., 210
Beck, B. R., 235, 284
Beck, D., 111, 172
Beck, M., 172
Beckert, K., 87
Behr, J. A., 148
Bergquist, J. C., 337
Berkeland, D. J., 337
Block, M., 393
Bluhm, R., 70, 133, 138
Blümel, R., 290, 329
Bollen, G., 111, 172
Bollinger, J. J., 295, 309
Borcea, C., 95
Bowe, P., 305
Brice, S. J., 143
Brodersen, C., 305
Brown, H. N., 52
Brunner, S., 125
Buchmann, L., 148
Budker, D., 175, 177
Bunce, G., 52
Bushan, K. G., 203
Byrne, J., 163

C

Carey, R. M., 52
Chu, X. Z., 329
Church, D. A., 235, 284
Crane, S. G., 143

Cruz, F. Z., 337
Cushman, P., 52

D

Danby, G. T., 52
D'Auria, J. M., 148
Dawber, P. G., 163
Debevec, P. T., 52
Dehmelt, H., 13, 261, 343
Deng, H., 52
Deninger, W., 52
de Saint Simon, M., 95, 111, 120
Deutsch, J., 148, 172
Dhawan, S. K., 52
Diederich, M., 43
Dilling, J., 148, 172
Dombsky, M., 148
Doubre, H., 95
Drewsen, M., 305
Driscoll, C. F., 277, 319
Druzhinin, V. P., 52
Dubé, P., 148
Duma, M., 95
Duong, L., 52

E

Earle, W., 52
Efstathiadis, E., 52
Eickhoff, H., 87
Eike, B., 194
Eisenbarth, U., 194
Engel, T., 125

F

Farley, F. J. M., 52
Farnham, D. L., 101
Fedotovich, G. V., 52
Franzke, B., 87

G

Gerlich, D., 80
Giesen, U., 148
Giron, S., 52
Goldschmidt, A., 143
Gorelov, A., 148

Gray, F., 52
Grieser, M., 194
Grimm, R., 194
Grosse Perdekamp, M., 52
Grossman, J. S., 155
Grossmann, A., 52
Gruber, L., 235, 284
Guckert, R., 143

H

Habs, D., 269
Haeberlen, U., 52
Häffner, H., 43
Hangst, J. S., 305
Hansen, K., 220
Hare, M., 52
Häusser, O., 148
Hazen, E. S., 52
Heber, O., 203
Helmcke, J., 348
Henry, S., 95, 120
Herfurth, F., 111
Hermanspahn, N., 43
Hernandez-Pozos, J.-L., 388
Hertzog, D. W., 52
Higaki, H., 324
Hime, A., 143
Höffges, J., 388
Holder, J. P., 235, 284
Hollmann, E. M., 277, 319
Holzscheiter, M. H., 65
Hood, C. J., 371
Hornek, L., 305
Houssin, M., 365
Huang, X.-P., 295, 309
Huesmann, R., 252
Hughes, V. W., 52
Hvelplund, P., 220

I

Illemann, J., 80
Immel, M., 43
Inabe, N., 210
Ioannou, I. I., 13
Itano, W. M., 295, 309, 337, 378
Iwasaki, M., 52

J

Jackson, K. P., 148
Jacotin, M., 95, 120

Jelenković, B. M., 295
Jennings, B., 148
Jentschura, U. D., 40
Jungmann, K., 52

K

Katayama, T., 210
Kawall, D., 52
Kawamura, M., 52
Képinski, J.-F., 95, 120
Khazin, B. I., 52
Kielpinski, D., 378
Kimble, H. J., 371
Kindem, J., 52
King, B. E., 378
Kluge, H.-J., 43, 111
Knoll, L., 185
Knoop, M., 365
Kohl, A., 111
Kostelecký, V. A., 70, 133, 138
Kretzschmar, M., 242
Krienen, F., 52
Kronkvist, I., 52

L

Lange, M., 185
Larsen, R., 52
Lauer, I., 194
Lebée, G., 95
Lee, Y. Y., 52
Leibfried, D., 378
Lenisa, P., 194
Le Scornet, G., 95
Levin, J., 185
Ley, R., 43
Lipphardt, B., 348
Liu, W., 52
Logashenko, I., 52
Luger, V., 194
Lunney, D., 95, 111, 120
Lynn, T., 371

M

Mabuchi, H., 371
Maleki, L., 357
Mann, R., 43
Martel-Bravo, I., 120
Maruyama, K., 210
Marx, G., 57
McDonald, J., 235

McNabb, R., 52
Melconian, D., 148
Meng, W., 52
Mi, J.-L., 52
Miller, D., 52
Miller, J. P., 52
Mitchell, T. B., 295, 309
Mittleman, R. K., 13
Mohr, P. J., 40
Monroe, C., 378
Monsanglant, C., 95
Moore, R. B., 111
Morse, W. M., 52
Mudrich, M., 194
Myatt, C. J., 378

N

Nagourney, W., 343
Neumayer, P., 52
Nolden, F., 87

O

Ohkawa, T., 210
Ohtomo, K., 210
Onderwater, G., 52
Orlov, Y., 52
Orozco, L. A., 155

P

Pai, C., 52
Pawletko, T., 365
Pearson, M., 155
Petrunin, V. V., 227
Phalet, T., 172
Plunien, G., 32
Podlech, C., 269
Polly, C., 52
Prestage, J. D., 357
Pretz, J., 52
Prieels, R., 172
Prigl, R., 52

Q

Quint, W., 43, 172

R

Rafac, R. J., 337
Rappaport, M., 203
Razvi, M. A. N., 329
Redin, S. I., 52
Reich, H., 87
Riehle, F., 348
Rind, O., 52
Roberts, B. L., 52
Russell, N., 70, 133, 138
Ryskulov, N., 52

S

Sanders, R., 52
Sapirstein, J., 3
Schark, E., 111
Schätz, T., 269
Scheffel, M., 185
Schiffer, J. P., 305
Schlemmer, S., 80
Schlitt, B., 87
Schmid, J., 148
Schmitt, A., 125
Schnatz, H., 348
Schneider, D., 235, 284
Schramm, U., 194, 269
Schützhold, R., 32
Schuurmans, P., 172
Schwalm, D., 185, 194
Schwarz, S., 111
Schweizer, T., 365
Schwinberg, P. B., 101
Sedykh, S., 52
Segal, D. M., 388
Seibert, P., 393
Semertzidis, Y. K., 52
Serednyakov, S., 52
Severijns, N., 172
Shabaev, V. M., 22
Shatunov, Y. M., 52
Simsarian, J. E., 155
Soff, G., 32, 40
Solodov, E., 52
Sossong, M., 52
Sprouse, G. D., 155
Stahl, S., 43
Stalgies, Y., 252
Steck, M., 87
Steiger, J., 235, 284
Steinmetz, A., 52
Sulak, L. R., 52
Swanson, T., 148
Szerypo, J., 111

T

Takanaka, M., 210
Tan, J. N., 295
Tanabe, T., 210
Tanaka, M., 52
Tanihata, I., 210
Thibault, C., 95
Thompson, R. C., 388
Timmermans, C., 52
Tjoelker, R. L., 357
Toader, C., 95
Tommaseo, G., 57
Torgerson, J., 343
Toschek, P. E., 252
Trebst, T., 348
Trinczek, M., 148
Trofimov, A., 52
Tupa, D., 143
Turchette, Q. A., 378

U

Urner, D., 52

V

van Dyck, Jr., R. S., 101
Vedel, F., 365
Vedel, M., 365
Verdú, J., 43
Vereecke, B., 172
Vernooy, D. W., 371
Versyck, S., 172
Vieira, D. J., 143
Vincent, J. R., 388

W

Wakasugi, M., 210
Walter, P. V., 52
Warburton, D., 52

Watanabe, S., 210
Wei, J., 269
Weidemüller, M., 194
Wellert, S., 80
Werth, G., 43, 57, 125, 329, 393
Wester, R., 185
Wineland, D. J., 295, 337, 378
Winkler, T., 87
Winn, D., 52
Wolf, A., 185
Wong, W., 148
Wood, C. S., 378

X

Xu, Q., 52

Y

Yamamoto, A., 52
Yano, Y., 210
Yashchuk, V., 177
Ye, J., 371
Yerokhin, V. A., 22
Yoshida, K., 210
Young, B. C., 337
Yu, N., 261

Z

Zafonte, S. L., 101
Zajfman, D., 203
Zhao, W. Z., 155
Zhao, X., 143
Zimmerman, D., 52
Zinner, G., 348
Zolotorev, M., 175, 177
Zschocke, S., 32
zu Putlitz, G., 52